VOLUME SIX

METHODS IN CHEMICAL PROCESS SAFETY

Methods to Assess and Manage Process Safety in Digitalized Process System

VOLUME SIX

METHODS IN CHEMICAL PROCESS SAFETY

Methods to Assess and Manage Process Safety in Digitalized Process System

Edited by

FAISAL KHAN
Mary Kay O'Connor Process Safety Center, Artie McFerrin Department of Chemical Engineering, Texas A&M University; TEES Mary Kay O'Connor Process Safety Center, Texas A&M University System, College Station, TX, United States

HANS PASMAN
Mary Kay O'Connor Process Safety Center, Artie McFerrin Department of Chemical Engineering, Texas A&M University; TEES Mary Kay O'Connor Process Safety Center, Texas A&M University System, College Station, TX, United States

MING YANG
Safety and Security Science Section, Department of Values, Technology, and Innovation, Faculty of Technology, Policy, and Management, Delft University of Technology, Delft, The Netherlands

ELSEVIER

ACADEMIC PRESS
An imprint of Elsevier

Academic Press is an imprint of Elsevier
50 Hampshire Street, 5th Floor, Cambridge, MA 02139, United States
525 B Street, Suite 1650, San Diego, CA 92101, United States
The Boulevard, Langford Lane, Kidlington, Oxford OX5 1GB, United Kingdom
125 London Wall, London, EC2Y 5AS, United Kingdom

First edition 2022

ISBN: 978-0-323-98897-1
ISSN: 2468-6514

Publisher: Zoe Kruze
Acquisitions Editor: Jason Mitchell
Developmental Editor: Naiza Ermin Mendoza
Production Project Manager: Vijayaraj Purushothaman
Cover Designer: Miles Hitchen

Typeset by STRAIVE, India

Contents

Contributors

Rouzbeh Abbassi
School of Engineering, Faculty of Science and Engineering, Macquarie University, Sydney, NSW, Australia

Sidum Adumene
School of Ocean Technology, Marine Institute Memorial University of Newfoundland, St. John's, NL, Canada

Md Alauddin
Centre for Risk, Integrity and Safety Engineering (C-RISE), Faculty of Engineering & Applied Science, Memorial University of Newfoundland, St. John's, NL, Canada

Md Tanjin Amin
Centre for Risk, Integrity and Safety Engineering (C-RISE), Faculty of Engineering & Applied Science, Memorial University of Newfoundland, St. John's, NL, Canada

Rajeevan Arunthavanathan
Centre for Risk, Integrity and Safety Engineering (C-RISE), Faculty of Engineering & Applied Science, Memorial University of Newfoundland, St. John's, NL, Canada

Mohsen Asaadi
Department of Automation, Tsinghua University, Beijing, China

Nikhil Bagalkot
Department of Mechanical, Electronics and Chemical Engineering, Oslo Metropolitan University, Norway

Yiming Bai
State Key Laboratory of Chemical Engineering, Department of Chemical Engineering, Tsinghua University, Beijing, China

Stewart Behie
TEES Mary Kay O'Connor Process Safety Center, Texas A&M University System, College Station, TX, United States

Guoming Chen
Centre for Offshore Engineering and Safety Technology (COEST), China University of Petroleum (East China), Qingdao, China

Valerio Cozzani
LISES - Laboratory of Industrial Safety and Environmental Sustainability - Department of Civil, Chemical, Environmental and Materials Engineering - DICAM, University of Bologna, Bologna, Italy

Ahmed Hamdy El-Kady
Mary Kay O'Connor Process Safety Center, Artie McFerrin Department of Chemical Engineering, Texas A&M University, College Station, TX, United States

Mihiran Galagedarage Don
Centre for Risk, Integrity and Safety Engineering (C-RISE), Faculty of Engineering & Applied Science, Memorial University of Newfoundland, St. John's, NL, Canada

Raymond G. Gosine
Centre for Risk, Integrity and Safety Engineering (C-RISE), Faculty of Engineering & Applied Science, Memorial University of Newfoundland, St. John's, NL; Innovation Policy Lab, Munk School of Global Affairs and Public Policy, University of Toronto, Toronto, ON, Canada

Syeda Halim
Mary Kay O'Connor Process Safety Center, Artie McFerrin Department of Chemical Engineering, Texas A&M University, College Station, TX, United States

Wenkai Hu
School of Automation, China University of Geosciences, Wuhan, China

Matteo Iaiani
LISES - Laboratory of Industrial Safety and Environmental Sustainability - Department of Civil, Chemical, Environmental and Materials Engineering - DICAM, University of Bologna, Bologna, Italy

Mohd Umair Iqbal
Department of Chemical Engineering, Indian Institute of Technology Gandhinagar, Gandhinagar, India

Arvind Keprate
Department of Mechanical, Electronics and Chemical Engineering, Oslo Metropolitan University, Norway

Faisal Khan
Mary Kay O'Connor Process Safety Center, Artie McFerrin Department of Chemical Engineering, Texas A&M University; TEES Mary Kay O'Connor Process Safety Center, Texas A&M University System, College Station, TX, United States

Xinhong Li
School of Resources Engineering, Xi'an University of Architecture and Technology, Xi'an, China

Yiliu Liu
Department of Mechanical and Industrial Engineering, Norwegian University of Science and Technology, Trondheim, Norway

Nicola Paltrinieri
Norwegian University of Science and Technology – NTNU, Trondheim, Norway

Hans Pasman
Mary Kay O'Connor Process Safety Center, Artie McFerrin Department of Chemical Engineering, Texas A&M University; TEES Mary Kay O'Connor Process Safety Center, Texas A&M University System, College Station, TX, United States

Payam Rahnamayiezekavat
School of Engineering, Design and Built Environment, Western Sydney University, Penrith, NSW, Australia

Mark Slezak
Occidental Petroleum Corporation, Houston, TX, United States

Babji Srinivasan
Department of Applied Mechanics; American Express Lab for Data Analytics, Risk & Technology, Indian Institute of Technology Madras, Chennai, India

Rajagopalan Srinivasan
Department of Chemical Engineering; American Express Lab for Data Analytics, Risk & Technology, Indian Institute of Technology Madras, Chennai, India

Hao Sun
Safety and Security Science Section, Department of Values, Technology, and Innovation, Faculty of Technology, Policy, and Management, Delft University of Technology, Delft, The Netherlands; College of Mechanical and Electronic Engineering, China University of Petroleum (East China), Qingdao, China

Mohammed Taleb-Berrouane
Centre for Risk, Integrity and Safety Engineering (C-RISE), Faculty of Engineering & Applied Science, Memorial University of Newfoundland, St. John's, NL, Canada

Alessandro Tugnoli
LISES - Laboratory of Industrial Safety and Environmental Sustainability - Department of Civil, Chemical, Environmental and Materials Engineering - DICAM, University of Bologna, Bologna, Italy

Thumeera R. Wanasinghe
Centre for Risk, Integrity and Safety Engineering (C-RISE), Faculty of Engineering & Applied Science, Memorial University of Newfoundland, St. John's, NL, Canada

Jiandong Wang
College of Electrical Engineering and Automation, Shandong University of Science and Technology, Qingdao, China

Zhen Wang
College of Electrical Engineering and Automation, Shandong University of Science and Technology, Qingdao, China

Shuaiyu Xiang
State Key Laboratory of Chemical Engineering, Department of Chemical Engineering, Tsinghua University, Beijing, China

Borui Yang
State Key Laboratory of Chemical Engineering, Department of Chemical Engineering, Tsinghua University, Beijing, China

Fan Yang
Department of Automation, Tsinghua University, Beijing, China

Ming Yang
Safety and Security Science Section, Department of Values, Technology, and Innovation, Faculty of Technology, Policy, and Management, Delft University of Technology, Delft, The Netherlands

Rioshar Yarveisy
C-RISE, Memorial University of Newfoundland, St. John's, NL, Canada

Mohammad Yazdi
Centre for Risk, Integrity, and Safety Engineering (C-RISE), Faculty of Engineering and Applied Science, Memorial University of Newfoundland, St. John's, NL, Canada; School of Engineering, Faculty of Science and Engineering, Macquarie University, Sydney, NSW, Australia

Esmaeil Zarei
Centre for Risk, Integrity, and Safety Engineering (C-RISE), Faculty of Engineering and Applied Science, Memorial University of Newfoundland, St. John's, NL, Canada

Aibo Zhang
Department of Advanced Design and Systems Engineering, City University of Hong Kong, Kowloon Tong, Hong Kong

Luyao Zhang
School of Resources Engineering, Xi'an University of Architecture and Technology, Xi'an, China

Yinong Zhang
College of Urban Rail Transit and Logistics, Beijing Union University, Beijing, China

Jinsong Zhao
State Key Laboratory of Chemical Engineering, Department of Chemical Engineering; Beijing Key Laboratory of Industrial Big Data System and Application, Tsinghua University, Beijing, China

Zeheng Zhao
State Key Laboratory of Chemical Engineering, Department of Chemical Engineering, Tsinghua University, Beijing, China

Preface

Volume 6 on the process safety aspects of digitalization of the process industry is the latest fruit of the *Methods of Chemical Process Safety* book series. Where previous volumes focused on established methods, including trends and new developments, the present volume has little to look back on in the past. In contrast, it is very future oriented. As mentioned in some chapter introductions of this volume, the Industry 4.0 revolution has caused a paradigm shift of methods. The term "Industry 4.0" was first coined in 2011, by Wolfgang Wahlster, Director and CEO of the German Research Center for Artificial Intelligence. The term obtained worldwide fame when Klaus Schwab, economist and engineer, founder and executive chairman of the World Economic Forum, produced in 2016 the book *The Fourth Industrial Revolution.* This was also the year the European branch of the Center for Chemical Process Safety organized its first conference in Frankfurt on big data and analytics. At that conference, the rough contours of what would come became visible. It didn't take long when the term "Industry 4.0" was followed by Safety 4.0. And that was not without reason. Prediction of hazards and risks for years had struggled with lack of data. Methods had been there, as systematic development of theory and experimental methods on process safety started after World War II, later followed by system-oriented methods, such as hazard and operability study, and, in the late 1970s, risk assessment. The latter stimulated field trials and development of computational methods, in particular computational fluid dynamics, to predict physical effects but also to support maintenance via risk-based inspection and other. However, the probability of possible scenarios, equipment failure, and human error remained difficult to predict. This was largely due to lack of possibility of making effective use of databases and the generation and processing of statistical data.

Independently, and starting even much earlier, the development of fundamental knowledge on how to treat observational data statistically focused on the classical Fisher approach and the more recent versatile Bayesian one. Later, in the 1960s and 1970s, the ambition of artificial intelligence (AI) with a view on robotic decision-making stimulated development that became increasingly stronger. It led to several products, such as expert systems and case-based reasoning, but also using graph theory and Pearl's cause-effect thinking in the Bayesian network. The latter caused an upswing in risk assessment as a result of enabling uncertainty to be quantified in a simulated complex causal structure leading to an unwanted event. Further, various data processing algorithms were developed, in particular over the last decade, enabling classification of data containing random noise with machine learning techniques and pattern recognition in all its different forms with the ever-further differentiating AI techniques. At the same time, computational capabilities (software, Internet of Things, cloud computing) and sensor availability increased tremendously, which enabled large-scale digitalization.

This sixth volume of the series provides, in a series of chapters, an overview of what this will mean for the process industry in plant design and operation, also with respect to effectiveness, efficiency, and resilience. With the new tools, the perspective of obtaining data enhances the accuracy of risk models, enabling dynamic monitoring of safety performance level and of equipment condition in view of maintenance—especially with respect to asset integrity and barrier effectiveness. And further, the tools are enabling fault detection and diagnosis, digital twin development that helps design, operation and operator training, and acquiring safety management information supporting risk-informed decision-making.

As every advantage is accompanied by some challenges, our strong reliance on the worldwide internet poses security threats such as data theft, abuse, misleading information and other interferences, and disruption. To prevent and negate hacking and intruder action, many new techniques are being developed. The interaction of cybersecurity with process safety is being analyzed.

We are grateful to all the authors who helped to create this volume. They were prepared to interrupt their research and educational work in this new and fascinating field to write a chapter. Many of them have worked under quite some stress to meet the deadline but, in the end, we made it. We also thank the Elsevier staff who have been responsive to our timing problem.

Faisal Khan
TEES Mary Kay O'Connor Process Safety Center
Texas A&M University System, College Station, TX, United States

Hans Pasman
TEES Mary Kay O'Connor Process Safety Center
Texas A&M University System, College Station, TX, United States

Ming Yang
Safety and Security Science Section, Department of Values, Technology, and Innovation, Faculty of Technology, Policy, and Management, Delft University of Technology, The Netherlands

CHAPTER ONE

Opportunities and threats to process safety in digitalized process systems—An overview

Hans Pasman[a,*], Hao Sun[b,c], Ming Yang[b,*], and Faisal Khan[a,*]
[a]TEES Mary Kay O'Connor Process Safety Center, Texas A&M University System, College Station, TX, United States
[b]Safety and Security Science Section, Department of Values, Technology, and Innovation, Faculty of Technology, Policy, and Management, Delft University of Technology, Delft, The Netherlands
[c]College of Mechanical and Electronic Engineering, China University of Petroleum (East China), Qingdao, China
*Corresponding authors: e-mail address: hjpasman@gmail.com; m.yang-1@tudelft.nl; fikhan@tamu.edu

Contents

1. Definition of digital process, digitalization, and safety

Several developments in the process industry, such as optimal use of energy, more complex processes and process conditions, and larger flexibility for product adaptation, with a work force that is more mobile, ask for a higher degree of automation. To adapt to this change and ensure production safety, digital technologies provide a potential way to improve production efficiency and reduce the likelihood and severity of industrial accidents.

The first push for digital computing in the process industry came in 1949 through IBM's industrial computing seminars (Lee, Cameron, &

Methods in Chemical Process Safety, Volume 6
ISSN 2468-6514
https://doi.org/10.1016/bs.mcps.2022.05.007

Hassall, 2019). The exponential increases in computing power and software development over the past 70 years have been the mainstay of the development of high-fidelity dynamic modeling of entire process plants (Lee et al., 2019). Fig. 1 shows the main development over the last 80 years.

Digital technologies, which are the bread and butter of Industry 4.0, have been used in many fields such as ecology, economy, engineering, and process system (Single, Schmidt, & Denecke, 2019). A clear definition of digital terms is the first step in determining its opportunities and threats in process systems.

Digital process refers to the basic process of transforming much complex and changeable information of a whole system into numbers and data that can be measured and then using these numbers and data to build an appropriate digital model, introducing them into the internal computer for unified processing (Porthin, Liinasuo, & Kling, 2020). Several authors have expressed their views in this regard in the available literature. However, there is no accepted definition in the academic community so far. There is a good example in process safety to describe the digital process. Single et al. (2019) developed an "semi-automated HAZOP" system to support a HAZOP team to identify potential hazards of a process plant. The information of Process Flow Diagrams (PFDs) and Piping and Instrumentation Diagrams (P&IDs) is exacted and used to represent the process plant. The nodes in PFDs and P&IDs can be modeled by a specific modeling language and a graphical editor. In the light of this, deviations in process variables can be automatically applied and propagated through the process system, detecting potential faults and hazardous events by employing inference methods. In this process, all information, including the information of the process plant and expert knowledge, is transformed into data processed by computers.

Digitization or *digitalization* is defined as the integration of digital technologies in process operations for greater efficiency and increased product quality (Kayikci, 2018; Khan, Amyotte, & Adedigba, 2021; Vaidya, Ambad, & Bhosle, 2018). Digitalization generally means encoding various information (e.g., images, voices, videos, data, etc.) into zeros and ones for better storage, processing, and transmission of information (Khan et al., 2021). The digitalization process comprises the increased use of robotics, automation solutions, and computerization, reducing costs, improving efficiency and productivity, and responding flexibly to changes.

The digitalization in process industries consists of two main parts: (i) physical digitalization (i.e., automation of equipment, holistic simulation

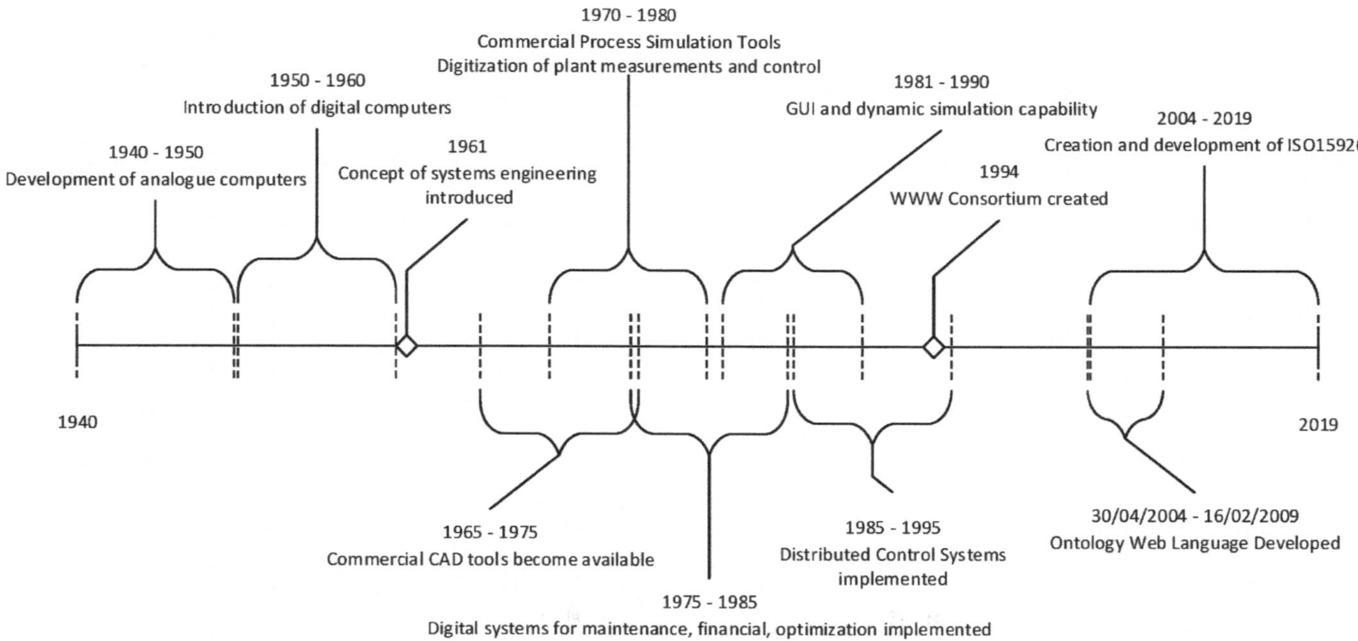

Fig. 1 The development of digitalization in process systems. *Adapted from Lee, J., Cameron, I., & Hassall, M. (2019). Improving process safety: What roles for Digitalization and Industry 4.0?* Process Safety and Environmental Protection, 132, *325–339. doi:10.1016/j.psep.2019.10.021.*

of process systems), (ii) information digitalization (i.e., automation of data collection, processing, and analysis). In the field of process safety, the latter is of greater interest (Lee et al., 2019). The essence of digitalization is to use the abundant data in systems to solve the uncertainty of complex systems. Digitalization enables digital twin, which focuses on the simulation of the overall system to optimize the system structure and verify the effectiveness of existing operating strategies, and collecting, storing, and processing a large amount of monitoring data and production data generated in systems. The main reasons for the growing interest in digitalization are exponential growth in technology development and the fact that data is becoming richer from more sources, real-time with higher data rates, more complex, and more useful.

Digitalization depends on data, and data relies on effective collection, and for reliable use, data should be keept cybersecure, data preparation and effective analysis is required to distil information from them, and in the end, their value would show up by useful application. The analysis focuses on identifying problems and analyzing them based on data and the main reasons behind the generation of such data to monitor the operations and status of process systems and develop strategies and measures for optimization.

Safety in the process industry can be defined as the ability or state of a system to be free from undesired accident risk during the production process (e.g., Hollnagel, 2008). Safety and risk are closely linked. Risk is a composite measure of the probability of an undesired accident and the corresponding consequences. From a conceptual point of view, reducing the accidents' probability and consequence can decrease sytem risk which will improve systems' safety. Early warning (Chang, Khan, & Ahmed, 2011; Schmitz, Swuste, Reniers, & van Nunen, 2020), fault detection and diagnosis (Fazai, Mansouri, Abodayeh, Nounou, & Nounou, 2019; Kopbayev, Faisal, Yang, & Halim, 2022), and safety barriers (e.g., physical barriers and non-physical barriers) can be used to improve the safety of process systems.

2. Brief history of process safety and reasons why digitalization can support process safety

2.1 A brief history of process safety

Process systems, and, in particular, chemical ones store and process a large amount of hazardous materials, which by explosion blast, radiant heat, or toxic substance exposure may lead to casualties, property damages, and ecological pollution (Abbassi, Khan, & Garaniya, 2015; Benson, Dimopoulos, & Argyropoulos, 2021; Khan, Wang, & Yang, 2016; Sun, Haiqing, Yang, & Reniers, 2021). Quite a few compilations of accidents that occurred are available, e.g., Marsh, 100 Largest Losses in the Hydrocarbon industry 1974–2019 (Marsh, 2020), while several countries maintain accident data bases, e.g., in Europe the JRC eMARS database. As a typical accident the Amuay refinery disaster in Venezuela can be mentioned causing more than 50 people dead, over 100 people injured, and about 1600 buildings destroyed, resulting in $1 billion in economic loss (Mishra, Wehrstedt, & Krebs, 2014). There have been many more, some more serious, others less.

Process safety plays a critical role in safety operations during production processes. It is identified as an integral part of process development and manufacturing rather than being viewed as an "add-on" to the process (Gibson, 1999). However, safety was not a high priority in the 1940s. At that time, the full-time safety worker were older foremen, retired army officers, and men (no women then) with non-technical backgrounds and experience (Kletz, 2012, see also, Swuste, Van Gulijk, & Zwaard, 2010). At British industrial pioneer Imperial Chemical Industries (ICI), in the late 1960s, only after several major safety incidents in a row, management decided that the safety work should be done by people with relevant knowledge and technology. In 1963, the method of Hazard and Operability (HAZOP) study started its development at ICI to identify the hazards and determine the potential equipment failures (Kletz, 1999, 2012). It is worth noting that late Trevor Kletz was the promotor of HAZOP within ICI and beyond. Besides, he advised and trained workers to conduct the HAZOP process. Since then, although the scale of process industries has doubled in the 1970s, the rate of fatal accidents in the process system has dropped significantly (Kletz, 1999), which can be seen in Fig. 2.

The origin of the term "process safety" and its evolution is related to the major accidents that occurred between 1960 and 1990 as a result of rapid industrialization and technological developments (Khan, Rathnayaka, & Ahmed, 2015). In other words, the driving force behind the movement to foster process safety and to regulate the industry is the unwavering occurrence of major accidents. For example, the Flixborough aerosol cloud explosion accident in the UK in June 1974 led to the death of 28 workers and injured 36 staff, besides destroying the plant and damaging nearby residential area. This disaster promoted in the same year the creation of the Advisory Committee on Major Hazards by the Health and Safety

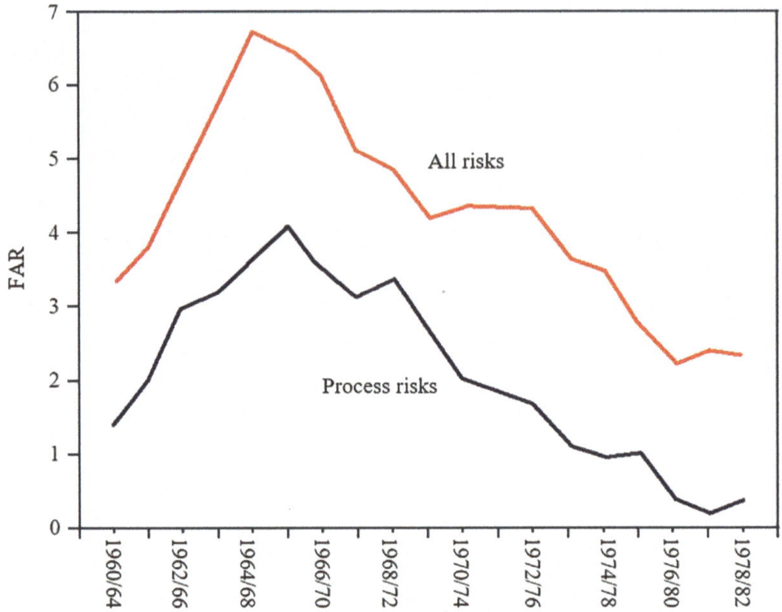

Fig. 2 Fatal accident rate from 1960 to 1982. *Adapted from Kletz, T. (2012). The history of process safety.* Journal of Loss Prevention in the Process Industries, 25, *763–765. doi:10.1016/j.jlp.2012.03.011.*

Committee in the United Kingdom, which itself was based on Health and Safety at Work etc. Act 1974.

Those developments were followed in continental Europe and the United States. But the engineering associations became already active earlier by organizing symposia to exchange experiences and to learn from each other. The earliest are the AIChE Loss Prevention symposia in the United States starting in 1967, while the IChemE 1971 Major Loss Prevention in Newcastle UK became the initiating gathering for the EFCE Loss Prevention symposia in Europe with the first one in 1974. In Table 1 a number of shocking accidents have been listed, which gave rise to changes in thinking and approaches to safety.

Economically the process industry was booming as the demand for its products grew as again seen in Japan, later in the early 1990s in the Middle-East followed by many Asian countries, notably China, and India. The hazardous nature of the substances (flammable, explosive, toxic) in the system makes it possible to cause severe consequences in the event of an accident. Especially, for toxic materials, in the event of a leakage, the consequences would be unacceptable. For instance, after the TCDD spreading

Table 1 The development of process safety influenced by major incidents.

Accident	Type	Time	Number of death and injury	Corresponding organization	Formed regulation	Reference
Flixborough, United Kingdom	Explosion	1974	28, 36	Advisory Committee on Major Hazards	New UK regulations for the control of industrial major accident hazards (CIMAH)	Mannan, Chowdhury, and Reyes-Valdes (2012)
Seveso, Italy	Loss of containment	1976	Severe pollution	The European Chemical Industry Council	Seveso I, II, and III	Hay (1977)
Bhopal, India	Loss of containment	1984	3000–20,000, hundreds of thousands were injured	Center for Chemical Process Safety (CCPS)	Environmental Policy Act; Air Act; Hazardous Waste (Management and Handling) Rules; Public Liability Insurance Act; Environmental Protection (Second Amendment) Rules	Mannan et al. (2005)
Pipe Alpha, United Kingdom	Loss of containment	1988	167, –	–	Safety case regulations; Permit-to-work system	Mannan et al. (2005)
Exxon Valdez Spill, United States	Loss of containment	1989	Severe pollution	–	Oil pollution Act	Khan (2007)

Continued

Table 1 The development of process safety influenced by major incidents.—cont'd

Accident	Type	Time	Number of death and injury	Corresponding organization	Formed regulation	Reference
Phillips 66, United States	Explosion	1989	23, hundreds	Chemical Safety and Hazard Investigation Board (CSB); Mary Kay O'Connor Process Safety Center (MKOPSC)	Clean Air Act Amendments, with the OSHA PSM regulation and later the RMP rule of EPA	Mannan et al. (2012)
BP Texas City, United States	Fires and explosions	2005	15, 200	Occupational Safety and Health Administration (OSHA)	Revision of the American Petroleum Institute's (API's) Recommended Practice (RP) 752	Kaszniak and Holmstrom (2008)
T2 Explosion, United States	Explosion	2007	4, 32	Accreditation Board for Engineering and Technology	Relevant evaluation criteria	Theis (2014)
Deepwater Horizon, United States	Explosion	2010	11, –	Bureau of Safety and Environmental Enforcement (BSEE); Bureau of Ocean Energy Management; Center for Offshore Safety; Ocean Energy Safety Institute	Relevant management criteria	Skogdalen, Khorsandi, and Vinnem (2012)
West Fertilizer Explosion, United States	Explosion	2013	15, 160	–	President Obama's Executive order to screen/improve regulation	Mannan, Reyes-Valdes, Jian, Tamim, and Ahammad (2016)

Seveso runaway accident, to avoid further dioxin contamination, more than 80,000 animals had to be exterminated, censored thousands of people in their activities, and allowed abortions based on mothers' decisions. To avoid the occurrence of major accidents, institutions were developed and regulations made. It can be seen in Table 1 that each major incident has facilitated the development of process safety, including promoting the establishment of relevant institutions and laws and regulations.

The focus of research on process safety has varied over time as systems and laws, and regulations have evolved. As mentioned above, in the 1960s, safety was not a concern until major accidents occurred. In order to prevent accidents, people with technical background were tasked to carry out relevant work. In the beginning, process safety focused on technical problems, which can be seen in Fig. 3. Accidents were seen as being caused by equipment failure. In quite a few cases the mechanism of a cause chain resulting in a severe consequence was not clear. Type denotations, such as reactor run-away and vapor cloud explosion, had to be invented yet. However, the call for proactive approach became stronger and Trevor Kletz's HAZOP, to identify already at the design stage potential technical failures of process systems became famous (Gowland, 2012). Also around 1974 following the example of the nuclear power industry, the concept of risk asessment emerged with the Canvey Island (UK, Canvey, 1978) and Rotterdam Rijnmond (Nl, COVO, 1982) studies. These studies recognized the problem of lack of knowledge on equipment failure data,

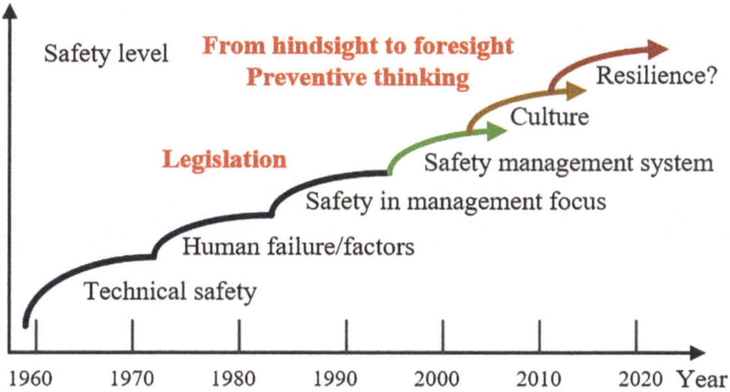

Fig. 3 Major contributions to the evolution of process safety presented in 1995 at the 8th European Loss Prevention symposium in Antwerp by Koos Visser, process safety pioneer at Shell, and since then stepwise expanded by Hans Pasman (Pasman & Fabiano, 2021).

consequence analysis models, and the importance of human factors in the safety performance of the process industry. A wave of activities in the 1980s and 1990s started to gather data by collecting equipment failure ones, by experiments and field trials, to develop models and to expand knowledge. Human factors led to significant developments in management and behavioral sciences, as well as advances in systems and cultural approaches (Khan et al., 2021). In this century, due to better insights and more complex process systems, more attention has been paid to socio-technical system approach, in which human factors, technical factors, and organizational and managerial factors interact. Note that since the Chernobyl nuclear reactor disaster in 1986, safety culture is recognized as important and over the years is taken into account in organizational and managerial factors. The Phillips Petroleum company 1989 dramatic vapor cloud explosion near Houston, US, gave the impetus to US Occupatioal Safety Administration (OSHA) to launch the Process Safety Management (PSM) requirements, which led further to safety management systems (SMS). The effect of these introductions of concepts appeared industry sometimes only many years later (Fig. 3). Nowadays, as digitalization proceeds, process systems belong to the cyber-physical system, which means more focus should be on managing process safety information and communicating these experiences as knowledge. Meanwhile, the impacts of digitalization on process safety require more discussion.

2.2 Why can digitalization support process safety?

To prevent accidents, scholars developed various risk assessment (RA) method variants to reduce the probability of an accident and mitigate accident consequences (Khan, Khan, & Veitch, 2020; Khakzad, 2019; Landucci, Argenti, Cozzani, & Reniers, 2017; Sultana, Anderson, & Haugen, 2019; Yang, 2018). RA plays an essential role in understanding the mechanism of accidents and ensuring system safety. Following Sam Mannan's thougths O'Connor, Pasman, and Rogers (2019) proposed three crucial elements of safety, namely, prevention, mitigation, and response, which can be integrated as a so-called safety triad. These three factors may seem simple, even intuitive. However, reports and investigations of accidents have proven that the main cause of accidents was the lack of these three factors. To determine their effectiveness methods, such as fault and event tree have been devised, which over the years became further developed. An example is bowtie which combines fault and event tree and shows besides scenarios also preventative and protective barriers. For quantitative evaluation of the probabilities of basic events to final major consequences bowties can be easily converted into Bayesian networks (see, e.g., Khakzad, Khan, & Amyotte, 2013). Because there may be interdependency of events and barriers, Ghosh, Ahmed, Khan, and Rusli (2020) utilized copula-based Bayesian network (BN), as well as traditional BN to assess the failure of the multivariable time-dependent system. Sun, Wang, Yang, and Reniers (2020) developed an integrated approach based on the window of opportunity and complex network to evaluate the risk of a process system.

The works described above show the significant progress on RA in process systems. Nevertheless, recurring accidents show that relying on conventional RA alone and following its recommendations is not enough to ensure system safety (Marsh, 2018). Marsh (2018) indicates the losses caused by accidents have not been reduced. In fact, semi-quantitative or quantitative risk assessment is not applied taking all possibilities and parameters into account, and moreover the used models and data contain large uncertainties and the analysts are limited in their knowledge and awareness as, e.g., well described by Rae, Alexander, and McDermid (2014). Nowadays, complex systems, which arise from non-linear interdependencies, are built rapidly to meet people's demand. The ensuing uncertainty, complex interaction, and interdependence between components (e.g., human, technical, and organizational elements) have become new risk factors in the process system.

These changes make it difficult to assess reliability, detect and diagnose faults, and determine the relationship between process parameters and system state. For example, deterioration and corrosion of equipment (such as valves) is a random process influenced by process parameters and "non-equipment factors" such as human factors, technical factors, environmental factors, etc. It is difficult to use the abovementioned traditional RA methods to ensure system safety under those conditions.

Automated and digital systems generate a large amount of data, leading to new opportunities for process safety and asset integrity assessment. Lee et al. (2019) suggest digitalization of chemical process industries could improve the mechanical issues by introducing early warning signaling, corrosion monitoring, remote sensing, increased connectivity, application of predictive models relying on real-time data, and machine learning. Moreover, they suggest digital systems could improve work processes, risk assessment, operator interfaces, and alarm management systems. In addition, digitization benefits from generating digital operational data and replacing manual operations with software that automates data collection, processing, and analysis for effective process monitoring.

Specifically, the benefits of digitalization are mainly reflected in five aspects: productivity, production quality, safety, efficiency, and flexibility (Hole, Hole, & MaFalone-Shaw, 2021). For example, digitalization can reduce labor and save costs. A high degree of automation can increase productivity while ensuring product quality and system safety. Digitalization provides novel and more efficient measures to improve the manufacturing process and safety system (e.g., safety barrier system). In practice, much information on site needs to be recorded by workers and fed back to the control room, which wastes a lot of time recording and feeding back, and increases the probability of human error. It is not conducive to safe and efficient production. Fortunately, digitalization can solve this problem. Besides, digitalization can help to improve control systems, which enables every process of production to be monitored effectively. This ensures the safety of the system and the quality of the product. For example, the boiled-off gas (BOG) is inevitably generated at the LNG receiving terminal during unloading, storage, and delivery. An effective control system can reduce the amount of BOG and thus save energy from liquefied BOG (Animah & Shafiee, 2020).

For processes, safety is a most critical concern. The benefits of digitalization for safety can ensure system safety by performing fault detection and diagnosis, risk assessment, and safety control. Reinartz, Kulahci, and Ole (2021) developed an extensive reference dataset, incorporating repeat simulations of healthy and faulty process data, additional measurements, and multiple magnitudes for all process disturbances. All six production modes of the Tennessee Eastman process (TEP) process control testbed as well as mode transitions and operating points in a region around the modes, are simulated. Besides, fault detection is conducted based on principal component analysis integrated with T^2 and Q charts using average run length as a performance metric to provide an initial benchmark for statistical process monitoring schemes for the presented data. Wu and Zhao (2021) presented a process topology convolutional network (PTCN) approach to conduct fault diagnosis of a chemical process system. The proposed method was validated by experiments on the benchmark Tennessee Eastman process (TEP), and results showed that PTCN improved the fault diagnosis accuracy. Cheded and Doraiswami (2021) developed a comprehensive approach, including model-free (MFA) and model-based (MBA) methods, for fault detection and isolation with in process system. Deng, Han, Cheng, et al. (2022)

proposed a real-time fault detection approach, which comprises space-time compressed matrix (STCM) and Naive Bayes (NB), to realize the fast learning and prediction in chemical process system. Experiments on the TEP show that the proposed approach reduces the sample size and feature size by 75% and 92%, respectively. Bi and Zhao (2021) presented a novel orthogonal self-attentive variational autoencoder (OSAVA) model, which includes orthogonal attention (OA) and variational self-attentive autoencoder (VSAE), to monitor the process system. OA is utilized to extract the correlations between different variables and the temporal dependency among different timesteps; VSAE is trained to detect faults through a reconstruction-based method, which employs self-attention mechanisms to comprehensively consider information from all timesteps and enhance detection performance. Arunthavanathan, Khan, Ahmed, and Imtiaz (2021) developed an early potential fault detection approach, including the Convolutional Neural Network (CNN) and Long Short-Term Memory (LSTM), to examine fault symptoms in chemical process systems. The performance of the proposed approach was validated by TEP again and indicated that the proposed method is efficient in finding out potential faults in chemical process systems. This will all be further detailed in Chapter 6.

3. Process safety treated in digitalized process systems

The benefits of digitalization in process systems are the generation of digitized operational data and the replacement of manual process operations with software that allows automated data collection for effective process monitoring (Khan et al., 2021). Digitalization can help systems predict uncertain disruptions, monitor, and control the process of systems, detect and diagnose faults, manage abnormal situations, and process massive amounts of data automatically. Therefore, the productivity, flexibility, and quality of process systems can be enhanced. However, new problems arise when most operations are digitally (e.g., sensor, controller, processor) dependent.

3.1 Human and management errors

The analysis of the causes of the 100 major accidents in the area of onshore oil, gas, and petrochemical from 1996 to 2015 was conducted by Jarvis and Goddard (2017). This study revealed major system failure causes: mechanical failures accounted for 43% of losses; operations, practices, and procedures made up 25%; work control accounted for 21%, etc. The above statistics illustrate the impact of management system failures on accident occurrence. It should be noted that nearly 60% of mechanical failures were due to management issues, such as inadequate inspection programs. Besides, hazard identification is involved in more than half of losses due to human errors.

Even though digitization is beneficial for reducing operators' workload and involvement time, it increases the complexity of human-computer interaction. This is because the screens, displayed data, and buttons in the control room will become more numerous and complex. Moreover, the system generates and automatically processes a large amount of data, and while most decisions are made automatically by the system, the decisions of the people in the control room remain also critical to ensuring the safety of the system. In addition, managers and operators are prone to laxity because they may believe all work can be done by digitalized equipment. Situation awareness, safety awareness, safety culture, management policy, etc., may decrease with the development of digitalization. This observation can be supported by an accident on December 11, 2005, at the Buncefield storage depot in the United Kingdom. The main cause of the overfilling of the gasoline (petrol) storage tank was the failure of the automatic tank measurement system while the control room was manned at the time (Paltrinieri, Øien, & Cozzani, 2012).

3.2 Process monitoring and control

Due to digitalization, equipment, including sensors, indicators, detectors, controllers, and valves of main process parameters (e.g., temperature, pressure, flow rate, etc.), are more automated. This means that fewer devices require human involvement. Although it reduces human errors and time, it lmay loose human as a vital back-up safety barrier. For example, there are also manual detectors in conventional process systems in addition to automatic detectors. When automatic detectors fail, regular manual detection can assist in the prevention of major accidents. However, digitalization may abandon most equipment that requires human involvement, which may increase systems risk if the reliability of the system is not sufficiently increased.

Process monitoring allows for fault detection and diagnosis by monitoring important process parameters, which is beneficial for ensuring system safety. However, relying entirely on automated monitoring cannot fully secure the system. For example, under normal conditions the safety instrumentation system (SIS) is dormant, and it only works when an accident occurs and demands it to function. Although faults can be signaled in part automatically, in particular of the electronics, for this type of equipment, human involvement of inspecting and testing will remain essential.

3.3 Massive data processing

With the process of Industry 4.0, process systems have become more automated and intelligent, which means that monitoring systems, control systems, and operation systems produce massive data (Pasman & Fabiano, 2021). Adequate use of those data is beneficial for early warning, fault detection, diagnosis of the system, etc. However, due to limited time and resources, it is challenging to analyze all of those data. To overcome the challenge, researchers paid more attention to developing a targeted approach to handling massive data. These methods are categorized as the data-driven method.

One of the representative methods among data-driven methods is machine learning. Machine learning consists of two types of methods: unsupervised learning and supervised learning.

Unsupervised learning is relevant when one wants to find out whether the data contain a pattern which at that moment is unknown. The technique includes clustering algorithms (e.g., K-means clustering) and dimension reduction algorithms (e.g., principal component analysis (PCA)). PCA is widely used to reduce the dimensionality of data by exacting the main features from the massive data. Ji, Jiao, Yuan, et al. (2021) employed PCA to analyze high dimensional data to assess the combustion risk of flammable liquids. Li, Hu, Gao, et al. (2021) Li, Jia, Zhang, et al. (2021), Li, Liu, Lin, et al. (2021), Li, Zhang, Khan, and Han (2021), Li, Zhou, and Wang (2021) developed a data-driven model, including PCA, artificial bee colony algorithm (ABC), and support vector regression (SVR). In this model, PCA is employed to reduce the dimension of corrosion influencing factors. The obtained primary components are selected as the input variables of the model. Kopbayev et al. (2022) combined Kernel principal component analysis (kPCA) and deep neural network (DNN) to perform fault detection and diagnosis in process system. In the proposed method, kPCA is used to reduce the dimensionality of the complex data. After this, the processed data is utilized for training DNN for detection and diagnosis. Amin, Khan, Ahmed, and Imtiaz (2021) presented a comprehensive method, including PCA and Bayesian network, for fault diagnosis. Wu, Li, and Li (2021) utilized the Principal Component Analysis (PCA) in combination with Support

Vector Machine (SVM) to classify the fault for complex process systems. The results show that the developed approach is efficient for fault detection and diagnosis. Harrou, Nounou, Nounou, et al. (2013) presented a PCA-based GLR fault detection method to detect faults in different process variables without a process model. Kaced, Kouadri, Baiche, et al. (2021) used the PCA to solve the problem of false alarms in a chemical process.

Supervised learning learns the underlying class distinctions or trends from a training data set. It comprises a classification algorithm and regression algorithm. K Nearest Neighbor is a typical classification method. Besides, classification algorithm includes Decision Tree, Support Vector Machine (SVM), Logistic Regression, Random Forest, etc. depending on whether the problme is linear or not and other. Meanwhile, the regression algorithm comprises Linear Regression, Least Square Regression, Artificial Neural Network (ANN), etc. In process safety, one of the most used methods is ANN. ANN is composed of three parts: neurons, layers, and networks. A typical ANN has three layers, the input layer, hidden layer, and output layer. The basic neurons are connected by weights between the input and hidden layers and the hidden layer and output. Put another way, the connection only exists between layers, and there is no connection within a layer since the information flows in one direction. It can be optimized based on new data and information. Due to these advantages of ANN, it is used in many research domains, such as risk analysis and fault detection and diagnosis. Adedigba, Khan, and Yang (2018) utilized multi-layer perceptron (MLP) and probability analysis to evaluate the safety of the process system. The non-linear relationships among process variables are determined by the MLP. Ayhan and Tokdemir (2019) developed a comprehensive approach, including latent class clustering analysis (LCCA), ANN, and case-based reasoning (CBR), for identifying potential accident scenarios. Sarbayev, Yang, and Wang (2019) utilized ANN to overcome the limitations of FT to quantify the risk of the process system. Li, Hu, et al. (2021), Li, Jia, et al. (2021), Li, Liu, et al. (2021), Li, Zhang, et al. (2021) and Li, Zhou, and Wang (2021) integrated computational fluid dynamics (CFD) with a general regression neural network (GRNN) to evaluate the rescue risk in explosion accidents. Li, Hu, et al. (2021), Li, Jia, et al. (2021), Li, Liu, et al. (2021), Li, Zhang, et al. (2021), and Li, Zhou, and Wang (2021) proposed a hybrid approach, which is composed of KPCA and BRANN, for predicting corrosion degradation of offshore oil pipelines. KPCA is utilized not only for decreasing the dimensionality of the factors affecting pipeline corrosion and for extracting principal features from massive data.

Meanwhile, BRANN is formed as a prediction model. Mamudu, Khan, Zendehboudi, and Adedigba (2021) proposed a comprehensive method, which consists of a multilayer perceptron–artificial neural network (ANN) and BN, to assess the risk of the system. Many of these methods will appear in several chapters of this volume.

Although massive data processing methods have undergone significant improvements globally over the past couple of decades, this has not translated into a substantial reduction in major accidents in process industries. The inherent characteristics of industrial system data create numerous challenges for effective process fault diagnosis (Khan et al., 2021). Those data are complex, nonlinear, highly time-variant, and non Gaussian in distribution. Therefore, it is also a challenge to extract important information from those complex data and to accurately perform fault detection and diagnosis.

3.4 Cyber-attacks

The rapid development of digitalization has brought about a strong increase of digitized process systems full of sensors, processing electronics, and often connected wireless via IIoT (the industrial internet of things), which inevitably creates new hazards and risks, like internet cyber-attacks. According to different attack motivations, attackers, hence hackers, can be divided into four categories: terrorists motivated by political gain and revenge, activists inspired by rebellion, disgruntled employees and contractors motivated by money and revenge, and criminals motivated by money (Iaiani, Tugnoli, Bonvicini, & Cozzani, 2021). Cyber-attacks are targeted, which means that they can bring about serious consequences. Moreover, due to the high level of automation and the ongoing digital transition (e.g., increased use of automated sensors, detectors, controllers, diagnostics, digital communications, wireless connections, the interconnection between control and safety instrumented systems, connections to external networks), incidents caused by cyber-attack are more frequent than before. The cybersecurity-related incidents in different industrial sectors are shown in Fig. 4. It is worth noting that the incidents in chemical and petrochemical account for 54.87%. This is because the cyber-attacks on chemical and petrochemical systems can result in severe impacts, which means it is more attractive for cyber-attacks than other areas. Therefore, there is a need to address the impact of cyber attacks on industrial sector facilities (such as nuclear power plants, water and food plants, chemical plants and oil refineries).

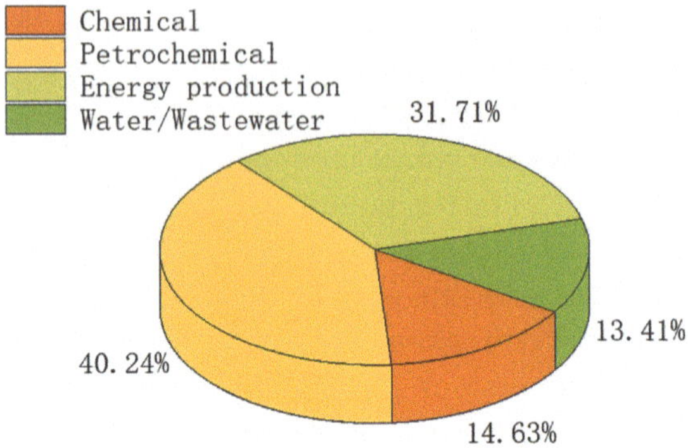

Fig. 4 The cybersecurity-related incidents in different industrial sectors.

4. Purpose and organization of this MCPS Volume 6

This book is the sixth volume of the *Methods in Chemical Process Safety* book series. This volume aims to provide state-of-art progress of digitalization and corresponding opportunities and threats in process safety. Chapter "State-of-the-art in process safety and digital system" by Amin et al. will do that in more detail. Chapter "Data-driven approaches: Use of digitized operational data in process safety" by Bai et al. details the application of data-driven approaches to process safety, including process monitoring, dynamic and operational risk assessment, and reliability modeling/predictive maintenance. Chapter "Industry 4.0 based process data analytics platform" by Wanasinghe et al. briefly presents Industry 4.0 based process data analytics platform, comprising data acquisition, IoT technologies, machine learning, and big data. Chapter "Digital process safety management" by Slezak et al. summarizes specific details on available process safety management (PSM) and safety management system. It discusses the advantages and disadvantages of digital (PSM and SMS). Chapter "Statistical approaches and artificial neural networks for process monitoring" by Alauddin et al. discusses statistical approaches and artificial neural networks for process monitoring for fault detection and diagnosis. Chapter "Alarm management techniques to improve process safety" by Yang et al. will be on alarm management, which in many cases is the start of abnormal situation management in digitalized process systems. It provides the details of conventional and advanced alarm system, including alarm signal processing, alarm prioritization and control, and alarm response procedures. Chapter "Performance evaluation of digitalized safety barriers" by

Zhang and Liu will be on smart safety instrumented system (SIS) and performance evaluation of digitalized safety barriers. Chapter "Dynamic operational risk assessment in process safety management" by Li et al. explains advantages of dynamic operational risk assessment in process safety management. Besides, in this chapter the methods of risk assessment and dynamic risk assessment are briefly discussed. Next, in chapter "Risk of cascading effects in digitalized process systems" by Iaiani et al. the risk of cascading effects and uncertainty modeling in risk assessment of digitalized process systems are provided. Chapter "Uncertainty modeling in risk assessment of digitalized process systems" by Yazdi et al. treats the uncertainty in risk assessment outcomes. Chapter "Human factors in digitalized process operations" by Srinivasan et al. describes the effect of human factors in digitalized process operations. Chapter "Safety assessment of complex socio-technical systems" by Paltrinieri explains the safety assessment of complex socio-technical systems, while chapter "Security of digitalized process systems" by El-Kady et al. will provide the state of play of security measures in digitalized process systems. Chapter "Integrated dynamic risk management in process plants" by Taleb-Berrouane and Pasman discusses aspects of the integrated dynamic risk management in process plants, including safety and security interactions and economic aspects. A more detailed treatment of application of digital twins in process safety management is provided in chapter "Use of digital twins for process safety management" by Keprate and Bagalkot. Chapter "Resilience analysis of digitalized process systems" by Yarveisy et al. discusses the details of resilience analysis of digitalized process systems, comprising various definitions and evolving methods. Finally, chapter "Risk assessment in Industry 4.0" by Amin and Khan shows and summarizes risk analysis in the Industry 4.0 time frame and concludes this volume.

References

Abbassi, R., Khan, F., Garaniya, V., et al. (2015). An integrated method for human error probability assessment during the maintenance of offshore facilities. *Process Safety and Environmental Protection*, *94*, 172–179. https://doi.org/10.1016/j.psep.2015.01.010.

Adedigba, S. A., Khan, K., & Yang, M. (2018). An integrated approach for dynamic economic risk assessment of process systems. *Process Safety and Environmental Protection*, *116*, 312–323.

Amin, M. J., Khan, F., Ahmed, S., & Imtiaz, S. (2021). A data-driven Bayesian network learning method for process fault diagnosis. *Process Safety and Environmental Protection*, *89*, 110–122. https://doi.org/10.1016/j.psep.2021.04.004.

Animah, I., & Shafiee, M. (2020). Application of risk analysis in the liquefied natural gas (LNG) sector: An overview. *Journal of Loss Prevention in the Process Industries*, *63*(103), 980. https://doi.org/10.1016/j.jlp.2019.103980.

Arunthavanathan, R., Khan, F., Ahmed, S., & Imtiaz, S. (2021). A deep learning model for process fault prognosis. *Process Safety and Environmental Protection, 154*, 467–479. https://doi.org/10.1016/j.psep.2021.08.022.

Ayhan, B. U., & Tokdemir, O. B. (2019). Safety assessment in megaprojects using artificial intelligence. *Safety Science, 118*, 273–287.

Benson, C., Dimopoulos, C., Argyropoulos, C., et al. (2021). Assessing the common occupational health hazards and their health risks among oil and gas workers. *Safety Science, 140*, 105284. https://doi.org/10.1016/j.ssci.2021.105284.

Bi, X. T., & Zhao, J. S. (2021). A novel orthogonal self-attentive variational autoencoder method for interpretable chemical process fault detection and identification. *Process Safety and Environmental Protection, 156*, 581–597. https://doi.org/10.1016/j.psep.2021.10.036.

Canvey. (1978). *An investigation of potential hazards form operations in the Canvey Island/Thurrock area, Health and Safety Executive, London, 1978.* Her Majesty's Stationary Office, ISBN:011883200X.

Chang, Y., Khan, F., & Ahmed, S. (2011). A risk-based approach to design warning system for processing facilities. *Process Safety and Environmental Protection, 89*, 310–316. https://doi.org/10.1016/j.psep.2011.06.003.

Cheded, L., & Doraiswami, R. (2021). A novel integrated framwork for fault diagnosis with application to process safety. *Process Safety and Environmental Protection, 154*, 168–188. https://doi.org/10.1016/j.psep.2021.08.008.

COVO. (1982). COVO Commission, risk analysis of six potentially hazardous industrial objects in the rijnmond area, a pilot study. In *A Report to the Rijnmond Public Authority, Central Environmental Control Agency, Schiedam, The Netherlands, 1981.* Dordrecht, The Netherlands: D. Reidel Publishing Co, ISBN:902771393 6.

Deng, Z. Y., Han, T., Cheng, Z. H., et al. (2022). Fault detection of petrochemical process based on space-time compressed matrix and Naive Bayes. *Process Safety and Environmental Protection, 158*, 146–158. https://doi.org/10.1016/j.psep.2022.01.048.

Fazai, R., Mansouri, M., Abodayeh, K., Nounou, H., & Nounou, M. (2019). Online reduced kernel PLS combined with GLRT for fault detection in chemical systems. *Process Safety and Environmental Protection, 128*, 228–243. https://doi.org/10.1016/j.psep.2019.05.018.

Ghosh, A., Ahmed, S., Khan, F., & Rusli, R. (2020). Process safety assessment considering multivariate non-linear dependence among process variables. *Process Safety and Environmental Protection, 135*, 70–80.

Gibson, N. (1999). Process safety—A subject for scientific research. *Transactions of the Institution of Chemical Engineering*, 77, 153–179.

Gowland, R. (2012). A journey into process safety with Trevor Kletz. *Journal of Loss Prevention in the Process Industries, 25*, 768–769.

Harrou, F., Nounou, M. N., Nounou, H. N., et al. (2013). Statistical fault detection using PCA-based GLR hypothesis testing. *Journal of Loss Prevention in the Process Industries, 26*, 129–139.

Hay, A. M. (1977). Tetrachlorodibenzo-p-dioxin release at Seveso. *Disasters, 4*, 289–308.

Hole, G., Hole, A. S., & MaFalone-Shaw, I. (2021). Digitalization in pharmaceutical industry: What to focus on under the digital implementation process? *Inteinational Journal of Pharmaceutics: X, 3*(100), 095. https://doi.org/10.1016/j.ijpx.2021.100095.

Hollnagel, E. (2008). Risk + barriers = safety? *Safety Science, 46*, 211–229. https://doi.org/10.1016/j.ssci.2007.06.028.

Iaiani, M., Tugnoli, A., Bonvicini, S., & Cozzani, V. (2021). Analysis of cybersecurity-related incidents in the process industry. *Reliability Engineering and System Safety, 209*(107), 485.

Jarvis, R., & Goddard, A. (2017). An analysis of common causes of major losses in the onshore oil, gas & petrochemical industries. *Loss Prevention Bulletin, 225*, 28–36.

Ji, C. X., Jiao, Z. R., Yuan, S., et al. (2021). Development of novel combustion risk index for flammable liquids based on unsupervised clustering algorithms. *Journal of Loss Prevention in the Process Industries, 70*(104), 422. https://doi.org/10.1016/j.jlp.2021.104422.

Kaced, R., Kouadri, A., Baiche, K., et al. (2021). Multivariate nuisance alarm management in chemical processes. *Journal of Loss Prevention in the Process Industries, 72*(104), 578.

Kaszniak, M., & Holmstrom, D. (2008). Trailer siting issues: BP Texas City. *Journal of Hazardous Materials, 159*, 105–111.

Kayikci, Y. (2018). Sustainability impact of digitization in logistics. *Procedia Manufacturing, 21*, 782–789.

Khakzad, N. (2019). System safety assessment under epistemic uncertainty: Using imprecise probabilities in Bayesian network. *Safety Science, 116*, 149–160.

Khakzad, N., Khan, F. I., & Amyotte, P. (2013). Dynamic safety analysis of process systems by mapping bow-tie into Bayesian network. *Process Safety and Environmental Protection, 91*, 46–53. https://doi.org/10.1016/j.psep.2012.01.005.

Khan, M. E. (2007). Environmental disasters as risk regulation catalysts? The role of Bhopal, Chernobyl, Exxon Valdez, Love Canal, and Three Mile Island in shaping U.S. environmental law. *Journal of Risk and Uncertainty, 35*, 17–43.

Khan, F., Amyotte, P., & Adedigba, S. (2021). Process safety concerns in process system digitalization. *Education for Chemical Engineers, 34*, 33–46. https://doi.org/10.1016/j.ece.2020.11.002.

Khan, B., Khan, F., & Veitch, B. (2020). A Dynamic Bayesian Network model for ship-ice collision risk in the Arctic waters. *Safety Science, 130*(104), 858.

Khan, F., Rathnayaka, S., & Ahmed, S. (2015). Methods and models in process safety and risk assessment: Past, present and future. *Process Safety and Environmental Protection, 98*, 116–147. https://doi.org/10.1016/j.psep.2015.07.005.

Khan, F., Wang, H., & Yang, M. (2016). Application of loss functions in process economic risk assessment. *Chemical Engineering Research and Design, 111*, 371–386. https://doi.org/10.1016/j.cherd.2016.05.022.

Kletz, T. (1999). The origins and history of loss prevention. *Process Safety and Environmental Protection*, 77(B), 109–116.

Kletz, T. (2012). The history of process safety. *Journal of Loss Prevention in the Process Industries, 25*, 763–765. https://doi.org/10.1016/j.jlp.2012.03.011.

Kopbayev, A., Faisal, K., Yang, M., & Halim, S. Z. (2022). Fault detection and diagnosis to enhance safety in digitalized process system. *Computers and Chemical Engineering, 158*(107), 609. https://doi.org/10.1016/j.compchemeng.2021.107609.

Landucci, G., Argenti, F., Cozzani, V., & Reniers, G. (2017). Assessment of attack likelihood to support security risk assessment studies for chemical facilities. *Process Safety and Environmental Protection, 110*, 102–114.

Lee, J., Cameron, I., & Hassall, M. (2019). Improving process safety: What roles for Digitalization and Industry 4.0? *Process Safety and Environmental Protection, 132*, 325–339. https://doi.org/10.1016/j.psep.2019.10.021.

Li, Z., Hu, S. P., Gao, G. P., et al. (2021). Decision-making on process risk of Arctic route for LNG carrier via dynamic Bayesian network modeling. *Journal of Loss Prevention in the Process Industries, 71*(104), 473. https://doi.org/10.1016/j.jlp.2021.104473.

Li, X. Y., Jia, R. C., Zhang, R. R., et al. (2021). A KPCA-BRANN based data-driven approach to model corrosion degradation of subsea oil pipelines. *Reliability Engineering & System Safety, 219*(108), 231. https://doi.org/10.1016/j.ress.2021.108231.

Li, X. Y., Liu, Y., Lin, Y. H., et al. (2021). A generalized petri net-based modeling framework for service reliability evaluation and management of cloud data centers. *Reliability Engineering & System Safety*, *207*(107), 381. https://doi.org/10.1016/j.ress.2020.107381.

Li, X. H., Zhang, L. Y., Khan, F., & Han, Z. Y. (2021). A data-driven corrosion prediction model to support digitization of subsea operations. *Process Safety and Environmental Protection*, *153*, 413–421. https://doi.org/10.1016/j.psep.2021.07.031.

Li, Q. Z., Zhou, S. N., & Wang, Z. Q. (2021). Quantitative risk assessment of explosion rescue by integrating CFD modeling with GRNN. *Process Safety and Environmental Protection*, *154*, 291–305.

Mamudu, A., Khan, F., Zendehboudi, S., & Adedigba, S. (2021). Dynamic risk modeling of complex hydrocarbon production systems. *Process Safety and Environmental Protection*, *151*, 71–84.

Mannan, M. S., Chowdhury, A. Y., & Reyes-Valdes, O. J. (2012). A portrait of process safety: From its start to present day. *Hydrocarbon Processing*, *91*, 55–62.

Mannan, M. S., Reyes-Valdes, O., Jian, P., Tamim, N., & Ahammad, M. (2016). The evolution of process safety: Current status and future direction. *Annual Review of Chemical and Biomolecular Engineering*, 7, 135–162.

Mannan, M. S., West, H. H., Krishna, K., Aldeeb, A. A., Keren, N., et al. (2005). The legacy ofBhopal: The impact over the last 20 years and future direction. *Journal of Loss Prevention in the Process Industries*, *18*, 218–224.

Marsh. (2018). *The 100 largest losses 1978–2017*. Accessed August 30, 2018. Retrieved from https://www.marsh.com/uk/insights/research/100-largestlosses-in-the-hydrocarbon-industry.html.

Marsh. (2020). *100 Largest losses in the hydrocarbon industry, 1974–2019* (26th ed.). Download from The 100 Largest Losses in the Hydrocarbon Industry (marsh.com).

Mishra, K. B., Wehrstedt, K. D., & Krebs, H. (2014). Amuay refinery disaster: The aftermaths and challenges ahead. *Fuel Processing Technology*, *119*, 198–203. https://doi.org/10.1016/j.fuproc.2013.10.025.

O'Connor, M., Pasman, H. J., & Rogers, W. J. (2019). Sam Mannan's safety triad, a framework for risk assessment. *Process Safety and Environmental Protection*, *129*, 202–209.

Paltrinieri, N., Øien, K., & Cozzani, V. (2012). Assessment and comparison of two early warning indicator methods in the perspective of prevention of atypical accident scenarios. *Reliability Engineering & System Safety*, *108*, 21–31. https://doi.org/10.1016/j.ress.2012.06.017.

Pasman, H. J., & Fabiano, B. (2021). The Delft 1974 and 2019 European Loss Prevention Symposia: Highlights and an impression of process safety evolutionary changes from the 1st to the 16th LPS. *Process Safety and Environmental Protection*, *147*, 80–91. https://doi.org/10.1016/j.psep.2020.09.024.

Porthin, M., Liinasuo, M., & Kling, T. (2020). Effects of digitalization of nuclear power plant control rooms on human reliability analysis—A review. *Reliability Engineering & System Safety*, *194*(106), 415.

Rae, A., Alexander, R., & McDermid, J. (2014). Fixing the cracks in the crystal ball: A maturity model for quantitative risk assessment. *Reliability Engineering and System Safety*, *125*, 67–81. https://doi.org/10.1016/j.ress.2013.09.008.

Reinartz, C., Kulahci, M., & Ole, R. (2021). An extended Tennessee Eastman simulation dataset for fault-detection and decision support systems. *Computers and Chemical Engineering*, *149*(107), 281.

Sarbayev, M., Yang, M., & Wang, H. Q. (2019). Risk assessment of process systems by mapping fault tree into artificial neural network. *Journal of Loss Prevention in the Process Industries*, *60*, 203–212.

Schmitz, P., Swuste, P., Reniers, G., & van Nunen, K. (2020). Mechanical integrity of process installations: Barrier alarm management based on bowties. *Process Safety and Environmental Protection*, *138*, 139–147.

Single, J., Schmidt, J., & Denecke, J. (2019). State of research on the automation of HAZOP studies. *Journal of Loss Prevention in the Process Industries*, *62*(103), 952. https://doi.org/10.1016/j.jlp.2019.103952.

Skogdalen, J. E., Khorsandi, J., & Vinnem, J. E. (2012). Evacuation, escape, and rescue experiences from offshore accidents including the Deepwater Horizon. *Journal of Loss Prevention in the Process Industries*, *25*(1), 148–158. https://doi.org/10.1016/j.jlp.2011.08.005.

Sultana, S., Anderson, B., & Haugen, S. (2019). Identifying safety indicators for safety performance measurement using a system engineering approach. *Process Safety and Environmental Protection*, *128*, 107–120.

Sun, H., Haiqing, W., Yang, M., & Reniers, G. (2021). Resilience-based approach to safety barrier performance assessment in process facilities. *Journal of Loss Prevention in the Process Industries*, *73*(104), 599. https://doi.org/10.1016/j.jlp.2021.104599.

Sun, H., Wang, H., Yang, M., & Reniers, G. (2020). On the application of the window of opportunity and complex network to risk analysis of process plants operations during a pandemic. *Journal of Loss Prevention in the Process Industries*, *68*(104), 322.

Swuste, P., Van Gulijk, C., & Zwaard, W. (2010). Safety metaphors and theories, a review of the occupational safety literature of the US, UK and The Netherlands, till the first part of the 20th century. *Safety Science*, *48*, 1000–1018.

Theis, A. (2014). Case study: T2 Laboratories explosion. *Journal of Loss Prevention in the Process Industries*, *30*, 296–300. https://doi.org/10.1016/j.jlp.2014.04.009.

Vaidya, S., Ambad, P., & Bhosle, S. (2018). Industry 4.0—A glimpse. *Procedia Manufacturing*, *20*, 233–238.

Wu, G. Y., Li, M. Y., & Li, Z. J. S. (2021). A Gene Importance based Evolutionary Algorithm (GIEA) for identifying critical nodes in Cyber–Physical Power Systems. *Reliability Engineering & System Safety*, *214*(107), 760.

Wu, D. Y., & Zhao, J. S. (2021). Process topology convolutional network model for chemical process fault diagnosis. *Process Safety and Environmental Protection*, *150*, 93–109. https://doi.org/10.1016/j.psep.2021.03.052.

Yang, M. (2018). Major process accidents: Their characteristics, assessment and management of the associated risks. *Process Safety Progress*, *37*, 268–275.

CHAPTER TWO

State-of-the-art in process safety and digital system

Md Tanjin Amin[a], Rajeevan Arunthavanathan[a], Md Alauddin[a], and Faisal Khan[b,*]

[a]Centre for Risk, Integrity and Safety Engineering (C-RISE), Faculty of Engineering & Applied Science, Memorial University of Newfoundland, St. John's, NL, Canada
[b]Mary Kay O'Connor Process Safety Center, Artie McFerrin Department of Chemical Engineering, Texas A&M University, College Station, TX, United States
*Corresponding author: e-mail address: fikhan@tamu.edu

Contents

1. Background

1.1 Definition

A digital system combines digital elements such as hardware, software, networks, strategies, and their use to improve its performance. It transforms analogue signals into a digital format to provide a digital solution (Kamalaldin, Sjödin, Hullova, & Parida, 2021). A modern industrial digital system is comprised of three key steps: digitization, digitalization, and digital transformation (Fig. 1). Nowadays, digital transformation has become one of the most widely used terms to enunciate the current progress and

Methods in Chemical Process Safety, Volume 6
ISSN 2468-6514
https://doi.org/10.1016/bs.mcps.2022.04.001

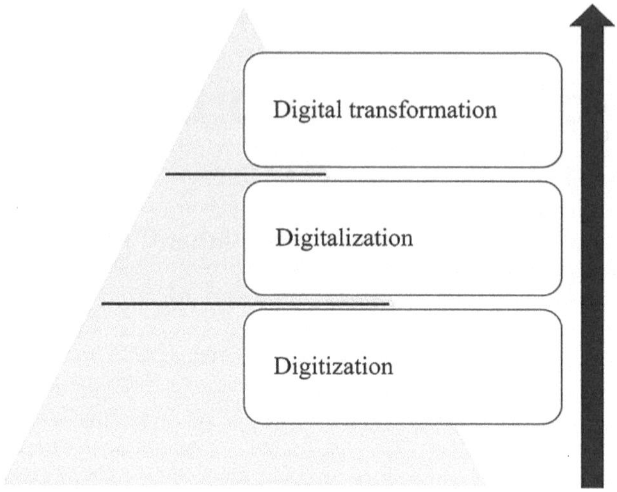

Fig. 1 Steps of building an industrial digital system.

required futuristic headway of the process industries. It impacts an industry in several ways, such as by enhancing safety, smoothness and rapidness of day to day operation and process optimization, and overall profitability, just to name a few (Devold, Graven, & Halvorsrød, 2017; He, Chen, Dong, Sun, & Shen, 2019; Jabłoński & Jabłoński, 2021; J. Lee, Cameron, & Hassall, 2019). High-reliability organizations, essential for enduring socio-economic growth and subject of interest to the research community, are adopting such a transformation at a rapid pace than ever to be competitive in the changing scenarios during the COVID-19 pandemic era (Kutnjak, 2021; Lee, Wang, Desouza, & Evans, 2021).

Perhaps, knowingly or unknowingly, we have enjoyed and are enjoying the benefits of digital evolution. Consider monitoring any intruder to properties sitting thousands of miles away with just a cellphone. This eliminates the necessity of recruiting dedicated security personnel. Meanwhile, an attempt to enter into an industrial site or personal property can be captured in real-time, and the consequence of such an attempt can be nullified or mitigated by synchronizing a security alarm with the local law enforcement department.

Although the past couple of decades have seen increasing usage of the term "digital transformation", some other fundamental terminologies like digitization and digitalization are often invariantly and interchangeably used to describe it without explaining the key differences among them from the conceptual and application perspectives. Nonetheless, there are notable distinctions that readers must know before jumping to any discussion on digital transformation or digital process systems. Fig. 1 shows the essential steps for developing an industrial digital system.

Digitization is the first step to a digital process; it refers to information digitizing. In simple words, digitization is converting an analogue signal into a digital one (Nelson & Ellis, 2019). Creating a scanned copy of a written document is one of the simplest examples of digitization in daily life. In process industries, the interpretation is slightly different. In earlier days, process operators used to manually measure different process variables, such as pressure, flow rate, volume, and temperature. This procedure was time-consuming and error-prone. With the advent of sensors, this practice was eliminated in many industrial facilities since the process variables' measurement could be converted to digital format. It eliminated the need for human interference. At the same time, it facilitated the data collection at a higher speed with more precision (Feise & Schaer, 2021; Urso, 2020). Eventually, it results in a significant reduction in operational costs (Shivajee, Singh, & Rastogi, 2019).

Digitalization is the way of enhancing the level of automation in a process system. It can be viewed as integrating digital technologies in process operations for better efficiency and product quality (Khan, Amyotte, & Adedigba, 2021). In addition to digitization, it offers the ability to analyze and store data for further utilization (Lee et al., 2019). Continuing from the example of converting analogue data to digital format, such data does not give the operators any valuable information about process abnormalities unless other technologies are used to extract valuable information from them. In reality, operators need crucial information about a process state that may help them to determine whether any action is required in a process operation. If only digitized data is used, human operators need to analyze those data manually. However, digital technologies based on artificial intelligence (AI) provide the mechanism to analyze these data automatically and contribute to a further level of automation.

Digital transformation relies on digitalization; however, it exceeds the latter's capacity (Vial, 2019). It offers enormous opportunities for an organization by integrating different digitalized segments and reasonably provides better outputs (de Beer & Depew, 2021). For example, with digitalization, a process abnormality can be captured. However, it may be valuable for only the maintenance department. The other departments (e.g., operation and supply chain departments) of a business cannot be highly benefitted.

Digital transformation is the means to interconnect different departments. It enables the supply chain department to know how much they will get affected due to an abnormality. Although the concepts of digitalization and digital transformation sound similar, these are indeed different. The first one is concerned with the automation of a part of a process. On the contrary, the latter focuses on transforming information to the whole process. Going beyond technologies, it involves digital business strategies, as well (Kane, Palmer, Phillips, Kiron, & Buckley, 2015). Securing an online presence instead of a paper or flyer-based marketing policy is an example of a strategic digital transformation that many businesses have adopted in recent years.

1.2 Drivers of digital transformation: A historical perspective

The current progress towards digital transformation has been possible due to several impetuses. The invention of modern analogue computers can be considered as the first step. Dumaresq is one of the earliest analogue computing machines in the modern era. Lieutenant John Dumaresq of the Royal Navy invented it in the early 1900s to handle a ship's fire control problem (Schleihauf, 2001). The AC network analyzers, FERMIAC, and MONIAC were other earliest analogue computers designed to solve specific problems (Haghighat, 2020; Stepney & Hickinbotham, 2018). These devices used either mechanical or electro-mechanical analogue signals for computation.

The need for digital computational machines was felt, and as a result of continual effort, ENIAC, the first electronic digital computer, was built during World War II (Haigh, Priestley, Priestley, & Rope, 2016). However, the first significant stimulation towards digital machines with high computational capacity was provided by IBM's Computation Seminar in 1949 (Lee et al., 2019). A total of 107 research engineers and scientists, experienced in applying mathematical methods for physical problems' computation, attended the seminar. This seminar displayed the then state-of-art progress and future needs in digital computing.

During the 1950s, research on digital sensor technology accelerated rapidly. A digital sensor is a device that produces a digital signal for an observed physical phenomenon (Gungor & Hancke, 2009). By the 1970s, many process industries shifted from manual measurement systems to digital sensors. It formulated an excellent ground for process digitization since a large number of process variables could be measured, avoiding error-prone manual systems and using sensors with high precision.

The advent of Process System Engineering (PSE) is a crucial factor behind process digitalization. PSE is a sub-discipline of Chemical Engineering. It is a System Engineering approach that integrates design, operation, control, and optimization of industrial process systems. According to Sargent (1983), "*Process systems engineering is all about the development of systematic techniques for process modelling, design and control*". "Process System Engineering" was first used during Kyoto's 1st Process Systems Engineering conference in 1982 (Pistikopoulos, Barbosa-Povoa, Lee, Misener, & Gani, 2021). Nonetheless, many of its sub-disciplines were well established before this conference (Stephanopoulos & Reklaitis, 2011).

The computational capacity bolstered PSE. Digital dynamic process simulators were developed using advanced computers. These enabled the generation of a virtual model of a process operation. As a result, several operational scenarios could be generated that helped in process control, optimization, and monitoring (de Beer & Depew, 2021; De Tommaso, Rossi, Moradi, Pirola, & Galli, 2020; Stephanopoulos, 2009). Simulators often provide faster solutions than real process dynamics, which is advantageous for process industries (Lee et al., 2019).

The introduction of distributed control system (DCS) around the 1980s was pivotal for process digitalization. Unlike the centralized control system, where the whole process is monitored by a central computer in a central control room, the DCS makes it possible to control the process in a distributed manner (Tang & Daoutidis, 2019). The significant advantage of using DCS is its ability to monitor complex processes with close loop control. Also, monitoring interconnected units becomes more manageable with the DCS (Xu & Bao, 2009).

Advanced sensor technologies and DCS made it possible to collect more plant data. It made the process plants a source of high dimensional datasets. Several software-based algorithms were developed for abnormal situation management and safety assessment. From the late 1980s, multivariate data-driven fault detection and diagnosis methods became widespread in process industries (Alauddin, Khan, Imtiaz, & Ahmed, 2018; Ge, Song, & Gao, 2013). These provided an alternative solution to model-based, limit checking, and univariate data-based tools.

Moreover, it was possible to extract meaningful information from the enormous amount of data generated in each instance. Consequently, process digitalization experienced a newer and increased pace due to the availability of software or algorithms to give meaning (i.e., safe or unsafe) to the digital signals collected by different sensors. In addition to fault detection and diagnosis (FDD) methods, several computer-aided methods have been developed for safety and risk analysis to facilitate process digitalization (Khan & Abbasi, 1999a, 1999b).

In recent years, the availability of large datasets, the internet, and the industrial internet of things (IIoT) speeded the adoption of digital technologies further. According to Boyes, Hallaq, Cunningham, and Watson (2018), "*IIoT is a system comprising networked smart objects, cyber-physical assets, associated generic information technologies and optional cloud or edge computing platforms, which enable real-time, intelligent, and autonomous access, collection, analysis, communications, and exchange of process, product and/or service information, within the industrial environment, so as to optimise overall production value. This value may include; improving product or service delivery, boosting productivity, reducing labour costs, reducing energy consumption, and reducing the build-to-order cycle.*" IIoT is an advanced version of DCS; it provides a higher degree of automation, optimization, and process control using cloud computing (Wang, Bi, & Xu, 2014).

Industry 4.0 or the fourth industrial revolution, originated from Germany, is the latest industrial revolution that the modern era is experiencing. It has many similarities with the IIoT. Both are focused on increasing the profitability of the manufacturing plants. Also, these are highly technology and human-driven. A seamless internet connection is essential for IIoT and Industry 4.0. Although these two concepts emphasize advanced digital technologies and automation, a skilled human resource who interprets and applies advanced data analytics is still required (Georgios, Kerstin, & Theofylaktos, 2019).

On the other hand, these are different from scope and action perspectives. IIoT is mainly concerned with the interconnectedness within a facility, while Industry 4.0 focuses beyond a plant only. Action-wise, Industry 4.0's scope is also broader. It uses many technologies such as IIoT and big data (Aazam, Zeadally, & Harras, 2018). Both IIoT and Industry 4.0 have contributed significantly towards digitalization and digital transformation (Belli, Davoli, Medioli, Marchini, & Ferrari, 2019; Ghobakhloo, 2020).

The latest driver behind the growing appeal towards digital transformation is the COVID-19 pandemic. This outbreak made the companies think about the traditional on-site work policy and harness the benefits of full-scale digital transformation. According to a survey by Ernst & Young, 90% of the executives from chemical industries expected a significant impact of digital technologies in their plants within the next 3 years (Jenner, 2021).

COVID-19 pandemic made people realize that some non-essential functions in an industry can be successfully conducted, being away from the plant using digital transformation in terms of remote operation (Gallacher & Hossain, 2020). For example, process safety engineers can get continuous information about an operation from the process monitoring modules. Also, any potential maintenance action can be suggested without staying on-site.

Another noteworthy change has been noticed in terms of strategic decisions. The conventional marketing strategies became ineffective since the consumers felt unsafe touching physical objects due to the highly contagious nature of COVID-19. Therefore, the industries realized the traditional marketing policy might lose their customers and stressed enhancing their online presence and digital marketing on social media (Hu & Olivieri, 2021). These are only a few examples of how the COVID-19 pandemic has increased enthusiasm towards the digital transformation of businesses worldwide.

1.3 Opportunities and challenges

Digital transformation is creating enormous opportunities. However, the scopes vary from industry to industry due to the pandemic. For example, the volume of customers for the petrochemical industries plummeted. On the contrary, pharmaceuticals saw a handsome rise in consumers and revenue. Digital transformation provides scopes to improve in all these plants. For instance, the pharmaceuticals are extending their business and opening new plants in new locations. Digital transformation is enabling them to operate these facilities under one umbrella. On the other hand, using digital technologies, the petrochemical industries are reducing waste and operational costs, meeting the safety requirements. The overall benefits are summarized below.

i. **Process optimization:** Digital transformation has reduced production and operational costs and increased profitability; it has been made possible using automated equipment, remote monitoring, and predictive maintenance (Pflaum & Gölzer, 2018).
ii. **Predictive maintenance:** The AI-based predictive analysis tools can indicate faults and failures early (Giunta, Bernasconi, Giro, & Cesari, 2020). Also, the safety analysis mechanisms can indicate predictive risk (Amin, Khan, Ahmed, & Imtiaz, 2021b). As a result, any equipment can undergo maintenance before failure, leading to reduced downtime and improved production efficiency. It is noteworthy to mention that

prescriptive maintenance, a new look into optimizing maintenance and production activities, is also remarkably getting advancements using the benefits of AI and digital transformation (Gordon, Burnak, Onel, & Pistikopoulos, 2020; Gordon & Pistikopoulos, 2021).

iii. **Remote operations:** During the pandemic, government protocols and safety regulations required the industries to be operated with fewer human resources. The digital communication system made it possible for different teams to communicate vertically and operate remotely. A survey indicates that the majority of the executives are planning to model a combination of on-site and work from home culture in the forthcoming years for safety and employee satisfaction (Delloite., 2021).

iv. **Waste reduction:** One of the significant challenges for process industries is maintaining product prices. Raw material prices see ups and downs. However, the customer prefers a fixed price from the manufacturer. Therefore, it is challenging for companies to meet consumer demand and maintain a certain profit level. Advanced data analytics provide an early indication of an anomaly in production processes and thereby help reducing waste due to a part failure and better utilization of raw materials (Borangiu, Trentesaux, Thomas, Leitão, & Barata, 2019).

v. **Supply chain management:** Market demand is usually volatile. Conventionally, the production team did not have any information about the real-time market demand. Digital transformation allows visualizing the current market demand, and then the data analytics techniques can be used to adjust the production rate to harmonize the demand (Centobelli, Cerchione, & Ertz, 2020). Also, the industries are trying to develop mechanisms to modify the product composition based on real-time market demand using digital transformation.

vi. **Safety and sustainability enhancement:** Chemical plants deal with a large volume of hazardous materials. Therefore, an equipment failure may result in a catastrophic accident (Khan, Rathnayaka, & Ahmed, 2015). Digital transformation has helped industries enhancing their safety profile by cutting emissions. Also, the AI-based predictive models help preventing a catastrophic failure by providing early warning.

In addition to these, the transformation of traditional processes to digital systems, minimizing technical errors, enhanced customer service, product quality improvement, and easier access to information are other notable benefits offered by a digital transformation (Hai, Van, & Tuyet, 2021; Nadkarni & Prügl, 2021).

Implementing digital transformation is not straightforward since it poses new challenges and threats. These can be listed as mentioned below.

i. **Leadership skill:** Business leaders need to work with a combination of digital and human workforces. The leaders need to know how to apply digital technologies. At the same time, they need to manage the human employees who may be reluctant to accept the new culture of digitalization. They need to bring unique ideas to solve this dilemma. More importantly, the leaders need to be ready and willing to embrace the new technologies (Vial, 2019).

ii. **Qualified workforce:** Although the ultimate goal of digital transformation is to use digital technologies and improve automation, a qualified workforce is required to implement the concept (Spöttl, 2017). Good command over data analytics is essential. However, process industries lack qualified personnel since the young talents who possess such skills often find other fields like data science more exciting.

iii. **Investment:** Lack of investment is another critical challenge. Unfortunately, many companies are yet to realize the importance and actual benefits of digital transformation. Several oil and gas industries were suspicious about the success of a full-scale digital transformation and reluctant to spend a sufficient amount behind it. However, the situation has changed due to COVID-19. The executives are willing to implement digital technologies to keep pace with the changing condition around the world.

iv. **Complex operation:** Digital transformation makes it possible to operate complex process operations. Nonetheless, it enhances the number of process variables to be monitored (Khan et al., 2021). Consequently, data-driven safety assessment, especially fault diagnosis, becomes difficult. Although the multivariate tools provide a reasonable solution, their performance is inconsistent. They often point towards a variable that is nowhere near the location of the actual root cause; it may add to process downtime.

v. **Digital threats:** As mentioned earlier, digital transformation produces a vast amount of data. These data contain a company's sensitive information. It is imperative to manage and store data correctly, especially when cyberattacks are increasing around the world. It was thought impossible for a process facility to be attacked through cyber means in earlier days. However, it is not an imagination anymore. Data theft and ransomware are two significant threats to cybersafety. The most alarming fact about cyberattacks is the ability to take control of process operations being far away. If such attacks go unnoticed, these can cause damage equal to any catastrophic accident. A known example is the Iranian nuclear facility that suffered a cyberattack through a unique worm named Stuxnet (Kerr, Rollins, & Theohary, 2011). It caused the reactor temperature to show abnormal behavior. Therefore, ensuring cybersecurity has become a significant challenge in digitalized process systems.

2. Current progress in digital process safety

This chapter uses a scientometric mapping or bibliometric analysis approach for investigating the current progress in digital process safety. Since Nalimov and Mul'chenko (1971) introduced the bibliometric analysis, it has become increasingly popular in the scientific domains due to its ability to capture the quantitative progress of a specific field. Several articles can be found in the current literature that has adopted such an approach to evaluate progress in different aspects of process safety (i.e., risk communication, pool fire, human reliability analysis, and so on).

The current work adopts the methodology used by Amin, Khan, and Zuo (2019). According to this method, several keywords are selected closely related to the field of interest, followed by an extensive search using different databases. Once bibliometric information is available, further search is conducted manually to select the articles that strongly suit the intended area. Finally, meaningful results are extracted from the refined documents using different software such as Microsoft Excel and VOSviewer.

2.1 Bibliographic information collection

First, several keywords related to a digital system and process safety were identified. Then the Web of Science Core Collection (WoS) database was used to search relevant documents. WoS was selected because of its enriched quality collection. However, it is noteworthy to mention that other databases such as Scopus, Compendex, and Google Scholar can also be used to extend the search horizon. This will result in detailed and exhaustive information; however, the entire process will be extensively time-consuming (Amin, Khan, & Amyotte, 2019). In this work, the documents were searched using the following keywords.

TOPIC: *digital* OR *digitalized* OR *digitalize* OR *digitalised* OR *digitalise* OR *digitalization* OR *digitalisation* OR *digitized* OR *digitize* OR *digitized* OR *digitize* OR *digitization* OR *digitisation* OR *industry 4.0* OR *internet of things*.

AND **TOPIC:** *chemical* OR *offshore* OR *process.*

AND **TOPIC:** *safety* OR *security* OR *cybersecurity* OR *intrusion* OR *risk* OR *monitoring* OR *fault* OR *hazard* OR *failure* OR *reliability* OR *integrity* OR *resilience* OR *vulnerability.*

WoS provided 44,586 documents from 200 different categories. However, most of the fields are not related to chemical process systems. Therefore, to reduce the number of documents and to get a list of documents closely related to digitalized process safety, the Engineering Multidisciplinary, Engineering Industrial, Chemistry Analytical, Operations Research Management Science, Chemistry Multidisciplinary, Engineering Chemical, Chemistry Physical, Mining and Mineral Processing, Engineering Marine, Engineering Ocean, Chemistry Applied, Engineering Petroleum, Automation Control Systems, Remote Sensing, and Energy Fuels categories were selected. As a result, 8713 documents were left that were further manually analyzed to find only the articles that matched digitalized process safety-related publications. This manual search found 338 documents for result generation. Among these documents, the number of journal articles, proceedings papers, reviews, meeting abstract, and editorial material are 201, 113, 22, 1, and 1, respectively.

2.2 Results

The results were interpreted in terms of annual publication volume and key areas. An annual publication trend gives an idea of how appealing a field is to the research community, while the latter shows what essential areas of this field are. Microsoft Excel and VOSviewer were used to generate the results.

2.2.1 Annual publication volume

The annual distribution pattern is shown in Fig. 2. It can be seen that the formal publications related to the digital chemical process started in the early 1970s. The progress till 2010 is steady. However, the field saw exponential growth from 2011. 288 out of the 338 studied documents have been published from 2011 to 2021. These statistics clearly indicate that although the research on digitalized chemical process safety was realized later, it has gained immense popularity in the past decade.

From the contents of these publications, a shift from digitization to digital transformation is also noticed. The very first document was a meeting abstract on digitization. It mentioned how manual pressure measurement could be made digital (Il, 1970). Most of the early documents (from the 1970s to the 1990s) are on the automation of control and measurement systems. However, this trend dramatically changed from the mid

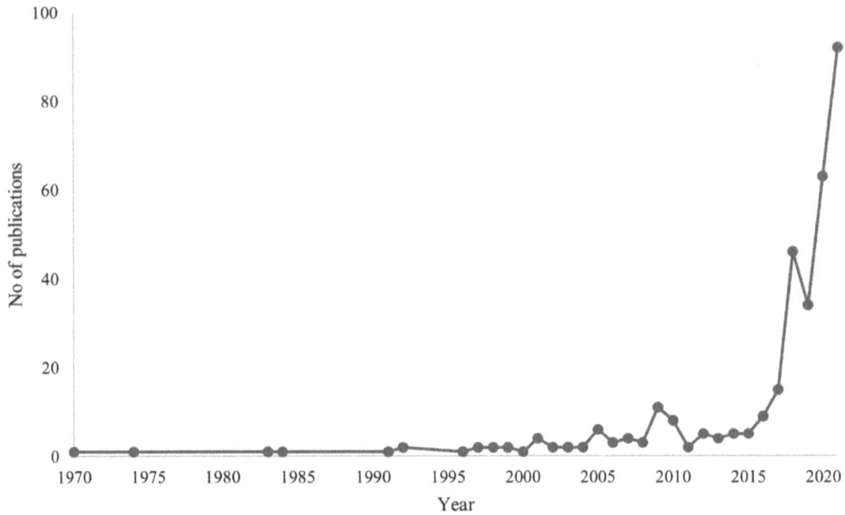

Fig. 2 Annual publication volume in digital process safety.

of 1990s, focusing on digitalization. The majority of the documents were related to developing algorithms for different aspects of process safety analysis (Kim, Kim, Shin, & Yoon, 2000; Sinnamon & Andrews, 1997).

Moreover, since the mid of 1990s, the support of electronic-based control systems such as programmable logic controllers (PLCs), supervisory control, and data acquisition systems (SCADA) ended a successful revolution in the process industry (Venkatasubramanian, 2001). With the help of automated control, process plants become highly complex dynamic processes comprised of many sensors and actuators. This revolution mainly adopts the system-level control except for plant management regarding emergency shutdowns and system safety. However, the advent of highly reliable electronics, compliant with the IEC 61508/61511 and ANSI/ISA 84 standards, produced a breakthrough. These sophisticated components' failure rates can be predicted and are used to build controllers and safety instrumented systems (SIS). Furthermore, recent research and development, continuous advances in computer and software technology, and reliable internet connections have enabled automated and intelligent solutions for Industry 4.0 evolution.

In addition to these, to prevent accidents, the need for integrating past accident repositories with computer-aided tools and digital control systems was demonstrated (Jefferson, Chung, & Kletz, 1997). Although process simulators are not new in process industries (Sikorski, 1991), these got renewed attention from the early 2010s due to the emergence of the concepts of

Industry 4.0 and IIoT. Also, several documents have reported the need for remote monitoring using digital transformation.

The majority of documents published in recent years focus on developing digital twin and advanced data-driven algorithms for predictive process monitoring and dynamic safety evaluation (Feng, Dai, Ji, Zhou, & Dang, 2021; Guo, Wan, Liu, Gao, & Yang, 2021; Kender, Kaufmann, Rößler, Wunderlich, & Klein, 2021; Rato & Reis, 2020; Topolski, 2021; Wattis, Murray, Baldan, Johnson, & Spitzenberger, 2021). Prognostic analysis and remaining useful life predictions using data-driven tools are becoming mainstream (Zhang, Yan, Shi, Li, & Yan, 2021). However, it is worth noting that the process industries still suffer from digitization due to a lack of understanding of the process or the difficulty to place a physical sensor, and therefore, process digitization remains a vital interest to many researchers (Li, Zhang, Khan, & Han, 2021; Villalpando-Hernandez, Munoz-Rodriguez, & Vargas-Rosales, 2021).

2.2.2 Key research areas

WoS provides indexed keywords by authors. These give a significant knowledge of the research areas. For instance, a work on IIoT is most likely to include the internet of things or IIoT in the keywords. VOSviewer was used to analyze these keywords. The key areas in digital process safety analysis are shown in Fig. 3.

A total of 1214 keywords were found from the 338 articles. However, the keywords that have been used at least five times (27 keywords) were included to develop the co-occurrence map. The influence of an area can be understood by the volume of its corresponding sphere, while the co-occurrence between two areas can be perceived by the width of the lines joining them. It can be seen that the conventional areas such as process safety analysis, risk assessment, safety management, process control, process monitoring, condition monitoring, fault diagnosis, and process optimization are vital in digital process safety.

In addition to these, Industry 4.0, internet of things, big data analysis, machine learning, digital twin, and security analysis are other areas that have got significant attention. Industry 4.0 and internet of things are the major areas of research, shown by the largest spheres. They have the strongest bibliographic coupling, followed by the link between industry 4.0 and big data. Industry 4.0 is part of the five most vital co-occurrence links. It is an indication that the recent research focus is entangling around the current industrial revolution, and process industries are adopting digital transformation to cope with this. Table 1 summarizes several works from distinct vital areas.

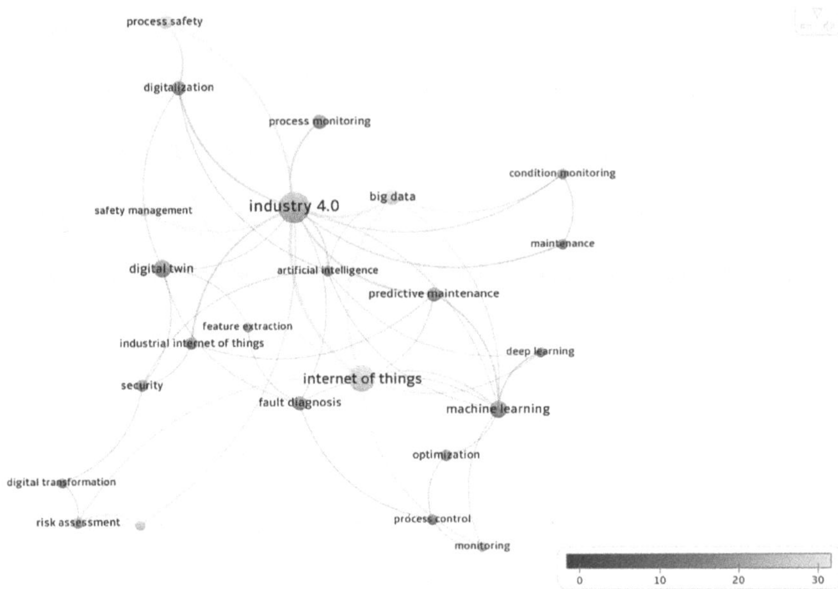

Fig. 3 Key research areas in digital process safety.

Table 1 A summary of publications from different key areas related to digital process safety.

Work	**Key feature(s)**
Arena, Florian, Zennaro, Orrù, and Sgarbossa (2022)	Presents a decision tree-based predictive maintenance framework.
Vrignat, Kratz, and Avila (2022)	Reviews the current growth in predictive maintenance and approaches used for prognostic failure analysis.
Lees and Johnstone (2021)	Emphasizes the need for implementing safety culture while adopting Industry 4.0 for sustainable growth.
Xu and Zhu (2021)	1. Proposes a dynamic auto-regressive latent variable model. 2. Captures data auto-correlations and cross-correlations. 3. Suitable for anomaly detection.

Table 1 A summary of publications from different key areas related to digital process safety.—cont'd

Work	Key feature(s)
Bi and Zhao (2021)	1. Proposes an orthogonal self-attentive variational autoencoder model for process monitoring. 2. Orthogonal attention is used for extracting variables' correlation. 3. Variational self-attentive autoencoder detects a fault.
Reis and Saraiva (2021)	1. Discusses the scopes and trends of Process System Engineering (PSE) evolution. 2. Emphasizes the use of data for expansion towards PSE 4.0.
Wu, Lind, Zhang, Pardhasaradhi, and Myllerup (2021)	Uses multilevel flow modeling for knowledge acquisition and representation to support process operations.
Feng et al. (2021)	Introduces bidirectional encoder representation from transformers (BERT) for analysis of hazard and operability (HAZOP) reports.
Amin, Khan, Ahmed, and Imtiaz (2021a)	Proposes a technique for developing Bayesian networks from process data to diagnose faults.
Wang and Wang (2021)	Presents an overview of big data theories, applications, and challenges in safety management.
Zhang et al. (2021)	Proposes a semi-empirical remaining useful life prediction model using Gamma process method, sequential Monte Carlo algorithm, and confidence interval quantization.
Medjek, Tandjaoui, Djedjig, and Romdhani (2021)	Describes a machine learning (ML)-based intrusion detection method for cybersecurity.
Andersen, Udugama, Gernaey, Khan, and Kulahci (2021)	Develops a graphical user interface-based Tennessee Eastman process simulator for big data generation, process control, and fault detection.
Abbasinejad, Hourfar, and Elkamel (2021)	Uses expected utility theory for optimum maintenance interval prediction.
Kender et al. (2021)	Uses a pressure-driven approach to develop a digital twin of distillation columns.

Continued

Table 1 A summary of publications from different key areas related to digital process safety.—cont'd

Work	Key feature(s)
Amin et al. (2021b)	Proposes an abnormal situation management method based on a dynamic risk profile.
Iaiani, Tugnoli, Bonvicini, and Cozzani (2021)	1. Develops a database of 82 cybersecurity-related incidents suffered by process industries. 2. Presents an analysis based on time trend, geographical distribution, distribution among the industrial sectors, impacts of the incidents, and nature of these cyber-attacks.
Benson, Argyropoulos, Dimopoulos, Mikellidou, and Boustras (2021)	Describes a conceptual risk assessment framework for digitalized process systems.
Niu and Yang (2021)	Proposes a one-dimensional convolutional neural network for fault diagnosis.
Hu, Wang, Su, Li, and Zhang (2021)	Proposes a digital twin architecture of petrochemical processes suitable for monitoring and fault diagnosis.
Progoulakis, Nikitakos, Rohmeyer, Bunin, and Karamperidis (2021)	1. Reviews current cybersecurity-related risk analysis and management approaches. 2. Presents results of a survey on cybersecurity for offshore oil and gas assets.
Amin et al. (2021b)	1. Reviews the existing methods for fault detection and diagnosis, abnormal situation management, and risk assessment. 2. Demonstrates future research direction in the contexts of Industry 4.0 and process safety.
Amin, Khan, Ahmed, and Imtiaz (2020)	Proposes an ML-based method for multivariate process monitoring and dynamic risk assessment.
Xie, Chen, Dong, and Zhang (2020)	Proposes a multi-sensor fusion approach-based risk assessment technique using cloud model, belief Jensen-Shannon divergence, and Dempster-Shafer evidence theory.

Table 1 A summary of publications from different key areas related to digital process safety.—cont'd

Work	Key feature(s)
Udugama, Gargalo, Yamashita, Taube, and Bayer (2020)	Provides an overview of operational requirements, availability and nature of data, and role of the control structure hierarchy in the context of big data.
Jung (2020)	Describes the importance of process simulators and machine learning algorithms in digital transformation.
Bajic, Rikalovic, Suzic, and Piuri (2020)	1. Differentiates between managerial and technological Industry 4.0 implementation challenges. 2. Classifies Industry 4.0 technologies into 10 categories. 3. Presents managerial challenges and opportunities for Industry 4.0 implementation.
Wassar, Mutlu, Franchek, and Hattab (2019)	Proposes an adaptive model-based approach for condition monitoring of an annular blowout preventer.
Patel, Bakhurji, Salloum, Kim, and Mubarak (2019)	Demonstrates how conventional oilfields can be benefitted by using advanced process control systems.
Aldemir, Guarro, Mandelli, Kirschenbaum, and Sun (2010)	Presents cell-to-cell mapping and dynamic flowgraph-based probabilistic risk assessment methods for digital instrumentation and control systems.

3. A future roadmap in digital process safety

The advances in digital technology in process systems have enabled process risk management to be enhanced by offering more insight into the systems' protection and prevention by controlling the primary process control and abnormal situation management (Arunthavanathan, Khan, Ahmed, & Imtiaz, 2021b). Adopting digital technologies in process systems may increase the use of sensors and actuators to control the basic process autonomously (Venkatasubramanian, 2001).

Industry 4.0 with a digitized automated process system will be the future trend in the process plant and industrial advancement. This implies the invention of cyber-physical process system in the upcoming process plant (Cohen & Singer, 2021; Ji, He, Xu, & Guo, 2016; Pivoto, de Almeida, da Rosa Righi, Rodrigues, & Alberti, 2021). However, adopting a cyber-physical process system by integrating a digital control process plant with cyber network communication and computational process will highly impact process safety. Along with the integration of cyber networks in process systems, cyber-security and process safety are considered key factors in cyber-physical process safety (Angelopoulos, Michailidis, Nomikos, Trakadas, & Zahariadis, 2020).

3.1 Safety in cyber-Physical process system

Digitalized process plant invention started with automating the basic control using computations (Venkatasubramanian, 2001). In process automation, safety instrumented systems (SISs) developed beyond the basic process control system as a protection level to protect the system from possible hazards (Stavrianidis & Bhimavarapu, 1998; Willey, 2014). Over the years, physical models, process data, and human knowledge have been used to manually monitor the process health (Alauddin et al., 2018; Tidriri, Chatti, Verron, & Tiplica, 2016). However, in recent years, safety decisions have been made automated by process controllers using process data with human interaction (Arunthavanathan, Khan, Ahmed, & Imtiaz, 2022; Çinar, Nuhu, Zeeshan, Korhan, & Safaei, 2020).

After many years, with the advancement and integration of digital communication and high computation in the physical plants, automated plants have been turned into cyber-physical systems (Ji et al., 2016; Prinsloo, Sinha, & von Solms, 2019). Cyber-physical systems connect resources, information, physical entities, and people through digital cyber embedding (Pivoto et al., 2021). In this Industry 4.0 era, cyber-physical system concepts are widely adopted in various manufacturing industries. Recent studies illustrate that nearly 73% of the process industries are willing to adopt Industry 4.0 in the next few years (Ji et al., 2016). However, with the adoption of Industry 4.0, it is also essential to be concerned about the adaptation of Safety 4.0 (Wang, 2021).

As shown in Fig. 4, in digitalized process plants, process safety relies on advanced technologies such as the internet of things (IoT) or industrial IoT (IIoT), big data, artificial intelligence (AI), and predictive maintenance (PM) for predicting remaining useful life (RUL).

Fig. 4 Process safety 4.0 framework.

3.1.1 Industrial internet of things (IIoT)

IIoT contributes to the process safety scenario through its concept and technologies to develop an industrial network composed of sensors and actuators connected to communication technologies that make it possible for the system to monitor and analyze the system by collecting and transmitting data quickly (Pivoto et al., 2021). Therefore, in the process safety, to monitor the process condition using data-driven models, IIoT plays a significant role in data acquisitions and transmission.

IIoT also generates significant benefits for the automated process plant, such as converging process plant operational technologies with information technologies (Angelopoulos et al., 2020). Although operational technologies, such as hardware and software, control the process plants, they were rarely connected to a network or a larger computerized system (Boyes et al., 2018). However, with emerging technologies, operational technologies components can interface directly with other machines and centralize servers, exchanging data via information technology and communication networks.

However, adopting the IIoT technologies to the process plant may increase the failure probability due to the dependency of technology and components. By implementing IIoT to operational technologies in process plants along with sensor, actuator, and process failure, communication network failures also need to be considered. Furthermore, when converging the operational and information technologies, either one of the failures may impact the other.

3.1.2 Big data/cloud computing

With the advancement of the digitization process plant to Industry 4.0, the volume of process plant resources and data flows with IIoT may significantly increase the systems' complexity. Cloud computing combined with Big Data can solve this complexity by improving system automation, operation, management, and data storage (Shah, Yuan, Lu, & Zerfos, 2018).

In addition to data storage, big data is used to visualize and understand the information. Therefore, big data requires data storage and management, hardware and software resources, appropriate domain knowledge, and new methods and technologies. Combining big data with analytics can provide a significant advantage of making an efficient process monitoring (Badri, Boudreau-Trudel, & Souissi, 2018).

Big data and cloud computing in Industry 4.0 can be used as a preliminary stage of data processing and monitoring process safety. Therefore, big data can be an initial and faster process monitoring level in process safety. Also, using big data is an advantage in maintaining quality and consistent data for further process monitoring.

3.1.3 Artificial intelligence

Implementing a cyber-physical process system is essential to monitoring the real-time process plant monitoring in Industry 4.0 (Jiang, Yin, & Kaynak, 2020). Autonomous fault detection and diagnosis are the initial actions to monitor the cyber-physical process plant safety. There are continuing efforts to develop fully automated fault detection and autonomous diagnosis methods without human interaction (Arunthavanathan et al., 2022; Çinar et al., 2020). AI will become a solution to develop self-learning process monitoring models with little or no human intervention (Arunthavanathan, Khan, Ahmed, Imtiaz, & Rusli, 2020; Chen, Tiňo, & Yao, 2014). The following sections present the research direction in developing AI-based models for the digitalization of process safety systems.

3.1.3.1 Development of AI-based algorithms

Numerous data-based methodologies such as principal component analysis (PCA) and partial least squares (PLS) (Kaspar & Harmon Ray, 1993; Kourti & MacGregor, 1995; Qin, 1998), independent component analysis

(ICA) (Stefatos & Hamza, 2010; Zhang & Qin, 2007; Zhang & Zhang, 2010), vine copula-based methods (Cui & Li, 2020; Ren, Zhu, Cai, & Li, 2017; Yu, Khan, & Garaniya, 2015), and Gaussian mixture model (GMM) (Li, Qin, & Yuan, 2016; Thissen, Swierenga, De Weijer, Wehrens, & Buydens, 2005; Yu, 2013; Yu & Qin, 2009) have been studied for FDD.

Many machine learning techniques such as artificial neural network (ANN) (Agatonovic-Kustrin & Beresford, 2000; Schmidhuber, 2015; Zhang, Bi, & Qiu, 2020; Zhao, Sun, & Jin, 2018), support vector machines (SVM) (Burges, 1998; Cortes & Vapnik, 1995; Widodo & Yang, 2007), case-based reasoning (CBR) (Zhao, Liu, Dong, Sun, & Ji, 2017), and classification and regression tree (CART) (Lawrence & Wright, 2001) have been successfully tested for fault detection and diagnostic methods for process systems. AI can capture the newly developed faults and be capable of classifying the known fault by the model. In recent years, integrating supervised and unsupervised learning in artificial intelligence has been applied to develop an autonomous and intelligent environment in process safety (Arunthavanathan et al., 2021b; Yu & Qi, 2021).

3.1.3.2 Tuning of the AI models using optimization algorithms

The structural optimization of distinct AI-based models is another area of research in enhancing the efficiency of a cyber-physical system. Many methods have been presented in improving the performance of data-driven methods by optimizing hyperparameters such as the number of hidden layers, number of neurons in each layer, the learning rate, and optimization techniques. Numerous approaches such as early stopping, regularization, dropout, and retraining have been devised for handling the overfitting and under-fitting of data-based models.

Evolutionary computing approaches have also been explored to exhilarate the training of artificial neural networks. Barakat, Druaux, Lefebvre, Khalil, and Mustapha (2011) developed a self-adaptive FDD based on artificial neural networks combined with an advanced signal processing method. Nawi, Khan, and Rehman (2013) combined Cuckoo Search (CS) and Levenberg Marquardt algorithm to train neural networks. Dehuri and Cho (2009) proposed a multi-objective Pareto-based particle swarm optimization (PSO) method for optimizing the architectural complexity and classification accuracy of a neural network.

Subudhi and Jena (2011a, 2011b) employed opposition-based differential evolution (ODE) for training a feed-forward neural network. Juang

(2004) trained a recurrent neural network with a hybrid GA-PSO algorithm using a PSO with a master-slave configuration. Subudhi and Jena (2011b) proposed a memetic algorithm-based differential evolution backpropagation (DEBP) for training a multilayer perceptron by exploiting local and global search spaces. Alauddin, Khan, Imtiaz, and Ahmed (2020) presented an ANN-based fault detection model that was tuned using a variable mosquito flying optimization (V-MFO) technique for maximizing fault detection rate (FDR) and minimizing false alarm rate (FAR).

3.1.3.3 Development of robust AI-based algorithms for handling uncertainties and low-quality data

The success of data-based models is utterly susceptible to the nature and availability of training data. The intelligent models rely on the assumption that the data is of high quality, sufficient granularity, and quantity.

Low quality, uncertainty and inconsistency are inevitable in process data. The process systems encounter various low data-quality problems such as incomplete data, outliers, missing information, imbalanced data, poorly labeled data, incorrectly mapped properties, noise, and inconsistency due to sensor breakdown, process shutdown, malfunctioning of equipment, random fluctuations, incorrect calibration, inconsistent sampling frequencies, and data entry errors due to human factors. Sparse data are also a growing concern for the efficient training of data-driven models. Usually, a large volume of data is available to represent the normal operating conditions, while faulty condition data is relatively scarce in industrial systems. The imbalanced data can lead to a biased outcome if not addressed in the modeling.

Handling low-quality data is crucial in the data-driven models of industry 4.0 in particular. Models based on the internet of things (IoT) where data come from multiple sources are susceptible to low quality and inconsistent data. Monte Carlo simulation and fuzzy logic can be used to capture the randomness and vagueness of the model parameters.

3.1.3.4 Development of hybrid models based on data-driven and first principle-based models

The AI-based models have the potential to significantly improve the efficiency, safety, and sustainability of process systems. The data-based models suffer from the lack of generalization ability and struggle when data is scarce or in extrapolation regimes (Fisher, Watson, Escrig, Witt, & Gomes, 2020; Pawar, San, Aksoylu, Rasheed, & Kvamsdal, 2021). Several studies reported improved generalization of the hybrid models. Moreover, the "black box" nature of the data-driven models, especially deep neural networks, is a barrier to large-scale adoption in industrial systems.

Furthermore, with the recent implementation of prognosis models using AI, it is also possible to detect the fault condition as early as possible by continually monitoring the fault symptoms (Arunthavanathan, Khan, Ahmed, & Imtiaz, 2021a). Furthermore, by combining AI models with plant equipment/components, degradation analysis or survival analysis can be used to estimate the remaining useful life (Li, Zhao, Zhang, & Zio, 2020; Tamssaouet, Nguyen, Medjaher, & Orchard, 2021).

3.1.4 Predictive maintenance

Predictive maintenance will be the emphasized ability in industry 4.0 with the use of integration of process system knowledge and process data knowledge through AI (Angelopoulos et al., 2020; Chen, Lu, Jiang, & Wang, 2021). By combining the advanced data accessing (i.e., big data), widened connectivity (i.e., IIoT), and advanced analytical tools (i.e., big data and cloud computing), predictive maintenance concept can be applied to maintain the process systems before a failure occurs (Zonta, da Costa, da Rosa Righi, de Lima, & Li, 2020).

However, predictive maintenance enables large amounts of data to be monitored and pushed into analytical tools to predict machine failure, downtime, overload, or other problems (Li, Wang, & Wang, 2017; Lv, Zhou, Li, & Li, 2021). Therefore, perspective to process safety with the advancing of digitized process system through cyber-physical process system, developing predictive maintenance will be a strength to monitor the fault to failure transmission in the system and investigating the failure condition before the system failure.

3.2 Security in cyber-physical process system

Cyber-security will be a major topic in process safety when Industry 4.0 is implemented with digital process systems by combining operational and information technologies in the process systems (Gu, Yin, Li, Su, & Shen, 2018). To control the process plants using operational technology, process controllers such as programmable logic controller (PLC), supervisory control and data acquisition systems (SCADA), and distributed control systems (DCS) are used in the implementation of basic process control. In addition, to implement safety of automated process systems, safety instrumentation systems are implemented to prevent the system from hazardous environment (Prinsloo et al., 2019). In combining with human-machine interaction frameworks, process controllers ensure plant safety by avoiding hazardous conditions. Therefore, most of the perspectives of the decision to the safety are made by skilled technicians, who can have exclusive access to the associated hardware and software (Kamble, Gunasekaran, & Gawankar, 2018).

The Maroochy Water Services cyber-attack (2000) incident is considered a prime example of a cyber-attack in the operational technology industrial plant (Slay & Miller, 2007). The lesson learned from the incident was that the operational controller's lack of process access control and cyber-security made it easy to access the process system controller and control the plant (Miller, Staves, Maesschalck, Sturdee, & Green, 2021). Since the process systems security condition is highly dependent on plant technicians, this emphasized one of the significant risks in process systems' security. However, it is complex and challenging to implement cyber-security with the embedded control system. Therefore, converging operational technologies with information technologies such as high computational devices, communication, and network protocols makes it possible to secure the embedded process system from cyber-attack.

However, with the use of advancing technologies in Industry 4.0, communicating data from operational technologies to information technologies, the probability of accessing the data by a third party will be higher. Therefore, in the upcoming process safety era, with the digitalization and implementation of Industry 4.0, it is necessary to balance the process system safety and information security.

Current research and development also focus on detecting and defusing the intruder threat on a component level. If process components, such as sensors and actuators connected to the IoT or IIoT or controllers, are made accessible from the internet, the threatening intruder level for the components will be higher. The layer of protection includes firewalls, intrusion detection systems, and anti-virus software's may protect the control systems such as SCADA, DCS, PLC, and remote telemetry units (Gu, Yin, Fan, & Xu, 2019). IoT components' security in process system mostly focuses on building security from the start, developing hardware impenetrable to the intruders, ensuring software security and firmware updates, and performing dynamic testing (Prinsloo et al., 2019). However, it is critical to weigh the cost of security against the risks before implementing an IoT advancement in process systems.

3.3 Integrating cybersecurity with process safety in cyber-physical process systems

Cyber-physical process systems must have both safety and security features. The goal of safety is to protect systems from accidental failures to avoid hazards, whereas security is to protect systems against intentional human attacks (Tantawy, Erradi, & Abdelwahed, 2019). However, the ultimate goal of safety and security analysis is the same, and they can be considered the two sides of the same coin. As the days will be progressing, the gap between safety and security analysis will be narrowed, and these will be captured in unified frameworks.

The International Society of Automation (ISA) and International Electrotechnical Commission (IEC) proposed two international standards: ISA 84/IEC 61511 and ISA 99/IEC 62443, to address cyber-physical system safety. ISA 84/IEC 61511 standards are well defined in process safety by inventing a safety instrumented system. This standard is generally practiced in automated process plants (Bukowski, 2006). Likewise, ISA 99/IEC 62443 standards are used in control system security.

With the advancement of cyber-physical process models, the integration of process control and cyber systems is happening (Fujdiak, Mlynek, Blazek, Barabas, & Mrnustik, 2018). In the future, ISA 84/IEC 61511 and ISA 99/IEC 62443 should be integrated to provide a safer and secure environment in process plants. Therefore, concerning the future in process safety, Industry 4.0 and Safety 4.0 development will cause urgent and deep changes in education curricula for engineers or technicians and with that also in research and development (Khan et al., 2021; Narayanamurti, 2022).

References

Aazam, M., Zeadally, S., & Harras, K. A. (2018). Deploying fog computing in industrial internet of things and industry 4.0. *IEEE Transactions on Industrial Informatics*, *14*(10), 4674–4682. https://doi.org/10.1109/TII.2018.2855198.

Abbasinejad, R., Hourfar, F., & Elkamel, A. (2021). Optimum maintenance interval determination for field instrument devices in oil and gas industries based on expected utility theory. *Computers & Chemical Engineering*, *152*, 107362. https://doi.org/10.1016/j.compchemeng.2021.107362.

Agatonovic-Kustrin, S., & Beresford, R. (2000). Basic concepts of artificial neural network (ANN) modeling and its application in pharmaceutical research. *Journal of Pharmaceutical and Biomedical Analysis*, *22*(5), 717–727. https://doi.org/10.1016/S0731-7085(99)00272-1.

Alauddin, M., Khan, F., Imtiaz, S., & Ahmed, S. (2018). A bibliometric review and analysis of data-driven fault detection and diagnosis methods for process systems. *Industrial and Engineering Chemistry Research*, *57*(32), 10719–10735. https://doi.org/10.1021/acs.iecr.8b00936.

Alauddin, M., Khan, F., Imtiaz, S., & Ahmed, S. (2020). A variable mosquito flying optimization-based hybrid artificial neural network model for the alarm tuning of process fault detection systems. *Process Safety Progress*, *39*(S1). https://doi.org/10.1002/prs.12122.

Aldemir, T., Guarro, S., Mandelli, D., Kirschenbaum, J., & Sun, X. (2010). Probabilistic risk assessment modeling of digital instrumentation and control systems using two dynamic methodologies. *Reliability Engineering & System Safety*, *95*(10), 1011–1039. https://doi.org/10.1016/j.ress.2010.04.011.

Amin, M. T., Khan, F., Ahmed, S., & Imtiaz, S. (2020). A novel data-driven methodology for fault detection and dynamic risk assessment. *The Canadian Journal of Chemical Engineering*, *1–20*. https://doi.org/10.1002/cjce.23760.

Amin, M. T., Khan, F., Ahmed, S., & Imtiaz, S. (2021a). A data-driven Bayesian network learning method for process fault diagnosis. *Process Safety and Environmental Protection*, *150*, 110–122. https://doi.org/10.1016/j.psep.2021.04.004.

Amin, M. T., Khan, F., Ahmed, S., & Imtiaz, S. (2021b). Risk-based fault detection and diagnosis for nonlinear and non-Gaussian process systems using R-vine copula. *Process Safety and Environmental Protection*, *150*, 123–136. https://doi.org/10.1016/j.psep.2021.04.010.

Amin, M. T., Khan, F., & Amyotte, P. (2019). A bibliometric review of process safety and risk analysis. *Process Safety and Environmental Protection*, *126*. https://doi.org/10.1016/j.psep.2019.04.015.

Amin, M. T., Khan, F., & Zuo, M. J. (2019). A bibliometric analysis of process system failure and reliability literature. *Engineering Failure Analysis*, *106*, 104152. https://doi.org/10.1016/j.engfailanal.2019.104152.

Andersen, E. B., Udugama, I. A., Gernaey, K. V., Khan, A. R., & Kulahci, M. (2021). An easy to use GUI for simulating big data using Tennessee Eastman process. *Quality and Reliability Engineering International*. https://doi.org/10.1002/qre.2975.

Angelopoulos, A., Michailidis, E. T., Nomikos, N., Trakadas, P., & Zahariadis, T. (2020). Tackling faults in the industry 4.0 era—A survey of machine-learning solutions and key aspects. *Sensors (Switzerland)*, *20*(1), 1–34. https://doi.org/10.3390/s20010109.

Arena, S., Florian, E., Zennaro, I., Orrù, P. F., & Sgarbossa, F. (2022). A novel decision support system for managing predictive maintenance strategies based on machine learning approaches. *Safety Science*, *146*, 105529. https://doi.org/10.1016/j.ssci.2021.105529.

Arunthavanathan, R., Khan, F., Ahmed, S., & Imtiaz, S. (2021a). A deep learning model for process fault prognosis. *Process Safety and Environmental Protection*, *154*, 467–479. https://doi.org/10.1016/j.psep.2021.08.022.

Arunthavanathan, R., Khan, F., Ahmed, S., & Imtiaz, S. (2021b). An analysis of process fault diagnosis methods from safety perspectives. *Computers & Chemical Engineering*, *145*, 107197. https://doi.org/10.1016/j.compchemeng.2020.107197.

Arunthavanathan, R., Khan, F., Ahmed, S., & Imtiaz, S. (2022). Autonomous fault diagnosis and root cause analysis for the processing system using one-class SVM and NN permutation algorithm. *Industrial & Engineering Chemistry Research*. https://doi.org/10.1021/ACS.IECR.1C02731. acs.iecr.1c02731.

Arunthavanathan, R., Khan, F., Ahmed, S., Imtiaz, S., & Rusli, R. (2020). Fault detection and diagnosis in process system using artificial intelligence-based cognitive technique. *Computers & Chemical Engineering*, *134*, 106697. https://doi.org/10.1016/J.COMPCHEMENG.2019.106697.

Badri, A., Boudreau-Trudel, B., & Souissi, A. S. (2018). Occupational health and safety in the industry 4.0 era: A cause for major concern? *Safety Science*, *109*, 403–411. https://doi.org/10.1016/J.SSCI.2018.06.012.

Bajic, B., Rikalovic, A., Suzic, N., & Piuri, V. (2020). Industry 4.0 implementation challenges and opportunities: A managerial perspective. *IEEE Systems Journal, 15*(1), 546–559. https://doi.org/10.1109/JSYST.2020.3023041.

Barakat, M., Druaux, F., Lefebvre, D., Khalil, M., & Mustapha, O. (2011). Self adaptive growing neural network classifier for faults detection and diagnosis. *Neurocomputing, 74*(18), 3865–3876. https://doi.org/10.1016/j.neucom.2011.08.001.

Belli, L., Davoli, L., Medioli, A., Marchini, P. L., & Ferrari, G. (2019). Toward industry 4.0 with IoT: Optimizing business processes in an evolving manufacturing factory. *Frontiers in ICT, 6*, 17. https://doi.org/10.3389/fict.2019.00017.

Benson, C., Argyropoulos, C. D., Dimopoulos, C., Mikellidou, C. V., & Boustras, G. (2021). Safety and risk analysis in digitalized process operations warning of possible deviating conditions in the process environment. *Process Safety and Environmental Protection, 149*, 750–757. https://doi.org/10.1016/j.psep.2021.02.039.

Bi, X., & Zhao, J. (2021). A novel orthogonal self-attentive variational autoencoder method for interpretable chemical process fault detection and identification. *Process Safety and Environmental Protection, 156*, 581–597. https://doi.org/10.1016/j.psep.2021.10.036.

Borangiu, T., Trentesaux, D., Thomas, A., Leitão, P., & Barata, J. (2019). Digital transformation of manufacturing through cloud services and resource virtualization. In *Computers in industry* (Vol. 108, pp. 150–162). Elsevier. https://doi.org/10.1016/j.compind.2019.01.006.

Boyes, H., Hallaq, B., Cunningham, J., & Watson, T. (2018). The industrial internet of things (IIoT): An analysis framework. *Computers in Industry, 101*, 1–12. https://doi.org/10.1016/j.compind.2018.04.015.

Bukowski, J. V. (2006). Incorporating process demand into models for assessment of safety system performance. *Proceedings of Annual Reliability and Maintainability Symposium, 00*(C), 577–581. https://doi.org/10.1109/RAMS.2006.1677435.

Burges, C. J. C. (1998). A tutorial on support vector machines for pattern recognition. *Data Mining and Knowledge Discovery, 2*(2), 121–167. https://doi.org/10.1023/A:1009715923555.

Centobelli, P., Cerchione, R., & Ertz, M. (2020). Agile supply chain management: Where did it come from and where will it go in the era of digital transformation? *Industrial Marketing Management, 90*, 324–345. https://doi.org/10.1016/j.indmarman.2020.07.011.

Chen, C., Lu, N., Jiang, B., & Wang, C. (2021). A risk-averse remaining useful life estimation for predictive maintenance. *IEEE/CAA Journal of Automatica Sinica, 8*(2), 412–422. https://doi.org/10.1109/JAS.2021.1003835.

Chen, H., Tiňo, P., & Yao, X. (2014). Cognitive fault diagnosis in Tennessee Eastman process using learning in the model space. *Computers and Chemical Engineering, 67*, 33–42. https://doi.org/10.1016/j.compchemeng.2014.03.015.

Çinar, Z. M., Nuhu, A. A., Zeeshan, Q., Korhan, O., & Safaei, B. (2020). Machine learning in predictive maintenance towards sustainable smart manufacturing in industry 4.0. *Sustainability (Switzerland), 12*(19). https://doi.org/10.3390/su12198211.

Cohen, Y., & Singer, G. (2021). A smart process controller framework for industry 4.0 settings. *Journal of Intelligent Manufacturing*. https://doi.org/10.1007/s10845-021-01748-5. 0123456789.

Cortes, C., & Vapnik, V. (1995). Support-vector networks. *Machine Learning, 20*(3), 273–297. https://doi.org/10.1023/A:1022627411411.

Cui, Q., & Li, S. (2020). Process monitoring method based on correlation variable classification and vine copula. *The Canadian Journal of Chemical Engineering*, 1–18. https://doi.org/10.1002/cjce.23702.

de Beer, J., & Depew, C. (2021). The role of process engineering in the digital transformation. *Computers & Chemical Engineering, 154*, 107423. https://doi.org/10.1016/j.compchemeng.2021.107423.

De Tommaso, J., Rossi, F., Moradi, N., Pirola, C., & Galli, F. (2020). Experimental methods in chemical engineering: Process simulation. *The Canadian Journal of Chemical Engineering, 98*(11), 2301–2320. https://doi.org/10.1002/cjce.23857.

Dehuri, S., & Cho, S. B. (2009). Multi-criterion Pareto based particle swarm optimized polynomial neural network for classification: A review and state-of-the-art. *Computer Science Review, 3*(1), 19–40. https://doi.org/10.1016/j.cosrev.2008.11.002.

Delloite. (2021). *The future of remote working and impact on employment tax*. https://www2.deloitte.com/us/en/pages/tax/articles/the-impact-of-work-from-home-survey.html (last accessed on 2021-12-21).

Devold, H., Graven, T., & Halvorsrød, S. O. (2017). Digitalization of oil and gas facilities reduce cost and improve maintenance operations. In *Offshore Technology Conference*. https://doi.org/10.4043/27788-MS.

Feise, H. J., & Schaer, E. (2021). Mastering digitized chemical engineering. *Education for Chemical Engineers, 34*, 78–86. https://doi.org/10.1016/j.ece.2020.11.011.

Feng, X., Dai, Y., Ji, X., Zhou, L., & Dang, Y. (2021). Application of natural language processing in HAZOP reports. *Process Safety and Environmental Protection, 155*, 41–48. https://doi.org/10.1016/j.psep.2021.09.001.

Fisher, O. J., Watson, N. J., Escrig, J. E., Witt, R., & Gomes, R. L. (2020). Considerations, challenges and opportunities when developing data-driven models for process manufacturing systems. *Computers and Chemical Engineering, 140*, 106881. https://doi.org/10.1016/j.compchemeng.2020.106881.

Fujdiak, R., Mlynek, P., Blazek, P., Barabas, M., & Mrnustik, P. (2018). Seeking the relation between performance and security in modern systems: Metrics and measures. In *2018 41st international conference on telecommunications and signal processing, TSP 2018*. https://doi.org/10.1109/TSP.2018.8441496.

Gallacher, G., & Hossain, I. (2020). Remote work and employment dynamics under COVID-19: Evidence from Canada. *Canadian Public Policy, 46*(S1), S44–S54. https://doi.org/10.3138/cpp.2020-026.

Ge, Z., Song, Z., & Gao, F. (2013). Review of recent research on data-based process monitoring. *Industrial & Engineering Chemistry Research, 52*(10), 3543–3562. https://doi.org/10.1021/ie302069q.

Georgios, L., Kerstin, S., & Theofylaktos, A. (2019). Internet of things in the context of industry 4.0: An overview. *International Journal of Entrepreneurial Knowledge*, 7(1), 4–19. https://doi.org/10.2478/ijek-2019-0001.

Ghobakhloo, M. (2020). Industry 4.0, digitization, and opportunities for sustainability. *Journal of Cleaner Production, 252*, 119869. https://doi.org/10.1016/j.jclepro.2019.119869.

Giunta, G., Bernasconi, G., Giro, R. A., & Cesari, S. (2020). Digital transformation of historical data for advanced predictive maintenance. In *Abu Dhabi international petroleum exhibition & conference*. https://doi.org/10.2118/202906-MS.

Gordon, C. A. K., Burnak, B., Onel, M., & Pistikopoulos, E. N. (2020). Data-driven prescriptive maintenance: Failure prediction using ensemble support vector classification for optimal process and maintenance scheduling. *Industrial & Engineering Chemistry Research, 59*(44), 19607–19622. https://doi.org/10.1021/acs.iecr.0c03241.

Gordon, C. A. K., & Pistikopoulos, E. N. (2021). Data-driven prescriptive maintenance toward fault-tolerant multiparametric control. *AIChE Journal*, e17489. https://doi.org/10.1002/aic.17489.

Gu, A., Yin, Z., Fan, C., & Xu, F. (2019). Safety framework based on blockchain for intelligent manufacturing cyber physical system. In *1st international conference on industrial artificial intelligence, IAI 2019, January*. https://doi.org/10.1109/IAI47267.2019.9085328.

Gu, A., Yin, Z., Li, Y., Su, B., & Shen, L. (2018). Functional safety assessment and security protection framework of cyber physical machine tool system. In *2018 IEEE 4th international conference on computer and communications, ICCC 2018, 2489-2494*. https://doi.org/10.1109/COMPCOMM.2018.8780915.

Gungor, V. C., & Hancke, G. P. (2009). Industrial wireless sensor networks: Challenges, design principles, and technical approaches. *IEEE Transactions on Industrial Electronics*, *56*(10), 4258–4265. https://doi.org/10.1109/TIE.2009.2015754.

Guo, K., Wan, X., Liu, L., Gao, Z., & Yang, M. (2021). Fault diagnosis of intelligent production line based on digital twin and improved random forest. *Applied Sciences*, *11*(16), 7733.

Haghighat, A. (2020). *Monte Carlo methods for particle transport* (2nd ed.). CRC Press.

Hai, T. N., Van, Q. N., & Tuyet, M. N. T. (2021). Digital transformation: Opportunities and challenges for leaders in the emerging countries in response to Covid-19 pandemic. *Emerging Science Journal*, *5*, 21–36. https://doi.org/10.28991/esj-2021-SPER-03.

Haigh, T., Priestley, P. M., Priestley, M., & Rope, C. (2016). *ENIAC in action: Making and remaking the modern computer*. MIT Press.

He, R., Chen, G., Dong, C., Sun, S., & Shen, X. (2019). Data-driven digital twin technology for optimized control in process systems. *ISA Transactions*, *95*, 221–234. https://doi.org/10.1016/j.isatra.2019.05.011.

Hu, L., & Olivieri, M. (2021). How to leverage digital marketing in B2B: A study on Italian SMEs during the Covid-19 outbreak. *Digital Marketing & ECommerce Conference*, *112–120*. https://doi.org/10.1007/978-3-030-76520-0_12.

Hu, S., Wang, S., Su, N., Li, X., & Zhang, Q. (2021). Digital twin based reference architecture for petrochemical monitoring and fault diagnosis. *Oil & Gas Science and Technology—Revue d'IFP Energies Nouvelles*, *76*, 9. https://doi.org/10.2516/ogst/2020095.

Iaiani, M., Tugnoli, A., Bonvicini, S., & Cozzani, V. (2021). Analysis of cybersecurity-related incidents in the process industry. *Reliability Engineering & System Safety*, *209*, 107485. https://doi.org/10.1016/j.ress.2021.107485.

Il, N. (1970). KMTS-1.6 digital process monitoring pressure gage. *Instruments and Experimental Techniques—USSR*, *5*, 1538.

Jabłoński, M., & Jabłoński, A. (2021). Shaping the safety culture of high reliability organizations through digital transformation. *Energies*, *14*(16), 4721. https://doi.org/10.3390/en14164721.

Jefferson, M., Chung, P. W. H., & Kletz, T. A. (1997). Learning the lessons from past accidents. *Institution of Chemical Engineers Symposium Series*, *141*, 217–226.

Jenner, F. (2021). *Chemical companies see a quantum leap through digitalization—From evolutionary to revolutionary and now disruptive*. https://www.ey.com/en_ca/advanced-manufacturing/how-long-before-the-digital-disruption-reaches-the-chemicals-industry.

Ji, X., He, G., Xu, J., & Guo, Y. (2016). Study on the mode of intelligent chemical industry based on cyber-physical system and its implementation. *Advances in Engineering Software*, *99*, 18–26. https://doi.org/10.1016/J.ADVENGSOFT.2016.04.010.

Jiang, Y., Yin, S., & Kaynak, O. (2020). Performance supervised plant-wide process monitoring in industry 4.0: A roadmap. *IEEE Open Journal of the Industrial Electronics Society*, *2*, 21–35. https://doi.org/10.1109/OJIES.2020.3046044.

Juang, C. F. (2004). A hybrid of genetic algorithm and particle swarm optimization for recurrent network design. *IEEE Transactions on Systems, Man, and Cybernetics, Part B: Cybernetics*, *34*(2), 997–1006. https://doi.org/10.1109/TSMCB.2003.818557.

Jung, N. (2020). The role of process engineering in the digital transformation. In *Computer aided chemical engineering* (Vol. 48, pp. 721–726). Elsevier. https://doi.org/10.1016/B978-0-12-823377-1.50121-X.

Kamalaldin, A., Sjödin, D., Hullova, D., & Parida, V. (2021). Configuring ecosystem strategies for digitally enabled process innovation: A framework for equipment suppliers in the process industries. *Technovation*, *102250*. https://doi.org/10.1016/j.technovation.2021.102250.

Kamble, S. S., Gunasekaran, A., & Gawankar, S. A. (2018). Sustainable industry 4.0 framework: A systematic literature review identifying the current trends and future perspectives. *Process Safety and Environmental Protection*, *117*, 408–425. https://doi.org/10.1016/J.PSEP.2018.05.009.

Kane, G. C., Palmer, D., Phillips, A. N., Kiron, D., & Buckley, N. (2015). Strategy, not technology, drives digital transformation. *MIT Sloan Management Review and Deloitte University Press*, *14*(1–25).

Kaspar, M. H., & Harmon Ray, W. (1993). Dynamic PLS modelling for process control. *Chemical Engineering Science*, *48*(20), 3447–3461. https://doi.org/10.1016/0009-2509(93)85001-6.

Kender, R., Kaufmann, F., Rößler, F., Wunderlich, B., & Klein, H. (2021). Development of a digital twin for a flexible air separation unit using a pressure-driven simulation approach. *Computers & Chemical Engineering*, *151*, 107349. https://doi.org/10.1016/j.compchemeng.2021.107349.

Kerr, P. K., Rollins, J., & Theohary, C. A. (2011). The Stuxnet computer worm: Harbinger of an emerging warfare capability. In *The Stuxnet computer worm and industrial control system security* (pp. 1–10). Congressional Research Service.

Khan, F. I., & Abbasi, S. A. (1999a). MAXCRED—A new software package for rapid risk assessment in chemical process industries. *Environmental Modelling & Software*, *14*(1), 11–25. https://doi.org/10.1016/S1364-8152(97)00031-5.

Khan, F. I., & Abbasi, S. A. (1999b). PROFAT: A user friendly system for probabilistic fault tree analysis. *Process Safety Progress*, *18*(1), 42–49. https://doi.org/10.1002/prs.680180109.

Khan, F., Amyotte, P., & Adedigba, S. (2021). Process safety concerns in process system digitalization. *Education for Chemical Engineers*, *34*, 33–46. https://doi.org/10.1016/j.ece.2020.11.002.

Khan, F., Rathnayaka, S., & Ahmed, S. (2015). Methods and models in process safety and risk management: Past, present and future. *Process Safety and Environmental Protection*, *98*, 116–147. https://doi.org/10.1016/j.psep.2015.07.005.

Kim, Y. H., Kim, K. H., Shin, D. P., & Yoon, E. S. (2000). Consequence analysis, emergency response and planning for an integrated risk management system: System design and implementation. *Institution of Chemical Engineers Symposium Series*, *148*, 611–620.

Kourti, T., & MacGregor, J. F. (1995). Process analysis, monitoring and diagnosis, using multivariate projection methods. *Chemometrics and Intelligent Laboratory Systems*, *28*(1), 3–21. https://doi.org/10.1016/0169-7439(95)80036-9.

Kutnjak, A. (2021). Covid-19 accelerates digital transformation in industries: Challenges, issues, barriers and problems in transformation. *IEEE Access*, *9*, 79373–79388. https://doi.org/10.1109/ACCESS.2021.3084801.

Lawrence, R. L., & Wright, A. (2001). Rule-based classification systems using classification and regression tree (CART) analysis. *Photogrammetric Engineering and Remote Sensing*, *67*(10), 1137–1142.

Lee, J., Cameron, I., & Hassall, M. (2019). Improving process safety: What roles for digitalization and industry 4.0? *Process Safety and Environmental Protection*, *132*, 325–339. https://doi.org/10.1016/j.psep.2019.10.021.

Lee, C.-H., Wang, D., Desouza, K., & Evans, R. (2021). Digital transformation and the new normal in China: How can enterprises use digital technologies to respond to COVID-19? *Sustainability*, *13*(18), 10195. https://doi.org/10.3390/su131810195.

Lees, M. J., & Johnstone, M. C. (2021). Implementing safety features of industry 4.0 without compromising safety culture. *IFAC—PapersOnLine*, *54*(13), 680–685. https://doi.org/10.1016/j.ifacol.2021.10.530.

Li, G., Qin, S. J., & Yuan, T. (2016). Data-driven root cause diagnosis of faults in process industries. *Chemometrics and Intelligent Laboratory Systems*, *159*, 1–11. https://doi.org/10.1016/j.chemolab.2016.09.006.

Li, Z., Wang, Y., & Wang, K. S. (2017). Intelligent predictive maintenance for fault diagnosis and prognosis in machine centers: Industry 4.0 scenario. *Advances in Manufacturing*, *5*(4), 377–387. https://doi.org/10.1007/s40436-017-0203-8.

Li, X., Zhang, L., Khan, F., & Han, Z. (2021). A data-driven corrosion prediction model to support digitization of subsea operations. *Process Safety and Environmental Protection*, *153*, 413–421. https://doi.org/10.1016/j.psep.2021.07.031.

Li, H., Zhao, W., Zhang, Y., & Zio, E. (2020). Remaining useful life prediction using multi-scale deep convolutional neural network. *Applied Soft Computing*, *89*, 106113. https://doi.org/10.1016/J.ASOC.2020.106113.

Lv, Y., Zhou, Q., Li, Y., & Li, W. (2021). A predictive maintenance system for multi-granularity faults based on AdaBelief-BP neural network and fuzzy decision making. *Advanced Engineering Informatics*, *49*, 101318. https://doi.org/10.1016/j.aei.2021.101318.

Medjek, F., Tandjaoui, D., Djedjig, N., & Romdhani, I. (2021). Fault-tolerant AI-driven intrusion detection system for the internet of things. *International Journal of Critical Infrastructure Protection*, *34*, 100436. https://doi.org/10.1016/j.ijcip.2021.100436.

Miller, T., Staves, A., Maesschalck, S., Sturdee, M., & Green, B. (2021). Looking back to look forward: Lessons learnt from cyber-attacks on industrial control systems. *International Journal of Critical Infrastructure Protection*, *35*, 100464. https://doi.org/10.1016/J.IJCIP.2021.100464.

Nadkarni, S., & Prügl, R. (2021). Digital transformation: A review, synthesis and opportunities for future research. *Management Review Quarterly*, *71*(2), 233–341. https://doi.org/10.1007/s11301-020-00185-7.

Nalimov, V. V., & Mul'chenko, Z. M. (1971). *Measurement of science. Study of the development of science as an information process*. National Technical Information Service.

Narayanamurti, V. (2022). *Technoscientific research—A missing term in R&D discourse | national academies*. National Academy of Engineering (NAE).

Nawi, N. M., Khan, A., & Rehman, M. Z. (2013). A new Levenberg Marquardt based back propagation algorithm trained with Cuckoo search. *Procedia Technology*, *11*(Iceei), 18–23. https://doi.org/10.1016/j.protcy.2013.12.157.

Nelson, G., & Ellis, S. (2019). The history and impact of digitization and digital data mobilization on biodiversity research. *Philosophical Transactions of the Royal Society B*, *374*(1763). https://doi.org/10.1098/rstb.2017.0391. 20170391.

Niu, X., & Yang, X. (2021). A novel one-dimensional convolutional neural network architecture for chemical process fault diagnosis. *The Canadian Journal of Chemical Engineering*. https://doi.org/10.1002/cjce.24087.

Patel, K. M., Bakhurji, A. S., Salloum, H. S., Kim, H. K., & Mubarak, S. (2019). Use of advanced process control for automating conventional oilfield operations. *SPE Production & Operations*, *34*(04), 678–692. https://doi.org/10.2118/192393-PA.

Pawar, S., San, O., Aksoylu, B., Rasheed, A., & Kvamsdal, T. (2021). Physics guided machine learning using simplified theories. *Physics of Fluids*, *33*(1), 011701. https://doi.org/10.1063/5.0038929.

Pflaum, A. A., & Gölzer, P. (2018). The IoT and digital transformation: Toward the data-driven enterprise. *IEEE Pervasive Computing*, *17*(1), 87–91. https://doi.org/10.1109/MPRV.2018.011591066.

Pistikopoulos, E. N., Barbosa-Povoa, A., Lee, J. H., Misener, R., & Gani, R. (2021). Process systems engineering—The generation next? *Computers & Chemical Engineering*, *147*, 107252. https://doi.org/10.1016/j.compchemeng.2021.107252.

Pivoto, D. G. S., de Almeida, L. F. F., da Rosa Righi, R., Rodrigues, J. J. P. C., & Alberti, A. M. (2021). Cyber-physical systems architectures for industrial internet of things applications in industry 4.0: A literature review. *Journal of Manufacturing Systems*, *58*, 176–192. https://doi.org/10.1016/J.JMSY.2020.11.017.

Prinsloo, J., Sinha, S., & von Solms, B. (2019). A review of industry 4.0 manufacturing process security risks. *Applied Sciences (Switzerland)*, *9*(23). https://doi.org/10.3390/app9235105.

Progoulakis, I., Nikitakos, N., Rohmeyer, P., Bunin, B., & Karamperidis, S. (2021). Perspectives on cyber security for offshore oil and gas assets. *Journal of Marine Science and Engineering*, *9*(2), 112. https://doi.org/10.3390/jmse9020112.

Qin, S. J. (1998). Recursive PLS algorithms for adaptive data modeling. *Computers and Chemical Engineering*, *22*(4–5), 503–514. https://doi.org/10.1016/s0098-1354(97)00262-7.

Rato, T. J., & Reis, M. S. (2020). An integrated multiresolution framework for quality prediction and process monitoring in batch processes. *Journal of Manufacturing Systems*, *57*, 198–216. https://doi.org/10.1016/j.jmsy.2020.09.007.

Reis, M. S., & Saraiva, P. M. (2021). Data-centric process systems engineering: A push towards PSE 4.0. *Computers & Chemical Engineering*, *155*, 107529. https://doi.org/10.1016/j.compchemeng.2021.107529.

Ren, X., Zhu, K., Cai, T., & Li, S. (2017). Fault detection and diagnosis for nonlinear and non-Gaussian processes based on copula subspace division. *Industrial and Engineering Chemistry Research*, *56*(40), 11545–11564. https://doi.org/10.1021/acs.iecr.7b02419.

Sargent, R. W. H. (1983). Advances in modelling and analysis of chemical process systems. *Computers & Chemical Engineering*, 7(4), 219–237. https://doi.org/10.1016/0098-1354(83)80013-1.

Schleihauf, W. (2001). The Dumaresq and the Dreyer: Part III. *Warship International*, *38*(3), 221–233.

Schmidhuber, J. (2015). Deep learning in neural networks: An overview. *Neural Networks*, *61*, 85–117. https://doi.org/10.1016/j.neunet.2014.09.003.

Shah, S. Y., Yuan, Z., Lu, S., & Zerfos, P. (2018). Dependency analysis of cloud applications for performance monitoring using recurrent neural networks. In *2017 IEEE international conference on big data (big data)* (pp. 1534–1543). https://doi.org/10.1109/bigdata.2017.8258087.

Shivajee, V., Singh, R. K., & Rastogi, S. (2019). Manufacturing conversion cost reduction using quality control tools and digitization of real-time data. *Journal of Cleaner Production*, *237*, 117678. https://doi.org/10.1016/j.jclepro.2019.117678.

Sikorski, M. (1991). Use of digital simulation in reliability analysis for the design of industrial process control systems. *Reliability Engineering & System Safety*, *31*(3), 281–295. https://doi.org/10.1016/0951-8320(91)90072-F.

Sinnamon, R. M., & Andrews, J. D. (1997). New approaches to evaluating fault trees. *Reliability Engineering & System Safety*, *58*(2), 89–96. https://doi.org/10.1016/S0951-8320(96)00036-1.

Slay, J., & Miller, M. (2007). Lessons learned from the Maroochy water breach. *IFIP International Federation for Information Processing*, *253*, 73–82. https://doi.org/10.1007/978-0-387-75462-8_6.

Spöttl, G. (2017). Skilled workers: Are they the losers of "Industry 4.0"? In *Advances in ergonomic design of systems, products and processes* (pp. 73–87). Springer. https://doi.org/10.1007/978-3-662-53305-5_5.

Stavrianidis, P., & Bhimavarapu, K. (1998). Safety instrumented functions and safety integrity levels (SIL). *ISA Transactions*, *37*(4), 337–351. https://doi.org/10.1016/S0019-0578(98)00038-X.

Stefatos, G., & Hamza, A. B. (2010). Dynamic independent component analysis approach for fault detection and diagnosis. *Expert Systems with Applications, 37*(12), 8606–8617. https://doi.org/10.1016/j.eswa.2010.06.101.

Stephanopoulos, G. (2009). Process systems engineering: From Solvay to the 21st century. A history of development, successes and prospects for the future. In *Computer aided chemical engineering* (Vol. 27, pp. 149–155). Elsevier. https://doi.org/10.1016/S1570-7946(09)70246-9.

Stephanopoulos, G., & Reklaitis, G. V. (2011). Process systems engineering: From Solvay to modern bio-and nanotechnology: A history of development, successes and prospects for the future. *Chemical Engineering Science, 66*(19), 4272–4306. https://doi.org/10.1016/j.ces.2011.05.049.

Stepney, S., & Hickinbotham, S. J. (2018). UCOMP roadmap: Survey, challenges, recommendations. In *Computational Matter* (pp. 9–31). Springer. https://doi.org/10.1007/978-3-319-65826-1_2.

Subudhi, B., & Jena, D. (2011a). A differential evolution based neural network approach to nonlinear system identification. *Applied Soft Computing Journal, 11*(1), 861–871. https://doi.org/10.1016/j.asoc.2010.01.006.

Subudhi, B., & Jena, D. (2011b). Nonlinear system identification using memetic differential evolution trained neural networks. *Neurocomputing, 74*(10), 1696–1709. https://doi.org/10.1016/j.neucom.2011.02.006.

Tamssaouet, F., Nguyen, K. T. P., Medjaher, K., & Orchard, M. E. (2021). Degradation modeling and uncertainty quantification for system-level prognostics. *IEEE Systems Journal, 15*(2), 1628–1639. https://doi.org/10.1109/JSYST.2020.2983376.

Tang, W., & Daoutidis, P. (2019). Distributed control and optimization of process system networks: A review and perspective. *Chinese Journal of Chemical Engineering, 27*(7), 1461–1473.

Tantawy, A., Erradi, A., & Abdelwahed, S. (2019). A modified layer of protection analysis for cyber-physical systems security. In *2019 4th international conference on system reliability and safety, ICSRS 2019* (pp. 94–101). https://doi.org/10.1109/ICSRS48664.2019.8987635.

Thissen, U., Swierenga, H., De Weijer, A. P., Wehrens, R., & Buydens, L. M. C. (2005). Multivariate statistical process control using mixture modelling. *Journal of Chemometrics, 19*(1), 23–31. https://doi.org/10.1002/cem.903.

Tidriri, K., Chatti, N., Verron, S., & Tiplica, T. (2016). Bridging data-driven and model-based approaches for process fault diagnosis and health monitoring: A review of researches and future challenges. *Annual Reviews in Control, 42*, 63–81. https://doi.org/10.1016/J.ARCONTROL.2016.09.008.

Topolski, M. (2021). Application of feature extraction methods for chemical risk classification in the pharmaceutical industry. *Sensors, 21*(17), 5753.

Udugama, I. A., Gargalo, C. L., Yamashita, Y., Taube, M. A., & Bayer, C. (2020). The role of big data in industrial (bio)chemical process operations. *Industrial & Engineering Chemistry Research, 59*(34), 15283–15297. https://doi.org/10.1021/acs.iecr.0c01872.

Urso, J. (2020). Remote operations. *Control Engineering, 67*(5), 14–15. link.gale.com/apps/doc/A628178247/AONE?u=anon~6692bcc9&sid=googleScholar&xid=d7e392c8.

Venkatasubramanian, V. (2001). Process fault detection and diagnosis: Past, present and future. *IFAC Proceedings Volumes, 34*(27), 1–13. https://doi.org/10.1016/s1474-6670(17)33563-2.

Vial, G. (2019). Understanding digital transformation: A review and a research agenda. *The Journal of Strategic Information Systems, 28*(2), 118–144. https://doi.org/10.1016/j.jsis.2019.01.003.

Villalpando-Hernandez, R., Munoz-Rodriguez, D., & Vargas-Rosales, C. (2021). Relational positioning method for 2D and 3D ad hoc sensor networks in industry 4.0. *Applied Sciences, 11*(19), 8907. https://doi.org/10.3390/app11198907.

Vrignat, P., Kratz, F., & Avila, M. (2022). Sustainable manufacturing, maintenance policies, prognostics and health management: A literature review. *Reliability Engineering & System Safety, 218*. https://doi.org/10.1016/j.ress.2021.108140, 108140.

Wang, B. (2021). Safety intelligence as an essential perspective for safety management in the era of safety 4.0: From a theoretical to a practical framework. *Process Safety and Environmental Protection, 148*, 189–199. https://doi.org/10.1016/J.PSEP.2020.10.008.

Wang, C., Bi, Z., & Xu, L. D. (2014). IoT and cloud computing in automation of assembly modeling systems. *IEEE Transactions on Industrial Informatics, 10*(2), 1426–1434. https://doi.org/10.1109/TII.2014.2300346.

Wang, B., & Wang, Y. (2021). Big data in safety management: An overview. *Safety Science, 143*. https://doi.org/10.1016/j.ssci.2021.105414, 105414.

Wassar, T., Mutlu, M., Franchek, M. A., & Hattab, O. (2019). Model-based health monitoring of annular blowout preventers. *SPE Drilling & Completion, 34*(4), 458–467. https://doi.org/10.2118/197041-PA.

Wattis, Z., Murray, P., Baldan, O., Johnson, M., & Spitzenberger, C. (2021). Asset digital twins: Integrating process safety risks. *Chemical Engineering Progress, 117*(11), 47–51. https://doi.org/10.3390/app11167733.

Widodo, A., & Yang, B. S. (2007). Support vector machine in machine condition monitoring and fault diagnosis. *Mechanical Systems and Signal Processing, 21*(6), 2560–2574. https://doi.org/10.1016/j.ymssp.2006.12.007.

Willey, R. J. (2014). Layer of protection analysis. *Procedia Engineering, 84*, 12–22. https://doi.org/10.1016/J.PROENG.2014.10.405.

Wu, J., Lind, M., Zhang, X., Pardhasaradhi, K., & Myllerup, C. M. (2021). Knowledge acquisition and representation for intelligent operation support in offshore fields. *Process Safety and Environmental Protection, 155*, 415–443. https://doi.org/10.1016/j.psep.2021.09.036.

Xie, S., Chen, Y., Dong, S., & Zhang, G. (2020). Risk assessment of an oil depot using the improved multi-sensor fusion approach based on the cloud model and the belief Jensen-Shannon divergence. *Journal of Loss Prevention in the Process Industries, 67*, 104214. https://doi.org/10.1016/j.jlp.2020.104214.

Xu, S., & Bao, J. (2009). Distributed control of plantwide chemical processes. *Journal of Process Control, 19*(10), 1671–1687. https://doi.org/10.1016/j.jprocont.2009.07.007.

Xu, B., & Zhu, Q. (2021). Concurrent auto-regressive latent variable model for dynamic anomaly detection. *Journal of Process Control, 108*, 1–11. https://doi.org/10.1016/j.jprocont.2021.10.012.

Yu, J. (2013). A new fault diagnosis method of multimode processes using Bayesian inference based Gaussian mixture contribution decomposition. *Engineering Applications of Artificial Intelligence, 26*(1), 456–466. https://doi.org/10.1016/j.engappai.2012.09.003.

Yu, H., Khan, F., & Garaniya, V. (2015). A probabilistic multivariate method for fault diagnosis of industrial processes. *Chemical Engineering Research and Design, 104*, 306–318. https://doi.org/10.1016/j.cherd.2015.08.026.

Yu, T., & Qi, Y. (2021). Intelligent fault diagnosis of chemical process based on twin convolution deep neural network of wavelet transform. In *2021 IEEE 10th Data driven control and learning systems conference (DDCLS)* (pp. 100–107). https://doi.org/10.1109/DDCLS52934.2021.9455551.

Yu, J., & Qin, S. J. (2009). Multiway gaussian mixture model based multiphase batch process monitoring. *Industrial and Engineering Chemistry Research, 48*(18), 8585–8594. https://doi.org/10.1021/ie900479g.

Zhang, S., Bi, K., & Qiu, T. (2020). Bidirectional recurrent neural network-based chemical process fault diagnosis. *Industrial and Engineering Chemistry Research*, *59*(2), 824–834. https://doi.org/10.1021/acs.iecr.9b05885.

Zhang, Y., & Qin, S. J. (2007). Fault detection of nonlinear processes using multiway kernel independent component analysis. *Industrial & Engineering Chemistry Research*, *46*(23), 7780–7787. https://doi.org/10.1021/ie070381q.

Zhang, S., Yan, Y., Shi, L., Li, L., & Yan, X. (2021). A semi-empirical model for underground gas storage injection-production string time series remaining useful life analysis in process safety operation. *Process Safety and Environmental Protection*, *154*, 1–17. https://doi.org/10.1016/j.psep.2021.08.003.

Zhang, Y., & Zhang, Y. (2010). Fault detection of non-Gaussian processes based on modified independent component analysis. *Chemical Engineering Science*, *65*(16), 4630–4639. https://doi.org/10.1016/j.ces.2010.05.010.

Zhao, H., Liu, J., Dong, W., Sun, X., & Ji, Y. (2017). An improved case-based reasoning method and its application on fault diagnosis of Tennessee Eastman process. *Neurocomputing*, *249*, 266–276. https://doi.org/10.1016/j.neucom.2017.04.022.

Zhao, H., Sun, S., & Jin, B. (2018). Sequential fault diagnosis based on LSTM neural network. *IEEE Access*, *6*, 12929–12939. https://doi.org/10.1109/ACCESS.2018.2794765.

Zonta, T., da Costa, C. A., da Rosa Righi, R., de Lima, M. J., & Li, G. P. (2020). Predictive maintenance in the industry 4.0: A systematic literature review. *Computers and Industrial Engineering*, *150*(August). https://doi.org/10.1016/j.cie.2020.106889.

Further reading

Li, X., Peng, L., Yao, X., Cui, S., & Chi, T. (2017). Long short-term memory neural network for air pollutant concentration predictions: Method development and evaluation. *Environmental Pollution*, *231*, 997–1004. https://doi.org/10.1016/j.envpol.2017.08.114.

Yu, L., Qu, J., Gao, F., & Tian, Y. (2019). A novel hierarchical algorithm for bearing fault diagnosis based on stacked LSTM. *Shock and Vibration*, *2019*. https://doi.org/10.1155/2019/2756284.

Zhou, D., Li, G., & Qin, S. J. (2010). Total projection to latent structures for process monitoring. *AIChE Journal*, *56*(1), 168–178. https://doi.org/10.1002/aic.11977.

CHAPTER THREE

Data-driven approaches: Use of digitized operational data in process safety

Yiming Bai[a], Shuaiyu Xiang[a], Zeheng Zhao[a], Borui Yang[a], and Jinsong Zhao[a,b,*]

[a]State Key Laboratory of Chemical Engineering, Department of Chemical Engineering, Tsinghua University, Beijing, China
[b]Beijing Key Laboratory of Industrial Big Data System and Application, Tsinghua University, Beijing, China
*Corresponding author: e-mail address: jinsongzhao@tsinghua.edu.cn

Contents

1. Importance of data in process safety

The continuous development of the economy and technology has caused an enormous increase in industry, which inevitably creates new hazards and increased risks. Against the background of complex failures, enhancing effective process safety is an effective approach to reduce the likelihood and severity of industrial accidents (Khan & Abbasi, 1999).

Methods in Chemical Process Safety, Volume 6
ISSN 2468-6514
https://doi.org/10.1016/bs.mcps.2022.04.002

Process safety plays a vital role in the strategies of hazard identification and analysis, risk assessment and evaluation, safety measures, and safe critical decision-making (Khan, Rathnayaka, & Ahmed, 2015).

The origin and progression of process safety is related to the major process accidents of the last century. Traditional process safety methods use qualitative or semi-quantitative approaches and models for hazard identification, risk assessment, and safety management (Khan et al., 2015). A widely used qualitative hazard identification method is hazard and operability analysis (HAZOP). HAZOP was developed to identify equipment failures leading to accidents (Kletz, 1999). Then an automation tool called CHECKOP was developed to optimize HAZOP to accomplish the goals of batch operation (Palmer & Chung, 2008, 2009). Multiple factors were introduced to HAZOP to achieve multi-task hazard analysis (Cagno, Caron, & Mancini, 2002; Kennedy & Kirwan, 1998). Additionally, HAZOP was integrated into other methods to further improve its capability to identify hazards (Wang & Gao, 2012). Then the exploitation of automatic HAZOP based on expert knowledge improved the efficiency of hazard identification, for example, HAZOPEX (Karvonen, Heino, & Suokas, 1990), optHAZOP (Khan & Abbasi, 1997), EXPERTOP (Khan & Abbasi, 2000a), and HAZOPExpert (Venkatasubramanian & Vaidhyanathan, 1994). Semi-quantitative analysis methods that produce approximate results are effective for process safety when process data cannot be measured directly. Layer of protection analysis (LOPA) is a widely applied semi-quantitative hazard analysis methodology that is typically performed in conjunction with other methods (Summers, 2003). Similarly, LOPA approaches that consider multiple factors and are based on expert systems have been intensively researched (Baybutt, 2002).

Quantitative analysis methods provide specific numerical estimates to enhance the understanding of faults, which helps operators to improve their responses to emergencies. Quantitative risk assessment (QRA), proposed in the aerospace industry in 1960, can be used to quantify the risk of hazardous chemical spills (Covello & Mumpower, 1985). To overcome the inability of QRA to change over time, Marseguerra et al. first applied artificial neural networks (ANNs) to it (Marseguerra, Ricotti, & Zio, 1995). Multiple failure scenarios have been accurately estimated and analyzed by building specific models, such as release and dispersion, fire, and explosion (Kukkonen & Nikmo, 1992; Matthias, 1990; Vandroux-Koenig & Berthoud, 1997). Hybrid analysis is a combination of qualitative and quantitative methods that can provide a comprehensive analysis of process safety. Risk assessment in

QRA has been determined using fault tree analysis and other further developed analysis methods, such as dynamic fault tree analysis (Čepin & Mavko, 2002), computer-aided fault tree analysis (Khan & Abbasi, 2000b), event tree analysis (Vílchez, Espejo, & Casal, 2011), bow-tie analysis (Markowski & Kotynia, 2011), and Bayesian network analysis (Kohda & Cui, 2007).

With increasing complexity and digitalization, the chemical industry has urgent requirements for process safety methods (Lee, Cameron, & Hassall, 2019). Additionally, researchers have also increased their focus on quantitative and dynamic issues in the field of process safety (Khan et al., 2015). And the availability and operability of large amounts of data used for process safety are becoming increasingly crucial (Dai, Cheng, Wu, Wu, & Zhao, 2020). Srinivasan and Zhao presented developments related to data in process safety for the following three aspects: deeper understanding of hazards, enhancing process safety management, and safer operations (Srinivasan & Zhao, 2022).

2. Data source

Data for the risk analysis of chemical processes are still largely derived from laboratories and safety management systems. Data processed for dynamic risk assessment, in most cases, originate from data monitored by sensors in automated systems.

Since the advent of the Big Data era, the means by which humans obtain data have been greatly enriched. In addition to the development of various monitoring instruments and sensors, the Internet of Things (IoT) has connected these data. As shown in Fig. 1, the information obtained through the IoT provides a perspective about the risk assessment process.

(1) Process data

At present, chemical processes are commonly equipped with distributed control systems (DCS) to improve quality and safety control. Consequently, the cost of acquiring data has significantly reduced, and process data can be easily and cheaply accessed from DCS.

(2) Alarm data

With the full deployment of DCS in chemical plants, the cost of acquiring and monitoring process information has significantly reduced, and it has become more pragmatic to set alarms for certain important monitoring variables. Alarm data are a collection of abnormal process data, which is a condensed form of data, and thus have become an important indicator in the risk assessment process.

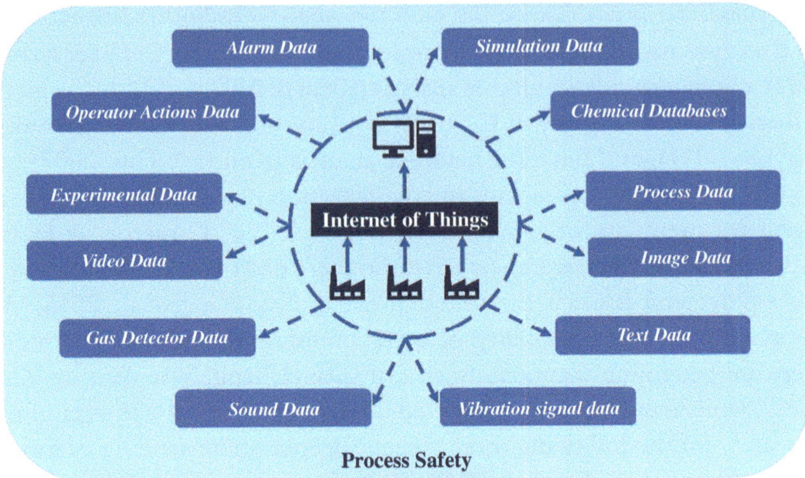

Fig. 1 Data from the IoT in a chemical plant for process safety.

(3) Experimental data

As laboratory information management systems are installed in many more modern laboratories and research facilities, and as an increasing number of institutions begin to make raw data from research publicly available, the process of obtaining experimental data for risk analysis is becoming practically feasible.

(4) Operator action data

In the event of significant changes in process conditions or unexpected scenarios, manual actions can replace automated actions to guarantee the safety and stability of the process. These operator actions and their results are also recorded in logs and used as information when responding to abnormal scenarios.

(5) Image data

With the creation of shared databases, image datasets for image recognition and inspection are increasingly being used in industrial scenarios. Typically, each large factory has tens to hundreds of cameras, and the amount of data collected by these cameras is substantial.

(6) Video data

To monitor the operation of plants and machines in real time, video streaming monitoring equipment is deployed throughout chemical plants. Helmet recognition and flame monitoring based on captured live video are gradually being used in various production scenarios. Higher accuracy can be achieved as each frame in the video actually contains temporal data.

(7) Gas detector data
Various flammable and toxic gas detectors are now available on the market. These gas detectors are used in laboratories and chemical plants to protect personnel and equipment. The data collected by these gas sensors can be accessed via the IoT and used to model and evaluate gas leakage conditions.
(8) Text data
Numerous text data have been available in plant logs and documents on chemical plant safety, such as risk analysis reports, historical accident reports, and near miss reports. These data are unstructured, but can be used to obtain information that can be useful for risk assessment using natural language processing (NLP).
(9) Chemical databases
The chemical databases currently in use store a large number of parameters for the properties of toxic and risky substances. By extracting and modeling these parameters, the risk of chemical systems can be appropriately assessed.
(10) Simulation data
Before a chemical plant can be set up, commercial software is used to account for and simulate the processes of the plant. Simulation data and scaled-up experimental data can then be used to assist in the subsequent safety assessment. With the development of virtual reality technology in the chemical industry, the availability of simulation data has gradually increased.
(11) Sound data
Malfunctioning chemical equipment emits characteristic audio signals, and different failure modes typically correspond to different characteristics. Therefore, the timing data of the audio signal can be analyzed to quickly determine the failure mode of the equipment and correct it instantly.
(12) Vibration signal data
In addition to airborne sound signal collectors, vibration signal collectors can be used to collect vibration information directly from the equipment. After processing using a certain algorithm, different failure patterns can be presented in different patterns, thus enabling the classification of failures.

3. Data preparation stages

3.1 Two stages of data preparation

Industrial process data, in most scenarios, are stored in relational databases. However, during the storage and updating process, the data can be mishandled or mistaken, leading to errors and flaws in the original data. Hence, it is necessary to perform two main stages of data preparation: error detection and error repairing.

3.1.1 *Error detection*

Good data records and data structures should satisfy integrity and consistency criteria. Numerous studies use integrity constraints (ICs) to capture data quality rules to which the database should conform. Error detection approaches can be classified according to whether and how humans are involved in the process. Error detection techniques can be completely automatic, such as detecting violations of functional dependencies (Bohannon, Fan, Flaster, & Rastogi, 2005), or human dependent, such as identifying duplicate records (Wang, Kraska, Franklin, & Feng, 2012). Chu et al. (Chu, Ilyas, Krishnan, & Wang, 2016) reviewed error detection methods, and Table 1 shows a sample.

Errors can occur anywhere in the process analysis. Unfortunately, while most error detection methods locate errors in the original database, some errors can only be noticed "a long time after the data preparation process has begun(Chalamalla et al., 2014). The role of error detection is not infinitesimal either. Small numerical errors and unstructured errors are not easily detected and may appear elsewhere. Checking and confirming that the data source is credible should be of high importance at all times.

3.1.2 *Error repairing*

An error repairing method or error repairing algorithm is the process of finding an improved version of the relational database that conforms to the data quality requirements. Similarly, error repairing methods can be divided into

Table 1 Example of error detection methods

Method	Classification	References
FDs value modification	Automatic	Bohannon et al. (2005)
Holistic data cleaning	Automatic	Chu, Ilyas, and Papotti (2013)
DBRx	Automatic	Chalamalla, Ilyas, Ouzzani, and Papotti (2014)
CrowdER	Human involved	Wang, Kraska, et al. (2012)
Corleone	Human involved	Gokhale et al. (2014)
Causality Analysis	Human involved	Meliou, Gatterbauer, Nath, and Suciu (2011)
Scorpion	Human involved	Wu and Madden (2013)

Table 2 Example of error repairing methods.

Method	Classification	References
FDs value modification	Automatic	Bohannon et al. (2005)
CFDs value modification	Automatic	Cong, Fan, Geerts, Jia, and Ma (2007)
LLUNATIC	Automatic	Geerts, Mecca, Papotti, and Santoro (2013)
Record matching and data repairing	Automatic	Fan, Ma, Tang, and Yu (2014)
Unified repair	Automatic	Chiang and Miller (2011)
Potter's Wheel	Human involved	Raman and Hellerstein (2001)
GDR	Human involved	Yakout, Elmagarmid, Neville, Ouzzani, and Ilyas (2011)
KATARA	Human involved	Chu et al. (2015)
Editing rules	Human involved	Fan, Li, Ma, Tang, and Yu (2012)

two categories: automatic and human involved. Table 2 illustrates some typical error repairing methods in both categories, which were also reviewed by Chu et al. (Chu et al., 2016).

Automatic error repairing is useful; however, in the field of process safety, it is often necessary to rely on expert knowledge or human experience to determine whether certain data are trustworthy or suspicious, and to use the experience of historical data processing for verification. The accuracy of the data has a significant effect on the accuracy of the subsequent data-driven models; hence, in addition to database-wise error repairing, the data should also be gated and processed in relation to production reality.

3.2 Preparation of four types of error

According to the source and manifestation, errors can be roughly divided into single/multisource errors and schema/instance-level errors. Generally, data cleaning can be performed in order from single to multiple errors, and from schema to instance errors (Rahm & Do, 2000; Ridzuan & Zainon, 2019). In the following, typical cases and repair methods for the four types of errors are described.

3.2.1 *Inconsistency error*

An inconsistency error is classified as a single-source and schema-level error. An inconsistency error may occur when the value of the data is inconsistent with the range of data (wrong range) or the constraint conditions between the data variables are not satisfied (wrong association). As an example of the wrong range, gender is a binary variable; therefore, other values imply an error. Additionally, the year of birth of people alive falling outside the last few decades erroneously may be considered as an error. As an example of wrong association, date of birth and age are not equal to the current year. ICs can be used to describe the constraints between these variables. These constraints are found in the analysis stage at the beginning of data cleaning and should be continuously improved thereafter. Simultaneously, whether to modify the constraints or modify the data, that is, be loyal to the data or loyal to the rules, should be considered as a trade-off in practical applications (Subramaniam, Palpanas, Papadopoulos, Kalogeraki, & Gunopulos, 2006).

3.2.2 *Incompleteness error*

An incompleteness error is classified as a single-source and instance-level error. The most significant problem with an incompleteness error is the filling in of missing values. It has always been a major concern because incomplete data lead to biased conclusions. There are many causes of missing values, such as various recording ages of data entries, difficulty in obtaining data, data being non-existent at first, data entry errors, and manual deletion. If the percentage of entries with missing values is small, deleting these entries will not have a big effect on the results of data analysis. If not, for numerical variables, statistics such as the median and average can be used as a substitute for real values. For more precise filling, the missing values can be filled according to the grouping of data (Wang et al., 2014). In some studies, researchers use Bayesian methods to infer missing values using other variables. For nominal variables, a new category can be created to represent missing values.

3.2.3 *Structural conflict*

A structural conflict is a multisource and schema-level error. Because of different schema designs, different data tables show different aspects of the data and use different names for variables. This means that there are overlaps between variables and the same variable may even have different names, whereas variables that share a name may have different meanings. Tables may also have different units and representations for the same variables, such as numerical variables expressed in currency. Schema translation and schema integration can resolve a structural conflict.

3.2.4 Duplication error

A duplication error is classified as a multisource and instance-level error. It is also known as an overlapping error. When there are multiple data tables, data entries may point to the same entity in each data table, and special attention should be paid to merging multiple tables. Because of the different data formats in the data tables and the differences in data collection, these duplicate entries are buried in the vast dataset and difficult to determine. Therefore, finding duplicates is typically performed in the last step after completing the merge and basic cleaning. By detecting highly unique variables, such as name, address, and phone number, a subset of the entries with exactly the same variables can be screened out. Furthermore, these variables can be fuzzy matched by calculating the similarity of the text (Chung, Mortensen, Binnig, & Kraska, 2018; Heise, Kasneci, & Naumann, 2014). The calculation of similarity can be multistep or simultaneous. For example, it can be filtered individually according to the order of the name, address, and phone number, or the similarity of these variables can be weighted and summed in one step. However, similarity-based detection requires a huge amount of calculation in a large-scale dataset. Because of the possibility of duplicate data, to facilitate the subsequent use of the dataset, the same number of new variables as the amount of original data tables should be added to the merged data table to represent the index of an entry in the original data table.

3.2.5 Tools

Data cleaning is traditionally a very labor-intensive and time-consuming task. Typically, 60% of data analysis time is spent on data cleaning. Although there are data preprocessing tools, particularly for data cleaning, the specific operations of data preprocessing vary greatly in different data tables, and the role of these tools in applications is relatively limited.

3.2.6 Data cleaning for big data

Big data has the characteristics of the 5Vs: value, volume, variety, velocity, and veracity (Chu et al., 2016). Therefore, it has problems that traditional data do not, and should be handled with special care. First, volume and variety mean that various problems may appear simultaneously, such as missing values and duplication. Additionally, traditional methods are not suitable for large-scale problems, as is stated before in text-similarity-based duplication detection. Second, increasing constraints makes it easier to find errors; however, it is more difficult to repair errors because other constraints may be violated after the repair. Thus, it is necessary to seek the smallest modification of the constraints. Third, big data contains a large amount of information; hence, expert knowledge is preferred for a deep understanding of the constraints. Typically, human intervention is required to acquire expert knowledge, and some recent software has specific solutions for certain disciplines.

4. Data-driven model for process safety

Since the beginning of the 21st century, the amount of data collected in chemical plants has increased rapidly with the improvement of sensor technology and computer processing power. The IoT produces an unimaginable amount of data, which contains deep hidden knowledge people were unable to master previously. The term "data" is the plural of "datum," which originated from the Latin word "datus." In the process safety domain, data are mainly the numerical values that indicate safety information, and are the results obtained through observation, experiment, or calculation, including numbers, text, images, and sound (Wen, Khan, Amin, & Halim, 2021).

4.1 Definition of the data-driven model

The data-driven model is fairly self-evident. The term generally refers to a model that is based on learning from the input and output data while disregarding explicit knowledge of chemical systems' physical features. It focuses on extracting knowledge from big and large datasets, and applies the knowledge and insights to solve problems in a wide range of applications.

The data-driven model collects a large amount of data using mobile Internet or other related software, organizes the data to form information, and then integrates and refines the relevant information, and forms an automated decision model after training and fitting based on the data. Therefore, the data-driven process is very complex, requires data and continuous input, and then the model needs to feed the deviation information into the machine learning algorithm after comparing the decision results with real data, and then improve itself in the process of continuous machine learning iteration.

Data-driven models are more powerful in mining this knowledge from big data than their counterpart, mechanism-driven models. Roughly speaking, data-driven models can be categorized into two main categories: statistical and artificial intelligence (AI).

4.2 Statistical models

In statistical models, statistical analysis is performed through variables. Statistical models focus on the "surface" of the data, such as statistical parameters (e.g., mean, median, mode, standard deviation, range, and interquartile range (Dai et al., 2020)), and exploratory analysis of the distribution, kurtosis, skewness, histogram, and control chart (Wen et al., 2021).

Statistical models can be further classified into two subcategories: univariate and multivariate models. The former requires the analysis of each variable individually, whereas the latter aims at projecting high-dimensional process data onto a lower-dimensional space, where then the data can be represented by a smaller set of latent variables (Dai et al., 2020). Univariate models are often used in specific cases, such as process variable monitoring (Harrou, Nounou, Nounou, & Madakyaru, 2015), threshold optimization (Yu, Wang, & Yang, 2017), and dead band setting (Afzal, Chen, Bandehkhoda, & Izadi, 2018).

In many scenarios, multivariate models occupy more weight as the amount of data and the demand for higher model accuracy increases. To improve process safety, a series of models and applications called multivariate statistical process monitoring (MSPM) were reviewed by Wang (Wang, Si, Huang, & Lou, 2018). Fig. 2 (Dai et al., 2020) shows the four typical categories of MSPM, that is, principal component analysis (PCA), partial least-squares (PLS), independent component analysis (ICA), Gaussian mixture model (GMM), and their modifications.

In addition to these models, hybrid and integrated models, such as multivariate exponentially weighted moving average PCA (Amin, Khan, Imtiaz, & Ahmed, 2019), dynamic Gaussian process regression-based PLS (Liu, Yang, Carlsson, Qin, & Yoo, 2019), co-integration/slow feature analysis (Zhao & Huang, 2018), and phase-based relevance vector machine (Zhu & Gao, 2018), have been proposed and used in process monitoring and analysis. These models are often executed on complex chemical process data with nonlinearity, non-Gaussianity, or multiple mode scenarios. Another series of multivariate models is the probabilistic model based on prior distributions. Two famous instances are Bayesian statistical models

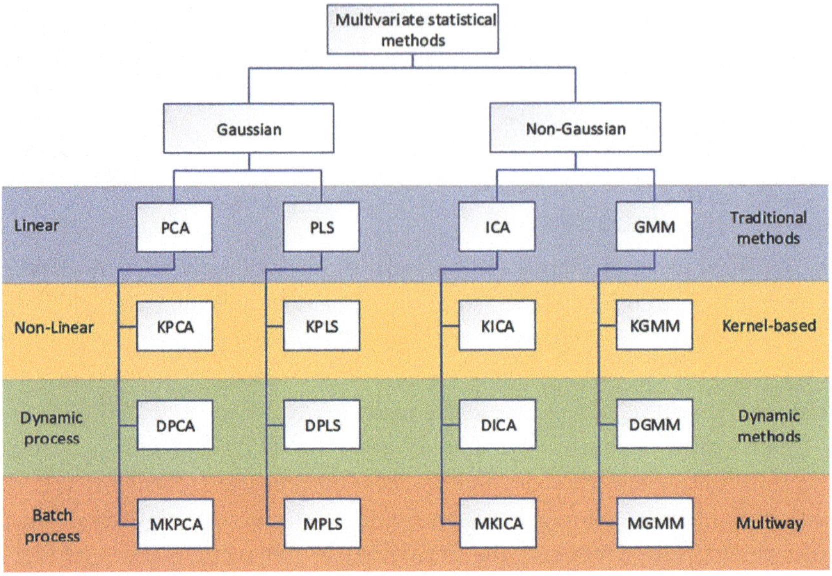

Fig. 2 Four categories of MSPM and their modifications (Dai et al., 2020).

and *n*-gram models which have been implemented in successful applications (Zhu, Wang, Li, Gao, & Zhao, 2016). Furthermore, models such as the probabilistic latent variable model are also helpful in process data analytics (Ge, 2018).

Statistical models have played a key role in traditional process safety analysis, and also enhance the physical model construction and validation process.

4.3 Artificial intelligence models

The traditional risk model and safety analysis (Stringfellow, Leveson, & Owens, 2010) are considered to be inadequate for handling the immense use of AI in safety-critical systems (Dey & Lee, 2021). With the development of computing power and algorithms, AI in chemical engineering entered a new era, and chemical engineers are excited about the potential of AI in industry (Venkatasubramanian, 2019). An example of an application of AI is machine learning, which aims at giving a system the ability to learn and improve from data automatically without being deliberately programmed (Bagheri, Akbari, & Mirbagheri, 2019). Based on the data characteristics, machine learning can be categorized into two parts: supervised and unsupervised learning models. The former requires labeled data to train the models, which means having a priori knowledge of the ground truth (what the result should be), and the latter, by contrast, tries to reconstruct such knowledge from scratch with minimal errors. Fig. 3 (Dai et al., 2020) illustrates the two categories and their detailed applications and corresponding typical models.

In the chemical engineering field, labeled data and supervised models are more commonly used, particularly for classification tasks. At the present time, the term machine learning is often associated with ANN. AI itself has a broader meaning, whereas ANNs, their variants, and derivatives are indeed the most classic data-driven models used for process safety. The typical architecture of a neural network is composed of an input layer, series of hidden layers, and output layer of neurons. Neurons between adjacent layers are associated with a transfer function using signals, weights, and biases. Advanced AI methods have more complicated structures to describe and learn from complex systems (Dai et al., 2020), such as convolutional neural networks (CNNs) (Wu & Zhao, 2018a), restricted Boltzmann machines (Zhang & Zhao, 2017), and residual networks (Amini & Zhu, 2021).

Deep neural networks can extract hidden features from big data, which have contributed a great deal to the development of AI models. Since 2005, significant progress has been made in terms of computing power and advanced models for AI applications. One example is ensemble learning (EL), which attempts to combine several submodels to increase the overall accuracy of the entire system (Zhou, Li, & Xiong, 2019). Multiple weak learners are used for building integrated models to reduce variance and prevent overfitting (Dai et al., 2020). A typical method is called bagging, which trains different learners on the same sample, and the outputs are used for voting (classification) or averaging (regression). Random forest is a famous bagging model (Bahl et al., 2019). Another EL model is boosting, which trains models iteratively, and each newly trained model attempts to fix previous

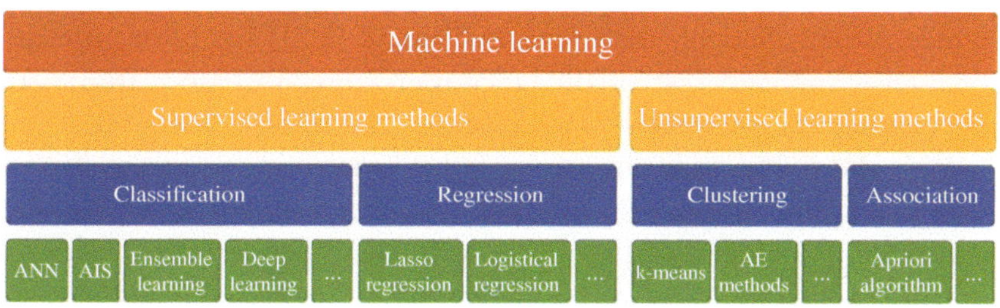

Fig. 3 Machine learning models and applications (Dai et al., 2020).

mistakes. A typical application is the gradient boosting decision tree (Cai, Gu, Ma, & Jin, 2019). Other EL models were reviewed by (Wang et al., 2022).

Regarding unlabeled data and unsupervised models, autoencoders (AEs) and *k*-means are the mostly commonly used algorithms, and are suitable for online application with minimal human intervention (Hinton & Salakhutdinov, 2006). An AE is an ANN capable of learning an efficient representation of the input data through unsupervised learning. This efficient representation of the input data, called encoding, is generally much smaller in dimension than the input data, which allows the AE to be used for dimensionality reduction. More importantly, self-encoders can be applied as powerful feature detectors for pre-training deep neural networks. Additionally, self-encoders can randomly generate data similar to the training data, which is called a generative model. An AE works by simply learning to copy the input to the output. This task may sound trivial, but it can be made extremely difficult by adding constraints to the neural network in different ways. For example, the size of the internal representation can be limited (which achieves dimensionality reduction), or noise can be added to the training data and the AE can be trained to recover the original data. These constraints prevent the AE from mechanically copying the input to the output and force it to learn an efficient representation of the data. The applications have been reported for state identification (Zheng & Zhao, 2018), data mining (Zheng & Zhao, 2018), and data preprocessing (Heinrich, Noering, Pruckner, & Jonas, 2021). Additionally, the short-time Fourier transform and categorical generative adversarial networks have also been used for unsupervised fault diagnosis (Tao, Wang, Chen, Stojanovic, & Yang, 2020).

Numerous other current AI models exist in the process safety field. For example, the artificial immune system applies immunology principles to generate responses against invading antigens (Artificial Immune System, 2016); real-time AI at the edge for vehicle detection/classification from a distance (Sabeti, Shoghli, Baharani, & Tabkhi, 2021) has been popular in computer vision; and a systematic literature review of AI safety from a software engineering perspective was conducted by (Dey & Lee, 2021).

With the further advancement of the Industry 4.0 process, and the increase of informationization and intelligence in modern society, data-driven AI models will play an increasingly important role in the safety of chemical systems. What the next high-reward research direction is for AI applications is worth thinking about, and controlling the boundaries of the AI jurisdiction is certainly meaningful.

5. Applications of data-driven models

5.1 Process monitoring in process safety

Process monitoring has become an essential aspect of chemical process safety because most modern chemical process systems are large-scale, highly complex, and complicated operations, with a large number of variables (Ayoubi & Isermann, 1997; Venkatasubramanian, Rengaswamy, Yin, & Kavuri, 2003). Process monitoring is achieved in a loop that includes fault detection, fault identification, fault diagnosis, and process recovery to continuously supervise the process (Chiang, Russell, & Braatz, 2000). Fault detection and diagnosis (FDD) is an essential part of the loop, and could improve the safety of plant operations and reduce production costs (Arunthavanathan, Khan, Ahmed, & Imtiaz, 2021a; Wang, Chai, Ding, & Brown, 2009).

The data-driven models of FDD apply multivariate statistical analysis and machine learning methods to a variety of industrial processes (Nor, Hassan, & Hussain, 2020). It is important that the specific measures of multivariate statistical variables in the process, such as Hotelling's T^2 statistic and Q statistic, could be used to detect and diagnose faults in the system (Chiang et al., 2000; Raich & Çinar, 1996). Many feature extraction methods, such as PCA, PLS and ICA have been effectively used in FDD, and many researchers have combined the development and characteristics of these methods (Nor et al., 2020; Park, Fan, & Hsu, 2020; Taqvi et al., 2021). Based on this, Jiang et al. proposed a fault identification model that provides typical variable analysis followed by slow feature analysis to monitor operating condition deviations and process dynamic anomalies (Jiang, Huang, Zhu, Yang, & Braatz, 2015; Shang et al., 2015). Because of the exponential growth in computational power, machine learning methods have become the most efficient approach to obtain knowledge from large amounts of process data. Various well-established neural network methods have performed well in FDD, such as ANN (Rostek, Morytko, & Jankowska, 2015), neuro-fuzzy (Pratama et al., 2013), SVM (Mahadevan & Shah, 2009), GMM (Chen & Zhang, 2010), kNN (Zhou, Wen, & Yang, 2016), BN (Zhang & Zhao, 2017), CNN (Wu & Zhao, 2018a), and AE (Cheng, He, & Zhao, 2019). Recently, Ajagekar and You used quantum computing to assist deep learning for FDD, and the method achieved excellent performance in both a continuous stirred tank reactor and Tennessee Eastman process (Ajagekar & You, 2020). Xiao et al. designed a variable selection strategy based on relevance and redundancy variable selection, which detects and solves minor abnormal conditions in time to ensure the long-term stable operation of a large-scale industrial process (Xiao et al., 2021). Yu and Yan proposed intensified iterative learning from a stacked autoencoder, which played an excellent role in industrial fault detection (Yu & Yan, 2021). Based on this, in recent studies, researchers have attempted to address the challenges crucial to achieving a smart process FDD.

Existing FDD methods have been successful in improving detection accuracy; however, they are only applicable to specific patterns when applied online. Therefore, how to make the model adaptive under different modes is of great value for practical applications. Studies on transfer learning (TL) have proved that deep neural networks can store knowledge gained while solving one problem and apply it to a different but related problem (Yosinski, Clune, Bengio, & Lipson, 2014). Wu and Zhao applied the method to achieve chemical FDD for multimode chemical processes that can accomplish the FDD task on the target domain by learning the data in the source domain, as shown in Fig. 4 (Wu & Zhao, 2020).

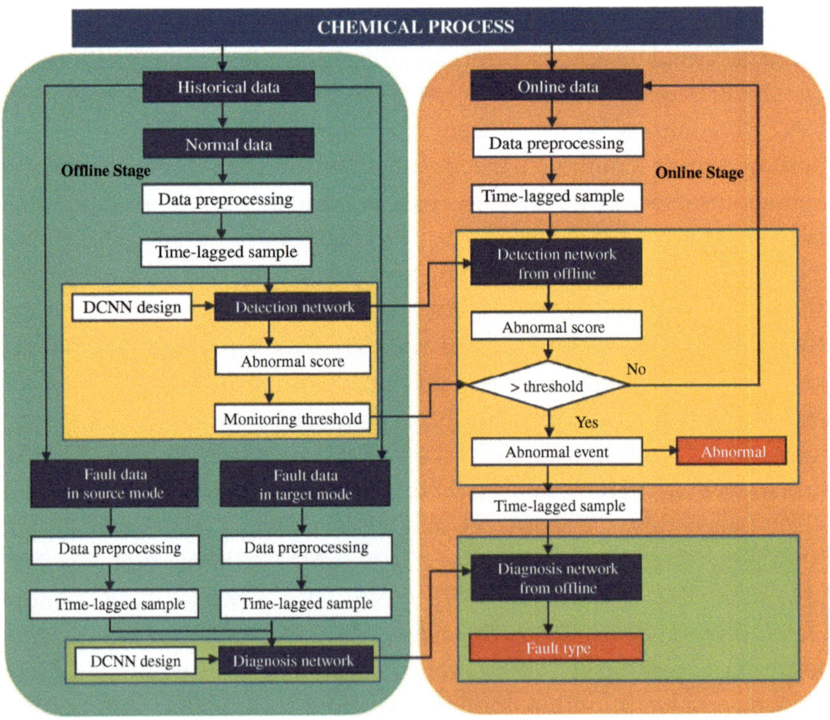

Fig. 4 Framework of the TL-based FDD method (Wu & Zhao, 2020).

Another challenge is that insufficient labeled data in the real world presents a huge challenge for supervised learning; hence, conducting unsupervised and semi-supervised research without labels or with fewer labels is of great importance in practical applications. Zheng and Zhao used historical data mining and knowledge discovery using unsupervised methods to further create labeled databases for building traditional supervised diagnostic models. They used pseudo labels for model training to improve the accuracy of the model through multiple rounds of iteration (Zheng & Zhao, 2020). Then they focused on the temporal characteristics of industrial process data and innovatively proposed a self-adaptive temporal-spatial self-training algorithm, which outperformed competing semi-supervised fault diagnosis methods (Zheng & Zhao, 2021).

As the accuracy of AI methods has increased, improving the interpretability of AI has become an important topic in the field of FDD, which is the same as other fields (Zhang & Zhu, 2018). In previous studies, the results of the FDD model were explained by statistical methods of feature extraction, Bayesian networks, and decision trees (Gharahbagheri, Imtiaz, & Khan, 2017; Jiarula, Gao, Gao, Jiang, & Wang, 2016). Wu and Zhao proposed the process topology convolutional network to integrate process knowledge when building the fault diagnosis model (Wu & Zhao, 2021). The calculation and inference process of this model is based on the linkage of process variables in the actual process, which allows for a better understanding of the computational process while ensuring its predictive performance (Mott, Zoran, Chrzanowski, Wierstra, & Rezende, 2019). In recent studies, the attention mechanism has been conspicuous in conferring interpretability to deep neural networks in many fields. As shown in Fig. 5, Bi and Zhao applied an orthogonal self-attentive variational autoencoder (OSAVA) to extract correlations between variables. When a fault is detected, the attention weights are used to identify the abnormal variables, thus providing an interpretable result (Bi & Zhao, 2021).

There are various challenges in FDD, and recent research has made some progress. However, each process monitoring method of FDD has advantages and disadvantages. The development of a hybrid FDD framework is an attempt to find an approach that meets the needs of a precise FDD system. A hybrid FDD framework system was proposed, where multiple methods perform collective problem-solving in a single FDD framework, which is also referred to as a multiagent method (Natarajan & Srinivasan, 2014). By contrast, implementing a hybrid fault diagnosis strategy in an FDD system, implies that the existence of two or more diagnostic methods

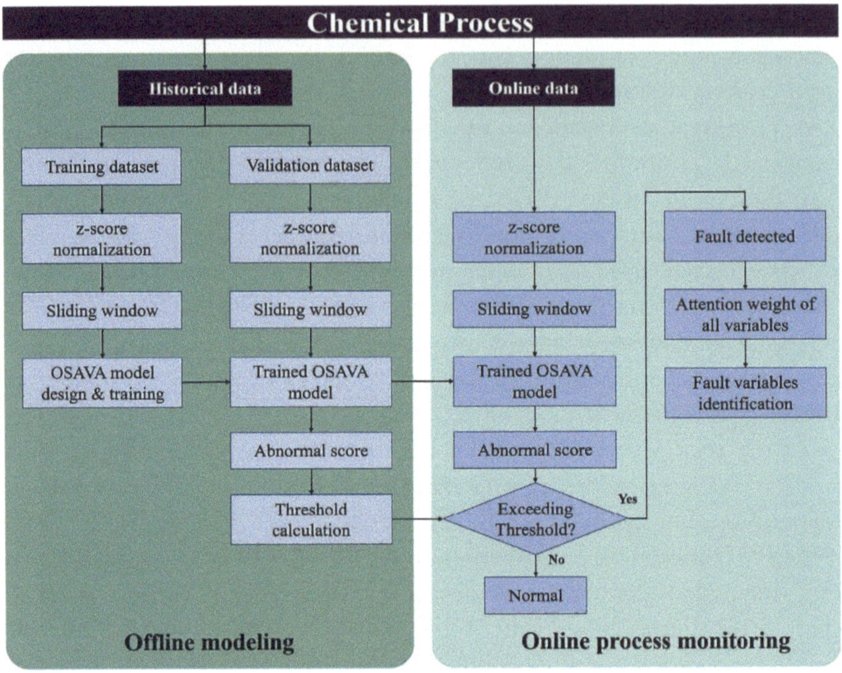

Fig. 5 Framework of the OSAVA-based FDD method (Bi & Zhao, 2021).

with multiple outputs may have potentially conflicting results (Chen, Ding, Luo, & Zhang, 2017). Methods such as averaging, majority voting, and Bayesian probability could effectively normalize the results to the final output (Uraikul, Chan, & Tontiwachwuthikul, 2007; Zhang, 2006; Zhang & Ge, 2015). The expanding size of process data increases the characteristics of irrelevance and redundancy, which would invert model complexity and reduce model performance (Ming & Zhao, 2017). Through previous studies, the ability of model FDD was effectively increased by adding discriminators, using causality, and applying Bayesian and other methods (Deng, Tian, Chen, & Harris, 2017; Jiang & Huang, 2016; Jiang, Zhu, Huang, & Braatz, 2015).

Although a great deal of research has been conducted and a wide range of methods have become available in process monitoring, most of them perform well in laboratory or simulation cases. However, some or most methods are not suitable for industrial application. Therefore, there is an urgent need to develop hybrid data-based model architectures, which could be effective for industrial examples, to allow FDD methods to play a vital role in chemical process safety.

5.2 Risk assessment in process safety

Risk assessment plays an important role in the entire life cycle of chemical process safety, including research, process design, operation, maintenance, and decommissioning. Traditional risk assessment methods use a structured approach to identify possible hazards, analyze their causes and consequences, and determine risk levels. QRA can provide more effective decision support than qualitative risk assessment.(Dai et al., 2020)

QRA consists of two main components: risk analysis and risk assessment. Risk analysis includes hazard identification and risk estimation. Based on the risk assessment results, the risks are compared with regulations and company-defined acceptable criteria, and then risk reduction measures are applied (Ahmadi, Mortazavi, Mahabadi, & Hosseinpouri, 2020). QRA has been effectively applied in several fields, such as risk-based design (Fadier & De la Garza, 2006), risk-based operations, quantitative assessment of domino scenarios (Cozzani, Antonioni, & Spadoni, 2006), and use of safety performance metrics to measure process safety (Khan, Abunada, John, & Benmosbah, 2010). However, there are still some limitations in the application of QRA. QRA methods cannot introduce new information or evidence. Additionally, probabilistic estimates are based on data obtained over decades and may introduce some uncertainty into the analysis (Villa, Paltrinieri, Khan, & Cozzani, 2016). By contrast, the QRA approach treats event relationships as primarily binary and cannot take into account the dependencies between contributing factors (Khan et al., 2015). To address risks in a changing environment and overcome the limitations of traditional methods, risk assessment methods have been dynamically developed to provide further risk assessment in conjunction with recent and accurate information.

The "domino" accident is the most common multihazard coupled disaster in the chemical industry, and refers to a technical accident that leads to one or more new technical accidents. The study of domino accidents has attracted a great deal of interest in the field of chemical process safety (Alileche, Olivier, Estel, & Cozzani, 2017). Most QRAs that respond to domino accidents do not consider the time interval between accidents, and are static approaches (Necci, Cozzani, Spadoni, & Khan, 2015). Abdolhamidzadeh proposed a QRA method based on Monte Carlo simulation (MCS), which has been an effective method to quantify the synergistic effects of domino accidents in chemical plants (Abdolhamidzadeh, Abbasi, Rashtchian, & Abbasi, 2010; Rad, Abdolhamidzadeh, Abbasi, & Rashtchian, 2014). Based on this, He and Wang proposed a domino accident QRA method based on field theory and MCS to perform QRA through a dynamic simulation approach (He & Weng, 2020). Recently, Liu et al. learned the potential relationship between variables using LSTM, and then obtained QRA based on variable prediction (Liu, Tian, Cui, Wei, & Li, 2021). Through these methods, the dynamic risk distribution was obtained, and thus the development pattern of domino accidents was proposed, which means that the results of the study might provide useful guidance for loss prevention, risk management, and personnel evacuation in the chemical industry.

Dynamic risk assessment (DRA) can be used to overcome the limitations of many QRA methods. DRA is defined as a method of updating the calculated primary risk number based on the performance of the control system, safety barriers, maintenance and inspection activities, human factors, and operational procedures (Khan et al., 2016). Swaminathan and Smidts first proposed a methodology to extend the application of an event sequence diagram to the modeling of dynamic scenarios and identification of missing accidental scenarios (Swaminathan & Smidts, 1999a, 1999b). The first complete dynamic risk assessment methodology for process facilities by estimating the dynamic probability of an accident, called dynamic failure assessment, was developed by Meel and Seider (Meel & Seider, 2008). DRA was further developed by Kalantarnia et al., who added accident analysis and a probability update (Kalantarnia, Khan, & Hawboldt, 2009; Kalantarnia, Khan, & Hawboldt, 2010).

Based on DRA, Yang and Mannan proposed the dynamic operational risk assessment (DORA) methodology for operational risk analysis in the oil/gas and chemical industries (Yang & Sam Mannan, 2010a, 2010b). The probabilistic modeling component of DORA integrates stochastic modeling and process dynamic modeling to assess operational risk. Stochastic system state trajectories are modeled based on the abnormal behavior or failure of each component. For each possible system state trajectory, a process dynamic evaluation is performed to examine the process variables.

Based on the DRA, the transition from traditional QRA to dynamic QRA (DQRA) is a natural evolution (Khan et al., 2015). Wang et al. proposed the use of loss functions to estimate operational losses in complex chemical processes, and the potential consequences depending on the scenario, which were used to quantify dynamic risks. They conducted a risk assessment by combining the probability of loss occurrence with the potential loss caused by unusual circumstances. Additionally, they monitored multiple variables simultaneously in the process of updating the loss probability

(Wang, Khan, Ahmed, & Imtiaz, 2016). Recent research on DQRA has placed increasingly great emphasis on the complex and nonlinear relationships that exist between risk elements in a process system. The interaction of risk factors can lead to the suddenness of an accident and further cause the escalation of risk factors. Based on dynamic and complex accident characteristics, Meng et al. proposed an integrated approach using DEMATEL-BN for the DQRA of offshore platform spills, covering both system vulnerability and dynamic risk perspectives (Meng, Chen, Zhu, & Zhu, 2019). Based on this, Ahmadi et al. developed a dynamic risk assessment method using fuzzy DEMATEL-BN and leading indicators that consider risk influencing factors, which update the calculated initial risk using leading indicators and Bayesian inference (Ahmadi et al., 2020). Determining the relationship between risk influencing factors using fuzzy DEMATEL, calculating the weights of risk influencing factors through fuzzy hierarchical analysis, and using fuzzy logic to determine the probabilities of starting events and barrier elements are the focuses of the method.

Multiple risk assessment methods for process safety have been effective in simulated cases and actual projects. As the complexity of industrial systems increases, the process safety of the system is further affected by system risk. In the field of chemical process safety, it will be industrially meaningful to use big data to construct a more accurate DQRA in future research.

With the development of NLP technology, the use of AI methods to process text to extract information has become an effective and speedy method in the field of process safety. NLP can be used in HAZOP analysis reports, not only to ensure the consistency of analysis results but also to help small chemical plants to conduct safety analysis (Feng, Dai, Ji, Zhou, & Dang, 2021). NLP can be used as a viable method to qualitatively and quantitatively explore the contributing factors of a particular accident. Wang and Zhao proposed an improved deep learning framework to systematically analyze the sources of confined space hazards and significantly improve the quality of risk assessment in confined space engineering (Wang & Zhao, 2022). Similarly, the NLP method can be used to efficiently and accurately classify and analyze the causes of various types of accidents (Jing et al., 2022). NLP can mine information from past text; however, unstructured text requires a large amount of labeled data, which poses many difficulties in the application of the method. Ahadh et al. proposed an automated, semi-supervised, domain-independent method to analyze incident reports that can yield accuracy comparable to supervised learning (Ahadh, Binish, & Srinivasan, 2021). The application of NLP in process safety is still in the exploratory stage, and more research is required to make NLP play an important role in better industrial applications of chemical process safety.

5.3 Process prognosis in process safety

As the complexity of chemical engineering systems has increased, chemical process safety has been challenged, to a large extent. Simply detecting and diagnosing a fault may be not sufficient in some cases, and it is necessary to prognose the fault ahead of time. Fault prognosis mainly aims at potential and future faults, which means the research object of fault prognosis is a future uncertain event (Smith & Randall, 2015; Zhang et al., 2011). The relationship between fault prognosis and fault diagnosis is shown in Fig. 6. Fault diagnosis determines the fault before the occurrence of secondary damage or the catastrophic failure of the system, subsystem, or component (Jiang, Li, Wu, & Zhou, 2018).

Current fault prognosis methods can be generally classified as three types: model-based methods, knowledge-based methods, and data-based methods (Zhong, Han, & Han, 2020). Model-based methods are approaches that build an accurate mathematical description of systems using physics or first principles. They tend to be better than other models when sufficient knowledge is available. However, they are generally unavailable for exact mathematical knowledge as industrial systems become increasingly complex. Knowledge-based methods are based on engineering experience and historical events, and the prognosis results obtained are more intuitive. They can be efficient when process knowledge has been readily accumulated. However, prediction accuracy and application scope are largely limited because of the combinatorial explosion problem. Additionally, data-based methods determine the status of the system within a certain period by analyzing previously observed data (Climente-Alarcon, Antonino-Daviu, Strangas, & Riera-Guasp, 2015). They can be applicable when data are abundant, and have a low operating cost because big data is playing an increasing role in the process industry.

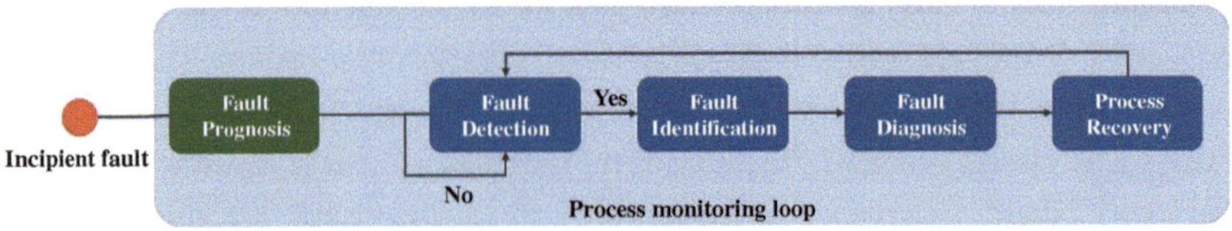

Fig. 6 Relationship between fault prognosis and process monitoring.

Data in industrial process systems have a large number of characteristics, and fault prediction methods have evolved a great deal with the changing characteristics of industrial data to better accommodate various data types. Dynamicity is an important characteristic of industrial process data, and mainly manifests as autocorrelation and crosscorrelation properties between different sampling points of the measurement variables (Tong & Shi, 2016). In recent years, many advanced methods have been proposed for dynamic system fault prognosis, such as the augmentation matrix, state estimation, and regression algorithm (Baruah & Chinnam, 2005; Li, Qin, Ji, & Zhou, 2010; Liu, Zhu, & Zeng, 2018; Wang, Sun, & Li, 2018; Zhao & Gao, 2015). The nonlinear characteristic is a manifestation of the process system controlled by nonlinear equations, which can greatly affect the accuracy of prediction. Kernel-based methods and ANN methods have been used to solve the problem (Daroogheh, Baniamerian, Meskin, & Khorasani, 2017). Non-Gaussian characteristics indicate that the measurement data cannot conform to a Gaussian distribution and fluctuate within a certain range, which is an important challenge for industrial process data. ICA separates data or signals into linear combinations of statistically independent and non-Gaussian sources using linear transformations (Ma, Wang, Li, & Zhou, 2014). Additionally, the kernel method can also be applied to ICA methods to solve nonlinear problems (Ma, Li, & Zhou, 2018). Another popular method for non-Gaussian system prognosis is the GMM, which assumes that process data in a complex industrial system can be described by some local linear models (Wang, Zarader, & Argentieri, 2012). Actual failures in industry processes may be of multiple types and may occur in multiple modes, which is a major difference from traditional fault prediction methods that assume that faults are single-mode and single-type. The most direct approach is to build multi-model approaches from existing knowledge (Verbert, De Schutter, & Babuška, 2018; Wen, Wu, & Yuan, 2017). When the system changes slowly, fault prognosis can be performed using adaptive and recursive methods (Prakash, Samantaray, Bhattacharyya, & Ghoshal, 2018). Additionally, isolation estimation allows fault prediction for multiple faults by building degenerate models for each model (Yu, Wang, & Luo, 2015). Recently, spatiotemporal attention was used to predict critical parameters in a chemical process, which significantly improved the accuracy of furnace tube temperature prediction (Yuan, Yang, Ling, Wu, & Li, 2021). Arunthavanathan et al. achieved early fault monitoring with an SVM using process data, which was forecasted by the CNN-LSTM model (Arunthavanathan, Khan, Ahmed, & Imtiaz, 2021b).

An important challenge in the practical application of fault prediction is to update the model and obtain the prediction results in real time. Bhadriraju and Khan proposed the operable adaptive sparse identification of systems to achieve fault prognosis (Bhadriraju, Kwon, & Khan, 2021). The sparse identification of nonlinear dynamics was used to fit nonlinear dynamics, and a deep neural network was used to match historical data with nonlinear relationships quickly. The model achieved the effect of predicting faults in advance in both simulated datasets, which passed certain robustness tests. However, there are still some drawbacks if the model needs to be applied in real production.

Remaining useful life (RUL) prediction is another important approach in process prediction. The accurate RUL prediction of critical assets is an important challenge in condition-based maintenance to improve reliability, and decrease the breakdown and cost of machines (Ben Ali, Chebel-Morello, Saidi, Malinowski, & Fnaiech, 2015). Bayesian methods can be used in the RUL prediction of critical components (Mosallam, Medjaher, & Zerhouni, 2013). He et al. proposed a trapezoidal network intelligent evaluation system and illustrated the effectiveness of the method using centrifugal pump RUL prediction as an example (He, Dai, Lu, & Mou, 2018).

Fault prognosis has yielded many impressive results in the field of process safety. However, there are still a number of research challenges that deserve continued in-depth study (Zhong et al., 2020). First, with the increase in database scale, and development of data collection and storage devices, fault prognosis in industrial systems is beginning to focus on big data analysis methods. High-dimensional data require more complex models that increase the computational effort, and redundant features can affect the accuracy of fault prediction. Effective feature extraction techniques play a more important role in fault prognosis. Additionally, the prediction step is one of the core issues regarding fault prognosis. Short-term prediction could increase the prediction accuracy, but cannot allow sufficient time for maintenance operations, which is an important reason that long-term forecasts need to be better studied. At present, several researchers are attempting to increase the accuracy of long-term predictions by incorporating effective machine learning models (Xiang, Bai, & Zhao, 2021). Some new models, such as Transformer (Vaswani et al., 2017), could also be introduced into the field of fault prognosis to improve the accuracy and timeliness of prediction. Finally, with the Industry 4.0 era, fault prognosis will become an important part of the general theme of process safety (Lee et al., 2019).

An important research direction for fault prediction development is how to combine fault prediction with FDD and risk assessment organically.

5.4 Emergency response in process safety

Emergency response is an integral part of process safety, and is vital to protect the health and safety of the public and workers, reduce environmental impact, and speed up the return to work and production (Hosseinnia, Khakzad, & Reniers, 2018). Most chemical plants are located in clusters because of factors such as environmental conditions, social motivations, and legal requirements, which results in major accidents in clusters that are more serious inside and outside the business premises than similar accidents in stand-alone, single plants (Reniers & Soudan, 2010). A comprehensive multi-plant emergency response consists of three main phases. The first phase consists of individual company site hazard identification, selection of plausible accident scenarios, and analysis of the subsequent results. The second phase is the creation of a standardized table of accident scenarios for bloodletting, and considering the likelihood and consequences of different accident scenarios. The third stage is to assess the level of urgency according to the criteria table and provide guidance corresponding to the subsequent actions (Hosseinnia et al., 2018).

Alarm systems play an important role in the emergency management of chemical plants, and alarm management is also an important component of process safety (Goel, Datta, & Mannan, 2017). In industrial systems, alarm flooding is a major factor in reducing the safety and reliability of alarm systems (Wang, Yang, Chen, & Shah, 2016). False alarms are most likely to occur during system state transitions. Zhu et al. proposed a dynamic alarm management strategy that effectively reduces alarm flooding in chemical process transitions (Zhu, Shu, Zhao, & Yang, 2014). Lai et al. responded to impending warning floods with online pattern matching and prediction (Lai, Yang, & Chen, 2017). Alarm flood sequence alignment is an important method to effectively ensure process safety against further losses, and methods such as matching-based, anomaly pattern mining and itemset mining methods have been proposed to accelerate sequence alignment (Lai, Yang, Chen, & Cao, 2019).

The big data era has created new challenges regarding correctly identifying failure scenarios. Chemical plants have a variety of detectors to identify hazardous scenarios, such as chemical spills or fires. Early detection can be effective in supporting decision-making and minimizing damage.

Cameras installed everywhere can help to detect hazardous scenes using computer vision methods. Wu et al. developed an intelligent fire detection method through a camera based on object detection and image classification methods (Wu, Wu, & Zhao, 2019). The method consisted of three steps: motion detection, fire detection, and area classification. First, moving objects were detected by the camera using a background differencing method. Then frames with moving objects were identified by a fire detection model that could output fire regions and their locations. Finally, a region classification model was used to identify whether a region was a on fire. Such an approach achieved a high detection rate for fires and facilitated the early detection of fires in industrial deployments. Smoke is an unusual signal in the pre-fire period, and correctly identifying smoke is more difficult and will be more effective for warning of a disaster. Similar image segmentation and recognition methods have been applied to the recognition of smoke, which required the accurate separation of smoke components in the background scene (Yuan et al., 2019). Additionally, infrared thermography could be used to determine whether liquid has leaked by detecting changes in temperature (Fahimipirehgalin, Trunzer, Odenweller, & Vogel-Heuser, 2019).

Real-time consequence estimation is an important aspect of the emergency response, and Wang et al. developed a fast prediction method that integrates gas detectors, ANN, and gas dispersion models (Wang, Chen, & Zhao, 2015). AI prediction models that combine mechanisms can also be used to estimate failures, such as leaks and fires (Zhang et al., 2017; Zhang, Collin, Moireau, Trouvé, & Rochoux, 2019). Jiao et al. developed a quantitative attribute-outcome relationship model based on deep learning for the rapid prediction of toxic diffusion to ensure safety (Jiao, Ji, Sun, Hong, & Wang, 2021). Chemical hazards are also an important factor in process safety. Shin et al. reasoned in real time to provide recommendations for an emergency response by building knowledge maps of major chemical elements (Shin, Yoo, Ju, & Shin, 2022).

Human behavior is a common cause of accidents and requires serious consideration in the emergency management of human behavior (Murphy & Paté-Cornell, 1996). Wearing a helmet is an effective and necessary safety precaution to reduce the risk of head injury; however, some workers do not always follow the rules. Wu and Zhao proposed a vision-based safety helmet identification method to accurately monitor whether a worker is wearing a helmet (Wu & Zhao, 2018b). Similarly, a range of systems ensures a safe working environment by detecting unsafe worker behavior or conditions, such as falls (Tsai, 2014) and climbing (Han, Lee, & Peña-Mora, 2013). In addition to human behavior in the field, the performance of control room operators is critical to process safety. Experiments have shown that simulator-based performance metrics provide a good estimate of operator reliability (Iqbal & Srinivasan, 2018). Additionally, the application of process synthesis methods to simultaneous intrinsic safety assessments reduces human error (Srinivasan, Srinivasan, Iqbal, Nemet, & Kravanja, 2019). The cognitive state of operators can be dynamically assessed through eye gaze behavior and electroencephalography to prevent accidents in the process industry (Shahab, Iqbal, Srinivasan, & Srinivasan, 2021).

The rapid growth and aggregation of industry has placed higher demands on emergency response, and the significant increase in data has created new opportunities for them in industry. The integrated and full use of data in all emergency response phases to resolve urgent matters is an important research direction for future emergency response industrial applications.

6. Summary and perspective

To summarize, process safety engineering for chemical processes can be applied across the entire life cycle. It includes process monitoring, risk assessment, fault prognosis, and emergency response that can be used in different phases of process safety. Data-driven approaches have been widely used for chemical process safety because they require less a priori knowledge than general methods and can dig into hidden information from large amounts of historical data. Statistical methods and AI-based methods are two of the most intensively studied data-driven approaches, and were presented separately in this chapter. The application of data-driven methods in the field of chemical process safety was elaborated in detail.

However, there is still a need to further develop data-driven approaches to more deeply exploit the variety of data from the IoT. Some areas for future research on data-driven approaches for process safety are as follows: (1) Most traditional data-driven approaches still focus on only one data type individually, which is not compatible with the multiple categories of data that are characteristic of Big Data. Therefore, advanced methods should be developed to handle different types of data in combined systems. (2) Process safety approaches have been greatly developed in past studies. Although they have achieved good results in various types of simulated datasets, they still need to be improved in applications to real industrial cases.

(3) Data are the basis of chemical process safety. However, data sharing is still an institutional problem because of confidentiality and differentiation between data. By contrast, many methods and models need specific data to function, and how to enhance the generality of the models using the migration model approach is also an important area of development for process safety.

References

Abdolhamidzadeh, B., Abbasi, T., Rashtchian, D., & Abbasi, S. A. (2010). A new method for assessing domino effect in chemical process industry. *Journal of Hazardous Materials, 182*, 416–426. https://doi.org/10.1016/j.jhazmat.2010.06.049.

Afzal, M. S., Chen, T., Bandehkhoda, A., & Izadi, I. (2018). Analysis and design of time-deadbands for univariate alarm systems. *Control Engineering Practice, 71*, 96–107. https://doi.org/10.1016/j.conengprac.2017.10.016.

Ahadh, A., Binish, G. V., & Srinivasan, R. (2021). Text mining of accident reports using semi-supervised keyword extraction and topic modeling. *Process Safety and Environmental Protection, 155*, 455–465. https://doi.org/10.1016/j.psep.2021.09.022.

Ahmadi, O., Mortazavi, S. B., Mahabadi, H. A., & Hosseinpouri, M. (2020). Development of a dynamic quantitative risk assessment methodology using fuzzy DEMATEL-BN and leading indicators. *Process Safety and Environmental Protection, 142*, 15–44. https://doi.org/10.1016/j.psep.2020.04.038.

Ajagekar, A., & You, F. (2020). Quantum computing assisted deep learning for fault detection and diagnosis in industrial process systems. *Computers & Chemical Engineering, 143*, 107119. https://doi.org/10.1016/j.compchemeng.2020.107119.

Alileche, N., Olivier, D., Estel, L., & Cozzani, V. (2017). Analysis of domino effect in the process industry using the event tree method. *Safety Science, Risk analysis and land use planning: managing safety on the short and long range, 97*, 10–19. https://doi.org/10.1016/j.ssci.2015.12.028.

Amin, M. T., Khan, F., Imtiaz, S., & Ahmed, S. (2019). Robust process monitoring methodology for detection and diagnosis of unobservable faults. *Industrial and Engineering Chemistry Research, 58*, 19149–19165. https://doi.org/10.1021/acs.iecr.9b03406.

Amini, N., & Zhu, Q. (2021). Fault detection and diagnosis with a novel source-aware autoencoder and deep residual neural network. *Neurocomputing*. https://doi.org/10.1016/j.neucom.2021.11.067.

Artificial Immune System. (2016). *Artificial immune system applications in Computersecurity* (pp. 1–25). John Wiley & Sons, Ltd. https://doi.org/10.1002/9781119076582.ch1.

Arunthavanathan, R., Khan, F., Ahmed, S., & Imtiaz, S. (2021a). An analysis of process fault diagnosis methods from safety perspectives. *Computers & Chemical Engineering, 145*, 107197. https://doi.org/10.1016/j.compchemeng.2020.107197.

Arunthavanathan, R., Khan, F., Ahmed, S., & Imtiaz, S. (2021b). A deep learning model for process fault prognosis. *Process Safety and Environmental Protection, 154*, 467–479. https://doi.org/10.1016/j.psep.2021.08.022.

Ayoubi, M., & Isermann, R. (1997). Neuro-fuzzy systems for diagnosis. *Fuzzy Sets and Systems, Application of Neuro-Fuzzy Systems, 89*, 289–307. https://doi.org/10.1016/S0165-0114(97)00011-0.

Bagheri, M., Akbari, A., & Mirbagheri, S. A. (2019). Advanced control of membrane fouling in filtration systems using artificial intelligence and machine learning techniques: A critical review. *Process Safety and Environmental Protection, 123*, 229–252. https://doi.org/10.1016/j.psep.2019.01.013.

Bahl, A., Hellack, B., Balas, M., Dinischiotu, A., Wiemann, M., Brinkmann, J., et al. (2019). Recursive feature elimination in random forest classification supports nanomaterial grouping. *NanoImpact, 15*, 100179. https://doi.org/10.1016/j.impact.2019.100179.

Baruah, P., & Chinnam, R. B. (2005). HMMs for diagnostics and prognostics in machining processes. *International Journal of Production Research, 43*, 1275–1293. https://doi.org/10.1080/00207540412331327727.

Baybutt, P. (2002). Layers of protection analysis for human factors (LOPA-HF). *Process Safety Progress, 21*, 119–129. https://doi.org/10.1002/prs.680210208.

Ben Ali, J., Chebel-Morello, B., Saidi, L., Malinowski, S., & Fnaiech, F. (2015). Accurate bearing remaining useful life prediction based on Weibull distribution and artificial neural network. *Mechanical Systems and Signal Processing, 56–57*, 150–172. https://doi.org/10.1016/j.ymssp.2014.10.014.

Bhadriraju, B., Kwon, J. S.-I., & Khan, F. (2021). Risk-based fault prediction of chemical processes using operable adaptive sparse identification of systems (OASIS). *Computers & Chemical Engineering, 152*, 107378. https://doi.org/10.1016/j.compchemeng.2021.107378.

Bi, X., & Zhao, J. (2021). A novel orthogonal self-attentive variational autoencoder method for interpretable chemical process fault detection and identification. *Process Safety and Environmental Protection, 156*, 581–597. https://doi.org/10.1016/j.psep.2021.10.036.

Bohannon, P., Fan, W., Flaster, M., & Rastogi, R. (2005). A cost-based model and effective heuristic for repairing constraints by value modification. In *Proceedings of the 2005 ACM SIGMOD International Conference on Management of Data, SIGMOD '05* (pp. 143–154). New York, NY, USA: Association for Computing Machinery. https://doi.org/10.1145/1066157.1066175.

Cagno, E., Caron, F., & Mancini, M. (2002). Risk analysis in plant commissioning: The multilevel Hazop. *Reliability Engineering & System Safety*, 77, 309–323. https://doi.org/10.1016/S0951-8320(02)00064-9.

Cai, L., Gu, J., Ma, J., & Jin, Z. (2019). Probabilistic wind power forecasting approach via instance-based transfer learning embedded gradient boosting decision trees. *Energies, 12*, 159. https://doi.org/10.3390/en12010159.

Čepin, M., & Mavko, B. (2002). A dynamic fault tree. *Reliability Engineering & System Safety, 75*, 83–91. https://doi.org/10.1016/S0951-8320(01)00121-1.

Chalamalla, A., Ilyas, I. F., Ouzzani, M., & Papotti, P. (2014). Descriptive and prescriptive data cleaning. In *Proceedings of the 2014 ACM SIGMOD International Conference on Management of Data, SIGMOD '14* (pp. 445–456). New York, NY, USA: Association for Computing Machinery. https://doi.org/10.1145/2588555.2610520.

Chen, Z., Ding, S. X., Luo, H., & Zhang, K. (2017). An alternative data-driven fault detection scheme for dynamic processes with deterministic disturbances. *Journal of the Franklin Institute, 354*, 556–570. https://doi.org/10.1016/j.jfranklin.2016.10.031.

Chen, T., & Zhang, J. (2010). On-line multivariate statistical monitoring of batch processes using gaussian mixture model. *Computers & Chemical Engineering, 34*, 500–507. https://doi.org/10.1016/j.compchemeng.2009.08.007.

Cheng, F., He, Q. P., & Zhao, J. (2019). A novel process monitoring approach based on variational recurrent autoencoder. *Computers & Chemical Engineering, 129*, 106515. https://doi.org/10.1016/j.compchemeng.2019.106515.

Chiang, F., & Miller, R. J. (2011). In *A unified model for data and constraint repair 2011 IEEE 27th International Conference on Data Engineering. Presented at the 2011 IEEE 27th International Conference on Data Engineering* (pp. 446–457). https://doi.org/10.1109/ICDE.2011.5767833.

Chiang, L. H., Russell, E. L., & Braatz, R. D. (2000). *Fault detection and diagnosis in industrial systems*. Springer Science & Business Media.

Chu, X., Ilyas, I. F., Krishnan, S., & Wang, J. (2016). Data cleaning: Overview and emerging challenges. In *Proceedings of the 2016 International Conference on Management of Data*ACM. https://doi.org/10.1145/2882903.2912574.

Chu, X., Ilyas, I. F., & Papotti, P. (2013). Holistic data cleaning: Putting violations into context. In *2013 IEEE 29th International Conference on Data Engineering (ICDE). Presented at the 2013 IEEE 29th International Conference on Data Engineering (ICDE)* (pp. 458–469). https://doi.org/10.1109/ICDE.2013.6544847.

Chu, X., Morcos, J., Ilyas, I. F., Ouzzani, M., Papotti, P., Tang, N., et al. (2015). KATARA: A data cleaning system powered by knowledge bases and crowdsourcing. In *Proceedings of the 2015 ACM SIGMOD International Conference on Management of Data, SIGMOD '15* (pp. 1247–1261). New York, NY, USA: Association for Computing Machinery. https://doi.org/10.1145/2723372.2749431.

Chung, Y., Mortensen, M. L., Binnig, C., & Kraska, T. (2018). Estimating the impact of unknown unknowns on aggregate query results. *ACM Trans. Database Syst.*, *43*. 3:1-3:37 https://doi.org/10.1145/3167970.

Climente-Alarcon, V., Antonino-Daviu, J. A., Strangas, E. G., & Riera-Guasp, M. (2015). Rotor-Bar breakage mechanism and prognosis in an induction motor. *IEEE Transactions on Industrial Electronics*, *62*, 1814–1825. https://doi.org/10.1109/TIE.2014.2336604.

Cong, G., Fan, W., Geerts, F., Jia, X., & Ma, S. (2007). Improving data quality: Consistency and accuracy. In *Proceedings of the 33rd International Conference on Very Large Data Bases, VLDB '07* (pp. 315–326). Vienna, Austria: VLDB Endowment.

Covello, V. T., & Mumpower, J. (1985). Risk analysis and risk management: An historical perspective. *Risk Analysis*, *5*, 103–120.

Cozzani, V., Antonioni, G., & Spadoni, G. (2006). Quantitative assessment of domino scenarios by a GIS-based software tool. *Journal of Loss Prevention in the Process Industries*, *19*, 463–477. https://doi.org/10.1016/j.jlp.2005.11.007.

Dai, Y., Cheng, F., Wu, H., Wu, D., & Zhao, J. (2020). Chapter Five - Data driven methods. In F. I. Khan, & P. R. Amyotte (Eds.), *Methods in chemical process safety, Advanced methods of risk assessment and management* (pp. 167–203). Elsevier. https://doi.org/10.1016/bs.mcps.2020.02.002.

Daroogheh, N., Baniamerian, A., Meskin, N., & Khorasani, K. (2017). Prognosis and health monitoring of nonlinear systems using a hybrid scheme through integration of PFs and neural networks. *IEEE Transactions on Systems, Man, and Cybernetics: Systems*, *47*, 1990–2004. https://doi.org/10.1109/TSMC.2016.2597272.

Deng, X., Tian, X., Chen, S., & Harris, C. J. (2017). Fault discriminant enhanced kernel principal component analysis incorporating prior fault information for monitoring nonlinear processes. *Chemometrics and Intelligent Laboratory Systems*, *162*, 21–34. https://doi.org/10.1016/j.chemolab.2017.01.001.

Dey, S., & Lee, S.-W. (2021). Multilayered review of safety approaches for machine learning-based systems in the days of AI. *Journal of Systems and Software*, *176*, 110941. https://doi.org/10.1016/j.jss.2021.110941.

Fadier, E., & De la Garza, C. (2006). Safety design: Towards a new philosophy. *Safety Science, Safety and Design*, *44*, 55–73. https://doi.org/10.1016/j.ssci.2005.09.008.

Fahimipirehgalin, M., Trunzer, E., Odenweller, M., & Vogel-Heuser, B. (2019). Automatic visual leakage inspection by using thermographic video and image analysis. In *2019 IEEE 15th International Conference on Automation Science and Engineering (CASE). Presented at the 2019 IEEE 15th International Conference on Automation Science and Engineering (CASE)* (pp. 1282–1288). https://doi.org/10.1109/COASE.2019.8842941.

Fan, W., Li, J., Ma, S., Tang, N., & Yu, W. (2012). Towards certain fixes with editing rules and master data. *The VLDB Journal*, *21*, 213–238. https://doi.org/10.1007/s00778-011-0253-7.

Fan, W., Ma, S., Tang, N., & Yu, W. (2014). Interaction between record matching and data repairing. J. Data and information. *Quality*, *4*, 16:1-16:38. https://doi.org/10.1145/2567657.

Feng, X., Dai, Y., Ji, X., Zhou, L., & Dang, Y. (2021). Application of natural language processing in HAZOP reports. *Process Safety and Environmental Protection, 155*, 41–48. https://doi.org/10.1016/j.psep.2021.09.001.

Ge, Z. (2018). Process data analytics via probabilistic latent variable models: A tutorial review. *Industrial and Engineering Chemistry Research, 57*, 12646–12661. https://doi.org/10.1021/acs.iecr.8b02913.

Geerts, F., Mecca, G., Papotti, P., & Santoro, D. (2013). The LLUNATIC data-cleaning framework. *Proceedings of the VLDB Endowment, 6*, 625–636. https://doi.org/10.14778/2536360.2536363.

Gharahbagheri, H., Imtiaz, S. A., & Khan, F. (2017). Root cause diagnosis of process fault using KPCA and Bayesian network. *Industrial and Engineering Chemistry Research, 56*, 2054–2070. https://doi.org/10.1021/acs.iecr.6b01916.

Goel, P., Datta, A., & Mannan, M. S. (2017). Industrial alarm systems: Challenges and opportunities. *Journal of Loss Prevention in the Process Industries, 50*, 23–36. https://doi.org/10.1016/j.jlp.2017.09.001.

Gokhale, C., Das, S., Doan, A., Naughton, J. F., Rampalli, N., Shavlik, J., et al. (2014). Corleone: Hands-off crowdsourcing for entity matching. In *Proceedings of the 2014 ACM SIGMOD International Conference on Management of Data, SIGMOD '14* (pp. 601–612). New York, NY, USA: Association for Computing Machinery. https://doi.org/10.1145/2588555.2588576.

Han, S., Lee, S., & Peña-Mora, F. (2013). Vision-based detection of unsafe actions of a construction worker: Case study of ladder climbing. *Journal of Computing in Civil Engineering, 27*, 635–644. https://doi.org/10.1061/(ASCE)CP.1943-5487.0000279.

Harrou, F., Nounou, M. N., Nounou, H. N., & Madakyaru, M. (2015). PLS-based EWMA fault detection strategy for process monitoring. *Journal of Loss Prevention in the Process Industries, 36*, 108–119. https://doi.org/10.1016/j.jlp.2015.05.017.

He, R., Dai, Y., Lu, J., & Mou, C. (2018). Developing ladder network for intelligent evaluation syste Case of remaining useful life prediction for centrifugal pumps. *Reliability Engineering & System Safety, 180*, 385–393. https://doi.org/10.1016/j.ress.2018.08.010.

He, Z., & Weng, W. (2020). A dynamic and simulation-based method for quantitative risk assessment of the domino accident in chemical industry. *Process Safety and Environmental Protection, 144*, 79–92. https://doi.org/10.1016/j.psep.2020.07.014.

Heinrich, F., Noering, F. K.-D., Pruckner, M., & Jonas, K. (2021). Unsupervised data-preprocessing for long short-term memory based battery model under electric vehicle operation. *Journal of Energy Storage, 38*, 102598. https://doi.org/10.1016/j.est.2021.102598.

Heise, A., Kasneci, G., & Naumann, F. (2014). Estimating the number and sizes of fuzzy-duplicate clusters. In *Proceedings of the 23rd ACM International Conference on Conference on Information and Knowledge Management, CIKM '14* (pp. 959–968). New York, NY, USA: Association for Computing Machinery. https://doi.org/10.1145/2661829.2661885.

Hinton, G. E., & Salakhutdinov, R. R. (2006). Reducing the dimensionality of data with neural networks. *Science, 313*, 504–507. https://doi.org/10.1126/science.1127647.

Hosseinnia, B., Khakzad, N., & Reniers, G. (2018). Multi-plant emergency response for tackling major accidents in chemical industrial areas. *Safety Science, 102*, 275–289. https://doi.org/10.1016/j.ssci.2017.11.003.

Iqbal, M. U., & Srinivasan, R. (2018). Simulator based performance metrics to estimate reliability of control room operators. *Journal of Loss Prevention in the Process Industries, 56*, 524–530. https://doi.org/10.1016/j.jlp.2017.10.011.

Jiang, Q., & Huang, B. (2016). Distributed monitoring for large-scale processes based on multivariate statistical analysis and Bayesian method. *Journal of Process Control, 46*, 75–83. https://doi.org/10.1016/j.jprocont.2016.08.006.

Jiang, B., Huang, D., Zhu, X., Yang, F., & Braatz, R. D. (2015). Canonical variate analysis-based contributions for fault identification. *Journal of Process Control, 26*, 17–25. https://doi.org/10.1016/j.jprocont.2014.12.001.

Jiang, Z., Li, Z., Wu, N., & Zhou, M. (2018). A petri net approach to fault diagnosis and restoration for power transmission systems to avoid the output interruption of substations. *IEEE Systems Journal, 12*, 2566–2576. https://doi.org/10.1109/JSYST.2017.2682185.

Jiang, B., Zhu, X., Huang, D., & Braatz, R. D. (2015). Canonical variate analysis-based monitoring of process correlation structure using causal feature representation. *Journal of Process Control, 32*, 109–116. https://doi.org/10.1016/j.jprocont.2015.05.004.

Jiao, Z., Ji, C., Sun, Y., Hong, Y., & Wang, Q. (2021). Deep learning based quantitative property-consequence relationship (QPCR) models for toxic dispersion prediction. *Process Safety and Environmental Protection, 152*, 352–360. https://doi.org/10.1016/j.psep.2021.06.019.

Jiarula, Y., Gao, J., Gao, Z., Jiang, H., & Wang, R. (2016). Fault mode prediction based on decision tree. In *2016 IEEE Advanced Information Management, Communicates, Electronic and Automation Control Conference (IMCEC). Presented at the 2016 IEEE Advanced Information Management, Communicates, Electronic and Automation Control Conference (IMCEC)* (pp. 1729–1733). https://doi.org/10.1109/IMCEC.2016.7867514.

Jing, S., Liu, X., Gong, X., Tang, Y., Xiong, G., Liu, S., et al. (2022). Correlation analysis and text classification of chemical accident cases based on word embedding. *Process Safety and Environmental Protection, 158*, 698–710. https://doi.org/10.1016/j.psep.2021.12.038.

Kalantarnia, M., Khan, F., & Hawboldt, K. (2009). Dynamic risk assessment using failure assessment and Bayesian theory. *Journal of Loss Prevention in the Process Industries, Quantitative Risk Analysis - Special issue devoted to Norberto Piccinini, 22*, 600–606. https://doi.org/10.1016/j.jlp.2009.04.006.

Kalantarnia, M., Khan, F., & Hawboldt, K. (2010). Modelling of BP Texas City refinery accident using dynamic risk assessment approach. *Process Safety and Environmental Protection, 88*, 191–199. https://doi.org/10.1016/j.psep.2010.01.004.

Karvonen, I., Heino, P., & Suokas, J. (1990). *Knowledge based approach to support Hazop-studies, Valtion teknillinen tutkimuskeskus. Tutkimuksia - research reports*. Espoo: VTT Technical Research Centre of Finland.

Kennedy, R., & Kirwan, B. (1998). Development of a Hazard and operability-based method for identifying safety management vulnerabilities in high risk systems. *Safety Science, 30*, 249–274. https://doi.org/10.1016/S0925-7535(98)00025-3.

Khan, F. I., & Abbasi, S. A. (1997). OptHAZOP—An effective and optimum approach for HAZOP study. *Journal of Loss Prevention in the Process Industries, 10*, 191–204. https://doi.org/10.1016/S0950-4230(97)00002-8.

Khan, F. I., & Abbasi, S. A. (1999). Major accidents in process industries and an analysis of causes and consequences. *Journal of Loss Prevention in the Process Industries, 12*, 361–378. https://doi.org/10.1016/S0950-4230(98)00062-X.

Khan, F. I., & Abbasi, S. A. (2000a). Towards automation of HAZOP with a new tool EXPERTOP. *Environmental Modelling & Software, 15*, 67–77. https://doi.org/10.1016/S1364-8152(99)00022-5.

Khan, F. I., & Abbasi, S. A. (2000b). Analytical simulation and PROFAT II: A new methodology and a computer automated tool for fault tree analysis in chemical process industries. *Journal of Hazardous Materials, 75*, 1–27. https://doi.org/10.1016/S0304-3894(00)00169-2.

Khan, F., Abunada, H., John, D., & Benmosbah, T. (2010). Development of risk-based process safety indicators. *Process Safety Progress, 29*, 133–143. https://doi.org/10.1002/prs.10354.

Khan, F., Hashemi, S. J., Paltrinieri, N., Amyotte, P., Cozzani, V., & Reniers, G. (2016). Dynamic risk management: A contemporary approach to process safety management. *Current Opinion in Chemical Engineering*, *14*, 9–17. https://doi.org/10.1016/j.coche.2016.07.006.

Khan, F., Rathnayaka, S., & Ahmed, S. (2015). Methods and models in process safety and risk management: Past, present and future. *Process Safety and Environmental Protection*, *98*, 116–147. https://doi.org/10.1016/j.psep.2015.07.005.

Kletz, T. A. (1999). The origins and history of loss prevention. *Process Safety and Environmental Protection, Developments in Process Safety*, 77, 109–116. https://doi.org/10.1205/095758299529938.

Kohda, T., & Cui, W. (2007). Risk-based reconfiguration of safety monitoring system using dynamic Bayesian network. *Reliability Engineering & System Safety, Special Issue on ESREL, 2005*(92), 1716–1723. https://doi.org/10.1016/j.ress.2006.09.012.

Kukkonen, J., & Nikmo, J. (1992). Modelling heavy gas cloud transport in sloping terrain. *Journal of Hazardous Materials*, *31*, 155–176. https://doi.org/10.1016/0304-3894(92)85003-J.

Lai, S., Yang, F., & Chen, T. (2017). Online pattern matching and prediction of incoming alarm floods. *Journal of Process Control*, *56*, 69–78. https://doi.org/10.1016/j.jprocont.2017.01.003.

Lai, S., Yang, F., Chen, T., & Cao, L. (2019). Accelerated multiple alarm flood sequence alignment for abnormality pattern mining. *Journal of Process Control*, *82*, 44–57. https://doi.org/10.1016/j.jprocont.2019.06.004.

Lee, J., Cameron, I., & Hassall, M. (2019). Improving process safety: What roles for digitalization and industry 4.0? *Process Safety and Environmental Protection*, *132*, 325–339. https://doi.org/10.1016/j.psep.2019.10.021.

Li, G., Qin, S. J., Ji, Y., & Zhou, D. (2010). Reconstruction based fault prognosis for continuous processes. *Control Engineering Practice*, *18*, 1211–1219. https://doi.org/10.1016/j.conengprac.2010.05.012.

Liu, Z., Tian, W., Cui, Z., Wei, H., & Li, C. (2021). An intelligent quantitative risk assessment method for ammonia synthesis process. *Chemical Engineering Journal*, *420*, 129893. https://doi.org/10.1016/j.cej.2021.129893.

Liu, H., Yang, C., Carlsson, B., Qin, S. J., & Yoo, C. (2019). Dynamic nonlinear partial least squares modeling using gaussian process regression. *Industrial and Engineering Chemistry Research*, *58*, 16676–16686. https://doi.org/10.1021/acs.iecr.9b00701.

Liu, T., Zhu, K., & Zeng, L. (2018). Diagnosis and prognosis of degradation process via hidden semi-Markov model. *IEEE/ASME Transactions on Mechatronics*, *23*, 1456–1466. https://doi.org/10.1109/TMECH.2018.2823320.

Ma, J., Li, G., & Zhou, D. (2018). Fault prognosis technology for non-gaussian and nonlinear processes based on KICA reconstruction. *The Canadian Journal of Chemical Engineering*, *96*, 515–520. https://doi.org/10.1002/cjce.23051.

Ma, J., Wang, Q., Li, G., & Zhou, D. (2014). Reconstruction-based fault prognosis for flue gas turbines with independent component analysis. *Asia-Pacific Journal of Chemical Engineering*, *9*, 205–213. https://doi.org/10.1002/apj.1760.

Mahadevan, S., & Shah, S. L. (2009). Fault detection and diagnosis in process data using one-class support vector machines. *Journal of Process Control*, *19*, 1627–1639. https://doi.org/10.1016/j.jprocont.2009.07.011.

Markowski, A. S., & Kotynia, A. (2011). "Bow-tie" model in layer of protection analysis. *Process Safety and Environmental Protection*, *89*, 205–213. https://doi.org/10.1016/j.psep.2011.04.005.

Marseguerra, M., Ricotti, M., & Zio, E. (1995). Approaching system evolution in dynamic PSA by neural networks. *Reliability Engineering & System Safety*, *49*, 91–99. https://doi.org/10.1016/0951-8320(95)00039-5.

Matthias, C. S. (1990). Dispersion of a dense cylindrical cloud in calm air. *Journal of Hazardous Materials, 24*, 39–65. https://doi.org/10.1016/0304-3894(90)80002-L.

Meel, A., & Seider, W. D. (2008). Real-time risk analysis of safety systems. *Computers & Chemical Engineering, 32*, 827–840. Festschrift devoted to Rex Reklaitis on his 65th Birthday. https://doi.org/10.1016/j.compchemeng.2007.03.006.

Meliou, A., Gatterbauer, W., Nath, S., & Suciu, D. (2011). Tracing data errors with view-conditioned causality. In *Proceedings of the 2011 ACM SIGMOD International Conference on Management of Data, SIGMOD '11* (pp. 505–516). New York, NY, USA: Association for Computing Machinery. https://doi.org/10.1145/1989323.1989376.

Meng, X., Chen, G., Zhu, G., & Zhu, Y. (2019). Dynamic quantitative risk assessment of accidents induced by leakage on offshore platforms using DEMATEL-BN. *International Journal of Naval Architecture and Ocean Engineering, 11*, 22–32. https://doi.org/10.1016/j.ijnaoe.2017.12.001.

Ming, L., & Zhao, J. (2017). Review on chemical process fault detection and diagnosis. In *2017 6th International Symposium on Advanced Control of Industrial Processes (AdCONIP). Presented at the 2017 6th International Symposium on Advanced Control of Industrial Processes (AdCONIP)* (pp. 457–462). https://doi.org/10.1109/ADCONIP.2017.7983824.

Mosallam, A., Medjaher, K., & Zerhouni, N. (2013). Bayesian approach for remaining useful life prediction. *Chemical Engineering Transactions, 33*, 139–144. https://doi.org/10.3303/CET1333024.

Mott, A., Zoran, D., Chrzanowski, M., Wierstra, D., & Rezende, D. J. (2019). Towards interpretable reinforcement learning using attention augmented agents. *arXiv*, (1906.02500 [cs, stat).

Murphy, D. M., & Paté-Cornell, M. E. (1996). The SAM framework: Modeling the effects of management factors on human behavior in risk analysis. *Risk Analysis, 16*, 501–515. https://doi.org/10.1111/j.1539-6924.1996.tb01096.x.

Natarajan, S., & Srinivasan, R. (2014). Implementation of multi agents based system for process supervision in large-scale chemical plants. *Computers & Chemical Engineering, 60*, 182–196. https://doi.org/10.1016/j.compchemeng.2013.08.012.

Necci, A., Cozzani, V., Spadoni, G., & Khan, F. (2015). Assessment of domino effect: State of the art and research needs. *Reliability Engineering & System Safety, Domino Effects in the Process Industries: fostering innovative approaches and advancing the State-of-the-art, 143*, 3–18. https://doi.org/10.1016/j.ress.2015.05.017.

Nor, N. M., Hassan, C. R. C., & Hussain, M. A. (2020). A review of data-driven fault detection and diagnosis methods: Applications in chemical process systems. *Reviews in Chemical Engineering, 36*, 513–553. https://doi.org/10.1515/revce-2017-0069.

Palmer, C., & Chung, P. W. H. (2008). A computer tool for batch hazard and operability studies. *Journal of Loss Prevention in the Process Industries, 21*, 537–542. https://doi.org/10.1016/j.jlp.2008.05.001.

Palmer, C., & Chung, P. W. H. (2009). An automated system for batch hazard and operability studies. *Reliability Engineering & System Safety, 94*, 1095–1106. https://doi.org/10.1016/j.ress.2009.01.001.

Park, Y.-J., Fan, S.-K. S., & Hsu, C.-Y. (2020). A review on fault detection and process diagnostics in industrial processes. *PRO, 8*, 1123. https://doi.org/10.3390/pr8091123.

Prakash, O., Samantaray, A. K., Bhattacharyya, R., & Ghoshal, S. K. (2018). Adaptive prognosis for a multi-component dynamical system of unknown degradation modes. In *Vol. 51. IFAC-PapersOnLine, 10th IFAC Symposium on Fault Detection, Supervision and Safety for Technical Processes SAFEPROCESS 2018* (pp. 184–191). https://doi.org/10.1016/j.ifacol.2018.09.576.

Pratama, M., Er, M. J., Li, X., Oentaryo, R. J., Lughofer, E., & Arifin, I. (2013). Data driven modeling based on dynamic parsimonious fuzzy neural network. *Neurocomputing, 110*, 18–28. https://doi.org/10.1016/j.neucom.2012.11.013.

Rad, A., Abdolhamidzadeh, B., Abbasi, T., & Rashtchian, D. (2014). FREEDOM II: An improved methodology to assess domino effect frequency using simulation techniques. *Process Safety and Environmental Protection, 92*, 714–722. https://doi.org/10.1016/j.psep.2013.12.002.

Rahm, E., & Do, H. (2000). Data cleaning: Problems and current approaches. *IEEE Data Engineering Bulletin, 23*, 3–13.

Raich, A., & Çinar, A. (1996). Statistical process monitoring and disturbance diagnosis in multivariable continuous processes. *AICHE Journal, 42*, 995–1009. https://doi.org/10.1002/aic.690420412.

Raman, V., & Hellerstein, J. M. (2001). Potter's wheel: An interactive data cleaning system. In *Proceedings of the 27th International Conference on Very Large Data Bases, VLDB '01* (pp. 381–390). San Francisco, CA, USA: Morgan Kaufmann Publishers Inc.

Reniers, G., & Soudan, K. (2010). A game-theoretical approach for reciprocal security-related prevention investment decisions. *Reliability Engineering & System Safety, 95*, 1–9. https://doi.org/10.1016/j.ress.2009.07.001.

Ridzuan, F., & Zainon, W. M. N. W. (2019). A review on data cleansing methods for big data. *Procedia Computer Science, 161*, 731–738. https://doi.org/10.1016/j.procs.2019.11.177.

Rostek, K., Morytko, Ł., & Jankowska, A. (2015). Early detection and prediction of leaks in fluidized-bed boilers using artificial neural networks. *Energy, 89*, 914–923. https://doi.org/10.1016/j.energy.2015.06.042.

Sabeti, S., Shoghli, O., Baharani, M., & Tabkhi, H. (2021). Toward AI-enabled augmented reality to enhance the safety of highway work zones: Feasibility, requirements, and challenges. *Advanced Engineering Informatics, 50*, 101429. https://doi.org/10.1016/j.aei.2021.101429.

Shahab, M. A., Iqbal, M. U., Srinivasan, B., & Srinivasan, R. (2021). Metrics for objectively assessing operator training using eye gaze patterns. *Process Safety and Environmental Protection, 156*, 508–520. https://doi.org/10.1016/j.psep.2021.10.043.

Shang, C., Yang, F., Gao, X., Huang, X., Suykens, J. A. K., & Huang, D. (2015). Concurrent monitoring of operating condition deviations and process dynamics anomalies with slow feature analysis. *AICHE Journal, 61*, 3666–3682. https://doi.org/10.1002/aic.14888.

Shin, E., Yoo, S., Ju, Y., & Shin, D. (2022). Knowledge graph embedding and reasoning for real-time analytics support of chemical diagnosis from exposure symptoms. *Process Safety and Environmental Protection, 157*, 92–105. https://doi.org/10.1016/j.psep.2021.11.002.

Smith, W. A., & Randall, R. B. (2015). Rolling element bearing diagnostics using the Case Western Reserve University data: A benchmark study. *Mechanical Systems and Signal Processing, 64–65*, 100–131. https://doi.org/10.1016/j.ymssp.2015.04.021.

Srinivasan, R., Srinivasan, B., Iqbal, M. U., Nemet, A., & Kravanja, Z. (2019). Recent developments towards enhancing process safety: Inherent safety and cognitive engineering. *Computers & Chemical Engineering, 128*, 364–383. https://doi.org/10.1016/j.compchemeng.2019.05.034.

Srinivasan, R., & Zhao, J. (2022). Editorial: Special issue on data analytics in process safety. *Process Safety and Environmental Protection, 159*, 625–626. https://doi.org/10.1016/j.psep.2022.01.039.

Stringfellow, M. V., Leveson, N. G., & Owens, B. D. (2010). Safety-driven Design for Software-Intensive Aerospace and Automotive Systems. *Proceedings of the IEEE, 98*, 515–525. https://doi.org/10.1109/JPROC.2009.2039551.

Subramaniam, S., Palpanas, T., Papadopoulos, D., Kalogeraki, V., & Gunopulos, D. (2006). Online outlier detection in sensor data using non-parametric models, in. In *Proceedings of the 32nd International Conference on Very Large Data Bases, VLDB '06* (pp. 187–198). Seoul, Korea: VLDB Endowment.

Summers, A. E. (2003). Introduction to layers of protection analysis. *Journal of Hazardous Materials, 104*, 163–168. https://doi.org/10.1016/S0304-3894(03)00242-5.

Swaminathan, S., & Smidts, C. (1999a). Identification of missing scenarios in ESDs using probabilistic dynamics. *Reliability Engineering & System Safety*, *66*, 275–279. https://doi.org/10.1016/S0951-8320(99)00024-1.

Swaminathan, S., & Smidts, C. (1999b). The event sequence diagram framework for dynamic probabilistic risk assessment. *Reliability Engineering & System Safety*, *63*, 73–90. https://doi.org/10.1016/S0951-8320(98)00027-1.

Tao, H., Wang, P., Chen, Y., Stojanovic, V., & Yang, H. (2020). An unsupervised fault diagnosis method for rolling bearing using STFT and generative neural networks. *Journal of the Franklin Institute*, *357*, 7286–7307. https://doi.org/10.1016/j.jfranklin.2020.04.024.

Taqvi, S. A. A., Zabiri, H., Tufa, L. D., Uddin, F., Fatima, S. A., & Maulud, A. S. (2021). A review on data-driven learning approaches for fault detection and diagnosis in chemical processes. *ChemBioEng Reviews*, *8*, 239–259. https://doi.org/10.1002/cben.202000027.

Tong, C., & Shi, X. (2016). Decentralized monitoring of dynamic processes based on dynamic feature selection and informative fault pattern dissimilarity. *IEEE Transactions on Industrial Electronics*, *63*, 3804–3814. https://doi.org/10.1109/TIE.2016.2530047.

Tsai, M.-K. (2014). Automatically determining accidental falls in field surveying: A case study of integrating accelerometer determination and image recognition. *Safety Science*, *66*, 19–26. https://doi.org/10.1016/j.ssci.2014.01.012.

Uraikul, V., Chan, C. W., & Tontiwachwuthikul, P. (2007). Artificial intelligence for monitoring and supervisory control of process systems. *Engineering Applications of Artificial Intelligence, Special Issue on Applications of Artificial Intelligence in Process Systems Engineering*, *20*, 115–131. https://doi.org/10.1016/j.engappai.2006.07.002.

Vandroux-Koenig, S., & Berthoud, G. (1997). Modelling of a two phase momentum jet close to the breach, in the containment vessel of a liquefied gas. *Journal of Loss Prevention in the Process Industries*, *10*, 17–29. https://doi.org/10.1016/S0950-4230(96)00038-1.

Vaswani, A., Shazeer, N., Parmar, N., Uszkoreit, J., Jones, L., Gomez, A. N., et al. (2017). Attention is all you need. In *Advances in neural information processing systems*. Inc: Curran Associates.

Venkatasubramanian, V. (2019). The promise of artificial intelligence in chemical engineering: Is it here, finally? *AICHE Journal*, *65*, 466–478. https://doi.org/10.1002/aic.16489.

Venkatasubramanian, V., Rengaswamy, R., Yin, K., & Kavuri, S. N. (2003). A review of process fault detection and diagnosis: Part I: Quantitative model-based methods. *Computers & Chemical Engineering*, *27*, 293–311. https://doi.org/10.1016/S0098-1354(02)00160-6.

Venkatasubramanian, V., & Vaidhyanathan, R. (1994). A knowledge-based framework for automating HAZOP analysis. *AIChE Journal (American Institute of Chemical Engineers); (United States)*, *40*(3). https://doi.org/10.1002/aic.690400311.

Verbert, K., De Schutter, B., & Babuška, R. (2018). A multiple-model reliability prediction approach for condition-based maintenance. *IEEE Transactions on Reliability*, *67*, 1364–1376. https://doi.org/10.1109/TR.2018.2825470.

Vílchez, J. A., Espejo, V., & Casal, J. (2011). Generic event trees and probabilities for the release of different types of hazardous materials. *Journal of Loss Prevention in the Process Industries, Frequency in Risk Assessment*, *24*, 281–287. https://doi.org/10.1016/j.jlp.2011.01.005.

Villa, V., Paltrinieri, N., Khan, F., & Cozzani, V. (2016). Towards dynamic risk analysis: A review of the risk assessment approach and its limitations in the chemical process industry. *Safety Science*, *89*, 77–93. https://doi.org/10.1016/j.ssci.2016.06.002.

Wang, H., Chai, T.-Y., Ding, J.-L., & Brown, M. (2009). Data driven fault diagnosis and fault tolerant control: Some advances and possible new directions. *Acta Automatica Sinica*, *35*, 739–747. https://doi.org/10.1016/S1874-1029(08)60093-2.

Wang, B., Chen, B., & Zhao, J. (2015). The real-time estimation of hazardous gas dispersion by the integration of gas detectors, neural network and gas dispersion models. *Journal of Hazardous Materials, 300*, 433–442. https://doi.org/10.1016/j.jhazmat.2015.07.028.

Wang, F., & Gao, J. (2012). A novel knowledge database construction method for operation guidance expert system based on HAZOP analysis and accident analysis. *Journal of Loss Prevention in the Process Industries, 25*, 905–915. https://doi.org/10.1016/j.jlp.2012.05.001.

Wang, H., Khan, F., Ahmed, S., & Imtiaz, S. (2016). Dynamic quantitative operational risk assessment of chemical processes. *Chemical Engineering Science, 142*, 62–78. https://doi.org/10.1016/j.ces.2015.11.034.

Wang, J., Kraska, T., Franklin, M. J., & Feng, J. (2012). CrowdER: Crowdsourcing entity resolution. *Proceedings of the VLDB Endowment, 5*, 1483–1494. https://doi.org/10.14778/2350229.2350263.

Wang, J., Krishnan, S., Franklin, M. J., Goldberg, K., Kraska, T., & Milo, T. (2014). A sample-and-clean framework for fast and accurate query processing on dirty data. In *Proceedings of the 2014 ACM SIGMOD International Conference on Management of Data, SIGMOD '14* (pp. 469–480). New York, NY, USA: Association for Computing Machinery. https://doi.org/10.1145/2588555.2610505.

Wang, Y., Si, Y., Huang, B., & Lou, Z. (2018). Survey on the theoretical research and engineering applications of multivariate statistics process monitoring algorithms: 2008–2017. *The Canadian Journal of Chemical Engineering, 96*, 2073–2085. https://doi.org/10.1002/cjce.23249.

Wang, Y., Sun, F., & Li, B. (2018). Multiscale neighborhood normalization-based multiple dynamic PCA monitoring method for batch processes with frequent operations. *IEEE Transactions on Automation Science and Engineering, 15*, 1053–1064. https://doi.org/10.1109/TASE.2017.2713800.

Wang, Z., Wen, H., Su, Y., Shen, W., Ren, J., Ma, Y., et al. (2022). Insights into ensemble learning-based data-driven model for safety-related property of chemical substances. *Chemical Engineering Science, 248*, 117219. https://doi.org/10.1016/j.ces.2021.117219.

Wang, J., Yang, F., Chen, T., & Shah, S. L. (2016). An overview of industrial alarm systems: Main causes for alarm overloading, research status, and open problems. *IEEE Transactions on Automation Science and Engineering, 13*, 1045–1061. https://doi.org/10.1109/TASE.2015.2464234.

Wang, Z., Zarader, J.-L., & Argentieri, S. (2012). A novel aircraft fault diagnosis and prognosis system based on gaussian mixture models. In *2012 12th International Conference on Control Automation Robotics Vision (ICARCV). Presented at the 2012 12th International Conference on Control Automation Robotics Vision (ICARCV)* (pp. 1794–1799). https://doi.org/10.1109/ICARCV.2012.6485422.

Wang, B., & Zhao, J. (2022). Automatic frequency estimation of contributory factors for confined space accidents. *Process Safety and Environmental Protection, 157*, 193–207. https://doi.org/10.1016/j.psep.2021.11.004.

Wen, H., Khan, F., Amin, M. T., & Halim, S. Z. (2021). Myths and misconceptions of data-driven methods: Applications to process safety analysis. *Computers & Chemical Engineering, 107639*. https://doi.org/10.1016/j.compchemeng.2021.107639.

Wen, Y., Wu, J., & Yuan, Y. (2017). Multiple-phase modeling of degradation signal for condition monitoring and remaining useful life prediction. *IEEE Transactions on Reliability, 66*, 924–938. https://doi.org/10.1109/TR.2017.2710319.

Wu, E., & Madden, S. (2013). Scorpion: Explaining away outliers in aggregate queries. *Proceedings of the VLDB Endowment, 6*, 553–564. https://doi.org/10.14778/2536354.2536356.

Wu, H., Wu, D., & Zhao, J. (2019). An intelligent fire detection approach through cameras based on computer vision methods. *Process Safety and Environmental Protection, 127*, 245–256. https://doi.org/10.1016/j.psep.2019.05.016.

Wu, H., & Zhao, J. (2018a). Deep convolutional neural network model based chemical process fault diagnosis. *Computers & Chemical Engineering, 115*, 185–197. https://doi.org/10.1016/j.compchemeng.2018.04.009.

Wu, H., & Zhao, J. (2018b). An intelligent vision-based approach for helmet identification for work safety. *Computers in Industry, 100*, 267–277. https://doi.org/10.1016/j.compind.2018.03.037.

Wu, H., & Zhao, J. (2020). Fault detection and diagnosis based on transfer learning for multimode chemical processes. *Computers & Chemical Engineering, 135*, 106731. https://doi.org/10.1016/j.compchemeng.2020.106731.

Wu, D., & Zhao, J. (2021). Process topology convolutional network model for chemical process fault diagnosis. *Process Safety and Environmental Protection, 150*, 93–109. https://doi.org/10.1016/j.psep.2021.03.052.

Xiang, S., Bai, Y., & Zhao, J. (2021). Medium-term prediction of key chemical process parameter trend with small data. *Chemical Engineering Science, 117361*. https://doi.org/10.1016/j.ces.2021.117361.

Xiao, B., Li, Y., Sun, B., Yang, C., Huang, K., & Zhu, H. (2021). Decentralized PCA modeling based on relevance and redundancy variable selection and its application to large-scale dynamic process monitoring. *Process Safety and Environmental Protection, 151*, 85–100. https://doi.org/10.1016/j.psep.2021.04.043.

Yakout, M., Elmagarmid, A. K., Neville, J., Ouzzani, M., & Ilyas, I. F. (2011). Guided data repair. *arXiv* (1103.3103 [cs].).

Yang, X., & Sam Mannan, M. (2010a). The development and application of dynamic operational risk assessment in oil/gas and chemical process industry. *Reliability Engineering & System Safety, 95*, 806–815. https://doi.org/10.1016/j.ress.2010.03.002.

Yang, X., & Sam Mannan, M. (2010b). An uncertainty and sensitivity analysis of dynamic operational risk assessment model: A case study. *Journal of Loss Prevention in the Process Industries, 23*, 300–307. https://doi.org/10.1016/j.jlp.2009.11.001.

Yosinski, J., Clune, J., Bengio, Y., & Lipson, H. (2014). *How transferable are features in deep neural networks?*.

Yu, M., Wang, D., & Luo, M. (2015). An integrated approach to prognosis of hybrid systems with unknown mode changes. *IEEE Transactions on Industrial Electronics, 62*, 503–515. https://doi.org/10.1109/TIE.2014.2327557.

Yu, Y., Wang, J., & Yang, Z. (2017). Design of Alarm Trippoints for univariate analog process variables based on alarm probability plots. *IEEE Transactions on Industrial Electronics, 64*, 6496–6505. https://doi.org/10.1109/TIE.2017.2682783.

Yu, J., & Yan, X. (2021). A new deep model based on the stacked autoencoder with intensified iterative learning style for industrial fault detection. *Process Safety and Environmental Protection, 153*, 47–59. https://doi.org/10.1016/j.psep.2021.07.002.

Yuan, Z., Yang, Z., Ling, Y., Wu, C., & Li, C. (2021). Spatiotemporal attention mechanism-based deep network for critical parameters prediction in chemical process. *Process Safety and Environmental Protection, 155*, 401–414. https://doi.org/10.1016/j.psep.2021.09.024.

Yuan, F., Zhang, L., Xia, X., Wan, B., Huang, Q., & Li, X. (2019). Deep smoke segmentation. *Neurocomputing, 357*, 248–260. https://doi.org/10.1016/j.neucom.2019.05.011.

Zhang, J. (2006). Improved on-line process fault diagnosis through information fusion in multiple neural networks. *Computers & Chemical Engineering, 30*, 558–571. https://doi.org/10.1016/j.compchemeng.2005.11.002.

Zhang, C., Collin, A., Moireau, P., Trouvé, A., & Rochoux, M. C. (2019). Front shape similarity measure for data-driven simulations of wildland fire spread based on state estimation: Application to the RxCADRE field-scale experiment. *Proceedings of the Combustion Institute, 37*, 4201–4209. https://doi.org/10.1016/j.proci.2018.07.112.

Zhang, F., & Ge, Z. (2015). Decision fusion systems for fault detection and identification in industrial processes. *Journal of Process Control, 31*, 45–54. https://doi.org/10.1016/j.jprocont.2015.04.004.

Zhang, C., Rochoux, M., Tang, W., Gollner, M., Filippi, J.-B., & Trouvé, A. (2017). Evaluation of a data-driven wildland fire spread forecast model with spatially-distributed parameter estimation in simulations of the FireFlux I field-scale experiment. In *Vol. 91. Fire Safety Journal, Fire Safety Science: Proceedings of the 12th International Symposium* (pp. 758–767). https://doi.org/10.1016/j.firesaf.2017.03.057.

Zhang, B., Sconyers, C., Byington, C., Patrick, R., Orchard, M. E., & Vachtsevanos, G. (2011). A probabilistic fault detection approach: Application to bearing fault detection. *IEEE Transactions on Industrial Electronics, 58*, 2011–2018. https://doi.org/10.1109/TIE.2010.2058072.

Zhang, Z., & Zhao, J. (2017). A deep belief network based fault diagnosis model for complex chemical processes. *Computers & Chemical Engineering, In honor of Professor Rafiqul Gani, 107*, 395–407. https://doi.org/10.1016/j.compchemeng.2017.02.041.

Zhang, Q., & Zhu, S.-C. (2018). Visual interpretability for deep learning: A survey. *arXiv*, (1802.00614 [cs).

Zhao, C., & Gao, F. (2015). Online fault prognosis with relative deviation analysis and vector autoregressive modeling. *Chemical Engineering Science, 138*, 531–543. https://doi.org/10.1016/j.ces.2015.08.037.

Zhao, C., & Huang, B. (2018). A full-condition monitoring method for nonstationary dynamic chemical processes with cointegration and slow feature analysis. *AICHE Journal, 64*, 1662–1681. https://doi.org/10.1002/aic.16048.

Zheng, S., & Zhao, J. (2018). States identification of complex chemical process based on unsupervised learning. In M. R. Eden, M. G. Ierapetritou, & G. P. Towler (Eds.), *Computer Aided Chemical Engineering, 13 International Symposium on Process Systems Engineering (PSE 2018)* (pp. 2239–2244). Elsevier. https://doi.org/10.1016/B978-0-444-64241-7.50368-2.

Zheng, S., & Zhao, J. (2020). A new unsupervised data mining method based on the stacked autoencoder for chemical process fault diagnosis. *Computers & Chemical Engineering, 135*, 106755. https://doi.org/10.1016/j.compchemeng.2020.106755.

Zheng, S., & Zhao, J. (2021). A self-adaptive temporal-spatial self-training algorithm for semi-supervised fault diagnosis of industrial processes. *IEEE Transactions on Industrial Informatics, 1–1*. https://doi.org/10.1109/TII.2021.3120686.

Zhong, K., Han, M., & Han, B. (2020). Data-driven based fault prognosis for industrial systems: A concise overview. *IEEE-CAA Journal of Automatica Sinica*, 7, 330–345. https://doi.org/10.1109/JAS.2019.1911804.

Zhou, Y., Li, S., & Xiong, N. (2019). Improved vine copula-based dependence description for multivariate process monitoring based on ensemble learning. *Industrial and Engineering Chemistry Research, 58*, 3782–3796. https://doi.org/10.1021/acs.iecr.8b04081.

Zhou, Z., Wen, C., & Yang, C. (2016). Fault isolation based on k-nearest neighbor rule for industrial processes. *IEEE Transactions on Industrial Electronics, 63*, 2578–2586. https://doi.org/10.1109/TIE.2016.2520898.

Zhu, J., & Gao, F. (2018). Improved nonlinear quality estimation for multiphase batch processes based on relevance vector machine with neighborhood component variable selection. *Industrial and Engineering Chemistry Research, 57*, 666–676. https://doi.org/10.1021/acs.iecr.7b03590.

Zhu, J., Shu, Y., Zhao, J., & Yang, F. (2014). A dynamic alarm management strategy for chemical process transitions. *Journal of Loss Prevention in the Process Industries, 30*, 207–218. https://doi.org/10.1016/j.jlp.2013.07.008.

Zhu, J., Wang, C., Li, C., Gao, X., & Zhao, J. (2016). Dynamic alarm prediction for critical alarms using a probabilistic model. *Chinese Journal of Chemical Engineering, 24*, 881–885. https://doi.org/10.1016/j.cjche.2016.04.017.

CHAPTER FOUR

Industry 4.0 based process data analytics platform

Thumeera R. Wanasinghe[a,*], Mihiran Galagedarage Don[a], Rajeevan Arunthavanathan[a,*], and Raymond G. Gosine[a,b]
[a]Centre for Risk, Integrity and Safety Engineering (C-RISE), Faculty of Engineering & Applied Science, Memorial University of Newfoundland, St. John's, NL, Canada
[b]Innovation Policy Lab, Munk School of Global Affairs and Public Policy, University of Toronto, Toronto, ON, Canada
*Corresponding authors: e-mail address: thumeerawa@mun.ca; rarunthavana@mun.ca

Contents

1. A brief overview for industrial revolution

Over the past centuries, industries have been evolving from muscle-driven (human-labor-driven) simple machinery to smart, fully automated machinery which cater to ever-changing human needs (Sharma & Singh, 2020a). The industrial world has undergone three distinct phases

Methods in Chemical Process Safety, Volume 6
ISSN 2468-6514
https://doi.org/10.1016/bs.mcps.2022.04.008

of the industrial revolution and is currently experiencing the fourth phase while laying the groundworks needed for the fifth phase. The first phase, sometimes referred to as Industry 1.0, mechanized some muscle-driven production activities using machines that were powered by water or steam. The second phase, referred to as Industry 2.0, introduced mass production assembly lines run by electric power. With advancements in information communication technologies (ICT) and micro-electromechanical systems (MEMS), the third phase of the industrial revolution, referred to as Industry 3.0, focussed on automating machinery and industrial processes. Modern digital technologies, including smart devices, the internet of things (IoT), big data analytics, artificial intelligence, cloud computing, robotics, and virtual-, augmented- or mixed-reality, are transforming the existing industrial ecosystem into connected cyber-physical systems (CPS). Industry 4.0 is the fourth industrial revolution and is widely referred to as the digital transformation (or digitalization) of the industrial ecosystem. As the past and current industrial revolutions increasingly automate and digitalize industrial processes and operations, many occupations, or at least manual tasks within occupations, face becoming obsolete. Unfortunately, this can lead to significant socioeconomic disruption. Therefore, the next (future) industrial revolution, referred to as Industry 5.0, will focus on implementing human-cyber-physical systems, which integrates cognitive computing capabilities of smart machines with human intelligence and resourcefulness to improve the productivity and safety of industrial processes while minimizing socioeconomic disruptions. In such implementation, human workers will be upskilled to operate advanced machinery installed in future workplaces.

2. Overview of process data analytic platforms & Industry 4.0

As of today, many industries (if not all), including the process industry,[a] are facing several challenges such as a desire for higher quality

[a] Note that the chemical industry is considered a subset of process industry. Since the plants deals with many chemical (e.g., refineries) typically referred to as process plant, but not as chemical plant, this chapter discusses process data analytic platforms for Industry 4.0 in the context of process industry and process plants.

and safety, ultra-short turnaround times, big crew changes, shortage of skilled workers, and the need to generate added value in rapidly changing industrial and socioeconomic environments. To address these challenges, Industry 4.0 has seen digital technologies used to acquire, process, analyze, visualize and store data to make data-driven decisions promptly. Data acquisition, processing, analytics, and data-driven decision-making are not new to the process industry. The process industry has been using supervisory control and data acquisition (SCADA) systems and wireless sensor networks (WSN) for decades to monitor assets (Nechibvute & Mudzingwa, 2013). However, deployment of industrial internet of things (IIoT) devices, cloud-based computing capabilities, advanced data analytics techniques (machine learning, deep learning, transfer learning, etc.), and advanced visualization techniques (virtual-, augmented- or mixed-reality) have been revolutionizing industrial process data analytic platforms.

Fig. 1 illustrates the general architecture for process data analytic platforms in the Industry 4.0 era, integrating enterprise software stack, data lake, knowledge lake, hardware (machines, sensors, actuators, etc.) modules, existing software tools (e.g., software tools for enterprise resource planning, operation management, product life-cycle management, etc.) systems into a cloud-based and IIoT enabled cyber-physical system. At an abstract level, such a platform can be divided into four layers:

1. *The hardware and sensing layer* is responsible for data acquisition and command execution.
2. *The data management layer* is responsible for storing, accessing, analyzing, and visualizing data and knowledge.
3. *The application layer* consists of all the business and enterprise-level software applications.
4. *The network layer* is responsible for securely relaying data between the above three layers.

The increased connectivity among devices and software tools in Industry 4.0 based process data analytic platforms significantly elevates cyber security threats, increasing the need for effective threat protection. Potential cybersecurity threats and remedies for the process data analytic platform are discussed in Section 7.

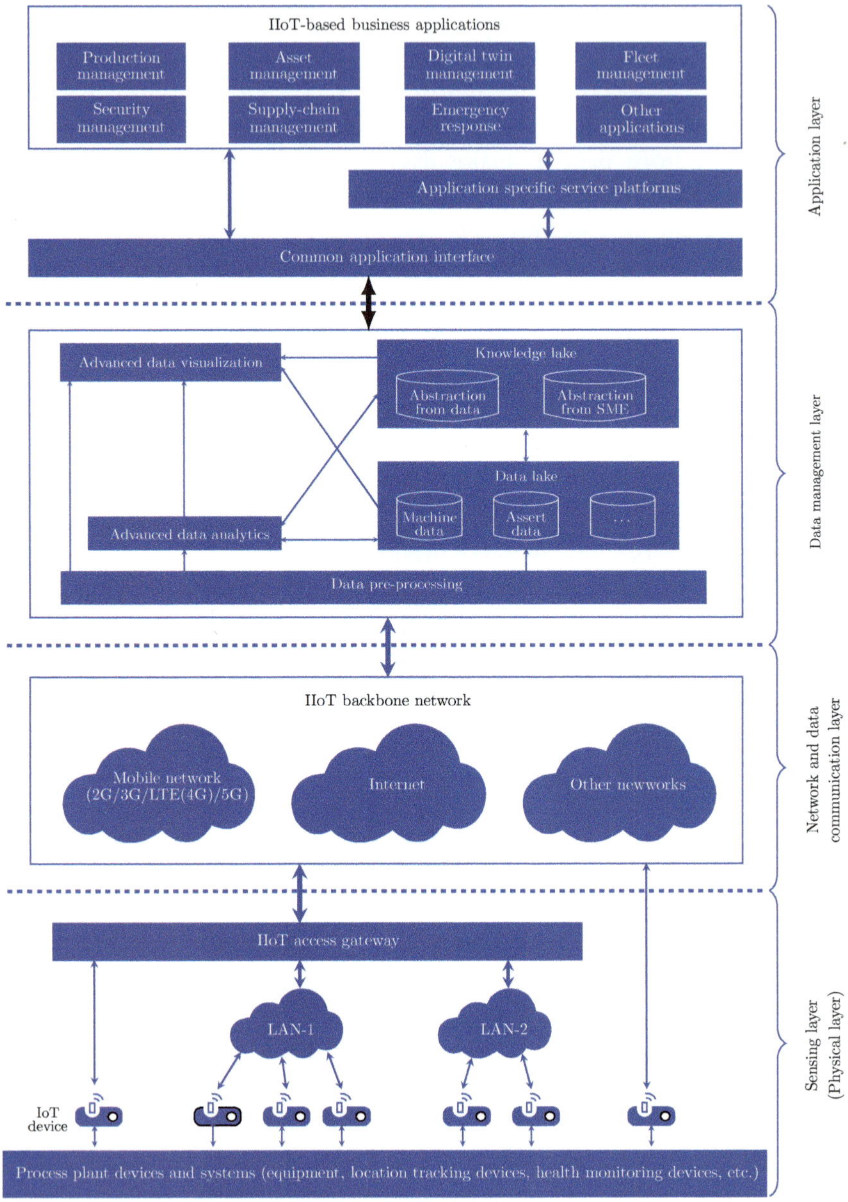

Fig. 1 High level system architecture for Industry 4.0 process data analytic platforms (SME-Subject Matter Experts).

3. Hardware and sensing layer

This layer contains all the physical components of the process plant, including machinery, sensors, actuators, and other infrastructure, such as buildings, pipelines, and valves. Sensors collect operations and health data associated with the physical components, while actuators allow operators to alter operation parameters to achieve their operation goals safely and effectively.

SCADA systems and WSNs have been the popular choices to enable machine-level connectivity and data acquisition for decades (Ashok & Rama, 2020; Bonastre, Capella, Ors, Peris, 2012; Gadhe, Bhojane, & Kawade, 2018; Mao, 2017; Patel, Patel, & Singh Bunet, 2015; Prabakaran, Sarma Dhulipala, Nirmal Kumar, & Ranjith, 2011; Sadiku, Musa, & Musa, 2018). However, these systems have several limitations. For instance, SCADA systems are typically not interoperable; their maintenance and equipment costs are relatively high and unscalable; changing their communication or operational protocols or software tools is highly challenging; and they have high latency (Khan, Aalsalem, Khan, Hossain, & Atiquzzaman, 2017; Shoja & Jalali, 2017). WSNs are not heterogeneous and typically lack coordinated communication and clarity within their operation areas and across processes (Shoja & Jalali, 2017). By addressing limitations associated with SCADA and WSNs, many industries, including process (Ley, Fitzpatrick, Ingham, & Nikbin, 2015; Wang, 2018), manufacturing (Yang, Kumara, Bukkapatnam, & Tsung, 2019), transportation (Zantalis, Koulouras, Karabetsos, & Kandris, 2019), mining (Aziz, Schelén, & Bodin, 2020), and oil and gas (Wanasinghe, Gosine, James, Mann, et al., 2020) industries, have now been deploying IIoT-based solutions for data acquisition, communication and processing. Similarly, applications such as smart cities (Talari, Shafie-Khah, Siano, Loia, et al., 2017), smart homes (Risteska Stojkoska & Trivodaliev, 2017), and smart grids (Kaur & Kalra, 2016) also utilize such IIoT-based solutions.

3.1 What is IoT/IIoT?

Internet of things (IoT) is a commonly used technical buzzword in the Industry 4.0 era. Many technological giants, small and medium-sized enterprises (SMEs), and start-up companies provide IoT-based solutions and services to many industries. Various definitions of IoT are often influenced by the vision of the proponent entity and the intended application (Minerva, Biru, & Rotondi, 2015).

Therefore, Minerva et al. reviewed a range of documents from standardization organizations, several research projects, international IoT initiatives, industrial activities related to IoT, white papers, and books (or book chapters) to develop a comprehensive definition for IoT. Their review led to two separate definitions of IoT, based on the level of complexity of the IoT network.

For the small environment scenario (also referred to as a low complexity scenario), IoT is defined as "*a network that connects uniquely identifiable 'Things' to the Internet. The 'Things' have sensing/actuation and potential programmability capabilities. Through the exploitation of unique identification and sensing, information about the 'Thing' can be collected, and the state of the 'Thing' can be changed from anywhere, anytime, by anything.*" For the large environment scenario (also referred to as high complexity scenario), IoT is defined as "*a self-configuring, adaptive, complex network that interconnects 'Things' to the Internet through the use of standard communication protocols. The interconnected things have physical or virtual representation in the digital world, sensing/actuation capability, a programmability feature, and are uniquely identifiable. The representation contains information including the Thing's identity, status, location, or any other business, social, or privately relevant information. The things offer services, with or without human intervention, through the exploitation of unique identification, data capture and communication, and actuation capability. The service is exploited through the use of intelligent interfaces and is made available anywhere, anytime, and for anything, taking security into consideration.*"

Based on these definitions, it is clear that the scope of IoT-based systems is not bounded for coordinated data collection like traditional WSNs, but it encompasses the power of intelligent machines, real-time intelligence, decision-making, and heterogeneity. Increased popularity and diverse application domains for IoT technology created two distinct groups of IoT devices: consumer IoT and industrial IoT (IIoT). A comparison between these two categories is summarized in Table 1.

3.2 Edge devices

IIoT devices that are (can be) deployed in the process industry can be divided into four main categories based on the intended applications. The first category consists of the intelligent sensor and actuators deployed in a process plant for operations and integrity monitoring. The second category consists

Table 1 Differences between consumer IoT and IIoT (Sharma & Singh, 2020b).

Prospect	Consumer IoT	IIoT
Linked things	User-level devices, basically not much expensive	Costly machines, sensors, systems, basically with high degree of difficulty
Service model	Human-based	Machine-based
Communication capacity	A smaller number of communication standards	A large range connectivity technologies and standards
Communication transportation	Typically, wireless	Both wired and wireless
Amount of Data	Medium to high	High to very high
Evaluative	Quite trivial	Becomes serious (timing, security, privacy, reliability)

of IIoT sensor nodes that are employed for location tracking purposes. This is not limited to personnel tracking systems but extends to supply chains, equipment, tools, machines, and any mobile asset tracking applications. IIoT sensors deployed for environmental monitoring, including temperature, hazardous level, radiation level, gas composition, belongs to the third category. The last category consists of IIoT nodes employed for physiological/biometric monitoring of plant workers. The connection between IIoT devices and an IIoT backbone network can be established in three ways: (1) directly, (2) through an IIoT access gateway, and (3) through a local area network followed by an IIoT access gateway, as shown in Fig. 1.

3.3 Location of data processing

The location where the data is processed is a crucial factor that governs the network's response time (i.e., the time taken from data collection to actuation). As shown in Fig. 2, there are three distinct paradigms associated with data processing: edge computing, fog computing, and cloud computing. More details on data processing platforms and techniques can be found in Section 4. Typical network latencies for edge, fog and cloud computing is given in Table 2. The response time can then be calculated as the summation of bi-directional network latency and the data processing time.

Edge processing (also known as edge computing) typically occurs directly on the IIoT or gateway device, which is physically adjacent to

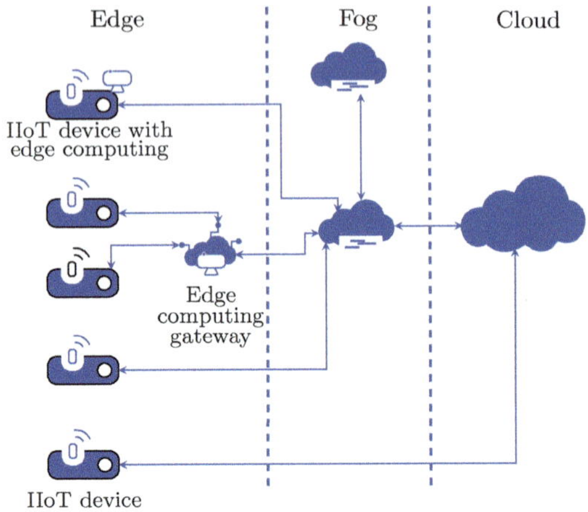

Fig. 2 Data processing location of IIoT.

Table 2 Typical network latency for edge, fog and cloud computing (Carroll & Chandramouli, 2019).

Location	Latency
On the device	Immediate
Edge gateway server	<1 ms
Fog (near edge)	1–4 ms
Fog (far edge)	3–6 ms
Cloud	50–55 ms

the IIoT sensor node. Fog computing occurs in processors (computers/servers) that are connected to the same local area network (LAN) as the IIoT device but may be physically distant from the sensors and actuators. Fog computing is sometimes referred to as the edge cloud. Cloud computing occurs at an immense computer cloud that can be located far away from the IIoT sensor node. Edge and fog computing essentially attempt to achieve one common goal. That is, both the approaches try to leverage the data processing capabilities within the LAN, minimizing the latency issue associated with the cloud-based approaches. For example, when edge and fog computing occur at the process plant, they enable faster data-driven decision-making about plant operations.

An IIoT sensor node has limited data processing capabilities and is thus suitable only for small computation tasks. For example, a smart temperature (or pressure) sensor can compare real-time temperature (or pressure) data with a predefined safety threshold and generate alarms or actuation commands when the measured value exceeds the safety threshold. In contrast, cloud computing offers a high processing and memory capacity, allowing the end-user to efficiently run advanced data analytic tools/algorithm. With its high processing and memory capacity, cloud computing is suited to complex data processing applications such as history mapping, 4D-simulations, pattern recognitions, high dimensional model generation and maintenance. An edge gateway, which route data from multiple edge devices to the fog, has higher processing capabilities than a sensor node, and fog computing processors have a higher processing and memory capacity than the edge gateway. Depending on the computational intensity, time criticalness, and computational power budget, IIoT system designers distribute computational tasks across the edge, fog, and cloud.

3.4 Technology enablers

Regardless of the architecture or the hardware and software platforms used, any end-to-end IoT solution should be able to identify IoT devices in the network uniquely; acquire data from IoT devices and actuate IoT devices; integrate heterogeneous objects and communicate data among the IoT devices; evaluate (process) the collected data; extract knowledge smartly; and finally offer high-level services, including identity-related services, information aggregation services, collaboration-aware services and ubiquitous services (Al-Fuqaha, Guizani, Mohammadi, Aledhari, & Ayyash, 2015). There is a range of technologies to empower these objectives. Some of these technologies, such as smart sensors, smart actuators, energy harvesting technologies, microprocessors, and RFID technologies, directly contribute to the development of an IoT system, while other technologies such as augmented reality, tangible user interfaces, and clean technologies, add value to IoT deployments (Sundmaeker, Guillemin, Friess, & Woelfflé, 2010). A comprehensive review of enabling technologies, associated challenges and the open research issues is presented in Čolaković and Hadžialić (2018), which categorizes the enabling technologies into four domains: the application domain, middleware domain, networking domain, and object domain. Table 3 summarizes the enabling technologies, samples, use cases, functionalities and utility factors (i.e., open issues and challenges that drive the research activities) for the four principal domains.

Table 3 Enabling technologies and functional blocks for IoT (Čolaković & Hadžialić, 2018).

Domain	Enabling technologies, samples and use cases		Functionalities	Utility factors												
				Standadization	Architecture	interoperability and integration	Availability and reliability	Data storage and processing	Contex awareness	Scalability	Management and configuration	Performance and QoS	Unique identification	Power and energy consumption	Security and privacy	Environmental issues
Application domain	IoT Application	Smart homes, smart cities, smart healthcare, IIoT, smart agriculture, smart oil fields, smart mines, smart farms, smart meters, environmental monitoring, etc.	Visualization, application development, system and device monitoring, control and management data and service management													
	Architectures	Software architectures, SOA, RESTFul, etc.														
	Software and APIs	OS (Contiki, FreeRTOS, LiteOS. Adroid, RIOT, Mbed, ...) APIs (JML, WebGL, RAML, ...), Embeded and custom apps built using IoT data, etc.														
Middleware domain	Cloud platforms	OpenIoT, Amazon, Google Cloud, Libelium, IBM Watson, Microsoft Azure, FIWARE, Arkadin, onePlatform, SensorCloud, SmartThings, ThingWorx, Oracle IoT, Nimbits, ThingSpeak, Xively, etc.	Data storage, data aggregation and processing, big data analytics, decision support (Expert and DSS systems), machine learning, AI													
	Data processing mechanisms	Data mining, Big Query, Cloud Datalab, Apache Hadoop, Apache Kafka, Apache Storm, RabbitMQ, SPARQL, SciDB, Semantic technologies (ISON, W3C, OWL, EXI,WSDL), etc.														
	Data storage	Storage infrastructure (public, private, hybrid), Database (MongoDB, Cassandra, Hadoop, Hbase, CouchDB, Redis, ...), Storage architectures, etc.														
Networking domain	Cloud platforms	Application (CoAP, MQTT, AMQP, XMPP, DDS, WebSocket, ...), Transport (TCP, UDP), Network (IPv4, IPv6), Routing (BPL), Service discovery (mDNS, DNS-SD, SSDP, SLP, ...), etc.	Seamless connectivity (anytime, anywhere by anyone and anything), data transfer (device-to-device, device-to-application, application-to-device)													
	Network interface	IEEE 802.11, IEEE 802.15.4, IEEE 802.15.1, IEEE 802.16, 3GPP, IEEE 802.15.6, WSN, Z-Wave, IEEE 802.3, RFID, NFC, UWB, IrDA, PLC, CAN, etc.														
	Adaption mechanisms	Adaption layer (6LoWPAN, 6TiSH, 6Lo, IEEE 1095.1, ...), Connectivity interfaces (RJ45, ODB2, RS-232, RS-4BS, PLC, USB, Modus, SPI, ...), Gateways (Advantech, ADLINK, ...), etc.														
Object domain	Hardware platforms	Arduino, Raspberry Pi, Intel Galileo, Intel Edison, BeagleBone Black, Broadcom, Netduino, Flutter, Marvell, Tessel 2, Practical.io, SmartThings, etc.	Identification, sensing, actuating, data processing and computation, power supply													
	Embeded objects	Embeded sensors, Actuators, RFID/NFC tags, identification (EPC, uCode, QR, ...), touch screan displys, firmwares, onboard softwares, etc.														
	Mechanical and electrical parts	Mechanical and electrical parts (batteries), processing units (microcontrollers, microprocessors), digital signal processing sensors, Peripheral controller chips, etc.														

4. Data management layer

This layer consists of a data lake, a knowledge lake, and a range of tools to analyze and visualize data. The data lake is a collection of connected databases (inhouse or cloud-based) that store all kinds of raw or processed data (e.g., machine data, asset health data, production data, maintenance data) associated with the process plant. In contrast, knowledge lake includes abstractions that are extracted from data or provided by subject matter experts. Note that these abstractions are not required to be generated by using the data from the process plant where the abstractions are going to be employed but can be from any facility with similar types of machines or operations. For example, abstractions generated by analyzing data from an automatic *fall detection sensor* installed in a safety helmet or jacket of a construction worker can be used for a fall detection application in a process plant. Concerning process data analytic platforms, a range of artificial intelligence (AI) and big-data analytics tools are available for analyzing the raw data to generate insights, abstractions, and digital twin (digital replica or models) for the process plant.

Industrial internet cloud platforms that hold data and knowledge lakes and perform data analytic and digital model building include Predix, MindSphere, Microsoft Azure, Amazon Web Services, IBM Cloud, and Google Cloud (Kabugo, Jämsä-Jounela, Schiemann, & Binder, 2020). With the deployment of IIoT enable devices and sensors, data processing can be done in a distributed manner where the data is partially processed at the IIoT devices (edge) and their immediate network (fog) while the advanced and resource-intensive algorithms run on a cloud.

4.1 Data preprocessing

With the deployment of IIoT sensors, modern process data analytics platforms capture an enormous amount of data from a given process plant. Unfortunately, this real-world data is often of poor quality, including noise, missing values, outliers, and inconsistencies. Such low-quality data leads to incorrect insights, negatively impacting the successive automated innovations and data-driven decisions. Therefore, it is essential to preprocess the real-world data before utilizing the data in data mining tasks (Alasadi & Bhaya, 2017; Fan, Chen, Wang, Wang, & Huang, 2021; Jane & Arockiam, 2021). A literature review on IoT data preprocessing presented in Jane and Arockiam (2021) categorized the data preprocessing techniques into four main categories: data cleaning, data integration, data reduction, and data transformation.

Data cleaning techniques attempt to address the missing value issue and noise associated with real-world data. The issue of missing values is addressed through deletion methods (i.e., variable-wise, pairwise, list-wise), imputation techniques (i.e., deductive imputations, mean/median/mode imputation, random sampling imputation, regression imputation, multiple imputation), and predictive and statistical models (i.e., linear regression, random forest, K-nearest neighbor, expectation-maximization, sensitivity analysis). Noise in a data set can be minimized through binning (i.e., smoothing by bin means, smoothing by bin medians), regression (i.e., linear regression, multiple linear regression), and outlier analysis (i.e., cluster-based outlier detection, support vector machine (SVM)-based outlier detection).

In a process plant, or other similar industrial settings, multiple IIoT sensors collect data on different parameters (e.g., temperature, pressure, composition) of the same process compound or processing equipment. Combining and visualizing the data from different sources as a single entity provides insight into the current status of the plant. The widely deployed approaches to integrating data include data consolidation, propagation, and virtualization. Data consolidation physically brings data together to one source through data warehousing. Data propagation creates data replicas, while data virtualization provides a real-time and unified view of data from multiple sources.

As mentioned previously, IIoT devices acquire enormous amounts of data at high rates. To analyze such high-volume data demands extensive memory and processing capabilities. Data analytic tasks may take longer to complete as a result. Therefore, various data reduction approaches have been employed to get a condensed representation of the data set, which is noticeably smaller in volume compared to the raw data set but yields the same analytic result. Commonly used data reduction techniques include data cube aggregation, data discretization, numerical reductions, concept hierarchy generation, attribute subset selection via statistical or computational methods, and dimensionality reduction via principal component analysis (PCA), backward feature elimination, forward feature construction, or discriminant methods.

Once the data cleaning, integration, and reduction has been performed, the resulting data set must be put into a form appropriate for data modeling. Strategies that enable data transformation include smoothing, aggregation, discretization, attribute construction, generalization, and normalization. Data normalization can use the min-max normalization or *Z*-score normalization approaches to transfer the data from its original format to the format suitable for data analytics tools.

4.2 Data analytics tools

Data acquired at different levels in the process industry are utilized in passive applications, such as process monitoring, and in active applications, such as

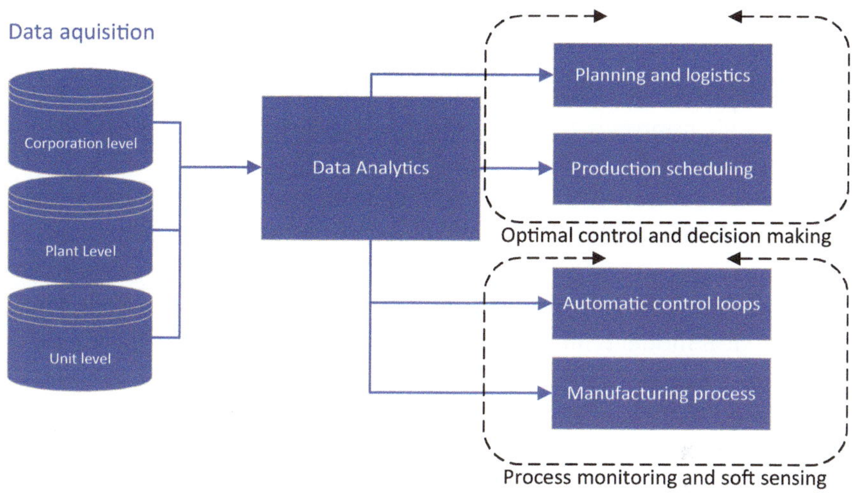

Fig. 3 The hierarchy of acquired data and their usage (Shang & You, 2019).

optimal controlling and decision making. As illustrated in Fig. 3, data analytics is an important step in a process data analytic platform (PDAPs) to cater to end-user applications.

Data analytics and machine learning have been well established and are employed to model, interpret, process, and utilize data leading to machine intelligence (Shang & You, 2019). The successful use of neural networks and back-propagation algorithms in process data analytics applications can be traced back to the late 1980s. Subsequently, principal component analysis (PCA), partial least squares (PLSs), and support vector machines (SVMs) have also become popular due to their inherent benefits, including strong statistical interpretation capability. The following sections present different software and hardware tools currently used in PDAPs in analyzing process data.

Representation learning is a new approach to data analytics and machine learning (Shang & You, 2019). The performance of supervised learning can be significantly improved using unsupervised learning algorithms with their feature detector ability, which is the core concept in representation learning. Further, deep learning is a subset of a larger family of artificial intelligent techniques based on representation learning and artificial neural networks (Schulz & Behnke, 2012). These techniques can be used in both active and passive applications in PDAPs. Soft sensing is another intelligent-sensing technology that can be developed to provide real-time estimation of key performance indicators (KPIs) of a process (Shardt,

Hao, & Ding, 2015). Soft sensing employs simple process variables to provide online estimates of difficult-to-measure but critical indices like product quality and other environmental indices. This technique is used to predict KPIs, which improves the decision-making process while avoiding time and money-sensitive experiments.

Optimal control of a process system using data analytics and machine learning can be taken as an active application of process data analytics. To determine the optimal control sequences in near future, model predictive control (MPC) can be employed. MPC usually contains a precisely known mathematical model which describes system behaviors (Chu & You, 2015). MPC has several limitations, however, including model mismatches, unmeasured disturbances, and random noise. It is possible to integrate data analytics and machine learning with MPC to address these limitations. Moreover, these model-based and data-driven techniques are utilized in optimization under uncertainty, leading to decision support.

Real-time, on-premises data analytics is becoming feasible with the development of edge computing. It is more efficient to use edge computing in comparison with the current automation systems as the plant devices can be directly connected to the IIoT cloud through a 5G network. Predix, MindSphere, and Sentience are some of the currently available IIoT platforms. Amazon Web Services, Microsoft Azure, Google Cloud, and IBM's Watson IoT provide a variety of multi-purpose IoT cloud platforms. This facilitates machine-to-machine communication (M2M) and real-time data processing. IIoT vendors such as Siemens and Honeywell are already developing devices to cater to the growing demand (Kabugo et al., 2020).

Process industries such as oil and gas are moving toward creating digital twins (DT) of their assets. As mentioned in Wanasinghe, Wroblewski, et al. (2020) the broadest definition of a digital twin is provided by the Defense Acquisition University's (DAU) glossary. It states that DT is an "integrated multi-physics, multi-scale, probabilistic simulation of an as-built system, enabled by digital thread, that uses the best available models, sensor information, and input data to mirror and predict activities/performance over the life of its corresponding physical twin." As illustrated in Fig. 4, a digital twin facilitates interactive optimization by acquiring data from the physical space and providing driving feedback to the physical space in return. The physical asset, as well as sensors and actuators, are all housed in the physical space. The virtual world serves as a digital mirror for the physical counterpart's high-fidelity simulation. Visualization services, product quality services, diagnostic services, model calibration services,

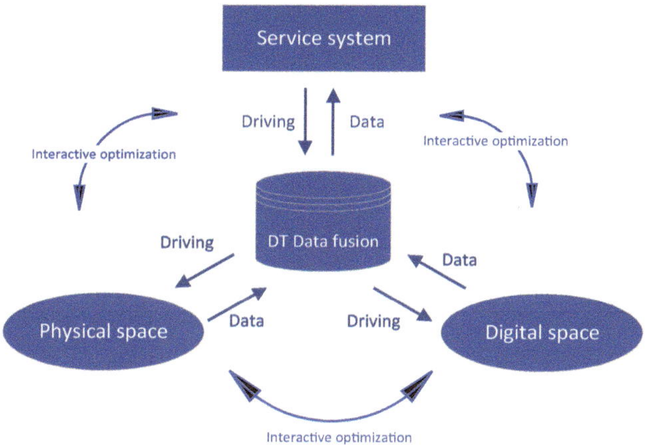

Fig. 4 A digital twin framework (Wanasinghe, Wroblewski, et al., 2020).

algorithm services, and numerous data services are all included in the service system. The DT data fusion model serves as a link between the real and virtual worlds and the service system. This module gathers information from physical world sensors, virtual world simulations and the service system. The DT data fusion model fuses and analyses the acquired data to provide actuating commands for the rest of the modules (Wanasinghe, Wroblewski, et al., 2020).

4.3 Process data analytics platforms

There are several cloud platforms, both commercial and open source, such as Amazon machine learning (Amazon ML), Microsoft Azure machine learning, Google machine learning and IBM Watson machine learning. There are sufficient provisions for data preprocessing, dimensionality reduction, predictive data analytics. The scalability and the availability of various machine learning algorithms are a few of the functionalities available in the aforementioned cloud platforms (Kabugo et al., 2020).

A real-time data analytics platform (RT-DAP) is introduced by Han, Gong, Nixon, Rotvold, et al. (2018) for large-scale industrial process monitoring and control. The data collected from heterogeneous resources are streamed, stored, processed, and visualized. The feedback is then sent to the control systems and the operators in real-time. These real-time data is processed by the run-time execution engine, which is a hybrid of a data flow system and a parallel real-time computing framework, that will conduct data analytics tasks in a parallel and resource-aware way. "Apache Kafka" and "Apache Storm" are examples of the data transportation and computing frameworks used in this type of application. The former is a data transportation medium with a latency of less than 10 ms used by more than 2000 firms worldwide. The latter is a free and open source distributed real-time computation system (Apache Software Foundation, 2022a, 2022b).

The scalable and adaptable IoT-enabled process data analytics pipeline "iProcess," allows analysts to ingest data from IoT devices, extract information from it, and connect it to process (execution) data (Beheshti, Schiliro, Ghodratnama, Amouzgar, et al., 2018). Although this was originally developed for providing real-time intelligence to organizations that can enhance knowledge-intensive processes, this has the potential to be adopted as a process data analytics platform, especially in diagnosing faults or determining the root causes for process-related accidents. As shown in Fig. 5, the data may be collected from IoT devices, media such as photographs, audio recordings, or directly from the workers and sent to the data lake. The data lake performs extensive analytics on massive hybrid collections of heterogeneous and largely unstructured data acquired through various methods, adding significant value to the data-in-motion. The primary purpose of the data lake is to preprocess the acquired data and store it until the data analyst decides how to use it. The concept of a knowledge lake, or a contextualized data lake, is developed to offer the foundation for big data analytics by automatically curating and preparing the raw data in the data lake for deriving insights. In the process narratives phase an online analytical processing (OLAP) style process data summary technique (Beheshti et al., 2018) is offered as an alternative to querying and analytical tools. As a result, the process analysts will be able to examine the process from numerous perspectives. Finally, an interface is developed linked with machine learning tools to facilitate the process analytics and decision-making phases.

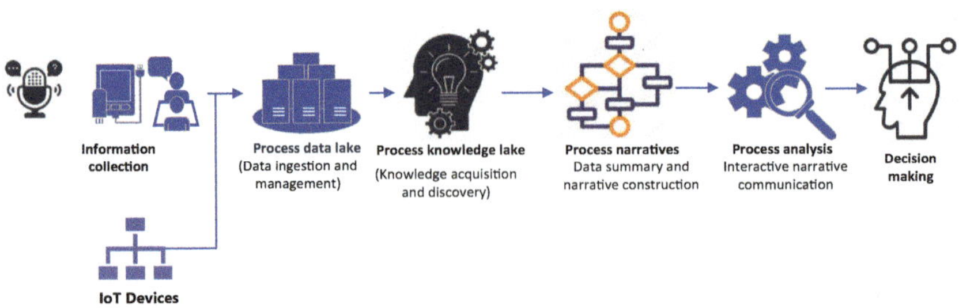

Fig. 5 IoT enabled process data analytics platforms (Beheshti et al., 2018).

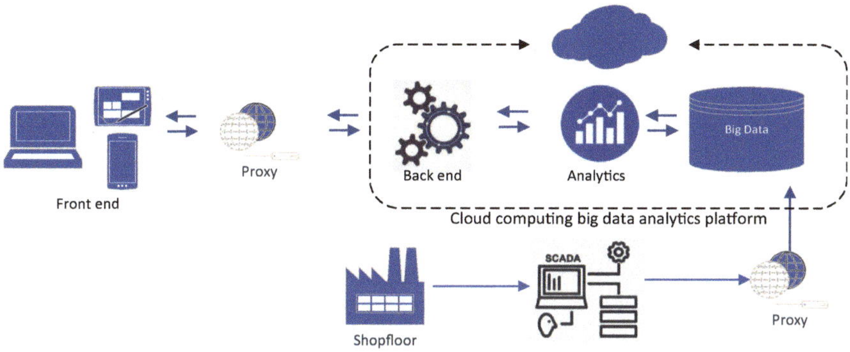

Fig. 6 A data analytic platform for SMEs (Bonnard et al., 2019).

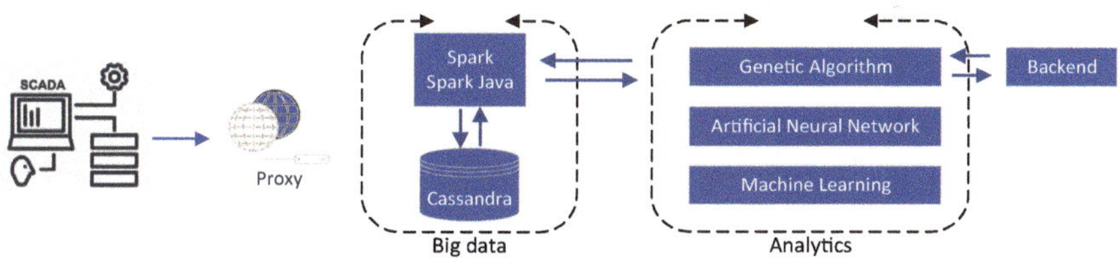

Fig. 7 The big data analytics architecture (Bonnard et al., 2019).

As a foundation for Industry 4.0, a conceptual data analytic platform is proposed by Bonnard, Magno, Vieira, Da, et al. (2019) to improve the efficiency, productivity and product quality of small and medium-sized enterprises (SMEs). This platform is illustrated in Fig. 6. In this platform, data is collected at the shop floor level using sensors and actuators which are connected to the company network through ethernet or wireless. Any devices that are not connectable to the network are bypassed to the big data cloud via application programming interface (API).

The big data analytics architecture of the same platform is presented in Fig. 7. The big data solution is implemented to store and process data coming from upstream. Java and Spark Java have been utilized for communicating between the big data and analytics solutions, while Apache Cassandra database was used for data storage. One of the drawbacks of this concept is the real-time control of systems that will require a data analysis and control system at the edge level instead of the cloud level unable to benefit from this form of data analytics architecture (Bonnard et al., 2019).

A cross-sectorial big data platform architecture for the process industry domain is presented and tested in aluminum and plastic molding industries (Sarnovsky, Bednar, & Smatana, 2018). The primary goal has been to create a scalable analytical platform that can handle data gathering, storage, and processing from various industries. The ability to be connected to the plant's existing environment and use the data gathered to construct predictive functions to improve production operations has been a major challenge. The transfer of knowledge across multiple domains is made possible by cross-sectorial sharing. Throughout the development process, a user-centered approach has been utilized to collect requirements from various stakeholders, which were then used to construct architectural models from various perspectives, ranging from contextual to deployment.

As depicted in Fig. 8, the cross-sectional data lab platform is connected to different industries, consisting of real-time operational platforms. These platforms send out predictive functions to each industry platform and receives their operational data. The predictive functions penetrate the enterprise, management, supervision, control, and field levels to provide operational data back to the platform.

A big data analytics platform for machining systems was developed by Woo, Shin, and Seo (2016), which is a conceptual architecture that specifies

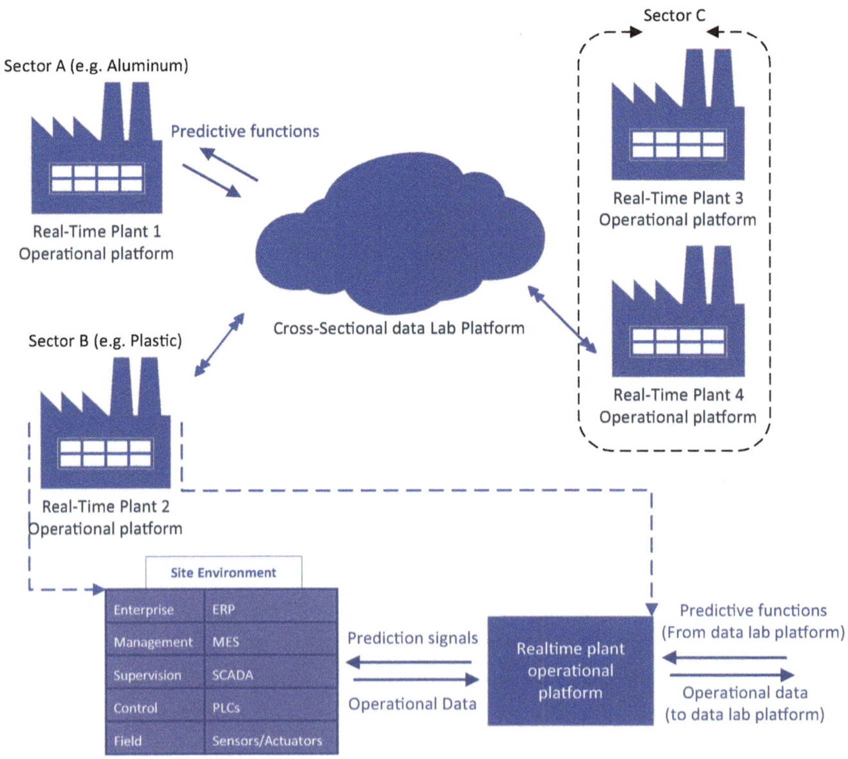

Fig. 8 High level architecture overview of the platform (Sarnovsky et al., 2018).

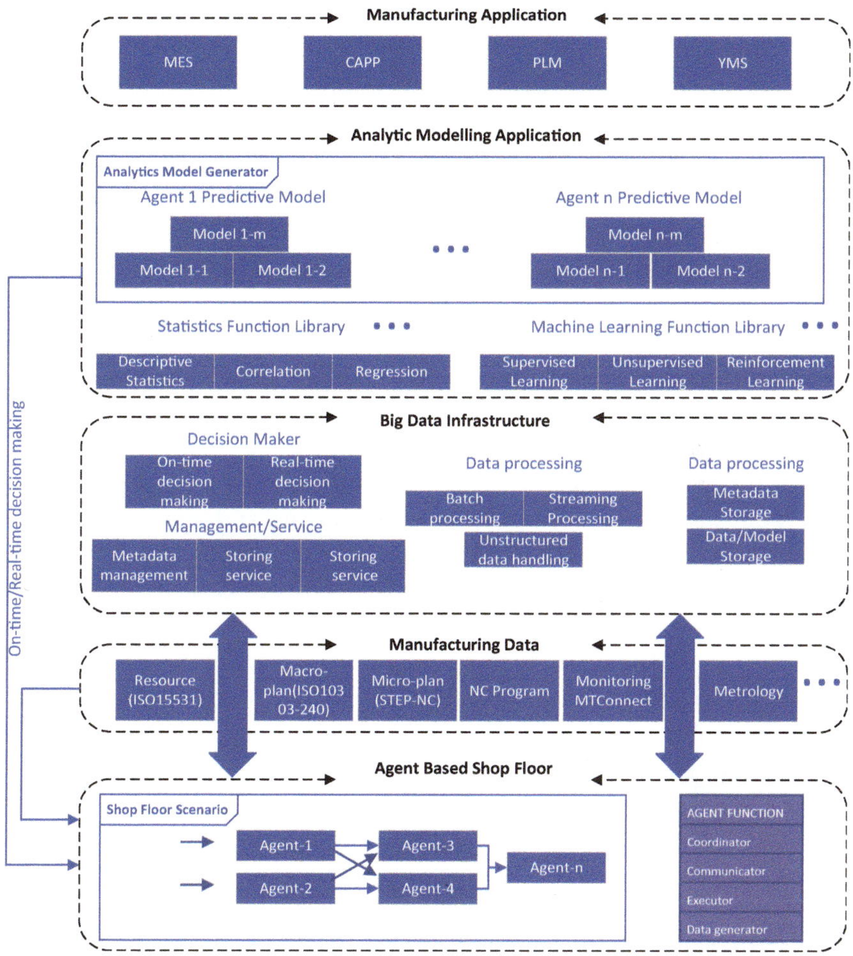

Fig. 9 A big data analytics platform for machining systems (Woo et al., 2016).

the operations needed to develop prediction models tailored to various machine tools using big data infrastructure. The schematic of the platform is illustrated in Fig. 9. The functions of the agent-based shop floor are representing and identifying shop floor objects, communicating with other objects across shop floors, collecting data from physical objects, and determining and controlling the actions of the objects, through the linkage with their virtual agents. The agent-based shop floor, the analytics modeling application, and the manufacturing applications are interfaced with the big data infrastructure. The infrastructure collects manufacturing data and associated configuration data generated on the shop floor. It uses the obtained data to monitor and observe the situation on the shop floor and then applies real-time control to the shop floor. The infrastructure also feeds manufacturing applications dynamics and operating conditions of the shop floor, allowing these applications to monitor and regulate KPIs, failure alerts, and shop floor activities. Manufacturing applications make short- and long-term planning choices and send them to the infrastructure and the shop floor based on the data and information given by the infrastructure. Finally, the analytics modeling application's goal is to allow manufacturers to develop analytics models based on their production data.

Numerous platforms and data analytics frameworks are available in the literature. A framework for integrating data from process and alarm databases and information on process connection is presented by Hu, Shah, and Chen (2018). Information gathered from various data sources may then be utilized for process and performance monitoring, such as alarm rationalization, root cause diagnosis of process defects, hazard and operability analysis, and safe and optimal process operation.

The RT-DAP real-time data analytics platform is developed by Han et al. (2018) to facilitate large-scale continuous data analytics in process industries. RT-DAP is built to stream, store, analyze, and display a high volume of real-time data flows gathered from a variety of plant resources and provide real-time feedback to the control system and operators. They have also developed a prototype on Microsoft Azure to demonstrate the validity of the platform in both the system and component levels.

4.4 Data processing hardware platforms

A comprehensive survey of various data processing hardware platforms that are currently available is presented in Singh and Reddy (2015). Data analysis can be done using vertical scaling or horizontal scaling. The former technique involves improving the performance of the available server, while the latter involves distributing the workload among many servers. The most suitable approach can be decided based on the application since the approaches each have advantages and disadvantages. Table 4 provides a qualitative comparison between different platforms based on the six criteria: scalability, data I/O performance, fault tolerance, real-time processing, data size support and the support for iterative tasks.

Table 4 A qualitative comparison between different hardware platforms (Singh & Reddy, 2015).

Scaling Type	Platforms (Communication Schemes)	System platform		Application algorithm			
		Scalability	Data I/O performance	Fault Tolerance	Real-Time processing	Data size supported	Iterative task support
Horizontal Scaling	Peer-to-peer (TCP/IP)	★★★★★	★	★	★	★★★★★	★★
	Virtual clusters (MapReduce/MPI)	★★★★★	★★	★★★★★	★★	★★★★	★★
	Virtual clusters (Spark)	★★★★★	★★★	★★★★★	★★	★★★★	★★★
Vertical Scaling	HPC Clusters (MPI/Mapreduce)	★★★	★★★★	★★★★	★★★	★★★★	★★★★
	Multicore (Multithreading)	★★	★★★★	★★★★	★★★	★★	★★★★
	GPU (CUDA)	★★	★★★★★	★★★★	★★★★★	★★	★★★★
	FPGA (HDL)	★	★★★★★	★★★★	★★★★★	★★	★★★★

5. Networking layer

As modern process data analytic platforms increasingly deploy IIoT devices, their core network typically follows the IIoT network framework (IIoTNF).[b] The existing IIoT platforms follow either the seven-layer open system interconnection (OSI) model or the four-layer TCP/IP model to implement their communication functionalities. In the OSI model, the IIoTNF is implemented utilizing the last three layers (i.e., physical, data link, and network layers) of the OSI model. In the TCP/IP model, the IIoTNF is implemented utilizing the last two layers (i.e., network interface and internet layers) of the TCP/IP model. In either case, IIoTNF is responsible for securely relaying data and commands among devices within the same layer (e.g., data exchange between edge sensor and actuator) or two different layers (e.g., data exchange between sensor and any data processor/storage device in the data management layer).

The network layer of the OSI model bridges data from source entities to destination entities through one or many networks. The core functionality of this layer includes addressing, routing and forwarding. The datalink layer provides the functions and procedural means for transferring data across network entities and provides the ability to detect and potentially fix faults in the physical layer. It mainly focuses on data exchange between devices in the same local area network (LAN). Its key functionalities include physical addressing, error control, flow control, time synchronization, scheduling/shaping, bandwidth reservation, security, framing, and frame replication & elimination. The physical layer is the physical media (i.e., wired or wireless) that connects two endpoints. It provides the mechanical, electrical, functional, and procedural mechanisms for activating, maintaining, and deactivating physical connections over a medium for bit transmission between datalink entities. The key functionality of this layer includes encoding/decoding, sequencing, bit-block multiplexing/de-multiplexing, fault detection and notifications, and link-status detecting and maintenance.

[b] More details and use-cases for IIoTNF can be found in Lou, Holler, Patel, Graf, and Gillmore (2021).

Following these models, many industrial networking technologies are available to enable the connectivity of devices in process data analytics platforms. A protocol stack-view of the industrial networking standards related to IIoTNF is given in Fig. 10. At the lower level, the communication stack can be divided into two categories: wireless and wired communication technologies. Internetworking technologies such as IPv4 and IPv6 are common elements for both categories. The wireless communication technologies used in the process data analytic platforms can be divided into four groups: mobile communication, low-power wide-area networks (LPWANs), short-range radio communication, and satellite communication. Due to the flexibility, wireless approaches are becoming a more popular option for intra-network/plant communication. Mobile communication technologies that are standardized under the third generation partnership project (3GPP) are now receiving more attention from many industries, including the process industry, as the networking solution for their plants. It is important to note that each mobile communication technology has unique characteristics and features (e.g., refer to Table 5). Moreover, different industrial applications have different network requirements, as given in Table 6. Therefore, data processing platforms designers must review these capabilities prior to selecting a communication mechanism for their platforms. In most cases, a mix of communication technologies (i.e., heterogeneous network architecture) is required to achieve uninterrupted connectivity across the devices, data management and application layers.

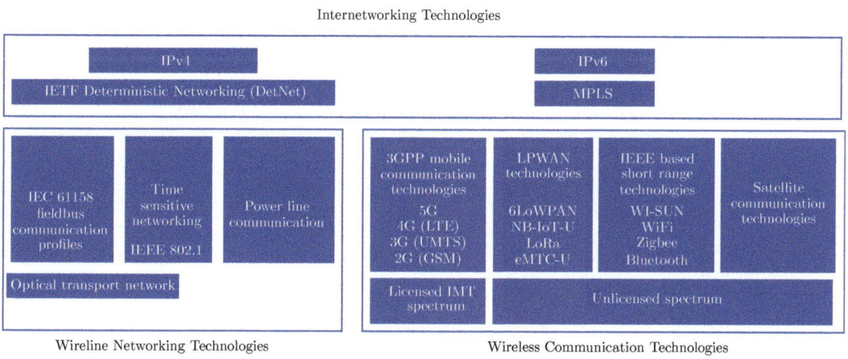

Fig. 10 A protocol stack-oriented view of the industrial networking standards (Lou et al., 2021).

Table 5 Summary of characteristics and example applications for 3GPP in IIoT (Lou et al., 2021).

	NB-IoT	Cat-M1	LTE	5G NR
Peak data rate	~100 Kbps	1 Mbps	25 Gbps (downlink) 12.9 Gbps (uplink)	37–38 Gbps (downlink) 17.9–18.9 Gbps (uplink)
Coverage	164 dB for 800 MHz band	160 dB for 800 MHz band	144 dB for 800 MHz band	–
Latency	<10 s	<10 s	<1 ms	<1 ms
Reliability	N/A	N/A	N/A	>99.999%
Battery	AA 10 years	10 years	N/A	N/A
Connection density	>10,000,000	>10,000,000	N/A	N/A
Typical applications	Track & trace env. Monitoring	Track & trace env. Monitoring	MTV, voice, mMTC	Critical MTC, AR/VR, eMBB

Table 6 Selected industrial applications in manufacturing and their network requirements (Lou et al., 2021).

	Condition monitoring	Motion control	Automated guided vehicles	Process automation	Augmented reality	Functional safety
Latency	20–100 ms	0.1–1 ms	10–20 ms	50 ms	10 ms	10 ms
Determinism	No	Yes	Yes	Yes	Yes	Yes
Time synchronization	No	No	Yes	No	Yes	No
Availability	99.99%	99.9999%	99.9999%	99.99%	99.99%	99.9999%
Redundancy	No	Yes	Yes	Yes	No	Yes
Mobility	Yes	No	Yes	Yes	Yes	Yes
Peak data rate	<1 Mbps	100 Mbps	100 Mbps	10 Mbps	1 Gbps	<1 Mbps
Battery power	Up to 10 years	Main power	Up to 8 h	Up to 10 years	2–3 h	Up to 10 years
Connection density (per 1000 m^2)	1000	100	200	1000	50	100

6. Application layer

This layer consists of a range of software tools (i.e., applications) to achieve various objectives of the process data analytic platform, including applications for predictive maintenance, automated KPI generations, automated error detection, remaining useful life calculation, supply-chain management. These applications can be provided by a third party or the industrial internet cloud platform provider. A sample of such applications is provided below.

6.1 Predictive maintenance

In past decades, reactive and proactive maintenance approaches have been commonly used in process systems. Reactive maintenance involves repairing or substituting a machine component only once it fails and can no longer operate (Lee, Ghaffari, & Elmeligy, 2011). A proactive maintenance strategy aims to identify and fix the cause of the system failure before it occurs (Wang & Rausand, 2014). The process industry is more likely to use a proactive approach to prevent the process plant before failure. Proactive maintenance is classified into three categories: preventive, condition-based, and predictive maintenance (Angelopoulos, Michailidis, Nomikos, Trakadas, et al., 2019; Peng, Dong, & Zuo, 2010). The preventive maintenance approaches perform maintenance and inspection at a regular interval. The discrete-time nature of such inspection/maintenance activities precludes identifying and preventing failures between two inspection/maintenance instances. As a result, preventive maintenance based on periodic inspection and replacement is also disappearing from industrial practice (ben Ali et al., 2015).

Condition-based maintenance (CBM) and predictive maintenance (PM) approaches have been widely targeted in Industry 4.0 system maintenance. These two approaches are highly dependent on the system's condition information (Biggio & Kastanis, 2020). Therefore, these two approaches rely on real-time sensor data and condition monitoring techniques. The main difference between these two methods is that PM combines sensor measurements with precise algorithms to predict the exact moment of the maintenance schedule. CBM takes maintenance action when a fault is detected, and PM extends the algorithm to predict the fault to failure transformation time to schedule the maintenance (Fink, Wang, Svensén, Dersin, et al., 2020).

PM and CBM methods can be classified into model-based, knowledge-based, and data-driven approaches. Model-based approaches rely on the physical system model, while knowledge-based approaches rely on expert knowledge algorithms and management systems. Finally, data-driven approaches rely on system data and data analytics methods such as statistical and artificial intelligence. Data-driven analysis has become more prevalent in decision-making in the process industry system maintenance in the Industry 4.0 era.

As illustrated in Fig. 11, maintenance based on condition monitoring depends on fault detection, localization/isolation, and diagnosis play a significant role in identifying the real-time process monitoring. Besides, predictive maintenance depends on failure prognosis and remaining useful life of an asset.

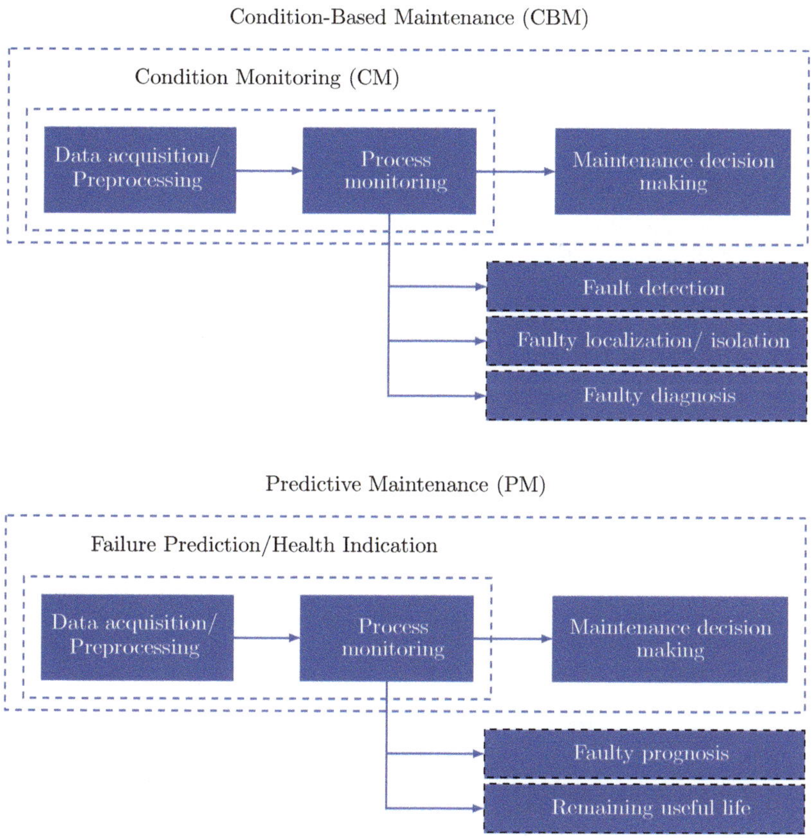

Fig. 11 Data-driven CBM vs PM process (Cohen & Singer, 2021).

6.2 Remaining useful life estimation

In recent years, estimating the remaining useful life (RUL) has played a significant role in plant failure prognosis, process health monitoring and predictive maintenance. The RUL estimation of the process systems is generally defined as the fault to failure degradation, as illustrated in Fig. 12. The RLU is estimated using the fault detection point as the base point. Therefore, RUL estimation based on the fault detection is formulated as:

$$RUL_{Fault\ condition} = t_f - t_d$$

In the process industry, RUL estimation techniques are used to predict the process systems failure condition as early as possible. This reduces maintenance costs, increases the machine's life cycle, and reduce risk (Li, Ding, & Sun, 2018; Wang, Chen, Cai, Gao, & Wang, 2020; Zhang, Yang, Xiu, Li, & Liu, 2021). Accurately predicting the RUL of equipment can guarantee failure prevention and allow scheduled maintenance of the equipment within a specific timeframe (Biggio & Kastanis, 2020).

The available RUL prediction methods are classified as physics-based and data-driven models (Ren, Dong, Wang, Meng, et al., 2021; Sayyad, Kumar, Bongale, Bongale, et al., 2021). Physics-based models are related to determining the mathematical relationship, such as the state-space models and analytical models. These include an observer-based method and parity

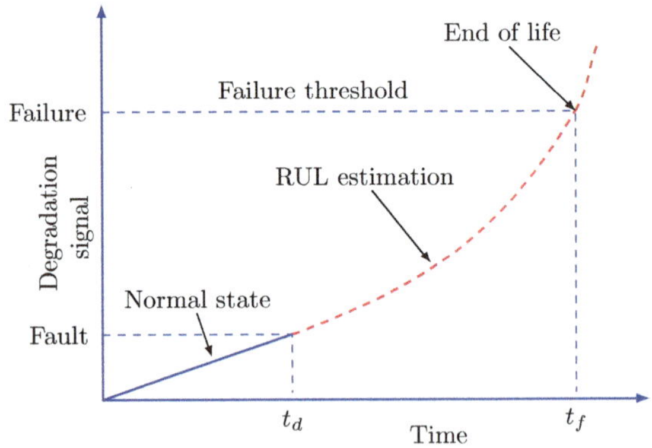

Fig. 12 RUL estimation based on fault to failure transformation (Witczak, Mrugalski, & Lipiec, 2021).

space to estimate the system fault condition (Gao, Li, Huang, & Wang, 2020). However, these models cannot monitor the system condition until a failure occurs or the system condition becomes observable. In contrast, data-driven models, such as traditional statistical analysis and machine learning models, define the degradation process as a functional relationship between system conditions and monitoring data (Ren et al., 2021).

Depending on the data availability, data-driven models are further classified as RUL estimation using survival, similarity, and degradation models (Zheng, 2019). System lifetime data proportional hazard and probabilistic distribution of the component failure time methods are utilized to estimate the RUL using the survival model (Si, 2015). RUL estimation using similarity models requires failure data from similar components or distinct components with similar behavior (Si, 2015; Si, Wang, Hu, & Zhou, 2011). Also, these models record the degradation profiles and compare them with process data. RUL estimations using degradation models predict the system health condition and analyze when the predictor crosses a predefined threshold (Si et al., 2011; Wang, Hu, Si, Pang, & Ren, 2019).

Process systems are constructed using several components in process industries, and each plant uses them for different purposes and unique operations. Collecting similar behavior failure data and lifetime data with failure conditions is a challenge and often impossible. Hence, degradation models are better for estimating the RUL in process systems since most process plants define the failure threshold for each component and parameter. RUL estimations using degradation models predict the system health condition and perform analysis when the predictor crosses a predefined threshold (Si et al., 2011; Wang et al., 2019). Also, since process plant systems are complex and combine several components, process system data are multivariate and highly correlated (Wu & Zhao, 2020). When applying degradation methods to estimate the RUL, the main challenge is to identify the condition indicator for the system condition deviation (Wang et al., 2019).

Also, process systems undergo modifications and changes in operating conditions with time, and the definition of normal and abnormal states can change with time (Arunthavanathan, Khan, Ahmed, & Imtiaz, 2021). Therefore, monitoring the normal to fault transition behavior and fault to failure estimation during the systems run time are essential. Also, in line with the Industry 4.0 concepts, online autonomous failure prediction and RUL estimation pose new challenges (Li, Wang, & Wang, 2017; Sajid, Haleem, Bahl, Javaid, et al., 2021).

6.3 Cyber physical systems

Cyber-physical systems (CPS) are embedded systems of integrating digitized process systems with digital communication primarily developed to monitor and control the physical devices in the systems. IoT and IIoT are the key technologies to network the physical process system. Intelligent systems are intended to control and monitor the system by self-learning models (An, Tao, Xu, el Mansori, & Chen, 2020; Jan, Lee, & Koo, 2021). Therefore, considering the risk assessment in the advanced process plant, safety and security are considered key properties.

As illustrated in Fig. 13, when applying CPS concepts into the process systems, following issues are considered from safety perspective: i. Traditional risk and safety assessment in process systems mainly consider sensor, actuator, controller equipment degradation, or component failure, while failure due to cyber-attacks is not part of the design process.

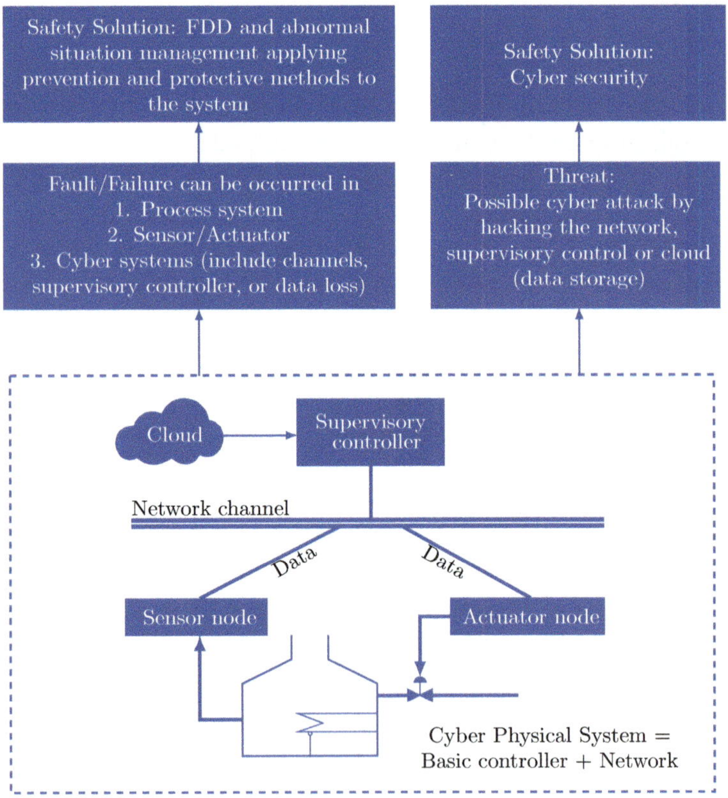

Fig. 13 CPS and its safety procedures (Tantawy, Erradi, & Abdelwahed, 2019).

Such attacks are extremely unlikely or even impossible in a traditional process plant (probability considered closer to zero.). ii. Cyber-physical interdependencies need to be considered in CPS where the system components, collaborative robots, and humans need to communicate for collective decision-making. Any exploitation or malfunction of one device may trigger a cascade of unexpected and often undesired changes of other devices and extend to the entire control network, introducing cascading failures.

7. Cybersecurity

Cybersecurity in the process industry began in the Industry 3.0 era with the invention and adaptation of supervision and data acquisition based on computer technology to basic process control parameters. Furthermore, introducing a LAN and further development in the communication protocols increased the vulnerability to cyber threats.

With Industry 4.0, it is common to divide the processing system into two parts, the operational technology (OT) and information technology (IT) systems. The OT system monitors and controls industrial equipment, assets, processes, and abnormal events. In contrast, the IT system includes all the process system hardware, software and peripheral equipment for storing, repossessing and communicating process data. Converging OT and IT with the use of IoT or IIoT raises cybersecurity concerns.

When integrating the process systems physical layer with the networking layer, it is important to focus on process system safety and security. Security focuses on protecting the systems from decisive attacks, while safety focuses on protecting the systems from accidental failures (Huang, Zhou, Yang, & Qin, 2015; Prinsloo, Sinha, & von Solms, 2019). Cybersecurity can be calculated using the detailed specification of the basic process control and safety instrumented system and their connectivity, including the embedded system hardware (i.e., PLC/SCADA based systems), operating system/system software, as well as the network connectivity (i.e., IoT/IIoT).

Recent global cybersecurity standards provide tools, policies, risk and safety management approaches, security countermeasures, and best practices to protect the cyber environment of a user or organization. Some cybersecurity standards in process industries are OT specifics, while some are IT specifics.

International Society of Automation (ISA) proposed two international standards: ISA 84 and ISA 99, to address cyber security for process systems. ISA 84 standards are well defined for process safety by including a safety instrumented system. This standard is generally practiced in automated process plants (Bukowski, 2006). Likewise, ISA 99 standards are used in control system security. With the advancement of cyber-physical process models, process control and cyber systems are integrated together into a single module (Fujdiak, Mlynek, Blazek, Barabas, & Mrnustik, 2018). Therefore, in the future, ISA 84 and ISA 99 should be integrated to provide a safer and secure environment in process plants.

8. Summary

The recent literature on Industry 4.0 adoption in process systems demonstrates extensive relevance and potential benefit to the process industry. This chapter discussed the overview of the Industry 4.0 revolution and its influence on process safety. The chapter was organized based on the process of systems architecture amendments due to advanced technologies in the systems' physical, communication, data management, and application layers. The impact on the process data collection due to the IoT/IIoT was reviewed in the physical layer. The required network protocols for inventing IoT/IIoT techniques were discussed in the network layer. Data preprocessing, analysis, and decision making using the Industry 4.0 technologies, such as cloud computing, big data, and AI, were discussed in the data management layer. The impact on the overall process system safety applications, such as cyber-physical model, remaining useful life, and predictive maintenance, were reviewed in the application layer. Finally, cybersecurity and its recent globalize standards were reviewed.

The review observes that Industry 4.0 research tends to focus more on the digital transformation of the process systems via integrating digitalized data acquisition, data transmission using communication protocols, and data-driven decision making. Therefore, it is important to unify several technological approaches, including IIoT (for data acquisition and transmission), Bigdata (for data preprocess and storage), artificial intelligence, and enhancing real-time monitoring and control (for data-driven decision making). Furthermore, Industry 4.0 in process plants focuses on routine operations and data analytics taking responsibility for smart decisions using statistical analysis and artificial intelligence. The merge between Industry 4.0 and data analytics implies transforming cyber-physical models in the advanced industry sector. Moreover, in Industry 4.0 era, data-driven decision-making in process plant perspectives to system maintenance or reliability focused on predictive maintenance and remaining useful time estimation. This can be achieved by preciously investigating the data and estimating the future.

The merging technologies in Industry 4.0 and data analytic platforms are upcoming streams in process industries. However, extensive knowledge is required from various disciplines, including information systems, engineering, and computer science, to establish the Industry 4.0 based data analytic platform in the process plant. Therefore, more efforts are required to transform the conceptual ideas into an operational context.

References

Alasadi, S. A., & Bhaya, W. S. (2017). Review of data preprocessing technique in data mining. *Journal of Engineering and Applied Science, 12*(16), 4102–4107.

Al-Fuqaha, A., Guizani, M., Mohammadi, M., Aledhari, M., & Ayyash, M. (2015). Internet of things: A survey on enabling technologies, protocols, and applications. *IEEE Communications Surveys and Tutorials, 17*(4), 2347–2376. https://doi.org/10.1109/COMST.2015.2444095.

An, Q., Tao, Z., Xu, X., el Mansori, M., & Chen, M. (2020). A data-driven model for milling tool remaining useful life prediction with convolutional and stacked LSTM network. *Measurement, 154*, 107461. https://doi.org/10.1016/J.MEASUREMENT.2019.107461.

Angelopoulos, A., Michailidis, E. T., Nomikos, N., Trakadas, P., ... Zahariadis, T. (2019). *Tackling faults in the industry 4.0 Era-A survey of machine-learning solutions and key aspects.* https://doi.org/10.3390/s20010109.

Apache Software Foundation. (2022a). *Apache Kafka.*

Apache Software Foundation. (2022b). *Apache Storm.*

Arunthavanathan, R., Khan, F., Ahmed, S., & Imtiaz, S. (2021). An analysis of process fault diagnosis methods from safety perspectives. *Computers & Chemical Engineering, 145*, 107197. https://doi.org/10.1016/J.COMPCHEMENG.2020.107197.

Ashok, M. L., & Rama, B. (2020). Real-time measurement and monitoring of industrial pharmaceutical chemical reactor process parameters (temperature, level and pressure) using PLC & SCADA system. *International Journal of Engineering Research & Technology, 9*(7), 1–6. www.ijert.org.

Aziz, A., Schelén, O., & Bodin, U. (2020). A study on industrial IoT for the mining industry: Synthesized architecture and open research directions. *IoT, 1*(2), 529–550. https://doi.org/10.3390/iot1020029.

Beheshti, A., Schiliro, F., Ghodratnama, S., Amouzgar, F., ... Motahari-Nezhad, H. R. (2018). iProcess: Enabling IoT platforms in data-driven knowledge-intensive processes. *Business Process Management Forum, 329*, 108–126. https://doi.org/10.1007/978-3-319-98651-7_7.

ben Ali, J., Chebel-Morello, B., Saidi, L., Malinowski, S., & Fnaiech, F. (2015). Accurate bearing remaining useful life prediction based on Weibull distribution and artificial neural network. *Mechanical Systems and Signal Processing, 56–57*, 150–172. https://doi.org/10.1016/J.YMSSP.2014.10.014.

Biggio, L., & Kastanis, I. (2020). Prognostics and health management of industrial assets: Current progress and road ahead. In *Vol. 3. Prognostics and health management of industrial assets: Current progress and road ahead* (pp. 1–24). https://doi.org/10.3389/frai.2020.578613.

Bonastre, A., Capella, J.v., Ors, R., & Peris, M. (2012). In-line monitoring of chemical-analysis processes using wireless sensor networks. *TrAC—Trends in Analytical Chemistry*, *34*, 111–125. https://doi.org/10.1016/j.trac.2011.11.009.

Bonnard, R., Magno, K., Vieira, M., Da, M., … Mattei, A. P. (2019). *EasyChair preprint a big data/analytics platform for industry 4.0 implementation in SMEs a BIG DATA/ANALYTICS PLATFORM FOR INDUSTRY 4.0 IMPLEMENTATION IN SMEs*. Easychair-Www.Easychair.Org https://easychair-www.easychair.org/publications/preprint_download/CKF1.

Bukowski, J.v. (2006). Incorporating process demand into models for assessment of safety system performance. In *Proceedings—Annual reliability and maintainability symposium* (pp. 577–581). https://doi.org/10.1109/RAMS.2006.1677435.

Carroll, K., & Chandramouli, M. (2019). *Scaling IoT to meet enterprise needs balancing edge and cloud computing*. Deloitte Insights.

Chu, Y., & You, F. (2015). Model-based integration of control and operations: Overview, challenges, advances, and opportunities. *Computers & Chemical Engineering*, *83*, 2–20. https://doi.org/10.1016/J.COMPCHEMENG.2015.04.011.

Cohen, Y., & Singer, G. (2021). A smart process controller framework for industry 4.0 settings. In *Journal of Intelligent Manufacturing: Vol. 32 (7)* (pp. 1975–1995). Springer. https://doi.org/10.1007/s10845-021-01748-5.

Čolaković, A., & Hadžialić, M. (2018). Internet of things (IoT): A review of enabling technologies, challenges, and open research issues. In *Vol. 144. Computer networks* (pp. 17–39). Elsevier B.V. https://doi.org/10.1016/j.comnet.2018.07.017.

Fan, C., Chen, M., Wang, X., Wang, J., & Huang, B. (2021). A review on data preprocessing techniques toward efficient and reliable knowledge discovery from building operational data. In *Frontiers in energy research: Vol. 9* (pp. 1–17). Frontiers Media S.A. https://doi.org/10.3389/fenrg.2021.652801.

Fink, O., Wang, Q., Svensén, M., Dersin, P., … Ducoffe, M. (2020). Potential, challenges and future directions for deep learning in prognostics and health management applications. *Engineering Applications of Artificial Intelligence*, *92*, 103678. https://doi.org/10.1016/J.ENGAPPAI.2020.103678.

Fujdiak, R., Mlynek, P., Blazek, P., Barabas, M., & Mrnustik, P. (2018, August 20). Seeking the relation between performance and security in modern systems: Metrics and measures. In *2018 41st international conference on telecommunications and signal processing, TSP 2018*. https://doi.org/10.1109/TSP.2018.8441496.

Gadhe, A., Bhojane, P., & Kawade, K. (2018). Chemical mixing for process industry using PLC & SCADA. *International Journal of Advanced Research in Electrical, Electronics and Instrumentation Engineering*, 7(4), 1784–1790. https://doi.org/10.15662/IJAREEIE.2018.0704045.

Gao, T., Li, Y., Huang, X., & Wang, C. (2020). *Data-driven method for predicting remaining useful life of bearing based on Bayesian theory*. https://doi.org/10.3390/s21010182.

Han, S., Gong, T., Nixon, M., Rotvold, E., … Ramamritham, K. (2018). RT-DAP: A real-time data analytics platform for large-scale industrial process monitoring and control. In *Proceedings—2018 IEEE international conference on industrial internet, ICII 2018, Icii* (pp. 59–68). https://doi.org/10.1109/ICII.2018.00015.

Hu, W., Shah, S. L., & Chen, T. (2018). Framework for a smart data analytics platform towards process monitoring and alarm management. *Computers and Chemical Engineering*, *114*, 225–244. https://doi.org/10.1016/j.compchemeng.2017.10.010.

Huang, S., Zhou, C. J., Yang, S. H., & Qin, Y. Q. (2015). Cyber-physical system security for networked industrial processes. *International Journal of Automation and Computing*, *12*(6), 567–578. https://doi.org/10.1007/s11633-015-0923-9.

Jan, S. U., Lee, Y. D., & Koo, I. S. (2021). A distributed sensor-fault detection and diagnosis framework using machine learning. *Information Sciences*, *547*, 777–796. https://doi.org/10.1016/J.INS.2020.08.068.

Jane, V., & Arockiam, L. (2021). Survey on IoT data preprocessing. *Turkish Journal of Computer and Mathematics Education*, *12*(9), 238–244.

Kabugo, J. C., Jämsä-Jounela, S. L., Schiemann, R., & Binder, C. (2020). Industry 4.0 based process data analytics platform: A waste-to-energy plant case study. *International Journal of Electrical Power and Energy Systems*, *115*. https://doi.org/10.1016/j.ijepes.2019.105508.

Kaur, M., & Kalra, S. (2016). A review on IOT based smart grid. *International Journal of Energy, Information and Communications*, 7(3), 11–22. https://doi.org/10.14257/ijeic.2016.7.3.02.

Khan, W. Z., Aalsalem, M. Y., Khan, M. K., Hossain, M. S., & Atiquzzaman, M. (2017). A reliable internet of things based architecture for oil and gas industry. In *International conference on advanced communication technology* (pp. 705–710). https://doi.org/10.23919/ICACT.2017.7890184.

Lee, J., Ghaffari, M., & Elmeligy, S. (2011). Self-maintenance and engineering immune systems: Towards smarter machines and manufacturing systems. *Annual Reviews in Control*, *35*(1), 111–122. https://doi.org/10.1016/J.ARCONTROL.2011.03.007.

Ley, S.v., Fitzpatrick, D. E., Ingham, R. J., & Nikbin, N. (2015). The internet of chemical things. *Beilstein Magazine*, *1*. https://doi.org/10.3762/bmag.2.

Li, X., Ding, Q., & Sun, J. Q. (2018). Remaining useful life estimation in prognostics using deep convolution neural networks. *Reliability Engineering & System Safety*, *172*, 1–11. https://doi.org/10.1016/J.RESS.2017.11.021.

Li, Z., Wang, Y., & Wang, K. S. (2017). Intelligent predictive maintenance for fault diagnosis and prognosis in machine centers: Industry 4.0 scenario. *Advances in Manufacturing*, *5*(4), 377–387. https://doi.org/10.1007/s40436-017-0203-8.

Lou, D., Holler, J., Patel, D., Graf, U., & Gillmore, M. (2021). *The industrial internet of things networking framework*.

Mao, C. (2017). Design and implementation of a WSN-based monitoring system for hazardous gas in chemical production. *Chemical Engineering Transactions*, *59*, 577–582. https://doi.org/10.3303/CET1759097.

Minerva, R., Biru, A., & Rotondi, D. (2015). *Towards definition of internet of things (IoT)*. IEEE Internet Initiative.

Nechibvute, A., & Mudzingwa, C. (2013). Wireless sensor networks for SCADA and industrial control systems. *International Journal of Engineering and Technology*, *3*(12), 1025–1035.

Patel, P. R., Patel, D. D., & Singh Bunet, A. (2015). Automation of chemical process plant based on PLC & SCADA. *International Journal of Innovative and Emerging Research in Engineering*, *2*(3), 46–49. www.ijiere.com.

Peng, Y., Dong, M., & Zuo, M. J. (2010). Current status of machine prognostics in condition-based maintenance: A review. *International Journal of Advanced Manufacturing Technology*, *50*(1–4), 297–313. https://doi.org/10.1007/s00170-009-2482-0.

Prabakaran, R., Sarma Dhulipala, V. R., Nirmal Kumar, K., & Ranjith, P. (2011). Future sensors and utilization of sensors in chemical industries with control of environmental hazards. In *2nd international conference in environmental science and development* (pp. 224–228). https://www.researchgate.net/publication/230659417.

Prinsloo, J., Sinha, S., & von Solms, B. (2019). *A review of industry 4.0 manufacturing process security risks*. https://doi.org/10.3390/app9235105.

Ren, L., Dong, J., Wang, X., Meng, Z., ... Deen, M. J. (2021). A data-driven auto-CNN-LSTM prediction model for lithium-ion battery remaining useful life. *IEEE Transactions on Industrial Informatics*, *17*(5), 3478–3487. https://doi.org/10.1109/TII.2020.3008223.

Risteska Stojkoska, B. L., & Trivodaliev, K.v. (2017). A review of internet of things for smart home: Challenges and solutions. In *Journal of Cleaner Production: Vol. 140* (pp. 1454–1464). Elsevier Ltd. https://doi.org/10.1016/j.jclepro.2016.10.006.

Sadiku, M. N. O., Musa, S. M., & Musa, O. M. (2018). SCADA in the chemical industry. *International Journal of Trend in Research and Developments, 5*(2), 123–125.

Sajid, S., Haleem, A., Bahl, S., Javaid, M., … Mittal, M. (2021). Data science applications for predictive maintenance and materials science in context to industry 4.0. *Materials Today: Proceedings, 45*, 4898–4905. https://doi.org/10.1016/J.MATPR.2021.01.357.

Sarnovsky, M., Bednar, P., & Smatana, M. (2018). Big data processing and analytics platform architecture for process industry factories. *Big Data and Cognitive Computing, 2*(1), 1–15. https://doi.org/10.3390/bdcc2010003.

Sayyad, S., Kumar, S., Bongale, A., Bongale, A., & Patil, S. (2021). *Estimating remaining useful life in machines using artificial estimating remaining useful life in machines using artificial intelligence: A scoping review*. University of Nebraska – Lincoln.

Schulz, H., & Behnke, S. (2012). Deep learning: Layer-wise learning of feature hierarchies. *KI - Kunstliche Intelligenz, 26*(4), 357–363. https://doi.org/10.1007/s13218-012-0198-z.

Shang, C., & You, F. (2019). Data analytics and machine learning for smart process manufacturing: Recent advances and perspectives in the big data era. *Engineering, 5*(6), 1010–1016. https://doi.org/10.1016/J.ENG.2019.01.019.

Shardt, Y. A. W., Hao, H., & Ding, S. X. (2015). A new soft-sensor-based process monitoring scheme incorporating infrequent KPI measurements. *IEEE Transactions on Industrial Electronics, 62*(6), 3843–3851. https://doi.org/10.1109/TIE.2014.2364561.

Sharma, A., & Singh, B. J. (2020a). Evolution of industrial revolutions: A review. *International Journal of Innovative Technology and Exploring Engineering, 9*(11), 66–73. https://doi.org/10.35940/ijitee.i7144.0991120.

Sharma, N., & Singh, R. (2020b). Evolution of IoT to IIoT: Applications & challenges. In *International conference on innovative computing and communication* (pp. 1–7). https://www.researchgate.net/publication/341495017.

Shoja, S., & Jalali, A. (2017). A study of the internet of things in the oil and gas industry. In *IEEE 4th international conference on knowledge-based engineering and innovation* (pp. 230–236).

Si, X. S. (2015). An adaptive prognostic approach via nonlinear degradation modeling: Application to battery data. *IEEE Transactions on Industrial Electronics, 62*(8), 5082–5096. https://doi.org/10.1109/TIE.2015.2393840.

Si, X. S., Wang, W., Hu, C. H., & Zhou, D. H. (2011). Remaining useful life estimation—A review on the statistical data driven approaches. *European Journal of Operational Research, 213*(1), 1–14. https://doi.org/10.1016/J.EJOR.2010.11.018.

Singh, D., & Reddy, C. K. (2015). A survey on platforms for big data analytics. *Journal of Big Data, 2*(1), 1–20. https://doi.org/10.1186/s40537-014-0008-6.

Sundmaeker, H., Guillemin, P., Friess, P., & Woelfflé, S. (2010). *Vision and challenges for Realising*. http://europa.eu/information_society.

Talari, S., Shafie-Khah, M., Siano, P., Loia, V., … Catalão, J. P. S. (2017). A review of smart cities based on the internet of things concept. In *Energies: Vol. 10, issue 4*. MDPI AG. https://doi.org/10.3390/en10040421.

Tantawy, A., Erradi, A., & Abdelwahed, S. (2019). A modified layer of protection analysis for cyber-physical systems security. In *2019 4th international conference on system reliability and safety, ICSRS 2019* (pp. 94–101). https://doi.org/10.1109/ICSRS48664.2019.8987635.

Wanasinghe, T. R., Gosine, R. G., James, L. A., Mann, G. K. I., … Warrian, P. J. (2020). The internet of things in the oil and gas industry: A systematic review. *IEEE Internet of Things Journal*, 7(9), 8654–8673. https://doi.org/10.1109/JIOT.2020.2995617.

Wanasinghe, T. R., Wroblewski, L., Petersen, B. K., Gosine, R. G., ... Warrian, P. J. (2020). Digital twin for the oil and gas industry: Overview, research trends, opportunities, and challenges. *IEEE Access*, *8*, 104175–104197. https://doi.org/10.1109/ACCESS.2020.2998723.

Wang, L. (2018). Environment supervision system for chemical Industry Park based on IoT. *Chemical Engineering Transactions*, *67*, 481–486. https://doi.org/10.3303/CET1867081.

Wang, Z., Chen, Y., Cai, Z., Gao, Y., & Wang, L. (2020). Methods for predicting the remaining useful life of equipment in consideration of the random failure threshold. *Journal of Systems Engineering and Electronics*, *31*(2), 415–431. https://doi.org/10.23919/JSEE.2020.000018.

Wang, X., Hu, C., Si, X., Pang, Z., & Ren, Z. (2019). An adaptive remaining useful life estimation approach for newly developed system based on nonlinear degradation model. *IEEE Access*, 7, 82162–82173. https://doi.org/10.1109/ACCESS.2019.2924148.

Wang, Y., & Rausand, M. (2014). Reliability analysis of safety-instrumented systems operated in high-demand mode. *Journal of Loss Prevention in the Process Industries*, *32*, 254–264. https://doi.org/10.1016/J.JLP.2014.09.007.

Witczak, M., Mrugalski, M., & Lipiec, B. (2021). *Remaining useful life prediction of MOSFETs via the Takagi-Sugeno Framework*. https://doi.org/10.3390/en14082135.

Woo, J., Shin, S.-J., & Seo, W. (2016). Developing a big data analytics platform for increasing sustainability performance in machining operations. In *Flexible automation and intelligent manufacturing, June* (pp. 1–8).

Wu, H., & Zhao, J. (2020). Fault detection and diagnosis based on transfer learning for multimode chemical processes. *Computers & Chemical Engineering*, *135*, 106731. https://doi.org/10.1016/J.COMPCHEMENG.2020.106731.

Yang, H., Kumara, S., Bukkapatnam, S. T. S., & Tsung, F. (2019). The internet of things for smart manufacturing: A review. *IISE Transactions*, *51*(11), 1190–1216. https://doi.org/10.1080/24725854.2018.1555383.

Zantalis, F., Koulouras, G., Karabetsos, S., & Kandris, D. (2019). A review of machine learning and IoT in smart transportation. In *Vol. 11, issue 4. Future internet* MDPI AG. https://doi.org/10.3390/FI11040094.

Zhang, Y., Yang, Y., Xiu, X., Li, H., & Liu, R. (2021). A remaining useful life prediction method in the early stage of stochastic degradation process. *IEEE Transactions on Circuits and Systems II: Express Briefs*, *68*(6), 2027–2031. https://doi.org/10.1109/TCSII.2020.3034393.

Zheng, Y. (2019). *Predicting remaining useful life based on Hilbert-Huang entropy with degradation model*. https://doi.org/10.1155/2019/3203959.

CHAPTER FIVE

Digital process safety management

Mark Slezak[a,*], Stewart Behie[b], and Hans Pasman[b,*]
[a]Occidental Petroleum Corporation, Houston, TX, United States
[b]TEES Mary Kay O'Connor Process Safety Center, Texas A&M University System, College Station, TX, United States
[*]Corresponding authors: e-mail address: mark_e_slezak@oxy.com; hjpasman@gmail.com

Contents

1. Introduction

In chapter "Opportunities and threats to process safety in digitalized process system—An overview" by Khan et al., a brief history of process safety is presented and reasons why digitalization can support process safety management. We revisit this subject here but focus more on some current, practical applications of digital technology as applied to a few of the elements that comprise a process safety management system. This chapter begins with a brief section, Basic Digital Concepts, to provide context as to what we mean by "digital." This is followed by a section The Case for Digital Process Safety Management to help the reader understand the magnitude of workflow processes associated with a process safety management program

Methods in Chemical Process Safety, Volume 6
ISSN 2468-6514
https://doi.org/10.1016/bs.mcps.2022.05.008

and how this supports the need to apply digital technologies for program improvements. The next section, Application of Digital Technologies in PSM, covers several PSM elements that require several workflow processes or have technical aspects that are amenable to implementing digital technologies. Several examples are provided for demonstrating how these digital advances in technology can be applied to improve process safety. The last section, Future Improvements in Process Safety through Digital Technology: Limited Literature Survey, references several recent papers that explore digital technologies that are evolving and introduces some concepts such as machine learning that may significantly change how future safety management systems will function.

In the context of managing processes and specifically chemical processes in a safe and controlled manner, the phrase "process safety management" is commonly used. Indeed, this seems a fitting description and it is also the title of an informational publication by the U.S. Occupational Safety and Health Administration (OSHA 3132, Process Safety Management) that provides an overview of U.S. OSHA regulation, 29 CFR § 1910.119 titled Process Safety Management of Highly Hazardous Chemicals. This regulation defines 14 elements of process safety and associated prescriptive requirements, and though it does not explicitly require a "management system," the regulation implies that a management system must be in-place for the overall integration of process safety elements. In the subsequent U.S. Environmental Protection Agency Risk Management Plan (RMP) rule, 40 CFR §68.15, similar process safety elements are specified, and additionally, it requires a management system for the development, implementation, and integration of the process safety elements. The U.S. Bureau of Safety and Environmental Enforcement (BSEE) regulation 30 CFR §250 Subpart S, applicable to offshore facilities, requires operators to develop and implement a Safety and Environmental Management Systems (SEMS) program. SEMS has process safety elements substantially similar to the OSHA process safety management regulation and EPA Risk Management Plan rule and requires a documented program. It is noted that other regulatory regimes outside of the U.S. often have similar safety management system (SMS) requirements. For example, the U.K. The Offshore Installations (Offshore Safety Directive) (Safety Case etc.) Regulations 2015, Section 8 requires a SEMS that includes process safety elements such as management of change, emergency planning and response, and audits and reviews, among others.

"Process safety management" or "PSM" is normally linked with "program" or "system," that is, a PSM program or PSM system. It is also commonly used in reference to the U.S. OSHA regulation such as the "PSM regulation." However, when used herein and unless noted otherwise, PSM, PSM program, or PSM system is intended to mean a generic management system having various elements that are applied to chemical or related processes to meet the requirements set forth by the company or applicable regulations to safely operate the processes. As previously noted, other nations as well as the U.S. offshore may more commonly refer to a PSM program as the SEMS or SMS.

2. Basic digital concepts

The term "digital" is commonly used to describe doing business in a way that uses new technology or data in more powerful ways that can lead to better and faster decision making. Organizations are said to develop digital strategies to drive business performance, or to reimagine the organization through digital transformation. In this context, digital is expressed as a desire of organizations to leverage new technology or a nascent technology in hopes of improving or transforming business processes, thereby making them more competitive and prosperous. These are worthy and perhaps necessary aspirations that can also apply to managing elements of process safety.

Before considering some of the esoteric digital technologies that are being developed and deployed, a more basic concept of "digital" and what it means is presented here in a more classical sense. Conceptually, digital or digitized can be thought of as chopping an analog signal such as an audio source into time-based thin slices and converting these slices into an electronic form that can be stored, processed, or displayed. The slices of electronic information, or digital data, consist of an ordered series of "ones" and "zeros" (bits) that comprise chains of data, typically referred to as bytes, longs, and words, among other expressions of digital information.

We contrast digital information with analog information. In the case of chemical process parameters like pressure and temperature, initial sensing and measurements are usually acquired by analog means. For instance, sensors used for temperature measurement such as resistance temperature detectors (RTD) and thermistors function by use of materials whose electrical resistance properties vary in response to temperature change. This unique property allows for these devices to output an analog electrical signal (current or voltage) that can be sent through an appropriately calibrated analog gauge, to be visually monitored and acted upon by operators, interfaced with process controllers, or sent to a storage device such as a magnetic tape-recording machine. This mostly outdated means of analog data storage is limited in usefulness compared with digital data because it does not readily allow for data processing, filtering, and other means used to analyze the data.

Alternatively, an analog electrical output from a temperature sensor can be sent to an analog-digital converter (ADC) device and, through use of a microprocessor, transformed into a digital electrical signal and then sent to a digital display device such an LED monitor (or digital device providing an analog representation) where it can be available for monitoring by process operators or other data users. Digital signals such as a temperature measurement can also be directly utilized for functions like process control, safety systems, and data retention. Digital storage devices provide the means to not only retain and retrieve large volumes of data, but also enables mathematical manipulation of the data (processing) in ways that can be useful for trending, analysis of inter-relationships between various parameters, and prediction of future values, among others.

When digital data that is derived from analog sources, the quality of data is dependent upon the sample rate and "bit depth" of the data, that is, number of bits that make up a sample. For example, the standard sample rate for CD-quality audio is 44.1 kHz and 16-bit depth. Inherently, the sampling process in digitalization results in some analog information being lost. When music was released on digital CDs, some audiophiles insisted that the sound quality was not as good as analog music recordings on phonographs (albums). The argument was valid as some data, albeit a very minor amount, was lost through digitizing the analog signals. In a purely technical sense, digital data taken from analog data will not be as good as the analog data. However, current digital sensor/converter resolution and processing power for many digital devices have progressed to an extent that data loss is generally negligible, making this a moot point.

Digital data lends itself to the implementation of systems that not only control processes, but also those systems necessary for managing the elements of process safety because this form of data can be compiled, parsed, and manipulated through mathematical functions in ways that facilitate the execution of data storage, retrieval, calculation, and display. On a more macro level, digital data in use today is integral to process system control as well as numerous workflows utilized for process safety management that are implemented in an electronic medium.

3. The case for digital process safety management

As mentioned in the introduction to this chapter, a PSM program is considered all of the systems and processes needed to manage and meet the requirements set forth by the company or applicable regulations to safely operate chemical processes. The PSM program defines aspects such as the organizational structure, roles, and responsibilities, training, reporting, and performance measures. For facilities located in the U.S. that are subject to the Occupational Safety and Health Administration (OSHA) process safety management regulation (29 CFR 1910.119), 14 elements of process safety management (PSM element) have been defined and that must be implemented. Each of these elements can be considered a discrete system to be managed in terms of executing the workflows and activities to meet the requirements of the PSM element, including the defined roles and responsibilities associated with the particular element, training, reporting, and defined system performance measures.

For consistent terminology used in this chapter, we consider the management of each PSM element to be at a systems level. Within each of the PSM elements, there are defined requirements. In the case of the Process Safety Information element, there are three higher level requirements as illustrated in Fig. 1, though at a more granular level, 17 specific requirements are defined. To fulfill these requirements, specific management processes are necessary, and, in the case of chemical hazard information, this may involve a process to maintain safety data sheets (SDS) as well as chemical reactivity data.

When extended to all process safety elements, there are numerous processes that must be managed just to maintain compliance. Further, some processes are supported by subprocesses. For example, the process to manage SDS may involve subprocesses such as new chemical approval and training on use and availability of SDS. Table 1 shows the OSHA PSM program elements, the number of discrete requirements for each program element, and a rough tally of the number of processes that may be needed to manage the program element.

As indicated in Table 1, a sizeable number of processes need to be managed to have well-functioning PSM element systems and in turn, a robust PSM program. In practice, some of the processes that are needed for a certain process safety element may overlap with other process safety elements, thereby lowering the number of management processes that are needed

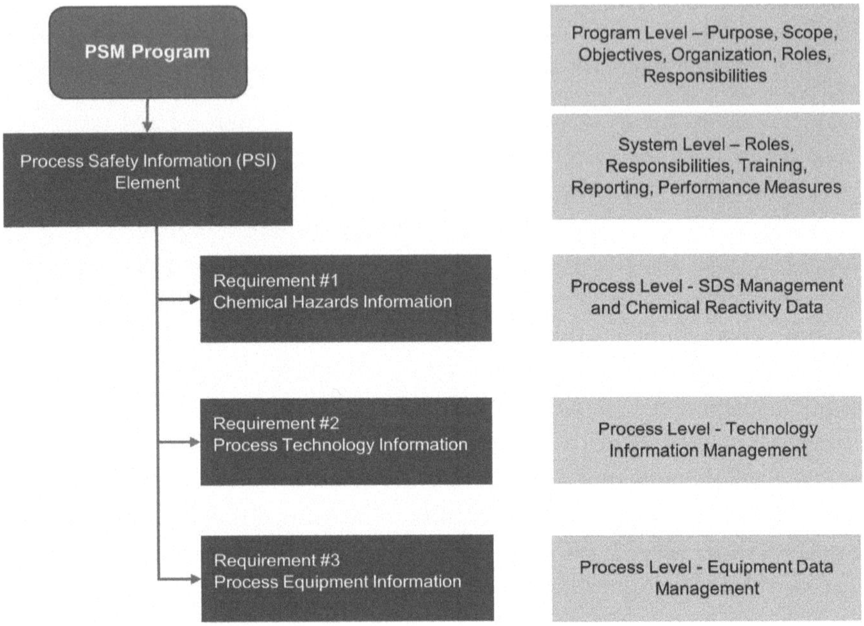

Fig. 1 General PSM program hierarchy for process safety information.

(e.g., process for training employees). Conversely, systems for managing process safety elements that are intended to be very robust and go beyond compliance may require additional processes to achieve a high level of effectiveness.

The management of each system or processes would typically involve development and maintenance of applicable standards, procedures, and guidelines and include sub-elements of roles and responsibilities, training requirements, data collection and management, and performance measures, among others. As any person having PSM program responsibilities can attest, considerable resources are needed to make the process safety elements highly compliant and sustainable. Market conditions, competitive pressures, and workforce changes can result in resource constraints that can erode PSM program effectiveness.

To combat these forces, companies have sought to make their systems and processes ever more efficient. Incremental improvements have naturally arisen from experience and creative thinking that serve to refine system's implementation. Without technological advances, these systems and processes would not likely progress and would be considered by many to be "as good as it gets." Fortunately, great technological progress has been made in recent decades and years that have opened the door to new opportunities

Table 1 OSHA PSM program element processes.

PSM program element	No. of requirements	Estimated no. of processes
Employee participation	3	2
Process safety information	17	12
Process hazard analysis	12	12
Operating procedures	5	5
Training	5	4
Contractors	6	5
Pre-startup review	2	4
Mechanical integrity	10	9
Hot work permit	4	3
Management of change	5	6
Incident investigation	7	5
Emergency planning and response	2	2
Compliance audits	5	3
Trade secrets	3	2
Total	86	74

for improvements to PSM programs in terms of both efficiency and effectiveness. Moreover, these advances are continuing at a rapid pace and to such an extent that "best available technology" today may be supplanted by better technology tomorrow.

Technological advances in the realm of digital data collection and analysis capabilities and has led to the common phrase "big data." Big data is commonly defined in terms of volume, velocity, and variety. That is, it involves data that is either so large, arrives at such high rates, or is of various data types/formats that it cannot be processed using traditional methods. While big data may not be directly applicable to process safety systems and even chemical process control systems in the true sense of the term, the large amounts of data that can be processed by current technology for these systems can provide stepwise improvements due to higher levels of digital data analysis and communication of this data. The remainder of this chapter highlights some examples of how digital technology can be applied to systems and processes used to manage the elements of process safety.

4. Application of digital technologies in PSM

In this section, several examples of systems using digital technology are discussed, along with the potential advantages and disadvantages, and possible future application of digital technology to elements of process safety. For many companies with mature PSM programs, these types of applications are likely to be in place and do not all represent novel ideas. However, for companies that have not made significant advancements in deploying digital technology, these examples may help in determining practical solutions for making improvements to their programs.

4.1 Process hazard analysis element

The intent of this element of process safety is to identify hazards associated with chemical processes, assess the risk associated with the identified hazards, and ensure that the hazards are appropriately controlled. OSHA's PSM regulation states: "The process hazard analysis (PHA) shall be appropriate to the complexity of the process and shall identify, evaluate, and control the hazards involved in the process and to use methodologies that are appropriate to determine and evaluate the hazards of the process being analyzed, along with other requirements pertaining to personnel, risk assessment, and addressing the findings and recommendations."

A high-level process (system) for which OSHA's PHA element can be implemented is illustrated in ISO 31000, Risk Management—Principles and Guidelines, Fig. 1 of that standard. The PHA element of PSM includes, among others, a risk assessment process that is defined as the overall process of risk identification (i.e., hazards as the source of risks in the context of physical harm), risk analysis, and risk evaluation. All of these processes or subprocesses involve information that is generated, analyzed, manipulated, or communicated and hence, are amenable to digitalization via software systems.

At least 12 processes have been identified that must be in place to manage this element of PSM, including processes for:

1. prioritizing the order of conducting PHAs
2. determining an appropriate PHA methodology
3. determining an appropriate team composition
4. assembling and organizing the information needed to conduct the PHA
5. identifying previous incidents in the workplace with potential for catastrophic consequences (or other relevant incidents)
6. training team members in the hazard analysis methodology being used

7. arranging meeting times, duration, location, and supporting activities
8. conducting and documenting PHAs
9. promptly addressing the team's findings and recommendations, including documentation and schedule of actions
10. communicating the actions taken to persons who may be affected by the recommendations or findings
11. updating and revalidating PHAs
12. retaining initial, updated, and revalidated PHAs and resolution to recommendations

Some of the above processes only involve static data that would not require workflows or sophisticated technology to manage. For instance, the priority order for conducting PHAs must be based on a rationale that considers factors such as the extent of the hazards, number of potentially affected employees, age of the process, and operating history of the [chemical] process. This process can be accomplished through means of a simple qualitative scoring table retained in commonly used software and, therefore, no need for more advanced digital technology nor for a workflow.

In contrast, several of the processes for managing the PHA element do require workflows as they would involve multiple steps and responsibilities of different people. Other processes, by their technical nature, lend themselves to more advanced digital systems such as for quantifying risk via quantitative risk analysis (QRA). For example, the process for conducting and documenting PHAs generally involves a workflow of several steps (or subprocesses) as well as technical processes. These are shown in Table 2, along with some of the digital means to perform the needed steps.

Digital technology as applied to the Process Hazard Analysis element is advancing primarily on two fronts: (1) technical safety software such as toxic, fire, explosion simulation and quantitative risk analysis, and (2) systems software for integrating PHA subsystems and linkage to other PSM elements. With respect to technical safety software, ever-increasing digital processing power available in mainstream computers allows for use of software that involves complex and intensive calculation such as required by computational fluid dynamics (CFD) and Monte Carlo simulations. Simulation software that previously utilized simplifying assumptions and approximation methods to conserve digital processing time can now use more refined analysis and higher complexity while not requiring undo calculation time. Additionally, more advanced simulation testing is possible for validating models through use of faster digital data acquisition and processing equipment such as might be used in large sensor networks to measure blast overpressures or to provide high-speed digital video recording.

Table 2 Digital applications in process hazard analyses.

Step	Digital applications
1. Delineate the specific sections of the process to be analyzed	• Computer-aided design and drafting software
2. Compile relevant process parameters	• Digital process information and drawing database • Digital equipment database • Safety data sheet (SDS) database
3. Analyzing and sections of the process	• Risk analysis software (What-If/Checklist, HAZID, HAZOP, LOPA, QRA, etc.) • Chemical release; gas dispersion, fire, explosion simulation software; computational fluid dynamics software • Consequence analysis software (fire, explosion, toxic hazards) • Computerized maintenance management systems (CMMS)
4. Documenting the analysis	• Risk methodology software; word processing and spreadsheet software
5. Reviewing, approving, distributing PHA reports	• Document management systems; secure online e-signatures and routing
6. Storage of PHA information	• Digital document storage systems • Network data storage; offsite data storage facilities; cloud-based data storage

These advancements have increased modeling granularity and led to a higher level of confidence that the use of these simulation tools will closely replicate real-world process incidents.

With respect to systems software for integrating PHA subsystems, consider common practices used within industry to conduct PHA studies, communicate results, and managing risks through addressing the recommendations. Software for documenting PHA studies using a methodology such as Hazard and Operability (HAZOP) have been commonly used for many years. The digitalization of this process and more advanced features has allowed for efficiency improvements in an otherwise time-consuming activity. A few such features include:

- Use of drop-down lists for any data fields that provide a selection
- Hyperlinks for quick navigation to data within the study or to external data sources

- Libraries to facilitate access to technical data involving equipment failure modes, equipment data, and checklists
- Configurable fields (e.g., naming conventions), data hierarchy, and reporting
- Use of "flags" to denote items such as scenarios needing review, special comments, and progress in revalidation of a study.

The output from PHA studies includes findings (e.g., risks) and recommendations to address the findings. These outputs are generally communicated via PHA report which itself may require a process for review approval, and communication. Another process may be used for assigning and tracking the recommendations to completion and yet another process may be needed for managing the risks that were identified in the study.

Digital advances have helped to drive efficiencies in these processes. For example, some PHA study software programs utilize internal report templates that can generate complete reports and output these in a common word processing format, thereby simplifying the process of compiling a report. The historically tedious and time-consuming process of physically "walking around" or routing of paper reports for review and approval signatures has been supplanted by online digital document routing and signature processes, a much more efficient means of completing this workflow.

PHA study software, like many others, has been shifting from locally installed applications specific to machine or user to online cloud-based applications. Cloud-based applications that are deployed enterprise-wide promotes consistency in practices and enhances data search capability. These also facilitate the management of PHAs in terms of licensing, standardization of studies, access to studies, transparency, and data security.

The process of tracking recommendations from within PHAs is often an integral feature of PHA study software. When deployed across a company business network or cloud-based environment, the actions to address PHA recommendations can receive visibility at higher levels of management and that can help drive accountability for timely completion. Also, features like email notification with hyperlinks for completing various steps in the action tracking workflow such as approving action item assignments, closing action items, and providing approval to close action items can make this workflow process very efficient. In addition, these action steps can be designed such that the user need not directly interface with the PHA software but accomplish their task via a simple online form. This is particularly useful feature as many persons who could potentially be assigned roles in the action tracking workflow do not need to access the PHA system and doing so would add to complexity and administrative burden.

However, facilities often generate action items from multiple sources such as compliance audits, new programs, inspection deficiencies, department goals, and many others in addition to PHAs. Hence, at a facility-level, a single action tracking system is desirable so that users only need to learn and interface with a single system. Moreover, a single action tracking system facilitates reporting and rolling-up of related performance metric. Therefore, having an action tracking system specific to PHA study software that is run in parallel would not be efficient. A better approach would be to build a data link to export actions from the PHA software to a facility action tracking system, thereby maintaining a single, simplified process.

Another output from PHAs are the findings and, specifically, the risks that are deemed to be at a level that warrants mitigation to a level that is within the organization's risk appetite. Generally, recommendations will be made to address these risks through more rigorous analysis or further engineering studies to confirm their risk level, or through additional controls. Within the risk management process, risks must be monitored and reviewed periodically and collectively to provide management with the information to make informed decisions. In a large organization, this process is needed at all levels and may involve lower risks that are managed at a single facility or business unit level to a corporate level. Therefore, the risk management process must be scaled, accordingly, with risk information accessible, rolled-up, and communicated at various organizational levels. To be effective, this necessitates an enterprise-wide system deployed over the organization's business network or through a cloud-based application. While standalone enterprise-wide risk management system can achieve the organization's objectives, an opportunity exists to leverage the risk data generated within an enterprise PHA study system and integrate this data directly with risk management functionality suitable for monitoring, reporting/communicating relevant information. A further opportunity exists to integrate data flow between a centralized PHA study system, action tracking system, and risk management system such that new or updated data in one system is reflected in the other systems. Systems integration such as this minimizes manual processes and contributes to improved efficiencies.

The concept of "dashboards" whereby users are presented screen views of information in real-time or near real-time has become commonplace. Dashboard views offer the advantage of presenting a "one stop" location for acquiring needed and useful information from various digital data sources and often contain hyperlinks to drill further down into the data. For instance, a dashboard relating to risks might present a chart showing the total number of risks for the organization. Through a filter selection, the user could drill down to a region or specific facility to view the number of risks. Further querying could present just the higher-level risks and hyperlinks would take the user to specific details such as the PHA study from which the risk originated and the current status of the action plans to mitigate the risk. An example dashboard the presents similar type of data is presented in Fig. 2.

The onset of the COVID pandemic in 2020 challenged companies in the process industries to maintain their operations. Along with this challenge, was continued implementation of process safety elements and in particular, the conduct of PHAs in which many employees were required to work from home and were restricted from gathering in-person for meetings. During this time, technologies quickly advanced that facilitated online meetings and allowed for successful continuation of PHAs. For U.S. facilities that are regulated under OSHA's process safety management regulation, this also meant they were able to maintain compliance with required periodic PHA revalidations. Following this period of restrictions, a question often asked was whether future PHAs should be conducted as historically done via in-person meetings or by remote online meetings. This will likely be decided by matter of preference based on individual experiences. The following is a short list of some advantages and disadvantages to conducting online meetings.

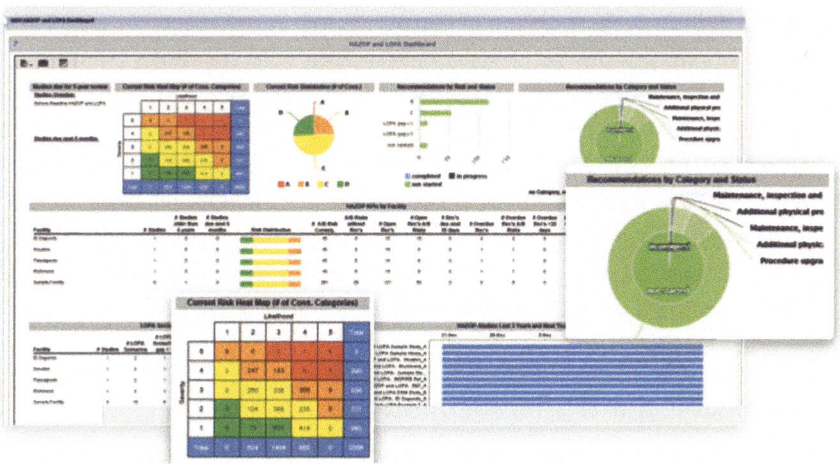

Fig. 2 Example risk studies and recommendations dashboard. *Courtesy of Sphera Solutions.*

Advantages:
- No need to schedule a meeting room.
- No need for activities necessary for bringing visitors on location (e.g., safety briefing, directing to meeting location, other logistics).
- No potential issues with connecting to video display equipment.
- No need for arrangement of meals or other customary practices for meetings.
- Software may permit users to enlarge or shrink their own video displays to suit their preference.
- A "raised hand" function can help the facilitator maintain order during discussions.
- Slight delay in audio transmission that is often experienced tends to encourage participants to pause before responding, promoting more orderly and respectful discussions.

Disadvantages:
- Lack of in-person presence and physical cues makes it harder for the facilitator to ensure team engagement and participation.
- Some team members may be uncomfortable with contributing without direct encouragement of the facilitator.
- Potential for limited audio-video bandwidth to maintain clear communication.
- Loss of in-person collaboration that might otherwise quickly resolve issues or disputes (e.g., question of instrument function or purpose as shown on a drawing).
- In some industries such as the upstream oil and gas sector, PHA team members may work in geographically separated locations by nature of their job function, making in-person PHA team meetings with all members is not always possible. Prior to the pandemic, remote participants typically joined PHA meetings via phone and by audio, only. Though feasible, the effectiveness of these types of meetings was not optimal. With rapid development of online meeting tools and greater user acceptance, coupled with greater digital communication bandwidth, online PHA meetings are viewed by many as a viable and effective option.

4.2 Mechanical integrity element

Mechanical Integrity (MI) programs are intended to prevent equipment failures and mainly, loss of primary containment (LOPC) events by detecting locations in process equipment and piping that have lost integrity, thereby allowing corrections to restore integrity be made before LOPC events occurs. Significant advances in tools used to detect various failure mechanisms such as corrosion, erosion, and cracking are allowing owners/operators to find and address deficiencies before LOPC can occur.

The techniques used to confirm mechanical integrity fall into two broad categories: Destructive testing (DT) and non-destructive testing (NDT). As the name implies, destructive testing renders a sample unsuitable for further use whereas NDT does not. DT methods such as tensile and hardness testing of material serve an important purpose for ensuring the integrity of materials used for constructing piping and equipment but are limited in usefulness following construction compared with NDT. Common NDT methods for confirming the ongoing integrity of piping and equipment are visual inspection, dye penetrant, leak testing, magnetic particle, ultrasonic, radiography, acoustic emission, and eddy current testing.

Most of these NDT methods have been and continue to be advanced through progress in digital technologies. For example, beyond the common pulse echo ultrasonic testing (UT) method, other types of UT have been developed such as shear wave, phased array ultrasonic testing, time of flight diffraction ultrasonic testing, and in applications such as automated UT. Even visual inspection methods have progressed through use of digital technology in the form of digital cameras having very high-resolution image sensors, large digital storage capacity, and deployment of these cameras by means of aerial drones that allow visual access where not easily accessible by humans.

In this section, two examples are presented of these technologies that have been extensively used in mechanical integrity applications: Digital radiography (DR) and use of unmanned aerial systems (UAS), more commonly referred to as drones, for visual inspections. These technologies have been utilized for many years but through advancements in sensor capabilities, increased digital processing power, digital storage capacity, and application refinements, they are providing increasingly higher levels of performance and cost savings to companies. The first example illustrates how DR has been applied in upstream oil and gas production.

4.2.1 Applying targeted corrosion approach through digital radiography

4.2.1.1 Introduction

As oil and gas production and well counts climbed over the past decade at one company, so did piping and equipment failures due to metal corrosion and erosion. Even with robust continuous chemical injection programs, production facilities experienced failures, primarily in wellhead piping, uncoated carbon steel water lines, and oil-gas-water separators. With stringent environmental regulations and high public visibility in certain locations, a more proactive solution was needed to prevent loss of primary containment (LOPC) events. For new production sites being constructed, potential failure mechanisms were addressed mostly by utilizing different piping materials, sizes, and flow paths. However, on already producing more mature sites, the challenge was finding the best method to avoid spill events. By minimizing these events, the company could benefit from avoiding environmental harm and cleanup costs, promoting their "license to operate" in a region, and minimizing unnecessary repair and replacement costs.

For facilities with a known corrosion risk, there are two extremes in approach: (1) Run the equipment to failure, or (2) Proactively replace all affiliated equipment. Fig. 3 illustrates how effective use of mechanical integrity monitoring reduces the cost of having unplanned down-time and the expense associated with adverse LOPC consequences likely to occur when equipment is run to failure. Likewise, effective use of mechanical integrity monitoring minimizes the added cost of unnecessarily replacing equipment whose integrity is sound when trying to avoid failures "at all costs."

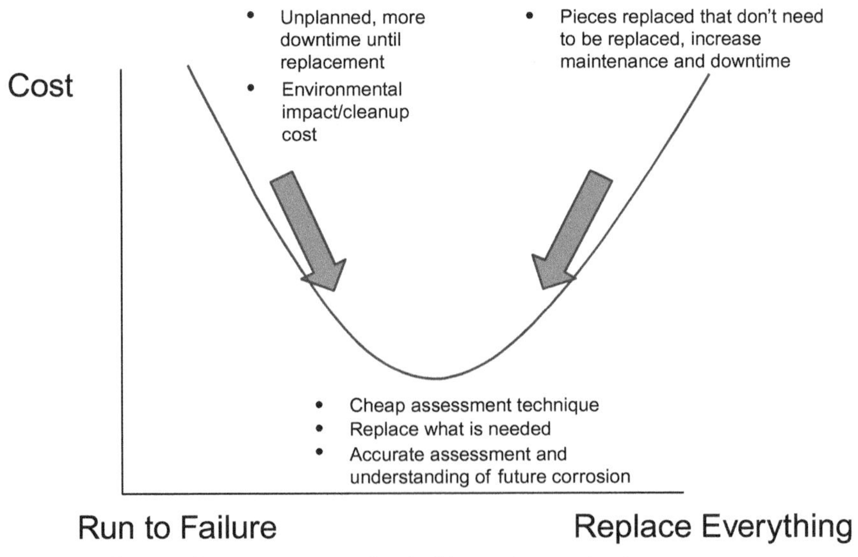

Fig. 3 Justification for proactive mechanical integrity monitoring.

Option 1 was deemed unacceptable due to the higher risk of environmental contamination, potential for prolonged downtime, and spill cleanup costs. Option 2 could result in unnecessary replacement spending as corrosion/erosion failure mechanisms may not have been present in the replaced equipment. Thus, the company sought to apply an effective mechanical integrity program for the various types of piping and equipment utilized in its operations that would be effective and minimize the overall cost of ownership.

An evaluation of various NDT methods was conducted with multiple field trials involving the different types of production equipment in use. Digital radiography (DR) proved to be most effective from a cost, downtime, safety, and assessment perspective. Digital radiography is a mechanical integrity assessment technique that utilizes X-rays (gamma rays) captured on a reusable flat panel detector. Utilizing traditional radiography equipment without the need for film and developing time, high quality DR images are captured in real time on the detector panel and wirelessly transferred to a computer or tablet device.

A common type of DR flat panel detector, or digital detector array (DDA), is based on a direct conversion process whereby x-rays striking a photoconductor layer like amorphous selenium cause a movement of electrons and resulting change in electrical conductivity. X-ray photons passing through a target such as a section of piping will be absorbed in relation to the target density, resulting in fewer photons reaching the photoconductor detector panel and corresponding change in conductivity and electrical charge output. A thin-film array, or TFT is then used to provide a visible readout of these electrical changes on the detector panel. The basic principle of DR operations is shown in Fig. 4.

One of the main benefits of digital radiography is that a single image can be used to evaluate a wide range of material densities and thicknesses.

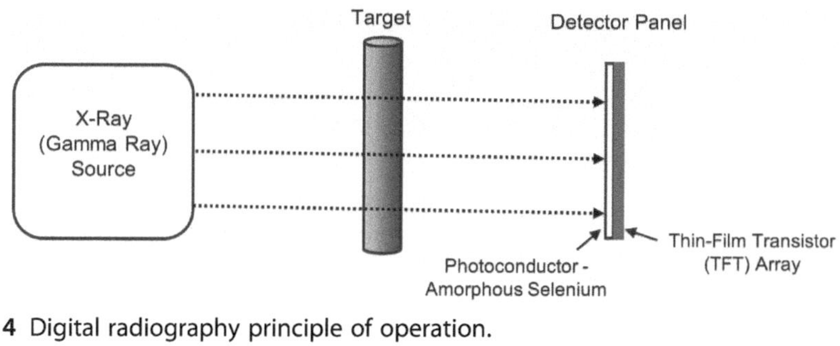

Fig. 4 Digital radiography principle of operation.

In contrast, conventional radiography could require multiple x-ray exposures at various distances and exposure times to capture the density of interest, potentially missing defects outside of the thickness range. Defects are easier to identify using DR through use of software editing functions and image filters to adjust parameters such as contrast. Accurate defect sizing and wall thickness determination can also be captured on the image, allowing for enhanced reporting and inspection detail. DR requires no downtime of equipment and does not require insulation removal. Limitations of DR include inability to scan below-ground piping and to inspect an entire above-ground piping section requires recording multiple images. Nonetheless, targeting DR to areas of known or suspected corrosion, or typical failure locations, has shown DR to be an excellent inspection method that is also relatively inexpensive when compared to some other NDT methods. Table 3 shows the company's experiential cost for several NDT methods per site assessed relative to DR. Another ultrasonic testing method for pipelines, Inline Inspection (ILI), is not directly comparable to inspection at a site but for a typical pipeline segment it is considerably more expensive.

4.2.1.2 Methodology

DR has been employed for inspection of a wide range of field equipment and piping. Extensive use of this digital technology has enabled creation of a risk profile in which various equipment piping components are ranked by the frequency and severity of corrosion locations. This information aids decision making with respect to prioritizing the areas to concentrate mechanical integrity inspections. By using this targeted approach, the likelihood of discovering areas of corrosion prior to an LOPC event is increased while making more effective use of mechanical integrity resources. Utilizing DR and as of this writing, the company has prevented a significant number of LOPC events over the past several years. Though DR has been applied to many types of equipment and piping, the focus areas have been on produced water lines, wellhead piping, and separator drain collar ports, the latter two types being discussed, following.

Table 3 Comparative costs of several NDT methods.

NDT method	Cost relative to DR
Digital radiography (DR)	1.0
Manual ultrasonic testing	0.1
Guided wave UT	5.0

4.2.1.3 Wellhead piping

The company's mechanical integrity team has performed more than 1000 wellhead piping inspections and, over time, adjusted the inspection workflow process to allow for a more targeted NDT approach. Initially, inspection resources were directed towards higher risk sites where LOPC events could have more adverse consequences. The team focused on piping tees and the motor valve as those are the primary locations where failures and corrosion have historically occurred (Fig. 5). The DR results and previous

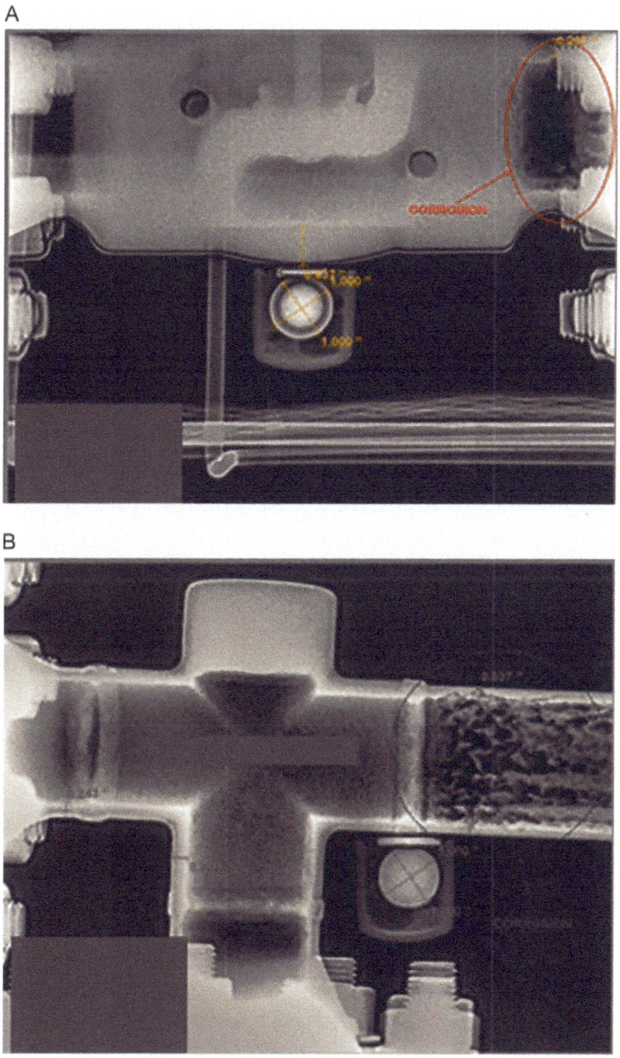

Fig. 5 DR images of wellhead motor valve and outlet tee.

failure data showed that higher average gas rates in the early life (0.25–0.75 years) of wells statistically increased the risk of failure (Fig. 6). With this information, the mechanical integrity team concentrated DR inspection activities to newer wells that met a specified average gas-rate criteria. Performing digital radiography in known issue areas of wellhead piping has been effective for identifying individual piping spool pieces that were at risk of failure. Moreover, the DR information has helped limit the scope of needed piping replacements which can range from single pipe spools to full wellhead assembly rebuilds. In the most current year of the company's proactive DR inspection and monitoring program, success was evident in not experiencing any wellhead piping LOPC events.

4.2.1.4 Separator drain collar ports

Separators that were manufactured from 2016 to 2018 with 1″ threaded drain collar ports proved to be a source of continuing issues, resulting in LOPC events that in one case, led to extensive downtime of a facility while excavation and soil analysis was conducted to ensure there was no environmental harm. After a series of drain collar port failures, the company's mechanical integrity team performed DR scans of these areas and based on wall loss, ranked them as high, medium, and minimal risk. The DR inspections confirmed that the oil drain collar ports were not corroding, allowing efforts to focus on the water drain collar ports where failures had occurred, resulting in further cost savings. DR images shown in Fig. 7 show minimal corrosion in the water drain collar port on the left, whereas significant corrosion is evident in the image on the right.

Based on the DR inspections, all high-risk water drain collar ports were replaced. Through the DR inspection program, the company is monitoring

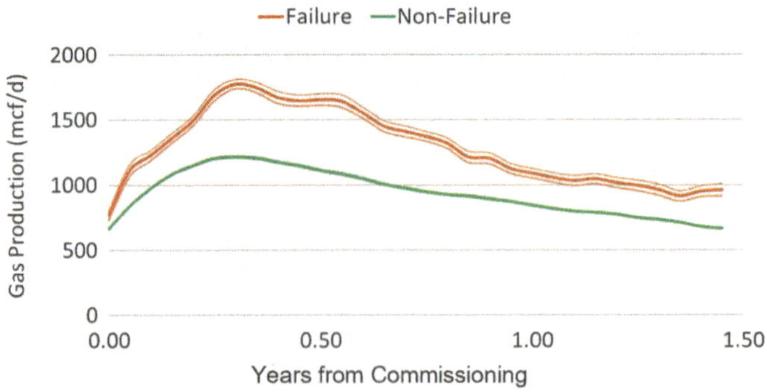

Fig. 6 Gas production for failure and non-failure subsets.

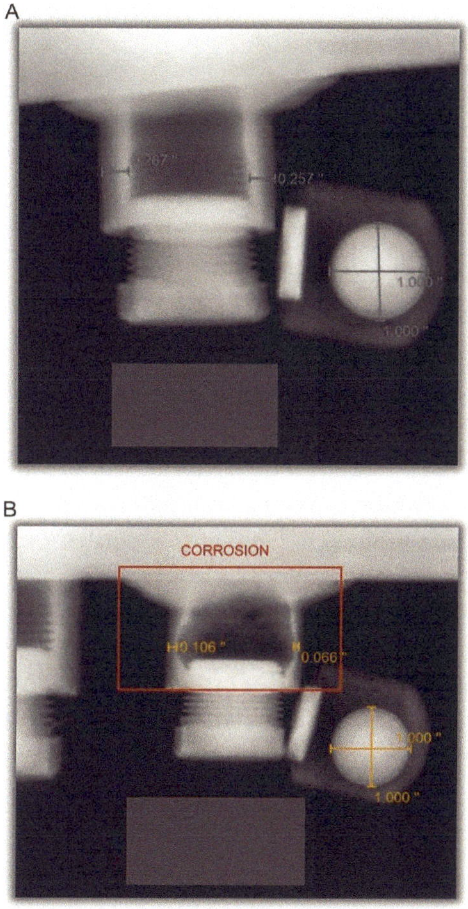

Fig. 7 DR images of threaded water drain collar ports.

the medium risk drain collar ports to determine the extent of corrosion progression, thereby providing the information needed for risk-based decision making moving forward.

4.2.1.5 Economics

As mentioned throughout this paper, digital radiography has allowed for a targeted approach to minimize LOPC events and avoid unnecessary replacement of piping. Consequently, when confronting corrosion or erosion issues on various types of equipment and piping configurations that may potentially have high spill consequences, DR has shown to be an effective NDT method that can minimize the overall cost of ownership.

In comparison to other inspection methods, especially with known areas of corrosion, digital radiography proved to be reasonably inexpensive. Since implementing and refining the DR program, LOPC events have been significantly reduced. Over a two-year time, costs associated with LOPC events involving the piping and equipment that the DR program focused on had decreased more than 10-fold and overall annual costs had decreased by more than half.

4.2.1.6 Conclusion

Digital radiography, an NDT method, was applied to assess corrosion of aboveground piping and equipment, better understand the corrosion risk profile, and reduce the overall cost of ownership. DR's advantages included its low cost, minimal operational interference (no downtime, no insulation removal), quality depiction of piping, and potential for direct quantifiability of wall corrosion losses. Since implementing a DR monitoring program for field production facilities, the rate of LOPC events is appreciably lower, and corrosion-related costs have decreased significantly. While every facility's value proposition is different, digital radiography DR can be a cost-effective NDT option for reducing the risk of LOPC events. Moreover, this method can help reduce the risk of experiencing LOPC events and promote environmental stewardship.

4.2.2 Tank internal inspection using aerial drones

4.2.2.1 Challenges

Atmospheric and low-pressure storage tanks are commonly used in chemical and petrochemical industries for holding various process fluids. Periodic inspections of tanks are performed to ensure that they are mechanically sound. One of the main challenges to conducting internal inspection of tanks is gaining access to high and complicated inspection locations, particularly in the tank roof and shell. Historically, the erection of scaffolding was the predominant way to access these areas. However, this method can require considerable time and high cost. The added time for constructing scaffolding often results in extending the total outage time.

Another option is to utilize ropes to reach elevated points for preliminary inspections. While the rope technique has proved to be faster and less expensive compared to scaffolding, this method comes with its own limitations and challenges such as a higher risk to humans and ability to comply fully to the defined acceptable distances for visual inspection.

4.2.2.2 Drone technology

Aerial drone technology, also referred to as unmanned aerial service (UAS), has improved dramatically in recent years, and it became more popular in inspecting points of limited or complex access like external tank roofs and flares. Initially, the company purchased several drones and used these for inspection of flare tips and appurtenances. Later, their use was expanded to inspection of equipment such as production headers in limited access areas, inspection of the roof of GTG air inlet ducts, and other remote or inaccessible locations. With recent developments in drone technology furthering its use in more complicated and congested areas, combining with artificial intelligence programs, the MI team considered its use for internal inspection of tanks a potential opportunity worth exploring for deployment in the MI program.

To evaluate the advantages of the technology and understand its limitation, the MI team opted to conduct a trial inspection in one of the larger tanks using a drone designed for this purpose such as the one shown in Fig. 8. The drone is relatively small in size, can maneuver in congested areas, and is equipped with soft and hard protections against collisions.

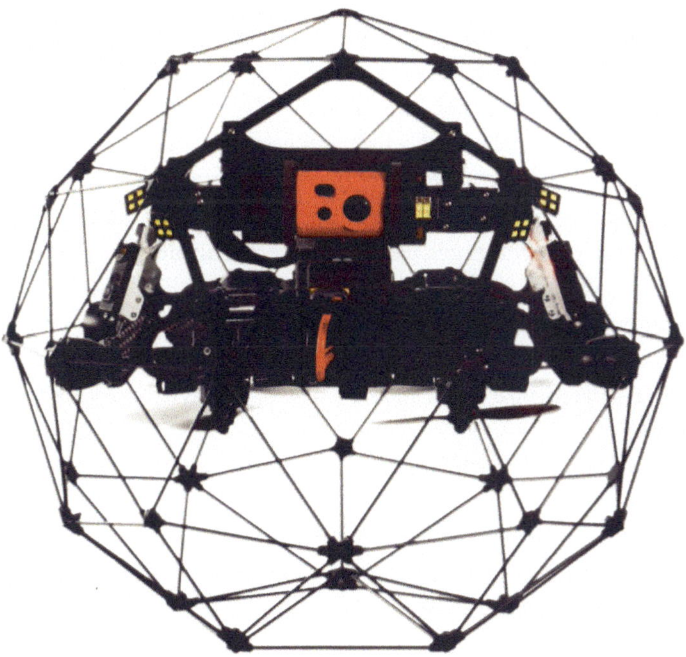

Fig. 8 Collision-resilient aerial drone. *Courtesy of FLYABILITY SA.*

The MI team worked with a third-party drone solutions company to conduct a trial tank internal inspection. The purpose of the trial was to evaluate the commercial and technical advantages of this technology in this application, and to understand its limitations and disadvantages compared to close physical inspection using scaffolding or rope access techniques.

4.2.2.3 Trial on sales oil tank

A sales oil tank was selected as a trial candidate since it was out of service and hydrocarbon free when the third-party drone company was ready to mobilize. The operation did not require confined space entry or lighting. The size of the drone was small enough to be inserted through the 36″ x48″ manway. Hence, it could be placed inside and operated without a need for the drone operator to be in the tank during inspection. The inspection of the entire tank took an average of 8 h a day over a period of 3 days. The trial was considered successful as it provided the expected data and achieved its goal to support decisions needed to move forward.

4.2.2.4 Trial outcome

Photos and videos taken by the drone were of high resolution and provided clear understanding on the general condition of the tank, allowing for an appropriate assessment and potential repair requirements. Fig. 9 shows several images taken from inside the tank in which a loose nut on a structural member and an anomaly in the internal coating were detected.

Based on the MI team members being involved with the inspection and reviewing the inspection report issued by the drone company, the main advantages and disadvantages of the drone technology are summarized as follows:

Advantages:

- Safe technique compared to rope access and eliminated the need for emergency response team standby associated with confined space entry activities
- High availability with minimum need for confined space trained personnel
- Reduced time for the full inspection of the tank
- Easy access to complicated and elevated internal inspection locations
- Reliable operation with multi-layer collision protection, soft and hard barriers
- Much more economical compared to scaffolding

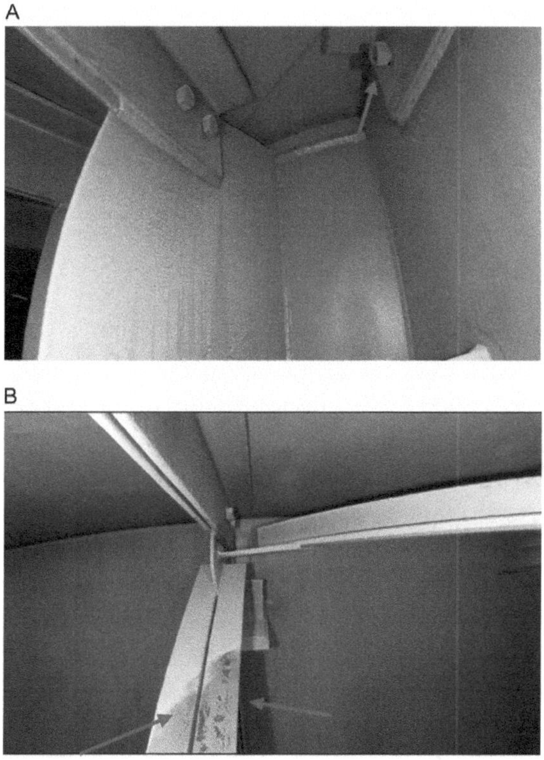

Fig. 9 Inside tank images showing anomalies.

- Can be used inside variety of field production equipment such as heat-recovery steam generators (HRSG), once-through steam generators (OTSG), vessels, pipe racks, and silos
- Clear, high-resolution images captured to evaluate general condition of inspected equipment
- Qualified inspector (not the pilot) does not need to be present at the site for the inspection; through live streaming, inspector can be anywhere in the world

Disadvantages:

- Scaffolding will still need to be erected if further cleaning/testing or repair activities are required; however, drone usage may reduce the amount of scaffolding needed and allow better planning
- Permits to conduct drone operations may be complicated and take substantial time to process, potentially causing delays

4.2.2.5 Next steps

The development of drone technology is likely to progress. Further enhancements are ongoing, such as linking drone images to artificial intelligence programs for troubleshooting and may suggest solutions for mechanical integrity aspects. With currently available and affordable technology, the MI team finds this technique sufficient for preliminary inspections, enabling quicker decisions on repair requirements or if there is need for more detailed inspection or repair plans.

4.2.2.6 Conclusion

Overall, the advantages of the drone internal tank inspection technique are that it is relatively inexpensive, and it provides quick and easy access to elevated, complicated, and congested inspection points. This technology can provide a preliminary inspection, or it may be combined with physical inspection via scaffolding and/or rope access as a solution for a specific inspection challenge. The trial supported including this technology in the MI program. A pending decision following the trial is whether to pursue a call off contract for this service (as needed) or to purchase an equivalent specialized drone and operate it using inhouse resources. In either case, the advantages and disadvantages are known, and these will aid in determining potential opportunities for further uses.

Note: Drone pilots typically need to be certified to fly the drone and other requirements may apply depending upon the local jurisdictional authority.

4.3 Management of change element

Management of change (MOC) is often regarding as one of the most important elements of process safety because it is intended to ensure that potential hazards, physical or administrative, that could be introduced by virtue of making a change to a process are identified and sufficiently addressed. This element of process safety interfaces with many of the other process safety elements and, if it is not implemented effectively, can cause other elements to be ineffective. This, in turn, increases the risk of having an unplanned event. Avoidance of incidents and their undesired consequences is a key motivator for companies to implement effective MOC systems. Management of Change done well, as noted by the author, simply means "no surprises."

The Management of Change process as applied to a single new facility or modification to a facility is commonly referred to within industry as a "change" or an "MOC," the latter being the term used here. Some of the common processes used within a Management of Change system are:

- Review and approval of MOC concept
- Review and approval of MOC detailed design
- Scheduling and coordinating of MOC hazard review
- Assignment and tracking of pre-startup and post-startup actions to closure
- Storage of MOC-related information
- Closure of MOCs
- Monitoring MOC system performance

An example of an MOC workflow that covers many of these processes is shown in Fig. 10.

In the execution of these MOC workflow processes, there may be many subprocesses needed because Management of Change, by its nature, involves other PSM elements, including Employee Participation, Process Safety Information, Process Hazard Analysis, Training, and Pre-Startup Review. Depending upon the type of modification, other PSM elements may be affected such as Operating Procedures, Contractors, Hot Work Permit, Mechanical Integrity, and Trade Secrets. As a result, a modification to equipment could trigger various subprocesses utilized for updating of:

- process documentation (e.g., piping and instrumentation diagrams)
- process hazard analysis documentation
- equipment registry and data files (specification, data sheets, vendor data)
- preventive maintenance and mechanical integrity systems
- operating procedures
- training program
- emergency equipment

Each of these MOC workflow processes and subprocesses can be implemented through manual steps and documented via hard-copy (paper) MOC-related forms and other information. This type of MOC system suffers several significant drawbacks: (1) inefficiencies introduced by nature of manual workflow processes and system performance monitoring activities, (2) potential for errors in performance of workflow steps due to dependence on human actions, and (3) potential for data integrity issues (loss, absence, insufficient, or incorrect data).

The complexity of Management of Change and its related processes provides the motivation for companies to implement digital (electronic) MOC systems. In these systems, workflow steps take place via data communication devices (desktop computers, tablets, mobile devices, etc.) and information is

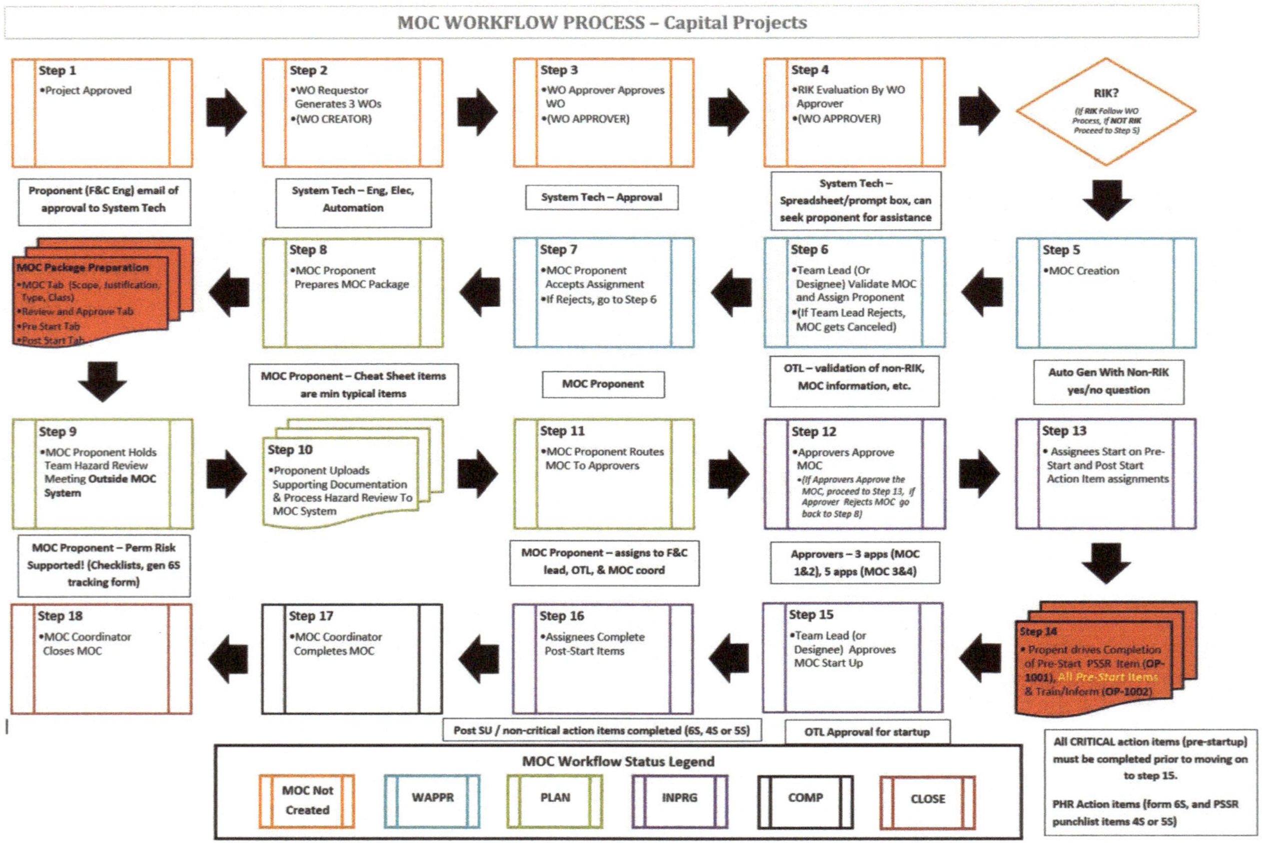

Fig. 10 Example MOC workflow process.

processed and stored in a digital data format. Some advantages of digital MOC systems include:

- Workflow steps are not bound by geographical distances, removing the need for in-person communications or activities such as signing forms.
- Workflow steps can be tailored to the type of change being made to allow for more efficient execution.
- Workflow steps such as MOC approvals can be programmed to proceed sequentially or in parallel paths. A sequential flow is often desirable when higher levels of management approval are required.
- Ability to define required data fields to ensure completeness as well as "gates" to ensure that the MOC workflow progresses in the desired manner.
- Data loss prevention through digital backups and offsite or "cloud" storage.
- Ability to automate data searches, queries, and filters to extract and report the desired information on a routine schedule.

In one such electronic MOC system, the home page presents the user with key data such as the number of MOCs that exist at various steps in the workflow, as well as the MOCs that are associated with a specific user. Hyperlinks can display lists of associated MOCs such as those awaiting approval. Several common types of search fields are provided to allow the user to find MOC data based on a unique identifier, MOC "owner," or activity. Keywords searches provide a powerful capability to quickly find information such as all modifications made in a particular facility/unit, or modifications involving a certain type of equipment. Another key feature of this and similar digital systems is the ability to restrict assess such that only information that is relevant to the users' areas of responsibility is visible, thus streamlining the search process for desired information.

It is desirable to monitor the MOC system performance not just to meet compliance requirements but to ensure that it is functioning effectively and efficiently. As indicated in the last bullet point, above, standard search, query, filtering, and reporting capabilities can be applied to enable MOC system improvement. Some common performance indicators that can be utilized to monitor and trend the health of an MOC system include:

- Percentage of audited MOCs that met all requirements
- Percentage of MOCs that remain open by more than NN days following implementation
- Number of temporary or emergency MOCs initiated during the month
- Number of post-startup action items that remain open

Extraction of data to monitor performance indicators can often be easily accomplished through setup of standardized and automated generation of reports within electronic MOC systems to minimize the clerical burden. As an example of how this can enhance process safety performance, consider the number of MOCs that are conducted on an emergency basis. That is, certain changes are deemed highly time-critical such that the risk of not making a modification immediately or in very short order is deemed to be greater than the risk of moving the proposed modification through the normal review and approval process. In general, emergency MOCs are not desirable as they are likely to introduce a higher level of risk than normal and, therefore, should be infrequent. The number of emergency MOCs that are conducted over a defined time period can be efficiently monitored through electronic MOC reporting to determine if this process is "out of control." If so, the digital MOC data can be further interrogated for possible causes or patterns such as: Is there a particular process area or unit where most of the emergency MOCs are originating or, more specifically, is there a particular individual who is responsible for more frequently generating emergency MOCs? If so, then underlying causes can be investigated such as a recurring process issue or perhaps an improper use of emergency MOCs to get modifications made more quickly.

Despite the advantages offered by electronic MOC systems, organizations should be aware of several challenges that can arise with these newer digital systems. As people must interact with the technology (human-machine interface) through various forms of data communication devices, they must be sufficiently proficient with operation of these devices. Transitioning from manual, paper-based systems to electronic systems can require keyboarding skills, among others, that may not be present to a high level within some workforces. This is becoming less of an issue as newer generations now develop these and other technical skills at early age, but it can still be a factor in some industries. Also, if the system is not user-friendly, difficult to use, or overly complex, then resistance to use of the system will inevitably result. Resistance can then lead to ineffective use and, ultimately, to loss of management support for the system. Another challenge is that the multitude of processes and subprocesses that exist within MOC, or which may interface with MOC and can lead to a very complex system that must be maintained. As such, changes to one system may have a negative impact on other systems. For instance, a data bridge may be built between the MOC system and a separate maintenance work order system to accumulate all of the work order costs for modifications that are made.

A bridged system may work fine if a common equipment reference number(s) is available to link the systems. However, if a delay or other issue arises in implementing a reference number in either system, then a system error may occur or possibly not generate an error, resulting in undetected missing information.

Many of the past obstacles to implementing robust electronic MOC systems such as data processing capability, bandwidth, storage capacity, human-machine interfaces, and costs are largely being overcome with advancements in digital technology. The use of digital data to monitor MOC system performance and the ability to interconnect other systems is helping to transform the MOC process.

4.4 Pre-startup review element

The Pre-Startup Review element is closely aligned with the Management of Change element as it is necessary whenever a modification is made to a facility that is significant enough to require a change in the process safety information, or when a new facility is installed and prior to introduction of hazardous materials to the process. This review is also referred to as Pre-Startup Safety Review (PSSR) in both the U.S. OSHA process safety management regulation and the EPA Risk Management Plan rule.

The PSSR serves as a final check to be sure the facility construction and equipment are in accordance with design specifications, relevant safety, operating, maintenance, and emergency procedures are in place, and other PHA, MOC, and training requirements are complete. Facility changes can range from the very simple such as adding a piping drain valve to very complex such as the addition of a new processing unit. As both of these changes would result in a change to the process safety information (process and instrument diagram or other technical documents), these would necessitate conducting a PSSR or even multiple PSSRs.

Due to the various types of PSSR checks that must be made, particularly with respect to confirming that construction and equipment are in accordance with design specifications, companies often employ checklists to document that they have confirmed the required PSSR items. While some types of changes may be sufficiently reviewed with a single-page checklist, more complex changes may warrant multi-disciplinary reviews using more specific checklists such as for safety and health, environmental, construction, mechanical integrity, electrical, and automation items, among others.

The nature of the PSSR information needing to be confirmed generally dictates an office-based review of documentation as well as a field-based review of the physical installation or modification. Therefore, the conduct of PSSRs may involve different persons, locations, and times. Checklist items that cannot be confirmed or that generate follow-up actions may also be documented on the PSSR form and input to an action tracking system, as applicable. For execution of the PSSR process safety element, at least four processes that involve information data are necessary:

- Process for coordinating and scheduling PSSRs
- Process for recording and storage of PSSR confirmation items
- Process for recording discrepancies/recommendations and tracking to completion
- Process for giving approval for startup and communicating to the appropriate persons

The execution of these processes has been done historically through use of PSSR checklists that exist as paper documents. Tracking of discrepancies and recommendations may be done on the same checklist or associated form with manual follow-up and communication as to documentation of when the items have been resolved. The same checklist may also include sign-off from an appropriate approving authority, signifying that the facility or equipment is ready for startup. When completed, common practice is to file the PSSR documentation with the associated project file or retain it in an MOC file if it is associated with a process modification. As such, PSSRs are considered static documents since they are generally "filed and forgotten," only to serve as proof that the PSSR requirements have been met. There is seldom further need of the information contained in a PSSR form apart from compliance audit evidence or in the event of an incident, in which case the investigation might probe for relevant PSSR evidence.

Management of paper based PSSRs can be challenging for reasons such as:

- potential for omissions and incorrect/inconsistent responses to checklist questions
- potential for misplaced checklists, particularly for facilities that are geographically distant from office locations
- geographic separation between persons needing to review, approve, and sign-off on PSSR actions
- coordinating various disciplines and collection of evidential information
- need for scanning paper copies into electronic files (e.g., PDFs)

All of the aforementioned PSSR processes can be implemented using digital technology and, in doing so, can offer significant benefits over paper-based systems. An obvious starting point for improving paper-based PSSR systems has been to convert hard-copy PSSR checklists into a digital format, typically PDF. This allows for retention in an electronic file repository that can be more broadly and easily accessible, should it be needed, and is generally more secure against loss such as could occur in the event of fire or natural catastrophe. However, converting PSSRs to digital format by itself does not appreciably enhance a PSSR system and it can add to the clerical burden slightly due to the added step of scanning paper documentation.

A more efficient system would utilize an electronic system in which all of the PSSR processes are digitized and interconnected. Such a system is illustrated in Fig. 11 in which a digital PSSR application was developed for deployment on mobile "smart" devices.

In this example application, PSSR processes are integrated into a common system that includes calendar, email, checklists, action tracking, and approval functions. Key benefits provided by this PSSR system include:

- Simple scheduling, notification, and means of confirming acceptance of PSSR meetings
- Required data fields to ensure all checklist questions are answered
- Assignment and completion of PSSR-related action items
- Remote/offline review of PSSRs, actions, and approvals
- Key-word search capabilities
- Linkage to MOC system for simple upload and data storage
- User app tutorial and help functionality

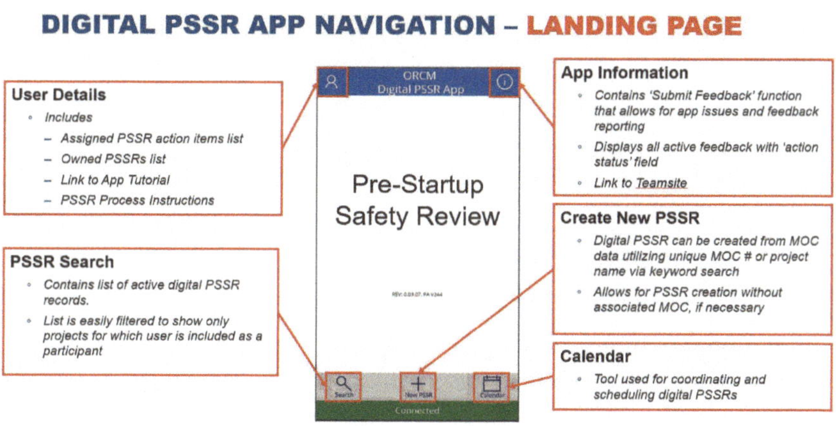

Fig. 11 Digital PSSR mobile application.

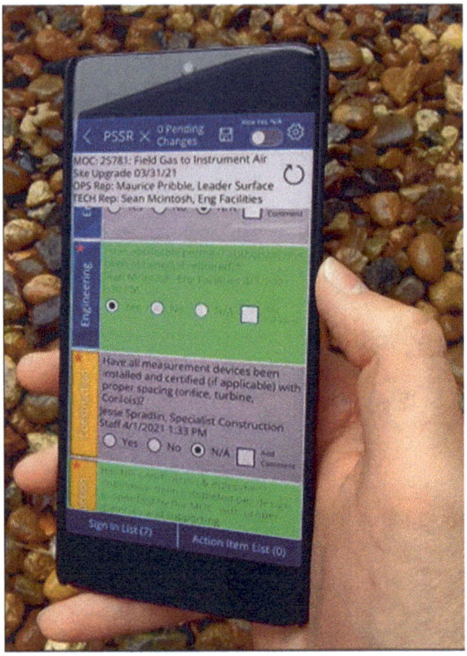

Fig. 12 Mobile phone PSSR application.

As shown in Fig. 12, this PSSR app has been deployed on a mobile phone, though it can also be utilized on tablets and other digital devices. Although this electronic means of conducting PSSRs offers key benefits over use of paper forms, this device in not intrinsically safe, meaning that it would not be permissible for use in certain hazardous class locations where flammable materials may be present. However, this limitation may be overcome with a built for purpose intrinsically safe mobile phone case. This application provides a linkage between the MOC and PSSR elements. Future advances may extend the integration of this app with the process safety information and operating procedures elements for additional capability.

5. Future improvements in process safety through digital technology: Limited literature survey

A small set of very recent papers has been scanned that relate to safety management and how this can be improved by digitalization as viewed by other authors. We shall start, though, with a recent paper of Behie, Halim, Efaw, O'Connor, and Quddus (2020) laying out a guideline for an effective process safety management system specifically for the oil and gas and chemical industry. The paper starts with considering the trends in process safety, which although not quite clear are not encouraging optimism either. To improve, several factors at play were analyzed: (1) Critical technical decisions taken at inappropriate management level, (2) Dynamic workforce changes, (3) Lack of process safety training/knowledge at all levels, (4). Reassessing process safety programs specific to aging plants, (5) Ineffective emergency response plans, and (6) External communication with the public to avid/mitigate mistrust with respect to the industry. Improvement may be made by: (1) strengthening the risk assessment function, (2) training for the entire workforce including the top, (3) vulnerability assessment and identification of best practices, (4) management of process safety of aging plants, (5) barrier (controls) health assessment, and (6) development of robust emergency response capability.

In the digital setting, the first review is a paper by Agnusdei, Elia, and Gnoni (2021a) on the support of safety management by the digital twin technology. After an introduction the paper presents a brief history of digital twin technology. Next, it reports a Scopus query using the search words "digital AND twin AND safety" over the period 2003–2021. This query yielded 25 papers in 2003 and 1279 in 2020; indicating exponential growth. The conclusion is that this technology will have implications for safety management, but implementation in industry is currently weak. In a second article, Agnusdei, Elia, and Gnoni (2021b) reviewed the literature with respect to applications of digital twin to strengthen risk assessment. In real life plants, the physical world and digital twin are communicating with each other through data, so that the technology enables mimicking the dynamics of real life by producing various kinds of data, depending on conditions. Processing the data may produce predictive information on human-machine interaction and on human and equipment-based risks which, in the end, will support safety management.

The work of Bevilacqua et al. (2019) referred to by Agnusdei et al. (2021a) describes the use of digital twin in process plants and defines conceptual guidelines for digital twin to predict and prevent risk in the oil & gas industry. Digital twin functioning and communicating with the physical space needs IoT, simulation technology, machine learning techniques, augmented and virtual reality, and cloud technology. The whole digital twin system also needs a user space, with user being human, device or system in which operational instructions (maintenance) and warnings and can be activated through devices, e.g., wearable systems. In a later paper, Costantino et al. (2021) stress that the new technology also introduces new and emerging hazards. The change in working environment, tools, and conditions and the need for new competencies will affect workers. A study was performed on what this could be and eight technologies were identified: (1) AM (additive manufacturing producing layered materials), (2) AGV (automatic guided vehicles with collision hazard), (3) the use of augmented reality and virtual reality (AV/VR) can also be a source of various hazards and discomfort, (4) digital twin itself did not generate hazards, (5) but the wrong use of exoskeletons do, (6) the use of robots and cobots (collaborative robots) can lead to hazardous situations, (7) smart wearables may cause allergies and discomfort, while continuous worker localizing may induce the fear of breaching privacy, and finally, (8) wireless communication technologies may cause heating hazards and interference between devices; long-term exposure to strong electro-magnetic radiation may lead to tissue and genetic damage. All this is based on a survey of about 50 literature references.

Zhao et al. (2021) worked out the support of IoT (Internet of Things) and a digital twin for maintaining operator safety in the case of a large warehouse storage system of fresh and refrigerated products. Human and machine tracking, and product positioning enabled monitoring the system. Man, machine, and materials (physical assets) are represented as real-time cyber entities on a location map, while their status will appear in color so that stakeholders (operator, supervisor, and manager) can continuously be informed. Data are stored (cloud) so that in case of an abnormal situation such as an operator being motionless due to accident, the person's position can be easily retrieved. An algorithm was developed to avoid false alarming of these situations.

Lee, Cameron, and Hassall (2019, 2022) first considered what roles digitalization and Industry 4.0 can have in improving process safety and, in their follow-on paper, they focus on information needs and challenges. In their first paper, they stress the importance of interactions of people, procedures, and plant. They describe how the ISO 15926 standardized language in ontological format facilitates the CAD design process as well as the follow-on building, start-up, and even operational stages of a plant. The ontology, rather than a data base, enables reasoning engines which will be helpful in hazard identification based on digital P&IDs. This, in combination with flowsheets, make use of CAPE (computer aided process engineering) software to develop a digital twin which can be used during the plant life cycle. Not only can such twins be used in operator training, it can also be most useful in the HAZOP process as it can mimic the dynamics of an operational process and thus, enables dynamic operational risk assessment which may even be done in real time. It will help to reduce incidents by effective risk-based inspection and reliability centered maintenance. An example of a PHA (process hazard analysis) is presented. Their 2022 paper considers DIKWA (data-information-knowledge-wisdom-action) cycle in relation to the Plant, People and Procedures (P3 model). To that end, the paper summarizes the sequence of data basics with its various aspects, deep learning using artificial neural networks, machine learning in its supervised and unsupervised modes and various techniques, artificial intelligence (sense, comprehend, decide/act, and learn), culminating in artificial wisdom. With respect to application for business gain and safety improvement, it is important to consider validation of data and information, leveraging affordances, and the effective management of risks, which are all further detailed. A literature search was conducted on the search terms "process safety" AND "big data" over the period 2016–2021 which, after reviewing, resulted in 16 relevant articles on big data and process safety. It showed that the structured approach of the framework concepts in DIKWA and P3 were still missing, a gap that needs to be filled. An example is given of high temperature hydrogen attack.

Mokhtarname et al. (2020) propose HAZOP 4.0 to deal with complexities such as "multiple failure, domino effects, time and amplitude dependent deviations etc." This is achieved by dynamic simulation with the aid of artificial intelligence algorithms and multi-variable process monitoring. This approach could enable a platform for sound elements of the process safety management system, such as "operating manuals, training and competency management, condition monitoring and predictive maintenance, management of change and pre-start-up review." The difference between conventional HAZOP and HAZOP 4.0 is shown for a batch styrene polymerization process.

The digitalization process will also affect human resource (HR) management. Fenech, Baguant, and Ivanov (2019) mentions a few brief literatures references how the human resource information system (HRIS) has brought change. A survey has been conducted that involved five HR managers who all stated that digitalization changed their world and even "revolutionized HR practices" such as training and development, health and safety, and culture.

Gnoni, Bragatto, Milazzo, and Setola (2020) describes how the Internet of Things and Bluetooth wireless technology enables the integration of signals from a variety of sensors, electronic wearables, RFIDs (radio frequency identification,) and PPE (personal protective equipment). This could be a means to improve safety management. Various concepts are mentioned, such as a digital platform to provide common data protocols, smart label applications for maintenance activities that can provide all mandatory information to stakeholders (operators, manager, inspection), a virtual aging application that uses a viewer to inform the supervisor or inspector on the state of equipment when walking through the plant, and worker location monitored via the safe work application. An illustrated example is described.

Sub-conclusions and reference to other chapters of *Methods in Chemical Process Safety*, this volume: According to a variety of authors, digitalization will support improvements in process safety by means of many different mechanisms. It will require, though, investment in expertise to establish and operate the technologically advanced tools. With these advancements, there are also downsides with respect to occupational safety that will need to be considered. Also not mentioned are the hazards due to cybersecurity breaches. These will be treated in chapter "Security of digitalized process systems" by El-Kady et al., while the interactions and interchanges of safety versus security is briefly described in a section of chapter "Integrated dynamic risk management in process plants" by Taleb-Berrouane and Pasman. More details on digital twins will be provided in chapter "Use of digital twins for process safety management" by Keprate and Bagalkot.

Acknowledgment

We gratefully acknowledge the contributions of Jacob Behling and Kolbee Hesseltine for information they provided on digital radiography, of Mohammed Al Kharusi for his contribution to the section on internal tank inspections using aerial drones, and of Benjamin Heasty for his contribution relating to a digital PSSR application.

References

Agnusdei, G. P., Elia, V., & Gnoni, M. G. (2021a). Is digital twin technology supporting safety management? A bibliometric and systematic review. *Applied Sciences*, *11*, 2767. https://doi.org/10.3390/app11062767.

Agnusdei, G. P., Elia, V., & Gnoni, M. G. (2021b). Review, a classification proposal of digital twin applications in the safety domain. *Computers & Industrial Engineering*, *154*(2021), 107137. https://doi.org/10.1016/j.cie.2021.107137.

Behie, S. W., Halim, S. Z., Efaw, B., O'Connor, T. M., & Quddus, N. (2020). Guidance to improve the effectiveness of process safety management systems in operating facilities. *Journal of Loss Prevention in the Process Industries*, *68*, 104257. https://doi.org/10.1016/j.jlp.2020.104257.

Bevilacqua, M., Bottani, E., Ciarapica, F. E., Costantino, F., Di Donato, L., Ferraro, A., et al. (2019). Digital twin reference model development to prevent operators' risk in process plants. *Sustainability*, *12*, 1088. https://doi.org/10.3390/su12031088.

Costantino, F., Falegnami, A., Fedele, L., Bernabei, M., Stabile, S., & Bentivenga, R. (2021). New and emerging hazards for health and safety within digitalized manufacturing systems. *Sustainability*, *2021*(13), 10948. https://doi.org/10.3390/su131910948.

Fenech, R., Baguant, P., & Ivanov, D. (2019). The changing role of human resource Management in an era of digital transformation. *International Journal of Entrepreneurship*, *22*(2), 166–175.

Gnoni, M. G., Bragatto, P. A., Milazzo, M. F., & Setola, R. (2020). Integrating IoT technologies for an "intelligent" safety management in the process industry. *Procedia Manufacturing*, *42*, 511–515. https://doi.org/10.1016/j.promfg.2020.02.040.

Lee, J., Cameron, I., & Hassall, M. (2019). Improving process safety: What roles for digitalization and industry 4.0? *Process Safety and Environmental Protection*, *132*, 325–339. https://doi.org/10.1016/j.psep.2019.10.021.

Lee, J., Cameron, I., & Hassall, M. (2022). Information needs and challenges in future process safety. *Digital Chemical Engineering*. https://doi.org/10.1016/j.dche.2022.100017.

Mokhtarname, R., Safavi, A. A., Urbas, L., Salimi, F., Zerafat, M. M., & Harasi, N. (2020). Toward HAZOP 4.0 approach for managing the complexities of the Hazard and operability of an industrial polymerization reactor. *IFAC PapersOnLine*, *53-2*, 13593–13600. http://creativecommons.org/licenses/by-nc-nd/4.0.

Zhao, Z., Shen, L., Yang, C., Wub, W., Zhang, M., & Huang, G. Q. (2021). IoT and digital twin enabled smart tracking for safety management. *Computers and Operations Research*, *128*, 105183. https://doi.org/10.1016/j.cor.2020.105183.

CHAPTER SIX

Statistical approaches and artificial neural networks for process monitoring

Md Alauddin[a], Rajeevan Arunthavanathan[a], Md Tanjin Amin[a], and Faisal Khan[b,*]

[a]Centre for Risk, Integrity and Safety Engineering (C-RISE), Faculty of Engineering & Applied Science, Memorial University of Newfoundland, St. John's, NL, Canada
[b]Mary Kay O'Connor Process Safety Center, Artie McFerrin Department of Chemical Engineering, Texas A&M University, College Station, TX, United States
*Corresponding author: e-mail address: fikhan@tamu.edu

Contents

1. Introduction

1.1 Background

Efficient process monitoring for detection and diagnosis of abnormalities at an early stage is of utmost importance in ensuring operational safety in process systems. Early detection of abnormal conditions gives a lead time to

Methods in Chemical Process Safety, Volume 6
ISSN 2468-6514
https://doi.org/10.1016/bs.mcps.2022.04.003

respond effectively to prevent or mitigate the damage due to the abnormal conditions. Thus, robust techniques of process monitoring, fault detection, identification, and diagnosis have become vital tools for the safety management of process systems.

The process monitoring tools can broadly be divided into two categories: model-based and data-based. Model-based or mechanistic, or first principle methods are based on rigorous mathematical formulations for representing conservation laws, domain knowledge, physical principles, and phenomenological behaviors (Liu, McDermid, & Chen, 2010; Venkatasubramanian, Rengaswamy, Yin, & Kavuri, 2003). These methods have extensively been used in process systems owing to their robustness and reliability. However, they are not easily implementable for early fault detection of complex processes since it becomes cumbersome to build a mechanistic model for complex process systems.

On the other hand, data-driven methods are based on process measurements, which do not require a priori quantitative or qualitative process knowledge (Isermann, 2005; Staroswiecki, 2000). The data-based FDD methods can further be categorized as statistical and machine learning methods. The statistical models exploit statistical parameters such as mean, median, mode, standard deviation, range, distribution, kurtosis, and skewness (Wen, Khan, Amin, & Halim, 2021). The machine learning-based methods, on the other hand, exploit the large volume of data to learn the hidden patterns representing the systems. Alauddin, Khan, Imtiaz, and Ahmed (2018) classified the evolution and development of research on FDD in four categories: formulation of basic data-driven algorithms, advancement of the algorithms, applications of data-based algorithms in process systems, and development and application of advanced hybrid techniques in process systems.

Experts predict that the 21st century will be known for the development of data-based systems that are revolutionizing the process design and monitoring of industrial systems. Many oil and gas companies are looking forward to adopting artificial intelligence-based models for their operations. Some of the recent studies on the survey of data-driven FDD includes: (Alauddin et al., 2018; Arunthavanathan, Khan, Ahmed, & Imtiaz, 2021; Dai & Gao, 2013; Gao, Cecati, & Ding, 2015; Ge, 2017; Qin, 2012; Tidriri, Chatti, Verron, & Tiplica, 2016; Yin, Ding, Haghani, Hao, & Zhang, 2012; Yin, Ding, Xie, & Luo, 2014). The following section describes commonly used data-based process monitoring models.

1.2 Common data-driven models

1.2.1 Principal component analysis and partial least squares-based FDD methods

Principal component analysis (PCA) and partial least squares (PLS) are the two leading multivariate fault detection and diagnostic approaches in process systems (Alauddin et al., 2018). These methods are based on data projection to a lower-dimensional space for achieving fault detection and diagnosis (Kourti & MacGregor, 1995). The PCA and PLS-based methods can handle a large number of highly correlated data, measurement errors, and missing data.

The dimensionality reduction feature of these algorithms makes them ideal candidates in combining with other algorithms for handling big data. The major drawback of PCA and PLS based methods is that they are not optimal for non-Gaussian and nonlinear data (Lawrence, 2005). Besides, Pearson's dependence on the covariance matrix makes PCA sensitive to outliers (Yu, Khan, & Garaniya, 2016). PLS does not perform well when input space has significant variation orthogonal to the output space.

Hence, numerous advanced data-based FDD algorithms have been developed over the last four decades such as, kernel principal component analysis (KPCA), multiway PCA, multi-block PCA, hierarchical PCA, recursive PCA, sparse PCA, and moving window KPCA. Similarly, some of the variants of PLS are as follows: orthogonal projections to latent structures (O-PLS), which removes variation from descriptor variables that are not correlated to property variables, total projection to latent structure (TPLS), which makes further decomposition on particular subspaces, multiway partial least squares (MPLS), dynamic PLS (DPLS), and recursive PLS (RPLS). Kaspar and Harmon Ray (1993) proposed DPLS for capturing the time-varying nature of process data. Qin (1998) formulated RPLS algorithms for online process modeling to adapt process changes and offline modeling to deal with huge data samples.

1.2.2 Independent component analysis-based FDD methods

Independent component analysis (ICA) was developed to handle non-Gaussian data that are frequently encountered in process systems. It is based on higher-order statistics (HOS), such as kurtosis and negentropy. Unlike PCA, ICA vectors are not orthogonal, and each component is equally important. Some of the key ICA algorithms include FastICA, simultaneous third-order tensor diagonalization (STOTD), joint approximate diagonalization of Eigen matrices (JADE), higher-order eigenvalue decomposition (HOEVD), maximal diagonality (MD), and maximum likelihood (ML) (De Lathauwer, De Moor, & Vandewalle, 2000).

The limitation of the conventional ICA is that the independent components cannot be ranked to the amount of variance explained by them (Chawla, 2011). Also, ICA is limited to linear systems and uses kernel-based mapping to deal with nonlinear systems that become computationally expensive with large samples. Moreover, fault identification and diagnosis are difficult due to the irreversible nature of kernel mapping (Deng, Tian, & Chen, 2013).

Many researchers came up with modifications in the existing algorithms to deal with these limitations. For instance, Zhang and Zhang (2010) introduced modified ICA based on particle swarm optimization (PSO). Stefatos and Ben Hamza (2010) used dynamic independent component analysis (DICA) to capture dynamic patterns. Zhang and Qin (2007) proposed a multiway kernel ICA (MKICA) method for dealing with nonlinearities in industrial systems. Rashid and Yu (2012) introduced the hidden Markov's model in ICA for handling multimodal problems of industrial processes. Cai and Tian (2014) and Cai, Tian, and Chen (2014) proposed robust ICA for handling outliers and noisy data. Tong, Palazoglu, and Yan (2014) suggested improved ICA based on ensemble learning and Bayesian inference. Du, Fan, Zhang, and Zhang (2017) introduced fault-related kernel independent component analysis (FKICA), which decomposes data into four subspaces and makes the algorithm more sensitive to specific faults.

1.2.3 Copula-based FDD methods

Vine copula-based methods are other potential tools for handling non-Gaussian data. Unlike PCA and PLS, which are based on dimensionality reduction, the vine copula-based FDD tools exploit features correlation to estimate copula density and joint probability distribution for reckoning the fault probability. Ren, Tian, and Li (2015) introduced C-vine copula in process monitoring. One of the C-vine-based method's intrinsic limitations is its inability to maximize the correlation structure. Cui and Li (2020) used a combination of C-vine and D-vine copula to capture both the stronger and weaker correlations by harnessing the maximum information coefficient (MIC) (Reshef, Reshef, Sabeti, & Mitzenmacher, 2014). Ren, Zhu, Cai, and Li (2017) proposed a copula subspace division-based fault diagnosis technique.

1.2.4 Gaussian mixture models of FDD

Mixture models are the other approach explored to deal with non-Gaussian data. The Gaussian mixture models (GMMs) are statistical methods based on the weighted sum of probability density functions of multiple Gaussian distributions. The number of Gaussian components is approximated by the expectation-maximization (EM) algorithm (Yu, 2013a, 2013b; Yu & Qin, 2009). The E-step determines the posterior probability, given all the parameters and the prior probabilities, while the M-step finds all the parameters. The maximum likelihood solution is obtained by repeating the E- and M-steps iteratively.

Thissen, Swierenga, De Weijer, Wehrens, et al. (2005) came up with GMM based on the Bayesian information criterion (BIC). Li, Qin, and Yuan (2016) devised a framework of root cause analysis for locating faults in stationary and dynamic processes. The authors proposed a dynamic time warping-based causality analysis for digging the causality relations of the candidates. Yu (2012a) devised a nonlinear kernel Gaussian mixture model (NKGMM) for handling nonlinearities. Yu and Qin (2008) proposed a finite Gaussian mixture model (FGMM) based on the Figueiredo-Jain (F-J) algorithm for optimizing Gaussian components and estimating statistical distributions. This formalism resulted in higher detection rates, lower false alarm rates, and shorter delays in fault detection.

Yu, Khan, and Garaniya (2015a) proposed a probabilistic multivariate method for fault diagnosis of industrial processes. The study employed a Gaussian copula based on rank correlation to model dependencies and nonlinearity of process variables. The technique is useful in handling nonlinearities; however, it requires a longer computational time for convergence. Similarly, Yu, Khan, and Garaniya (2015b) devised a three-layered nonlinear Gaussian belief network (NLGBN) for fault diagnosis in industrial systems by introducing G-index, a novel parameter for process monitoring based on expectation maximization.

1.2.5 Bayesian network and hidden Markov models of FDD

Bayesian network (BN) and hidden Markov models (HMM) are tools for handling uncertainty using probabilistic reasoning. The HMM is an efficient tool to estimate probability distributions of state transitions and unobservable states of a continuous process in a datastream. It can be applied in diagnostics, prognostics, and fault detection. Smyth (1994) employed the HMM for continuous monitoring of complex dynamic systems. Yu (2012b) proposed the multiway discrete hidden Markov model (MDHMM) for fault detection and classification for handling uncertainty.

Bayesian networks (BNs) are a commonly used technique for fault diagnosis. A BN can also be used to model hierarchical levels of multiple causes and effects. A dynamic Bayesian network (DBN) for process monitoring was implemented by Yu and Rashid (2013). Zhang and Dong (2014) proposed a multiple time-slice DBN-based FDD for handling dynamic nature and missing data. Amin, Khan, Imtiaz, and Ahmed (2019) proposed a dynamic Bayesian anomaly index (DBAI)-based thresholding technique for fault detection and diagnosis.

1.2.6 Machine learning-based models for FDD

Machine learning (ML) refers to techniques aiming to program computers to learn from experience. ML methods are divided into supervised, unsupervised, and semi-supervised learning methods (Lei, Yang, Jiang, Jia, & Nandi, 2020). Supervised learning is a process of adjusting the model parameters based on supervisory labels. The historical process data containing the anomalies is used to train fault detection systems. Many machine learning techniques have successfully been tested for fault detection and diagnostic methods of process systems. The artificial neural network (ANN) (Agatonovic-Kustrin & Beresford, 2000; Schmidhuber, 2015), support vector machines (SVM) (Burges, 1998; Cortes & Vapnik, 1995; Widodo & Yang, 2007), case-based reasoning (CBR) (Zhao, Liu, Dong, Sun, & Ji, 2017), classification and regression tree (CART) (Lawrence & Wright, 2001) are some of the prominent methods in this category (see Table 1).

The applications of ML algorithms in chemical health and safety studies can date back to the mid-1990s. However, the rapid advancement of computational resources and sophisticated algorithms accelerated the neural network domain's research over the last decade. ANN has been inspired by biological neural systems. A standard ANN consists of many processors, also called neurons, which generate a sequence of real-valued activations (Agatonovic-Kustrin & Beresford, 2000; Schmidhuber, 2015). A neural network is a universal approximator and can approximate any function given sufficient structural complexity (Hornik, Stinchcombe, & White, 1989). ANN consists of a network of input, hidden and output layers.

Depending on the configuration of neural structure, several deep learning-based formulations, such as multilayer perceptron (MLP), deep autoencoder (DAE) networks, and convolutional neural networks (CNNs), have been

Table 1 Summary of statistical and machine learning methods of FDD and process monitoring.

Methods/ algorithms	Principles	Advantages	Limitations	Variants/derivatives	Studies
Principal component analysis and partial least squares	Data projection to a lower-dimensional space for achieving effective process monitoring and fault detection and diagnosis indices	• Can handle large numbers of highly correlated data, measurement errors, and missing data • Leads to simple monitoring charts, consistent with the philosophy of statistical process control • Ideal candidates in combining with other algorithms for handling big data	• Not optimal for non-Gaussian and nonlinear data	Kernel principal component analysis, multiway PCA, multi-block PCA, hierarchical PCA, recursive PCA, sparse PCA, moving window KPCA, orthogonal projections to latent structures (O-PLS), total projection to latent structure (TPLS), multiway PLS, dynamic PLS, and recursive PLS	Gajjar, Kulahci, and Palazoglu (2018), Gharahbagheri, Imtiaz, and Khan (2017), Kaspar and Harmon Ray (1993), Kourti and MacGregor (1995), Lawrence (2005), Nomikos and MacGregor (1994, 1995), Portnoy, Melendez, Pinzon, and Sanjuan (2016), Qin (1998), Rato and Reis (2013), Trygg and Wold (2002), Wise, Veltkamp, and Davis (1988), Yu et al. (2016), Zass and Shashua (2007)
Independent component analysis	Based on higher-order statistics (HOS), such as kurtosis and negative entropy	• Can handle non-Gaussian data that are frequently encountered in process systems • Provide latent variables that are independent to each other • Able to extract high-order data information	• Independent components cannot be ranked to the amount of variance explained by them • ICA is limited to linear systems and uses kernel-based mapping to deal with nonlinear systems that become computationally expensive with large samples • The monitoring performance of the ICA-based method is unstable, difficult to determine the control limit and selecting the number of independent components	*FastICA, simultaneous third-order tensor diagonalization (STOTD), joint approximate diagonalization of Eigen matrices (JADE), higher-order eigenvalue decomposition (HOEVD), maximal diagonality (MD), and maximum likelihood (ML), dynamic independent component analysis, multiway kernel ICA, and ICA based on ensemble learning and Bayesian inference*	Cai et al. (2014), Du et al. (2017), Hyvärinen and Oja (2000), Kano, Tanaka, Hasebe, Hashimoto, and Ohno (2003), Lee, Yoo, and Lee (2004), Stefatos & Ben Hamza (2010)m Tong et al. (2014), Zhang, An, and Zhang (2013), Zhang, Li, and Teng (2012), Zhang and Qin (2007), Zhang and Zhang (2010)

Continued

Table 1 Summary of statistical and machine learning methods of FDD and process monitoring.—cont'd

Methods/ algorithms	Principles	Advantages	Limitations	Variants/derivatives	Studies
Copula-based methods	Exploit features correlation to estimate copula density and joint probability distribution	• Can handle non-Gaussian data	• Inability to maximize the correlation structure	*C-vine copula, D-vine copula, and R-vine copula*	Amin, Khan, Ahmed, and Imtiaz, (2021a, 2021b), Cui and Li (2020), Ren et al. (2017), Wan and Li (2019), Zhou and Li (2018)
Gaussian mixture model	Based on the weighted sum of probability density functions of multiple Gaussian distributions. The number of Gaussian components is approximated by the expectation-maximization (EM) algorithm	• Can handle non-Gaussian data	• Slow convergence	Nonlinear kernel Gaussian mixture model, finite Gaussian mixture model, nonlinear Gaussian belief network, parsimonious Gaussian mixture models, and adaptive Gaussian mixture models	Cai, Tian, and Chen (2017), Cai and Tian (2014), Du et al. (2017), Nakamura, Palhares, Caminhas, Menezes, et al. (2017), Ren, Song, Yang, and Ren (2015), Thissen et al. (2005), Xie and Shi (2012), Yu et al. (2015a, 2015b), Yu (2012a, 2012b, Yu and Qin (2009)
Bayesian network	Handling uncertainty by using probabilistic reasoning	• A white box model as the graphical representation makes it intuitively easy for the user to understand the interaction between the model variables • Robust prediction based on the prior knowledge • Can also be used to model hierarchical levels of multiple causes and effects	• Challenging task of construction of the structure of all interacting variables and parameters • Misleading priors can degrade the performance	Static Bayesian network dynamic Bayesian network hierarchical Bayesian network, discrete-time Bayesian network, continuous-time Bayesian network, conditional Gaussian network, and dynamic Bayesian network with fuzzy sets theory	Amin, Khan, Ahmed, and Imtiaz (2020), Chen and Ge (2020), Gharahbagheri et al. (2017), Jiang and Yan (2015), Mehranbod, Soroush, and Panjapornpon (2005), Rojas-Guzman and Kramer (1993), Xu and Deng (2016), Yu et al. (2015b), Yu (2013b), Yu and Rashid (2013), Zhang, Wu, Zheng, and Fan (2018), Zhang and Dong (2014)

Hidden Markov models	An efficient tool to estimate probability distributions of state transitions and unobservable states	• Convenient to build a unified PHM model and increase model efficiency • Facilitates modeling with unobservable states	• Sensitive to parameters initialization • Expensive in terms of memory and compute time	Hidden Markov chain model, hidden semi-Markov model, hidden Markov model-based independent component analysis, machine learning-based hidden Markov model, discrete HMM, hierarchical hidden Markov model, and mixture-of-Gaussian hidden Markov model	de Almeida and Park (2012), Galagedarage Don and Khan (2019a, 2019b), Lee, Cho, and Choi (2021), Lee, Li, and Ni (2010), Lemos, Caminhas, and Gomide (2013), Mondal, Ghalyan, Ray, and Mukhopadhyay (2019), Rebello, Yu, and Ma (2019), Smyth, (1994), Yu (2010)
ANN and Deep Learning	Inspired by biological neural systems. A standard ANN consists of many processors, also called neurons, which generate a sequence of real-valued activations	• Can model nonlinearities and complexities • No need to specify the parameter and network structure like in the case of BN	• Prone to overfitting, i.e., the relationships between the inputs and outputs are generalized to a given set of data • The "black box" methods and they cannot explicitly identify possible causal relationships between input and output variables • Easy to converge at local minimum and leads to unstable results for each run; not able to adapt to process drifts	Feed-forward neural network, recurrent neural networks (RNNs), gated recurrent unit (GRU), Long short-term memory (LSTM) bidirectional recurrent neural networks, vanilla LSTM, convolutional neural networks, concurrent convolution neural network (CeCNN), autoencoder, variational autoencoders, restricted Boltzmann machine (RBM) stacked autoencoders (SAE), deep Boltzmann machine (DBM), and generative adversarial network (GAN)	Agatonovic-Kustrin and Beresford (2000), AL-Qutami, Ibrahim, Ismail, and Ishak (2018), Alauddin, Khan, Imtiaz, and Ahmed (2020), Arunthavanathan, Khan, Ahmed, Imtiaz, and Rusli (2020), Barakat, Druaux, Lefebvre, Khalil, and Mustapha (2011), Cheng, Zhu, Wu, and Shao (2019), Deng et al. (2013), Heo and Lee (2018), Hoskins and Himmelblau (1988), Hu, Li, Xia, and Luo (2018), Kang (2020), Li et al. (2017), Schmidhuber (2015), Su, Tao, Jin, Wang, and Wang (2020), Uraikul, Chan, and Tontiwachwuthikul (2007), Wang, Li, Zhao, and Gao (2020), Wang, Pan, Yuan, Yang, and Gui (2020), Wu and Zhao (2018, 2020), Xie, Yang, Li, and Ji (2019), Yu, Gomm, and Williams (1999), Yu, Qu, Gao, and Tian (2019), Yu and Qi (2021), Zhang, Zhang, Zou, and Fan (2020), Zhang, Bi, and Qiu (2020), Zhang and Zhao (2017), Zhao, Sun, and Jin (2018)
Support vector machines (SVM)	A kernel-based method based on optimal fitting of hyperplanes to achieve the intended function	• Good generalization with limited data • Can handle both the linear and nonlinear process data	• Selection of kernel is tedious • Not able to adapt to process drifts	Support vector data description (SVDD), semi-supervised SVM (S3VM), and safe semi-supervised SVM (S4VM)	Burges (1998), Cortes and Vapnik (1995), Li, Wang, and Yuan (2006), Md Nor, Hussain, and Che Hassan (2017), Monroy, Escudero, and Graells (2009), Onel, Kieslich, Guzman, and Pistikopoulos (2018), Sheibat-Othman, Laouti, Valour, and Othman (2014), Widodo and Yang (2007), Yang and Hou (2016), Yu (2012c), Zhang, Hu, and Zhang (2016), Zhang (2008)

Continued

Table 1 Summary of statistical and machine learning methods of FDD and process monitoring.—cont'd

Methods/ algorithms	Principles	Advantages	Limitations	Variants/derivatives	Studies
Expert system and fuzzy logic	Heuristic Algorithm/rules	• Fuzzy rules are easy to understand • Modularity rules can be added and removed as needed • Efficient control of machines • Fuzzy systems can handle vague, distorted or imprecise data	• Selection of membership functions and basic rules is challenging • Compromised accuracy • Fuzzy-based systems may not scale well to large or complex problems	Fuzzy classifier, fuzzy logic system, fuzzy-genetic algorithm, fuzzy rules, and genetic algorithm	Basila, Stefanek, and Cinar (1990), Gomm (1996), Jokinen (1992), Lau, Ghosh, Hussain, and Che Hassan (2013), Musulin, Yélamos, and Puigjaner (2006), Neumann, Deerberg, and Schlüter (1999), Orantes, Kempowsky, Le Lann, and Aguilar-Martin (2008), Ruiz, Nougués, and Puigjaner (2001), Da Silva, Saxena, Balaban, and Goebel (2012), Szafnicki, Narce, and Bourgois (2005), Theilliol, Weber, Ghetie, and Noura (1995), Tudón-Martínez and Morales-Menendez (2015), Zumoffen and Basualdo, 2008)
Decision tree random forest, and ensemble learning	Based on tree structure with multiple judgment nodes/training multiple base learners as ensemble members for improved performance on intended function(s)	• High accuracy	• Slow convergence • Overfitting	Classification and regression tree (CART), bagging, boosting, XGBoost, LightGBM, and adaptive LightGBM	Auret and Aldrich (2010), Chen and Ge (2019), Dorgo, Palazoglu, and Abonyi (2021), Gordon and Pistikopoulos (2021), Liu and Ge (2019), Marins, Barros, Santos, Barrionuevo, et al. (2021), Svetnik, Wang, Tong, Liaw, and Song (2005), Zhao et al. (2017)
Unsupervised/ clustering	Aims to divide n observations into clusters, in which the observation belongs to a particular cluster with stated criteria	• Simple to implement • Do not require many labeled data	• Sensitive to the choice of the initial cluster centers	*K-means algorithm, fuzzy C-means (FCM), hierarchical clustering method, and Gaussian mixture model*	Auret and Aldrich (2010), Bahrampour, Moshiri, and Salahshoor (2011), Bhushan and Romagnoli (2008), Detroja, Gudi, and Patwardhan (2006), Elshenawy, Mahmoud, and Chakour (2020), Escobar, Kaneko, and Funatsu (2016), Eslamloueyan (2011), He, Qin, and Wang (2005), Khediri, Weihs, and Limam (2012), Lemos et al. (2013), Monroy, Benitez, Escudero, and Graells (2010), Petković, Rapaić, Jeličić, and Pisano (2012), Yu (2012c), Ziaei-Halimejani, Nazemzadeh, Zarghami, Gernaey, et al. (2021)

explored for improved accuracy and precision. Recurrent neural networks (RNNs) and their variants have been developed to deal with complex sequential problems of time series data. Long short-term memory (LSTM) has been employed for fault detection and diagnosis for several processing applications (Li et al., 2017; Zhao et al., 2018). Bidirectional recurrent neural networks (Zhang, Zhang, et al., 2020) and convolutional neural networks (Wu & Zhao, 2018) have also been successfully exploited for the fault diagnosis of chemical processes. A comprehensive review of the application of machine learning and deep learning methods have in chemical safety is presented in (Arunthavanathan et al., 2021; Jiao, Hu, Xu, & Wang, 2020).

Support vector machine (SVM) is a kernel-based ML method used in regression and classification tasks. The main idea is to find two parallel hyperplanes to separate two sets of data in a multi-dimensional space and maximize the margin width between the hyperplanes. SVM is an excellent technique in modeling both linear and nonlinear relationships. The availability of large training datasets is a challenge in machine learning; however, SVM tends to generalize well even with a limited amount of training data.

Many ensemble learning techniques have been devised for handling complex systems of chemical processing facilities. Ensemble machine learning formalism aims at developing effective and reliable models with higher accuracy than single machine learning. Conventional ensemble learning techniques include bagging, boosting, and the random subspace method.

One of the major drawbacks of supervised learning is that labeled data are required to train the model. However, generating data with such information is a challenge in the process industries. Consequently, in recent years, the unsupervised learning approach has become popular in machine learning FDD. Some widely used unsupervised techniques include the K-means algorithm, fuzzy C-means (FCM), hierarchical clustering method, and Gaussian mixture model.

1.2.7 Hybrid data-driven FDD models

Many hybrid methods have been formulated for efficient process monitoring and FDD by exploiting the diagnosis capability of Bayesian. Alauddin et al. (2018) illustrated how the Bayesian network has evolved as the choice of recombinant and hybrid algorithms to detect and diagnose process faults. Mallick and Imtiaz (2013) combined PCA with BN for improving diagnostic performance. Gharahbagheri et al. (2017) proposed a data-driven BN learning algorithm using Granger causality and transfer entropy. Amin, Imtiaz, and Khan (2018) presented a hybrid approach integrating the

PCA and T^2 statistics with a Bayesian network (BN) model. A combination of the hidden Markov model (HMM) and a BN was put forward by Galagedarage Don and Khan (2019a) and Galagedarage Don and Khan (2019b) to detect and diagnose process abnormalities. Rebello et al. (2019) proposed an integrated approach based on the hidden Markov model and Bayesian network to estimate the safety threshold for complex industrial processes.

Numerous hybrid algorithms have been asserted to exhilarate the training of artificial neural networks. Barakat et al. (2011) developed a self-adaptive FDD based on artificial neural networks combined with an advanced signal processing method. Nawi, Khan, and Rehman (2013) combined Cuckoo Search (CS) and Levenberg Marquardt algorithm to train neural networks. Dehuri and Cho (2009) proposed a multi-objective Pareto-based particle swarm optimization (PSO) method for optimizing the architectural complexity and classification accuracy of a neural network. Subudhi and Jena (2011a) employed opposition-based differential evolution (ODE) for training a feed-forward neural network.

Juang (2004) trained a recurrent neural network with a hybrid GA-PSO algorithm using a PSO with a master-slave configuration. Subudhi and Jena (2011b) proposed a memetic algorithm-based differential evolution backpropagation (DEBP) for training a multilayer perceptron by exploiting local and global search spaces. Alauddin et al. (2020) presented an ANN-based fault detection model that was tuned using a variable mosquito flying optimization (V-MFO) technique for maximizing fault detection rate (FDR) and minimizing false alarm rate (FAR). Van Lith, Betlem, and Roffel (2003) augmented a simple physical framework with fuzzy logic. Lee and Kang (2007) formulated a modified backpropagation neural network (BPN) based meta-model that ensures constraint feasibility of the approximate optimal solution.

Data-driven, also known as empirical models, are generic and can adapt to meet the complex systems' requirements of the modern era. Data-based models have the potential to significantly improve the efficiency, safety, and sustainability of process systems. The data-based models, especially the ANN, can approximate complex functions. Machine learning models can be very good at fitting observations; yet, predictions may be contradictory with the laws of physics or even physically nonsensical (e.g., negative density estimations) (Svendsen, Piles, Munoz-Mari, Luengo, et al., 2021). Data-based models assume that the data is of sufficient granularity and quantity.

Although today's smart sensors and sophisticated technologies have paved the way for the availability of tremendous data for cyber-physical industrial systems, selecting legitimate data for training in process systems is still a daunting task. The model will fail at making a reliable assessment if the training data is only representative of a subsection of the processing system (Alvarez, Luengo, & Lawrence, 2009). They also suffer from the lack of generalization ability and struggle when data is scarce or in extrapolation regimes (Fisher, Watson, Escrig, Witt, et al., 2020; Pawar, San, Aksoylu, Rasheed, & Kvamsdal, 2021).

Moreover, the "black box" nature of some data-driven models, especially deep neural networks, is a barrier to large-scale adoption in industrial systems. The data-based models can be improved by conforming mechanistic science, expert knowledge, and meta-learning insights. The inclusion of the first principle models, which are robust and have a higher degree of generalization, can compensate for the lack of high-quality data (Fisher et al., 2020).

Many efforts are made to bridge the data-based and mechanistic models. Hybrid models combine the flexibility and scalability of machine learning while respecting the physics of the underlying systems. Muralidhar, Islam, and Marwah (2018) illustrated how incorporating domain constraints directly into the loss function can improve the learning of parameters. The hybrid models can result in improved prediction accuracy, better calibration properties, enhanced extrapolation properties, and better interpretability (Chaffart & Ricardez-Sandoval, 2018; Psichogios & Ungar, 1992; Reichstein, Camps-Valls, Stevens, Jung, et al., 2019; Svendsen et al., 2021; Thompson & Kramer, 1994; Van Lith et al., 2003; von Stosch, Oliveira, Peres, & Feyo de Azevedo, 2014).

The synergy between the two approaches has been gaining attention, by either redesigning the model's architecture, augmenting training datasets with simulated data, or including physical principles as constraints in the cost function. The hybrid models are also referred to as a gray-box model (GBM) (Acuña, González, Curilem, & Cubillos, 2013), integrated neural network (Mavkov, Forgione, & Piga, 2020), semi-mechanistic (Lima & Saraiva, 2007), semi-parametric (McBride, Sanchez Medina, & Sundmacher, 2020; von Stosch et al., 2014), informed machine learning (Vonrueden, Mayer, Beckh, Georgiev, et al., 2021), domain adaptive (Liu, Zheng, Zong, Lu, & Tang, 2020; Schuld, Sinayskiy, & Petruccione, 2014), theory-guided data science (Colwell, Collet, & Downton, 2020; Karpatne, Atluri, Faghmous, Steinbach, et al., 2017), physics-guided neural network

(Hu, Hu, Verma, & Zhang, 2021; Murphy & Kerekes, 2021; Wang, Li, et al., 2020; Wang, Pan, et al., 2020), and process dynamic-guided neural network (Alauddin, 2021).

Integrated mechanistic knowledge in the form of physics-based priors has been rendered in many data-based models, including kernel machines (Acuña et al., 2013), Bayesian non-parametric approaches such as Gaussian Process (GP) regression (Camps-Valls, Martino, Svendsen, Campos-Taberner, et al., 2018; Raissi, Perdikaris, & Karniadakis, 2017; Svendsen et al., 2021) and latent force models (LFMs) (Alvarez et al., 2009). Many studies demonstrated improved performance of neural networks by imparting knowledge from the first principles and field experience. The graph neural network (GNN) can exploit the structural relation of process variables in the training of the neural network.

Hendriks, Jidling, Wills, and Schön (2020) presented a neural network-based model that can explicitly satisfy known linear operator constraints. Wu and Zhao (2020) proposed three distinct hybrid formalisms based on a recurrent neural network to integrate physical knowledge in data-driven models. Wu and Zhao (2021) presented a process topology convolutional network (PTCN) for harnessing the simple graphical relationship of process variables. Several comprehensive reviews on hybrid methods applied to process systems have been presented (Bikmukhametov & Jäschke, 2020; Sansana, Joswiak, Castillo, Wang, et al., 2021; von Stosch et al., 2014; Xiong & Jutan, 2002).

Table 1 presents a brief overview of statistical and machine learning methods of FDD and process monitoring. Usually, machine learning-based models offer higher accuracy, while statistical approaches directly handle the overfitting problem.

2. Statistical process monitoring

This section discusses the selection criteria and applications of statistical process monitoring (SPM) tools to industrial process systems. First, a generalized representation of how SPM-based methods work in process industries is discussed. Then a comparative performance evaluation of some commonly used tools such as the PCA, KPCA, ICA, and KICA to a binary distillation column in two fault scenarios is presented.

2.1 Generalized framework for statistical process monitoring

While applying any SPM tool, several issues, such as data characteristics, data pre-treatment, proper model selection, online operation of the model, and

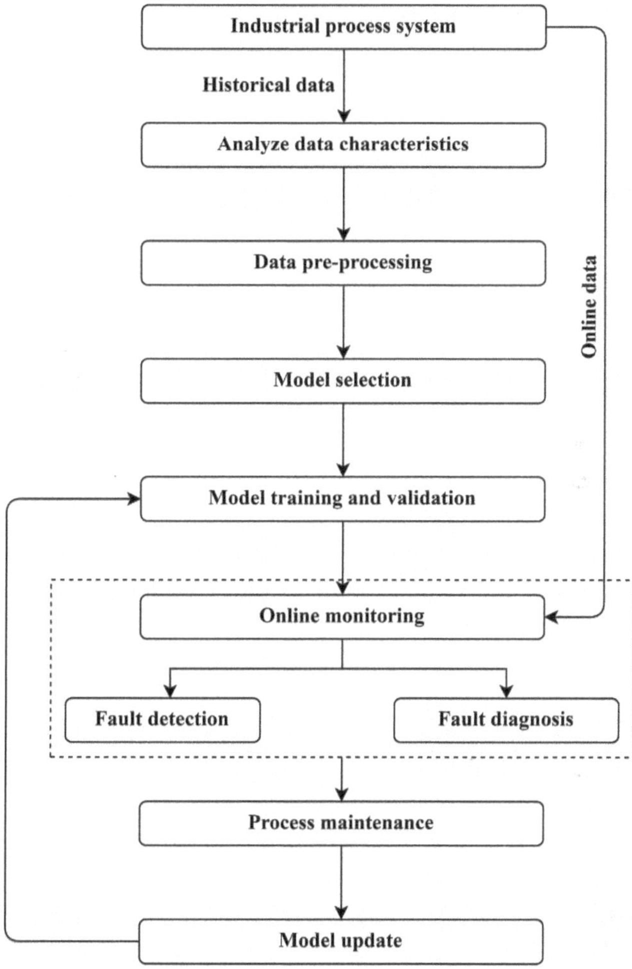

Fig. 1 A generalized SPM framework.

model tuning, need to be considered for efficient performance and tight quality control. Fig. 1 shows a generalized framework for applying SPM tools to process plants. Different components of this framework are discussed in the following sub-sections.

2.1.1 Data characterization

In modern days, large datasets provide valuable information about process operations. Thanks to the revolutionary progress in sensor technologies and process digitization. However, even if the operational datasets are absent, the process simulators provide great flexibility in generating a wide range of realistic scenarios and gathering valuable data (Nazir & Manca, 2015). The historical process data, collected either from an industrial database or a simulator, can be utilized to develop the monitoring model using an SPM algorithm.

Dimension of the dataset is the first thing that needs careful examination. It is reasonably expected that there will be digitized data for several observations; however, the data can be of different dimensions. Consider two process systems, where 2 and 20 process variables need to be monitored, respectively. The SPM models for these two cases will be different. The first one can be monitored with a couple of control charts only, while the latter can be monitored with 20 control charts. Nonetheless, it may not provide a feasible solution if multiple charts need to be monitored simultaneously.

The next task is to examine the relationships (i.e., linear and nonlinear) among the process variables and the distribution (i.e., Gaussian and non-Gaussian) these variables are following. The type of relationship is an essential aspect of model selection and development. If the variables exhibit linear relationships, the remaining procedures to develop an SPM model become relatively more straightforward than if they follow a nonlinear relationship. Fig. 2 shows how linear and nonlinear relationships look for two variables. In a linear relationship, a variable maintains a linear trend with the increase or decrease of another variable. On the other hand, this trend is nonlinear when the variables do not follow a linear relationship. The Pearson correlation coefficient can quantify the strength of the linear relationship between two variables, while the Spearman and Kendall rank correlation coefficient can be applied to nonlinear variables.

Distribution assumption is also crucial. Gaussianity and non-Gaussianity are the key factors here (see Fig. 3). Most of the earlier FDD tools are developed considering the Gaussian distribution assumption. However, this may not be a valid assumption in real industrial processes. Hence, the control charts developed using the Gaussianity assumption, may fail when process data follow the non-Gaussian distribution. The Akaike information criterion (AIC) (Akaike, 1998) and Bayesian information criterion (BIC) (Schwarz, 1978) can be used to find the most suitable distribution that the variables follow.

Due to time correlation and feedback controllers used in process operations, data often show similar characteristics after a certain period (see Fig. 4). It can be noticed that the variable is exhibiting a similar trend frequently; this is known as autocorrelation. It may result in biased parametric estimation. Enhancing the sampling frequency may provide a solution to

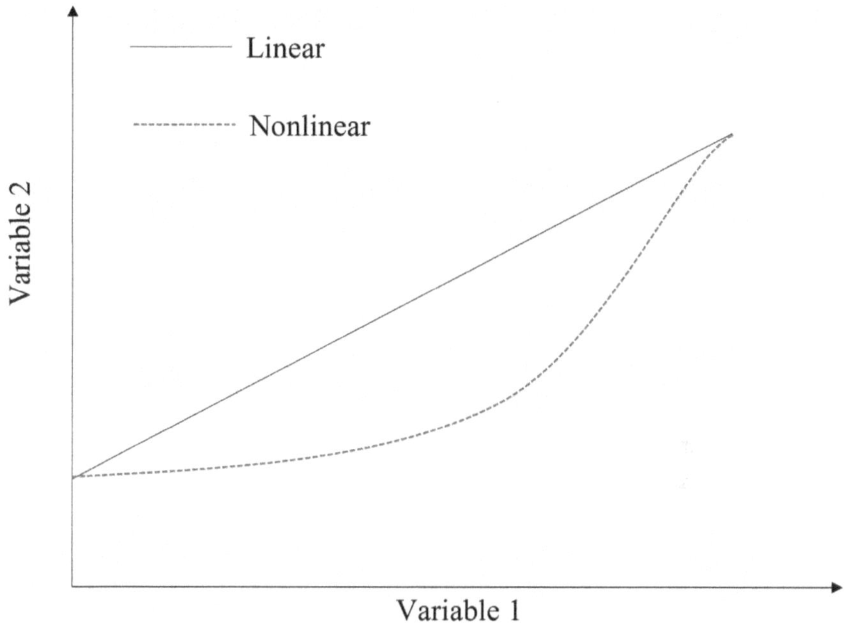

Fig. 2 Bivariate linear and nonlinear relationships.

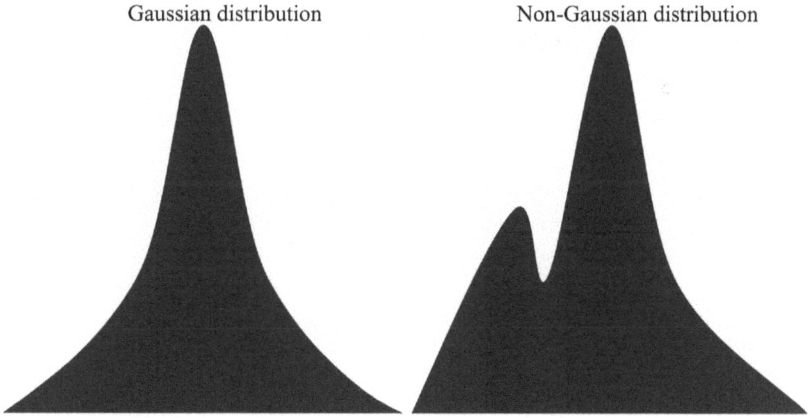

Fig. 3 Gaussian and non-Gaussian distributions.

autocorrelated datasets. However, at the same time, the datasets may lose valuable information.

Identifying random unacceptable noise and missing data (Fig. 5) are two other crucial issues. Noise may result from external disturbances. Suppose a sensor is measuring the temperature of a room. Suddenly an operator opens

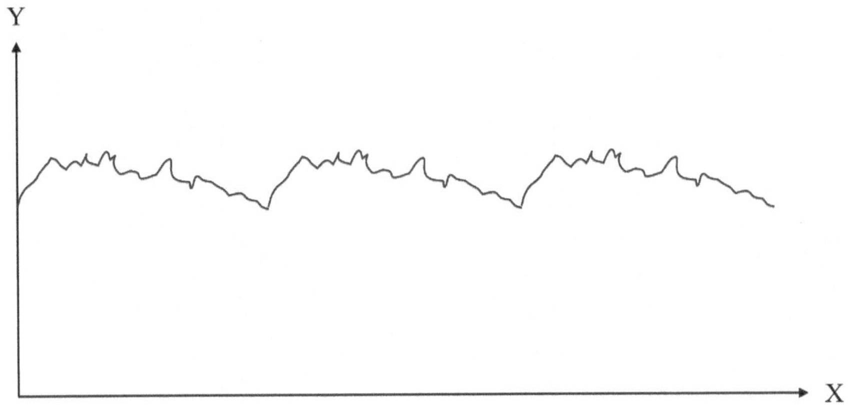

Fig. 4 Autocorrelated process data.

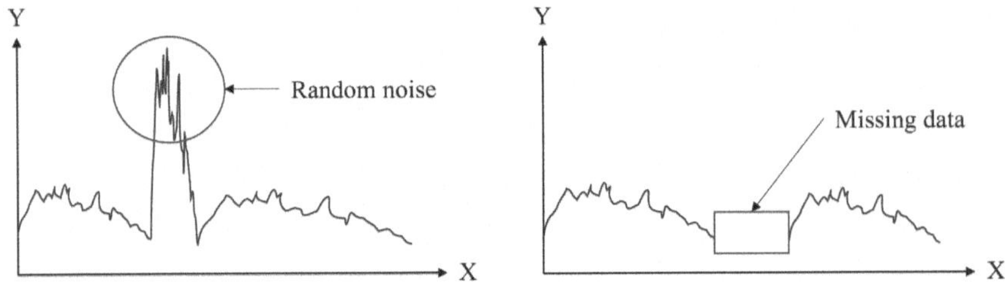

Fig. 5 Random noise vs missing data.

the door for a while. This action may work as a disturbance, and a larger spike in temperature profile may be noticed. On the other hand, missing data may result from malfunctioning hardware. If the same sensor fails to work for a period, the measurements for this span will be missing.

2.1.2 Data pre-processing

The studied dataset needs to be represented properly for efficient performance. For example, if the most widely used data-driven method PCA is used for monitoring a process, the dataset needs to be auto-standardized. It means all the observations need to be normalized to zero mean and unit variance. This pre-treatment allows scaling the effect of different variables. As a result, the monitoring algorithm does not show bias to any random variable.

Eliminating the extremely noisy samples are another critical data pre-processing task before applying a monitoring model. An inappropriate value may be noticed if such data are considered while computing mean and standard deviation that are the foundations of many SPM algorithms. However, eliminating a window may mean losing information too. Filters can also be used to suppress the effect of noise. Data pre-filtering may provide a robust solution for early fault detection since it may make feature extraction easier (Amin, Khan, & Imtiaz, 2019; Amin, Khan, Imtiaz, & Ahmed, 2019; Bin Shams, Budman, & Duever, 2011). The maximum likelihood estimation-based techniques such as the expectation-maximization (EM) algorithm can be used to predict the missing data (Dempster, Laird, & Rubin, 1977). Another simple way to handle this issue is by replacing the missing values with mean or median values.

2.1.3 Model selection

The performance of an SPM tool largely depends on selecting a suitable model to monitor the process. Since SPM tools are developed using process data, selecting an optimal model relies on data characteristics. Suppose a process needs to be monitored with only a couple of variables; the conventional univariate Shewhart chart will be sufficient. However, since this control chart cannot capture data correlation, the other univariate control charts, such as the exponentially weighted moving average (EWMA) and cumulative sum (CUSUM) charts, can be utilized to handle the correlative nature of process data.

With the advancement of sensor technologies and feedback control, most process operations contain a high dimensional dataset. Therefore, the multivariate statistical process monitoring (MSPM) tools provide more flexibility. If a process with 20 variables needs to be monitored, the univariate tools require 20 control charts; this burdens the operators. Fortunately, the same process can be monitored with a couple of control charts using the MSPM tools.

Nonlinearity and non-Gaussianity are the following factors that need to be accounted for. If process data is found to be following a linear relationship and Gaussian distribution, the conventional PCA and PLS-based monitoring models can provide optimal performance (Kresta, Macgregor, & Marlin, 1991). When there is a deviation from any of these two characteristics, PCA and PLS may not provide suitable performance (Ge, Song, & Gao, 2013). The ICA and GMM can be used for monitoring non-Gaussian processes (Kano et al., 2003; Yu & Qin, 2008).

Data nonlinearity can be addressed using kernel tricks (Lee, Qin, & Lee, 2007). Kernel functions provide better results in nonlinear process systems (Chouaib, Mohamed-Faouzi, & Messaoud, 2015; Fan & Wang, 2014).

KPCA, KPLS, KICA, and KGMM are some tools that have integrated the kernel function with the conventional MSPM tool. Kernel functions map low dimensional data to a high dimensional feature space. Selecting the proper kernel function and tuning parameters is a challenge and requires a healthy amount of time and operator experience. Linear approximation is an alternative solution to kernel tricks in the context of nonlinear datasets. Several local models describe the nonlinear space in this approach (Kerschen & Golinval, 2002).

In addition to these, dynamic and multimodal process natures are other crucial issues that require attention. The majority of the abovementioned MSPM tools consider a single operating condition and static process operation that may not be practical in many process operations where slow drift and frequent operating condition changes are noticed (Huang & Yan, 2015; Ku, Storer, & Georgakis, 1995; Yu, 2010). Although numerous methods have been proposed over the past three decades to capture different aspects of data characteristics, unfortunately, a ubiquitous rule is yet to be established to select an optimal model.

2.1.4 Model training and validation

Once a model is selected, the historical process data is used to train the model. The training steps vary depending on the used algorithm. However, the principal aim of any SPM algorithm is to estimate monitoring indexes and their control limits. For instance, PCA uses Hotelling's T^2 and Q or SPE statistics to reduce the large dimensional process data to be monitored with only two control charts. These indexes are sensitive to a breakdown of data correlation.

The developed model's performance needs to be evaluated before using it for online applications. The false alarm rate (FAR) and fault detection rate (FDR) are two key performance metrics to examine a developed tool's efficacy. The historical normal and faulty datasets can be used in this regard. The ultimate goal is to develop a model in such a way that it gives no false alarm and detects all the faulty samples. However, in reality, it is not always possible to reach such an ideal condition. Therefore, a trade-off between FAR and FDR is compromised.

2.1.5 Online monitoring

In this step, a data-based SPM tool decides about the possibility of a fault using the online samples. When online data comes, these are converted into monitoring indexes (e.g., T^2 and Q values). If there is any violation of the predefined control limits, an alarm is generated; it indicates a faulty condition in process operation. The next task is to diagnose the root cause of this fault. Using univariate charts provides an easy solution in the context of root cause diagnosis since the variable that exceeds the threshold can be identified as the cause of possible process upset.

However, fault diagnosis is not a straightforward task while MSPM tools are used. The multivariate contribution plots are used for finding the variable that is the reason behind an alarm. The variable with the highest contribution in the multivariate contribution plot is selected as the root cause. Nevertheless, due to the smearing effect, this diagnosis result may often lead to a false diagnosis (Liu, 2012). Still, the contribution plots provide helpful information that can be used to predict the faulty part of a process.

2.1.6 Process maintenance

Once an abnormality has been detected, the process needs to be returned to normal operating conditions. The diagnosis information is vital in this context. The operators can use the diagnosis results provided by the SPM tools and take necessary actions for process recovery. Once the faulty part of a process is repaired, the process maintenance part is complete.

2.1.7 Model update

If no other conditions such as operating point change, slow drift, and process modification do not occur, the developed monitoring model requires no update. However, all these conditions change in industrial operation. As a result, the performance of the developed SPM model deteriorates over time. This issue can be solved by periodically inspecting the efficacy and updating the model. One of the simplest ways is to update process variables' mean and standard deviation on a window basis if no fault is reported. It dynamically updates the model parameters (Amin et al., 2021b). Also, adaptive and moving-window based updating techniques are utilized to keep the model up-to-date with the most current process condition (Jeng, 2010; Wang, Kruger, & Irwin, 2005).

2.2 A comparative performance analysis

This sub-section provides a comparative efficiency of a few widely used MSPM tools. The binary distillation column (Fig. 6), a common sub-system in several process operations, is considered. This process model consists of 40 stages and produces a product with 96% purity. The detailed process description can be found in the work by Skogestad (1997).

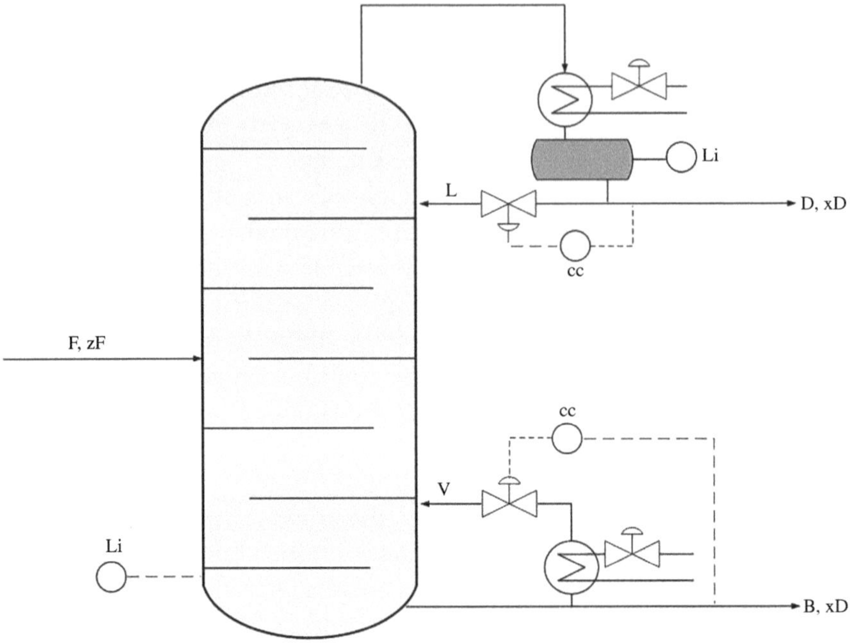

Fig. 6 A binary distillation column. *Modified and redrawn from Skogestad, S. (1997). Dynamics and control of distillation columns: A tutorial introduction.* Chemical Engineering Research and Design, *75(6), 539–562. doi:10.1205/026387697524092.*

This chapter considers five variables from the distillation column model: feed rate (F), feed composition (z_F), reflux flowrate (L), top composition (x_D), and bottom composition (x_B). The first three are the inputs to the system, while the remaining two are the outputs.

A total of two case studies are considered. In the first case study, a random variation in the reflux flowrate occurs. The latter case study consists of a scenario where a step change occurs in the feed rate. Both these faults happen in the input to the system. Hence, the outputs: top composition and bottom composition, get upset, and it becomes incredibly arduous to isolate the root cause. A total of 500 data samples are collected with a sampling time of 1 s. Both faults occur after 400 s of normal operating conditions. PCA, KPCA, ICA, and KICA were applied to these scenarios, and the results are summarized in Table 2.

The performance of PCA, KPCA, ICA, and KICA is compared in terms of FAR, FDR, detection delay (DD), and accurate diagnosis. Detection delay is the time required for a model to detect a fault. The univariate SPM tools are excluded from the comparison since these are not suitable

Table 2 A comparative performance analysis of different MSPM tools applied to a binary distillation column (Amin et al., 2021a).

Fault scenario no	Performance criteria	PCA		KPCA		ICA		KICA	
		T^2	SPE	T^2	SPE	I^2	SPE	I^2	SPE
1	FAR (%)	0	2.50	0	0.50	1.25	1.50	1.25	1.50
	DR (%)	0	60	0	13	100	100	100	100
	DD (s)	NA	0	NA	27	0	0	0	0
	Accurate diagnosis	NA	No	No	No	No	Yes	No	Yes
2	FAR (%)	0	2.50	0	0.50	1.25	1.50	1.25	1.50
	DR (%)	45	100	44	90	100	100	100	100
	DD (s)	55	0	59	7	0	0	0	0
	Accurate diagnosis	No	Yes	No	No	Yes	No	Yes	No

to monitor large scale process systems. T^2 statistics with PCA and KPCA provides the worst performance from detection and diagnostic perspectives. PCA-T^2 fails to distinguish a random variation in reflux flowrate from a random process noise. The diagnosis performance is also unsatisfactory. It fails to diagnose the root causes in both cases. Compared to the PCA-T^2-based monitoring, the PCA-SPE-based monitoring provides a better result. It can detect the fault earlier and diagnose the root cause in the second fault scenario.

It can be observed that ICA provides superior performance compared to PCA and KPCA. It can immediately detect faults with 100% FDR. Also, the FAR is lower than the PCA. It indicates the presence of non-Gaussian data in the studied process operation. Another noticeable fact is the same level of performance for ICA and KICA; it implies the dataset does not exhibit strong nonlinearity. It should be noted that the Gaussian kernel function has been used in both KPCA and KICA models.

ICA-I^2 can accurately diagnose the root cause in the second fault scenario. On the contrary, ICA-SPE provides an accurate diagnosis in the first case study. Despite simultaneously detecting the abnormalities, the diagnostics of the two charts are contradictory. This can create a dilemma from a maintenance perspective. It will be difficult to decide which contribution plot to be trusted. In essence, it can be concluded that while the detection performance of MPSM tools is remarkable, these tools severely suffer from accurate diagnosis capacity.

3. Artificial neural network for process monitoring

This section describes the ANN-based methods and models that have been developed to monitor the process systems. Process industries have been using ANNs for decades in various applications, and they are one of the most powerful and versatile artificial intelligence tools. An ANN is a computer tool that provides a simplified mathematical model of brain functions.

3.1 Neural network basic model

The first conceptual model of an ANN was devised by neuroscientist Warren S. McCulloch and logician Walter Pitts (Pospíchal & Kvasnička, 2015). In their model, an ANN was proposed as a computational model based on the brain that was created to solve specific types of problems. Based on their concept, one of the most important characteristics of a neural network is learning ability (McCulloch & Pitts, 1990).

A neural network (NN) is a sophisticated adaptive system, which means it may modify its internal structure in response to the information it receives (Abiodun, Jantan, Omolara, Dada, et al., 2018; Hunt, Sbarbaro, Zbikowski, & Gawthrop, 1992). Typically, this is achieved by altering the weights of the components. A connection between two neurons is shown by each line in Fig. 7, which represents the pathway for information to pass between two neurons. Each connection between the neurons has a weight, which is a number that affects the signal between two neurons. This weight will be adjusted based on the output accuracy. Moreover, the activation function used in each connection allows the model to create a complex mapping between input and output nodes. According to the different activation functions, nonlinear transformation can be done on the input before sending it to the next layer.

Binary step, linear, and nonlinear activation functions are categorized as neural network activation functions used in recent models (Apicella, Donnarumma, Isgrò, & Prevete, 2021; Laudani, Lozito, Fulginei, & Salvini, 2015). The binary step function is conditional on a threshold value that determines whether or not to activate a neuron (Zhao & Xiao, 2021). Linear activation function, often called the identity function, is a function in which the activation is proportionate to the input. Nonlinear activation functions are implanted to address the following drawbacks of linear activation functions (Ertuğrul, 2018; Xiao, Li, Tan, Zhang, et al., 2019):

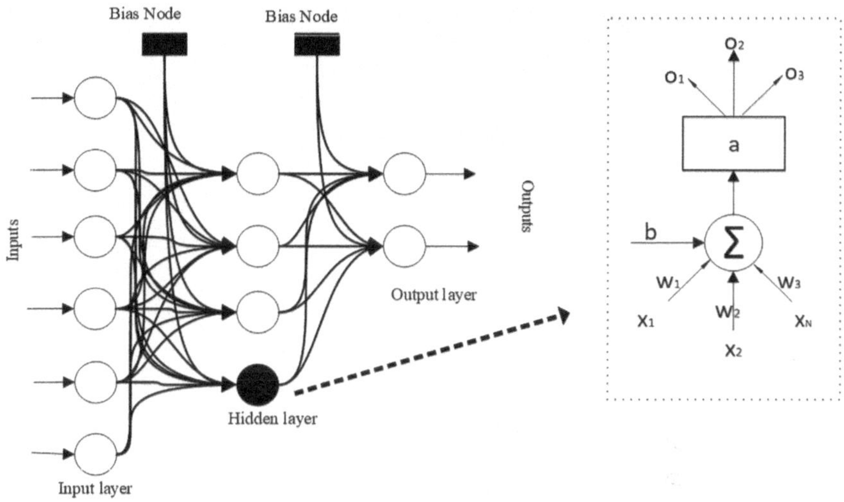

Fig. 7 Neural network basic architecture. *Adapted from Arunthavanathan, R., Khan, F., Ahmed, S., Imtiaz, S., & Rusli, R. (2020). Fault detection and diagnosis in process system using artificial intelligence-based cognitive technique.* Computers and Chemical Engineering, *134. doi:10.1016/j.compchemeng.2019.106697.*

- Backpropagation is possible with nonlinear activation functions because they have a derivative function connected to the inputs.
- To form a deep neural network, nonlinear activation functions allow numerous layers of neurons to be "stacked." To learn complicated data sets with high degrees of accuracy, multiple hidden layers of neurons are needed.

The most popular nonlinear functions are tanh, sigmoid, rectified linear unit (ReLU), exponential linear unit (ELUs), and Softmax (Apicella et al., 2021).

In process monitoring, supervised and unsupervised learning strategies are used to train the neural network models depending on the available data. Supervised learning is predicated on a large amount of faulty and normal data being collected in the past (Dai & Gao, 2013; Hoskins & Himmelblau, 1988). As a result, all data pertaining to different process conditions must be gathered at the system level to build the model.

Unsupervised machine learning models are trained on historical data that does not include all the process condition classification labels (Arunthavanathan et al., 2021; Dike, Zhou, Deveerasetty, & Wu, 2019). These models, on the other hand, employ an algorithm to select classification tags by recognizing a hidden pattern in data. Clustering and anomaly detection are two often used unsupervised model behaviors in process monitoring.

3.2 Neural network for supervised learning

In ANN modeling, the backpropagation algorithm (BPA) based on the delta rule is widely used as a supervised training method to optimize the ANN parameters such as weights and bias (Sung, 1998; Wong, McDonald, & Palazoglu, 2001). The BPA's popularity in supervised training of ANN models is largely due to its simplicity of comprehension and execution. The first step in adopting this strategy is to make an assumption about the weights and biases' initial values (Buscema, 2009). It is necessary to iteratively update the starting values of the weight and bias. The iterative training method is repeated until the correlation coefficient satisfies the threshold criterion (Li, Cheng, Shi, & Huang, 2012). The BPA based on delta rules is defined as:

$$E = (y - \hat{y})^2 \rightarrow \frac{\partial E}{\partial y_j^{(l)}} \tag{1}$$

$$w_{ij}^{new(l)} = w_{ij}^{old(l)} + lr\left(-\frac{\partial E}{\partial w_{ij}^{(l)}}\right) \tag{2}$$

In the above equation, E is average error estimation, y and $\hat{y}$ are real and simulated outputs. Further, $w_{ij}^{(l)}$ is the weight for linking the ith neuron to the jth neuron in the lth layer, and lr is the learning rate between the value of 0 and 1.

Backpropagation algorithm for supervised learning:

Algorithm 1. Supervised NN

Offline train model/online Update

1a. Input: Labeled data.

2a. Output: NN model.

3a. Train feed-forward: initializing weight, bias and activation function.

$$o^{(l)} = a\left(w^{(l)}.o^{(l-1)} + b\right) \tag{3}$$

4a. For each layer, weights and bias matrix connecting to a new layer, obtained using the backpropagation algorithm by Eq. (2), set the learning rate.

5a. for (1: length_data):

6a. Perform forward propagation and backpropagation to optimize the w and b.

7a. Verify the backpropagation using Cost Function and determine the appropriate learning rate for the optimization algorithm.
8a. NN_Model is developed for label prediction.
Online/real-time classification
1a. **Input**: Data
2a. **Output**: Classification/Label
3a. Develop/updated NN model (input sample)
4a. **for** (1: length_input sample)**:**
5a. Predict the classification for each sample.
Using classified label, define the test accuracy.

3.2.1 Procedure to apply ANN for supervised learning

The procedure for applying the approximation function is explained in the three steps: (i) data pre-processing, (ii) choosing the best ANN model, and (iii) validation of chosen ANN model.

Step 1: Data pre-processing for NN

After identifying the input data as predictors, data pre-processing methods are implemented to improve the neural network training process, stability, and modeling performance (Kotsiantis & Kanellopoulos, 2006). Choosing an appropriate pre-processing strategy improves the accuracy of the output and reduces the learning process's computing cost. The widely used pre-processing techniques are explained below,

- Normalization using min-max techniques uses linear interpolation to rescale input values that frequently fall within the range of [0, 1], as illustrated in Eq. (4).

$$\widehat{x} = \frac{x - min_x}{max_x - min_x} \tag{4}$$

  where x is the value of input data and $\widehat{x}$ is normalized data.
- Standardization using Z-score uses the mean and standard deviation for each set of input data to generate a set of standardized data. The standardization Z-score formula is illustrated in Eq. (5).

$$\widehat{x} = \frac{x - \mu_x}{\sigma_x} \tag{5}$$

  where x is the value of input data, and $\widehat{x}$is the standardized data. μ_x and σ_x are the mean and standard deviation of the input data, respectively.

According to the needs of pre-processing, standardization and normalization can be selected. In min-max techniques, all features have the same scale; however, outliers are poorly handled. On the other hand, in Z-score standardization, outliers are handled; however, the provided data do not have the same magnitude (Iliou, Anagnostopoulos, Stephanakis, & Anastassopoulos, 2017). Furthermore, to reduce the feature dimension, PCA, KPCA, or KICA, as described in Section 2, can be used before processing the ANN.

Step 2: Select the best ANN architecture

As shown in Fig. 7, the number of hidden layers, number of neurons in each layer (input, hidden, and output), particular transfer (activation) functions, connection patterns between layers, and learning techniques are all factors to consider while developing an ANN architecture (Kotsiantis, 2007). The ANN model's architecture controls how a network converts its input into an output. Similarly, different neuron connectivity topologies exhibit ANN functioning. The most common supervised learning networks in process monitoring are multilayer fully connected NN and multilayer deep NN, including recurrent neural network (RNN), convolutional neural network (CNN), and radial basis function (RBF) (Abiodun et al., 2018). Depending on the complexity of the supervised data, a network model can be selected for the specific operations. When modeling an ANN, it is a common practice to develop a model with a simple network and further modified to complex models to achieve expected accuracy (Wang, Liu, & Griffin, 2000).

An essential part of ANN modeling is network training, which is utilized as a mapping procedure to calibrate the chosen neural network. Underfitting and overfitting can occur in the training process for any ANN design, and the calibration procedure can help reduce this problem (Abiodun et al., 2018). It is possible to reduce the negative consequences of overfitting and underfitting in the model training stage by using a well-designed network architecture that considers the trade-off between bias and variance (Kotsiantis, 2007). Nevertheless, the optimal number of neurons and hidden layers is obtained by comparing the error in the training and testing procedures of the ANN model.

Step 3: Postprocessing of data

ANN models are tested using postprocessing approaches. Performance evaluation in training and testing phases with statistical measures is used to examine the correctness of the neural network models (Kotsiantis, 2007). Following statistical metrics are commonly used to evaluate the ANN models.

- Mean absolute error (MAE) is used to determine the average deviations in a series of predictions without regard for their direction. It is the average of the absolute differences between prediction and actual observation over the test sample, where all individual deviations are given equal weight.

$$MAE = \frac{1}{n}\sum_{j=1}^{n}\left|y_j - \widehat{y_j}\right| \tag{6}$$

- The root mean square error (RMSE) is a quadratic scoring rule that determines the average deviation. It is the square root of the average of the squared discrepancies between the predicted and observed values.

$$RMSE = \sqrt{\frac{1}{n}\sum_{j=1}^{n}\left(y_j - \widehat{y_j}\right)^2} \tag{7}$$

The RMSE will always be greater than or equal to the MAE; the greater the gap between them, the more variance in the sample's error. If the root mean square error equals the mean absolute error, all errors are of the same magnitude (Pham, Nguyen, Bui, Prakash, et al., 2019). MAE and RMSE both have a range of 0 to infinity. They are descriptive scores, with lower values being preferable (Sorsa & Koivo, 1993).

Network architecture must be updated through an iterative process to reduce the estimation error between simulated and observed data to the acceptable level if postprocessing approaches fail to discover good results. Iteratively adjusting the number of hidden layers, the number of neurons in each layer, the transfer functions, and initial predictions for biases and weights are all part of the technique (Gomm, 1996; Heo & Lee, 2018).

3.2.2 An overview of the application of supervised neural network modeling in process monitoring

This section demonstrates the application of ANN in process monitoring of the benchmark Tennessee Eastman (TE) chemical process for various studies, including fault diagnosis considered. A detailed description of the TE process was introduced in Downs and Vogel (1993). To evaluate the performance of ANN, results from (Arunthavanathan et al., 2020; Sheta, Braik, & Al-Hiary, 2019; Zhang, Qian, Chai, Li, & Liu, 2015) are used in this section.

Step 1: Data pre-processing using normalization or standardization and feature reduction (if applicable).
Step 2: Combine and shuffle the normal and abnormal data groups and divide them into train, test, and validate datasets.
Step 3: In this step, the NN architecture is determined. This includes selecting the number of nodes in input and output layers depending on the number of features and the number of classification groups in the data. Table 3 illustrates the NN model accuracy with 0, 1, 2, 3, and 4 hidden layers. It can be seen that increasing the number of hidden layers will not be beneficial without the proper evaluation of other hyperparameters.
Step 4: The model is trained as described in Algorithm 1. When training the NN model, proper learning rate as described in Eq. (2), appropriate activation function can be selected by iterating the model with different values. Table 4 shows the model training accuracy with various activation functions. Nevertheless, for different applications, these results can be varied.
Step 5: Validate the model by and define the model accuracy using simulated and observed data.
Step 6: Test the model using test data.

3.3 Neural network for unsupervised learning

Unsupervised learning models are trained with historical data without any classification labels. However, by finding a hidden pattern from data, these models select classification tags using the algorithm. The clustering and anomaly detection are widely used unsupervised model behaviors in

Table 3 Fault detection accuracy of ANN with different number of hidden layers (Heo & Lee, 2018),

Number of hidden layers	0	1	2	3	4
Accuracy (%)	80.76	83.24	97.25	97.6	97.6

Table 4 ANN model performance based on different activation functions (Zhang et al., 2015).

	Activation function			
Model training accuracy	Linear	Poly	RBF	Sigmoid
	85.79%	97.43%	98.89%	80%

ANN. Clustering is a process of aligning the set of data as a smaller group and is used for multiclass classification. Anomaly detection is a binary classification to determine the outlier.

A training algorithm for the unsupervised clustering methods has been developed by (Kohonen and Somervuo (1998), which is based on the processing of input data with modification of weights and biases and resulting in some output classes. In recent years, one class neural network has become more popular in unsupervised classification. One class neural network is a type of ANN and similar to one class SVM in that it separates all data points from the origin. Arunthavanathan et al. (2020), Chalapathy, Menon, and Chawla (2018) described one class NN model illustrated in Fig. 8.

In the above neural network, "γ" is the scalar matrix from the hidden to the output layer, "α" is the weight matrix from the input to the hidden layer, F is the feature space, "β" is the bias matrix, a is the NN activation function, and N is the number of samples from the origin.

The one class neural network objective is formulated as,

$$\min_{\gamma,\alpha,\beta} \frac{1}{2}\|\gamma\|_2^2 + \frac{1}{2}\|\alpha\|_F^2 + \frac{1}{\alpha}\cdot\frac{1}{N}\left[\sum_{n=1}^{N} \max\left(0, \beta - \langle\gamma, a(\alpha \mathbf{X}_{n:})\rangle\right) - \beta\right] \quad (8)$$

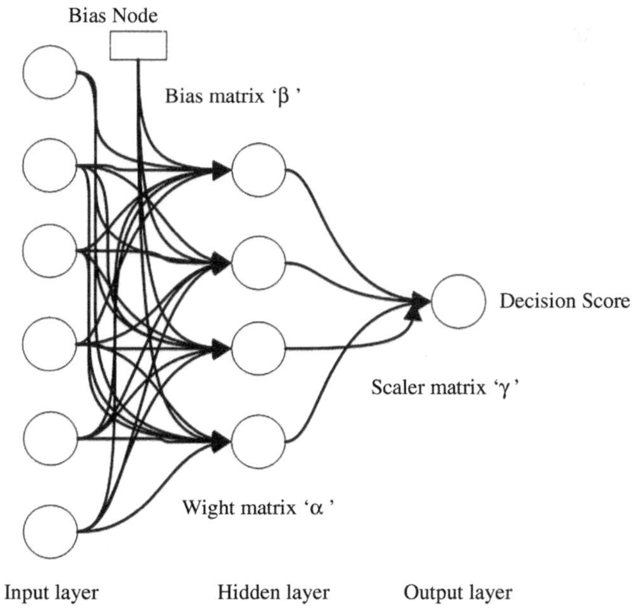

Fig. 8 One class neural network model (Arunthavanathan et al., 2020).

From Eq. (8), bias "β," weight matrix "α," and the output matrix "γ" can be optimized by a minimization approach. Chalapathy et al. (2018) described the optimization algorithm for one class neural network by optimizing "α" and "γ" separately while updating the "β" based on optimized "α" and "γ."

$$\underset{\gamma,\alpha}{argmin}\ \frac{1}{2}\|\gamma\|_2^2 + \frac{1}{2}\|\alpha\|_F^2 + \frac{1}{\alpha}\cdot\frac{1}{N}\left[\sum_{n=1}^{N} \max\left(0, \left(\beta - [\gamma, g(\alpha \boldsymbol{x}_{n:})]\right)\right)\right] \quad (9)$$

Similarly, the optimization for β,

$$\underset{\beta}{argmin}\left[\frac{1}{N\alpha}\cdot\sum_{n=1}^{N} \max\left(0, \left(\beta - [< \gamma, a(\alpha \boldsymbol{x}_{n:}) >]\right) - \beta\right].\right.$$

The input layer of a one-class neural network algorithm feeds the set of multivariate samples and produces the data sample decision score for each data sample using a single output.

Eq. (8) illustrates the optimization of "α" and "γ" using the "β" value. Similarly, Eq. (9) illustrates the optimization of "β" by using the optimized "α" and "γ" values. Initial decision scores are computed for each data sample based on the optimized γ, α, and β values. By using the decision score, normal and abnormal data samples are classified. Finally, with the optimized model, positive decision score data will identify as a normal condition, and a negative decision score will be defined as an abnormal condition. One class algorithm is summarized in Algorithm 2.

Algorithm 2. Unsupervised one class NN

1. **Input**: Data points $\boldsymbol{x}_{n:}$ $\boldsymbol{n = 1: N}$
2. **Output**: Set of decision scores (for one data frame)/update One class NN
3. **while** (1: "n" sample window):
4. Define "β" value
5. **for** (1: No of epochs)
6. Update ("α" and "γ"); optimize the NN weights and output layer
7. Optimize the r-value using updated "α" and "γ."
8. **end**
9. Compute the decision score for each sample: d(n)
10. **if** decision score ($d(n) \geq 0$) then
11. $\boldsymbol{x}_n$:Normal data point
12. **else**

13. x_n:Anomaly data point

14. end

Arunthavanathan et al. (2020) proposed a TE process monitoring based on the one-class NN. In the application, one class NN is used to detect the number of anomalies from the dataset.

4. Conclusions

This chapter discusses various widely applied statistical and artificial neural network methods in process system monitoring. Statistical process monitoring models have gained popularity in the past few decades due to their ease of construction and reliable fault detection performance. These methods can monitor and visualize the natural variation of the process efficiently. However, statistical methods lack accurate root cause diagnosis capacity and often require human knowledge for successful fault diagnosis.

In recent years, with the integration of unsupervised and supervised learning methods, autonomous process monitoring have sufficiently developed. However, artificial neural networks and statistical process monitoring are not competing data-driven methodologies. Statistical methodologies can directly be applied to the neural network in several ways, including data preprocessing, criteria estimation, parameter optimization, and confidence interval estimation. Therefore, properly integrated statistical process monitoring and artificial neural network methods would be beneficial in automated and digitalized process monitoring.

References

Abiodun, O. I., Jantan, A., Omolara, A. E., Dada, K. V., … Arshad, H. (2018). State-of-the-art in artificial neural network applications: A survey. *Heliyon*, *4*(11), e00938. https://doi.org/10.1016/J.HELIYON.2018.E00938.

Acuña, G., González, J., Curilem, M., & Cubillos, F. (2013). A SVM gray-box model for a solid substrate fermentation process. *Chemical Engineering Transactions*, *35*, 961–966.

Agatonovic-Kustrin, S., & Beresford, R. (2000). Basic concepts of artificial neural network (ANN) modeling and its application in pharmaceutical research. *Journal of Pharmaceutical and Biomedical Analysis*, *22*(5), 717–727. https://doi.org/10.1016/S0731-7085(99)00272-1.

Akaike, H. (1998). *Information theory and an extension of the maximum likelihood principle* (pp. 199–213). https://doi.org/10.1007/978-1-4612-1694-0_15.

Alauddin, M. (2021). *Data-driven models for safety management of complex systems* (Unpublished PhD thesis). Memorial University of Newfoundland.

Alauddin, M., Khan, F., Imtiaz, S., & Ahmed, S. (2018). A bibliometric review and analysis of data-driven fault detection and diagnosis methods for process systems. *Industrial and Engineering Chemistry Research*, *57*(32), 10719–10735. https://doi.org/10.1021/acs.iecr.8b00936.

Alauddin, M., Khan, F., Imtiaz, S., & Ahmed, S. (2020). A variable mosquito flying optimization-based hybrid artificial neural network model for the alarm tuning of process fault detection systems. *Process Safety Progress*, *39*(S1), e12122. https://doi.org/10.1002/prs.12122.

AL-Qutami, T. A., Ibrahim, R., Ismail, I., & Ishak, M. A. (2018). Virtual multiphase flow metering using diverse neural network ensemble and adaptive simulated annealing. *Expert Systems with Applications*, *93*, 72–85. https://doi.org/10.1016/j.eswa.2017.10.014.

Alvarez, M., Luengo, D., & Lawrence, N. D. (2009). Latent force models. *Journal of Machine Learning Research*, *5*, 9–16.

Amin, M. T., Imtiaz, S., & Khan, F. (2018). Process system fault detection and diagnosis using a hybrid technique. *Chemical Engineering Science*, *189*, 191–211. https://doi.org/10.1016/j.ces.2018.05.045.

Amin, M. T., Khan, F., Ahmed, S., & Imtiaz, S. (2020). A novel data-driven methodology for fault detection and dynamic risk assessment. *Canadian Journal of Chemical Engineering*, *98*(11), 2397–2416. https://doi.org/10.1002/cjce.23760.

Amin, M. T., Khan, F., Ahmed, S., & Imtiaz, S. (2021a). A data-driven Bayesian network learning method for process fault diagnosis. *Process Safety and Environmental Protection*, *150*, 110–122. https://doi.org/10.1016/j.psep.2021.04.004.

Amin, M. T., Khan, F., Ahmed, S., & Imtiaz, S. (2021b). Risk-based fault detection and diagnosis for nonlinear and non-Gaussian process systems using R-vine copula. *Process Safety and Environmental Protection*, *150*, 123–136. https://doi.org/10.1016/j.psep.2021.04.010.

Amin, M. T., Khan, F., & Imtiaz, S. (2019). Fault detection and pathway analysis using a dynamic Bayesian network. *Chemical Engineering Science*, *195*, 777–790. https://doi.org/10.1016/j.ces.2018.10.024.

Amin, M. T., Khan, F., Imtiaz, S., & Ahmed, S. (2019). Robust process monitoring methodology for detection and diagnosis of unobservable faults. *Industrial and Engineering Chemistry Research*, *58*(41), 19149–19165. https://doi.org/10.1021/acs.iecr.9b03406.

Apicella, A., Donnarumma, F., Isgrò, F., & Prevete, R. (2021). A survey on modern trainable activation functions. *Neural Networks*, *138*, 14–32. https://doi.org/10.1016/J.NEUNET.2021.01.026.

Arunthavanathan, R., Khan, F., Ahmed, S., & Imtiaz, S. (2021). An analysis of process fault diagnosis methods from safety perspectives. *Computers & Chemical Engineering*, *145*. https://doi.org/10.1016/j.compchemeng.2020.107197, 107197.

Arunthavanathan, R., Khan, F., Ahmed, S., Imtiaz, S., & Rusli, R. (2020). Fault detection and diagnosis in process system using artificial intelligence-based cognitive technique. *Computers and Chemical Engineering*, *134*. https://doi.org/10.1016/j.compchemeng.2019.106697.

Auret, L., & Aldrich, C. (2010). Unsupervised process fault detection with random forests. *Industrial and Engineering Chemistry Research*, *49*(19), 9184–9194. https://doi.org/10.1021/ie901975c.

Bahrampour, S., Moshiri, B., & Salahshoor, K. (2011). Weighted and constrained possibilistic C-means clustering for online fault detection and isolation. *Applied Intelligence*, *35*(2), 269–284. https://doi.org/10.1007/s10489-010-0219-2.

Barakat, M., Druaux, F., Lefebvre, D., Khalil, M., & Mustapha, O. (2011). Self adaptive growing neural network classifier for faults detection and diagnosis. *Neurocomputing*, *74*(18), 3865–3876. https://doi.org/10.1016/j.neucom.2011.08.001.

Basila, M. R., Stefanek, G., & Cinar, A. (1990). A model—Object based supervisory expert system for fault tolerant chemical reactor control. *Computers & Chemical Engineering*, *14*(4–5), 551–560. https://doi.org/10.1016/0098-1354(90)87026-L.

Bhushan, B., & Romagnoli, J. A. (2008). Self-organizing self-clustering network: A strategy for unsupervised pattern classification with its application to fault diagnosis. *Industrial & Engineering Chemistry Research*, *47*(12), 4209–4219. https://doi.org/10.1021/ie071549a.

Bikmukhametov, T., & Jäschke, J. (2020). Combining machine learning and process engineering physics towards enhanced accuracy and explainability of data-driven models. *Computers and Chemical Engineering*, *138*, 106834. https://doi.org/10.1016/j.compchemeng.2020.106834.

Bin Shams, M. A., Budman, H. M., & Duever, T. A. (2011). Fault detection, identification and diagnosis using CUSUM based PCA. *Chemical Engineering Science*, *66*(20), 4488–4498. https://doi.org/10.1016/j.ces.2011.05.028.

Burges, C. J. C. (1998). A tutorial on support vector machines for pattern recognition. *Data Mining and Knowledge Discovery*, *2*(2), 121–167. https://doi.org/10.1023/A:1009715923555.

Buscema, M. (2009). Back propagation Neural Networks, *33*(2), 233–270. doi:https://doi.org/10.3109/10826089809115863.

Cai, L., & Tian, X. (2014). A new fault detection method for non-Gaussian process based on robust independent component analysis. *Process Safety and Environmental Protection*, *92*(6), 645–658. https://doi.org/10.1016/j.psep.2013.11.003.

Cai, L., Tian, X., & Chen, S. (2014). A process monitoring method based on noisy independent component analysis. *Neurocomputing*, *127*, 231–246. https://doi.org/10.1016/j.neucom.2013.07.029.

Cai, L., Tian, X., & Chen, S. (2017). Monitoring nonlinear and non-gaussian processes using gaussian mixture model-based weighted kernel independent component analysis. *IEEE Transactions on Neural Networks and Learning Systems*, *28*(1), 122–135. https://doi.org/10.1109/TNNLS.2015.2505086.

Camps-Valls, G., Martino, L., Svendsen, D. H., Campos-Taberner, M., … García-Haro, F. J. (2018). Physics-aware Gaussian processes in remote sensing. *Applied Soft Computing Journal*, *68*, 69–82. https://doi.org/10.1016/j.asoc.2018.03.021.

Chaffart, D., & Ricardez-Sandoval, L. A. (2018). Optimization and control of a thin film growth process: A hybrid first principles/artificial neural network based multiscale modelling approach. *Computers and Chemical Engineering*, *119*, 465–479. https://doi.org/10.1016/j.compchemeng.2018.08.029.

Chalapathy, R., Menon, A. K., & Chawla, S. (2018). *Anomaly detection using one-class neural networks* (pp. 19–23).

Chawla, M. P. S. (2011). PCA and ICA processing methods for removal of artifacts and noise in electrocardiograms: A survey and comparison. *Applied Soft Computing Journal*, *11*(2), 2216–2226. https://doi.org/10.1016/j.asoc.2010.08.001.

Chen, G., & Ge, Z. (2019). SVM-tree and SVM-forest algorithms for imbalanced fault classification in industrial processes. *IFAC Journal of Systems and Control*, *8*. https://doi.org/10.1016/j.ifacsc.2019.100052, 100052.

Chen, G., & Ge, Z. (2020). Robust Bayesian networks for low-quality data modeling and process monitoring applications. *Control Engineering Practice*, *97*(June 2019), 104344. https://doi.org/10.1016/j.conengprac.2020.104344.

Cheng, Y., Zhu, H., Wu, J., & Shao, X. (2019). Machine health monitoring using adaptive kernel spectral clustering and deep long short-term memory recurrent neural networks. *IEEE Transactions on Industrial Informatics*, *15*(2), 987–997. https://doi.org/10.1109/TII.2018.2866549.

Chouaib, C., Mohamed-Faouzi, H., & Messaoud, D. (2015). New adaptive kernel principal component analysis for nonlinear dynamic process monitoring. *Applied Mathematics & Information Sciences*, *9*(4), 1833–1845.

Colwell, T., Collet, O., & Downton, J. (2020). Theory-guided data science-based for reservoir characterization. In *1st EAGE conference on Machine Learning Americas: Applications in the O and G industry, challenges and opportunities* (pp. 1–5). https://doi.org/10.3997/2214-4609.202084013.

Cortes, C., & Vapnik, V. (1995). Support-vector networks. *Machine Learning*, *20*(3), 273–297. https://doi.org/10.1023/A:1022627411411.

Cui, Q., & Li, S. (2020). Process monitoring method based on correlation variable classification and vine copula. *The Canadian Journal of Chemical Engineering*, 1–18. https://doi.org/10.1002/cjce.23702.

Da Silva, J. C., Saxena, A., Balaban, E., & Goebel, K. (2012). A knowledge-based system approach for sensor fault modeling, detection and mitigation. *Expert Systems with Applications*, *39*(12), 10977–10989. https://doi.org/10.1016/j.eswa.2012.03.026.

Dai, X., & Gao, Z. (2013). From model, signal to knowledge: A data-driven perspective of fault detection and diagnosis. *IEEE Transactions on Industrial Informatics*, *9*(4), 2226–2238. https://doi.org/10.1109/TII.2013.2243743.

de Almeida, G. M., & Park, S. W. (2012). Fault detection in continuous industrial chemical processes: A new approach using the hidden Markov modeling. Case study: A boiler from a Brazilian cellulose pulp mill. In *Vol. 7435 LNCS. Lecture notes in computer science (including subseries lecture notes in artificial intelligence and lecture notes in bioinformatics)* (pp. 743–752). https://doi.org/10.1007/978-3-642-32639-4_88.

De Lathauwer, L., De Moor, B., & Vandewalle, J. (2000). An introduction to independent component analysis. *Journal of Chemometrics*, *14*(3), 123–149. https://doi.org/10.1002/1099-128X(200005/06)14:3<123::AID-CEM589>3.0.CO;2-1.

Dehuri, S., & Cho, S. B. (2009). Multi-criterion pareto based particle swarm optimized polynomial neural network for classification: A review and state-of-the-art. *Computer Science Review*, *3*(1), 19–40. https://doi.org/10.1016/j.cosrev.2008.11.002.

Dempster, A. P., Laird, N. M., & Rubin, D. B. (1977). Maximum likelihood from incomplete data via the EM algorithm. *Journal of the Royal Statistical Society. Series B (Methodological)*, 1–38. https://doi.org/10.1111/j.2517-6161.1977.tb01600.x.

Deng, X., Tian, X., & Chen, S. (2013). Modified kernel principal component analysis based on local structure analysis and its application to nonlinear process fault diagnosis. *Chemometrics and Intelligent Laboratory Systems*, *127*, 195–209. https://doi.org/10.1016/j.chemolab.2013.07.001.

Detroja, K. P., Gudi, R. D., & Patwardhan, S. C. (2006). A possibilistic clustering approach to novel fault detection and isolation. *Journal of Process Control*, *16*(10), 1055–1073. https://doi.org/10.1016/j.jprocont.2006.07.001.

Dike, H. U., Zhou, Y., Deveerasetty, K. K., & Wu, Q. (2019). Unsupervised learning based on artificial neural network: A review. In *2018 IEEE International Conference on Cyborg and Bionic Systems, CBS 2018* (pp. 322–327). https://doi.org/10.1109/CBS.2018.8612259.

Dorgo, G., Palazoglu, A., & Abonyi, J. (2021). Decision trees for informative process alarm definition and alarm-based fault classification. *Process Safety and Environmental Protection*, *149*, 312–324. https://doi.org/10.1016/j.psep.2020.10.024.

Downs, J. J., & Vogel, E. F. (1993). A plant-wide industrial problem process. *Computers & Chemical Engineering*, *17*(3), 245–255. https://doi.org/10.1016/0098-1354(93)80018-I.

Du, W., Fan, Y., Zhang, Y., & Zhang, J. (2017). Fault diagnosis of non-Gaussian process based on FKICA. *Journal of the Franklin Institute*, *354*(6), 2573–2590. https://doi.org/10.1016/j.jfranklin.2016.11.012.

Elshenawy, L. M., Mahmoud, T. A., & Chakour, C. (2020). Simultaneous fault detection and diagnosis using adaptive principal component analysis and multivariate contribution analysis. *Industrial & Engineering Chemistry Research*, *59*(47), 20798–20815. https://doi.org/10.1021/acs.iecr.0c04572.

Ertuğrul, Ö. F. (2018). A novel type of activation function in artificial neural networks: Trained activation function. *Neural Networks*, *99*, 148–157. https://doi.org/10.1016/J.NEUNET.2018.01.007.

Escobar, M. D. S., Kaneko, H., & Funatsu, K. (2016). Data visualization & clustering: Generative topographic mapping similarity assessment allied to graph theory clustering. In *Vol. 1222. ACS Symposium series* (pp. 175–210). https://doi.org/10.1021/bk-2016-1222.ch010.

Eslamloueyan, R. (2011). Designing a hierarchical neural network based on fuzzy clustering for fault diagnosis of the Tennessee-Eastman process. *Applied Soft Computing Journal, 11*(1), 1407–1415. https://doi.org/10.1016/j.asoc.2010.04.012.
Fan, J., & Wang, Y. (2014). Fault detection and diagnosis of non-linear non-Gaussian dynamic processes using kernel dynamic independent component analysis. *Information Sciences, 259*, 369–379. https://doi.org/10.1016/j.ins.2013.06.021.
Fisher, O. J., Watson, N. J., Escrig, J. E., Witt, R., … Gomes, R. L. (2020). Considerations, challenges and opportunities when developing data-driven models for process manufacturing systems. *Computers and Chemical Engineering, 140*. https://doi.org/10.1016/j.compchemeng.2020.106881, 106881.
Gajjar, S., Kulahci, M., & Palazoglu, A. (2018). Real-time fault detection and diagnosis using sparse principal component analysis. *Journal of Process Control, 67*, 112–128. https://doi.org/10.1016/j.jprocont.2017.03.005.
Galagedarage Don, M., & Khan, F. (2019a). Dynamic process fault detection and diagnosis based on a combined approach of hidden Markov and Bayesian network model. *Chemical Engineering Science, 201*, 82–96. https://doi.org/10.1016/j.ces.2019.01.060.
Galagedarage Don, M., & Khan, F. (2019b). Process fault prognosis using hidden Markov model–Bayesian networks hybrid model. *Industrial & Engineering Chemistry Research, 58*(27), 12041–12053. https://doi.org/10.1021/acs.iecr.9b00524.
Gao, Z., Cecati, C., & Ding, S. X. (2015). A survey of fault diagnosis and fault-tolerant techniques—Part I: Fault diagnosis with model-based and signal-based approaches. *IEEE Transactions on Industrial Electronics, 62*(6), 3757–3767.
Ge, Z. (2017). Review on data-driven modeling and monitoring for plant-wide industrial processes. *Chemometrics and Intelligent Laboratory Systems, 171*, 16–25. https://doi.org/10.1016/j.chemolab.2017.09.021.
Ge, Z., Song, Z., & Gao, F. (2013). Review of recent research on data-based process monitoring. *Industrial & Engineering Chemistry Research, 52*(10), 3543–3562. https://doi.org/10.1021/ie302069q.
Gharahbagheri, H., Imtiaz, S. A., & Khan, F. (2017). Root cause diagnosis of process fault using KPCA and Bayesian network. *Industrial & Engineering Chemistry Research, 56*(8), 2054–2070. https://doi.org/10.1021/acs.iecr.6b01916.
Gomm, J. B. (1996). On-line learning for fault classification using an adaptive neuro-fuzzy network. *IFAC Proceedings Volumes, 29*(1), 6494–6499. https://doi.org/10.1016/S1474-6670(17)58724-8.
Gordon, C. A. K., & Pistikopoulos, E. N. (2021). Data-driven prescriptive maintenance toward fault-tolerant multiparametric control. *AIChE Journal*. https://doi.org/10.1002/aic.17489.
He, Q. P., Qin, S. J., & Wang, J. (2005). A new fault diagnosis method using fault directions in Fisher discriminant analysis. *AICHE Journal, 51*(2), 555–571. https://doi.org/10.1002/aic.10325.
Hendriks, J., Jidling, C., Wills, A., & Schön, T. (2020). *Linearly constrained neural networks*. http://arxiv.org/abs/2002.01600.
Heo, S., & Lee, J. H. (2018). Fault detection and classification using artificial neural networks. *IFAC-PapersOnLine*. https://doi.org/10.1016/j.ifacol.2018.09.380.
Hornik, K., Stinchcombe, M., & White, H. (1989). Multilayer feedforward networks are universal approximators. *Neural Networks, 2*(5), 359–366. https://doi.org/10.1016/0893-6080(89)90020-8.
Hoskins, J. C., & Himmelblau, D. M. (1988). Artificial neural network models of knowledge representation in chemical engineering. *Computers and Chemical Engineering, 12*(9–10), 881–890. https://doi.org/10.1016/0098-1354(88)87015-7.
Hu, X., Hu, H., Verma, S., & Zhang, Z. L. (2021). Physics-guided deep neural networks for power flow analysis. *IEEE Transactions on Power Systems, 36*(3), 2082–2092. https://doi.org/10.1109/TPWRS.2020.3029557.

Hu, G., Li, H., Xia, Y., & Luo, L. (2018). A deep Boltzmann machine and multi-grained scanning forest ensemble collaborative method and its application to industrial fault diagnosis. *Computers in Industry*, *100*, 287–296. https://doi.org/10.1016/j.compind.2018.04.002.

Huang, J., & Yan, X. (2015). Dynamic process fault detection and diagnosis based on dynamic principal component analysis, dynamic independent component analysis and Bayesian inference. *Chemometrics and Intelligent Laboratory Systems*, *148*, 115–127. https://doi.org/10.1016/j.chemolab.2015.09.010.

Hunt, K. J., Sbarbaro, D., Zbikowski, R., & Gawthrop, P. J. (1992). Neural networks for control systems—A survey. *Automatica*, *28*(6), 1083–1112. https://doi.org/10.1016/0005-1098(92)90053-I.

Hyvärinen, A., & Oja, E. (2000). Independent component analysis: Algorithms and applications. *Neural Networks*, *13*(4–5), 411–430. https://doi.org/10.1016/S0893-6080(00)00026-5.

Iliou, T., Anagnostopoulos, C. N., Stephanakis, I. M., & Anastassopoulos, G. (2017). A novel data pre-processing method for boosting neural network performance: A case study in osteoporosis prediction. *Information Sciences*, *380*, 92–100. https://doi.org/10.1016/J.INS.2015.10.026.

Isermann, R. (2005). Model-based fault-detection and diagnosis—Status and applications. *Annual Reviews in Control*, *29*(1), 71–85. https://doi.org/10.1016/j.arcontrol.2004.12.002.

Jeng, J.-C. (2010). Adaptive process monitoring using efficient recursive PCA and moving window PCA algorithms. *Journal of the Taiwan Institute of Chemical Engineers*, *41*(4), 475–481. https://doi.org/10.1016/j.jtice.2010.03.015.

Jiang, Q., & Yan, X. (2015). Nonlinear plant-wide process monitoring using MI-spectral clustering and Bayesian inference-based multiblock KPCA. *Journal of Process Control*, *32*, 38–50. https://doi.org/10.1016/j.jprocont.2015.04.014.

Jiao, Z., Hu, P., Xu, H., & Wang, Q. (2020). Machine learning and deep learning in chemical health and safety: A systematic review of techniques and applications. *Journal of Chemical Health and Safety*, *27*(6), 316–334. https://doi.org/10.1021/acs.chas.0c00075.

Jokinen, P. A. (1992). Non-linear network model for continuously learning expert systems. Part II. Application. *Process Control and Quality*, *2*(2), 155–168.

Juang, C. F. (2004). A hybrid of genetic algorithm and particle swarm optimization for recurrent network design. *IEEE Transactions on Systems, Man, and Cybernetics, Part B: Cybernetics*, *34*(2), 997–1006. https://doi.org/10.1109/TSMCB.2003.818557.

Kang, J.-L. (2020). Visualization analysis for fault diagnosis in chemical processes using recurrent neural networks. *Journal of the Taiwan Institute of Chemical Engineers*, *112*, 137–151. https://doi.org/10.1016/j.jtice.2020.06.016.

Kano, M., Tanaka, S., Hasebe, S., Hashimoto, I., & Ohno, H. (2003). Monitoring independent components for fault detection. *AICHE Journal*, *49*(4), 969–976. https://doi.org/10.1002/aic.690490414.

Karpatne, A., Atluri, G., Faghmous, J. H., Steinbach, M., … Kumar, V. (2017). Theory-guided data science: A new paradigm for scientific discovery from data. *IEEE Transactions on Knowledge and Data Engineering*, *29*(10), 2318–2331. https://doi.org/10.1109/TKDE.2017.2720168.

Kaspar, M. H., & Harmon Ray, W. (1993). Dynamic PLS modelling for process control. *Chemical Engineering Science*, *48*(20), 3447–3461. https://doi.org/10.1016/0009-2509(93)85001-6.

Kerschen, G., & Golinval, J.-C. (2002). Non-linear generalization of principal component analysis: From a global to a local approach. *Journal of Sound and Vibration*, *254*(5), 867–876. https://doi.org/10.1006/jsvi.2001.4129.

Khediri, I. B., Weihs, C., & Limam, M. (2012). Kernel k-means clustering based local support vector domain description fault detection of multimodal processes. *Expert Systems with Applications*, *39*(2), 2166–2171. https://doi.org/10.1016/j.eswa.2011.07.045.

Kohonen, T., & Somervuo, P. (1998). Self-organizing maps of symbol strings. *Neurocomputing*, *21*(1–3), 19–30. https://doi.org/10.1016/S0925-2312(98)00031-9.

Kotsiantis, S. B. (2007). Supervised machine learning: A review of classification techniques. *Informatica*, *31*, 249–268.

Kotsiantis, S. B., & Kanellopoulos, D. (2006). Data pre-processing for supervised leaning. *International Journal of Computer Science*, *1*(2), 1–7. https://doi.org/10.1080/02331931003692557.

Kourti, T., & MacGregor, J. F. (1995). Process analysis, monitoring and diagnosis, using multivariate projection methods. *Chemometrics and Intelligent Laboratory Systems*, *28*(1), 3–21. https://doi.org/10.1016/0169-7439(95)80036-9.

Kresta, J. V., Macgregor, J. F., & Marlin, T. E. (1991). Multivariate statistical monitoring of process operating performance. *The Canadian Journal of Chemical Engineerging*, *69*(1), 35–47. https://doi.org/10.1258/phleb.2011.010101.

Ku, W., Storer, R. H., & Georgakis, C. (1995). Disturbance detection and isolation by dynamic principal component analysis. *Chemometrics and Intelligent Laboratory Systems*, *30*(1), 179–196. https://doi.org/10.1016/0169-7439(95)00076-3.

Lau, C. K., Ghosh, K., Hussain, M. A., & Che Hassan, C. R. (2013). Fault diagnosis of Tennessee Eastman process with multi-scale PCA and ANFIS. *Chemometrics and Intelligent Laboratory Systems*, *120*, 1–14. https://doi.org/10.1016/j.chemolab.2012.10.005.

Laudani, A., Lozito, G. M., Fulginei, F. R., & Salvini, A. (2015). On training efficiency and computational costs of a feed forward neural network: A review. *Computational Intelligence and Neuroscience*, *2015*, 818243. https://doi.org/10.1155/2015/818243.

Lawrence, N. (2005). Probabilistic non-linear principal component analysis with Gaussian process latent variable models. *Journal of Machine Learning Research*, *6*, 1783–1816.

Lawrence, R. L., & Wright, A. (2001). Rule-based classification systems using classification and regression tree (CART) analysis. *Photogrammetric Engineering and Remote Sensing*, *67*(10), 1137–1142.

Lee, J., Cho, W., & Choi, J. (2021). Fault detection for IoT hydrogen refueling station system using a combined hidden Markov model mixed with Gaussian. In *2021 International Conference on Electrical, Computer, Communications and Mechatronics Engineering (ICECCME)* (pp. 1–6). https://doi.org/10.1109/ICECCME52200.2021.9590853.

Lee, J., & Kang, S. (2007). GA based meta-modeling of BPN architecture for constrained approximate optimization. *International Journal of Solids and Structures*, *44*(18–19), 5980–5993. https://doi.org/10.1016/j.ijsolstr.2007.02.008.

Lee, J. M., Yoo, C. K., & Lee, I. B. (2004). Statistical process monitoring with independent component analysis. *Journal of Process Control*, *14*(5), 467–485. https://doi.org/10.1016/j.jprocont.2003.09.004.

Lee, S., Li, L., & Ni, J. (2010). Adaptive anomaly detection using a hidden Markov model. In *Vol. 2, 2. ASME 2010 international manufacturing science and engineering conference* (pp. 599–606). https://doi.org/10.1115/MSEC2010-34169.

Lee, J.-M., Qin, S. J., & Lee, I.-B. (2007). Fault detection of non-linear processes using kernel independent component analysis. *The Canadian Journal of Chemical Engineering*, *85*(4), 526–536. https://doi.org/10.1002/cjce.5450850414.

Lei, Y., Yang, B., Jiang, X., Jia, F., & Nandi, A. K. (2020). Applications of machine learning to machine fault diagnosis: A review and roadmap. *Mechanical Systems and Signal Processing*, *138*, 106587. https://doi.org/10.1016/J.YMSSP.2019.106587.

Lemos, A., Caminhas, W., & Gomide, F. (2013). Adaptive fault detection and diagnosis using an evolving fuzzy classifier. *Information Sciences*, *220*, 64–85. https://doi.org/10.1016/j.ins.2011.08.030.

Li, J., Cheng, J. H., Shi, J. Y., & Huang, F. (2012). Brief introduction of Back propagation (BP) neural network algorithm and its improvement. In D. Jin, & S. Lin (Eds.), *Vol. 169. Advances in computer science and information engineering. Advances in intelligent and soft computing*. Berlin, Heidelberg: Springer. https://doi.org/10.1007/978-3-642-30223-7_87.

Li, G., Qin, S. J., & Yuan, T. (2016). Data-driven root cause diagnosis of faults in process industries. *Chemometrics and Intelligent Laboratory Systems*, *159*, 1–11. https://doi.org/10.1016/j.chemolab.2016.09.006.

Li, X., Peng, L., Yao, X., Cui, S., Hu, Y., You, C., et al. (2017). Long short-term memory neural network for air pollutant concentration predictions: Method development and evaluation. *Environmental Pollution*, *231*, 997–1004. https://doi.org/10.1016/j.envpol.2017.08.114.

Li, Y., Wang, Z., & Yuan, J. (2006). On-line fault detection using SVM-based dynamic MPLS for batch processes. *Chinese Journal of Chemical Engineering*, *14*(6), 754–758. https://doi.org/10.1016/S1004-9541(07)60007-X.

Lima, P. V., & Saraiva, P. M. (2007). A semi-mechanistic model building framework based on selective and localized model extensions. *Computers and Chemical Engineering*, *31*(4), 361–373. https://doi.org/10.1016/j.compchemeng.2006.07.006.

Liu, J. (2012). Fault diagnosis using contribution plots without smearing effect on non-faulty variables. *Journal of Process Control*, *22*(9), 1609–1623.

Liu, Y., & Ge, Z. (2019). Deep ensemble forests for industrial fault classification. *IFAC Journal of Systems and Control*, *10*. https://doi.org/10.1016/j.ifacsc.2019.100071, 100071.

Liu, S., McDermid, J. A., & Chen, Y. (2010). A rigorous method for inspection of model-based formal specifications. *IEEE Transactions on Reliability*, *59*(4), 667–684. https://doi.org/10.1109/TR.2010.2085571.

Liu, J., Zheng, W., Zong, Y., Lu, C., & Tang, C. (2020). Cross-corpus speech emotion recognition based on deep domain-adaptive convolutional neural network. *IEICE Transactions on Information and Systems*, *E103D*(2), 459–463. https://doi.org/10.1587/transinf.2019EDL8136.

Mallick, M. R., & Imtiaz, S. A. (2013). A hybrid method for process fault detection and diagnosis. *IFAC Proceedings Volumes*, *46*(32), 827–832. https://doi.org/10.3182/20131218-3-IN-2045.00099.

Marins, M. A., Barros, B. D., Santos, I. H., Barrionuevo, D. C., … Netto, S. L. (2021). Fault detection and classification in oil wells and production/service lines using random forest. *Journal of Petroleum Science and Engineering*, *197*, 107879. https://doi.org/10.1016/j.petrol.2020.107879.

Mavkov, B., Forgione, M., & Piga, D. (2020). Integrated neural networks for nonlinear continuous-time system identification. *IEEE Control Systems Letters*, *4*(4), 851–856. https://doi.org/10.1109/LCSYS.2020.2994806.

McBride, K., Sanchez Medina, E. I., & Sundmacher, K. (2020). Hybrid semi-parametric modeling in separation processes: A review. *Chemie-Ingenieur-Technik*, *92*(7), 842–855. https://doi.org/10.1002/cite.202000025.

McCulloch, W. S., & Pitts, W. (1990). A logical calculus of the ideas immanent in nervous activity (reprinted from bulletin of mathematical biophysics, vol 5, pg 115-133, 1943). *Bulletin of Mathematical Biology*, *52*(1–2), 99–115.

Md Nor, N., Hussain, M. A., & Che Hassan, C. R. (2017). Fault diagnosis and classification framework using multi-scale classification based on kernel Fisher discriminant analysis for chemical process system. *Applied Soft Computing*, *61*, 959–972. https://doi.org/10.1016/j.asoc.2017.09.019.

Mehranbod, N., Soroush, M., & Panjapornpon, C. (2005). A method of sensor fault detection and identification. *Journal of Process Control*, *15*(3), 321–339. https://doi.org/10.1016/j.jprocont.2004.06.009.

Mondal, S., Ghalyan, N. F., Ray, A., & Mukhopadhyay, A. (2019). Early detection of thermoacoustic instabilities using hidden Markov models. *Combustion Science and Technology*, *191*(8), 1309–1336. https://doi.org/10.1080/00102202.2018.1523900.

Monroy, I., Benitez, R., Escudero, G., & Graells, M. (2010). A semi-supervised approach to fault diagnosis for chemical processes. *Computers and Chemical Engineering*, *34*(5), 631–642. https://doi.org/10.1016/j.compchemeng.2009.12.008.

Monroy, I., Escudero, G., & Graells, M. (2009). Anomaly detection in batch chemical processes. *Computer Aided Chemical Engineering*, *26*, 255–260. https://doi.org/10.1016/S1570-7946(09)70043-4.

Muralidhar, N., Islam, M. R., & Marwah, M. (2018). Incorporating prior domain knowledge into deep neural networks. *IEEE Big Data*, 36–45. http://people.cs.vt.edu/naren/papers/PID5657885.pdf.

Murphy, C. P., & Kerekes, J. P. (2021). Physics-guided neural network for predicting chemical signatures. *Applied Optics*, *60*(11), 3176. https://doi.org/10.1364/ao.420688.

Musulin, E., Yélamos, I., & Puigjaner, L. (2006). Integration of principal component analysis and fuzzy logic systems for comprehensive process fault detection and diagnosis. *Industrial & Engineering Chemistry Research*, *45*(5), 1739–1750. https://doi.org/10.1021/ie050695i.

Nakamura, T. A., Palhares, R. M., Caminhas, W. M., Menezes, B. R., ... Lemos, A. P. (2017). Adaptive fault detection and diagnosis using parsimonious Gaussian mixture models trained with distributed computing techniques. *Journal of the Franklin Institute*, *354*(6), 2543–2572. https://doi.org/10.1016/j.jfranklin.2016.11.024.

Nawi, N. M., Khan, A., & Rehman, M. Z. (2013). A new Levenberg Marquardt based Back propagation algorithm trained with cuckoo search. *Procedia Technology*, *11*(Iceei), 18–23. https://doi.org/10.1016/j.protcy.2013.12.157.

Nazir, S., & Manca, D. (2015). How a plant simulator can improve industrial safety. *Process Safety Progress*, *34*(3), 237–243. https://doi.org/10.1002/prs.11714.

Neumann, J., Deerberg, G., & Schlüter, S. (1999). Early detection and identification of dangerous states in chemical plants using neural networks. *Journal of Loss Prevention in the Process Industries*, *12*(6), 451–453. https://doi.org/10.1016/S0950-4230(98)00059-X.

Nomikos, P., & MacGregor, J. F. (1994). Monitoring batch processes using multiway principal component analysis. *AICHE Journal*, *40*(8), 1361–1375. https://doi.org/10.1002/aic.690400809.

Nomikos, P., & MacGregor, J. F. (1995). Multi-way partial least squares in monitoring batch processes. *Chemometrics and Intelligent Laboratory Systems*, *30*(1), 97–108. https://doi.org/10.1016/0169-7439(95)00043-7.

Onel, M., Kieslich, C. A., Guzman, Y. A., & Pistikopoulos, E. N. (2018). Simultaneous fault detection and identification in continuous processes via nonlinear support vector machine based feature selection. *Computer Aided Chemical Engineering*, *44*, 2077–2082. https://doi.org/10.1016/B978-0-444-64241-7.50341-4.

Orantes, A., Kempowsky, T., Le Lann, M.-V., & Aguilar-Martin, J. (2008). A new support methodology for the placement of sensors used for fault detection and diagnosis. *Chemical Engineering and Processing: Process Intensification*, *47*(3), 330–348. https://doi.org/10.1016/j.cep.2007.01.024.

Pawar, S., San, O., Aksoylu, B., Rasheed, A., & Kvamsdal, T. (2021). Physics guided machine learning using simplified theories. *Physics of Fluids*, *33*(1), 011701. https://doi.org/10.1063/5.0038929.

Petković, M., Rapaić, M. R., Jeličić, Z. D., & Pisano, A. (2012). On-line adaptive clustering for process monitoring and fault detection. *Expert Systems with Applications*, *39*(11), 10226–10235. https://doi.org/10.1016/j.eswa.2012.02.150.

Pham, B. T., Nguyen, M. D., Bui, K. T. T., Prakash, I., ... Bui, D. T. (2019). A novel artificial intelligence approach based on multi-layer perceptron neural network and biogeography-based optimization for predicting coefficient of consolidation of soil. *CATENA*, *173*, 302–311. https://doi.org/10.1016/J.CATENA.2018.10.004.

Portnoy, I., Melendez, K., Pinzon, H., & Sanjuan, M. (2016). An improved weighted recursive PCA algorithm for adaptive fault detection. *Control Engineering Practice*, *50*, 69–83. https://doi.org/10.1016/j.conengprac.2016.02.010.

Pospíchal, J., & Kvasnička, V. (2015). 70th anniversary of publication: Warren McCulloch & walter pitts—A logical calculus of the ideas immanent in nervous activity. *Adv. Intell. Syst. Comput.*, *316*, 1–10. https://doi.org/10.1007/978-3-319-10783-7_1.

Psichogios, D. C., & Ungar, L. H. (1992). A hybrid neural network-first principles approach to process modeling. *AICHE Journal*, *38*(10), 1499–1511. https://doi.org/10.1002/aic.690381003.

Qin, S. J. (1998). Recursive PLS algorithms for adaptive data modeling. *Computers and Chemical Engineering*, *22*(4–5), 503–514. https://doi.org/10.1016/s0098-1354(97)00262-7.

Qin, S. J. (2012). Survey on data-driven industrial process monitoring and diagnosis. *Annual Reviews in Control*, *36*(2), 220–234. https://doi.org/10.1016/j.arcontrol.2012.09.004.

Raissi, M., Perdikaris, P., & Karniadakis, G. E. (2017). Machine learning of linear differential equations using Gaussian processes. *Journal of Computational Physics*, *348*, 683–693. https://doi.org/10.1016/j.jcp.2017.07.050.

Rashid, M. M., & Yu, J. (2012). Hidden markov model based adaptive independent component analysis approach for complex chemical process monitoring and fault detection. *Industrial and Engineering Chemistry Research*, *51*(15), 5506–5514. https://doi.org/10.1021/ie300203u.

Rato, T. J., & Reis, M. S. (2013). Fault detection in the Tennessee Eastman benchmark process using dynamic principal components analysis based on decorrelated residuals (DPCA-DR). *Chemometrics and Intelligent Laboratory Systems*, *125*, 101–108. https://doi.org/10.1016/j.chemolab.2013.04.002.

Rebello, S., Yu, H., & Ma, L. (2019). An integrated approach for real-time hazard mitigation in complex industrial processes. *Reliability Engineering & System Safety*, *188*, 297–309. https://doi.org/10.1016/j.ress.2019.03.037.

Reichstein, M., Camps-Valls, G., Stevens, B., Jung, M., ... Prabhat. (2019). Deep learning and process understanding for data-driven Earth system science. *Nature*, *566*(7743), 195–204. https://doi.org/10.1038/s41586-019-0912-1.

Ren, S., Song, Z., Yang, M., & Ren, J. (2015). A novel multimode process monitoring method integrating LCGMM with modified LFDA. *Chinese Journal of Chemical Engineering*, *23*(12), 1970–1980. https://doi.org/10.1016/j.cjche.2015.09.007.

Ren, X., Tian, Y., & Li, S. (2015). Vine copula-based dependence description for multivariate multimode process monitoring. *Industrial and Engineering Chemistry Research*, *54*(41), 10001–10019. https://doi.org/10.1021/acs.iecr.5b01267.

Ren, X., Zhu, K., Cai, T., & Li, S. (2017). Fault detection and diagnosis for nonlinear and non-Gaussian processes based on copula subspace division. *Industrial and Engineering Chemistry Research*, *56*(40), 11545–11564. https://doi.org/10.1021/acs.iecr.7b02419.

Reshef, Y. A., Reshef, D. N., Sabeti, P. C., & Mitzenmacher, M. (2014). Theoretical foundations of equitability and the maximal information coefficient. *ArXiv* (Preprint ArXiv:1408.4908).

Rojas-Guzman, C., & Kramer, M. A. (1993). Comparison of belief networks and rule-based expert systems for fault diagnosis of chemical processes. *Engineering Applications of Artificial Intelligence*, *6*(3), 191–202. https://doi.org/10.1016/0952-1976(93)90062-3.

Ruiz, D., Nougués, J. M., & Puigjaner, L. (2001). Fault diagnosis support system for complex chemical plants. *Computers & Chemical Engineering*, *25*(1), 151–160. https://doi.org/10.1016/S0098-1354(00)00638-4.

Sansana, J., Joswiak, M. N., Castillo, I., Wang, Z., ... Reis, M. S. (2021). Recent trends on hybrid modeling for industry 4.0. *Computers and Chemical Engineering*, *151*, 107365. https://doi.org/10.1016/j.compchemeng.2021.107365.

Schmidhuber, J. (2015). Deep learning in neural networks: An overview. *Neural Networks*, *61*, 85–117. https://doi.org/10.1016/j.neunet.2014.09.003.

Schuld, M., Sinayskiy, I., & Petruccione, F. (2014). Quantum computing for pattern classification. *Lecture Notes in Computer Science (Including Subseries Lecture Notes in Artificial Intelligence and Lecture Notes in Bioinformatics)*, *8862*, 208–220. https://doi.org/10.1007/978-3-319-13560-1.

Schwarz, G. (1978). Estimating the dimension of a model. *The Annals of Statistics*, *6*(2), 461–464. https://doi.org/10.2307/2958889.

Sheibat-Othman, N., Laouti, N., Valour, J.-P., & Othman, S. (2014). Support vector machines combined to observers for fault diagnosis in chemical reactors. *The Canadian Journal of Chemical Engineering*, *92*(4), 685–695. https://doi.org/10.1002/cjce.21881.

Sheta, A., Braik, M., & Al-Hiary, H. (2019). Modeling the Tennessee Eastman chemical process reactor using bio-inspired feed-forward neural network (BI-FF-NN). *International Journal of Advanced Manufacturing Technology*, *103*(1–4), 1359–1380. https://doi.org/10.1007/s00170-019-03621-5.

Skogestad, S. (1997). Dynamics and control of distillation columns: A tutorial introduction. *Chemical Engineering Research and Design*, *75*(6), 539–562. https://doi.org/10.1205/026387697524092.

Smyth, P. (1994). Hidden Markov models for fault detection in dynamic systems. *Pattern Recognition*, *27*(1), 149–164. https://doi.org/10.1016/0031-3203(94)90024-8.

Sorsa, T., & Koivo, H. N. (1993). Application of artificial neural networks in process fault diagnosis. *Automatica*, *29*(4), 843–849. https://doi.org/10.1016/0005-1098(93)90090-G.

Staroswiecki, M. (2000). Quantitative and qualitative models for fault detection and isolation. *Mechanical Systems and Signal Processing*, *14*(3), 301–325. https://doi.org/10.1006/mssp.2000.1293.

Stefatos, G., & Ben Hamza, A. (2010). Dynamic independent component analysis approach for fault detection and diagnosis. *Expert Systems with Applications*, *37*(12), 8606–8617. https://doi.org/10.1016/j.eswa.2010.06.101.

Su, Y., Tao, F., Jin, J., Wang, T., & Wang, L. (2020). Failure prognosis of complex equipment with multistream deep recurrent neural network. *Journal of Computing and Information Science in Engineering*, *20*(2). https://doi.org/10.1115/1.4045445/1068285.

Subudhi, B., & Jena, D. (2011a). A differential evolution based neural network approach to nonlinear system identification. *Applied Soft Computing Journal*, *11*(1), 861–871. https://doi.org/10.1016/j.asoc.2010.01.006.

Subudhi, B., & Jena, D. (2011b). Nonlinear system identification using memetic differential evolution trained neural networks. *Neurocomputing*, *74*(10), 1696–1709. https://doi.org/10.1016/j.neucom.2011.02.006.

Sung, A. H. (1998). Ranking importance of input parameters of neural networks. *Expert Systems with Applications*, *15*(3–4), 405–411. https://doi.org/10.1016/s0957-4174(98)00041-4.

Svendsen, D. H., Piles, M., Munoz-Mari, J., Luengo, D., … Camps-Valls, G. (2021). Integrating domain knowledge in data-driven earth observation with process convolutions. *IEEE Transactions on Geoscience and Remote Sensing*, 1–15. https://doi.org/10.1109/TGRS.2021.3059550.

Svetnik, V., Wang, T., Tong, C., Liaw, A., & Song, Q. (2005). Boosting: An ensemble learning tool for compound classification and QSAR modeling. *Journal of Chemical Information and Modeling*, *45*(3), 786–799. https://doi.org/10.1021/ci0500379.

Szafnicki, K., Narce, C., & Bourgois, J. (2005). Towards an integrated tool for control, supervision and operator training—Application to industrial wastewater detoxication plants. *Control Engineering Practice*, *13*(6), 729–738. https://doi.org/10.1016/j.conengprac.2004.07.002.

Theilliol, D., Weber, P., Ghetie, M., & Noura, H. (1995). A hierarchical fault diagnosis method using a decision support system applied to a chemical plant. In *1995 IEEE international conference on systems, man and cybernetics. Intelligent systems for the 21st century, 3* (pp. 2205–2210). https://doi.org/10.1109/ICSMC.1995.538108.

Thissen, U., Swierenga, H., De Weijer, A. P., Wehrens, R., … Buydens, L. M. C. (2005). Multivariate statistical process control using mixture modelling. *Journal of Chemometrics, 19*(1), 23–31. https://doi.org/10.1002/cem.903.

Thompson, M. L., & Kramer, M. A. (1994). Modeling chemical processes using prior knowledge and neural networks. *AICHE Journal, 40*(8), 1328–1340. https://doi.org/10.1002/aic.690400806.

Tidriri, K., Chatti, N., Verron, S., & Tiplica, T. (2016). Bridging data-driven and model-based approaches for process fault diagnosis and health monitoring: A review of researches and future challenges. *Annual Reviews in Control, 42*, 63–81. https://doi.org/10.1016/J.ARCONTROL.2016.09.008.

Tong, C., Palazoglu, A., & Yan, X. (2014). Improved ICA for process monitoring based on ensemble learning and Bayesian inference. *Chemometrics and Intelligent Laboratory Systems, 135*, 141–149. https://doi.org/10.1016/j.chemolab.2014.04.012.

Trygg, J., & Wold, S. (2002). Orthogonal projections to latent structures (O-PLS). *Journal of Chemometrics, 16*(3), 119–128. https://doi.org/10.1002/cem.695.

Tudón-Martínez, J. C., & Morales-Menendez, R. (2015). Actuator fault diagnosis in a heat exchanger based on classifiers—A comparative study. *IFAC-PapersOnLine, 48*(21), 1210–1215. https://doi.org/10.1016/j.ifacol.2015.09.691.

Uraikul, V., Chan, C. W., & Tontiwachwuthikul, P. (2007). Artificial intelligence for monitoring and supervisory control of process systems. *Engineering Applications of Artificial Intelligence, 20*(2), 115–131. https://doi.org/10.1016/j.engappai.2006.07.002.

Van Lith, P. F., Betlem, B. H. L., & Roffel, B. (2003). Combining prior knowledge with data driven modeling of a batch distillation column including start-up. *Computers and Chemical Engineering, 27*(7), 1021–1030. https://doi.org/10.1016/S0098-1354(03)00067-X.

Venkatasubramanian, V., Rengaswamy, R., Yin, K., & Kavuri, S. N. (2003). A review of process fault detection and diagnosis, part I quantitative model-based methods. *Computers & Chemical Engineering, 27*(3), 293–311. https://doi.org/10.1016/S0098-1354(02)00160-6.

von Stosch, M., Oliveira, R., Peres, J., & Feyo de Azevedo, S. (2014). Hybrid semi-parametric modeling in process systems engineering: Past, present and future. *Computers and Chemical Engineering, 60*, 86–101. https://doi.org/10.1016/j.compchemeng.2013.08.008.

Vonrueden, L., Mayer, S., Beckh, K., Georgiev, B., … Schuecker, J. (2021). Informed machine learning—A taxonomy and survey of integrating prior knowledge into learning systems. *IEEE Transactions on Knowledge and Data Engineering*, 1. https://doi.org/10.1109/TKDE.2021.3079836.

Wan, J., & Li, S. (2019). Modeling and application of industrial process fault detection based on pruning vine copula. *Chemometrics and Intelligent Laboratory Systems, 184*, 1–13. https://doi.org/10.1016/j.chemolab.2018.11.005.

Wang, X., Kruger, U., & Irwin, G. W. (2005). Process monitoring approach using fast moving window PCA. *Industrial & Engineering Chemistry Research, 44*(15), 5691–5702. https://doi.org/10.1021/ie048873f.

Wang, J., Li, Y., Zhao, R., & Gao, R. X. (2020). Physics guided neural network for machining tool wear prediction. *Journal of Manufacturing Systems, 57*, 298–310. https://doi.org/10.1016/j.jmsy.2020.09.005.

Wang, Z., Liu, Y., & Griffin, P. J. (2000). A combined ANN and expert system tool for transformer fault diagnosis. 2000 IEEE power engineering society. *Conference Proceedings, 2*, 1261–1269. https://doi.org/10.1109/PESW.2000.850127.

Wang, Y., Pan, Z., Yuan, X., Yang, C., & Gui, W. (2020). A novel deep learning based fault diagnosis approach for chemical process with extended deep belief network. *ISA Transactions*, *96*, 457–467. https://doi.org/10.1016/j.isatra.2019.07.001.

Wen, H., Khan, F., Amin, M. T., & Halim, S. Z. (2021). Myths and misconceptions of data-driven methods: Applications to process safety analysis. *Computers & Chemical Engineering*. https://doi.org/10.1016/j.compchemeng.2021.107639, 107639.

Widodo, A., & Yang, B. S. (2007). Support vector machine in machine condition monitoring and fault diagnosis. *Mechanical Systems and Signal Processing*, *21*(6), 2560–2574. https://doi.org/10.1016/j.ymssp.2006.12.007.

Wise, B., Veltkamp, D., & Davis, B. (1988). Principal components analysis for monitoring the west valley liquid fed ceramic melter. In *Waste management '88: Symposium on radioactive waste management* (pp. 811–818). Tucson, AZ, USA, 28 Feb–3 Mar.

Wong, J. C., McDonald, K. A., & Palazoglu, A. (2001). Classification of abnormal plant operation using multiple process variable trends. *Journal of Process Control*, *11*(4), 409–418. https://doi.org/10.1016/S0959-1524(00)00011-1.

Wu, H., & Zhao, J. (2018). Deep convolutional neural network model based chemical process fault diagnosis. *Computers and Chemical Engineering*, *115*, 185–197. https://doi.org/10.1016/j.compchemeng.2018.04.009.

Wu, H., & Zhao, J. (2020). Fault detection and diagnosis based on transfer learning for multimode chemical processes. *Computers and Chemical Engineering*, *135*. https://doi.org/10.1016/j.compchemeng.2020.106731.

Wu, D., & Zhao, J. (2021). Process topology convolutional network model for chemical process fault diagnosis. *Process Safety and Environmental Protection*, *150*, 93–109. https://doi.org/10.1016/j.psep.2021.03.052.

Xiao, L., Li, K., Tan, Z., Zhang, Z., … Li, S. (2019). Nonlinear gradient neural network for solving system of linear equations. *Information Processing Letters*, *142*, 35–40. https://doi.org/10.1016/J.IPL.2018.10.004.

Xie, X., & Shi, H. (2012). Dynamic multimode process modeling and monitoring using adaptive gaussian mixture models. *Industrial and Engineering Chemistry Research*, *51*(15), 5497–5505. https://doi.org/10.1021/ie202720y.

Xie, Z., Yang, X., Li, A., & Ji, Z. (2019). Fault diagnosis in industrial chemical processes using optimal probabilistic neural network. *Canadian Journal of Chemical Engineering*, *97*(9), 2453–2464. https://doi.org/10.1002/cjce.23491.

Xiong, Q., & Jutan, A. (2002). Grey-box modelling and control of chemical processes. *Chemical Engineering Science*, *57*(6), 1027–1039. https://doi.org/10.1016/S0009-2509(01)00439-0.

Xu, Y., & Deng, X. (2016). Fault detection of multimode non-Gaussian dynamic process using dynamic Bayesian independent component analysis. *Neurocomputing*, *200*, 70–79. https://doi.org/10.1016/j.neucom.2016.03.015.

Yang, C., & Hou, J. (2016). Fed-batch fermentation penicillin process fault diagnosis and detection based on support vector machine. *Neurocomputing*, *190*, 117–123. https://doi.org/10.1016/j.neucom.2016.01.027.

Yin, S., Ding, S. X., Haghani, A., Hao, H., & Zhang, P. (2012). A comparison study of basic data-driven fault diagnosis and process monitoring methods on the benchmark Tennessee Eastman process. *Journal of Process Control*, *22*(9), 1567–1581.

Yin, S., Ding, S. X., Xie, X., & Luo, H. (2014). A review on basic data-driven approaches for industrial process monitoring. *IEEE Transactions on Industrial Electronics*, *61*(11), 6418–6428. https://doi.org/10.1109/TIE.2014.2301773.

Yu, J. (2010). Hidden Markov models combining local and global information for nonlinear and multimodal process monitoring. *Journal of Process Control*, *20*(3), 344–359. https://doi.org/10.1016/j.jprocont.2009.12.002.

Yu, J. (2012a). A nonlinear kernel Gaussian mixture model based inferential monitoring approach for fault detection and diagnosis of chemical processes. *Chemical Engineering Science, 68*(1), 506–519. https://doi.org/10.1016/j.ces.2011.10.011.

Yu, J. (2012b). A particle filter driven dynamic gaussian mixture model approach for complex process monitoring and fault diagnosis. *Journal of Process Control, 22*(4), 778–788. https://doi.org/10.1016/j.jprocont.2012.02.012.

Yu, J. (2012c). Multiway discrete hidden Markov model-based approach for dynamic batch process monitoring and fault classification. *AICHE Journal, 58*(9), 2714–2725. https://doi.org/10.1002/aic.12794.

Yu, J. (2013a). A new fault diagnosis method of multimode processes using Bayesian inference based Gaussian mixture contribution decomposition. *Engineering Applications of Artificial Intelligence, 26*(1), 456–466. https://doi.org/10.1016/j.engappai.2012.09.003.

Yu, J. (2013b). A support vector clustering-based probabilistic method for unsupervised fault detection and classification of complex chemical processes using unlabeled data. *AICHE Journal, 59*(2), 407–419. https://doi.org/10.1002/aic.13816.

Yu, L., Qu, J., Gao, F., & Tian, Y. (2019). A novel hierarchical algorithm for bearing fault diagnosis based on stacked LSTM. *Shock and Vibration, 2019*(12), 1–10. https://doi.org/10.1155/2019/2756284.

Yu, D. L., Gomm, J. B., & Williams, D. (1999). Sensor fault diagnosis in a chemical process via RBF neural networks. *Control Engineering Practice*, 7(1), 49–55. https://doi.org/10.1016/S0967-0661(98)00167-1.

Yu, H., Khan, F., & Garaniya, V. (2015a). A probabilistic multivariate method for fault diagnosis of industrial processes. *Chemical Engineering Research and Design, 104*, 306–318. https://doi.org/10.1016/j.cherd.2015.08.026.

Yu, H., Khan, F., & Garaniya, V. (2015b). Nonlinear Gaussian belief network based fault diagnosis for industrial processes. *Journal of Process Control, 35*, 178–200. https://doi.org/10.1016/j.jprocont.2015.09.004.

Yu, H., Khan, F., & Garaniya, V. (2016). An alternative formulation of PCA for process monitoring using distance correlation. *Industrial and Engineering Chemistry Research, 55*(3), 656–669. https://doi.org/10.1021/acs.iecr.5b03397.

Yu, J., & Qin, S. J. (2008). Multimode process monitoring with bayesian inference-based finite Gaussian mixture models. *AICHE Journal, 54*(7), 1811–1829. https://doi.org/10.1002/aic.11515.

Yu, J., & Qin, S. J. (2009). Multiway Gaussian mixture model based multiphase batch process monitoring. *Industrial and Engineering Chemistry Research, 48*(18), 8585–8594. https://doi.org/10.1021/ie900479g.

Yu, J., & Rashid, M. M. (2013). A novel dynamic Bayesian network-based networked process monitoring approach for fault detection, propagation identification, and root cause diagnosis. *AICHE Journal, 59*(7), 2348–2365. https://doi.org/10.1002/aic.14013.

Yu, T., & Qi, Y. (2021). Intelligent fault diagnosis of chemical process based on twin convolution deep neural network of wavelet transform. In *2021 IEEE 10th data driven control and learning systems conference (DDCLS)* (pp. 100–107). https://doi.org/10.1109/DDCLS52934.2021.9455551.

Zass, R., & Shashua, A. (2007). Nonnegative sparse PCA. *Advances in Neural Information Processing Systems, 19*, 1561–1568.

Zhang, L., Wu, S., Zheng, W., & Fan, J. (2018). A dynamic and quantitative risk assessment method with uncertainties for offshore managed pressure drilling phases. *Safety Science, 104*(December), 39–54. https://doi.org/10.1016/j.ssci.2017.12.033.

Zhang, X., Hu, J., & Zhang, L. (2016). Performance assessment of fault classifier of chemical plant based on support vector machine. In *2016 12th International conference on natural computation, fuzzy systems and knowledge discovery (ICNC-FSKD)* (pp. 46–51). https://doi.org/10.1109/FSKD.2016.7603149.

Zhang, Y. (2008). Fault detection and diagnosis of nonlinear processes using improved kernel independent component analysis (KICA) and support vector machine (SVM). *Industrial & Engineering Chemistry Research, 47*(18), 6961–6971. https://doi.org/10.1021/ie071496x.

Zhang, Y., An, J., & Zhang, H. (2013). Monitoring of time-varying processes using kernel independent component analysis. *Chemical Engineering Science, 88*, 23–32. https://doi.org/10.1016/j.ces.2012.11.008.

Zhang, S., Bi, K., & Qiu, T. (2020). Bidirectional recurrent neural network-based chemical process fault diagnosis. *Industrial and Engineering Chemistry Research, 59*(2), 824–834. https://doi.org/10.1021/acs.iecr.9b05885.

Zhang, Z., & Dong, F. (2014). Fault detection and diagnosis for missing data systems with a three time-slice dynamic Bayesian network approach. *Chemometrics and Intelligent Laboratory Systems, 138*, 30–40. https://doi.org/10.1016/j.chemolab.2014.07.009.

Zhang, Y., Li, S., & Teng, Y. (2012). Dynamic processes monitoring using recursive kernel principal component analysis. *Chemical Engineering Science, 72*, 78–86. https://doi.org/10.1016/j.ces.2011.12.026.

Zhang, K., Qian, K., Chai, Y., Li, Y., & Liu, J. (2015). Research on fault diagnosis of Tennessee Eastman process based on KPCA and SVM. In *Proceedings–2014 7th international symposium on computational intelligence and design, ISCID 2014, 1* (pp. 490–495). https://doi.org/10.1109/ISCID.2014.234.

Zhang, Y., & Qin, S. J. (2007). Fault detection of nonlinear processes using multiway kernel independent component analysis. *Industrial & Engineering Chemistry Research, 46*(23), 7780–7787. https://doi.org/10.1021/ie070381q.

Zhang, Y., & Zhang, Y. (2010). Fault detection of non-Gaussian processes based on modified independent component analysis. *Chemical Engineering Science, 65*(16), 4630–4639. https://doi.org/10.1016/j.ces.2010.05.010.

Zhang, Q., Zhang, J., Zou, J., & Fan, S. (2020). A novel fault diagnosis method based on stacked LSTM. *IFAC-PapersOnLine, 53*(2), 790–795. https://doi.org/10.1016/j.ifacol.2020.12.832.

Zhang, Z., & Zhao, J. (2017). A deep belief network based fault diagnosis model for complex chemical processes. *Computers and Chemical Engineering, 107*, 395–407. https://doi.org/10.1016/j.compchemeng.2017.02.041.

Zhao, H., Liu, J., Dong, W., Sun, X., & Ji, Y. (2017). An improved case-based reasoning method and its application on fault diagnosis of Tennessee Eastman process. *Neurocomputing, 249*, 266–276. https://doi.org/10.1016/j.neucom.2017.04.022.

Zhao, H., Sun, S., & Jin, B. (2018). Sequential fault diagnosis based on LSTM neural network. *IEEE Access, 6*, 12929–12939. https://doi.org/10.1109/ACCESS.2018.2794765.

Zhao, Y., & Xiao, J. (2021). An adiabatic method to train binarized artificial neural networks. *Scientific Reports, 11*(1), 1–8. https://doi.org/10.1038/s41598-021-99191-2.

Zhou, N., & Li, S. (2018). Nonlinear and non-Gaussian process monitoring based on simplified R-Vine copula. *Industrial and Engineering Chemistry Research, 57*(22), 7566–7582. https://doi.org/10.1021/acs.iecr.8b00701.

Ziaei-Halimejani, H., Nazemzadeh, N., Zarghami, R., Gernaey, K. V., … Mostoufi, N. (2021). Fault diagnosis of chemical processes based on joint recurrence quantification analysis. *Computers & Chemical Engineering, 155*, 107549. https://doi.org/10.1016/j.compchemeng.2021.107549.

Zumoffen, D., & Basualdo, M. (2008). From large chemical plant data to fault diagnosis integrated to decentralized fault-tolerant control: Pulp mill process application. *Industrial & Engineering Chemistry Research, 47*(4), 1201–1220. https://doi.org/10.1021/ie071064m.

Further reading

Joe Qin, S. (1998). Recursive PLS algorithms for adaptive data modeling. *Computers & Chemical Engineering*, *22*(4–5), 503–514. https://doi.org/10.1016/S0098-1354(97)00262-7.

Liu, R., Yang, B., Zio, E., & Chen, X. (2018). Artificial intelligence for fault diagnosis of rotating machinery: A review. *Mechanical Systems and Signal Processing*, *108*, 33–47. https://doi.org/10.1016/J.YMSSP.2018.02.016.

Onel, M., Kieslich, C. A., & Pistikopoulos, E. N. (2019). A nonlinear support vector machine-based feature selection approach for fault detection and diagnosis: Application to the Tennessee Eastman process. *AICHE Journal*, *65*(3), 992–1005. https://doi.org/10.1002/aic.16497.

Zhou, D., Li, G., & Qin, S. J. (2010). Total projection to latent structures for process monitoring. *AICHE Journal*, *56*(1), 168–178. https://doi.org/10.1002/aic.11977.

CHAPTER SEVEN

Alarm management techniques to improve process safety

Fan Yang[a,*], Jiandong Wang[b], Mohsen Asaadi[a], Wenkai Hu[c], Zhen Wang[b], and Yinong Zhang[d]

[a]Department of Automation, Tsinghua University, Beijing, China
[b]College of Electrical Engineering and Automation, Shandong University of Science and Technology, Qingdao, China
[c]School of Automation, China University of Geosciences, Wuhan, China
[d]College of Urban Rail Transit and Logistics, Beijing Union University, Beijing, China
*Corresponding author: e-mail address: yangfan@tsinghua.edu.cn

Contents

1. Philosophy of alarm management

Process facilities cost billions of dollars. They are some of the most complex systems ever developed by humankind. As evidenced, for example, by the tragic explosion at BP's refinery in Texas City in 2005 (CSB, 2007), the costs in terms of economics and casualties when things go wrong are staggering (Marsh, 2019; just property loss about 300 M$, but liabilities, fines

Methods in Chemical Process Safety, Volume 6
ISSN 2468-6514
https://doi.org/10.1016/bs.mcps.2022.04.009

and settlement payments went up to a few billion), To decrease the mentioned losses, alarm management plays an important role. Alarm management is to effectively design, implement, operate, and maintain alarms in industrial manufacturing/process plants. Alarms alert/warn for an abnormal value of a process variable signaling an abnormal and perhaps potential hazardous occurrence. Based on what alarm, frequency, and alarm flood the underlying cause should be traced and corrected. Such response, called abnormal situation management (Shu, Ming, Cheng, Zhang, & Zhao, 2016), relies on the data enabling fault detection and diagnosis (chapter "Data-driven approaches: Use of digitized operational data in process safety" by Bai et al. and chapter "Statistical approaches and artificial neural networks for process monitoring" by Alauddin et al.).

In practice, alarm management does not always fulfill its task due to incorrect design, inadequate documentation, changes made without a thorough review, or a failure to convey sufficient and correct information to the operator. A dependable alarm system is a critical component of a safe and reliable process. Ineffective alarm systems may have catastrophic consequences and result in alarm flooding, many standing alarms, low alarm prioritizing, and inappropriate or no alarm response. Alarm management is required in a process plant managed by an operator through a control system such as a distributed control system (DCS). Alarm management projects have been carried out to improve the performance of alarm systems to handle various alarm tags in such a manner that the operator receives an accurate alarm showing the root cause. Standards establish guidelines for properly designing alarm systems and managing alarm datasets in this field (ANSI/ISA-18.2, 2016; EEMUA-191, 2013; International Electrotechnical Commission, 2014). We begin with a widely used and relevant standard in the alarm management sector, namely, ANSI/ISA-18.2, which is concerned with alarm systems integrated into modern control systems, such as DCSs, SCADA systems, programmable logic controllers, or safety systems. It applies to plants where operators react to alarms shown on a computer-type display or an annunciator. ISA-18.2 is written with a life cycle structure comprised of 10 stages shown in Fig. 1.

(1) Alarm philosophy

Because alarm philosophy is the initial step in the alarm management lifecycle (ANSI/ISA-18.2, 2016), it is critical to create a proper alarm philosophy document based on the project's goals and requirements. Some critical functions that the alarm philosophy document plays in the alarm management project's outcomes are as follows:

- Developing and maintaining a successful alarm system over time involves a systematic alarm philosophy.

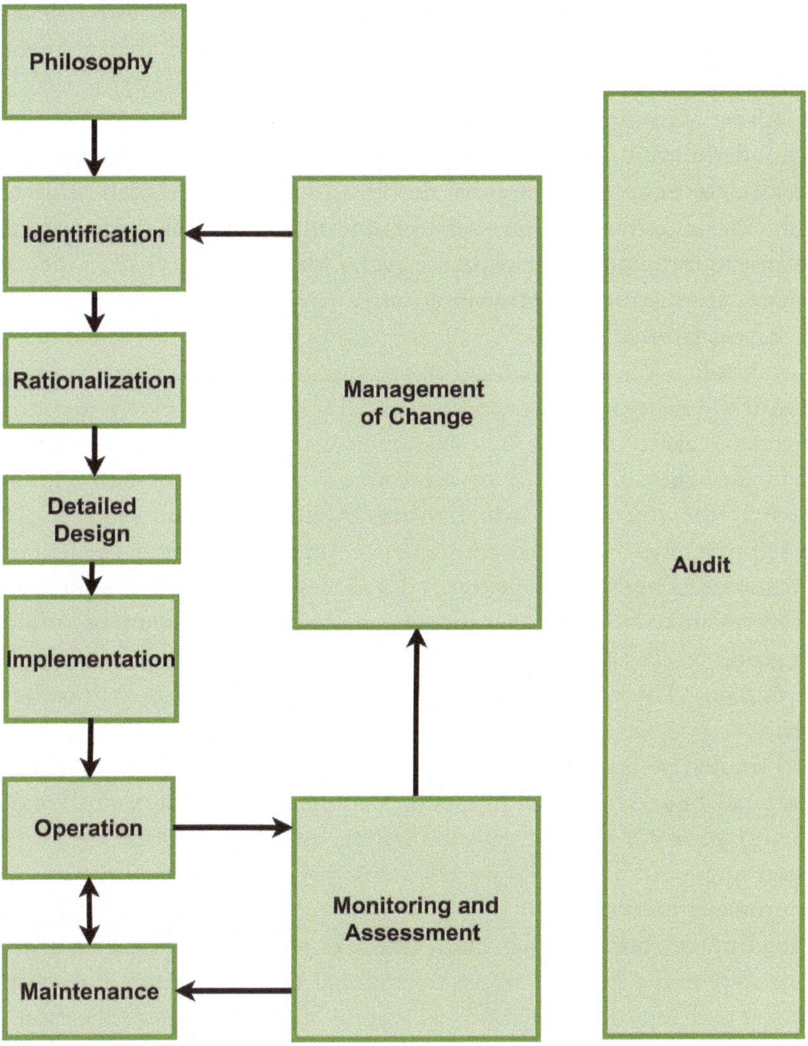

Fig. 1 Alarm management lifecycle.

- Develop a shared knowledge of alarm management.
- Define how alarm management will be done in the workplace.
- Describe how to deal with alarm management issues.
- Establish alarm rationalization rules.
- Define responsibilities for managing the alarm.

Creating an alarm philosophy document is the cornerstone for the development and sustainability of effective alarm management. The alarm philosophy document establishes the guidelines for addressing all aspects of alarm management at a site, including roles and responsibilities, rationalization, design, operations, maintenance, testing, training, and management of change.

(2) Alarm identification

Alarm identification is a conventional name for the many techniques available for determining the potential need for an alarm or a modification to an alarm. The identification stage serves as the entry point for suggested alarms or alarm adjustments in the alarm life cycle. Alarms that have been identified provide information for rationalization.

(3) Alarm rationalization

Alarm justification, documentation, prioritizing, and categorization are just a few of the many essential actions that fall under this category. Existing or prospective alarms are carefully examined against the criteria for alarms set out in the alarm philosophy to determine whether or not they should be raised. If the suggested alarm fits the requirements, the kind of alarm, the setpoint, the cause, the consequence, and the operator's reaction are all recorded. The alarm is prioritized and classed based on the philosophy of the organization. The classification includes the assignment of alarms to a group or class that defines particular administrative requirements, such as fire alarms. These actions are often grouped to form a single rationalization activity.

(4) Detailed design

A number of additional alarm qualities are defined and created during the design stage in accordance with the criteria established during the rationalization step.

(5) Implementation

At the implementation stage, all of the tasks required to install an alarm or alarm system and bring it up to operational level are done. The system's physical and logical installation, as well as its functional verification, are all part of the implementation process.

(6) Operation

During the operation stage, the alarm or alarm system is operational and fulfills the purpose for which it was designed. This level includes refresher training on both the alarm philosophy and the purpose of each alarm, as well as the purpose of each alarm. To put it another way, the alarm system is functional; operators diagnose and react to alarms using the tools that are accessible to them (e.g., shelving and alarm response processes).

(7) Maintenance

Alarms are taken out of service for repair and replacement at this point, based on alarm monitoring reports and alarm philosophy.

(8) Monitoring and assessment
Alarm system performance is measured and compared against key performance indices from the philosophy.
(9) Management of change
Changes are accepted and recorded in the Management of Change process. These changes occur at every level of the process, from identification through implementation.
(10) Audit
The Audit stage entails completing evaluations that may discover areas of opportunity for improvement that are not obvious during ordinary monitoring.

2. Conventional alarm systems vs. advanced alarm systems

An alarm system is a collection of hardware and software that detects, communicates and annunciates alarms to the operator. Based on the ISA standard (ANSI/ISA-18.2, 2016), "an alarm system is the collection of hardware and software that detects an alarm state, communicates the indication of that state to operators, and records changes in the alarm state." Process health is expected to be accurately communicated by an alarm system that has been properly developed. When it comes to alarm systems, there are two elements to consider. The first is alarm generation algorithms that may range from something as basic as limit checking logic on a raw process variable to something as complex as employing machine learning technologies for fault categorization to get the desired result. The second element of the alarm system is the clear and easy announcement of a malfunction, allowing the operator to take prompt action to correct the problem before it becomes worse. The design of the Human Machine Interface (HMI) occurs to be a component of the second element. This task aims to evaluate the overall performance of the engineering parts of the alarm system (Kondaveeti, Izadi, Shah, & Black, 2010). In the past decades, alarm systems have developed continuously. Conventional alarm systems refer to those which the alarm generation mechanism is threshold-based. An alarm is triggered when the process variable is higher or lower than the threshold chosen based on the safety considerations or optimization of the key performance indices. Advanced alarm systems refer to the systems with advanced and mostly data-driven based alarm generation mechanisms. In a nutshell, advanced alarm systems take more aspects of the process monitoring, including the strategies based on multivariate statistics, performance deterioration, pattern matching in alarm floods, fault propagation, and risk assessment into account to generate an alarm. Relatively few individual alarms (e.g., 10–20) often produce a large percentage of the total alarm system load (e.g., 20–80%). The most frequent alarms should be reviewed regularly (e.g., daily, weekly, or monthly). Substantial performance improvement can be made by addressing the most frequent alarms (ANSI/ISA-18.2, 2016; EEMUA-191, 2013).

Therefore, the first and relatively easy step is to remove nuisance alarms; and this is the primary goal of conventional alarm systems. The more difficult step, however, is to look beyond the bad actors, which requires using all available resources: process data, alarm data, and connectivity data. Commercial alarm management software uses alarm data. However, the rationalization process is based on process data and connectivity information. Automated rationalization and regular alarm data monitoring is a key factor for a safe and reliable process operation. This need is motivated by competitiveness and loss avoidance. Advanced alarm systems are the logical next step in our progression towards operating these facilities safely, in reliability, and cost-effectively.

In the next sections, univariate alarm processing strategies (used in conventional alarm systems) and the improved univariate alarm processing strategies (used in advanced alarm systems) will be explained briefly.

3. Conventional alarm management techniques with improvements

3.1 Methods of data analytics for alarm management

There are several ways to analyze binary data sets, emphasizing alarm data sets to manage them; the primary objective is to find duplicate alarms in a simple manner. The most frequently utilized procedures in the industry are listed below. Two unique techniques for visualizing alarm data are presented: (1) The High Density Alarm Plot (HDAP) visualizes the most critical alarms over a specified period, and (2) the Alarm Similarity Color Map (ASCM) visually shows related and redundant alarms (Kondaveeti, Izadi, Shah, Black, & Chen, 2012). Yang, Shah, Xiao, and Chen (2012) proposed a novel technique for assessing several alarm series to detect duplication (Yang et al., 2012). Statistical procedures such as improved correlation and singular value analysis are employed after converting binary data to continuous (pseudo) data using a unique Gaussian kernel approach. The ASCM tool can show the numerical findings by appropriately arranging and grouping alarm tags. It demonstrates that the clustered correlation map of pseudo data has the following primary advantages: (1) it is robust to missed, false, and chattering alarms; (2) the correlation of pseudo data provides directional (positive or negative) information in addition to similarity; (3) the pseudo data used to generate the color map can also be used in other statistical analyses, providing additional potential; and (4) there is no knowledge of delays between correlated alarms. Hu, Al-Dabbagh, Chen, and Shah (2018) conducted a detailed evaluation of the literature on graphical and visualization plots. The classification of the instruments enables us to see their capabilities clearly. Their attributes were compared in terms of event types, dimensions, online/offline applications, and data volume. Additionally, novel visualization approaches have been developed on the basis of conventional visualization plots or in response to the unique needs of industrial alarm systems.

3.2 Recognition of nuisance alarms

Nuisance alarms refer to the occurred alarms not associated with any actual abnormalities. Nuisance alarms do not mean those process variables deviate from normal conditions in reality and are not true alarms corresponding to abnormal conditions.

The essence of nuisance alarms recognition requires detecting whether the production process is in the normal or abnormal conditions during alarm occurrences or clearances, so that nuisance alarms are detected as false alarms occurring in normal conditions. However, detecting normal or abnormal conditions generally depends on the definition of normal or abnormal conditions and related information such as process knowledge, operating experience, and recent data of monitoring variables. There are various causes resulting in nuisance alarms, so automatically determining the normal or abnormal conditions is often tricky. Chattering alarms and oscillating alarms are common nuisance alarms with typical characteristics. Therefore, the existing and emerging nuisance alarms can be automatically recognized according to these characteristics.

3.2.1 Chattering alarms

Chattering alarms are nuisance alarms that repeatedly and rapidly make transitions between alarm and normal states in a short time period (Wang & Chen, 2013). Chattering alarms have short alarm durations or intervals when process variables are in normal or abnormal conditions. Thus, if alarm durations or alarm intervals are less than a certain value of T_c, then corresponding alarms are considered to be chattering alarms.

As a rule of thumb in the industrial field, when alarms occur or clear more than three times per minute, it is considered that chattering alarms are present. Hence, the default value of T_c is 20 s. The value of T_c is different for process variables with different speeds of change. For example, some process variables such as flow and pressure change rapidly, so that T_c usually takes 15 s; by contrast, other process variables such as level and temperature change slowly, and the value of T_c is usually 60 s.

3.2.2 Oscillating alarms

Oscillating alarms, also known as cycling alarms, are the alarms making repeatedly transitions between alarm and non-alarm states within a fixed time period (Rothenberg, 2009). They are often caused by oscillations presented in process variables, which are the phenomena frequently observed in process industries (Thornhill & Hägglund, 1997). Hence, if alarm durations or alarm intervals are equal to or close to a constant value, then oscillating alarms are present.

The hypothesis test is usually used to judge whether alarm durations or intervals are constant. Taking alarm durations as an example, the main idea of the hypothesis test is that if the standard deviation of alarm durations is small with respect to the mean of alarm durations, then alarm durations are claimed to be constant. In the hypothesis test, the null hypothesis is that the standard deviation of alarm durations is equal to the mean, and the alternative hypothesis is that the mean of alarm durations is greater than the standard deviation. If alarm durations or intervals fail to pass the hypothesis test, i.e., the null hypothesis is rejected, it is considered that there are oscillating alarms.

3.2.3 Recognition of nuisance alarms based on the comparison of the sample mean and alarm threshold

The alarm threshold is usually the basic process knowledge to distinguish normal and abnormal conditions. However, due to the influence of noises and disturbances, comparing a single sample of process variables with the alarm threshold to determine normal and abnormal conditions often leads to erroneous results. Therefore, multiple adjacent sample points of process variables can be automatically divided into data segments with the same statistical distribution. The sample means of data segments are compared with the alarm threshold. Doing so can accurately determine the normal and abnormal conditions, and the occurrence of nuisance alarms can be recognized.

The recognition of nuisance alarms in this part mainly includes two steps (Xu, Wang, Izadi, & Chen, 2011). The first step is to divide multiple adjacent sample points of a process variable into data segments with the same statistical distribution. The second step is to compare the sample mean and alarm threshold based on the hypothesis test of each data segment, determine normal or abnormal conditions, and recognize nuisance alarms. For example, if a data segment is determined in normal conditions, the alarms occurring in the data segment are all nuisance alarms.

3.3 Recognition of stale alarms

Stale alarms are alarm events that stay in the alarm state for a long time. Some closely-related definitions of stale alarms are: "alarms in the alarm state continuously for more than 24 h" (Bill & Eddie, 2011), "an alarm that remains in the alarm state for an extended period of time (e.g., 24h)" (ANSI/ISA-18.2, 2016), "the former as the alarms acknowledged but uncleared for 8–12h" (Rothenberg, 2009), and "any alarm active for a full operating shift or longer" (Brown, 2003). These definitions have a common feature of having large alarm durations, but are different in alarm duration thresholds.

The causes of stale alarms are summarized as follows:

(1) The alarm is indicative of an operational problem that the operator should do something about or of which the operator should be aware.
(2) The alarm is indicative of a problem about which others (typically maintenance staff) should do something.
(3) The alarm is indicative of a problem that cannot be resolved in the short term (for example, it may have to wait until an annual overhaul).
(4) The alarm is spurious because it requires no action and does not indicate an operational problem (i.e., it is not a true alarm caused by an actual abnormality).

Stale alarms caused by the last reason are one type of nuisance alarms because a well-accepted criterion to distinguish nuisance alarms is that an informative alarm must require an operator response (ANSI/ISA-18.2, 2016). By contrast, stale alarms due to the first three reasons are not nuisance alarms because they require corrective actions. Therefore, stale alarms deserve more attention due to the last reason as they need to be recognized and removed. The main causes of stale alarms caused by the last reason are identified as follows.

The first cause of stale alarms is that alarm variables should not be configured with alarms. In modern computerized monitoring systems such as the DCS, alarm variables are very easily realized in a technical sense by clicking a mouse and entering alarm threshold values. Such a quick alarm configuration is regarded as "free" without any cost, and is often implemented without careful study. Hence, some alarm variables, which should not be configured with alarms, may lead to stale alarms.

The second cause of stale alarms is that the design of alarm variables is inconsistent with discrete-valued operating states. The daily operation of industrial plants usually involves several states such as startup, half-rate, and full-rate operations, each of which has different demands for equipment. Some equipment is intentionally stopped after industrial plants transit from one state to another. These alarm variables are usually designed only for the running status of the equipment; thus, stale alarms arise when the equipment is turned off.

The third cause of stale alarms is that alarm thresholds are inconsistent with operating states represented by continuous-valued process variables. Alarm variables are generated by comparing the measurements of process variables with the high and/or low alarm thresholds. In contemporary alarm systems, alarm thresholds usually take constant values. However, process variables may experience large-scale variations at different operating states that are represented by some continuous-valued process variables. The constant alarm thresholds are usually suitable only for one operating state, and hardly agree with other operating states so that stale alarms may arise.

Stale alarms have large alarm durations or intervals. Thus, if alarm durations or intervals are greater than the threshold value T_p, then the corresponding alarms are considered to be stale alarms. A general guideline in the industrial field is that if alarm durations or intervals are longer than 8–24 h, then the occurring alarms are treated as stale alarms. Since industrial operators are generally in an 8-h shift team, the default value of T_p is 8 h.

3.4 Univariate alarm processing strategies

The primary goal of using alarm processing strategies is to accurately trigger an alarm and notify industrial operators of occurring abnormalities. Fig. 2 shows the important role of alarm strategies on the alarm system

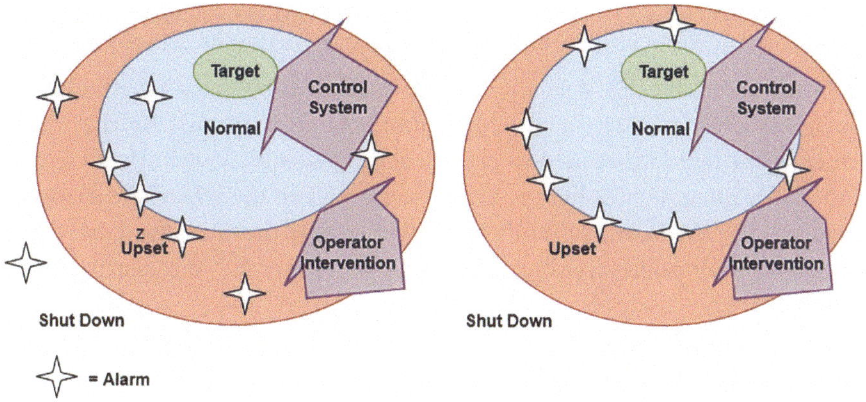

Fig. 2 Left: Operation zones without the proper alarm processing strategies. Right: Operation zones associated with the proper alarm processing strategies.

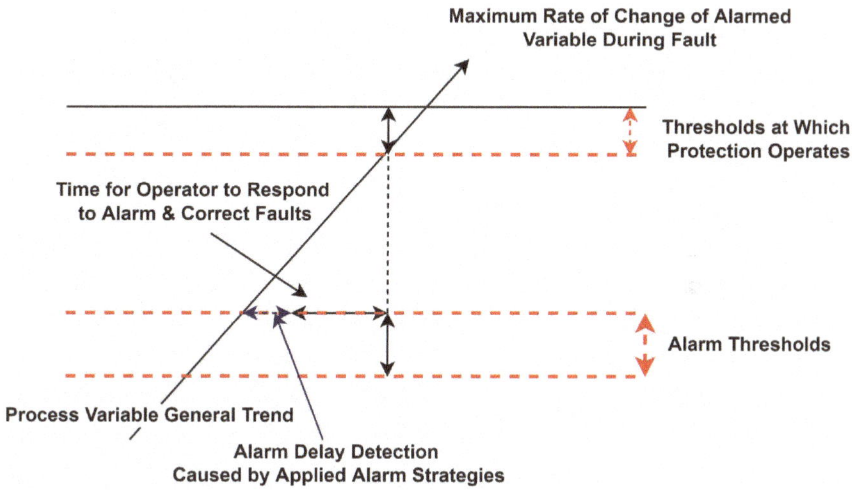

Fig. 3 Given time to operator to take the best decision.

performance. When the process variable exceeds the normal zone and enters the upset zone, an alarm must be arisen and industrial operators should take the best reaction to the occurring abnormality based on the remained time left for the variable to remain in the normal zone. The length of this period is directly affected by the strategies utilized by the alarm system designer.

Fig. 3, which is based on EEMUA-191 (2013), illustrates the typical behavior of a faulty process variable and the amount of time available to the operator to make the optimum reaction. As can be seen, the time allotted to the operator is influenced by the Alarm Detection Delay (ADD), which is the time taken by the alarm system to notify the operator of a fault. The designer is responsible for the ADD immediately impacted by the policy. Overall, the techniques used to process alarm variables have both advantages and disadvantages, and there is a trade-off between the effect of the applied strategies on the alarm system's performance and the time available to the operator to respond appropriately. The more precise the alarm system, the greater the delay.

To this point, we have discussed the general influence of alarm processing strategies on an alarm system's performance. The ensuing paragraphs summarize and describe the most frequently used alarm processing strategies.

3.4.1 Filtering

Filtering is a commonly used technique for lowering noise, flattening curves, and modifying distributions; it is used in signal and image processing, time series analysis, and process industries. Slight noise may cause chattering when process variables are operating near their trip points. Filtration has previously been used in the process sector to resolve these issues (Azad, Ude, Dillmann, & Cheng, 2004). Generally, control systems use a range of filter techniques to improve the signal quality of noisy process variable signals. We are interested in the element of filters that affects statistical distributions. The moving average and exponentially weighted moving average filters are the two most often used filters in industry. When calculated with a given periodicity length, it is also commonly used to separate frequency components, extract characteristics, and eliminate periodic effects. The use of a moving average filter considerably decreases the noise in process data, allows the trip point to be more effective in generating an alarm with increased accuracy. However, filtering resulted in a longer detection latency. A filter is most effective when it improves accuracy while minimizing false alarm rate and missed alarm rate (Adnan & Izadi, 2013).

3.4.2 Deadband

Another approach, known as "alarm deadband," is often utilized in industry to create an effective alarm system. When a deadband is applied to the process variable, the alarm is triggered when the process variable passes the alarm trip point. However, when the process variable is inside the same trip point,

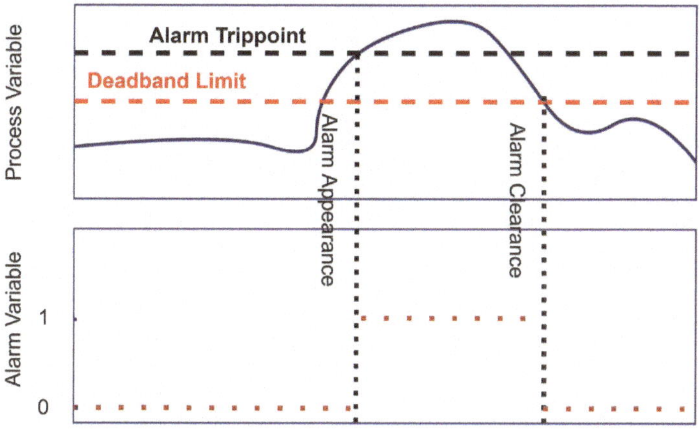

Fig. 4 Alarm deadband mechanism.

the alarm is not removed; instead, a smaller limit is utilized to clear the alarm. Indeed, the process variable should be brought inside stricter boundaries in order to clear the generated alarms. In other words, two distinct alarm limitations are used to trigger and clear an alarm (Fig. 4).

The alarm deadband is usually measured as a portion of the process variable range, or the alarm trip point. On the one hand, an alarm deadband effectively reduces alarm chattering; on the other hand, it increases the time of alarm clearance (Wang, Yang, Chen, & Shah, 2015).

3.4.3 Delay timers

A delay-timer is a simple but efficient approach that may greatly minimize the amount of false, nuisance and missed alarms which may significantly minimize missed alarm rate and false alarm rate. If a variable is set with a delay timer, an alarm is triggered if n sequential samples exceed the alarm threshold. This is alluded to as on-delay. Similarly, if an alarm is triggered, it will persist until m successive samples fall below the threshold, a condition known as off-delay. If the system returns to its regular operational condition during the intermediate states in the on-delay scenario, the alarm is not triggered; in the off-delay situation, the alarm is activated. More precisely, the alarm delay-timer separates the alarm and nonalarm states into many sub-states in the sequence specified. The direction in which sub-states are transferred is determined by the value of the real-time variable.

Fig. 5 shows the alarm generation mechanism using n-m-order delay timer. The delay timer alarming strategy classifies the alarm status and the

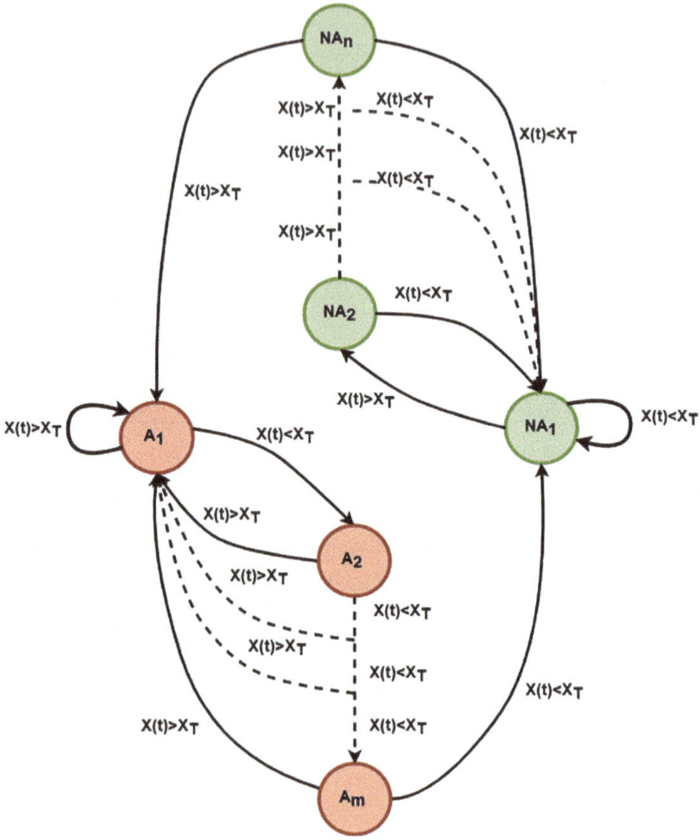

Fig. 5 Alarm state transition diagram of a *n*-*m*-order alarm delay timer.

non-alarm status into n and m sub-statuses, respectively. When the value of a process variable in normal process condition consecutively exceeds the predefined setpoint X_t for n times, the system status switches from non-alarm status NA_1 to alarm status A_1 via sub-statuses of NA_2, ..., NA_n. If the variable value drops within the alarm limit at any time instant in this process, the system status returns to NA_1 immediately. On the other hand, when the value of a process variable in the abnormal condition is consecutively within X_t for n times the system status switches from alarm status A_1 to non-alarm status NA_1 via sub-statuses of A_2, ..., A_n. If the variable value drops outside the alarm limit at any time instant in this process, the system status jumps back to alarm statusA_1.

3.5 Improvement of univariate alarm processing techniques

In Section 3.4, we attempted to enumerate the most essential univariate strategies in the simplest form possible, as is the nature of research in general, with each technique or method advancing daily. This basic rule applies to univariate alarm processing strategies as well. Prior to the explanation of the improvement of univariate alarm processing strategies, we should indicate that the ingredients for any changes and improvements are the result of either process variable model justification or adjustment of the basic alarm processing techniques. There is no single solution for improving univariate alarm processing strategies; however, several methods for optimizing univariate alarm processing strategies can be applied in a variety of areas, including multivariate process monitoring, state-based priority setting, model-based process monitoring, threshold design, and data processing (Ahnlund, Bergquist, & Spaanenburg, 2003; Asaadi, Aslansefat, Izadi, & Yang, 2022; Asaadi, Izadi, Hassanzadeh, & Yang, 2022; Izadi, Shah, & Chen, 2010; Izadi, Shah, Shook, Kondaveeti, & Chen, 2009; Su et al., 2018). Methods that use moving window approaches or multi-setpoint settings are intended to fulfill stricter alarm requirements. Moving window alarming settings indicate that the alarm state is determined by a function transformation of some previously observed values and alarm statuses. The alarm delay-timer and the alarm filter are examples of such methods (Adnan, Cheng, Izadi, & Chen, 2013; Kondaveeti, Izadi, Shah, & Chen, 2011; Xu et al., 2011). Fig. 6 shows two typical alarm processing strategies "reset" (Xu et al., 2011) and "n_1 − out of −n" (Adnan et al., 2013) situations for building delay timers. Assume that n is the duration of the window

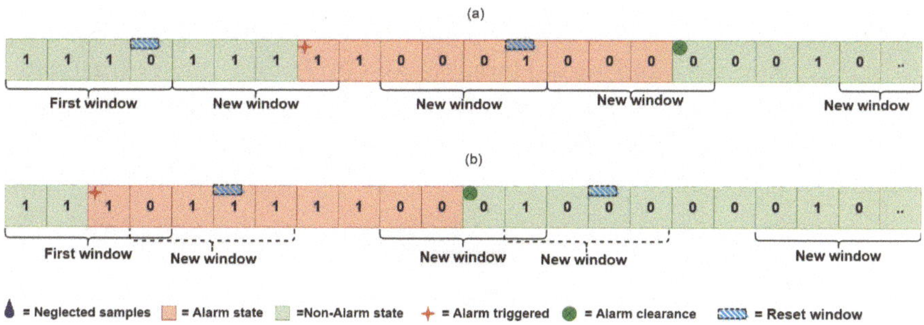

Fig. 6 Scenario for raising and clearing alarms (A) in reset scenario for $n = 4$ and (B) in 3-out of-4 scenario.

during which the alarm may be raised or removed. Additionally, suppose that "1" and "0" represent a certain sample of process data that is larger or less than x_{tp}. If n successive samples of the process data become "1" in the reset scenario, an alarm is sounded. If n successive samples of the process data become "0," the alarm is cleared. If a unique "0" happens between n consecutive "1 s" or a unique "1" occurs between n consecutive "0 s," the window is reset.

In the "n_1-out of-n" scenario, if n_1 number of "1 s" occur inside a window of length, an alarm is triggered. The primary distinction between the "reset" and "n_1-out of-n" situations is that in the reset scenario, n consecutive "1 s" must occur to trigger an alarm, but in the "n_1-out of-n" scenario, n_1 consecutive "1 s" are not required. In Fig. 6B, as can be seen, if n_1 equals n, both situations have the same outcomes. In other words, the reset scenario is a subset of the "n_1-out of-n" scenario, which occurs when n_1 equals n. There is another extension of the above methods with a new parameter named "penalty". In a scenario with n-sample delay, penalty must be varying from 0 to $n - 1$. In this strategy, if a single "0" occurs among n number of "1 s," the moving window will be reset but starts from "i" steps backward. Fig. 7 shows and example of 4-sample on/off delay with penalty $i = 2$.

The term "multi-setpoint" refers to the fact that the alarm annunciation and alarm clearing of an alarm status correspond to distinct setpoints (Su et al., 2018); a common example is the alarm deadband, which is often formed by two setpoints (Adnan et al., 2013).

The multi-setpoint delay-timer alarm processing strategy makes use of extra setpoints to segment the domain of the process variable value into several periods. Fig. 8 depicts the probability density functions for process values under normal and abnormal states, as well as five setpoints, namely X_{NL}, X_{AL}, X_T, X_{NH}, and X_{AH}. Here, NL and NH signify the normal state's low and high alarm limits, whereas AL and AH denote the abnormal state's low and high alarm limits.

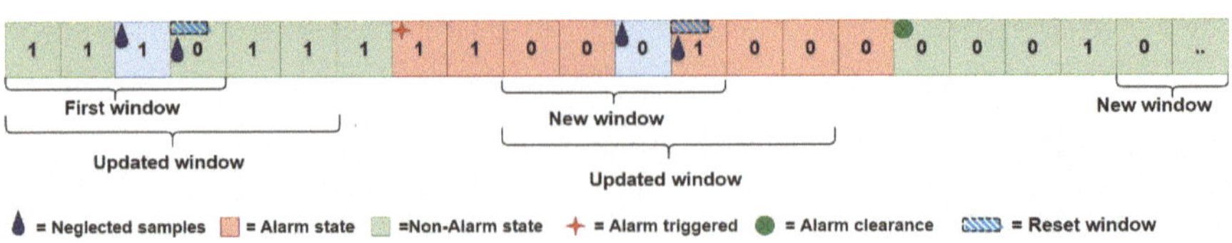

Fig. 7 Penalty scenario for case $n = 4$ and $i = 2$.

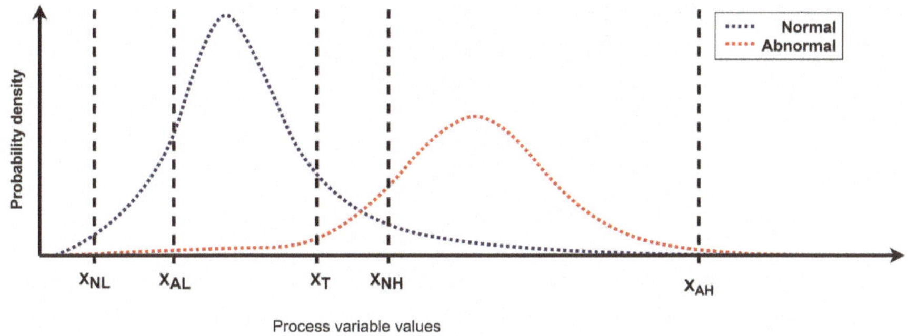

Fig. 8 Probability distributions of the process variable in normal and abnormal conditions and the multi-setpoint thresholds.

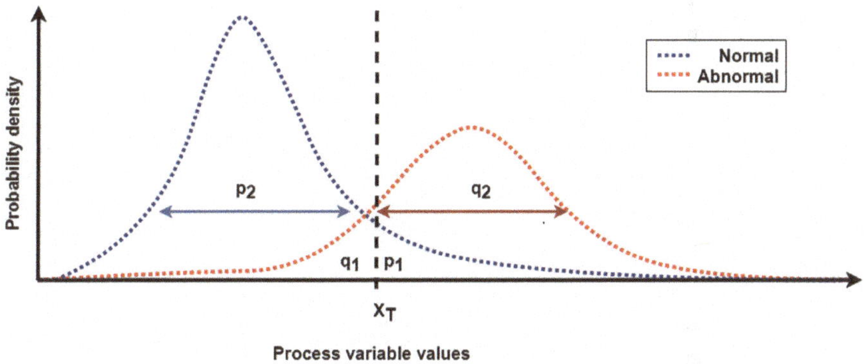

Fig. 9 Probability distributions of the process variable in normal and abnormal conditions.

The triggering mechanism caused by delay timer strategies are investigated thoroughly in the previous section. In here, the Markov process of the improved strategies are discussed based on the stationary model considered for the process variable. Fig. 9 shows the stationary probability densities of the process variable, the Markov processes uses the transition probabilities of normal to abnormal condition transitions calculated based on probability densities of process variable.

According to Fig. 9, p_1 is the probability of false alarm occurrence while the process is operating in its normal state, and q_1 is the probability of missed alarm occurrence while the process is operating in abnormal state.

Markov process of n_1-out of-m scenario is depicted in the Fig. 10 for the case of 2-out of-3 for both on and off delay timers, where A_n are the alarm

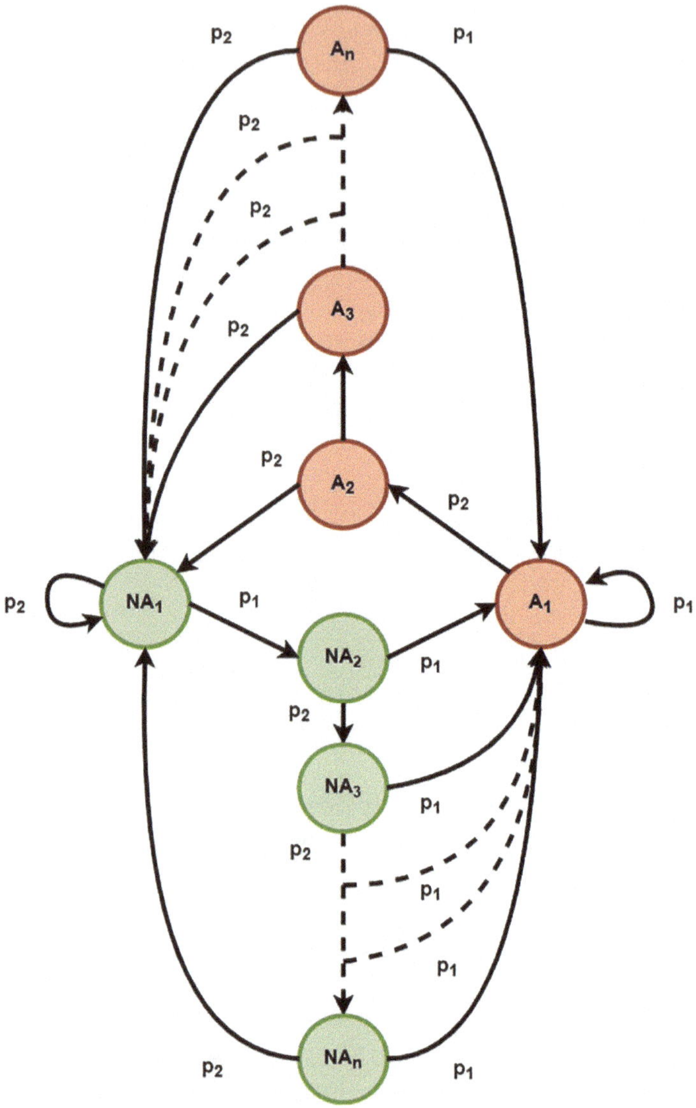

Fig. 10 Markov chain diagram of a 2-out of-3 alarm delay-timer for both on and off delay timer.

states, and NA_n are the non-alarm states. We use the same terms for Markov process states of other improved delay timer strategies. It is assumed that the process is operating in the normal operation zone, which means, the triggered alarms are false, for the abnormal operation zone, one must change p_1 to q_2 and p_2 to q_1 and the non-alarm states indicate missed alarms.

Fig. 11 shows the Markov process of delay timer scenario $n = m = 4$ associated with penalty $i = 2$ for both on and off delay timers (in abnormal operation zone).

For the scenario of multiset point, the Markov process of the case in Fig. 8 is shown in Fig. 12.

As we have more setpoints in this scenario, there are different transition probabilities; where, each of which concerns a special interval of process variable indicated by setpoints; P_1, P_2, P_3, and P_4 indicate the probability of x falling in intervals $[X_{NL}, X_{AL}]$, $[X_{AL}, X_T]$, $[X_T, X_{NH}]$ and $[X_{NH}, X_{AH}]$, respectively. When the transition probabilities are calculated as the sum of two probability values, they represent the likelihood of falling inside the union of two corresponding intervals. For example, $P_1 + P_2$ is the probability of falling in the interval of $[X_{NL}, X_{AL}] \cup [X_{AL}, X_T]$.

The model selection is a critical part of alarm system design; thus far, the process variable models in the area of alarm management mostly have used stationary model to model the behavior of an abrupt fault. While, a great portion of fault occurrences are of intermittent and incipient faults (IF, and INF respectively). The behavior of IF, and INF are different from what is considered to design alarm systems till now. In Asaadi, Aslansefat, et al. (2022) and Asaadi, Izadi, et al. (2022) for the first time the behavior of an intermittent fault is models by a nonstationary statistical model to design and assess an alarm system. The model is a time-variant finite mixture model which changes the way we used to evaluate the alarm system by KPIs. To put it in a nutshell, it defines different zones of a faulty process's operation. Fig. 13 shows the general idea of the work and it could be extended to Fig. 14.

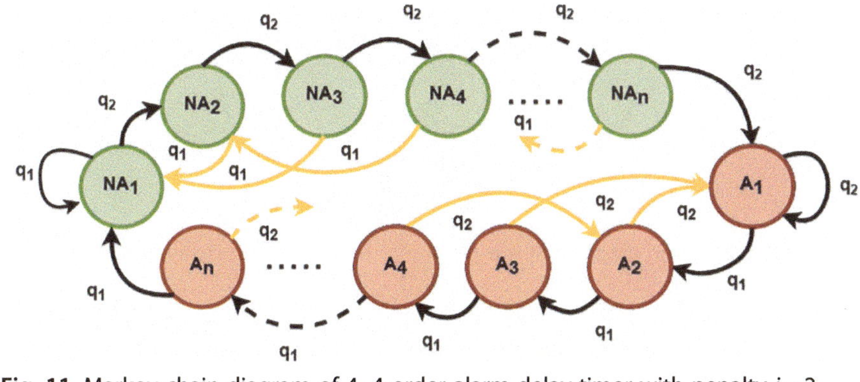

Fig. 11 Markov chain diagram of 4–4-order alarm delay timer with penalty $i = 2$.

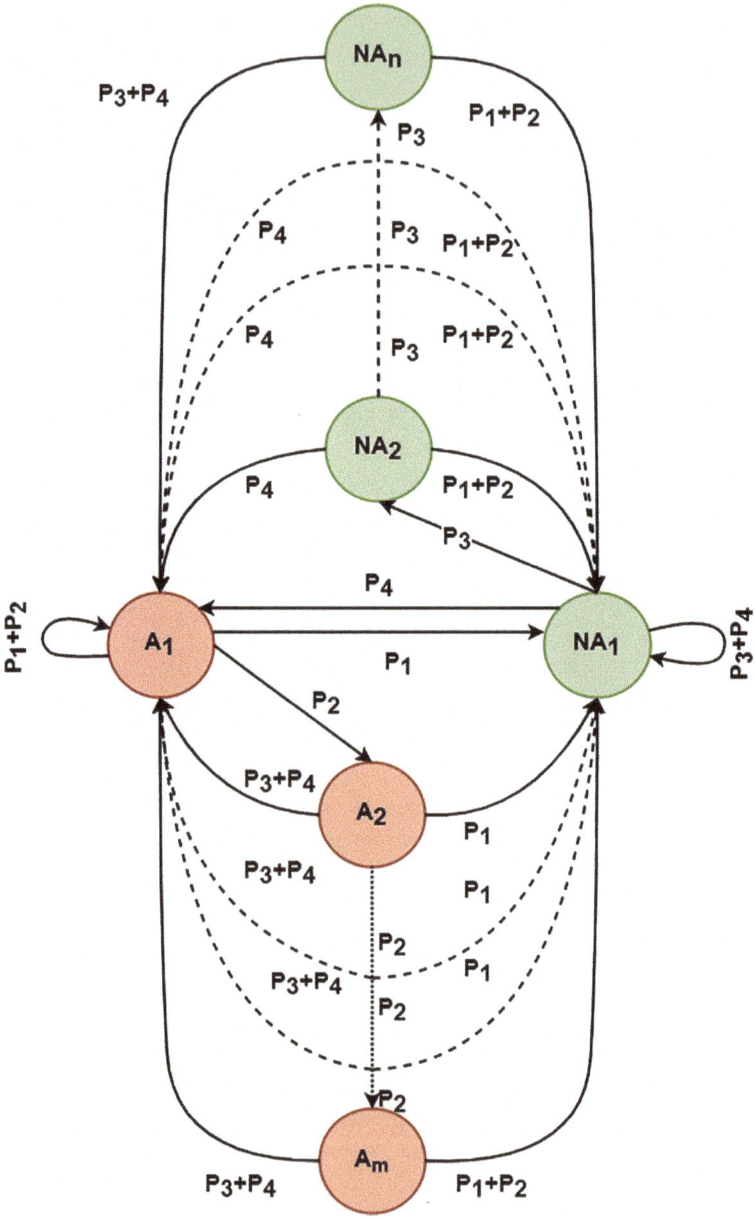

Fig. 12 Markov chain diagram of the *n*-*m*-order multi-setpoint delay-timer.

The fault occurred in Fig. 13 is not an intermittent fault but clearly shows the idea. There is another operation zone of a faulty process variable which overlooked by researches that is Return to Normal zone depicted in Fig. 14.

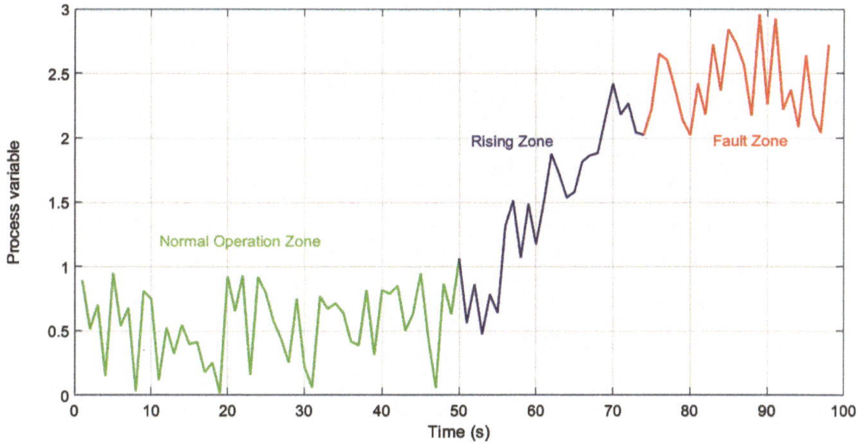

Fig. 13 Different operation zones of a faulty process variable.

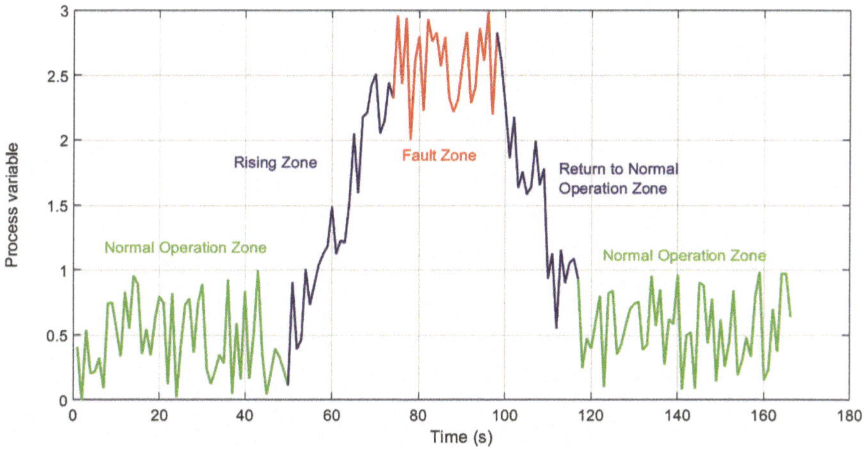

Fig. 14 Generalized operation zones of a faulty process variable.

This behavioral modeling, bring us more detailed information of the process variable behavior and can help us improve the univariate alarm processing strategies. This literacy of fault explanation may change the methods have been used to assess alarm systems.

All univariate alarm processing strategies have an online version due to the real-time assessment mechanism used to optimize the univariate alarm system's performance during operation. The general concept behind this method consists of two stages: the first stage involves designing an

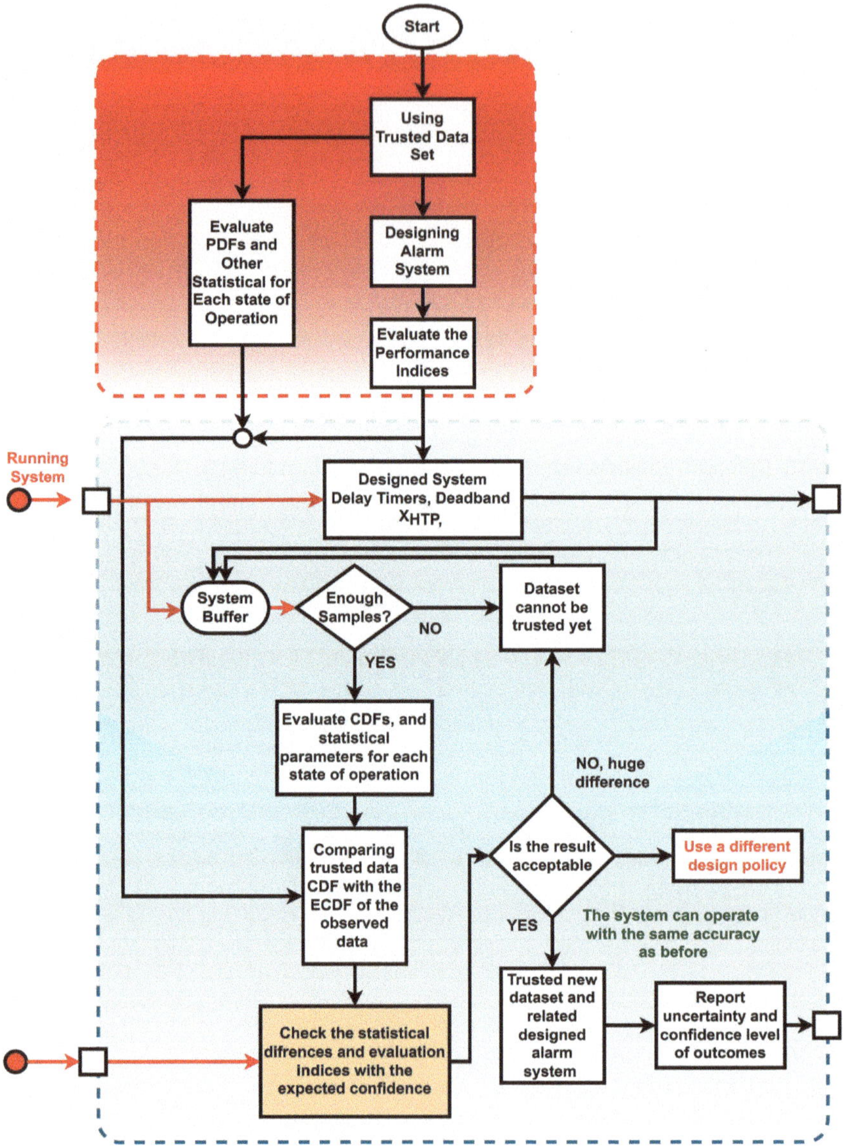

Fig. 15 Proposed real-time evaluation algorithm in Asaadi, Aslansefat, et al. (2022) and Asaadi, Izadi, et al. (2022).

alarm system based on historical data, and the second stage involves monitoring the alarm system to ensure that it is performing as well as it did during the designing stage, which is accomplished through the use of statistical difference measures. Fig. 15 depicts the method used to evaluate an alarm system online. In other words, this technology enables us to re-engineer the alarm system while the observed process is operating.

4. Advanced alarm management techniques

4.1 State-based and dynamic alarming

This section illustrates a dynamic state-based alarm system to avoid the appearance of stale alarms caused by the inconsistency between the alarm design and discrete-valued operating states, which is the second main cause of long-standing alarms in Section 3.2.

The alarm generation mechanism of the dynamic state-based alarm system is that when state variables are in the on state, alarms are generated or eliminated by comparing samples with alarm threshold; when state variables are in the off state, and the alarm system is in the non-alarm state at the previous moment, then the alarm system is in the non-alarm state, i.e.,

$$x_{a,s}(t) = \begin{cases} 0, & x_s(t) \& x_{a,s}(t-1) = 0 \\ x_a(t), & \text{otherwise} \end{cases} \tag{1}$$

Here, the state variable, denoted as $x_s(t)$, indicates the discrete-valued operating states; thus, it is usually a digital variable with binary values "0" and "1." Without loss of generality, let $x_s(t)=0$ stand for the state that does not affect the design of alarm variables. The alarm variable $x_a(t)$ is comparing samples of the process variable with the alarm threshold. State variables $x_s(t)$ generally represent the operating state of equipment or devices related to $x_a(t)$. Thus, the key to designing dynamic state-based alarm systems is to select appropriate state variables.

The qualification rule for state variables is formulated as:

Rule 1: If an alarm variable $x_a(t)$ is in the non-alarm state at time instant (t_0-1), then the value change of the state variable $x_s(t)$ from "0" to "1" at the next time instant $t_0+\tau$ is always accompanied by an alarm occurrence, i.e., $x_a(t)$ switches to the alarm state at $t_0+\tau$.

Owing to the potential dynamics between $x_a(t)$ and $x_s(t)$ as well as the presence of stochastic noises/disturbances, there may be a random delay τ between $x_a(t)$ and $x_s(t)$. The physical meaning of this rule is that when the alarm variable is in the non-alarm state, every switch of the state variable from "0" to "1" will always be accompanied by a switch of the alarm variable from the non-alarm state "0" to the alarm state "1" after a period of time.

This rationale of Rule 1 is based on the following observations on the relation between alarm and state variables:

(1) Whenever some equipment is switched off, the related alarm variable always runs into the alarm state, if the alarm variable previously is in the non-alarm state.

(2) The alarm variable may have already been in the alarm state, when the equipment is switched off.

(3) When a piece of equipment is switched on, the alarm variable does not necessarily switch to the non-alarm state.

(4) When the equipment is staying at the running or shutdown status, the alarm variable could be either in the alarm or non-alarm state.

In order to select state variables that meet the Rule 1, relevant state-on variables $x_s'(t)$ is defined as: when $x_s(t)$ changing from "0" to "1" and $x_a(t-1)$ is in non-alarm state "0," $x_s'(t)$ changes from "0" to "1"; otherwise, $x_s'(t)$ is "0," i.e.,

$$x_s'(t) = \begin{cases} 1, & x_a(t-1) = 0, x_s(t-1) \& x_s(t) = 1 \\ 0, & \text{otherwise} \end{cases} \quad (2)$$

Meanwhile, the relative of $x_a(t)$, referred to as the alarm-on variable, takes the value "1" only at the time instant when $x_a(t)$ switches from the non-alarm state to the alarm state, i.e.,

$$x_a'(t) = \begin{cases} 1, & x_a(t-1) = 0 \& x_a(t) = 1 \\ 0, & \text{otherwise} \end{cases} \quad (3)$$

To tolerate such a random delay τ, additional "1 s" are appended in front of each "1" of $x_a'(t)$, e.g., $x_a'(t) = \cdots, 0, \underset{\uparrow}{1}, 0, \cdots$ becomes

$$x_{a,L}'(t) = \cdots, 0, \underbrace{1, \cdots, 1}_{L}, \underset{\uparrow}{1}, 0, \cdots \quad (4)$$

Here L is the number of additional "1s," and should be smaller than $L_{\max}$ that is the minimum value of "alarm durations" of $x_s(t)$. The consistency ratio ρ becomes

$$\rho(L) = \frac{\sum_{t=1}^{T_{\max}} x_s'(t) \cdot x_{a,L}'(t)}{\sum_{t=1}^{T_{\max}} x_s'(t)} \quad (5)$$

where $x_s(t)$ is the selected state variable by changing $L=0, 1, 2, \cdots, L_{max}$ to satisfy the requirement $\rho=1$. The dynamic state-based alarm system can reduce mostly stale alarms when using this state variable.

4.2 Pattern recognition of alarm floods

Alarm floods are variable-duration periods of alarm activity with annunciation rates likely to exceed the operator response capability. They typically occur due to the presence of faults or the changes of the process states (e.g., from normal to shut down, state A to state B, or normal to abnormal conditions). In an alarm flood, the alarm system is likely to be ineffective in assisting the operator. The ISA-18.2 standard defines an alarm flood as "a condition during which the alarm rate is greater than the operator can effectively manage" and suggests 10 alarms per 10 min for each operator as the benchmark threshold to identify alarm floods (ANSI/ISA-18.2, 2016). The duration of alarm flood should be short, and the total number of alarms in an alarm flood should be low. As a recommended target, the total alarm flood duration should be less than 1% of the reporting period. However, alarm floods in actual processes are far worse than the targets.

To reduce alarm floods and mitigate their negative effects, pattern recognition methods have been proposed in literature to identify similar alarm floods and detect alarm patterns. The results can be used to predict incoming alarm floods or help plant operators to handle alarm floods and mitigate their negative effects. The following subsections introduce the detection of alarm floods first, and then present several alarm flood pattern recognition methods.

4.2.1 Detection of alarm floods

Such an alarm flood event is characterized by a sequence of alarms over a certain time period. Fig. 16 presents an example of the alarm burst plot, where alarm flood periods can be identified by comparing the burst alarm rate (the alarm count over a duration of 10 min) with the threshold (black horizontal line in the figure). The dashed blue curve and solid red curve represent the variations of burst alarm rates calculated based on the raw alarm data and the alarm data with chattering alarms reduced by an off delay-timer. The largest burst alarm rates are around 300 and 100 alarms/10 min, respectively. It can be observed that after reducing chattering alarms, most peaks of the curve disappear and the rest presents true alarm flood periods.

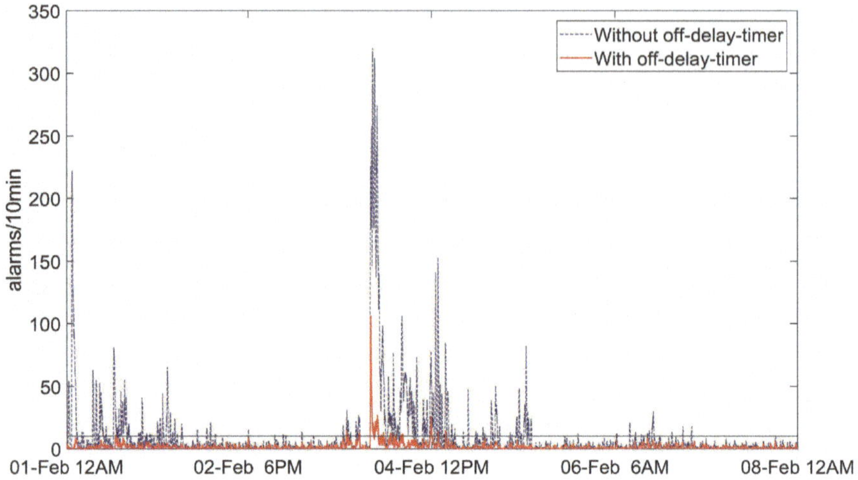

Fig. 16 An example of alarm burst plot.

Burst alarm rate is a common key index to evaluate the performance of alarm systems, and the presence of alarm floods can be detected by comparing the alarm rate with the benchmark thresholds. An alarm flood sequence consists of a set of alarms annunciated in a chronological order. Given the burst alarm rate $\gamma(t)$ that denotes the alarm count over a period of 10 min from t-599 s to t, an indexing variable $\varphi(t)$ is then defined as (Hu, Chen, & Shah, 2018)

$$\varphi(t)=\begin{cases}1 & \text{if } \gamma(\mathrm{t})\geq 10, \text{and } \varphi(t-1)=0\\ 0 & \text{if } \gamma(\mathrm{t})<5, \text{and } \varphi(t-1)=0\\ \varphi(t-1) & \text{otherwise}\end{cases} \tag{6}$$

Then, the start and end timestamps of each alarm flood can be easily found based on an indexing variable $\varphi(t)$. Formally, an alarm flood is said to begin at a time instant t_s if

$$\varphi(t_s)=1, \text{and } \varphi(t_s-1)=0 \tag{7}$$

which indicates the alarm rate reaching the threshold of 10 alarms over a period of 10 min, and end at time instant t_e if

$$\varphi(t_e)=0, \text{and } \varphi(t_e-1)=1 \tag{8}$$

which indicates the alarm rate dropping below the threshold of 5 alarms over a period of 10 min. Thus, given an alarm flood starting and ending at t_s and t_e, respectively, the corresponding alarm flood sequence is represented by $F = [a_1, a_2, \ldots, a_i, \ldots, a_n]$, where a_i denotes the ith alarm and N is the length of the sequence.

4.2.2 Pairwise sequence alignment of alarm floods

A major technique for pattern recognition of alarm floods is the sequence alignment, which compares two alarm flood sequences to measure the similarity and identify the common subsequence. Alarm flood pattern recognition is helpful for root cause analysis of historical floods and prediction of incoming floods. Since an alarm pattern is usually associated with a specific potential exception, all alarms in a particular pattern can be temporarily suppressed when a corresponding exception is detected.

Fig. 17 presents a general framework of alarm flood sequence alignment. The first step is to extract the alarm flood sequences from the historical data. Then, the similarities between the sequences are measured in pairs. Last, similar alarm floods are clustered to form groups consisting of flood sequences resembling each other. Similar alarm floods usually relate to the same kind of underlying abnormality. By clustering similar alarm floods, experts can find flood pattern candidates and root causes of alarm floods.

In the framework of similarity-based techniques for sequence-based anomaly detection, a key step is the pairwise sequence alignment of alarm floods. There are two types of typical sequence alignment algorithms, namely, the global and local sequence alignment. The global one aligns the whole sequence from end to end, which is characterized by sequences of roughly equal size; a typical global alignment method is the Needleman-Wunsch Algorithm (Needleman & Wunsch, 1970). The local alignment compares similar regions; a typical local alignment method is the Smith-Waterman Algorithm (Smith & Waterman, 1981).

(1) The Needleman-Wunsch Algorithm (NWA) was originally proposed in 1970. It provides optimal global alignment, namely, the end-to-end alignment between two sequences (Smith & Waterman, 1981). It starts

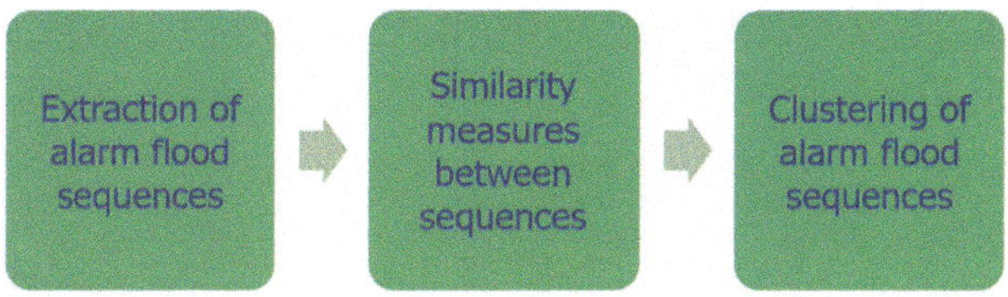

Fig. 17 Framework of alarm flood sequence alignment.

by initializing a score matrix, and then calculates the scores based on the matches, mismatches, and gaps between two sequences. Last, the optimal alignment is found by tracing back from the last element of the matrix to the first one. An example of sequence alignment using NWA is shown in Fig. 18. The NWA was applied in Charbonnier, Bouchair, and Gayet (2016) for comparison of industrial alarm flood sequences.

(2) The Smith-Waterman Algorithm (SWA) was proposed by Smith and Waterman (Smith & Waterman, 1981). Its objective is to find a pair of segments from two sequences, such that there is no other pair of segments with greater similarity. The SWA also calculates the similarity score between two sequences and forms a similarity matrix. Then, it applies a backward path search based on the similarity table to locate the common subsequence. An example of sequence alignment using SWA is shown in Fig. 19. The SWA was applied in Cheng, Izadi, and Chen (2013) for comparison of industrial alarm flood sequences.

The main differences between the pairwise sequence alignment algorithms (SWA and NWA) for similarity analysis of alarm flood sequences are summarized as follows (Parvez, Hu, & Chen, 2020):

(1) Manners to find the optimal alignment are different: The SWA belongs to local alignment and it finds the optimal local subsequence pair between two alarm flood sequences, while the NWA belongs to optimal global alignment and finds the end-to-end alignment.

(2) The ways of initialization are different: The SWA initializes the scoring matrix by assigning zero values to the first row and first column, while the NWA adds a gap penalty as it moves to the right in the first row and down in the first column.

(3) Similarity scoring strategies are different: The negative similarity score is set to zero in the SWA, while the similarity score between the alarm tags can be either positive or negative in the NWA.

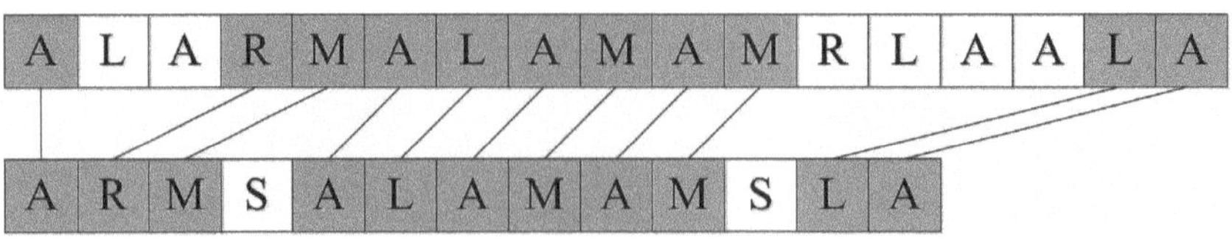

Fig. 18 An example of sequence alignment based on the Needleman-Wunsch algorithm.

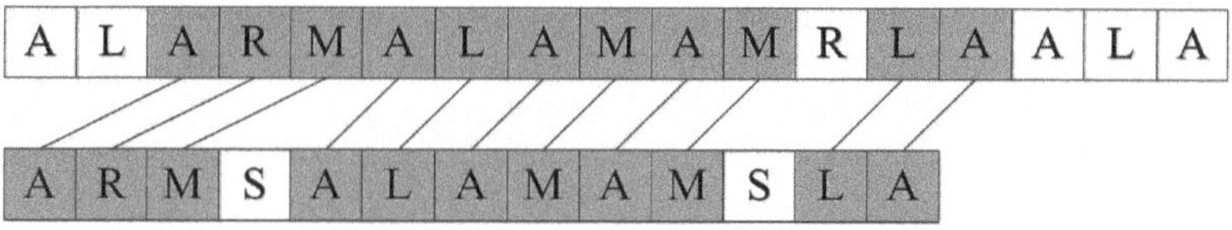

Fig. 19 An example of sequence alignment based on the Smith–Waterman algorithm.

(4) Lengths of optimal alignment are different: For pairwise sequence alignment of alarm floods, the optimal local alignment length using SWA is always no greater than the optimal global alignment length using NWA.

(5) Back tracing procedures are different: To find the best local alignment using SWA, the backtracking starts at the maximum value and ends with the element equal to zero in the score matrix. To find optimal global alignment using NWA, backtracking starts at the lower-right element of the score matrix and ends at the upper-left element.

If a unique symbol is used to represent each alarm type, the similarity index and common segment of the alarm flood sequence are found to be very similar to the approximately matched target. Particularly, a good alarm sequence alignment algorithm should have the following properties (Cheng et al., 2013):

(1) Insensitive to some irrelevant alarms occurring in one or both of the alarm sequences.

(2) Somewhat tolerant to the ambiguity of order; the closer the time stamps of two alarms are, the less important the order of them is.

The standard approximate matching algorithm considering the gap and mismatch can meet the first requirement well. As for the second one, it is possible that several strongly connected alarms occur almost simultaneously, but their order in the alarm sequence varies in cases. In addition, an alarm may appear after its cascade of alarms due to random detection delays. Therefore, when the timestamps of two alarms are close, the order of the two alarms is less important. Thus, the sequence alignment incorporates a time ambiguity tolerance strategy, that utilizes the time stamp information of the sequence to blur the order of alarms when they occur close to each other (Cheng et al., 2013). Such a strategy can improve the accuracy of pattern recognitions of alarm floods.

4.2.3 Accelerated method for alarm flood sequence alignment

Sequence alignment algorithms are useful to find similar alarm floods in the historical alarm and event database. However, the existing Smith-Waterman

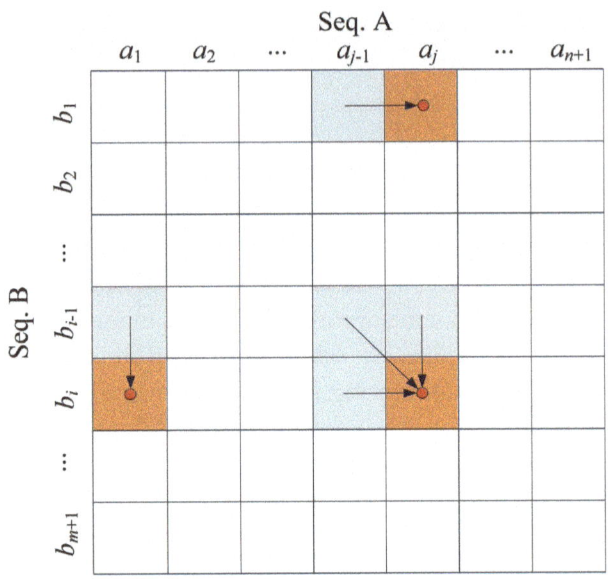

Fig. 20 The calculation of the score matrix using NWA and SWA.

algorithm and Needleman-Wunsch algorithm are typical dynamic programming approaches. A major defect of the two algorithms lies in the high computational complexity, which may prevent their applications within a tolerable computational time period. Both the two methods require the calculation of a score matrix (as shown in Fig. 20). In the computation, each cell of the matrix needs to be proceeded and accordingly the computational complexity is very high especially for long alarm sequences and comparison of a large amount of alarm floods.

In the biological sequence alignment, an efficient tool called Basic Local Alignment Search Tool (BLAST) was formulated in Altschul et al. (1997)) and Altschul, Gish, Miller, Myers, and Lipman (1990). It locates similar segments between a query sequence and object sequences by a seeding step and an extending step. In the first step, the sequences are broken into short words of a fixed size, and a lookup table is built for all possible short combinations of elements. Then, similar segments of the query and object sequences are quickly located by indexing short words in the lookup table. The pairs of words with relatively high similarity scores are taken as seeds. Afterwards, each seed is extended in two directions via sequence alignment in the extending step until the similarity score falls below a threshold. The segment pairs with the highest score are obtained as the most similar segments.

Comparing with the NWA and SWA, the BLAST does not align all elements in sequences and thus achieves fast alignments by pruning most search space.

To adapt the BLAST for alignment of alarm flood sequences, some significant modifications were made to achieve high computational efficiency and better alignment results (Hu, Wang, & Chen, 2016):

(1) Prior to conducting sequence alignment between two alarm flood sequences, it is necessary to compare their alarm sets, which is a much easier way for comparison. If two alarm floods share no common alarms, it should not proceed to sequence alignment steps. Therefore, such a set based matching mechanism can be used as a pre-step to exclude irrelevant alarm floods.

(2) The BLAST requires a predefined lookup table comprised of all possible short words of a fixed size. However, the number of unique alarm tags is usually large in an alarm system, making it almost infeasible to construct such a lookup table, which could be very huge. Therefore, in the seeding stage, it searches all matched pairs of different sizes rather than find short words from a lookup table. As a result, similar regions between two sequences are found quickly.

(3) In an alarm system, the priority is assigned for each alarm to indicate its importance. Since alarms with higher priorities are less likely to occur but more important, it is necessary to make the sequence alignment more sensitive to alarms of higher priorities. This can be achieved through a priority-based scoring scheme that assign higher scores to matches of alarms with higher priorities.

(4) In the extending stage, a time ambiguity tolerance strategy can be used to make the sequence alignment less sensitive to orders of alarms occurring almost simultaneously. An example of extending based on a seed comprised by three matched components is presented in Fig. 21.

Eventually, a large number of computation cells are avoided. This method is fast in computation and achieves high alignment accuracy, owing to three novelties, namely, the priority-based similarity scoring strategy, the set-based pre-matching mechanism, and the modified seeding-extending steps.

In addition to the above method, the Match-based Accelerated Alignment (MAA) is also an effective way to generate insightful and informative alarm sequence alignments with reduced computation (Guo, Hu, Lai, Yang, & Chen, 2017). MAA focuses mainly on alarm match analysis and outperforms other methods in terms of robustness towards nuisance alarms and improved computational efficiency. More importantly, the alarm

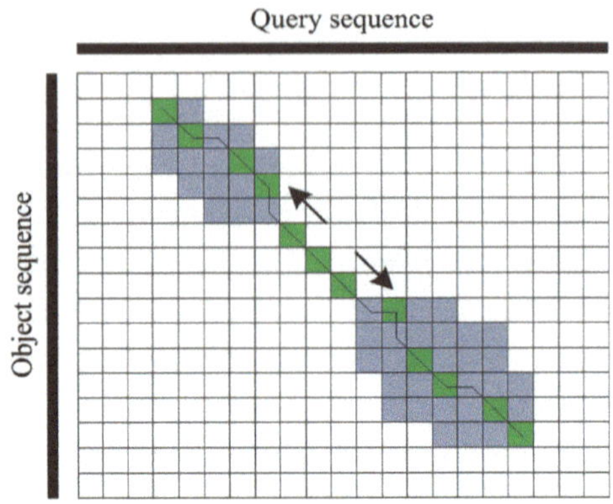

Fig. 21 Graphical illustration of the extending stage.

time information is well considered and explored in MAA, rendering its alignment results capable of revealing the real similarity of alarm floods.

The main idea of the MAA method is that the alignment of two similar alarm flood sequences should be mainly composed of alarm matches, as the objective of alarm flood sequence alignment is to find the way of aligning two sequences that corresponds to a high similarity score, and only matches increase the score while mismatches and gap-matches decrease it. The MAA method works by firstly finding all the matches and using them to assemble the sequence alignment. The main steps of MAA are as follows:

(1) Potential match selection: MAA is based on the analysis of matches, and thus efforts are paid to selecting matches that may remain in the alignment. Given two alarm flood sequences, it firstly finds out potential matches that are highly possible to be present in the alignment.

(2) Conflicting match alignment: In two typical situations, potential matches are considered as conflicting matches that need further processing. The first one is about the multiple matches sharing mutual alarms, which could be caused by the repeating of one alarm in an alarm flood; thus, they may align multiple times with the same alarm in the other flood. The second one is that if alarms are announced close enough, their occurrence order may switch but is of little importance in alignment. Therefore, these conflicting matches should be removed in the alignment.

(3) Gap-match insertion: After conflicting matches are detected and removed, the non-conflicting potential matches are connected with the remaining conflicting matches based on the orders of alarms in the sequences. The match-only alignment is then generated and all the matches contained in this alignment are alignment matches. Subsequently, the other alarms that are not involved in the match-only alignment are aligned to gaps and the obtained gap-matches are inserted in the alignment. Then, the alignment is completed.

4.2.4 Multiple sequence alignment of alarm floods

Pairwise sequence alignment algorithms compare alarm flood sequences and measure their similarities. Then, groups of similar alarm floods are discovered by clustering. In order to find out how the alarm floods in a cluster resemble each other and identify the common subsequence shared by multiple alarm flood sequences, the multiple sequence alignment can be conducted. The idea to align multiple sequences is similar to pairwise sequence alignment, but extends the calculation of the score matrix and the back tracing into a higher dimension space.

An algorithm that extends the modified SWA in Cheng et al. (2013) to align multiple alarm flood sequences was proposed (Lai, Yang, Chen, & Cao, 2019). The main steps of multiple sequence alignment of alarm floods are summarized as follows with an example of three sequences (Lai et al., 2019):

(1) The basic similarity scores are defined in a multi-way scoring function that incorporates the alarms from multiple sequences. The score is calculated based on the match, mismatch, and gap penalty. In order to tolerate the order ambiguity of alarms with close time stamps, the time distance and weight vectors are calculated in the same way (Cheng et al., 2013).

(2) In the dynamic programming, a score matrix is calculated in multiple directions according to the number of alarm flood sequences in comparison. In the case with three floods, the score matrix is in the form of a cuboid. An example is presented in Fig. 22 to illustrate the difference between the calculation of score matrices for two sequences and three sequences. In Fig. 22A, the score (red dot) in a cell of the score matrix is obtained based on three candidates (pink dots) for two sequences X and Y. In Fig. 22B, the score (red dot) is obtained based on seven candidate objects (pink dots) in each step of dynamic programming for three sequences X, Y, and Z.

(3) Given the calculated score matrix, back tracing is carried out on the three-dimensional space. Fig. 23 presents an example of back tracking

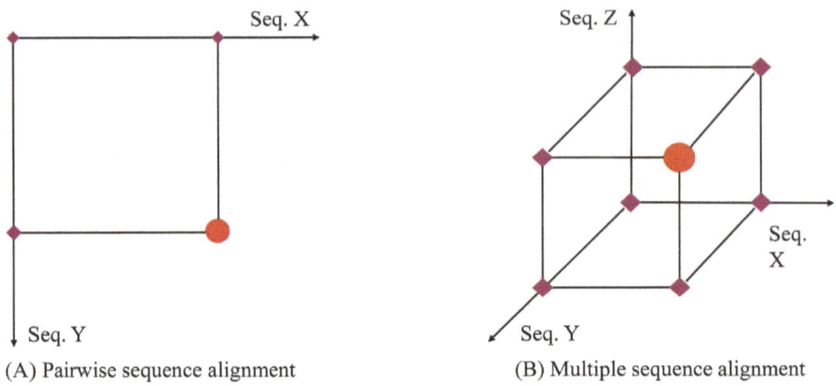

Fig. 22 Illustration of similarity score calculation for two and three sequences.

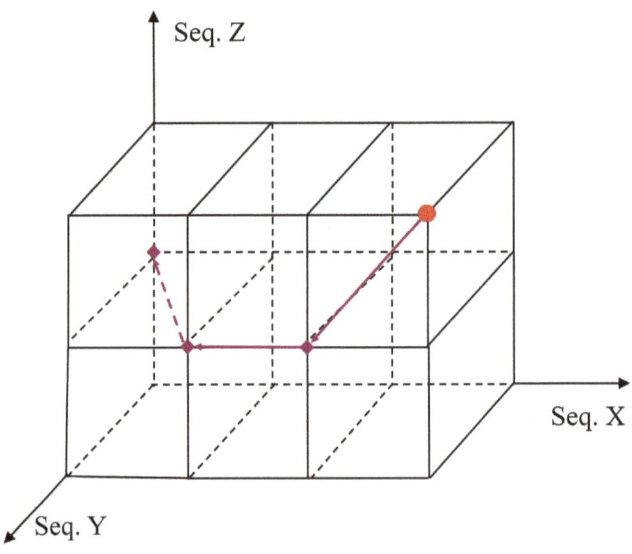

Fig. 23 An example of back tracing of a case with three alarm flood sequences.

on the three-dimensional space for sequences X, Y, and Z (Lai & Chen, 2017). Analogous to SWA, the back tracing starts from the position of the largest score in the cuboid, and end at the position with a zero score.

The same as the traditional pairwise sequence alignment, the above multiple sequence alignment also suffers from high computational complexity. Thus, an accelerated multiple pattern mining algorithm was proposed in Lai and Chen (2017) to find a suitable alignment of multiple alarm flood sequences and obtain the pattern for a selected alarm flood cluster. The algorithm incorporates a generalized pairwise sequence alignment method and a progressive multiple sequence alignment approach, which significantly reduce the computational complexity.

Generalized sequence alignment consists of three elements, time distance and weight vectors, generalized similarity score, and generalized dynamic programming procedure. The difference between the generalized and the standard dynamic programming program is that the former treats sequences from the perspective of tuples, which makes sequences either alarm sequences or the alignment of multiple alarm sequences. Thus, the standard dynamic programming procedure can be seen as a special case of the generalized one. Then, a progressive multiple sequence alignment method based on the generalized dynamic programming is used to find the optimal sequence alignment. First, all the pairwise similarity scores between the alarm sequences are calculated. Then, a dendrogram is built based on these pairwise similarity scores using UPGMA (Unweighted Pair Group Method with Arithmetic Mean). Next, based on the dendrogram, the primitive alignment is obtained by progressively aligning all the sequences, from the most similar pair to the most dissimilar ones.

4.2.5 Online sequence alignment of alarm floods

The above techniques for sequence alignment of alarm floods are conducted offline. In practice, it is a more critical and challenging issue to assist operators during the ongoing alarm flood situation. Conducting online similarity analysis of alarm floods is supposed to provide decision supports for plant operators to judge the root cause, predict incoming alarms, and make correct responses (Parvez, Hu, & Chen, 2022).

An online algorithm was designed (Lai, Yang, & Chen, 2017) to provide early prediction of an incoming alarm flood by matching the online alarm sequence with a pattern database and measuring their similarities iteratively. The key to extend the offline sequence alignment to an online one lies in the calculation of the score matrix via dynamic programing. Instead by calculating a score matrix in one time, the online alignment conducts incremental dynamic programming to increase the size of the score matrix based on the new alarms that appear in the online alarm flood sequence (Lai et al., 2017). To show how to incrementally calculate the score when new alarms are raised and added into the online sequence, an example is presented below. Given the pattern sequence <(1,3), (2, 26), (3, 28), (4, 293)> and the online sequence <(1, 4), (8, 26), (9, 105)>, when a new alarm

(A) Previous score matrix

	Δ	(1,3)	(2,26)	(3,28)	(4,293)
Δ	0	0	0	0	0
(1,4)	0	1	0.8	0.72	0.52
(8,26)	0	0.8	0.6	0.52	0.32
9,105	0	0.6	0.4	0.32	0.12

(B) Updated score matrix

	Δ	(1,3)	(2,26)	(3,28)	(4,293)
Δ	0	0	0	0	0
(1,4)	0	1	0.8	0.72	0.52
(8,26)	0	0.8	0.6	0.52	0.32
(9,105)	0	0.6	0.4	0.32	0.12
(2,122)	**0**	**0.4**	**1.6**	**1.52**	**1.32**

Fig. 24 An illustrative example of calculating the score matrix incrementally.

(2,122) is raised and included into the online sequence, the score matrix can be updated by augmenting the old one (Fig. 24A) with a new row, as shown in Fig. 24B. In each bracket, the first number indicates an alarm tag, while the second one represents its timestamp. In the dynamic programming to calculate the score matrix in Fig. 24, the time ambiguity order tolerance strategy is applied.

When the online sequence is identified to be similar to an offline alarm pattern, the prediction of incoming alarms will be provided based on the record of the sequence window filter and the pattern in the database. The alarm messages contained in the pattern but not included in the online sequence over a window filter will be served as predictions. Fig. 25 shows an example of a prediction procedure. As common alarms are identified for the online sequence and the offline pattern, the other alarms in the pattern can be taken as the prediction of incoming alarms for the online sequence.

In the incremental calculation of the above method, the similarity of the sequence to be compared is continuously calculated, and the sequence with a large similarity difference is discarded. However, this algorithm ignores the fact that there may be several relatively similar sequence patterns in practice, and the sequence pattern is updated in real-time. Therefore, a real-time pattern matching and ranking algorithm was designed in Parvez et al. (2022) for online sequence alignment of alarm floods, and the results are derived as a sorted list of historical alarm flood sequences. The approach consists of the following main steps:

(1) As a key pre-matching step, plant unit-based screening can be used to screen out irrelevant alarm floods, so as to avoid unnecessary sequence alignment computation in later steps and thus can reduce the computational time a lot. This step can narrow the comparison down to a smaller set of historical alarm floods that have alarms coming from the same plant units as those in the online alarm flood. Meanwhile, the plant unit-based similarity S_{unit} is calculated for each historical alarm flood sequence and the online one.

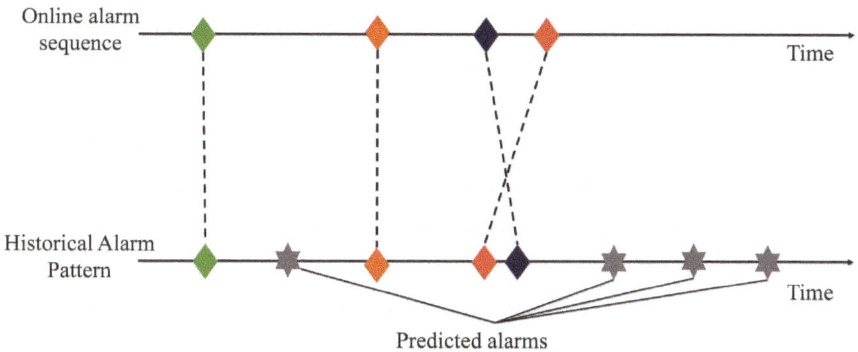

Fig. 25 An illustrative example of predicting incoming alarms.

(2) It should be noticed that two alarm flood sequences cannot be similar to each other if their alarm sets are not similar. Pre-matching based on alarm sets can be used to find and exclude historical alarm flood sequences that have little similarities with the online sequence in terms of alarm sets. Meanwhile, the alarm set based similarity S_{set} is calculated. This pre-matching step can further screen out a large number of irrelevant alarm floods.

(3) To measure the similarities between the ongoing alarm flood sequence with the historical sequences after screening, incremental sequence alignment is conducted. This step exploits the SWA or NWA to calculate an initial score matrix at the triggering time instant of the online alarm flood, and update the score matrix with the addition of new incoming alarms. Meanwhile, the alarm sequence-based similarity S_{seq} is calculated. Such an incremental sequence alignment strategy can avoid comparing the complete sequences again whenever a new alarm occurs.

(4) Given the calculated three similarity scores, including S_{unit}, S_{set}, and S_{seq}, historical alarm floods are ranked based on a certain criterion to export a ranking list, such that the most similar historical alarm flood should appear at the top. The historical alarm floods are ranked in a descending order of S_{seq}; if two historical floods hold the same S_{unit}, their ranking will be based on S_{set}; if two historical floods hold the same S_{seq} and the same S_{set}, their ranking will be based on the S_{unit}. The ranking list is updated when new alarms are raised in the ongoing alarm flood along with time. Dynamic ranking of the most similar historical alarm floods can help operators to observe all the similar sequences and recall the root causes, and thus make corrective responses to handle alarm floods.

4.2.6 Cross-process alarm flood sequence alignment

In a complex large-scale industrial facility, there may exist many fault types and a huge number of alarm patterns, which pose great challenges for analysis of alarm floods. Modern industrial facilities are usually composed of a series of processes or units that may have the same types of equipment or have similar functionalities, such as the pump stations in an oil and gas pipeline system. The abnormalities that happen in these similar processes or units could be in the same types, and the generated alarm floods may contain analogous series of consequential alarms. If these alarm floods are identified and grouped, the resulting information could facilitate root cause investigation and lead to more generalized solutions. However, the alarms in different processes or units are assigned with different tag names, making it rather difficult to analyze these alarm sequences across processes or units. Thus, a generalized method for pattern matching of alarm floods across different processes or units was proposed (Zhou, Hu, Brown, & Chen, 2020). The framework of the method is presented in Fig. 26 and summarized as follows:

(1) Alarm flood sequences are extracted from the historical Alarm & Event (A&E) log by considering the process unit information. As a result, alarm floods in each process unit form an individual set. Later, in the similarity analysis of alarm floods, it should not only compare alarm floods within each set, but also make comparisons between alarm floods in different sets.

(2) Generalization of alarm representations is conducted to reconstruct abstracted alarm descriptors based on key words distilled from detailed alarm descriptions. An example is shown in Fig. 27. The three alarms have different tag names, but indicate the same type of abnormality. Based on key words shared in the alarm descriptions of three alarms, the phrase "Pump Pressure High" is reconstructed as the abstracted alarm descriptor that can represent the three alarms identically. A word processing method can be used to achieve this purpose. The generalization of alarm representations is conducted in three steps, namely, the distillation of key words, the determination and removal of useless words, and the reconstruction of abstracted alarm descriptors based on the purified set of words.

(3) Similarity analysis of alarm floods is then conducted to find clusters of similar alarm floods across different processes or units. Here, the alarm set based pre-matching can be used to exclude dissimilar alarm floods. Then, alarm floods with similarity scores passing a threshold in the

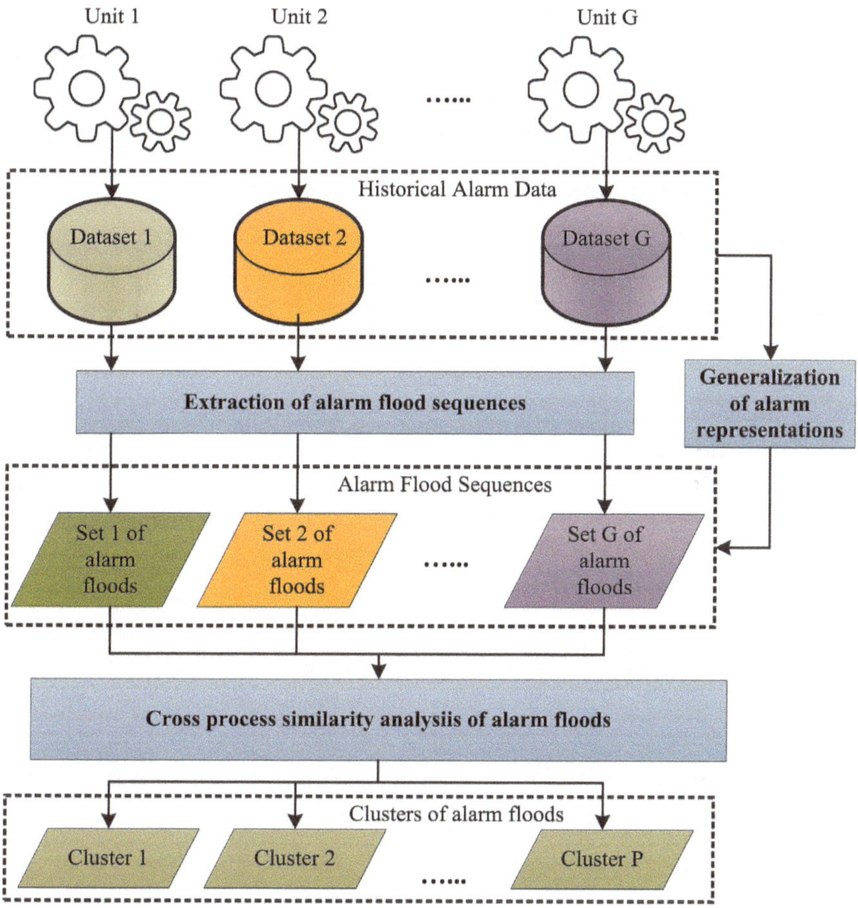

Fig. 26 Framework for the cross-process alarm flood sequence alignment.

Alarm Tag	Alarm Description
U1P1.PVHI	Pump pressure U1P1 in Unit 1 exceeds high alarm limit
U1P2.PVHI	Pump pressure U1P2 in Unit 1 exceeds high alarm limit
U2P2.PVHI	Pump pressure U2P2 in Unit 2 exceeds high alarm limit

↑ Generalize Alarm Representations ↓ Words Distillation

Abstracted Alarm Descriptor: "Pump Pressure High"

Fig. 27 An example for reconstruction of an abstracted alarm descriptor (Zhou et al., 2020).

set-based pre-matching are compared through sequence alignment. The major difference between the cross-process sequence alignment and the traditional one lies in the scoring function. In the cross-process sequence alignment, the scoring function is based on the abstracted alarm descriptors. According to the similarities obtained in cross-process sequence alignment, groups of similar alarm floods are obtained by clustering.

Fig. 28 presents an industrial example in Zhou et al. (2020). Two alarm flood sequences A and B are from two different units, namely, U1 and U3. The tag names of the alarms in the two sequences are distinct. If directly comparing the sequences represented by these alarm tags, it will find that two alarm floods are not similar. Using the cross-process sequence alignment, the two alarm floods are identified to share a significant subsequence represented by abstracted alarm descriptors. In fact, although the two alarm floods are from two different units, they are associated with the same problem, namely, the unusual value operation mode.

In conclusion, Fig. 28 reveals that the proposed method can discover similar alarm floods across different processes without the restriction of common tag names, whereas the traditional sequence alignment, such as that in Cheng et al. (2013), is based on exact alarm tag names and thus is not able to identify these similar alarm floods. The discovered clusters of similar floods can help users to find common root causes and seek generalized

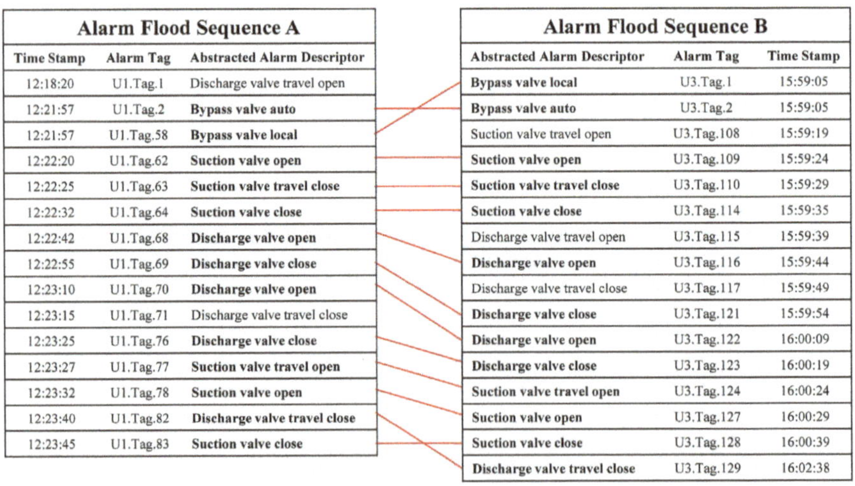

Fig. 28 Sequence alignment of two alarm floods from two different units (Zhou et al., 2020).

solutions to address alarm floods from different processes or units all at once. As a result, the efficiency of alarm flood analysis can be significantly improved.

4.3 Alarm prioritization and alarm response procedure

Alarm prioritization is a procedure to assign priorities to alarms based on a certain alarm philosophy. Alarm priorities are commonly used in alarm systems to indicate the relative importance of alarms and can assist operators in selecting alarm response procedures with different urgency levels and allowable response time. Usually, operators can distinguish the importance of alarms based on their priority attributes, and pay more attention to alarms of higher priorities in rea-time monitoring, as these alarms may relate to more serious consequences. The ISA-18.2 standard provides relevant provisions for priority division rules: Effective prioritization should be such that high priorities occur less frequently than low priorities, and most alarms should be given the lowest priority (ANSI/ISA-18.2, 2016). Reasonable alarm prioritization can ensure good performance of an alarm system.

This section mainly introduces the basic knowledge of alarm priorities in industrial alarm systems, presents the methods for alarm prioritization based on industrial regulations and literatures, and outlines alarm response procedures.

4.3.1 Basics of alarm priorities

In ANSI/ISA-18.2 (2016), alarm priority is defined as the relative importance assigned to an alarm within the alarm system to indicate the urgency of response (e.g., seriousness of consequences and allowable response time) (ANSI/ISA-18.2, 2016). In the design and rationalization of alarms, each alarm is assigned with a priority attribute to indicate its relative importance, such that the operator can decide the order of responding to certain alarms based on their priorities. This is helpful especially during periods of high alarm rates.

In the DCSs, it usually allows for multiple alarm priorities. An appropriate allocation in practice is to classify alarms into three priorities. Four or five priorities are also possible. Four priority allocation often includes an additional highest priority for a very few selected alarms. Additional special-purpose priorities may be useful, such as the lowest priority for instrument malfunction or diagnostic alarms with very limited and prescribed operator actions (ANSI/ISA-18.2, 2016). But more than that may affect the perception of operators, thus causing confusions rather than being helpful (Bill & Eddie, 2011; Bransby & Jenkinson, 1998).

It is well established through industrial studies and the industrial standards that the distribution of alarm priorities for all configured alarms should be assigned like Tables 1 and 2 (ANSI/ISA-18.2, 2016; Bill & Eddie, 2011). For higher priorities, the associated alarms should be relatively rare, where the percentage of alarms with lower priorities should be greater. Alarms with the lowest priorities usually account for the largest.

The alarm priority distribution guidelines shown above should be the natural outcome of a sound and consistent alarm rationalization process, not an artificially forced assignment simply to meet best practice guidelines (Bill & Eddie, 2011). These numbers are commonly referred to as the 80%/15%/5% distribution for three priority allocation, and distributions with deviating from these percentages may compromise the value of prioritization and make operator confused.

To distinguish alarms in different priorities within an alarm system, a combination of indications can be used, including visual and audible indications (ANSI/ISA-18.2, 2016; Bill & Eddie, 2011). Indications are illustrated as follows:

(1) Color indications: Each alarm priority must be associated with a unique color indication, and thus color must be reserved and not used for other purposes in the HMI of an alarm system. A suggestion of color choices is that red for Highest Priority, yellow for High Priority, orange for Medium Priority, and magenta for Low Priority (Bill & Eddie, 2011).

(2) Symbol indications: Same as color indications, each alarm priority should be associated with a unique symbol to reinforce color coding, such as a certain shape or text designation.

Table 1 Alarm distribution for three priorities.

	Priority 1	**Priority 2**	**Priority 3**
Alarm priority	High	Medium	Low
Distribution	5%	15%	80%

Table 2 Alarm distribution for four priorities.

	Priority 1	**Priority 2**	**Priority 3**	**Priority 4**
Alarm priority	Highest	High	Medium	Low
Distribution	<1%	5%	15%	80%

(3) Audible indications: In some environments, an audible indication is also assigned for each alarm priority. Adjusting the sounds should guarantee that operators can easily distinguish alarms in different priorities from their hearing.

(4) Message indications: The alarm messages provide further clarification of alarms and may also include part of the alarm responses.

4.3.2 Alarm prioritization methods

Alarm prioritization is defined as a process of assigning a level of operational importance to an alarm in the (ANSI/ISA-18.2, 2016). It is an important step of rationalization and a critical way for improving the discrimination between alarms of different importance. In assigning priorities to alarms, it needs to make sure that alarms in high priorities are not very frequent while alarms in low priorities should predominate in the alarm distribution. There exist quite a few studies on alarm prioritization in different areas, such as oil refinery (Foong, Sulaiman, Rambli, & Abdullah, 2009), petroleum (Chen, Wang, An, & Huang, 2015), chemical process (Dai & Qiu, 2019; Parsa, Hassall, & Naderpour, 2021), pharmaceutical plant (Agapito, Prasetyo, Persada, & Redi, 2021), and cellular networks (García, Toril, Oliver, Luna-Ramírez, & Ortiz, 2020).

In Clause 6.2.9 of the ISA-18.2 standard, basic alarm prioritization methods are suggested. Specific elements that shall be considered in alarm prioritization include (ANSI/ISA-18.2, 2016):

(1) The basis for alarm prioritization (e.g., time to respond, the severity of consequence, etc.);

(2) The metrics for alarm configuration (e.g., alarm count, priority distribution);

(3) Any impact of classification on prioritization.

Effective alarm prioritization typically results in most of the alarms assigned to the lowest priority and the fewest alarms to the highest priority. The assigned alarm priority should be aligned with the severity of consequence if the alarm is failed and the allowable time to respond to the alarm. Alarms in the highest priority should be associated with potentially high consequences in terms of safety of personnel, protection of equipment, environment, and maintaining and optimizing production. In addition, highest alarm priority implies the shortest allowable response time, such that alarms in the highest priority require immediate attentions or actions within the limited time to avoid consequences.

A grid-based system for alarm prioritization in Bill and Eddie (2011) is a simple yet effective way to allocate priorities to alarms, and has been proven in a number of successful projects. There are three grids to build for alarm prioritization (Bill & Eddie, 2011):

(1) The grid for areas of impact and severity of consequences is presented in Table 3. The contents of each row in the table must be customized for each plant, as any single scenario may have different severities in the different impact categories.

(2) The grid for maximum time available for response and correction is presented in Table 4. The maximum time to respond is the time duration within which the operators can take action(s) to prevent or mitigate the undesired consequence(s) associated with the alarm. Usually, shorter time to respond implies the higher priory of the alarm.

(3) The grid for severity of consequences and time to respond for alarm priority determination is presented in Table 5. This grid determines

Table 3 Grid for areas of impact and severity of consequence (Bill & Eddie, 2011).

Impact Category	Severity			
	None	**Minor**	**Major**	**Severe**
Public or Environment	No effect	• Local environmental effect • Does not cross fence line • Contained release • Little, if any, clean up • Negligible financial consequence • Internal or routine reporting requirements only	• Contamination causes some non-permanent damage • Single complaint • Single exceedance of statutory or prescribed limit • Reporting required at the local or state agency level	• Limited or extensive toxic release • Crosses fence line • Impact involving the community • Repeated exceedances • Uncontained release of hazardous materials with major environmental impact and 3rd party impact • Extensive cleanup measures and financial consequences • Reporting required at the state or federal agency level
Costs / Production Loss / Down-time / Quality	No loss	• < $10,000 • Reporting required only at the department level	• Between $10,000 and $100,000 • Reporting required at the site level	• > $100,000 • Reporting required above the site level
Personnel Safety	No injury or health effect	• For any alarm, if taking operator action is the primary way to avoid harming people, the highest priority should be assigned.		

Table 4 Grid for maximum time available for response and correction (Bill & Eddie, 2011).

	Maximum time for alarm response	
Text	**Option 1**	**Option 2**
Re-configure alarm for urgency	>30 min	>30 min
Promptly	Between 10 and 30 min	Between 15 and 30 min
Rapidly	Between 3 and 10 min	Between 5 and 15 min
Immediately	<3 min	<5 min

Table 5 Grid for severity of consequences and time to respond for alarm priority determination (Bill & Eddie, 2011).

Maximum Time to Respond	Consequence Severity		
	Minor	Major	Severe
> 30 minutes	Re-configure alarm for urgency		
Between 10 and 30 min	Priority 3	Priority 3	Priority 3
Between 3 and 10 min	Priority 3	Priority 2	Priority 2
< 3 minutes	Priority 2	Priority 1	Priority 1

the appropriate priority for the alarms based on the results from the previous two grids. Priority 1, Priority 2, and Priority 3 are three different priorities in a descending order based on the relative importance. Given three levels of alarm priority, they are typically defined along the lines of (Bill & Eddie, 2011; Bransby & Jenkinson, 1998):

Priority 1: urgent/emergency that requires immediate operator attention and the related consequences are severe;

Priority 2: warning/caution that requires operator attention, but the consequences of not responding to the alarms will either not be very severe, or will not occur immediately;

Priority 3: information/advisory that implies a potential problem nonsignificant, but requires operator's investigation.

4.3.3 Alarm response procedures

Given an annunciated alarm in an alarm system, the operator needs to take actions to respond to the alarm, so as to avoid consequence. For alarms in different priorities, the requirements on attentions and allowable response

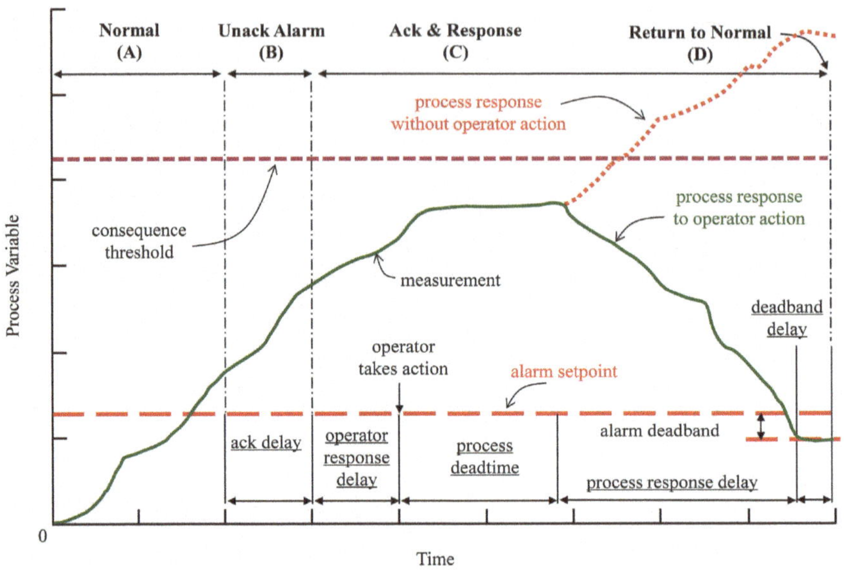

Fig. 29 Alarm timeline.

time may vary, as shown in Table 4. Regardless of the types and priorities, the lifecycle of alarms is certain. Fig. 29 presents the alarm timeline from a normal condition to an abnormal condition and the two possible scenarios based on whether the operator takes the corrective action or not. It is usually that the operator will firstly acknowledge the alarm after a certain delay of the alarm occurrence. Then, if correct actions are taken to respond to the alarm and handle the abnormality, the alarm will return to the normal state; otherwise, the problem may get worse without operator's correct response and lead to serious consequences.

An alarm response is a human cognitive process involving thought and analysis. ANSI/ISA-18.2 (2016) recommends a series of alarm response procedures, including the alarm type, alarm setpoint, potential causes, consequence of deviation, corrective action, allowable response time, and alarm class. In Bill and Eddie (2011), it is suggested the following steps for the operator to respond to an alarm. It is noteworthy that the above steps are usually not taken simultaneously. Instead, some steps can be performed in a sequential order, while they could also be in parallel in some scenarios.

(1) Detecting the alarm.
(2) Silencing and/or acknowledging the alarm.

(3) Navigating to the appropriate screen to obtain contextual information from the process of which the alarm is a part.
(4) Verifying that the alarm is valid and not a malfunction.
(5) Analyzing the process situation to determine the alarm's cause, and deciding on the proper action(s) to take in response to the alarm.
(6) Implementing the chosen action(s), generally through manipulation of the control system, contacting and directing other people to perform tasks, leaving the console to take action that cannot be accomplished without doing so, or a combination of all of these.
(7) Continuing to monitor the system to ensure the action(s) performed correctly the situation causing the alarm.

According to Rothenberg (2009) and Hu, Al-Dabbagh, et al. (2016) and Hu, Wang, and Chen (2016), actions for the operator to respond to an alarm are categorized into two major types, namely, primary actions and secondary actions. Primary actions are usually deterministic, while secondary actions are uncertain and usually performed along with primary actions. Moreover, primary actions can be easily historized and used for online analysis, while secondary actions are considered as non-standard and non-documented actions and usually cannot be historized by computerized control and information systems. Table 6 presents some typical operator actions in two categories related to alarm responses.

Table 6 Examples of primary and secondary operator actions.

Type	Examples	Characteristics
Primary operator actions	(1) Acknowledge, shelve, and suppress alarms; (2) Manipulate pumps, valves, motors or fans; (3) Set and adjust parameters and alarm limits; (4) Change operating states and control modules.	Deterministic; Can be easily historized
Secondary operator actions	(1) Communicate with other engineers or operators; (2) Schedule tasks contemporaneously; (3) Check operation manuals or notes; (4) Think about causes and seek for solutions	Uncertain and usually performed together with primary actions; Non-standard and hard to get historized

The responses and procedures to handle alarms may vary. Operators usually take strict training to learn how to manage different situations. In practice, alarm response manuals a related and documents are helpful for operators to correctly respond to alarms and address the abnormalities behind. However, such a manual might not always be up-to-date and updating manuals is rather complex and resource intensive. Especially, the procedures taken by skilled operators to respond to alarms could be better solutions but not recorded in the manuals. To recover such experience and use them to guide operators for correct alarm responses, it can mine operational patterns from historical alarm and event data and construct workflow models that represent the actions in response to alarms. The main idea is that in many alarm systems, the operator actions (mainly primary actions) are historized together with alarm events. Then, techniques, such as process discovery approaches, can be applied to capture relationships between operator actions and alarms (Hu, Al-Dabbagh, et al., 2016; Hu, Al-Dabbagh, Chen, & Shah, 2018; Hu, Al-Dabbagh, Li, & Chen, 2018; Hu, Wang, & Chen, 2016). As a result, workflow models for alarm responses are constructed and visualized. Such workflow models obtained from historical alarm and event data can be used to supplement the alarm response manuals and thus assist operators in decision making.

5. Applications of alarm management

This section presents the results of alarm management applications to a complex industrial facility of a 300 MW thermal power generation unit at Shandong Province in China.

5.1 Steps of applications

Many techniques are presented for a single alarm variable in the previous sections. In practice, there are usually hundreds and thousands of alarm variables in an industrial alarm system. For such a large number of alarm variables, the techniques are applied in the following steps:

Step 1. Collect the basic information for all alarm variables. The basic information includes names (or tagnames) of alarm variables and related process variables, alarm thresholds, and alarm generation mechanisms.

Step 2. Prepare historical data samples of all alarm variables and related process variables. It is a common practice that historical data samples of all process variables are stored in industrial plant databases. Some alarm variables take the same values of related process variables.

Step 3. Find the alarm variables suffering from nuisance alarms. First, nuisance alarms can be recognized according to Section 3.2; an alternative way is to calculate the number of alarm occurrences for each alarm variable. Attentions should be given to the alarm variable with a large number of alarm occurrences. After screening nuisance alarms, Rule 1 can be exploited to detect stale alarms.

Step 4. Design advanced alarm management techniques for the alarm variables suffering from too many nuisance alarms. For example, the corresponding delay timers can be designed (Xu et al., 2011). The delay timer is an integrated module in most industrial alarm systems, so that the deployment of the designed delay timers is easy in the technical sense.

Step 5. Repeat Steps 2–4 as a regular maintenance task in a fixed time period (e.g., 1 day or 1 week). The performance of an industrial alarm system is expected to be improved greatly after the first-time implementation of Steps 1–4. However, some alarm variables may experience high severities of nuisance alarms, when an industrial plant experience some operating status or external environment changes. By repeating the above steps regularly, the industrial alarm system can preserve a good performance in a long term.

5.2 Basics of the power generation unit

The main operation of the power generation unit is to convert the heat energy of combustion by burning coal into the thermal energy of high-pressure high-temperature steam, and to generate the electricity via an electrical generator from the mechanical energy provided by a steam turbine. Two major subsystems, namely, the coal-burning subsystem and the water–steam cycling subsystem, are schematically shown in Figs. 30 and 31. The DCS of the power generation unit involves 24,079 process variables. There are 8145 analog process variables, among which 208 variables are configured with alarm variables. There are also 2018 out of 15,934 digital process variables, being configured with alarm variables. The basic form of comparison with alarm threshold is used to generate all 2226 alarm variables in the current alarm system (CAS) of the DCS.

5.3 Comparison of the current and advanced alarm systems

We studied the performance of the CAS since January 2014 and developed a so-called advanced alarm system (AAS) that had been running in parallel to the CAS based on the same 2226 alarm variables.

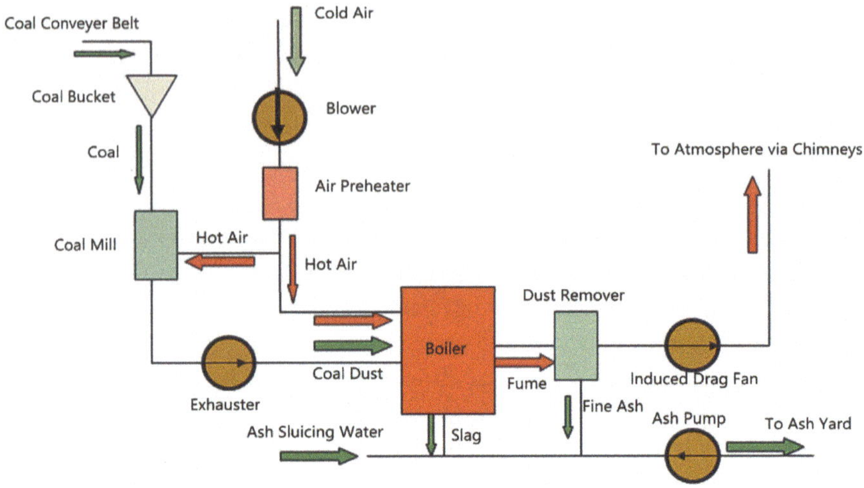

Fig. 30 The schematic diagram of coal-burning subsystem.

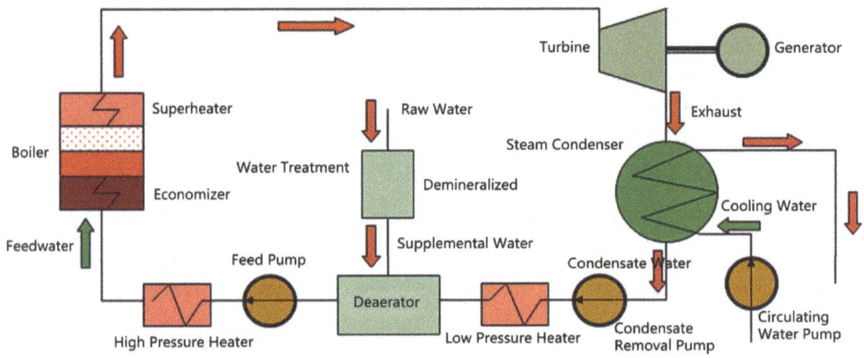

Fig. 31 The schematic diagram of water–steam cycling subsystem.

The performances of the CAS and AAS in the year of 2015 are compared here. Table 7 compares three metrics of alarm systems in the EEMUA benchmark (Brown, 2003), the power industry survey (Rothenberg, 2009), the CAS and AAS here. The three metrics of the CAS are much worse than the EEMUA benchmarks, while the counterparts of the AAS are quite close to the EEMUA benchmarks. The performances of the CAS and AAS in each month have also been investigated. Table 8 compares the numbers of alarm occurrences in the CAS and AAS for 12 months of 2015.

In particular, the average number of alarm occurrences per day in the CAS is enormous up to 18,280, while the counterpart in the AAS is only

Table 7 Performance metrics of industrial alarm systems.

	EEMUA	Power	CAS	AAS
Average alarms/day	144	2000	18,280	359
Peak alarms/10 min	10	350	405	43
Average alarms/10 min	1	8	127	2.5

Table 8 The numbers of alarm occurrences in the CAS and AAS for 12 months in 2015.

	CAS	AAS	% of reduction
January	582,615	6424	98.90%
February	667,302	8740	96.69%
March	393,220	10,420	97.35%
April	341,794	10,533	96.92%
May	724,686	7915	98.91%
June	575,079	6737	98.83%
July	705,110	8825	98.75%
August	560,846	12,744	97.73%
September	527,927	15,358	97.09%
October	583,834	12,725	97.82%
November	588,066	15,884	97.30%
December	421,565	14,708	96.51%

359; thus, about 98.04% alarm occurrences in the CAS has been removed. Even though there is still some space for improvements in the AAS, the severity of alarm overloading phenomenon in the power generation unit had been greatly alleviated.

6. Conclusions

In this chapter, alarm management techniques are overviewed with applications. Compared to conventional alarm management techniques that mainly focus on alleviation of alarm overloading, advanced alarm management targets on smart monitoring of abnormal situations and has demonstrated much better performance.

References

Adnan, N. A., Cheng, Y., Izadi, I., & Chen, T. (2013). Study of generalized delay-timers in alarm configuration. *Journal of Process Control, 23*(3), 382–395.

Adnan, N. A., & Izadi, I. (2013). On detection delays of filtering in industrial alarm systems. In *21st Mediterranean conference on control and automation* (pp. 113–118). IEEE.

Agapito, A. M., Prasetyo, Y. T., Persada, S. F., & Redi, A. P. (2021). Quality assessment support system for alarm prioritization in pharmaceutical plant operations. In *Proceedings of the IEEE 8th International Conference on Industrial Engineering and Applications (ICIEA), Chengdu, China* (pp. 390–393). IEEE.

Ahnlund, J., Bergquist, T., & Spaanenburg, L. (2003). Rule-based reduction of alarm signals in industrial control. *Journal of Intelligent Fuzzy Systems, 14*(2), 73–84.

Altschul, S. F., Gish, W., Miller, W., Myers, E. W., & Lipman, D. J. (1990). Basic local alignment search tool. *Journal of Molecular Biology, 215*(3), 403–410.

Altschul, S. F., Madden, T. L., Schäffer, A. A., Zhang, J., Zhang, Z., Miller, W., et al. (1997). Gapped BLAST and PSI-BLAST: A new generation of protein database search programs. *Nucleic Acids Research, 25*(17), 3389–3402.

ANSI/ISA-18.2. (2016). *Management of alarm systems for the process industries*. Durham, NC USA: ISA (International Society of Automation).

Asaadi, M., Aslansefat, K., Izadi, I., & Yang, F. (2022). Real-time safety evaluation of alarm systems design policies. In *Submitted for the 7th International Symposium on Advanced Control of Industrial Processes,* Vancouver, BC, Canada, August 7–9, 2022.

Asaadi, M., Izadi, I., Hassanzadeh, A., & Yang, F. (2022). Assessment of alarm systems for mixture processes and intermittent faults. *Journal of Process Control, 114*, 120–130. accepted.

Azad, P., Ude, A., Dillmann, R., & Cheng, G. (2004). A full body human motion capture system using particle filtering and on-the-fly edge detection. In *Vol. 2. 4th IEEE/RAS international conference on humanoid robots* (pp. 941–959).

Bill, R. H., & Eddie, H. (2011). *Alarm management: A comprehensive guide* (2nd ed.). International Society of Automation.

Bransby, M. L., & Jenkinson, J. (1998). *The management of alarm systems* (pp. 193–195). Sudbury: HSE Books.

Charbonnier, S., Bouchair, N., & Gayet, P. (2016). Fault template extraction to assist operators during industrial alarm floods. *Engineering Applications of Artificial Intelligence, 50*, 32–44.

Chen, T., Wang, H., An, Y., & Huang, Q. (2015). Study on multi-dimensional method for quantified setting alarm priority in petroleum industry. *China Safety Science Journal, 25*(01), 165–170.

Cheng, Y., Izadi, I., & Chen, T. (2013). Pattern matching of alarm flood sequences by a modified Smith–Waterman algorithm. *Chemical Engineering Research and Design, 91*(6), 1085–1094.

CSB. (2007). *Investigation report refinery explosion and fire (15 killed, 180 injured), BP, Texas City, Texas, March 23, 2005*. U.S. Chemical Safety and Hazard Investigation Board. Report no. 2005-04-I-TX.

Dai, Y., & Qiu, Y. (2019). Risk matrix and event tree based half quantitative alarm priority analysis for alarm systems. *Process Integration and Optimization for Sustainability, 3*(2), 159–166.

EEMUA-191. (2013). *Alarm systems: A guide to design, management and procurement*. London: EEMUA (Engineering Equipment and Materials Users' Association).

Foong, O., Sulaiman, S., Rambli, D. R. B. A., & Abdullah, N. (2009). ALAP: Alarm prioritization system for oil refinery. In *Vol. 2. Proceedings of the world congress on engineering and computer science* (p. 2009).

García, A. J., Toril, M., Oliver, P., Luna-Ramírez, S., & Ortiz, M. (2020). Automatic alarm prioritization by data mining for fault management in cellular networks. *Expert Systems with Applications*, *158*, 113526.

Guo, C., Hu, W., Lai, S., Yang, F., & Chen, T. (2017). An accelerated alignment method for analyzing time sequences of industrial alarm floods. *Journal of Process Control*, *57*, 102–115.

Hu, W., Al-Dabbagh, A. W., Chen, T., & Shah, S. L. (2016). Process discovery of operator actions in response to univariate alarms. *IFAC-PapersOnLine*, *49*(7), 1026–1031.

Hu, W., Al-Dabbagh, A. W., Chen, T., & Shah, S. L. (2018). Design of visualization plots of industrial alarm and event data for enhanced alarm management. *Control Engineering Practice*, *79*, 50–64.

Hu, W., Al-Dabbagh, A. W., Li, D., & Chen, T. (2018). Extraction and graphical representation of operator responses to multivariate alarms in industrial facilities. *IFAC-PapersOnLine*, *51*(18), 25–30.

Hu, W., Chen, T., & Shah, S. L. (2018). Detection of frequent alarm patterns in industrial alarm floods using itemset mining methods. *IEEE Transactions on Industrial Electronics*, *65*(9), 7290–7300.

Hu, W., Wang, J., & Chen, T. (2016). A local alignment approach to similarity analysis of industrial alarm flood sequences. *Control Engineering Practice*, *55*, 13–25.

International Electrotechnical Commission. (2014). *IEC 62682 Management of alarm systems for the process industries* (pp. 24–26). Geneva, Switzerland: IEC.

Izadi, I., Shah, S. L., & Chen, T. (2010). Effective resource utilization for alarm management. In *49th IEEE conference on decision and control (CDC)* (pp. 6803–6808).

Izadi, I., Shah, S. L., Shook, D. S., Kondaveeti, S. R., & Chen, T. (2009). A framework for optimal design of alarm systems. *IFAC Proceedings Volumes*, *42*(8), 651–656.

Kondaveeti, S. R., Izadi, I., Shah, S. L., & Black, T. (2010). Graphical representation of industrial alarm data. *IFAC Proceedings Volumes*, *43*(13), 181–186.

Kondaveeti, S. R., Izadi, I., Shah, S. L., Black, T., & Chen, T. (2012). Graphical tools for routine assessment of industrial alarm systems. *Computers & Chemical Engineering*, *46*, 39–47.

Kondaveeti, S. R., Izadi, I., Shah, S. L., & Chen, T. (2011). On the use of delay timers and latches for efficient alarm design. In *19th Mediterranean conference on control & automation (MED)* (pp. 970–975). IEEE.

Lai, S., & Chen, T. (2017). A method for pattern mining in multiple alarm flood sequences. *Chemical Engineering Research and Design*, *117*, 831–839.

Lai, S., Yang, F., & Chen, T. (2017). Online pattern matching and prediction of incoming alarm floods. *Journal of Process Control*, *56*, 69–78.

Lai, S., Yang, F., Chen, T., & Cao, L. (2019). Accelerated multiple alarm flood sequence alignment for abnormality pattern mining. *Journal of Process Control*, *82*, 44–57.

Marsh. (2019). *100 Largest losses in the hydrocarbon industry 1974–2019* (26th ed.). Marsh JLT Specialty.

Needleman, S. B., & Wunsch, C. D. (1970). A general method applicable to the search for similarities in the amino acid sequence of two proteins. *Journal of Molecular Biology*, *48*(3), 443–453.

Parsa, K., Hassall, M., & Naderpour, M. (2021). Enhancing alarm prioritization in the alarm management lifecycle. *IEEE Access*, *10*, 99–111.

Parvez, M. R., Hu, W., & Chen, T. (2020). Comparison of the Smith-Waterman and Needleman-Wunsch algorithms for online similarity analysis of industrial alarm floods (pp. 1–6). IEEE Electric Power and Energy Conference (EPEC).

Parvez, M. R., Hu, W., & Chen, T. (2022). Real-time pattern matching and ranking for early prediction of industrial alarm floods. *Control Engineering Practice*, *120*, 105004.

Rothenberg, D. H. (2009). *Alarm management for process control: A best-practice guide for design, implementation, and use of industrial alarm systems.* Momentum Press.

Shu, Y., Ming, L., Cheng, F., Zhang, Z., & Zhao, J. (2016). Abnormal situation management: Challenges and opportunities in the big data era. *Computers and Chemical Engineering*, *91*, 104–113. https://doi.org/10.1016/j.compchemeng.2016.04.011.

Smith, T. F., & Waterman, M. S. (1981). Identification of common molecular subsequences. *Journal of Molecular Biology*, *147*(1), 195–197.

Su, J., Guo, C., Zang, H., Yang, F., Huang, D., Gao, X., et al. (2018). A multi-setpoint delay-timer alarming strategy for industrial alarm monitoring. *Journal of Loss Prevention in the Process Industries*, *54*, 1–9.

Thornhill, N. F., & Hägglund, T. (1997). Detection and diagnosis of oscillation in control loops. *Control Engineering Practice*, *5*(10), 1343–1354.

Wang, J., & Chen, T. (2013). An online method for detection and reduction of chattering alarms due to oscillation. *Computers & Chemical Engineering*, *54*, 140–150.

Wang, J., Yang, F., Chen, T., & Shah, S. L. (2015). An overview of industrial alarm systems: Main causes for alarm overloading, research status, and open problems. *IEEE Transactions on Automation Science and Engineering*, *13*(2), 1045–1061.

Xu, J., Wang, J., Izadi, I., & Chen, T. (2011). Performance assessment and design for univariate alarm systems based on FAR, MAR, and AAD. *IEEE Transactions on Automation Science and Engineering*, *9*(2), 296–307.

Yang, F., Shah, S. L., Xiao, D., & Chen, T. (2012). Improved correlation analysis and visualization of industrial alarm data. *ISA Transactions*, *51*(4), 499–506.

Zhou, B., Hu, W., Brown, K., & Chen, T. (2020). Generalized pattern matching of industrial alarm flood sequences via word processing and sequence alignment. *IEEE Transactions on Industrial Electronics*, *68*(10), 10171–10179.

CHAPTER EIGHT

Performance evaluation of digitalized safety barriers

Aibo Zhang[a,*] and Yiliu Liu[b,*]
[a]Department of Advanced Design and Systems Engineering, City University of Hong Kong, Kowloon Tong, Hong Kong
[b]Department of Mechanical and Industrial Engineering, Norwegian University of Science and Technology, Trondheim, Norway
[*]Corresponding authors: e-mail address: aibozhang0121@gmail.com; yiliu.liu@ntnu.no

Contents

1. Introduction

When digitalized safety barriers are installed in the process industries, or they are operated to reduce the frequency of hazardous events and mitigate their consequence, it is critical to know the situation of such protection systems and evaluate whether their performance is good enough for their safety functions. This chapter will deal with the measures, numerical analysis methods, and modeling approaches in the performance evaluation of safety barriers.

In performance evaluation, we consider whether and how well the actual performance of a function fits with the expectation or the target value. The function associated with a safety barrier is called as barrier function in this

Methods in Chemical Process Safety, Volume 6
ISSN 2468-6514
https://doi.org/10.1016/bs.mcps.2022.04.006

chapter, and it can be described as "verb + noun (related to negation or deviation)," for example, release pressure, reduce high temperature, and stop gas flow. If the actual performance of a barrier function is within the acceptable limits of deviation from the expectation (target value), the barrier is regarded as functioning. If the actual performance is beyond the acceptable limits, the barrier is not as effective as expected. It can fully fail without any contribution to risk control, or it can be in a degraded state, with lower capability in risk control.

With the development of sensing and digital technologies, more and more data on the operational conditions of safety barriers can be collected, and accurate predictions on the failure time and residually useful life (RUL) of barriers can be expected. This chapter will thus discuss the challenges and opportunities brought by new technologies.

The remainder of this chapter is organized as follows: Section 2 will introduce the failure modes, operational modes, and strategies of digitalized safety barriers, as the preparation of performance evaluation. Section 3 is for presenting the classification of main measures, followed by Section 4 with modeling approaches for simple barrier systems. Analytical formulas and some modeling examples of safety barrier systems are discussed in Section 5. Section 6 presents some new methods with the digital twin principle for evaluating conditions and planning for future operations.

2. Failure and operational analysis of safety barriers

Safety barriers are versatile, from organizational, cultural, signal, physical, and technical. Digitalized safety barriers, in our understanding, belong to technical barriers, where (digital) technologies are involved. A digital barrier system functionally consists of three subsystems: sensing-, analyzing-, and executing subsystems. The sensing subsystem is used to collect data with the purpose of detecting the abnormal situations in the asset of interest, for example, pressure transmitters and level sensors in the process industry. The analyzing subsystem deals with the collected data, extracting useful information, and makes decisions on what kind of response is needed and whether barrier functions should be performed. In traditional safety-instrumented systems, this subsystem is called the logic solver, but for digitalized safety barriers, the data analysis function will be enhanced. Decisions will be made not only based on simple and clear logic or rules but based on comprehensive evaluation with higher requirements on the computational capability. The executing subsystem is the element that really conducts a safety or barrier function, and it often needs to integrate the digital technologies with specific mechanical, electric, or hydraulic components. The state of these components also can be monitored, and some of their failure and degradation information can be diagnosed and be feedbacked to the analyzing subsystem.

The operational mode of safety barriers is different from that of normal production systems. Technical safety barriers are designed and deployed to respond to demands, which are the negations or deviations in the definition of barrier function. Demands can be external shocks and other events resulting in damages to the asset of interests. When there is no demand, the executing subsystem of a safety barrier is in a dormant state, meaning that it is not activated or withstands any workload. Diagnosis modules can be installed to detect whether degradations and failures occur on the mechanical, electric, or hydraulic components of the executing subsystem, but not all failures and degradations can be revealed. For example, the barrier function of a shutdown valve is to stop gas flow in case high pressure is found in the downstream tank of a pipeline. When the valve closes, a failure can be the leakage of gas from the gate valve. Such kind of failure is difficult to be found only with diagnosis before the valve really finishes its action for the barrier function.

Some failures of barriers are not directly related to its barrier function or the safety of the asset of interests. Still for the example of the shutdown valve, after it closes, another possibility is the failure to open again. Such a failure will result in some loss of production time but not serious damage. Thus, it is necessary to classify the failures of safety barriers according to their consequences and detectability (Rausand, 2014), as:

- Dangerous undetected (DU) failures: Failure to perform barrier function, not detectable by self-diagnosis, but can be revealed by proof-testing or the occurrence of real demand, such as the leakage (through the valve) in the closed position (LCP) for the shutdown valve.
- Dangerous detected (DD) failures: Failure to perform barrier function, but detectable immediately after occurring, such as signal loss, signal out of range, and final element in the wrong position. The detection can be via self-diagnosis modules and by human operators. Normally DD failures are expected to be fixed in a short time.
- Safe undetected (SU) failures: Failure not directly related to the barrier function, not detectable by self-diagnosis, such as failure to open(FTO).
- Safe detected (SD) failures: Failure not directly related to the barrier function, and detectable immediately after occurring, such as a spurious trip.

When dealing with safety barriers, a focus is to achieve successful safety performance. A frequently mentioned term is "safety critical failure," which Selvik and Signoret (2017) interpret as "A final failure triggering an unsafe situation and making a hazardous event possible". The term may vary from different aspects but express a consensus, referring to the failure event has a significant effect on the system protection and may lead to unacceptable risk. Considering the four categories mentioned earlier, these DU failures that might occur in digitalized safety barriers deserve more attention in operation and management.

DU failures can be found in a real demand or a simulation of demand in a proof test. Proof tests are conducted regularly to examine all functions of a barrier, and their frequency cannot be too high because they need to stop production or interrupt the normal operation of the asset that barriers are protecting. With a certain delay (e.g., a fixed period of τ) between two sequential proof tests, a failure can occur and bring the barrier to an unavailable state. Then, as shown in Fig. 1, if a demand arrives, the asset of interest will be exposed to a hazard without any safeguard or mitigation, and an accident will occur.

Therefore, it can be found that DU failures are the main contributor to accidents from the perspective of safety barriers. In the remainder of this chapter, we will pay most attention to the evaluation related to DU failures.

If there is, fortunately, no demand between a DU failure of the barrier and the next proof test, the fault can be revealed in the test, and the failed barrier can be repaired. In many cases, the repair work is assumed as perfect, and the barrier is then restored to an as-good-as-new condition, given that degradation is not taken into account. This is the normal operation and maintenance strategy. However, for digitalized safety barriers, with more

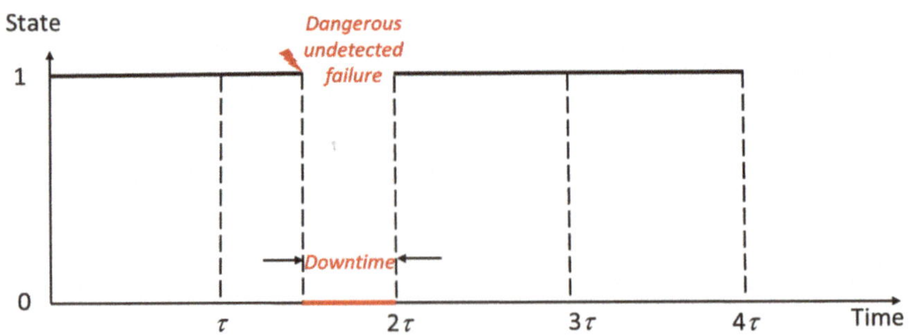

Fig. 1 The state variable X(t) of a periodically tested barrier with respect to DU failure.

information about the barrier condition that can be collected, preventive maintenances and condition-based maintenances are more reasonable to take actions before real failures and reduce the risk to the asset of interests. We will briefly discuss such considerations in the performance evaluation of safety barriers at the end of this chapter.

3. Measures for performance evaluation

According to the study by Johansen and Rausand, (2015), performance of safety barriers can be evaluated from many different perspectives, including specificity, functionality, reliability, response time, capacity, durability, robustness, auditability, and independence, some of which are design properties (such as specificity, functionality, auditability, and independence) while some are easier to be qualified and used as measures in performance evaluation. In the review paper of Liu (2020), performance measures of safety barriers are generally summarized as effectiveness and availability. In the process industries,

- Effectiveness can be regarded as the functions of a barrier and its capability in reducing accidental risks. Specified response time to a hazardous event and durability of a barrier in withstanding a demand are also contributors of effectiveness.
- Availability is used to reflect the ability of a barrier to be effective in performing its safety function at a certain moment or in a certain period. Since we often assume that safety barriers are repairable, it makes more sense to use availability to indicate the combined effects of the internal reliability of a barrier and its test and maintenance strategy.

Detailed discussions on more classifications and types of performance measures for safety barriers also can be found in Prashanth, Fernandez, Sunder, and Boardman (2017) and Liu (2020).

In this chapter, we mainly consider technical barriers where digital technologies are applied for monitoring and controlling. In both the generic standard IEC61508 (2010) and the specific standard IEC61511 (2016) for the process industries, availability or safety integrity is used as the measure of performance of safety-instrumented systems, as the safety barriers using electrical, electronic or programmable electronic techniques. ISO13702 (2015) then implements the terminology of integrity for all safety barriers. Digitalized safety barriers always integrate with electrical and electronic techniques, and thus it is natural to employ the measures in these standards.

Table 1 Determination of SIL by PFD and PFH.

SIL	PFD	PFH
4	$\geq 10^{-5}$ to $<10^{-4}$	$\geq 10^{-9}$ to $<10^{-8}$
3	$\geq 10^{-4}$ to $<10^{-3}$	$\geq 10^{-8}$ to $<10^{-7}$
2	$\geq 10^{-3}$ to $<10^{-2}$	$\geq 10^{-7}$ to $<10^{-6}$
1	$\geq 10^{-2}$ to $<10^{-1}$	$\geq 10^{-6}$ to $<10^{-5}$

Adopted from IEC61508. (2010). *Functional safety of electrical/electronic/programmable electronic safety-related systems*. Geneva, Switzerland: International Electrotechnical Commission.

To facilitate risk management and communication, discrete levels (normally 4) are used to specify different requirements or performance of safety barriers. As shown in Table 1, safety integrity level 4 (SIL 4) means the strictest requirement or highest performance, while SIL 1 is the lowest.

SIL should be determined based on the calculation results of the average probability of failure on demand (PFD_{avg}) or average frequency of failures (PFH, from the old name of "probability of having a dangerous failure per hour"). The previous measure is for barriers in the low-demand mode (demand frequency is low and demands come seldomly), while the latter is for barriers operating in the high-demand or continuous mode (demand frequency is high and demands come often). In fact, both of these two are measuring the unavailability of a safety barrier.

For PFD of a barrier element by a certain time before the first proof test, namely PFD(t), it is simply equal to the failure probability of this item because there is no conducted repair. PFD_{avg} is the average value of PFD(t) in the first test interval, which is the period from the time 0 when the item is put into operation to the moment just before a proof test. In mathematics,

$$\mathrm{PFD}_{\mathrm{avg}} = \frac{1}{\tau}\int_0^{\tau} \mathrm{PFD}(t)dt = \frac{1}{\tau}\int_0^{\tau} F(t)dt \tag{1}$$

where τ is the proof test interval, and $F(t)$ is the failure probability by time t.

Suppose we can assume that a proof test can detect all faults and the following maintenance is perfect, meaning that the barrier element can be restored to an "as-good-as-new" state. In that case, each proof test can be regarded as a renewal of the item, and Eq. (1) can be applied for any interval. It can be found that not only item reliability (reflected by $F(t)$) but the operational strategy (by determining the length of τ) can influence the value of PFDavg.

When a safety barrier is running continuously or with frequent demands, PFH can be used for evaluating its performance. In such cases, by time t we have

$$\text{PFH} = \frac{F(t)}{t} = \frac{E(N(0,t))}{t} \tag{2}$$

where $E(N(0,t))$ is the expected number of failures occurring in the period of $(0,t)$. According to Eq. (2), test interval is not an influencing factor of PFH because demands will always arrive before the next proof test, and any failures of safety barriers will probably result in an accident.

PFH is equivalent to the rate of occurrence of failures (ROCOF) which is a common reliability measure of normally operational items. The analysis methods for ROCOF can be used for calculating PFH. Therefore, in the rest of this chapter, we will focus more on low-demand safety barriers and use PFD as their performance measure.

4. Modeling approaches

Several approaches can be used in the modeling of safety barriers for their system analysis and performance evaluation. When we talk about a modeling approach, it always includes two elements: the modeling framework, which is used to represent a system; and the algorithm for performance analysis, which is behind the representation.

In this chapter, our aim is not to introduce each approach in detail but only to mention some of them briefly and highlight their benefits and limitations. Readers can find more information about these modeling methods in many textbooks in reliability engineering. The approximation formulas based on Eqs. (1) and (2) are not included in this section but in Section 5 specific for integrity analysis. It is also noted that many terminologies in this section are not strictly defined and used; a textbook of reliability theory is recommended if the readers have no relevant background.

4.1 Fault tree analysis (FTA) and reliability block diagram (RBD)

Fault tree analysis (FTA) and reliability block diagram (RBD) are the most important tools in system reliability analysis and evaluation. We introduce these two methods together since their models can be interchanged in many applications.

FTA is a failure-oriented approach to describe how a system failure occurs due to different causes, for example, component failures. This method is based on a top-down logic diagram like a tree, as shown in Fig. 2A.

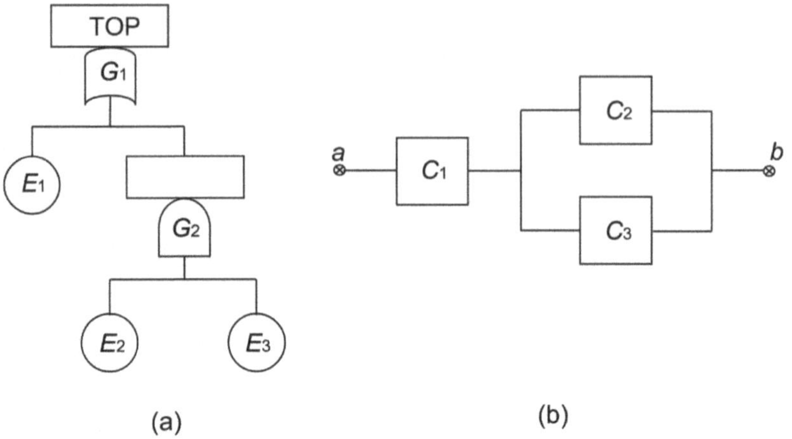

Fig. 2 Introductory examples of fault tree (A) and reliability block diagram (B).

RBD is, however, a success-oriented approach to describe how a specified system function is fulfilled with the connections of subfunctions, e.g., functions of components. RBD is a kind of network with functional blocks (rectangles) and connections (lines) between blocks, as shown in Fig. 2B.

In a fault tree, the system failure is at the top, called as the TOP event, shown as a rectangle node. The fault tree is then established by identifying the causes of this event, which can be human errors, component failures, or combinations of component failures. They are the lower-level events, and then they can be decomposed further until basic events E_i, which are some root causes or individual component failures, shown as circulars in Fig. 2A. In the simplest FTA, upper-level events and lower-level events are connected with two types of logic gates: OR-gate (as G_1 in Fig. 2A) and AND-gate (as the G_2 in Fig. 2A). The OR-gate means that the output event of the gate (upper-level event) will occur if any of the input events (lower-level events) occurs, and the AND-gate means that the output event of the gate can occur only if all of the input events occur. The fault tree is often used as a Boolean model, meaning that for each node, no matter it is the TOP event or basic event, it only can be in one of the two states: occurring or not occurring. If we only consider an event as a failure, the two states can be failed and functioning.

In RBD, a system is considered as functioning if it is possible to pass through all the blocks on the path from one endpoint to the other endpoint (e.g., from *a* to *b* in Fig. 2B). It should be noted RBD is the logical connections that are required to function, rather than the physical layout of a system.

A basic RBD is also a Boolean model, with each block having two states: functioning and failed, and the two ends of the block are connected if it is functioning, otherwise, the two ends are disconnected. An RBD model can have two basic structures: series structure and parallel structure. In a series structure, the system is functioning if and only if all of the n components within the structure are functioning, while in the parallel structure, the system is functioning if at least one of the n components within the structure are functioning.

It can be found that a parallel structure of RBD is equivalent to the AND-gate in a fault tree, and a series structure of RBD is equivalent to the OR-gate in a fault tree. In this case, we can regard that C_i in RBD denotes component i, and E_i is the failure of component i.

Both models are effective in the modeling of large-scale systems and structural analysis, and they support both qualitatively causal analysis and quantitative performance and evaluation.

Identification of minimal cut sets in a fault tree can be a starting point of quantitative evaluation. A cut set is a set of basic events whose simultaneous occurrence ensures that the TOP event occurs, and a cut set is said to be minimal if the set cannot be reduced without losing its status as a cut set. For the example in Fig. 2A, the minimal cut sets of the fault tree include: $\{E_1\}$, $\{E_2, E_3\}$. This implies that when the event E_1 occurs, or events E_2 and E_3 occur at the same time, the TOP event occurs. In mathematics, the occurrence of basic event E_i at time t is a random variable $X_i(t)$ with two possible values of 1 (occur) and 0 (not occur), and then $q_i(t)$ can be used to denote the occurrence probability of the event (failure probability of component i by time t), namely, $q_i(t) = \Pr[X_i(t) = 1]$.

If the basic event i is one of n input events of an AND-gate, the occurrence probability of the output event $q_j(t)$ of the gate is

$$q_j(t) = \prod_{i=1}^{n} q_i(t) \tag{3}$$

While if the basic event i is one of n input events of an OR-gate, the occurrence probability of the output event $q_j(t)$ of the gate is

$$q_j(t) = 1 - \prod_{i=1}^{n} (1 - q_i(t)) \tag{4}$$

We can consider the fault tree in Fig. 2A as a model for parts of a technical safety barrier, where E_1 means the failure of the logic solver, and E_2 and E_3 are for the failure of two redundant shutdown valves, respectively. Then based on Eqs. (3) and (4), we can have

$$\text{PFD}(t) = q_{\text{TOP}}(t) = q_1(t) \cdot (q_2(t) + q_3(t) - q_2(t)q_3(t))$$

For such a safety barrier, FTA is effective in obtaining the instantaneous unavailability, namely PFD(t), but it should be noted that we cannot use values of PFD_{avg} of basic events to calculate the PFDavg of the barrier system, in consideration of the impacts of tests.

For RBD, it is needed to discover the paths that can link the two endpoints of the model. A set of nodes (components) that, by connecting (functioning), ensure that the system is functioning is called a path set, and it is said to be minimal if the path set cannot be reduced without losing its status. In Fig. 1B, the sets $\{E_1, E_2\}$ and $\{E_1, E_3\}$ are two minimal path sets.

In a quantitative analysis, the state of component i at time t can be regarded as a random variable $Y_i(t)$ with two possible values: 0 (failed) and 1 (functioning), and $r_i(t)$ is the probability that component i is functioning at time t, so $r_i(t) = \Pr[Y_i(t) = 1]$. $r_i(t)$ is the reliability or survivor probability. For a series structure j with n component, its probability of functioning is

$$r_j(t) = \prod_{i=1}^{n} r_i(t) \tag{5}$$

While for a parallel structure with n components, its probability of functioning is

$$r_j(t) = 1 - \prod_{i=1}^{n} (1 - r_i(t)) \tag{6}$$

Since these two methods are based on the Boolean algorithm, it is a bit difficult to apply them for degradation modeling or analysis for systems with more complex behaviors.

Both FTA and RBD are recommended in IEC61508 (2010). More symbols and information about FTA can be found in the standard IEC61025 (2006), and more details about RBD models can be found in the latest version IEC61078 (2016).

4.2 Bayesian network

A Bayesian network, or Bayesian belief network (BBN), is a graphical model consisting of consists of two parts: A directed acyclic graph (DAG) with a finite set of nodes and a set of directed edges, and conditional probability tables (CPTs) associated with nodes. A BBN can be written as

$N = <(V, E), P>$, where (V, E) represents the nodes and edges of DAG, and P is a probability distribution of each node.

BBN can illustrate the causal relationships between causes and outputs of a system by using directed edges between the nodes. As shown in Fig. 3, the node TOP has two inputs nodes E_1 and E_4, which are called as the parent nodes of TOP, denoting the causes of the TOP event. TOP is called as the child node of E_1 and E_4. Node E_4 also has two parent nodes E_2 and E_3. Nodes without parent nodes are called as root nodes, as E_1, E_2, and E_3 in this case. Nodes without child nodes are called leaf nodes, as the TOP node in Fig. 3, and all the others are intermediate nodes. Two CPTs are associated with nodes E_4 and TOP, and those root nodes have no CPT.

In Fig. 3, the causal logic between E_4 and its parent node is just like an AND-gate in a fault tree. Based on the CPT, we can see that only when both events E_2 and E_3 occur, E_4 will occur. Similarly, it can be found that the causal logic between TOP and parent nodes is like an OR-gate. A fault tree model can thus be translated into a BBN model without losing any details. The algorithm for FTA also can be used for BBN.

If we generally consider a BBN with n nodes, the probability that $X_i(t) = x_i$ (for $i = 1, 2, \ldots, n$) is

$$\begin{aligned}\Pr(X_1(t) = x_1 \cap X_2(t) = x_2 \cap \cdots \cap X_n(t) = x_n) \\ = \prod_{i=1}^{n} \Pr(X_i(t) = x_i | Parent(X_i))\end{aligned} \tag{7}$$

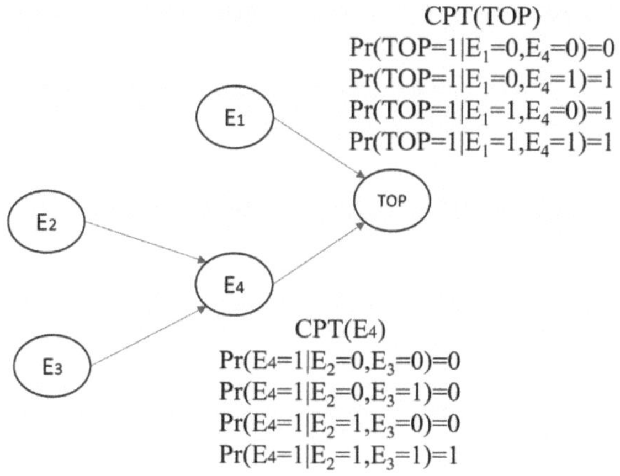

Fig. 3 An introductory example of BBN.

BBN is an intuitive representation of causal relationships, and it is very powerful in analyzing and evaluating complex systems with uncertainties. BBN can replace FTA in many applications with more flexibilities. We can find several studies for technical safety barriers, such as Wu, Zhang, Lundteigen, Liu, and Zheng (2018). However, BBN requires a lot of work for data collection and establishment of CPTs, and this workload increases exponentially when the number of nodes increases. In the following illustrative examples of performance evaluation of safety barriers in this chapter, we do not use BBN.

4.3 Markov method

It is difficult to use FTA or RBD to conduct the performance evaluation for systems with dynamic behaviors, such as proof tests that are regularly performed on safety barriers, degradation of barriers, and associated maintenances. State transition models are therefore introduced, and one of the most widely used is the Markov method.

We can use a Markov diagram, as shown in Fig. 4, to illustrate discrete states of a barrier system and the transitions between these states. When the system performance is time-dependent, the process behind the model can be called as a continuous Markov process.

In Fig. 4, the circles represent system states. For example, we can set state 0 as the functioning state of the system and state 1 as the faulty state. The transition from state 0 to state 1 is the failure of the system with a rate of λ, and the transition from state 1 to state 0 is the restoration of the system with a rate of μ. When we want to evaluate the system performance, we are interested in the sojourn time or sojourn probabilities of the system in state 0 and state 1.

The calculation behind a Markov diagram is always based on the assumption that the state transitions are stochastic processes following the Markov property. According to the textbook by Rausand, Barros, and Hoyland (2020), the Markov property means that the future state of a process is only

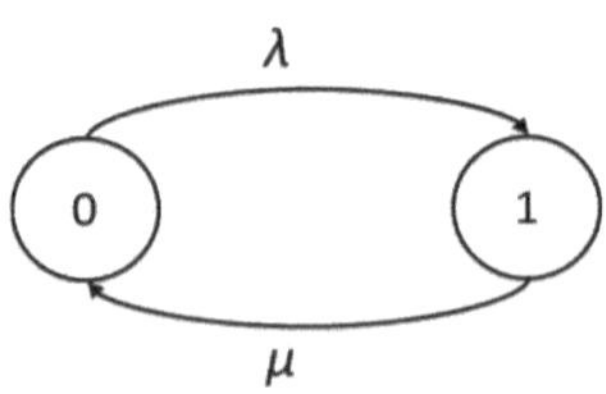

Fig. 4 A simple Markov diagram.

dependent on the present state and is not influenced by how the process arrived in the present state. In mathematics,

$$\Pr[X(t+s)=j|X(s)=i, X(u)=x(u), 0\leq u<s] = Pr[X(t+s)=j|X(s)=i]$$
$$\text{for all possible } x(u) \tag{8}$$

A stochastic process with such a property is also called as the process with stationary transition probability or time-homogeneous process, meaning that a Markov process is exponentially distributed in terms of transition time.

A Markov model can be quantitatively analyzed. Consider a Markov process $\{X(t), t\geq 0\}$ with the state space $\Omega=\{0,1,2,\ldots,n\}$. The transition probability from state i to state j by time t is

$$p_{ij}(t) = \Pr[X(t)=j|X(0)=i] \tag{9}$$

For a system with n states, a transition matrix can be set up as

$$\boldsymbol{P}(t) = \begin{pmatrix} p_{00}(t) & p_{01}(t) & \cdots & p_{0n}(t) & \\ p_{10}(t) & p_{11}(t) & & p_{1n}(t) & \\ \vdots & & \ddots & \vdots & \\ p_{n0}(t) & & p_{n1}(t) & \cdots & p_{nn}(t) \end{pmatrix} \tag{10}$$

And we also know when a process is in state i at time 0, it must be either in the same state or have entered a different state at time t, so that

$$\sum_{j=0}^{N} p_{ij}(t) = 1 \text{ for all } i\in\Omega \tag{11}$$

Based on these basic considerations and the exponential distribution assumption, it is possible to calculate the steady-state probability of a system in the long term. The concrete calculation procedure is skipped here, and readers can reference Chapter 11 in the textbook by Rausand et al. (2020) if they have an interest in the details of the analytical process of the Markov model.

One benefit of the Markov method lies in its clear illustration of the dynamics of a system, and thus it will be mainly used in the rest of this chapter for describing what is considered in the safety barrier performance evaluation. Markov model is flexible and is effective in modeling complex behaviors of a small system, but its application to a large system is difficult considering the rapid increase of model size when more items and more states are included. IEC61508 (2010) has included this method in the integrity analysis of technical safety barriers.

Besides the steady-state probability calculation, the multi-phase Markov method also has been developed to capture system performance since the system state is observed and known through proof tests. The main idea is to reallocate the distribution of system states at each proof test time. More details can be found in research, e.g., Innal, Lundteigen, Liu, and Barros (2016), Wu, Zhang, Lundteigen, et al. (2018), Srivastav, Barros, and Lundteigen (2020), and Zhang, Srivastav, Barros, and Liu (2021), but no further discussion here.

4.4 Petri net

Petri net (PN) is another alternative of the state transition model, especially when the assumptions on the Markov property are not valid for reality. According to the terminology specified in the standard IEC62551 (2013) for PN, such a model includes two types of static elements: places (circles in Fig. 5) and transitions (rectangles in Fig. 5), and they are connected by directed arcs (with arrows in Fig. 5). The model also includes dynamic elements, namely tokens that are used to denote the movable resources in a system. Tokens can reside in the places, and the distribution of tokens in the places is regarded as a system state. Thus, the system state shown in Fig. 5 can be written as $X = \langle p_1(1), p_2(0) \rangle$.

Tokens move via arcs, and each arc is assigned a multiplicity as its token deliverability. When the number of tokens in all input places of a transition are at least equal to the multiplicities of the associate arcs to the transition, the transition is enabled. For example, in the current state shown in Fig. 5, transition t_1 is enabled. Then, the transition can be fired to change the distribution of tokens in places. Time from being enabled to being fired can have different distributions, called as firing time, reflected by different symbols of transitions in the model. According to IEC62551 (2013), a blank

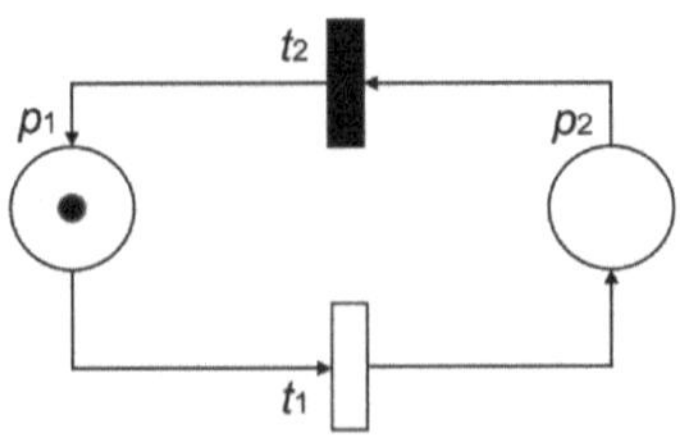

Fig. 5 An introductory example of Petri net.

rectangle (as t_1 in Fig. 5) means that the time follows the exponential distribution, and a black rectangle (as t_2 in Fig. 5) means a deterministic delay.

We can use Fig. 5 to model the failure and repair of a system. When a token is in place p_1, the system is functioning and t_1 is enabled, meaning that the system is subject to fail. If t_1 is fired, the token in p_1 is removed, and then a new token is released to p_2, since the two arcs p_1 to t_1 and t_1 to p_2 have a multiplicity of 1 (1 is default in drawing if no specific digit is added). The new state is that one token in p_2 and no token in p_1, indicating the system fails and t_2 is enabled. If we suppose that the failed system will be restarted after 1 week when everything is ready, transition t_2 is fired with the deterministic delay of 1 week.

When all transitions in a PN have zero firing time or exponentially distributed firing time, the model is actually equivalent to a Markov process, and the numerical analysis approaches suitable for Markov models can be adopted. While deterministic transitions and other non-exponential transitions are involved like in Fig. 5, Monte Carlo simulation is needed to capture the useful information.

Petri net has many extensions. In IEC61508 (2010), a reliability block diagram (RBD) driven PN is recommended, where two formulas are added in a PN:

- Predicate (identified by "?" or "??")—to control the enabling condition of a transition;
- Assertion (identified by "!" or "!!")—to update one variable when the transition is fired.

PN is an excellent tool for modeling complex behaviors and compared with the Markov model, it is not necessary to identify all potential states of a system when modeling. PN is also very effective in modeling some parallel events and conditions, and thus it has been adopted in studies about the impacts of different test and maintenance strategies on technical safety barriers, more details can be found in Liu (2014) and Rausand (2014), etc. However, this method is not widely used in industries since it needs specific training before it can be understood by engineers without relevant background. Targeting to improve the readability and understandability of PN modeling for large systems, both for PN designers and readers, Signoret et al. (2013) has conducted the research by introducing predicates and assertions to develop the sub-PNs, furtherly open the way to build RBD driven-PNs by using the above sub-PNs. This method has been verified with strong capability in the performance assessment of safety systems. More details of basic Petri net models can be found in IEC62551 (2013), and details of RBD Petri net are in part 6 of IEC61508 (2010).

5. Integrity analysis and evaluation of safety barriers

We start from the integrity analysis of a single safety barrier element and barriers systems where different components are independent of each other, meaning that the failure probability of one component is not affected by the failure of others.

The state variable $X(t)$ of an item with respect to DU failure, as shown in Fig. 1, is

$$Xt = \begin{cases} 1 & \text{if the item is able to function as a barrieri.e., no DU failure is present} \\ 0 & \text{if the item is not able to function as a barrieri.e., DU failure is present} \end{cases}$$

After a test(repair), the item is considered to be as-good-as-new if a DU failure mode is not present.

Generally, a technical safety barrier consists of three main subsystems: (1) a sensor subsystem, (2) a logic solver subsystem, and (3) a final element subsystem, as illustrated in Fig. 6.

The safety barrier fails on demand if any of the subsystem fails. When the three subsystems are independent and have high reliability, indicating the probability that two or three subsystems fail at the same time is negligible, the average probability that the safety barrier fails on demands, PFD(SIF) avg., can be determined by adding the PFD_{avg}'s for the three subsystems, as

$$\text{PFD}_{\text{avg}}^{(\text{SIF})} \approx \text{PFD}_{\text{avg}}^{(\text{S})} + \text{PFD}_{\text{avg}}^{(\text{LS})} + \text{PFD}_{\text{avg}}^{(\text{FE})} \tag{12}$$

where PFD(S) avg., PFD(LS) avg., PFD(FE) avg. denote the PFD_{avg} of the sensor, logic solver, and final element subsystem, respectively.

To improve barrier reliability, generally, each subsystem is allocated as a *k*-out-of-*n* (koon) voted structure, meaning that the subsystem is functioning when at least k of its n channels are functioning, e.g., 1oo2, 2oo2, 2oo3, etc. The simplified formulas and Markov models as the widely used quantitative methods in barrier performance evaluation are thus introduced here.

Following the Markov models, the barrier can be in a finite state space. In some states, the barrier is able to perform its safety function; in other states, it is not. Let D denote the set of the states where the barrier is 'down' and not

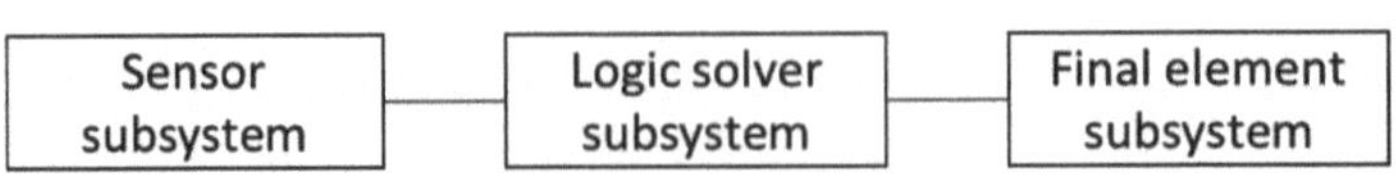

Fig. 6 The three subsystems of a safety barrier.

able to perform its function. The probability that the barrier is not able to perform its safety function at time t, the PFD(t) is given by

$$\mathrm{PFD}(t) = \sum_{i \in D} P_i(t) \tag{13}$$

Then PFDavg over a test interval (0, τ) is

$$\mathrm{PFD}_{\mathrm{avg}} = \frac{1}{\tau}\int_0^{\tau} \sum_{i \in D} P_i(t)\,dt \tag{14}$$

In the cases where it is relevant to assume a steady-state situation, we get $\mathrm{PFD}_{\mathrm{avg}}$ of the barrier as

$$\mathrm{PFD}_{\mathrm{avg}} = \sum_{i \in D} P_i \tag{15}$$

where P_i is the steady-state probability in state i.

Considering a 1oo1 configuration with constant DU failure rate λ, and it is proof-tested at regular intervals of length τ. The survivor function of the configuration is $R(t) = e^{-\lambda t}$ and the $\mathrm{PFD}_{\mathrm{avg}}$ is from Eq. (1)

$$\mathrm{PFD}_{\mathrm{avg}}^{1oo1} = \frac{1}{\tau}\int_0^{\tau} \mathrm{PFD}(t)\,dt = 1 - \frac{1}{\tau}\int_0^{\tau} R(t)\,dt \tag{16}$$

When $\lambda\tau$ is small (i.e., <0.10), by replacing $e^{-\lambda t}$ with its Maclaurin series, the $\mathrm{PFD}_{\mathrm{avg}}$ can be approximated as

$$\mathrm{PFD}_{\mathrm{avg}}^{1oo1} \approx \frac{\lambda\tau}{2} \tag{17}$$

The 1oo1 barrier can have two possible states (Table 2):

A Markov model for the 1oo1 barrier is shown in Fig. 4.

The parameter μ stands for the repair rate, which can be given by the reverse of the mean downtime (MDT).

$$\mu = \frac{1}{\mathrm{MDT}} = \frac{1}{\frac{\tau}{2} + \mathrm{MRT}} \tag{18}$$

Table 2 possible states the 1oo1 barrier with DU failures.

State	State description
0	The barrier is able to function (i.e., does not have a DU fault)
1	The barrier has a DU fault

If DD failure is considered, then the 1oo1 barrier can have three possible states (Table 3):

A Markov model for the barrier is shown in Fig. 7. In this model it is tacitly assumed that a DU fault cannot develop into a DD fault.

Assuming a configuration 2oo2 with two dependent units with failure rate λ_1 and λ_2, respectively, with respect to DU failures, it can be regarded as a series structure where both units have to function for the structure to function. The survivor function for the structure is

$$R(t) = R_1(t)R_2(t) = e^{-(\lambda_1+\lambda_2)t} \tag{19}$$

The 2oo2 configuration $\mathrm{PFD}_{\mathrm{avg}}$ then can be approximated as

$$\mathrm{PFD}_{\mathrm{avg}}^{2oo2} = 1 - \frac{1}{\tau}\int_0^{\tau} R(t)dt \approx \frac{(\lambda_1 + \lambda_2)\tau}{2} \tag{20}$$

For the 1oo2 configuration with identical units with failure rate λ, it means that it is sufficient if one unit is functioning for the voted group to function. It can be regarded as a parallel structure. Then survivor function for the voted group is

$$R(t) = 1 - (1 - R_1(t))(1 - R_2(t)) = 2e^{-\lambda t} - e^{-2\lambda t} \tag{21}$$

Table 3 Possible states for the 1oo1 barrier with DD and DU failures

State	State description
0	The barrier is able to function (i.e., does not have a DU fault)
1	The barrier has a DD fault
2	The barrier has a DU fault

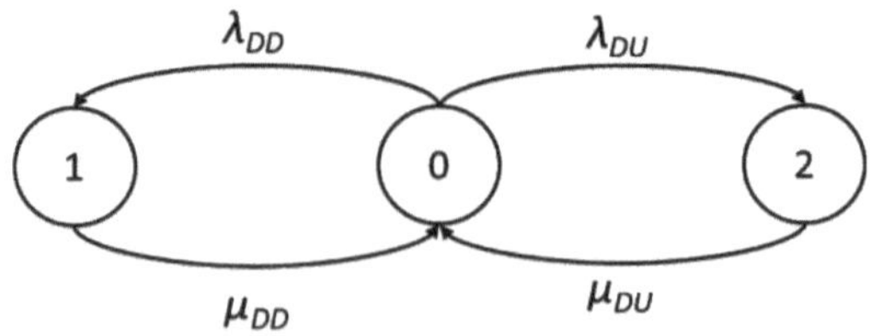

Fig. 7 Markov model (state transition diagram) for the 1oo1 barrier with DD and DU failures.

And when $\lambda\tau$ is small, the system PFD_{avg} can be approximated by

$$PFD_{avg}^{1oo2} = 1 - \frac{1}{\tau}\int_0^{\tau} R(t)dt \approx \frac{(\lambda\tau)^2}{3} \quad (22)$$

For a barrier with independent and identical units, there are three possible states (Table 4):

The 1oo2 barrier is functioning as long as it is in state 0 or state 1 and fails when it is in state 2. Because P_2 is the long-term or average probability that the barrier is in state 2 in steady-state approach, then $PFD_{avg} = P_2$. While for the 2oo2 barrier, the barrier is functioning only in state 0 and fails in either state 1 or 2, then $PFD_{avg} = P_1 + P_2$.

Here, we take the 1oo2 barrier as an example to explain the evaluation method based on the Markov model. The corresponding transition rate matrix for the 1oo2 barrier, as shown in Fig. 8, can be written as

$$A = \begin{bmatrix} -2\lambda & 2\lambda & 0 \\ \mu_1 & -(\mu_1 + \lambda) & \lambda \\ \mu_2 & 0 & -\mu_2 \end{bmatrix} \quad (23)$$

Table 4 Possible states for the barrier of two independent and identical units, where only DU failures are considered.

State	State description
0	Both units are able to function (i.e., do not have a DU fault)
1	One unit is able to function and the other has a DU fault
2	Both units have DU faults

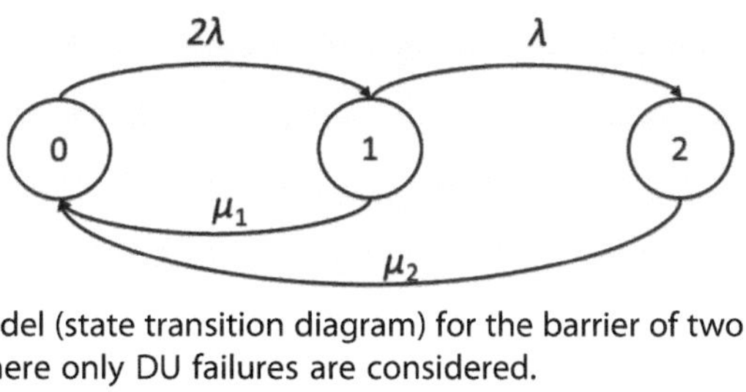

Fig. 8 Markov model (state transition diagram) for the barrier of two independent and identical units, where only DU failures are considered.

Then, the steady-state probabilities can be determined from three equations as

$$\begin{aligned} P_0 + P_1 + P_2 &= 1 \\ 2\lambda P_0 - (\mu_1 + \lambda)P_1 &= 0 \\ \lambda P_1 - \mu_2 P_2 &= 0 \end{aligned} \tag{24}$$

As the DU failure rate is small compared to with the repair rates, P_2 can be obtained and approximated by solving these equations,

$$P_2 = \frac{2\lambda^2}{2\lambda^2 + 3\lambda\mu_2 + \mu_1\mu_2} \approx \frac{2\lambda^2}{\mu_1\mu_2} \tag{25}$$

For example, if the failure rate of the two units is 1×10^{-6}, MRT$=$ 10 h and test interval $\tau = 8760$ h. Then the exact value for barrier PFD_{avg} can be calculated as 2.54×10^{-5}, while the approximated value is 2.57×10^{-5}. And the approximated result by the simplified formulas is 2.56×10^{-5}.

For the 2oo3 configuration with identical units with failure rate λ with respect to DU failures, the three units are proof-tested at the same time with test interval τ. The survivor function for the voted group is

$$R(t) = 3e^{-2\lambda t} - 2e^{-3\lambda t} \tag{26}$$

The PFD_{avg} of 2oo3 voted group is

$$\text{PFD}_{\text{avg}}^{2\text{oo}3} = 1 - \frac{1}{\tau}\int_0^{\tau} R(t)dt \approx (\lambda\tau)^2 \tag{27}$$

Assuming the 2oo3 barrier of three independent and identical units, without common cause failures. The barrier can have four states (Table 5 and Fig. 9):

The 2oo3 barrier is functioning when it is in state 0 and state 1 and fails in state 2 or 3, then $\text{PFD}_{\text{avg}} = P_2 + P_3$. Barrier performance then can be obtained.

Table 5 Possible states for 2oo3 of three independent and identical units, where only DU failures are considered.

State	State description
0	Three units are able to function (i.e., do not have a DU fault)
1	Two units are able to function and one has a DU fault
2	One unit is able to function and two have DU faults
3	Three units have DU faults

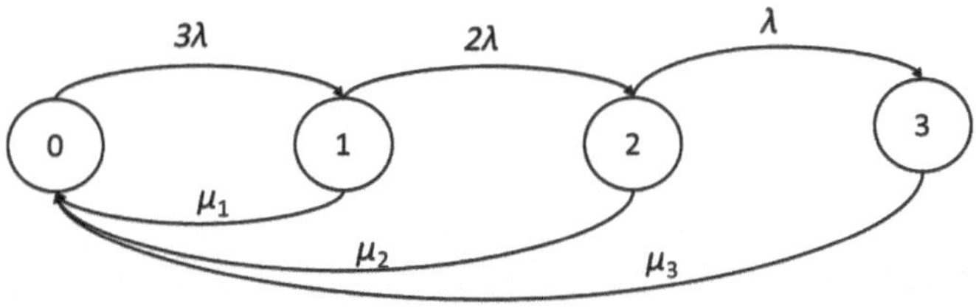

Fig. 9 Markov model (state transition diagram) for 2oo3 of three independent and identical units, where only DU failures are considered.

6. Digital twin approach for condition-based evaluation and prognosis

Currently, diagnostic tests are automatedly performed on many safety barriers in their operational phase, but only a ratio of faults can be found immediately since these technical barriers are often subject to several failure mechanisms. Regular proof tests are still the main approach to ensure safety barrier high availability and EUC safety (IEC61508, 2010; IEC61511, 2016) by revealing those undetected faults and then restoring the system to an as-good-as-new condition if necessary.

In practice, the availability and performance of safety barriers depend on multiple factors, e.g., demands, structure, and chronic degradations due to mechanisms of corrosion, wear-out, etc. These factors challenge the sufficiency of failure rate-based methods, namely binary system states, in safety barrier performance evaluation, as shown in Fig. 10.

The recent advances in digitalization provide more efficient approaches for barrier performance evaluation. In digital twin, a more representative picture of the current barrier status shall be available based on the measured data at the testing time. Generally, the full trip proof tests including several tasks, such as examining alarm functions, measuring response time for a specific SIF, and operating all input devices (Rausand, 2014). Prognostics can evaluate the degradation of the barrier so as to detect incipient deviations, relying on certain barrier health indicators, e.g., leakage rate, the closing time for valves (Nystad, Gola, Hulsund, & Roverso, 2010).

For low-demand safety barriers, it might not always be worthwhile proof test periodically, especially if the shutdown and restart of the process are costly. It implies that the test of barriers is a two-fold consideration from safety and economic perspectives. As safety barriers are designed and installed to protect the EUC, the first principle is to meet the required

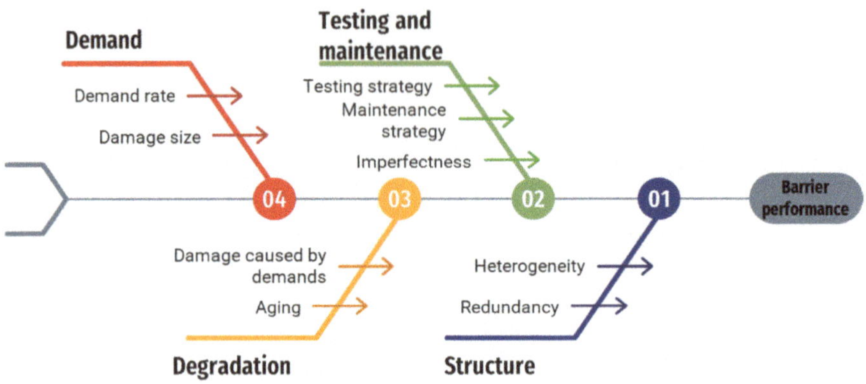

Fig. 10 Influence factors of barrier performance in the operational phase. *Adopted from Zhang, A. (2021).* Prognostics and health management of safety-instrumented systems: Approaches of degradation modeling and decision-making. *Norwegian University of Science and Technology. Trondheim. https://hdl.handle.net/11250/2726046.*

performance. Briefly, the main issues regarding the safety barrier management in the operational phase can broadly be summarized into two categories "Is the safety barrier (still) sufficient to implement the predefined functions based on the current state information? (short-term)," "How should activities be scheduled and coordinated to ensure that the safety barrier remains at the required integrity level? (long-term)."

A digital twin can make barrier performance influencing factors more visible and convenient for health management. Based on the capability of estimation of remaining useful lifetime (RUL), it can help practitioners to monitor the real-time performance of barriers, and then make plans for testing and maintenance not only keeping the SIS effective but reducing the impact on EUC to the lowest. Also, with the digital twin, the probability of failure on demand of barriers will be optimized to be more suitable for the dynamic working conditions.

In this case, the barrier management can be shifted from the time-based to the performance-based scheme. The actual safety barrier performance can be quantified as the deviation from the designed target value. If the actual performance is within the acceptable limits, marked as threshold L, which refers to the deviation from the desired performance(target value), the performance of the safety barrier is acceptable, and the safety barrier can be qualified as functioning. If the actual performance is beyond the acceptable limit, then the barrier is not effective any longer in terms of the required function for risk control, as described in Fig. 11.

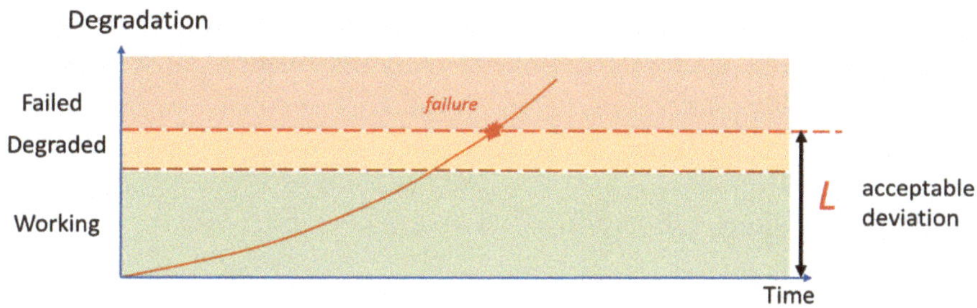

Fig. 11 Degradation, acceptable deviation and failure.

Table 6 System state definition.

State	Status	State description
0	Working	The barrier is functioning as specified
1	Degraded	The barrier has a degraded performance but functioning
2	Failed	The system has a fault

Adopted from Zhang, A., Srivastav, H., Barros, A., & Liu, Y. (2021). Study of testing and maintenance strategies for redundant final elements in SIS with imperfect detection of degraded state. *Reliability E-ngineering & System Safety*, *209*, 107393. https://doi.org/10.1016/j.ress.2020.107393

Based on the actual performance, we can consider the barrier with three different states: working, degraded, and failed, as shown in Table 6. It should be noted that the degradation still can exist in state 0, but it can be accepted without any maintenance action.

Conditioning on the revealed system state in the tests and the expensive proof tests, three options of follow-ups are possible: (1) no action if the barrier in test in working well, referring to the estimation of failure probability of the barrier by the next test is rather low based on the observed situation; (2) preventive maintenance (PM) if a certain degradation has been identified. In this way, the degradation is mitigated but is not eliminated considering costly perfect maintenance; (3) repair or replacement of the barrier if it is failed.

The degradation process in barriers can be obtained by historical data and expert knowledge. The typical process in modeling these mechanisms includes the Gamma process, Wiener process, etc. The barrier starts to work with a new state and generally is subject to a monotonical degradation process modeled as $X(t)$. Then the instantaneous $PFD(t)$, $t < \tau$, of the barrier in the first test interval $[0, \tau]$ is

$$\mathrm{PFD}(t) = 1 - \Pr(X(t) < L) \tag{28}$$

From the second interval, the prior test result acts as the condition to estimate barrier PFD, taking the collected barrier information μ ($\mu < L$) in tests as input, for $\tau < t < 2\tau$:

$$\text{PFD}(t) = 1 - \Pr(X(t) < L|X(\tau) = \mu) \tag{29}$$

Similarly, we can get *PFD*(t), $(i-1)\ \tau < t < i\tau$,

$$\text{PFD}(t) = 1 - \Pr(X(t) < L|X((i-1)\tau)) \tag{30}$$

Accordingly, we can calculate the barrier PFD_{avg}. The time-dependent barrier degradation and PFD_{avg} can be described in Fig. 12.

The degradation process-based performance evaluation model can represent the related working conditions and tolerable performance requirements with different thresholds. Also, from the performance perspective, the as-good-as-new assumption after each proof test is far from practical even though the system is qualified. The barrier performance can be evaluated in a more dynamic way.

Redundant structures are often used in safety barriers to improve the performance availability and so to enhance safety, e.g., two shutdown valves are installed on the pipeline to stop flow when the downstream pressure is too high. When one of them cannot be activated, the process, namely EUC, is still safe if the other valve work. Such kind of configuration is called 1-out-of-2 (1oo2). The digital twin can record the activation history of the barriers,

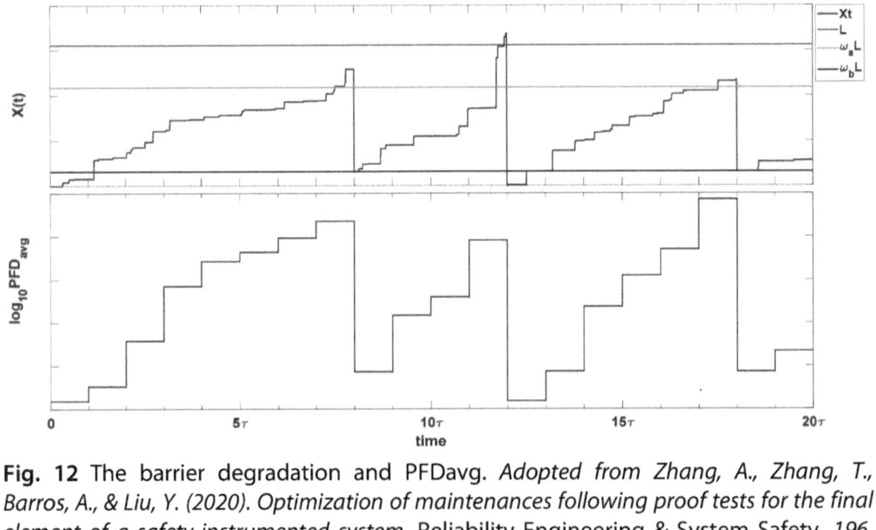

Fig. 12 The barrier degradation and PFDavg. *Adopted from Zhang, A., Zhang, T., Barros, A., & Liu, Y. (2020). Optimization of maintenances following proof tests for the final element of a safety-instrumented system.* Reliability Engineering & Safety System, *196, 106779. https://doi.org/10.1016/j.ress.2019.106779.*

referring to the demands on EUC. These demands can lead to internal stress given the occasional high pressure in the valves, consequently intensifying the degradation. As shown in Fig. 10, these factors can be integrated into the performance evaluation with the assistance of the digital twin.

The instantaneous unavailability of the safety barrier within the first proof test interval (Zhang, Barros, & Liu, 2019) is

$$\mathrm{PFD}(t) = \Pr(X_1(t) > L \cap X_2(t) > L) \tag{31}$$

Similar to the single element, the performance information in prior tests can be used as the input to estimate the condition barrier performance in the upcoming test interval.

In the long run, for the lifecycle management of safety barriers, the digital twin can provide useful information during the design phase and guild for decision supporting and operation control throughout the operation phase. In the operation phase, the real-time dynamic performance of barriers can also be transferred in digital twin for the barrier health management, and the real working conditions are also beneficial for the design of the new barriers systems.

7. Closure and further reading

In this chapter, we have discussed the considerations, models, and some analytical formulas for digitalized safety barriers. Due to the page limits, many important topics about safety barriers are just simply mentioned or totally skipped in this chapter. We provide some references for further reading if readers have interests in the following topics:

- Condition monitoring on degradation and performance evaluation: Zhang et al. (2019), Srivastav et al. (2020), Zhang et al. (2022) and Zhang et al. (2021).
- Impact of test strategies on performance evaluation: IEC61508 (2010), Liu (2014), Liu and Rausand (2016), Wu, Zhang, Barros, Zheng, and Liu (2018), and Zhang et al. (2021).
- Data-driven performance evaluation approaches for safety barriers: Selvik and Abrahamsen (2017), and Xie, Håbrekke, Liu, and Lundteigen (2019).
- Performance evaluation of safety barriers with common cause failures: Lundteigen and Rausand (2007) and Hauge, Hokstad, Håbrekke, and Lundteigen (2016).

- Performance evaluation of safety barriers with cascading failures: Xie, Lundteigen, and Liu (2020) and Xie, Lundteigen, and Liu (2021).
- Performance evaluation on the digitalized prognosis of safety barriers: Zhang et al. (2021).

References

Hauge, S., Hokstad, P., Håbrekke, S., & Lundteigen, M. A. (2016). Common cause failures in safety-instrumented systems: Using field experience from the petroleum industry. *Reliability Engineering & System Safety, 151*, 34–45. https://doi.org/10.1016/j.ress.2015.09.018.

IEC61025. (2006). *Fault tree analysis (FTA)*. Geneva, Switzerland: International Electrotechnical Commission.

IEC61078. (2016). *Reliability block diagram*. Geneva, Switzerland: International Electrotechnical Commission.

IEC61508. (2010). *Functional safety of electrical/electronic/programmable electronic safety-related systems*. Geneva, Switzerland: International Electrotechnical Commission.

IEC61511. (2016). *Functional safety - safety instrumented systems for the process industry sector*. Geneva, Switzerland: International Electrotechnical Commission.

IEC62551. (2013). *Analysis techniques for dependability - Petri net techniques*. Geneva, Switzerland: International Electrotechnical Commission.

Innal, F., Lundteigen, M. A., Liu, Y., & Barros, A. (2016). PFDavg generalized formulas for SIS subject to partial and full periodic tests based on multi-phase Markov models. *Reliability Engineering & System Safety, 150*, 160–170. https://doi.org/10.1016/j.ress.2016.01.022.

ISO13702. (2015). *Petroleum and natural gas industries- Control and mitigation of fires and explosions on offshore production installations-Requirements and guidelines*. Geneva, Switzerland: International Organization for Standardization.

Johansen, I. L., & Rausand, M. (2015). Barrier management in the offshore oil and gas industry. *Journal of Loss Prevention in the Process Industries, 34*, 49–55. https://doi.org/10.1016/j.jlp.2015.01.023.

Liu, Y. (2014). Optimal staggered testing strategies for heterogeneously redundant safety systems. *Reliability Engineering & System Safety, 126*, 235–243. https://doi.org/10.1016/j.ress.2014.01.013.

Liu, Y. (2020). Safety Barriers: Research advances and new thoughts on theory, engineering and management. *Journal of Loss Prevention in the Process Industries, 67*, 104260. https://doi.org/10.1016/j.jlp.2020.104260.

Liu, Y., & Rausand, M. (2016). Proof-testing strategies induced by dangerous detected failures of safety-instrumented systems. *Reliability Engineering & System Safety, 2016*(145), 366–372. https://doi.org/10.1016/j.ress.2015.06.016.

Lundteigen, M. A., & Rausand, M. (2007). Common cause failures in safety instrumented systems on oil and gas installations: Implementing defense measures through function testing. *Journal of Loss Prevention in the Process Industries, 20*(3), 218–229. https://doi.org/10.1016/j.jlp.2007.03.007.

Nystad, B. H., Gola, G. E., Hulsund, J., & Roverso, D. (2010). Technical condition assessment and remaining useful life estimation of choke valves subject to erosion. In *Annual Conference of the PHM Society*.

Prashanth, I., Fernandez, G. J., Sunder, R. G., & Boardman, B. (2017). Factors influencing safety barrier performance for onshore gas drilling operations. *Journal of Loss Prevention in the Process Industries, 49*, 291–298. https://doi.org/10.1016/j.jlp.2017.07.009.

Rausand, M. (2014). *Reliability of safety-critical systems: theory and applications.* John Wiley & Sons.

Rausand, M., Barros, A., & Hoyland, A. (2020). *System reliability theory: Models, statistical methods, and applications* (3rd ed.). Wiley-Blackwell.

Selvik, J. T., & Abrahamsen, E. B. (2017). How to classify failures when collecting data for safety-instrumented systems in the oil and gas industry. *Journal of Risk Research, 20*(7), 952–962. https://doi.org/10.1080/13669877.2015.1121910.

Selvik, J. T., & Signoret, J. P. (2017). How to interpret safety critical failures in risk and reliability assessments. *Reliability Engineering & System Safety, 161*, 61–68. https://doi.org/10.1016/j.ress.2017.01.003.

Signoret, J. P., Dutuit, Y., Cacheux, P. J., Folleau, C., Collas, S., & Thomas, P. (2013). Make your Petri nets understandable: Reliability block diagrams driven Petri nets. *Reliability Engineering & System Safety, 113*, 61–75. https://doi.org/10.1016/j.ress.2012.12.008.

Srivastav, H., Barros, A., & Lundteigen, M. A. (2020). Modelling framework for performance analysis of SIS subject to degradation due to proof tests. *Reliability Engineering & System Safety, 2020*(195), 106702. https://doi.org/10.1016/j.ress.2019.106702.

Wu, S., Zhang, L., Barros, A., Zheng, W., & Liu, Y. (2018). Performance analysis for subsea blind shear ram preventers subject to testing strategies. *Reliability Engineering & System Safety, 169*, 281–298. https://doi.org/10.1016/j.ress.2017.08.022.

Wu, S., Zhang, L., Lundteigen, M. A., Liu, Y., & Zheng, W. (2018). Reliability assessment for final elements of SISs with time dependent failures. *Journal of Loss Prevention in the Process Industries, 51*, 186–199. https://doi.org/10.1016/j.jlp.2017.12.007.

Xie, L., Håbrekke, S., Liu, Y., & Lundteigen, M. A. (2019). Operational data-driven prediction for failure rates of equipment in safety-instrumented systems: A case study from the oil and gas industry. *Journal of Loss Prevention in the Process Industries, 60*, 96–105. https://doi.org/10.1016/j.jlp.2019.04.004.

Xie, L., Lundteigen, M. A., & Liu, Y. (2020). Reliability and barrier assessment of series-parallel systems subject to cascading failures. *Proceedings of the Institution of Mechanical Engineers, Part O: Journal of Risk and Reliability, 234*(3), 455–469. https://doi.org/10.1177/1748006X19899235.

Xie, L., Lundteigen, M. A., & Liu, Y. (2021). Performance assessment of K-out-of-N safety instrumented systems subject to cascading failures. *ISA Transactions, 118*, 35–43. https://doi.org/10.1016/j.isatra.2021.02.015.

Zhang, A., Barros, A., & Liu, Y. (2019). Performance analysis of redundant safety-instrumented systems subject to degradation and external demands. *Journal of Loss Prevention in the Process Industries, 62*, 103946. https://doi.org/10.1016/j.jlp.2019.103946.

Zhang, A., Srivastav, H., Barros, A., & Liu, Y. (2021). Study of testing and maintenance strategies for redundant final elements in SIS with imperfect detection of degraded state. *Reliability Engineering & System Safety, 209*, 107393. https://doi.org/10.1016/j.ress.2020.107393.

Zhang, A., Wu, S., Fan, D., Xie, M., Cai, B., & Liu, Y. (2022). Adaptive testing policy for multi-state systems with application to the degrading final elements in safety-instrumented systems. *Reliability Engineering & System Safety, 108360.* https://doi.org/10.1016/j.ress.2022.108360.

CHAPTER NINE

Dynamic operational risk assessment in process safety management

Xinhong Li[a,*], Luyao Zhang[a], Faisal Khan[b], and Guoming Chen[c]
[a]School of Resources Engineering, Xi'an University of Architecture and Technology, Xi'an, China
[b]Mary Kay O'Connor Process Safety Center, Artie McFerrin Department of Chemical Engineering, Texas A&M University, College Station, TX, United States
[c]Centre for Offshore Engineering and Safety Technology (COEST), China University of Petroleum (East China), Qingdao, China
[*]Corresponding author: e-mail addresses: safety_lxh@163.com; lixinhong@xauat.edu.cn

Contents

1. Necessity of dynamic risk assessment

To ensure the safety of process operations, considerable efforts were made on identification of the potential hazards that can cause severe consequences. QRA is a popular tool that can quantify the risk associated with a particular process activity (Kalantarnia, Khan, & Hawboldt, 2009; Li et al., 2021). The industry 4.0 environment brings the availability of a large

Methods in Chemical Process Safety, Volume 6
ISSN 2468-6514
https://doi.org/10.1016/bs.mcps.2022.04.004

volume of the data on the operational information of process systems due to the automation and digitalization of process system. The conventional QRA is limited since it is unable to use the accident precursor information to revise the risk profile. Therefore, the improved methodology that can handle continuous monitoring data for dynamic risk assessment is required in the digitalized age of process operations urgently. Dynamic risk assessment methodology can reach more efficient and accurate management of process safety, and assist the process digitalization (Khan, Amyotte, & Adedigba, 2021).

The conventional risk assessment methods including Fault tree (FT), Event tree (ET) and Bow-tie (BT) were widely adopted in the previous studies. These methods are known as static and fail to capture the variation of risk as change occurs in the operation and environment (Abimbola, Khan, & Khakzad, 2014). In addition, conventional methods are unable to account for the various states of a variable, common cause failures and conditional dependency among contributing events of an accident (Khakzad, Khan, & Amyotte, 2013; Li, Chen, & Zhu, 2016; Yuan, Khakzad, Khan, & Amyotte, 2015). BN was proven to be useful to overcome the limitations of conventional methods and achieve dynamic risk assessment (Li et al., 2018). The conditional dependency among various variables can be considered through directed arcs and conditional probability tables (CPTs). Applying Bayes theorem, BN can update the initial beliefs when the new information about the system is available. In the digitalized age, the digitalization of process operations now sites as an indispensable, enabling philosophy and technology that embraces industrial change (Wang, 2021). A better understanding of process operations needs a digitization to establish appropriate monitoring safety measures. As computer technology rises, developing dynamic risk assessment approaches using data-driven techniques is required by the digitization of process operations, which is critical to improve the intelligent level of risk management of process operations. Digitalization can provide early-warnings (Schmitz, Swuste, Reniers, & van Nunen, 2020), fault detection (Arunthavanathan, Khan, Ahmed, & Imtiaz, 2021; Fazai, Mansouri, Abodayeh, Nounou, & Nounou, 2019), and dynamic diagnosis (Nan, Khan, & Iqbal, 2008). The digitalization of process operations can foster better process safety to keep the operational environment healthy for the workers and their operational facilities safe (Romero et al., 2018).

This chapter presents two dynamic risk assessment methods for process operations, namely Bayesian approach and data-driven approach, which aims to contribute to the development of overall better risk management and improve the safety of process operations in harsh environments.

The structure of this chapter is as follows: Section 2 introduces dynamic risk assessment methods, including a Bayesian approach and a data-driven approach. Case studies is presented in Sections 3 and 4, respectively to demonstrate the feasibility and effectiveness of the proposed dynamic risk assessment methods.

2. Dynamic risk assessment methodology

2.1 A Bayesian approach

Fig. 1 presents the flowchart of the Bayesian approach for dynamic risk assessment of process operations. The main steps of the methodology include:

- Accident scenario modeling;
- Information collection;
- Probability estimation of basic events;
- Dynamic risk assessment using precursor data.

2.1.1 Accident scenario modeling

Developing accident scenario is the basic but essential for risk assessment. There are a series of accident scenario modeling methods including conventional and advanced methods. BT is a synthesis of FT and ET, which is a graphical method representing an accident from causes to consequences. This study uses BT method to identify the potential causes, safety barriers, and consequences. The subsequent scenario evolution process of accidental leakage of subsea pipelines is obtained. However, BT is known to be static, failing to capture the variation of risk as changes in operation or environmental condition takes place. Also, primary events and safety barriers in BT model are assumed to be independent, which is not consistent with reality and introduces certain uncertainty into results (Abimbola et al., 2014; Li et al., 2016). To relax the limitations of BT and address the model uncertainty, previous researchers usually couple BT with Bayes' theorem or map BT into BN (Ale et al., 2009). In the present study, the developed BT mode is transformed into BN to perform a quantitative analysis.

BN is a probabilistic graphical based network that is represented as a directed acyclic graph, mainly for describing knowledge uncertainty (He et al., 2018; Ping, Wang, Kong, & Chen, 2018). A typical BN includes nodes, arcs and CPTs. Nodes represent random variables, and arcs represent

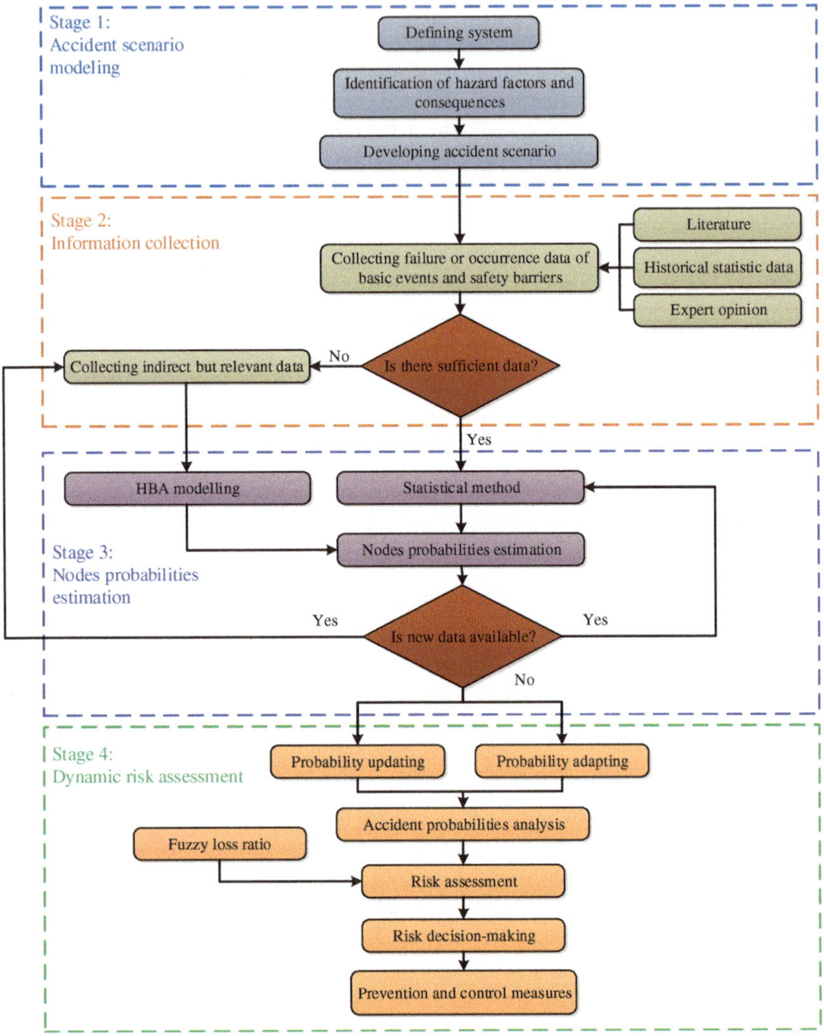

Fig. 1 Flowchart of Bayesian approach for dynamic risk assessment.

the conditional dependencies among linked nodes. CPTs describe the type and strength of these dependencies among nodes. Fig. 2 presents a simple example of BN comprising random variables X_1-X_4.

Based on the conditional dependencies of variables, the joint probability distribution $P(U)$ of variables $U=\{A_1, \ldots, A_n\}$ can be represented as:

$$P(U) = \prod_{i=1}^{n} P(A_i|Pa(A_i)) \tag{1}$$

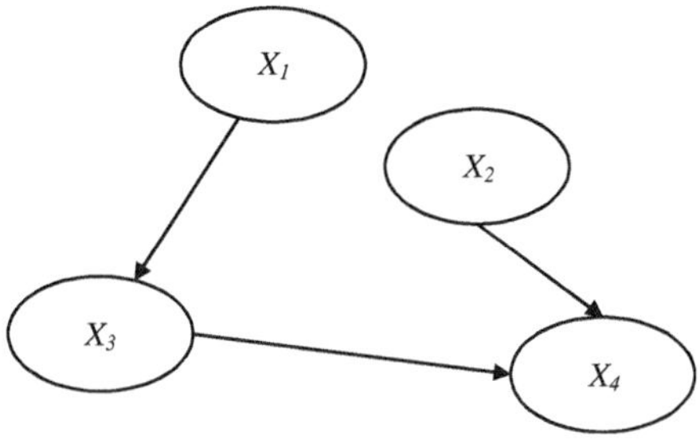

Fig. 2 An example of BN (Arzaghi et al., 2018).

where $Pa(A_i)$ is the parent set of variables A_i.

Based on Bayes theorem and new observation or evidence E, the prior probabilities of variables are updated to be the posterior probabilities:

$$P(U|E) = \frac{P(U,E)}{P(E)} = \frac{P(U,E)}{\sum_U P(U,E)} \tag{2}$$

2.1.2 Information collection

The purpose of this stage is to collect the direct or indirect but relevant data of basic events, safety barriers and consequences in the accident scenario model. Collecting information about them is vital to conduct a quantitative risk assessment. The information needs to be collected are: (1) Occurrence number within specific time intervals; (2) Direct probabilities for some events which is usually from literature or failure database, e.g., OREDA; (3) Loss caused by each consequence type once accident happens, and total investment of a project.

During information collection, failure data of basic events and safety barriers can be found from literature, report of historical statistics. Furthermore, experts' opinion is also an important approach to obtain the expected failure data, which is helpful for the events that have no direct or indirect relevant data. When direct data are not sufficient, failure data can be collected from different sources including different regions, operational conditions, and different industries. If the data relevant to subsea pipelines leak are scarce in the OOG sector, secondary data can be collected from the nuclear, chemical, and mining sectors; which is mainly due to (1) Within each sector, installation, equipment, and procedures are similar; and (2) Each sector has a rich operational data history (Yang, Khan, & Lye, 2013).

2.1.3 Nodes' probabilities estimation

For the events or safety barriers with enough direct relevant data, their probabilities can be determined by the statistical method. For the events or safety barriers with limited even scarce direct or indirect relevant data, their probabilities can be derived from HBA modeling.

HBA is commonly used to consider the uncertainty induced by data collected from different sources. In general, a multi-stage prior distribution is utilized in HBA framework, which is very complicated to be solved numerically and has large computational requirements. However, the availability of Markov Chain Monte Carlo (MCMC) sampling technique facilitates the application of HBA (Kelly & Smith, 2009). The main advantages of HBA are as follows: (1) HBA can address the population variability of the data from different sources; and (2) make full use of scarce data and borrow strength from other indirect but relevant data (Yang et al., 2013).

In the HBA framework, the data from different sources are used to establish an informative prior distribution for the parameter of interest θ. To build an informative prior distribution θ is considered to follow a generic distribution with its own parameters α and β (hyper parameters), i.e., $p\ (\theta \mid \alpha, \beta)$. This is the first-stage prior representing the uncertainty in θ. The hyper parameters are also uncertain and usually considered to follow diffusive or non-informative distributions, such as uniform distribution and Jeffreys priors (Khakzad, Khakzad, & Khan, 2014; Yang, Khan, Lye, & Amyotte, 2015). This is the second-stage prior representing the uncertainty in α and β.

The second stage prior can be updated given generic data D. The posterior distribution of hyper parameters $p(\alpha, \beta \mid D)$ can be written as follows (Yu, Khan, & Veitch, 2017):

$$p(\alpha,\beta|D) = \frac{p(\alpha,\beta)l(D|\alpha,\beta)}{\iint p(\alpha,\beta)l(D|\alpha,\beta)d\alpha d\beta} \tag{3}$$

where $p(\alpha, \beta \mid D)$ is the posterior distribution of hyper parameters. The likelihood function of hyper parameters, i.e. $l(D \mid \alpha, \beta)$ is shown as follows:

$$l(D|\alpha,\beta) = \int l(D|\theta)p(\theta|\alpha,\beta)d\theta \tag{4}$$

There are a lot of methods to build $l(D \mid \alpha, \beta)$ using precursor data. After obtaining the posterior distribution of hyper parameters, the first stage prior can be updated to get an informative prior distribution for the parameter of interest θ by averaging over values of hyper parameters as follows:

$$p(\theta|D) = \iint p(\theta|\alpha,\beta)p(\alpha,\beta|D)d\alpha d\beta \quad (5)$$

This informative prior distribution represents the source to source uncertainty in θ. Given more case-specific data D^*, $p(\theta \mid D)$ can be further updated as follows (Kelly & Smith, 2009):

$$p(\theta|D^*,D) = \frac{p(\theta|D)l(D^*|\theta)}{\int p(\theta|D)l(D^*|\theta)d\theta} \quad (6)$$

where $p(\theta \mid D^*, D)$ is the case-specific posterior distribution of the parameter of interest. The complete HBA procedure for obtaining the case-specific posterior distribution of parameter of interest θ is illustrated in Fig. 3.

This stage focuses on clarifying the application of HBA in probabilities estimation of nodes in accident scenario model.

Conjugate pairs are commonly used in HBA framework. In the present study, the Poisson likelihood-Gamma prior pair is employed to predict the occurrence number of events in the next time interval. The Binomial likelihood-Beta prior is utilized to estimate the probabilities of nodes that could be defined as rare events. The aggregated data from different source

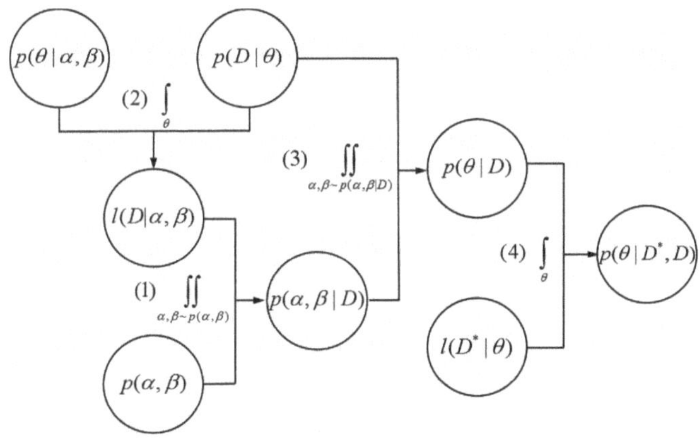

Fig. 3 Graphic illustration of HBA procedure. *Adapted from Yu, H., Khan, F., & Veitch, B. (2017). A flexible hierarchical Bayesian modeling technique for risk analysis of major accidents.* Risk Analysis, *37(9), 1668–1682.*

are the basis of constructing likelihood functions. In HBA, safety expert's feedback or plant-specific data are used to determine the parameters of hyper-priors.

Once the HBA model is established, the posterior probability distribution of parameters of interest can be obtained to solve the model equations based on aggregated data. The data uncertainty due to different sources is considered in this distribution. As can be seen, given the new data, this distribution can also be used as an informative prior distribution. Fig. 4 depicts the application of HBA to establish a case-specific posterior distribution for the nodes in the BN model.

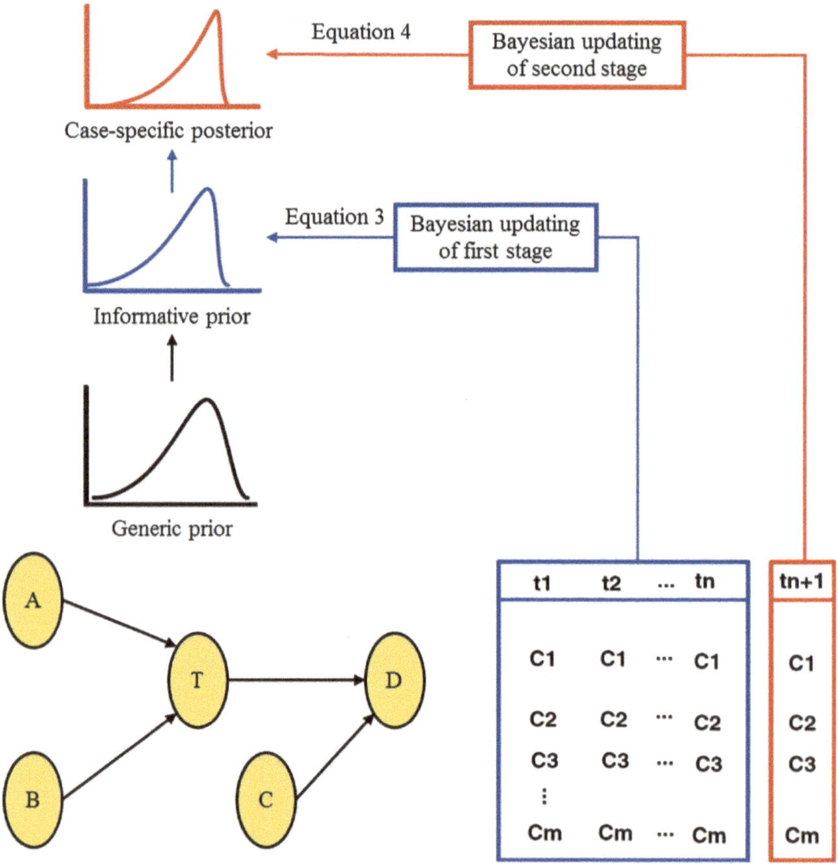

Fig. 4 Illustration of the HBA procedure to establish a case-specific posterior distribution for nodes in BN. *Adapted from Khakzad, N., Khan, F., & Paltrinieri, N. (2014). On the application of near accident data to risk analysis of major accidents.* Reliability Engineering & System Safety, 126, *116–125.*

2.1.4 Dynamic risk assessment

Through the above procedures, the probability of nodes in BN model are determined. Thus, risk assessment of process operations could be achieved in BN framework. The occurrences probabilities of all consequence types induced by subsea pipelines leak are derived by forward inference. In addition, the probability updating and adapting can also be performed in BN. Introducing a new observation, the probability updating is conducted by backward inference to find the most probable causes and evolution paths of consequence type of interest. Probability adapting could use precursor data including near accident data and accident data to get the probabilities of basic events and safety barriers at the different time. The dynamic failure probability of the system over time can be obtained from probability adapting approach. The concept of fuzzy loss ratio is utilized to assess the severity of each consequence type induced by subsea pipelines leak.

The concept of the loss ratio of consequence is defined as the ratio of monetary loss value induced by accident (including casualties, direct assets loss, environment pollution and delay in project period) and the total investment of the project, shown as Eq. (7) (Zhou, 2009).

$$T_i = W_i/W \tag{7}$$

where W is the total investment, W_i is the total monetary loss of accident consequence i. The loss ratio intervals are shown in Table 1. There are totally 5 grades for accident consequence loss. Note that the loss ratio T_i may exceed 1 for some specific accident such as BP Deepwater Horizon accident. In such case, the loss ratio can be taken as 1.

The median values of intervals in Table 1 can be expressed by a vector as follows:

$$R_{cv} = [0.5 \;\; 0.05 \;\; 0.005 \;\; 0.0005 \;\; 0.00005] \tag{8}$$

Table 1 Loss ratio intervals (Tan, Chen, Chang, & Fu, 2013).

Loss grade	Intervals of T_i	Median values of intervals	Description
5	$0.1 \sim 1$	0.5	Catastrophic
4	$0.01 \sim 0.1$	0.05	Extremely serious
3	$0.001 \sim 0.01$	0.005	Serious
2	$0.0001 \sim 0.001$	0.0005	Needs to be considered
1	$0.00001 \sim 0.0001$	0.00005	Negligible

Actually, the same accident scenario may result in different grades of consequence loss. There is a certain likelihood for consequence loss ratio locating in a specific interval. A loss ratio probability vector q_i shown as Eq. (9) is used to show the likelihood of consequence loss ratio at different intervals.

$$q_i = [q_{i1} \quad q_{i2} \quad q_{i3} \quad q_{i4} \quad q_{i5}] \tag{9}$$

where i is the type of accident consequence, q_{i1} is the probability that loss ratio of consequence i reaches the grade 1 in Table 1, q_{i2}, q_{i3}, q_{i4}, q_{i5} can be reasoned and so on. Statistical data or expert opinion is often used to determine the loss ratio probability vector.

Regarding a specific accident consequence i, its loss grade can be presented using the fuzzy loss ratio which is a concept that integrating the loss ratio with its probability, as shown in Eq. (10).

$$c_i = q_i \bullet R_{cv}^T \tag{10}$$

where c_i is the fuzzy loss ratio of consequence type i; q_i is the likelihood of loss ratio at different intervals; $RT\ cv$ is the transposed form of R_{cv}.

Based on the above, the fuzzy loss ratio of specific consequence is determined by loss ratio probability vector and median values vector of loss ratio intervals. In general, loss ratio probability vector is usually determined by experts' opinion and statistic data.

Eventually, the risk assessment of a major accident is achieved by integrating probabilities with fuzzy loss ratio of consequences. A risk decision-making can be made based on assessed results. Furthermore, some prevention and control measures can be developed for the specific accident whose risk exceeds the acceptable level in the OOG sector.

2.2 Data-driven approach

Fig. 5 presents the flowchart of data-driven approach for dynamic risk assessment of process operations. Here the discussions are presented using subsea pipeline corrosion rate prediction as an example. The main steps of the approach include:

- Selecting model input and output variables;
- Dividing sample sets for model training and testing;
- Selection and optimization of model parameters;
- Model training and testing.

In the proposed approach framework, Support vector regression (SVR) is the baseline model for developing the improved approach. SVR is a

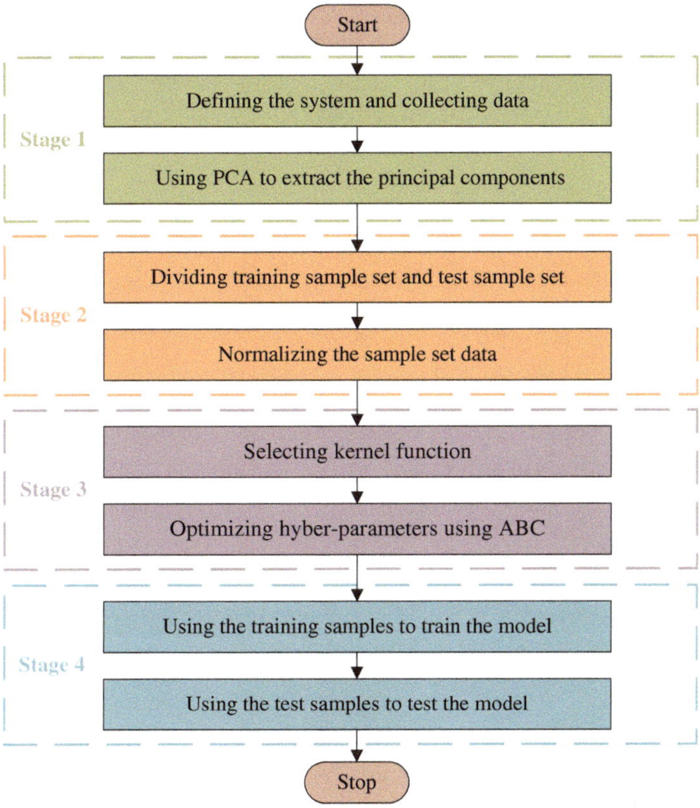

Fig. 5 Framework of the proposed data-driven approach.

machine learning technique for solving regression problems, which is promoted by Support Vector Machine (SVM). Based on the principle of structural risk minimization, SVR can more effectively cope with over-fitting and under-fitting to achieve accurate data prediction (Güner & Çomak, 2011; Safavi & Esmikhani, 2013).

In the SVR framework, first suppose that the sample set is as follows:

$$D = \{(x_i, y_i) | x_i \in R^n, y_i \in R, i = 1, 2, \cdots m\} \tag{11}$$

where x_i is the input variable; y_i is the output variable; R^n is the n-dimensional input space; R is the output space; m is the number of samples.

The SVR function can be described by Eq. (12).

$$f(x) = \omega^T \varphi(x) + b \tag{12}$$

where ω is the weight vector; $\varphi(x)$ is the nonlinear mapping function; b is the offset vector.

Then, to ensure optimization, the slack variable constraints θ and θ^* are used in Eq. (13) (Güven & Şimşir, 2020).

$$\varphi(\omega, \theta, \theta^*) = \frac{1}{2}\|\omega\|^2 + C\left(\Sigma_{i=1}^{i}\theta_i + \Sigma_{i=1}^{i}\theta_i^*\right)$$
$$S.T.\begin{cases} y_i - \omega^T\varphi(x_i) - b \leq \varepsilon + \theta_i, i = 1, 2, \cdots N \\ \omega^T\varphi(x_i) + b - y_i \leq \varepsilon + \theta_i^*, i = 1, 2, \cdots N \\ \theta_i, \theta_i^* \geq 0 \end{cases} \tag{13}$$

where C is the penalty factor; ε is the insensitive loss parameter.

Next, in order to solve the SVR, Lagrangian function constraints are introduced to transform the nonlinear problem into a dual programming, i.e.,

$$\max\left[-\frac{1}{2}\Sigma_{i=1}^{n}\Sigma_{i=1}^{n}(\alpha_i - \alpha_i^*)\left(\alpha_j - \alpha_j^*\right)K\left(x_i, x_j\right) + \Sigma_{i=1}^{n}\alpha_i(y_i - \varepsilon)\right.$$
$$\left. - \Sigma_{i=1}^{n}\alpha_i^*(y_i - \varepsilon)\right] \tag{14}$$
$$S.T.\begin{cases} \Sigma_{i=1}^{n}(\alpha_i - \alpha_i^*) = 0 \\ 0 \leq \alpha_i \leq C0 \leq \alpha_i^* \leq C \end{cases}$$

where α_i and α_i^* are Lagrange multipliers; $K(x_i, x_j)$ is the kernel function.

Finally, the SVR function is expressed by using Eq. (15).

$$f(x) = \sum_{i=1}^{n}(\alpha_i - \alpha_i^*)K(x_i, x) + b \tag{15}$$

Fig. 6 presents a schematic diagram of the SVR.

2.2.1 Selecting model input and output variables

Selecting model input and output variables is a basic but essential step for corrosion rate prediction of subsea pipelines. In the established prediction model, the output variable is the corrosion rate of the subsea pipelines, and the input variables are determined by the influencing factors of the pipeline corrosion. Information/Data regarding the influencing factors of subsea pipelines corrosion is collected through inspection, measurement and fluid

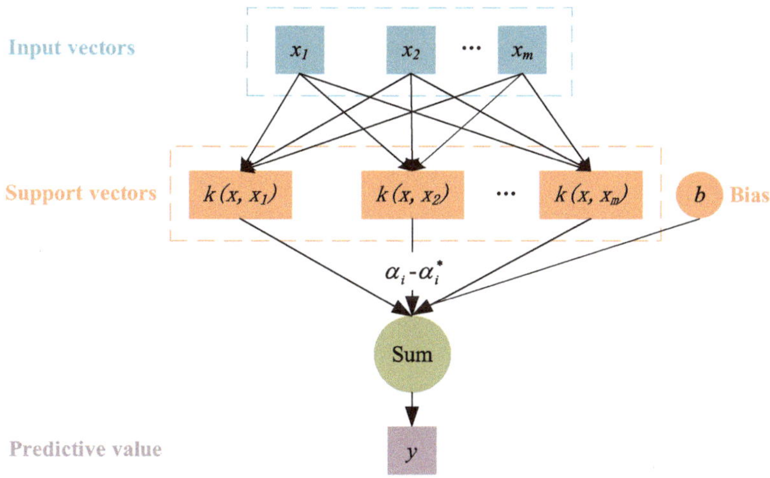

Fig. 6 The SVR model diagram.

analysis. However, it is known that there are too many influencing factors of pipeline corrosion. Due to the strong correlation between corrosion influencing factors, there may be redundant variables in these factors. The existence of redundant variables will make the training easy to fall into the local minimum value point, and ultimately lead to the poor generalization ability of the model. Principal component analysis (PCA) is a commonly used technology of data dimensionality reduction (Wang, Liu, Bao, & Zhang, 2019). Its basic principle is to select a set of uncorrelated principal components from a large number of linearly related original variables, and make them retain as much information of the original variables as possible (He & Zhang, 2018; Lei, Shen, Wang, & Chen, 2019). The specific steps of PCA are as follows:

Firstly, the original matrix is standardized to eliminate the influence of different dimensions, and the standardized matrix $\boldsymbol{Y}$ obtained is shown as follows:

$$Y = \left(Y_1 Y_2 \cdots Y_p\right) = \begin{pmatrix} y_{11} & y_{12} & \cdots & y_{1p} \\ y_{21} & y_{22} & \cdots & y_{2p} \\ \vdots & \vdots & & \vdots \\ y_{n1} & y_{n2} & \cdots & y_{np} \end{pmatrix} \tag{16}$$

Secondly, the correlation matrix $\boldsymbol{R}$ of the standardized matrix $\boldsymbol{Y}$ is calculated by Eq. (17):

$$r_{ij} = \frac{\Sigma_{k=1}^{n}(y_{ki} - \overline{y}_i)\left(y_{kj} - \overline{y}_j\right)}{\sqrt{\Sigma_{k=1}^{n}(y_{ki} - \overline{y}_i)^2 \Sigma_{k=1}^{n}\left(y_{kj} - \overline{y}_j\right)^2}} (i, j = 1, 2, \cdots, p) \tag{17}$$

where $i = 1, 2, \ldots, n$; $j = 1, 2, \ldots, p$; $\overline{y}_i$ represents the average value of the ith row; $\overline{y}_j$ represents the average value of the jth row.

Subsequently, the eigenvalues and eigenvectors of the correlation matrix $\boldsymbol{R}$ are calculated by Eq. (18), and the eigenvalues are arranged in descending order, namely $\lambda_1 \geq \lambda_2 \geq \ldots \geq \lambda_p \geq 0$.

$$\left|R - \lambda_j E\right| = 0 \tag{18}$$

where $j = 1,2,\ldots, p$; E is the identity matrix.

Eventually, the contribution rate and cumulative contribution rate of the principal components are calculated by using Eqs. (19) and (20):

$$l_j = \frac{\lambda_j}{\Sigma_{k=1}^{p} \lambda_j} \tag{19}$$

$$K = \frac{\Sigma_{j=1}^{m} \lambda_j}{\Sigma_{j=1}^{p} \lambda_j} \tag{20}$$

where l_j is the contribution rate of the jth principal component; K represents the cumulative contribution rate.

When the cumulative contribution rate reaches 85%–98%, m principal components are extracted, and m principal components are uncorrelated. Generally, the 1th, 2th, …, mth principal components correspond to the eigenvalues λ_1 , λ_2 , … λ_m in turn.

In the present study, PCA is adopted to reduce the dimensions of the corrosion influencing factors, and eliminate the redundant variables. Subsequently, the obtained principal components are used as input variables of the prediction model.

2.2.2 Dividing sample sets for model training and testing

The purpose of this process is to divide sample sets for the prediction model training and testing. The division of sample sets is vital to model training and testing. The collected sample data were divided into two sets for training and testing of the model respectively. In addition, before inputting the sample data sets into the prediction model, it needs to be normalized to eliminate the adverse effects of the singular sample data (Li, Wang, Xu, Yi, & Wang, 2019).

2.2.3 Selection and optimization of model parameters

This stage determines the structure of the model. In the prediction model, the penalty factor C controls the generalization ability and training error of the model. Another parameter to be optimized depends on the choice of the kernel function. Table 2 presents the common kernel functions. Radial Basis Function (RBF) is utilized in this paper because of its strong applicability (Feng, Zhang, Yang, Li, & Zhang, 2019; Wauters & Vanhoucke, 2014). And RBF depends on the regularization parameter γ.

The selection of hyper-parameters (C, γ) plays a crucial role in obtaining the prediction model with superior performance (Vanhoucke & Vandevoorde, 2007). In order to improve the prediction accuracy of the model, the artificial bee colony (ABC) algorithm is used in this work to optimize the above two hyper-parameters. ABC is a meta-heuristic algorithm based on bee foraging behavior. It was proposed by Karaboga (2005), including three types of bees, namely employed bees, onlooker bees and scout bees (Zhang, Tan, Zhu, Chen, & Liu, 2019). In the ABC algorithm, the bees perform different activities according to their respective division of labor, and share swarm information to determine the global optimal solution of the problem (Cheng, Cao, & Tran, 2014). The specific steps of the ABC algorithm are as follows:

Step 1: According to the ABC algorithm, an initial population of NP nectar sources (solutions) randomly distributed can be created in the search space. The initial position of each nectar source is generated by using Eq. (21).

$$x_{i,j} = x_{\min,j} + rand[0,1] \times \left(x_{\max,j} - x_{\min,j}\right) \tag{21}$$

where $x_{min,j}$ represents the lower bound of the search space; $x_{max,j}$ represents the upper bound of the search space; $j=\{1,2,\ldots, D\}$, D is the number of parameters to be optimized, in the present study, $D=2$; rand [0,1] is a random number in the range of [0,1].

Table 2 Common kernel functions.

Kernel name	Formula	Parameter
Linear	$K(x_i, x_j) = x_i x_j$	
Polynomia	$K(x_i, x_j) = (\gamma x_i x_j + f_{coe})^d$	γ, d, f_{coe}
Radial Basis	$K(x_i, x_j) = \exp(-\gamma \lVert x_i - x_j \rVert^2)$	γ
Sigmoidal	$K(x_i, x_j) = \tanh(\gamma x_i x_j + f_{coe})$	γ, f_{coe}

The nectar content of the nectar source is the fitness value of the solution. The fitness value is calculated as follows:

$$fitness_i = \frac{1}{1 + Obj.Fun._i} \tag{22}$$

where $Obj.Fun._i$ represents the objective function, which is defined as follows (Fei & He, 2015):

$$Obj.Fun._i = \frac{1}{\nu} \sum_{q=1}^{\nu} \left| \frac{y_q - \widehat{y}_{q,i}}{y_q} \right| \tag{23}$$

where ν represents the number of training samples in the training sample set; y_q is the actual value; $\widehat{y}_{q,j}$ is the validation value.

Step 2: The employed bee searches for a new nectar source *xnew i,j*in the neighborhood of the initial nectar source $x_{i,j}$ by using Eq. (24).

$$x_{i,j}^{new} = x_{\min,j} + rand[-1, 1] \times \left(x_{i,j} - x_{l,j}\right) \tag{24}$$

where $l \neq i$; rand $[-1,1]$ is a random number in the range of $[-1,1]$.

By comparing the fitness values of the new nectar source *xnew i,j* and the initial nectar source $x_{i,j}$, the nectar source with a better fitness value is retained.

Step 3: After all the employed bees complete the search, they share the nectar source information to the onlooker bees, the onlooker bees evaluate the fitness of all nectar sources, and select the nectar source according to the probability P_i calculated by Eq. (25). After that, the onlooker bee updates the selected nectar source by using Eq. (24), and calculates the fitness value of the updated nectar source to further retain the better nectar source.

$$P_i = \frac{fitness_i}{\Sigma_{i=1}^{NP} fitness_i} \tag{25}$$

where $fitness_i$ denotes the fitness value of the solution; *NP* is the number of nectars.

Step 4: If there is a nectar source for multiple iterations without improvement, the nectar source is abandoned, and a new nectar source is randomly generated by the scout bee using Eq. (21) to replace the abandoned nectar source.

When the first cycle is completed, the algorithm will store the optimal solution so far, and repeat steps 2 to 4 until the maximum number of iterations is reached to output the optimal solution.

2.2.4 Model training and testing

The final phase is to use the training sample set to train the prediction model and to apply the test sample set to evaluate the performance of the prediction model. Obviously, criterions for measuring the performance of the prediction model need to be proposed. This paper adopts the mean absolute error (*MAE*), the root mean square error (*RMSE*) and coefficient of determination (R^2) as evaluation indices to measure the prediction accuracy of the model, which are defined as follows:

$$MAE = \frac{1}{n}\sum_{i=1}^{n}\left|y^i - \hat{y}^i\right| \tag{26}$$

$$RMSE = \sqrt{\frac{1}{n}\sum_{i=1}^{n}\left(y^i - \hat{y}^i\right)^2} \tag{27}$$

$$R^2 = 1 - \frac{\sum_{i=1}^{n}\left(y^i - \hat{y}^i\right)^2}{\sum_{i=1}^{n}\left(\overline{y}^i - \hat{y}^i\right)^2} \tag{28}$$

where n represents the number of test samples; y^i, $\hat{y}^i$ and $\overline{y}^i$ represent the actual values, the predicted values and the average values of corrosion rate, respectively. It is generally considered that models with smaller *MAE* and *RMSE* values are more accurate, while models with larger R^2 values are more stable.

3. Case Study 1: Dynamic risk assessment of subsea pipeline leak

3.1 Scenario development

This section presents a case study to demonstrate the proposed dynamic approach. Subsea pipelines leak incidents are rare but can induce major consequences, such as fatalities, capital losses, and irreversible damage to the environment. A BT model for subsea pipelines leak is shown in Fig. 7, including potential causes and consequences due to subsequent failure of safety barriers. The causes of subsea pipelines leak include external and internal factors. External factors involve corrosion, natural disaster, external loads and spanning vibration, whereas internal factors comprise welding seam and material defects, as well as attachments failure. Detailed bottom factors are presented in Tables 3 and 4. These are some safety barriers to prevent leak escalating into more severe consequences, including (*S1*) leakage detection,

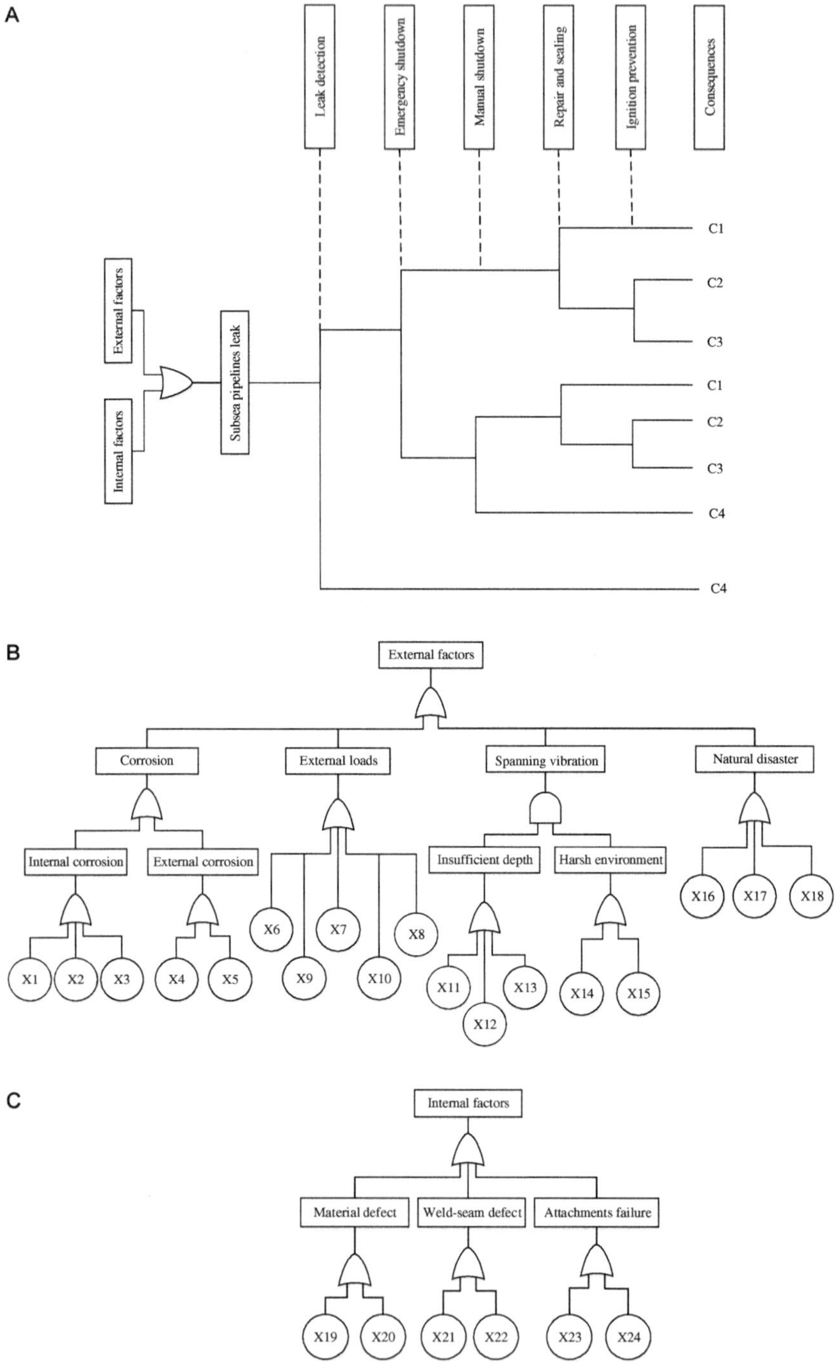

Fig. 7 (A) BT model for subsea pipelines leak. (B) Sub-FT for external factors in BT model. (C) Sub-FT for internal factors in BT model.

Table 3 Basic events in BT and their probabilities from literature (Li et al., 2016; Participants, 2002).

Index	Description	Prior probability
X1	Without removing corrosive gas and impurities	1.0E-03
X3	Not pigging regularly	2.0E-04
X14	Strong current and wave	2.0E-05
X19	Design defect of material	8.4E-04
X20	Construction defect of material	9.7E-04
X21	Design defect of weld-seam	2.3E-04
X22	Construction defect of weld-seam	6.5E-04
X23	Attachments aging	3.2E-03
X24	Design defect of attachments	1.0E-05

Table 4 Basic events in BT and their probabilities from HBA.

Index	Description	Prior probability	97.5% Confidence interval
X2	Without adding a corrosion inhibitor	8.3E-03	(1.97E-03, 1.82E-02)
X4	Anticorrosive coating failure	3.3E-03	(2.69E-03, 9.84E-03)
X5	Cathodic protection failure	7.8E-03	(1.69E-03, 1.75E-02)
X6	Dropped objects	8.3E-03	(1.61E-03, 1.91E-02)
X7	Anchoring work	8.9E-03	(2.76E-03, 1.82E-02)
X8	Fishing gear interaction	9.4E-03	(2.40E-03, 1.99E-02)
X9	Oil stolen by drilling hole	4.4E-03	(7.15E-05, 1.41E-02)
X10	Offshore engineering construction	6.7E-03	(1.46E-03, 1.51E-02)
X11	Insufficient buried depth due to design	5.2E-03	(2.72E-05, 1.83E-02)
X12	Insufficient buried depth due to operation	4.4E-03	(3.52E-04, 1.19E-02)
X13	Not treating timely	3.9E-03	(9.89E-05, 1.19E-02)
X15	Seabed soil are eroded easily	4.2E-03	(7.56E-07, 1.63E-02)
X16	Subsea earthquake	2.7E-03	(1.27E-05, 9.41E-03)
X17	Seabed movement	6.1E-03	(8.98E-04, 1.47E-02)
X18	Typhoon	3.3E-03	(1.07E-04, 9.99E-03)

(*S2*) emergency shutdown, (*S3*) manual shutdown, (*S4*) repair and sealing and (*S5*) ignition prevention. Since the subsequent failure of safety barrier, the initial leakage incident can induce a series of severe consequences and even develop into a major accident, such as fire, explosion or oil spill in the large area. The end states in Fig. 7 consists of four conditions: (*C1*) near miss; (*C2*) a small vapor cloud and environmental pollution; (*C3*) a small fire and environment pollution; (*C4*) a huge vapor cloud and environmental pollution.

In this study, BT is only utilized to develop the initial accident scenario. As discussed in previous sections, BT is known as a static method that induces certain uncertainty into results. To relax its limitations, BT is transformed into a BN (Fig. 8). As can be seen, the conditional dependencies between safety barriers S2 and S3, as well as S4 and S5 are considered. Also, the uncertainties among father and child nodes are also through an amending CPT where a set of uncertain values, e.g. 0.99 or 0.01, are used. The expert experience is used to determine the use of these uncertain values in CPTs.

3.2 Nodes' probabilities estimation

In reality, direct data on failure probabilities of basic events and safety barriers are limited and insufficient. Therefore, the root nodes in BN model are divided into two types. The first type includes the nodes have the directly

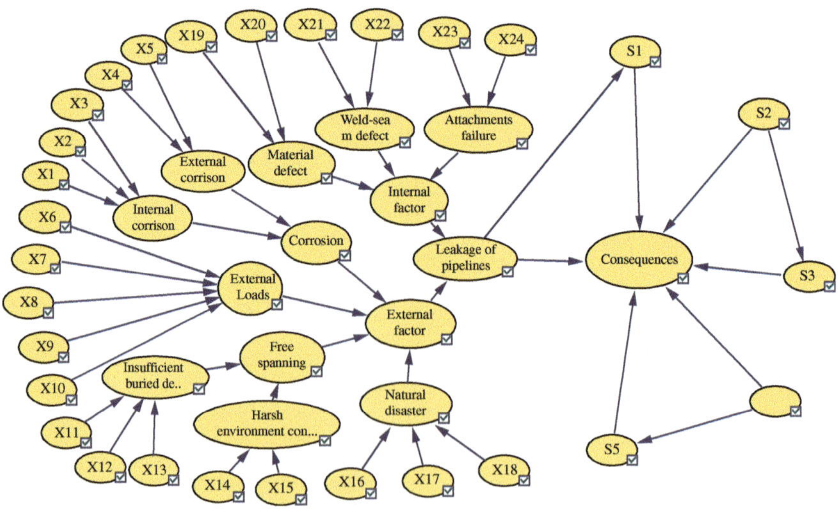

Fig. 8 BN model mapping from BT.

available failure probabilities that can be found in literature or other sources, whereas the other type embodies the nodes without the available probabilities. Table 3 presents the basic events with directly available probabilities. Table 4 shows the basic events that have no direct probabilities, and the probabilities of these events can be derived from the HBA method.

Available indirect but relevant data pooled from different sources can be incorporated into the HBA framework to estimate the probabilities of basic events in Table 4. For the illustrative purpose, Table 5 gives the hypothetical precursor data of basic events and consequence states. It is assumed that these data are observed over the course of 10 years and collected from 170 pipelines. Based on these data, the precursor data of safety barriers can be

Table 5 Precursor data for basic events and consequences.

Year	1	2	3	4	5	6	7	8	9	10
X2	0	1	2	2	1	2	0	1	2	3
X4	1	0	1	0	0	0	1	1	0	1
X5	1	2	2	3	0	1	2	1	0	1
X6	1	0	3	2	0	1	1	2	3	1
X7	2	1	2	1	0	2	3	2	1	1
X8	2	3	1	0	2	2	0	3	1	2
X9	0	1	0	0	2	1	1	0	0	2
X10	1	2	0	1	1	0	2	1	2	1
X11	0	0	1	0	0	1	2	0	3	1
X12	0	1	1	1	0	1	0	2	1	0
X13	0	0	1	2	0	0	1	1	0	1
X15	0	2	0	0	1	0	2	1	0	0
X16	0	0	1	0	1	0	1	0	1	0
X17	1	0	1	2	0	2	2	1	0	1
X18	1	0	1	1	0	0	1	0	0	1
C1	3	2	1	2	2	0	0	1	1	0
C2	1	1	0	1	1	3	3	1	0	0
C3	0	1	1	1	0	2	0	0	0	0
C4	1	0	0	1	1	0	1	0	0	1

reasoned from the logic structure of event tree model. The occurrence number of leak incident can also be obtained which is the sum of the occurrence number of all consequence states. Note that the dynamic risk in this case means the risk variation on an annual scale.

Using the precursor data in Table 5, the hierarchical Bayesian model can be utilized to predict the number of leakage and different consequences. HBA is executed using OpenBUGS, which is a freely available Bayesian inference software (OpenBUGS).

The dataset of pipeline leak and consequences can be modeled from data in Table 5. The occurrence number of pipeline leak and consequence states can be modeled using a Poisson distribution with the parameter of interest λ. λ is the occurrence rate and is considered to follow the conjugate prior gamma distribution with hyperparameters α, β. α and β follow the independent diffusive distribution. Fig. 9 gives the predictive distribution for the occurrence rate of a pipeline leak. Likewise, Fig. 10 presents the posterior predictive distributions for the occurrence number of different consequence states. These posterior predictive distributions can be employed to predict the occurrence number of leak incident and the corresponding consequence states in next interval, that is, 11th year in this paper.

Table 6 presents the expected values of these posterior predictive distributions shown in Figs. 9 and 10, they are the predicted occurrence number of leak incident and consequence state in next time interval. The 97.5% confidence interval represents the uncertainty in predicted values. Furthermore, the occurrence rate of leak and different consequences is also estimated by merely taking the 10 years' averages. The results in Table 6 demonstrate the difference between the two estimators. As for the leak, *C2*, *C3* and *C4*, the occurrence rates estimated by HBA is higher than the ones that obtained from taking averages, while the occurrence rate of *C1* from HBA approach is lower than the one derived from taking the average. The reason is that the sensitivity of average to data points introduces certain bias in results.

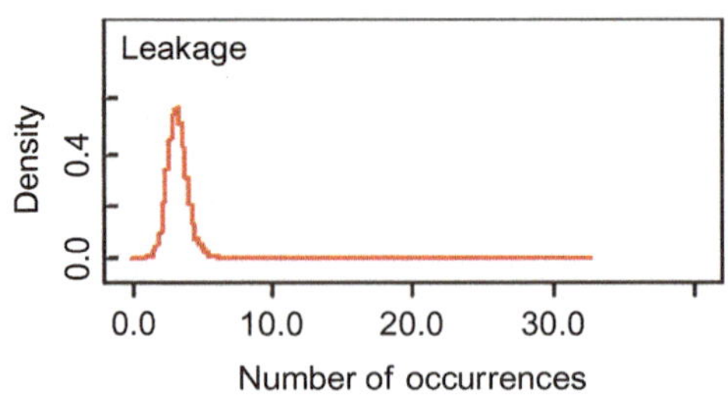

Fig. 9 Predictive distribution for occurrence rate of a pipeline leak.

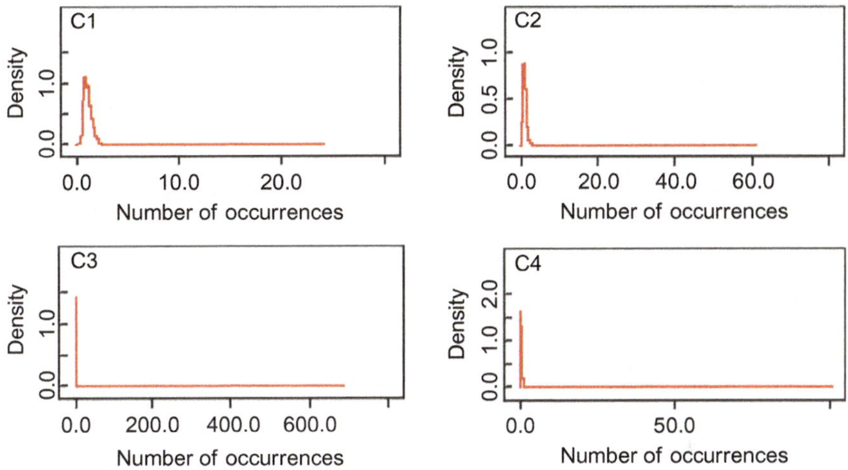

Fig. 10 Predictive distribution for occurrences rate of different consequence states.

Table 6 Occurrence rate of leak incident and different consequence states from HBA approach.

Events	Occurrence rate	97.5% Confidence interval	Ten years' average
Pipeline leak	3.321	(1.809, 5.249)	3.3
C1	1.162	(0.3397, 2.471)	1.2
C2	1.123	(0.2079, 2.57)	1.1
C3	0.553	(0.02719, 1.31)	0.5
C4	0.531	(0.09026, 1.182)	0.5

In addition to the prediction of occurrence rate, the other important application of HBA is to calculate the probabilities of basic events and safety barriers with insufficient data. As for the basic events without direct available probabilities data, their failure probabilities can be derived from the HBA approach using relevant precursor data shown in Table 5. A binomial distribution with parameter p is assigned to each basic event to model their failures number per time interval, in which p is considered to follow the beta distribution with hyperparameters a and b. An independent diffusive distribution is assigned to hyperparameters a and b. Fig. 11 presents the predictive distributions for failure probabilities of basic events, whereas Table 4 gives the expected values with 97.5% confidence interval.

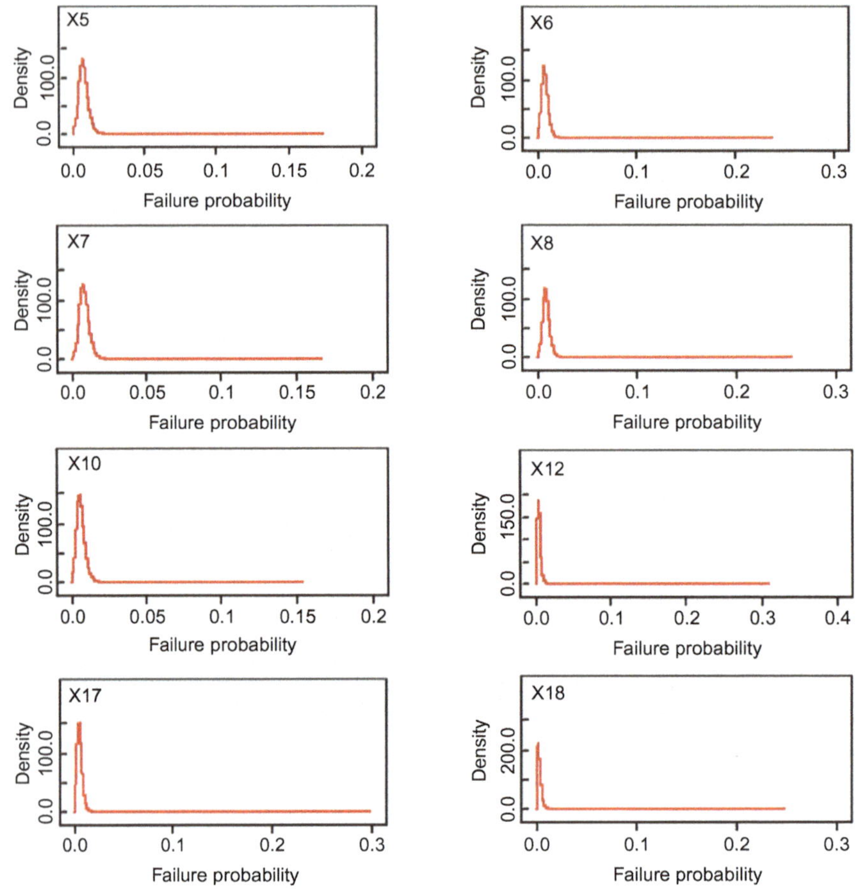

Fig. 11 Predictive distributions for failure probabilities of basic events.

The precursor data of safety barriers are reasoned from event tree structure and precursor data of different consequence states. The same HBA procedure is utilized to calculate the failure probabilities of safety barriers. Fig. 12 presents the posterior predictive distributions for failure probabilities of safety barriers. The expected values of these distributions and 97.5% confidence intervals are summarized in Table 7. The 97.5% confidence intervals show the uncertainties among estimated failure probabilities of safety barriers. By HBA approach, the uncertainty among precursor data from different sources has been addressed.

3.3 Probability updating

Based on the probabilities in Tables 3, 4 and 7, the occurrence probabilities of pipeline leak and consequence states are estimated via forward inference.

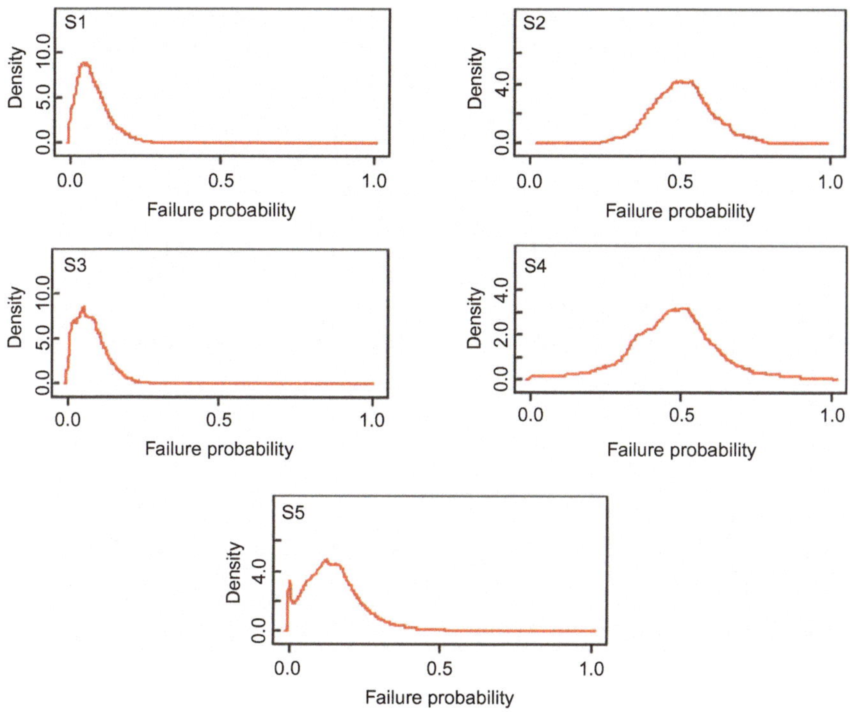

Fig. 12 Predictive distributions for failure probabilities of safety barriers.

Table 7 Safety barriers and failure probabilities.

Index	Description	Prior probability	97.5% Confidence interval
S1	Leakage detection	8.28E-02	(5.23E-03, 2.36E-01)
S2	Emergency shutdown	5.06E-01	(3.07E-01, 7.19E-01)
S3	Manual shutdown	8.23E-02	(7.46E-03, 2.20E-01)
S4	Repair and sealing	4.76E-01	(1.34E-01, 8.23E-01)
S5	Ignition prevention	1.58E-01	(2.78E-03, 4.44E-01)

The probability of pipeline leak is estimated to be 6.87E-02. Table 8 shows the occurrence probabilities of different consequence states. The occurrence probabilities of four consequence states from high to low are *C1*, *C2*, *C4*, and *C3*, respectively, *C1* (*Near miss*) is the most probable consequence state. The severities of them increase from *C1* to *C4*. *C4* (*Huge vapor cloud and environment pollution*) is the most dangerous consequence state which may induce catastrophic impact on corporation reputation and environment. In the present study, *C4* is chosen as the target incident to mitigate in safety management of subsea pipelines, and used for the subsequent analysis.

Table 8 Consequence states and the estimated probabilities.

Index	Description	Probability
C1	Near miss	3.29E-02
C2	Small vapor cloud and environment pollution	2.52E-02
C3	Small fire and environment pollution	4.72E-03
C4	Huge vapor cloud and environment pollution	5.95E-03

The main advantage of BN is that it could perform a probability updating given a evidence. This makes BN be superior to conventional risk analysis methods such as BT. Probability updating can help to find the most probable events contributing to the occurrence of evidence (Li et al., 2018). this study uses consequence state *C4* as a new evidence to obtain the updated probabilities of basic events, i.e., P(*Xi* | consequence = *C4*).

Fig. 13 presents the comparison of prior probabilities and updated probabilities of basic events. As can be seen, the updated probabilities of some basic events increase significantly, meaning their critical role in causing a disastrous consequence. These basic events include *X2* (Without adding corrosion inhibitor), *X4* (Anticorrosive coating failure), *X6* (Dropped objects), *X7* (Anchoring work), *X8* (Fishing gear interaction), *X9* (Oil stolen by drilling hole), *X10* (Offshore engineering construction), *X16* (Subsea earthquake), *X17* (Seabed movement), *X18* (Typhoon) and *X23* (Attachments aging), indicating their critical role in causing occurrences of leakage and *C4*. According to updated probability, these basis events should be paid enough attention to reduce their failure probability or strengthen protective measures.

Fig. 14 shows the comparison of prior and updated failure probabilities of safety barriers. It could be found that the updated probabilities of *S1* (leak detection) and *S2* (Emergency shutdown) increase. This could indicate that *S1* and *S2* are critical factors for the occurrence of *C4*. Mainly, *S1* is the most critical safety barrier to prevent the escalation of leakage incident. When leak incident is not perceived by the SCADA system or patrol workers successfully, the occurrence of *C4* will be caused directly. Also, although the leakage is detected successfully, the oil and gas will still release out of control if the emergency shutdown does not be performed in time. Therefore, the reliability of *S1* and *S2* is required to improve safety management for reducing the risk of leakage escalating into a major accident.

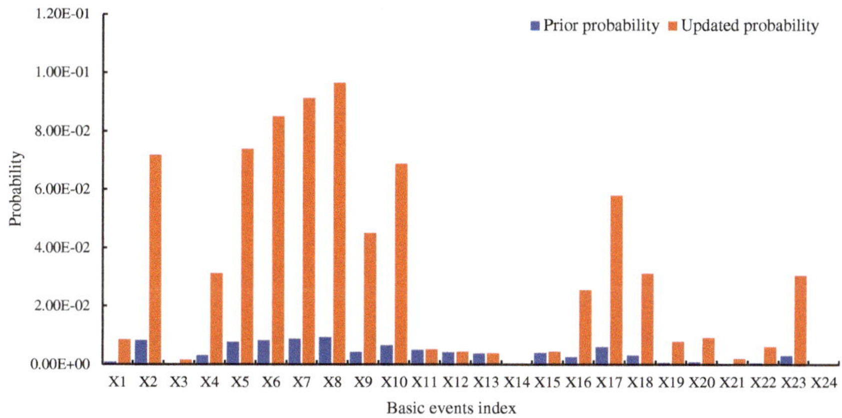

Fig. 13 Comparison of prior and updated probabilities of basic events.

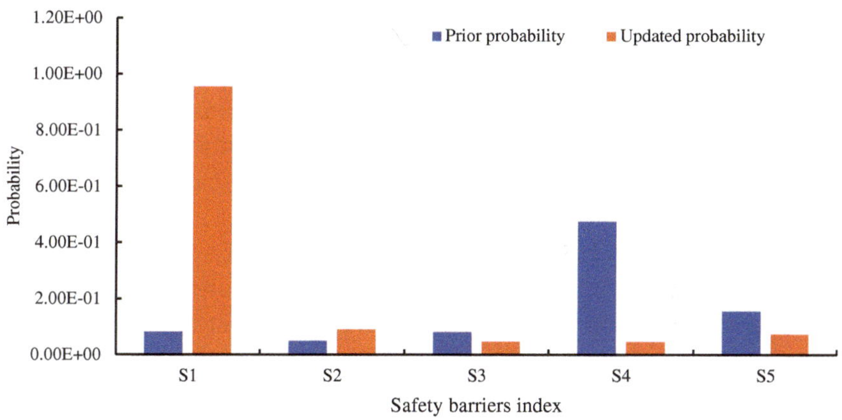

Fig. 14 Comparison of prior and updated failure probabilities of safety barriers.

3.4 Probability adapting

Probability adapting is also known as sequential learning or experience learning (Kjaerulff & Madsen, 2008; Yuan et al., 2015). It is an important application of BN in risk analysis field, and particularly suitable for dynamic risk modeling of the system. Probability adapting is able to utilize new information that were recorded over time to revise the failure probability of the system. Thus, failure probability of the system could be adapted to be more case-specific and reflect the system characteristics (Li et al., 2016). In general, new information involves the observations of near misses, mishaps, incidents and root events during the life cycle of the system. The form of the cumulative number of the observations is utilized to develop likelihood function. Then the prior probabilities of nodes in BN can be updated over time by Bayes theorem shown as Eq. (29). Propagating new information through the BN, the probability of accident, its consequences and safety barriers will change over time.

$$P(n|data) = \frac{P(n)L(data|n)}{\sum P(n)L(data|n)} \tag{29}$$

where $P(n)$ is the nodes' prior probabilities, $L(n \mid data)$ is the likelihood function that is constructed using precursor data, $P(n \mid data)$ is the nodes' updated probabilities after adapting.

Based on the precursor data in Table 4, the probabilities of critical basic events and safety barriers identified in probability updating, are updated over time. The dynamic probabilities of them can be propagated in BN. Thus, dynamic probabilities of different consequence states are produced. Fig. 15 presents the probability adapting results for the top event (subsea pipelines leak). In the figure, year 0 denotes the priors. As can be seen, given the appearance of precursor data, the occurrence probability of pipeline leak is increasing over a 10-year period, which indicates that the likelihood of subsea pipelines leak is increasing over time.

Fig. 16 depicts the dynamic failure probabilities of safety barriers in the appearance of precursor data. As can be seen, the failure probabilities of all safety barriers, in general, are increasing over time, which are in agreement

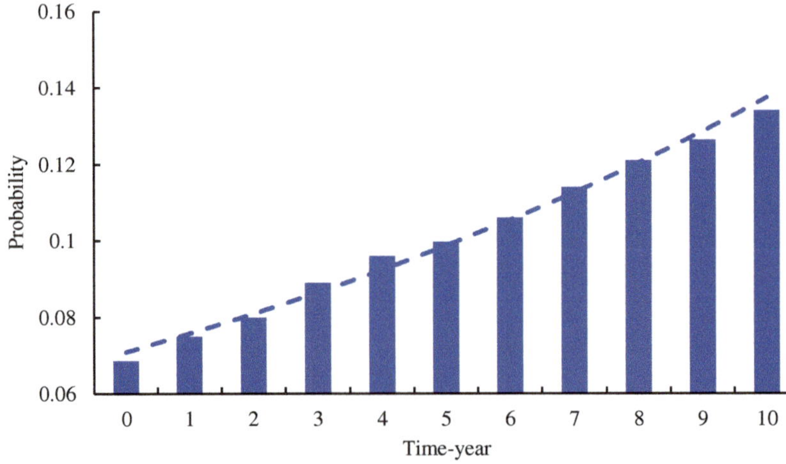

Fig. 15 Dynamic occurrence probability of subsea pipelines leak.

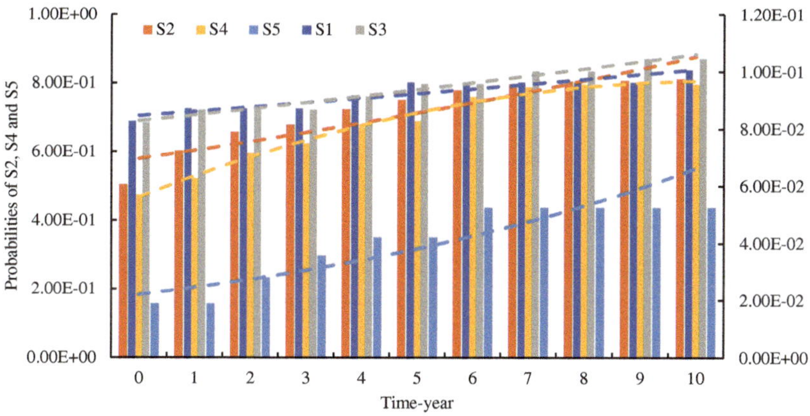

Fig. 16 Dynamic failure probabilities of safety barriers.

with the probability of the top event. However, the failure probabilities of S1, S2, S4 and S5 stand at a steady value or have a slight drop from the 6th year. This is due to the infrequency of precursor data relating to these safety barriers. Dynamic failure probabilities of safety barriers could help to identify the weak links in the safety system of subsea pipelines and give useful guidance for further improvement of the safety system.

Integrating precursor data with probability adapting of BN could revise generic prior probabilities of basic events, safety barriers and consequence states to be more case-specific updated probabilities. Such dynamic probabilities of consequence states can be used to predict the development trend of the consequence of interest and render more practical guidance for risk prevention and control.

3.5 Risk assessment

Based on the definition of risk, the consequence is the other important aspect of risk. In this study, the consequence severity of subsea pipelines leak is measured by the concept of the fuzzy loss ratio. The loss ratio probability vector q_i for consequence state i must be estimated before determining the fuzzy loss ratio. Experts' judgment is employed to obtain the loss ratio probability vector, and the weighted average method is used to aggregate m experts' knowledge. The nth element in q_i from the aggregation of m experts' judgment is shown as Eq. (30) (Ferdous, Khan, Sadiq, Amyotte, & Veitch, 2012).

$$\overline{q}_{in} = \frac{\sum_{j=1}^{m} w_j q_{in-j}}{\sum_{j=1}^{m} w_j} \tag{30}$$

where $q_{in\text{-}j}$ is the *nth* element in vector q_i elicited from expert *j*, *m* is the number of experts, w_j is the weighting factor of expert *j*, $\overline{q}_{in}$ is the aggregated value of $q_{in\text{-}j}$.

After obtaining the loss ratio probability vector q_i, the fuzzy loss ratio of consequence state *i* can be determining by Eq. (10). Estimated loss ratio probability vectors and fuzzy loss ratios for all consequences states are summarized in Table 9.

In terms of subsea pipelines leak, there are four potential consequences states (*C1*, *C2*, *C3* and *C4*). Therefore, the risk of subsea pipelines leak is the sum of risk value for all consequences states, shown as Eq. (31) (Tan et al., 2013).

$$R = \sum_{i=1}^{4} c_i p_{Ci} = c_1 p_{C1} + c_2 p_{C2} + c_3 p_{C3} + c_4 p_{C4} \tag{31}$$

where *R* is the risk value of all consequence states, c_i is the fuzzy loss ratio of consequence state *i*, P_{Ci} is the occurrence probability of consequence state *i*.

Fig. 17 presents the dynamic risk profile of subsea pipelines leak, indicating that the total risk of subsea pipelines leak is increasing over time. Based on the result presented here, the oil corporation managers can know the overall risk trend of subsea pipelines leak, and conduct risk decision-making given some critical risk limits. In general, the definition of critical risk limits varies in different oil corporations, depending on their production scale, management level, safety culture and risk tolerance, etc. Using the critical risk values, the risk early warning can be implemented to prevent and control unexpected incidents, and the managers can also make a decision

Table 9 Fuzzy loss ratio for different consequences states.

Symbol of consequences states	Loss ratio probability vector q_i	Fuzzy loss ratio c_i
C1	[0,0,0,0,1]	5.0E-05
C2	[0, 0.1, 0.2, 0.6, 0.1]	6.3E-03
C3	[0, 0.3, 0.5, 0.2, 0]	1.76E-02
C4	[0.3, 0.6, 0.1, 0, 0]	1.81E-01

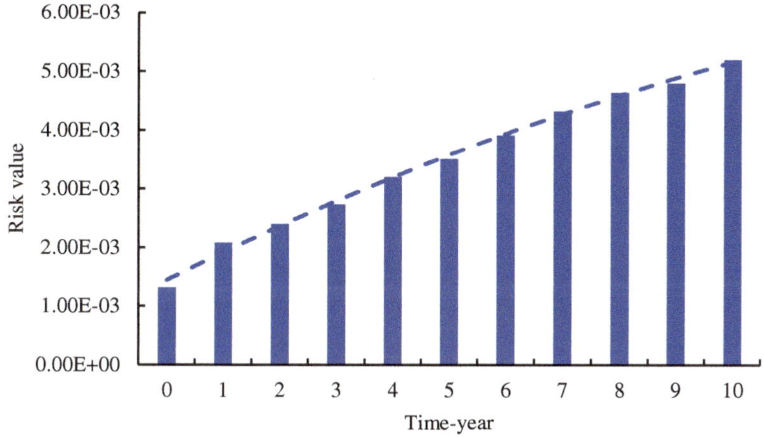

Fig. 17 Dynamic risk profile of subsea pipelines leak.

on the further expenses to reduce the risk. Also, the dynamic risk profile can be used to predict the future risk status of subsea pipelines leak.

4. Case Study 2: Dynamic prediction of subsea pipelines corrosion degradation

A case study is provided in this section to illustrate the proposed approach for dynamic prediction of subsea pipelines corrosion degradation. The descriptive statistics involving corrosion comes from a pipeline in the South China Sea. The pipeline has a total length of 4.8 km and was put into use in 2008 with a design life of 20 years. And the pipe material is X80 steel.

4.1 Index selection

4.1.1 Corrosion influencing factors of subsea pipelines

The influencing factors for the corrosion rate of subsea pipelines include:

(1) Water content (X_1): For subsea pipelines, the proportion of water content is an important factor affecting pipeline corrosion. There is a positive correlation between water content and pipelines corrosion rate. As the water content increases, the corrosion rate increases significantly compared with the low water content. What's more, the liquid phase water quickens the dissolution of acid gases, which provides favorable conditions for the occurrence of corrosion.

(2) H_2S content (X_2): In the process of pipelines transportation, H_2S dissolved in water is highly corrosive to pipelines. On the one hand, the ionized H^+ diffuses into the inside of the pipelines steel material, forming high-pressure hydrogen gas, which causes hydrogen damage of the pipelines steel material. On the other hand, because the pipeline is exposed to the environmental medium containing H_2S, a layer of corrosion product film will be formed on the inner surface of the pipe. FeS is the main component of the corrosion product film. When FeS is densely attached to the surface of the pipelines, it will hinder corrosion. On the contrary, it will promote the corrosion reaction.

(3) CO_2 content (X_3): CO_2 corrosion is one of the main reasons for the corrosion failure of subsea pipelines. It reacts with water to form H_2CO_3, which then eventually produces corrosion product $FeCO_3$. However, due to the poor compactness of $FeCO_3$ film, serious local corrosion is often induced at a certain temperature. In addition, it is worth noting that in a system where CO_2 and H_2S coexist, the partial pressure ratio of the two also affects the corrosion process of the pipeline.

(4) Temperature (X_4): In different temperature ranges, the tightness and adhesion of the corrosion product film are different, which will affect the corrosion behavior of subsea pipelines. When the temperature is about 100°C, the corrosion rate of the pipeline is the largest.

(5) Pressure (X_5): Pressure is also a significant parameter affecting the corrosion rate of subsea pipelines. As the pressure increases, the corrosion rate of the pipeline increases.

(6) pH value (X_6): pH is the standard for measuring the concentration of hydrogen ions. The corrosion rate of the pipeline is greater at lower pH value.

(7) Cl ion content (X_7): Due to the strong penetrating ability of Cl ions, it could be preferentially adsorbed to the surface of the pipelines steel and interact with it to form soluble compounds, which causes the aggravation of pipelines corrosion.

(8) Ca and Mg ions content (X_8): The corrosion mechanism of Ca and Mg ions is more complicated. On the one hand, Ca and Mg ions will cause scaling of the pipelines, thus inhibiting corrosion; On the other hand, Ca and Mg ions are prone to under-deposit corrosion. Regarding the pipelines system, under-deposit corrosion is one of the most destructive forms of corrosion, which seriously threatens the safe operation of the pipelines.

(9) Flow rate (X_9): The increase of the flow rate will destroy the corrosion product film on the surface of the pipelines steel, leading to the acceleration of the corrosion process. Thus, the corrosion rate of the pipelines increases as the flow rate increases.

(10) Dissolved oxygen content (X_{10}): The increase of dissolved oxygen content will increase the corrosion rate of subsea pipelines. Because the increase of the dissolved oxygen content will accelerate the oxygen depolarization corrosion process of the cathode, thereby promoting the corrosion of the pipelines.

The corrosion rate of the pipeline is measured by a professional organization. A total of 100 groups is used for the illustrative purpose. Table 10 presents the statistics (mean, std., min, max) of the collected data of these factors.

4.1.2 Principal component analysis of corrosion influencing factors

The correlation analysis of the above 10 influencing factors has been carried out through SPSS statistical software. Table 11 presents the obtained correlation coefficients. The larger the correlation coefficient between the two influencing factors, the stronger the correlation between them. As can be observed from Table 11, there is a strong correlation between the corrosion influencing factors of subsea pipelines, thereby leading to redundant variables in these factors. The existence of redundant variables will affect the prediction accuracy of the prediction model.

To eliminate redundant variables, PCA was carried out for the 10 influencing factors of corrosion rate for subsea pipelines. Table 12 presents the PCA results. It can be observed that the cumulative contribution rate of the first four principal components has reached 95.71%, which indicates that most information of the original variables is retained. Therefore, in this study, the first four principal components were selected as input variables of the prediction model.

4.2 Division of sample sets

The samples were randomly sorted by using Randperm function in MATLAB, and the first 80% of the samples constituted the training sample set while the remaining 20% samples constituted the testing sample set. It is noteworthy that the sample data needs to be normalized before being used. Data normalization can eliminate the influence of unit and dimension of the dataset of corrosion factors, and improve their comparability. Besides, it can also help to improve the speed to get optimal solution and prevent data

Table 10 Descriptive statistics sheet.

	X_1	X_2	X_3	X_4	X_5	X_6	X_7	X_8	X_9	X_{10}	Corrosion rate
No.	**%**	**mg/m^3**	**%**	**°C**	**MPa**	****	**mg/L**	**mg/L**	**m/s**	**mg/L**	**mm/a**
Mean	67.2	202.05	5.57	30.53	4.72	5.81	9308	442.39	0.061	1.98	1.15
Std	12.11	205.85	2.38	3.74	0.65	0.87	4695	158.99	0.024	1.01	0.64
Min	51.51	14.31	3.24	22.52	3.61	3.89	5016	235.43	0.036	0.81	0.48
Max	94.14	542.69	10.99	38.25	6.06	7.09	19,369	785.74	0.121	3.93	2.34

Table 11 Correlation analysis of influencing factors.

	X_1	X_2	X_3	X_4	X_5	X_6	X_7	X_8	X_9	X_{10}
X_1	1.00	0.77	−0.28	−0.15	0.29	−0.01	0.94	0.12	0.08	0.92
X_2	0.77	1.00	−0.25	0.10	0.01	−0.53	0.74	0.57	−0.31	0.81
X_3	−0.28	−0.25	1.00	−0.18	−0.17	0.14	−0.29	−0.48	−0.20	−0.49
X_4	−0.15	0.10	−0.18	1.00	0.31	−0.27	−0.09	0.14	−0.74	−0.05
X_5	0.29	0.01	−0.17	0.31	1.00	0.52	0.48	−0.54	−0.19	0.35
X_6	−0.01	−0.53	0.14	−0.27	0.52	1.00	0.10	−0.84	0.44	−0.06
X_7	0.94	0.74	−0.29	−0.09	0.48	0.10	1.00	0.00	0.02	0.95
X_8	0.12	0.57	−0.48	0.14	−0.54	−0.84	0.00	1.00	−0.16	0.24
X_9	0.08	−0.31	−0.20	−0.74	−0.19	0.44	0.02	−0.16	1.00	0.06
X_{10}	0.92	0.81	−0.49	−0.05	0.35	−0.06	0.95	0.24	0.06	1.00

Table 12 PCA results.

Principal component	**Contribution rate**	**Cumulative contribution rate**
M_1	39.58%	39.58%
M_2	26.75%	66.33%
M_3	18.62%	84.95%
M_4	10.76%	95.71%
M_5	1.88%	97.59%
M_6	0.95%	98.54%
M_7	0.67%	99.21%
M_8	0.56%	99.77%
M_9	0.13%	99.90%
M_{10}	0.09%	100.00%

explosion. There are two common methods used for data normalization. One is min max normalization, and this method can scale the data into 0 to 1 according to a certain ratio. The other is standard deviation normalization, and this method can scale the dataset with the mean of 0, and the variance of 1. This paper uses the first method, i.e., min max normalization, to normalize the dataset of pipeline corrosion. Mapminmax function in MATLAB was adopted to normalize the sample data to the interval of [0, 1].

4.3 Comparison of predicted results

4.3.1 Compared with SVR and PCA-SVR

In this section, the prediction results of PCA-ABC-SVR are compared with the SVR and PCA-SVR models.

Fig. 18 presents the predicted results of the SVR model. It can be observed that the predicted value of corrosion rate differs greatly from the actual value, indicating that the prediction accuracy of the SVR model is low. Fig. 19 presents the predicted results of the PCA-SVR model. As can be observed, after selecting the principal components of corrosion influencing factors as input variables, the prediction accuracy of the model is greatly improved. However, at some testing sample points, there is a large error between the predicted value and the actual value.

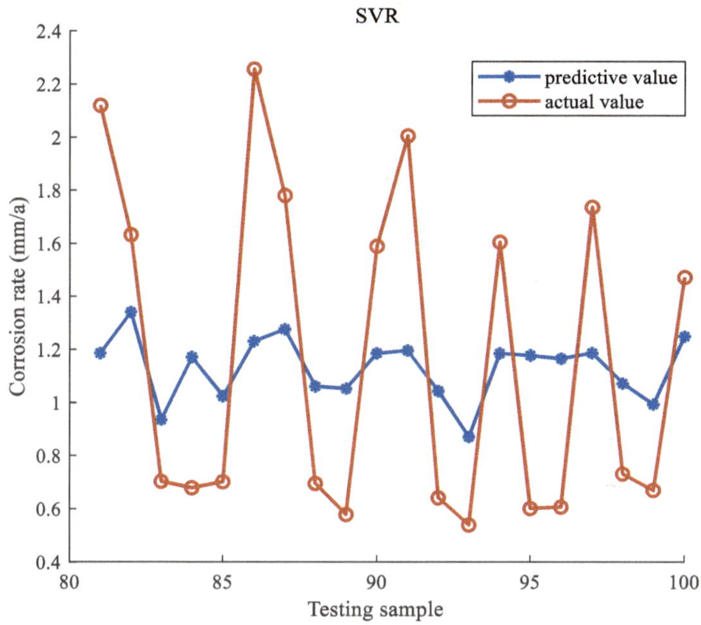

Fig. 18 SVR model prediction results.

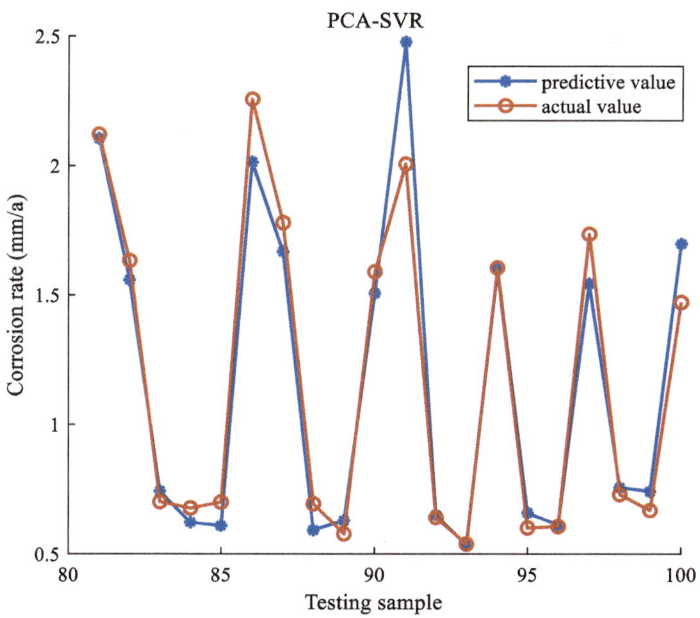

Fig. 19 PCA-SVR model prediction results.

It can be observed that PCA-SVR model has a better prediction accuracy than SVR alone. The reason is that the use of PCA takes the redundant information out of the corrosion influencing factors and eliminate their correlation. The dataset after dimension reduction contains the main information of all corrosion factors. This contributes to the improvement of prediction accuracy. Besides, the use of PCA also reduces the computational time compared to SVR alone. The computational times for SVR alone and PCA-SVR model are 12 and 8 s, respectively.

Fig. 20 presents the predicted results of the PCA-ABC-SVR model. It can be seen that using the ABC algorithm to optimize the model, compared with the PCA-SVR model, the prediction value of corrosion rate is closer to the actual value. And currently, the values of the hyper-parameters (C, γ) are 3.87 and 79.186, respectively.

Table 13 illustrates the prediction accuracy of the above three models. Among these three models, PCA-ABC-SVR shows better performance based on evaluation indices. Specifically, the PCA-ABC-SVR model has the smallest *MAE* and *RMSE* values, which are 7.10% and 9.19%, respectively, and the R^2 value is the largest, equal to 0.976. This proves that PCA is

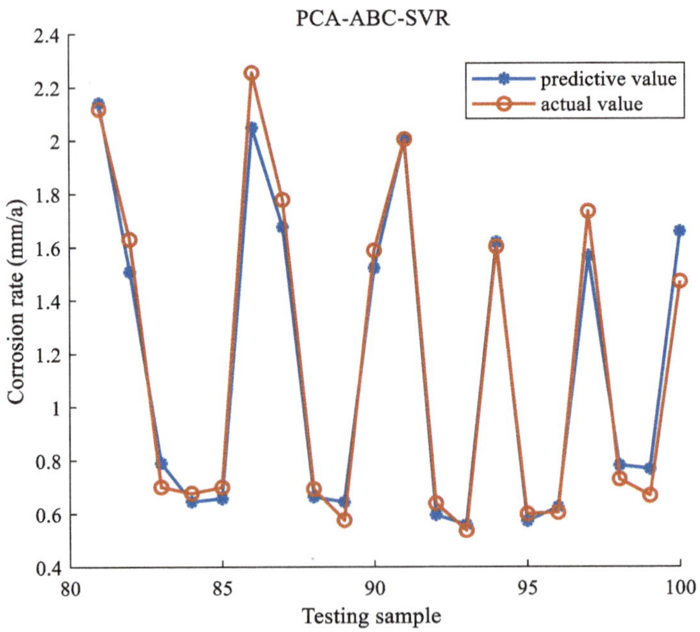

Fig. 20 PCA-ABC-SVR model prediction results.

Table 13 Error comparison between SVR, PCA- SVR, PCA-ABC-SVR.

	MAE	*RMSE*	R^2
SVR	47.91%	52.47%	0.231
PCA-SVR	9.72%	14.68%	0.939
PCA-ABC-SVR	7.10%	9.19%	0.976

used to select the principal components of corrosion influencing factors, and ABC algorithm is employed to optimize the model, which is beneficial to improve the prediction accuracy of the model. Therefore, the improvement of prediction accuracy benefits from both PCA and ABC. PCA reduces the dimension of dataset, while ABC optimizes hyper-parameters. Both PCA and ABC contribute to the better prediction accuracy. However, PCA-ABC-SVR model requires more computational time compared to SVR alone and PCA-SVR model. The computational time of the case using PCA-ABC-SVR model is 17 s.

The over-fitting problem is a common problem faced by most machine learning algorithms. To examine the over-fitting possibility of the established PCA-ABC-SVR model, this study presents both training and testing results of PCA-ABC-SVR model, as shown in Figs. 20 and 21. Table 14 summarizes the errors for training and testing processes of the model. It can be found that PCA-ABC-SVR model has the excellent goodness of fitting in both training and testing processes, and the training and testing errors are relatively low. This observation can demonstrate that that the established PCA-ABC-SVR model does not over-fit.

4.3.2 Compared with PCA-GA-SVR and PCA-PSO-SVR

In this section, the prediction results of the PCA-ABC-SVR model are compared with those of the PCA- Genetic algorithm (GA)-SVR and PCA-Particle swarm optimization (PSO)-SVR models, as shown in Fig. 22 and Table 15.

According to Fig. 22 and Table 15, it is obvious that the PCA-ABC-SVR model can provide more accurate results for the prediction of pipelines corrosion rate when compared to PCA-GA-SVR and PCA-PSO-SVR models. This indicates that PCA-ABC-SVR model has the better performance in predicting corrosion rate of subsea pipelines.

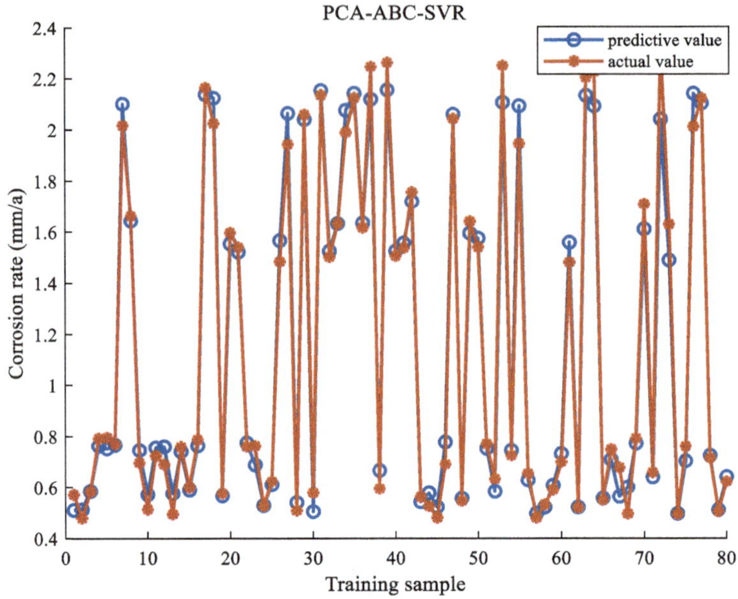

Fig. 21 PCA-ABC-SVR model prediction results (training error).

Table 14 Error of PCA-ABC-SVR model.

	MAE	*RMSE*	R^2
Training error	4.94%	6.91%	0.989
Testing error	7.10%	9.19%	0.976

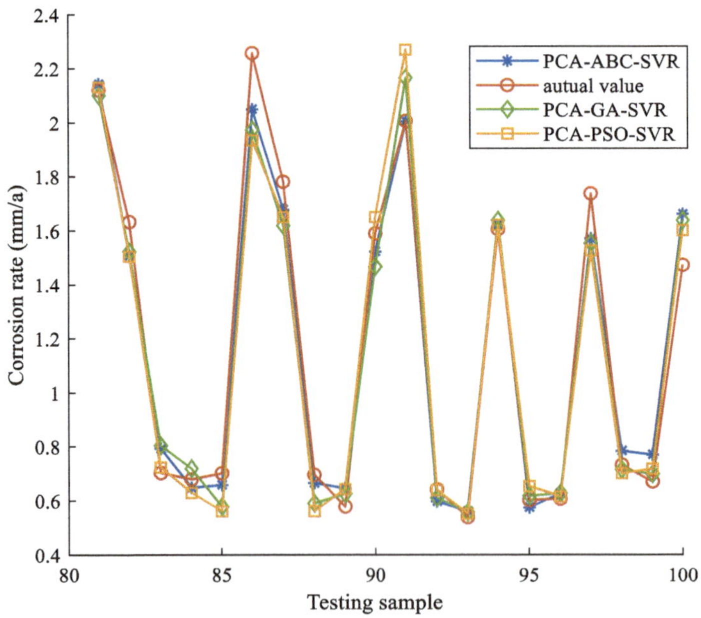

Fig. 22 Comparison of prediction results between models.

Table 15 Error comparison of different models.

	MAE	*RMSE*	R^2
PCA-ABC-SVR	7.10%	9.19%	0.976
PCA-GA-SVR	8.93%	11.51%	0.963
PCA-PSO-SVR	9.17%	12.68%	0.955

5. Conclusions

This chapter illustrates two dynamic risk assessment methods for process operations, including a Bayesian approach and a data-driven approach. The Bayesian approach consists of four steps: accident scenario modeling, information collection, probability estimation of basic events, and dynamic risk assessment using precursor data. The data-driven approach consists of four steps, namely selecting model input and output variables, dividing sample sets for model training and testing, selection and optimization of model parameters, model training and testing. The proposed dynamic risk assessment methods can be an important part of process safety management to support the digitization of process operations.

References

Abimbola, M., Khan, F., & Khakzad, N. (2014). Dynamic safety risk analysis of offshore drilling. *Journal of Loss Prevention in the Process Industries*, *30*, 74–85.

Ale, B. J., Bellamy, L. J., Van der Boom, R., Cooper, J., Cooke, R. M., Goossens, L. H., et al. (2009). Further development of a causal model for air transport safety (CATS): Building the mathematical heart. *Reliability Engineering & System Safety*, *94*(9), 1433–1441.

Arunthavanathan, R., Khan, F., Ahmed, S., & Imtiaz, S. (2021). An analysis of process fault diagnosis methods from safety perspectives. *Computers & Chemical Engineering*, *145*, 107197.

Arzaghi, E., Abbassi, R., Garaniya, V., Binns, J., Chin, C., Khakzad, N., et al. (2018). Developing a dynamic model for pitting and corrosion-fatigue damage of subsea pipelines. *Ocean Engineering*, *150*, 391–396.

Cheng, M. Y., Cao, M. T., & Tran, D. H. (2014). A hybrid fuzzy inference model based on RBFNN and artificial bee colony for predicting the uplift capacity of suction caissons. *Automation in Construction*, *41*, 60–69.

Fazai, R., Mansouri, M., Abodayeh, K., Nounou, H., & Nounou, M. (2019). Online reduced kernel PLS combined with GLRT for fault detection in chemical systems. *Process Safety and Environmental Protection*, *128*, 228–243.

Fei, S. W., & He, Y. (2015). Wind speed prediction using the hybrid model of wavelet decomposition and artificial bee colony algorithm-based relevance vector machine. *International Journal of Electrical Power & Energy Systems*, *73*, 625–631.

Feng, Y., Zhang, P., Yang, M., Li, Q., & Zhang, A. (2019). Short term load forecasting of offshore oil field microgrids based on DA-SVM. *Energy Procedia*, *158*, 2448–2455.

Ferdous, R., Khan, F., Sadiq, R., Amyotte, P., & Veitch, B. (2012). Handling and updating uncertain information in bow-tie analysis. *Journal of Loss Prevention in the Process Industries*, *25*(1), 8–19.

Güner, N., & Çomak, E. (2011). Predicting performance of first year engineering students in calculus by using support vector machines. *Pamukkale University Journal of Engineering Sciences*, *17*(2), 87–96.

Güven, İ., & Şimşir, F. (2020). Demand forecasting with color parameter in retail apparel industry using artificial neural networks (ANN) and support vector machines (SVM) methods. *Computers & Industrial Engineering*, *147*, 106678.

He, R., Li, X., Chen, G., Wang, Y., Jiang, S., & Zhi, C. (2018). A quantitative risk analysis model considering uncertain information. *Process Safety and Environmental Protection*, *118*, 361–370.

He, F., & Zhang, L. (2018). Prediction model of end-point phosphorus content in BOF steelmaking process based on PCA and BP neural network. *Journal of Process Control*, *66*, 51–58.

Kalantarnia, M., Khan, F., & Hawboldt, K. (2009). Dynamic risk assessment using failure assessment and Bayesian theory. *Journal of Loss Prevention in the Process Industries*, *22*(5), 600–606.

Karaboga, D. (2005). *An idea based on honey bee swarm for numerical optimization. Vol. 200* (pp. 1–10). Erciyes University, Engineering Faculty, Computer Engineering Department. Technical report-tr06.

Kelly, D. L., & Smith, C. L. (2009). Bayesian inference in probabilistic risk assessment—The current state of the art. *Reliability Engineering & System Safety*, *94*(2), 628–643.

Khakzad, N., Khakzad, S., & Khan, F. (2014). Probabilistic risk assessment of major accidents: Application to offshore blowouts in the Gulf of Mexico. *Natural Hazards*, *74*(3), 1759–1771.

Khakzad, N., Khan, F., & Amyotte, P. (2013). Quantitative risk analysis of offshore drilling operations: A Bayesian approach. *Safety Science*, *57*, 108–117.

Khan, F., Amyotte, P., & Adedigba, S. (2021). Process safety concerns in process system digitalization. *Education for Chemical Engineers*, *34*, 33–46.

Kjaerulff, U. B., & Madsen, A. L. (2008). *Bayesian networks and influence diagrams. Vol. 200* (p. 114). Springer Science + Business Media.

Lei, Z., Shen, J., Wang, Q., & Chen, Y. (2019). Real-time weld geometry prediction based on multi-information using neural network optimized by PCA and GA during thin-plate laser welding. *Journal of Manufacturing Processes*, *43*, 207–217.

Li, X., Chen, G., Jiang, S., He, R., Xu, C., & Zhu, H. (2018). Developing a dynamic model for risk analysis under uncertainty: Case of third-party damage on subsea pipelines. *Journal of Loss Prevention in the Process Industries*, *54*, 289–302.

Li, X., Chen, G., & Zhu, H. (2016). Quantitative risk analysis on leakage failure of submarine oil and gas pipelines using Bayesian network. *Process Safety and Environmental Protection*, *103*, 163–173.

Li, F., Wang, W., Xu, J., Yi, J., & Wang, Q. (2019). Comparative study on vulnerability assessment for urban buried gas pipeline network based on SVM and ANN methods. *Process Safety and Environmental Protection*, *122*, 23–32.

Li, X., Zhang, Y., Abbassi, R., Yang, M., Zhang, R., & Chen, G. (2021). Dynamic probability assessment of urban natural gas pipeline accidents considering integrated external activities. *Journal of Loss Prevention in the Process Industries*, *69*, 104388.

Nan, C., Khan, F., & Iqbal, M. T. (2008). Real-time fault diagnosis using knowledge-based expert system. *Process Safety and Environmental Protection*, *86*(1), 55–71.

Participants, O. (2002). *OREDA offshore reliability data handbook*. DNV.

Ping, P., Wang, K., Kong, D., & Chen, G. (2018). Estimating probability of success of escape, evacuation, and rescue (EER) on the offshore platform by integrating Bayesian network and fuzzy AHP. *Journal of Loss Prevention in the Process Industries, 54*, 57–68.

Romero, D., Mattsson, S., Fast-Berglund, Å., Wuest, T., Gorecky, D., & Stahre, J. (2018). In *Digitalizing occupational health, safety and productivity for the operator 4.0 IFIP international conference on advances in production management systems* (pp. 473–481). Cham: Springer.

Safavi, H. R., & Esmikhani, M. (2013). Conjunctive use of surface water and groundwater: Application of support vector machines (SVMs) and genetic algorithms. *Water Resources Management, 27*(7), 2623–2644.

Schmitz, P., Swuste, P., Reniers, G., & van Nunen, K. (2020). Mechanical integrity of process installations: Barrier alarm management based on bowties. *Process Safety and Environmental Protection, 138*, 139–147.

Tan, Q. L., Chen, G. M., Chang, Y. J., & Fu, J. M. (2013). Dynamic risk analysis of high-sulfur wellhead gas-liquid separator based on bayesian method. *Journal of China University of Petroleum (Edition of Natural science), 37*(6), 129–134.

Vanhoucke, M., & Vandevoorde, S. (2007). A simulation and evaluation of earned value metrics to forecast the project duration. *Journal of the Operational Research Society, 58*(10), 1361–1374.

Wang, B. (2021). Safety intelligence as an essential perspective for safety management in the era of safety 4.0: From a theoretical to a practical framework. *Process Safety and Environmental Protection, 148*, 189–199.

Wang, Y., Liu, M., Bao, Z., & Zhang, S. (2019). Stacked sparse autoencoder with PCA and SVM for data-based line trip fault diagnosis in power systems. *Neural Computing and Applications, 31*(10), 6719–6731.

Wauters, M., & Vanhoucke, M. (2014). Support vector machine regression for project control forecasting. *Automation in Construction, 47*, 92–106.

Yang, M., Khan, F. I., & Lye, L. (2013). Precursor-based hierarchical Bayesian approach for rare event frequency estimation: A case of oil spill accidents. *Process Safety and Environmental Protection, 91*(5), 333–342.

Yang, M., Khan, F., Lye, L., & Amyotte, P. (2015). Risk assessment of rare events. *Process Safety and Environmental Protection, 98*, 102–108.

Yu, H., Khan, F., & Veitch, B. (2017). A flexible hierarchical Bayesian modeling technique for risk analysis of major accidents. *Risk Analysis, 37*(9), 1668–1682.

Yuan, Z., Khakzad, N., Khan, F., & Amyotte, P. (2015). Risk analysis of dust explosion scenarios using Bayesian networks. *Risk Analysis, 35*(2), 278–291.

Zhang, M., Tan, Y., Zhu, J., Chen, Y., & Liu, H. (2019). Modeling and simulation of improved artificial bee colony algorithm with data-driven optimization. *Simulation Modelling Practice and Theory, 93*, 305–321.

Zhou, H. B. (2009). Risk assessment of deep foundation pit by using fuzzy comprehensive evaluation method based on Bayesian networks. *Journal of Shanghai Jiaotong University, 43*(9), 1473–1479.

CHAPTER TEN

Risk of cascading effects in digitalized process systems

Matteo Iaiani, Alessandro Tugnoli, and Valerio Cozzani*
LISES - Laboratory of Industrial Safety and Environmental Sustainability - Department of Civil, Chemical, Environmental and Materials Engineering - DICAM, University of Bologna, Bologna, Italy
*Corresponding author: e-mail address: valerio.cozzani@unibo.it

Contents

1. Introduction

Chemical and process plants are highly interconnected and interdependent systems, and are vital installations for the society, being part of critical energy and materials supply chains (Landucci & Reniers, 2019). At the same time, they frequently present an inherent hazard due to the processing and storage of hazardous materials (e.g., flammable and toxic materials) that can lead to adverse events with severe consequences on humans, assets, and the environment (Mannan, 2012), i.e., the so-called major events such as energy and hazardous material releases, fires and toxic dispersions. Among these adverse events, cascading events (also defined as "chain of accidents" (Khan & Abbasi, 1998), "knock-on effects" (Reniers & Dullaert, 2008), and "domino effects" (Reniers & Cozzani, 2013)) have received increasing attention in the scientific and technical literature since the 1990s (Bagster & Pitblado, 1991), and a high number

Methods in Chemical Process Safety, Volume 6
ISSN 2468-6514
https://doi.org/10.1016/bs.mcps.2022.04.010

of such events has been recorded in chemical and process plants (Abdolhamidzadeh, Abbasi, Rashtchian, & Abbasi, 2011; Casal & Darbra, 2013; Darbra, Palacios, & Casal, 2010; Delvosalle, 1996; Iaiani, Casson Moreno, Reniers, Tugnoli, & Cozzani, 2021; Kourniotis, Kiranoudis, & Markatos, 2000; Rasmussen, 1996).

Cascading events are phenomena where a primary major event propagates to nearby units causing one or more secondary incidents, leading to an overall consequence more severe than that of the primary event (Cozzani, Gubinelli, Antonioni, Spadoni, & Zanelli, 2005).

Cascading events can be triggered by either unintentional (safety-related) events such as random failure of components, human errors, natural disasters (NaTech), or intentional (security-related) events such as physical or cyber terrorist attacks (Chen, Reniers, & Khakzad, 2020; Iaiani, Casson Moreno, Reniers, et al., 2021, Iaiani, Tugnoli, Bonvicini, & Cozzani, 2021a; Landucci, Argenti, Tugnoli, & Cozzani, 2015; Landucci, Reniers, Cozzani, & Salzano, 2015; Reniers & Cozzani, 2013). In the latter case, cascading events may be deliberately triggered by adversaries intending to generate catastrophic events (i.e., the adversary is specifically aimed at generating a cascading event), or can be indirectly induced due to the interaction between the target damaged installation and the nearby installations (i.e., the adversary is not aimed at specifically generating a cascading event, but the cascading event occurs) (Chen, Reniers, & Khakzad, 2019).

Compared with cascading events caused by safety-related causes, intentional cascading events may generate more severe consequences due to the simultaneous damage of installations induced by multiple target attacks. For example, in 1993 in Warrington (UK), undefined adversaries crossed the site fences overnight and placed and detonated explosives on the roof of a gas-holder causing a fireball. The fragments projected perforated a second gasholder causing a jet-fire (cascading event). No injuries and fatalities followed the incident, but huge economic losses were suffered by the targeted company (eMARS database, 2021).

The operations of the chemical and process industry (CPI), either continuous or batch, are typically interdependent and constitute a connected process system with its own dynamics, which is typically monitored and controlled by Industrial Automation and Control Systems (IACSs) (Stouffer, Pillitteri, Lightman, Abrams, & Hahn, 2015). IACSs present themselves as a hierarchy of subsystems (see Fig. 1), which are commonly divided in the Information Technology (IT) system (corporate network and local network) and the Operational Technology (OT) system (Basic Process Control System

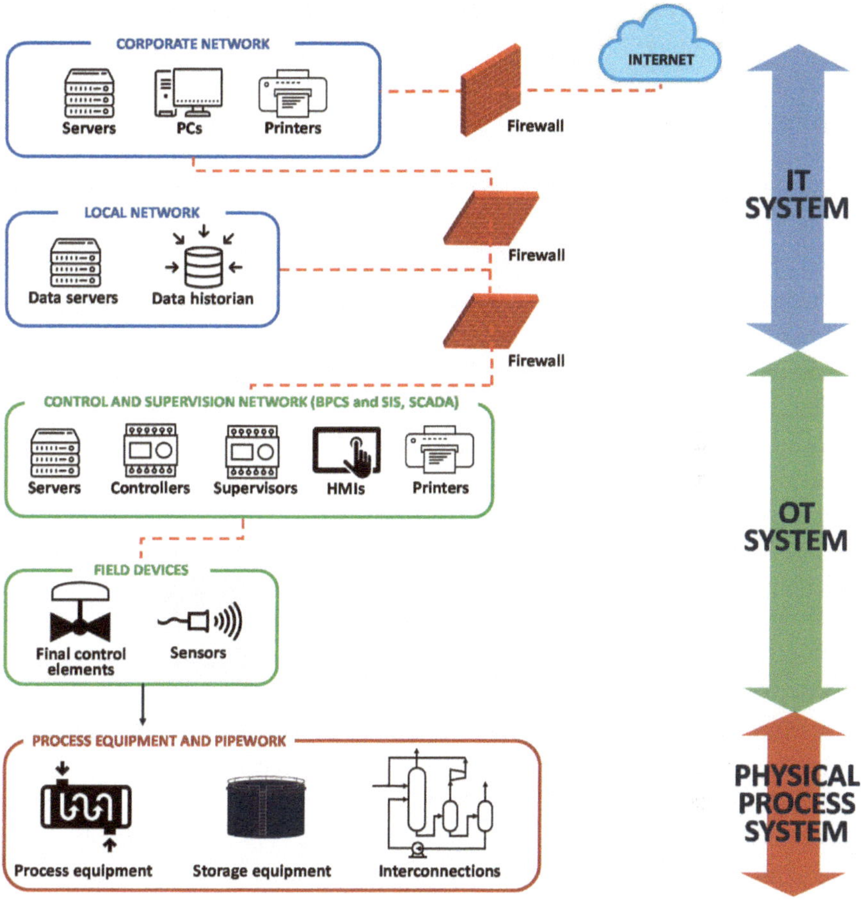

Fig. 1 Typical structure of IACS (IT system and OT system) and physical process system. *Adapted from Iaiani, M., Tugnoli, A., Bonvicini, S., & Cozzani, V. (2021a). Analysis of cybersecurity-related incidents in the process industry.* Reliability Engineering and System Safety, 209, *107485. doi:10.1016/j.ress.2021.107485.*

(BPCS) and Safety Instrumented System (SIS), or Supervisory Control And Data Acquisition (SCADA) system): while the IT system includes all hardware, software and peripheral equipment for storing, retrieving and transmitting data or information (Rennie & Law, 2019), the OT system provides the direct monitoring and control of industrial equipment, assets, processes and events (Stouffer et al., 2015).

The interdependence between a considerable number of subsystems (e.g., corporate network, local network, BPCS, SIS, physical process plant, see Fig. 1), along with the still ongoing digital transition that is taking place in the chemical and process industry toward even higher levels of automation and of interconnection (e.g., digital sensors, improved network protocols, increased interconnection with external networks, etc.) (Lezzi, Lazoi, & Corallo, 2018; Nicholson, Webber, Dyer, Patel, & Janicke, 2012), makes cyber-related cascading events hazards that CPI can no longer disregard (Crucitti, Latora, Marchiori, & Rapisarda, 2004; Srivastava & Gupta, 2010; Wang, Jiang, & Qian, 2014; Wang, Rong, Zhang, & Zhang, 2008). In fact, despite digital transition ensures several advantages thanks to the possibility to analyze the huge amount of data coming from industrial processes (e.g., reduction of total machine downtime due to predictive maintenance and remote monitoring, or increase of the labor productivity due to automation of manual work) (Corallo, Lazoi, & Lezzi, 2020), the adoption of IACSs by CPI has given rise to some criticisms with regard to functional safety issues and cyber-attacks possibility, as they are software-based and may therefore be vulnerable to random failures, cyber threats, and other external circumstances (Cherdantseva et al., 2016; Iaiani, Tugnoli, Bonvicini, & Cozzani, 2021a; Kjaerland, 2006; Kornecki & Zalewski, 2010; Ogie, 2017; Willems, 2011).

Functional safety is the part of the overall safety of plant and equipment that depends on the correct operation of safety-related systems and other risk reduction measures such as BPCS, SIS, SCADA, and alarm system, and thus it mainly concerns the OT system part of the typical structure of a CPI facility (see Fig. 1) (International Electrotechnical Commission, 2010). Functional safety in the CPI is addressed by standard IEC 61511 ("Functional safety - Safety instrumented systems for the process industry sector"), standard BS 6739 ("Code of practice for instrumentation in process control systems: installation design and practice"), standard IEC 62682 ("Management of alarm systems for the process industries"), and HSE Operational Guidance (Health and Safety Executive, 2021).

On the other hand, cybersecurity is the ability to protect or defend the use of cyberspace (IT and OT systems in the case of CPI) from cyber threats, intended as any circumstance or event with the potential of adversely impacting organizational operations, organizational assets, or individuals through an information system via unauthorized access, destruction, disclosure, modification of information, and/or denial of service (Paulsen & Byers, 2019). Due to the particular nature of the CPI, a full cybersecurity analysis requires two complementary parts: the first concerns the IT and OT systems (see Fig. 1) and consists in identifying and managing the vulnerabilities that can be exploited by the adversaries to gain access and starting the infection.

The second concerns the physical process system (see Fig. 1) and consists in identifying and analyzing the possible impacts of a cyber-attack mediated by the control system (e.g., major events with potential escalation of the consequences), and the physical action mechanisms by which adversaries can trigger such adverse events. Cybersecurity issues for IT systems are addressed by the ISO/IEC 27000 family of standards on information security management system, while those of OT systems are addressed by the ISA/IEC 62443 family of standards on cybersecurity of IACSs, and by HSE Operational Guidances (Corallo et al., 2020; Health and Safety Executive, 2021).

Based on the panorama outlined, the following sections are aimed at characterizing cascading events in digitalized process systems, with particular focus on the CPI, pointing out their generic definition (Section 2), their specific causes (Section 3), the prevention and mitigation strategies (Section 4), as well as the main features of both unintentional (Section 5) and intentional (Section 6) cascading events.

2. Definition of cascading event

Cozzani, Gubinelli and Salzano (2006) define a cascading event as "an accident in which a primary event propagates to nearby equipment, triggering one or more secondary events resulting in overall consequences more severe than those of the primary event," and identify three common features in the sequence of events which are shown by any cascading event:

i. a primary event, which initiates the domino sequence;

ii. the propagation of the primary event, due to an "escalation vector" generated by the physical effects of the primary event, that results in the damage of at least one secondary equipment item;

iii. one or more secondary events (e.g., fire, explosion, toxic dispersion, soil contamination), involving the damaged equipment items (the number of secondary events is usually the same of the damaged equipment items).

It is important to remark that the overall severity of the cascading event should be higher than that of the primary event taken alone, thus for a relevant cascading event to take place, an "escalation" of the primary event should occur, triggering one or more secondary events.

Escalation vectors are defined as the physical effects (e.g., heat radiation, fire impingement, overpressure, fragment impact) which can cause damage to nearby equipment items and initiate/maintain a cascading event depending on the type of primary event, the distance between the equipment items, and the type of nearby equipment items (Cozzani et al., 2006).

Escalation vectors should be above some predefined threshold values to determine which equipment items are impacted and take part of the cascading event: threshold values for different escalation vectors and target equipment items are reported in several studies by Cozzani and co-workers (Cozzani, Gubinelli, & Salzano, 2006; Cozzani & Salzano, 2004a, 2004b, 2004c).

Cascading events can be triggered by either unintentional causes (safety-related, e.g., failure of equipment item, failure of IACS, human error, NaTech event), or intentional causes (security-related, e.g., physical attack, cyber-attack, cyber-physical attack). In the latter case, adversaries may intend to specifically generate a cascading event, or it occurs as a result of a physical and/or cyber-attack not aimed at generating it (Chen et al., 2020).

The comparison of the main characteristics between unintentional and intentional cascading events is reported in Table 1. Besides the nature of

Table 1 Comparison of the main characteristics of unintentional and intentional cascading events.

	Unintentional cascading event	Intentional cascading event
Cause of primary event	Unintentional	Intentional
Location of primary event	Usually occurring at plant equipment items (process/ storage)	Usually occurring at plant equipment items (process/ storage) or in any position within the plant boundaries or outside the area nearby (only in case of physical attacks)
Sources of hazards	Hazardous materials in plant equipment items and hazardous material from loading and unloading vehicles	Hazardous materials in plant equipment items and external hazardous material carried by adversaries (only in case of physical attacks)
Main escalation vectors	Heat radiation, fire impingement, overpressure, fragment impact	Heat radiation, fire impingement, overpressure, fragment impact
Simultaneous primary scenarios	Usually involving a single plant equipment item	Multiple plant equipment items can be involved due to multiple target attacks

Adapted from Chen, C., Reniers, G., & Khakzad, N. (2020). Cost-benefit management of intentional domino effects in chemical industrial areas. *Process Safety and Environment Protection 134*, 392–405. doi:10.1016/j.psep.2019.10.007.

the cause (unintentional or intentional), the main difference consists in the fact that intentional cascading events may induce more severe consequences due to simultaneous damage of equipment items induced by multiple target attacks (e.g., as in the incident occurred in Warrington in 1993 described above). Moreover, in case of physical attacks, also the hazardous material carried by the adversaries (e.g., explosive or incendiary material) can be a source of hazard that can trigger a cascading event due to the effects of the blast wave of explosives or the heat load of incendiary substances. However, both unintentional and intentional cascading events show the sequence of events reported above.

In the following, the specific causes of cascading events will be described in detail, with a specific focus on digitalized process systems, reporting also examples from past events occurred worldwide in the last decades.

3. Specific causes of cascading events in digitalized process systems

In this section, the specific causes of cascading events in digitalized process systems such as chemical and process facilities are discussed.

According to the NIST Special Publication 800–82 ("Guide to Industrial Control Systems (ICS) Security") (Stouffer et al., 2015), four different categories of causes are possible, all discussed in the following: adversarial, accidental, structural, and environmental. In order to exemplify the cause categories with real cases, some past events occurred in the chemical and process industry and similar sectors (i.e., chemical, petrochemical, pipelines, and energy production sectors, with the latter including also nuclear power plants) and initiated by actions attributable to each cause category, are briefly described in Table 2. However, it is important to remark that, even if no cascading events have been documented in some of the past events reported in the table, they confirm that the considered causes have the potential to generate primary events which can lead to an escalation of the consequences, and thus to trigger cascading events.

Adversarial causes include individuals, groups, organizations, or states that seek to exploit the organization dependence on cyber resources (e.g., information in electronic form, information and communications technologies, and the communications and information-handling capabilities provided by those technologies). Therefore, this category includes both accidental and intentional (internal or external) cyber-attacks. Accidental cyber-attacks are not directed toward a specific target but infect any

Table 2 Example of past events occurred in the CPI and in similar sectors initiated by adversarial, accidental, structural, and environmental causes according to definitions reported in Stouffer et al. (2015).

Cause category	Specific cause	Year, location—industrial sector	Description of the past event
Adversarial	Intentional cyber-attack from outsiders	2008, Turkey—Pipeline	Adversaries were able to exploit vulnerabilities of the camera communication software in order to access the OT system of the BTC (Baku-Tbilisi-Ceyhan) pipeline and compromise the leak detection system, the automated pressure reliefs, the alarm servers, the input traffic from field devices and the PLCs (Programmable Logic Controllers) operating on the pipeline valves. Adversaries were able to pressurize the pipeline and misreport pressure values back to the control room (man-in-the-middle attack). The attack resulted in an explosion, in the release of more than 30,000 barrels of oil in an area above a water aquifer, in a fire lasting more than 2 days, and in losses for BP and its partners of $5 million a day (Lee, Assante, & Conway, 2014; The Repository of Industrial Security Incidents, 2015)
	Intentional cyber-attack from outsiders	2010, Iran—Energy production	Adversaries infected the BPCS and SIS (OT system) of the Natanz nuclear power plant with Stuxnet worm, starting from accessing the IT system. The worm consisted of a layered attack against three different systems: the Windows operating system, industrial software applications that run on Windows, and the software operating on the PLCs. Stuxnet was also able to misreport process variable values back to the control room (man-in-the-middle attack). The infection started with an infected USB stick. The spinning speed of thousands of centrifuges was fluctuating without the monitors detected any malfunction, causing their damage. The uranium enrichment was stopped for about 1 week, causing huge economic losses (Firoozjaei, Mahmoudyar, Baseri, & Ghorbani, 2022; The Repository of Industrial Security Incidents, 2015)

Table 2 Example of past events occurred in the CPI and in similar sectors initiated by adversarial, accidental, structural, and environmental causes according to definitions reported in Stouffer et al. (2015).—cont'd

Cause category	Specific cause	Year, location—industrial sector	Description of the past event
	Intentional cyber-attack from outsiders	2017, Saudi Arabia—Petrochemical	Adversaries (state-sponsored attack) gained remote access to a SIS engineering workstation of the Saudi Arabian Oil Company (Saudi Aramco) and deployed the Triton malware which was able to insert malicious firmware that changed the logic of the final control elements, potentially leading to a plant disaster. A near miss occurred as some SIS controllers entered a failsafe mode as the hackers attempted to reprogram them, causing the shut down of the related processes, but allowing the plant to spot the attack (Firoozjaei et al., 2022; Johnson et al., 2017)
	Accidental cyber-attack	2004, Chad—Petrochemical	A serious near miss occurred due to malwares accidentally entering and infecting several servers of the OT system. The malware infection caused server and communications failure throughout the system from the wells and manifolds to the Floating Production Offshore Platform (FPSO) without more severe consequences (The Repository of Industrial Security Incidents, 2015)
Accidental	Erroneous action of personnel	2005, Unknown—Petrochemical	A natural gas facility hired an IT security consulting organization to conduct penetration testing on its IT corporate network. The consulting organization carelessly ventured into a part of the network that was directly connected to the SCADA system (OT system). The penetration test locked up the SCADA system causing a serious near miss (Stouffer et al., 2015)

Continued

Table 2 Example of past events occurred in the CPI and in similar sectors initiated by adversarial, accidental, structural, and environmental causes according to definitions reported in Stouffer et al. (2015).—cont'd

Cause category	Specific cause	Year, location—industrial sector	Description of the past event
Accidental and Adversarial	Erroneous action of personnel + accidental cyber-attack	2003, United Kingdom—Petrochemical	A user installed a software in his/her own laptop, unaware that it included an unpatched version of Microsoft SQL. Sometime later, the user connected the laptop to the Internet (in violation of company policy) to access email via an ISP (Internet Service Provider). SQL Slammer worm infected the internet connected machine. The user then brought the infected machine into the office and connected it to the network, causing the outbreak of the malware within the IT corporate network. An unfirewalled data acquisition server and the BPCS became infected causing its shutdown. SQL Slammer was then removed by IT security personnel preventing more severe consequences (The Repository of Industrial Security Incidents, 2015)
Structural	Failure of SCASA system	1999, United States—Pipelines	237,000 gal (900,000 L) of gasoline leaked from the Washington Gasoline Pipeline and ignited 1.5 h later causing 3 deaths, 8 injuries, and extensive property damage. The pipeline failure was exacerbated by control systems not able to perform control and monitoring functions: in fact, immediately before and during the incident, the SCADA system exhibited poor performance that inhibited the pipeline controllers from seeing and reacting to the development of an abnormal pipeline operation (Stouffer et al., 2015)
	Failure of pumping shutdown system	2005, United Kingdom—Petrochemical	The failure of the pumping shutdown system did not prevent the overfilling of unleaded petrol in one of the storage tanks at the Buncefield oil depot that led to a release and the consequent formation of a flammable vapor cloud which dispersed inside the plant and among the surrounding facilities. As soon as the

Table 2 Example of past events occurred in the CPI and in similar sectors initiated by adversarial, accidental, structural, and environmental causes according to definitions reported in Stouffer et al. (2015).—cont'd

Cause category	Specific cause	Year, location—industrial sector	Description of the past event
			vapor cloud came into contact with an ignition source (it is believed to have been in the fire pumphouse, in the generator cabin, or from a car engine), a VCE of unexpected strength was generated. The explosion was followed by a huge fire, which engulfed 20 fuel storage tanks (cascading event). There were also other smaller explosions, independent of the first one (probably, inside tanks affected by fire). (Casal & Darbra, 2013; Paltrinieri, Dechy, Salzano, Wardman, & Cozzani, 2012)
Environmental	Tsunami	2011, Japan—Energy production	The Great East Japan Earthquake on 11 March 2011 struck off the coast of Japan, sending a massive tsunami inland toward the Fukushima Daiichi Nuclear Power Plant. The tsunami compromised the plants seawall, flooding much of the plant including the location housing the emergency generators. This emergency power was critical to operate the control rooms and also to provide coolant water for the reactors. The loss of coolant caused the reactor cores to overheat to the point where the fuel zirconium cladding reacted with water, releasing hydrogen gas and fueling large explosions in three of the four reactor buildings. This resulted in large-scale radiation leakage that has impacted plant employees, nearby citizens, and the local environment. Post event analysis found that the plant emergency response center had insufficient secure communication lines to provide other areas of the plant with information on key safety related instrumentation. (The American Nuclear Society Special Committee on Fukushima, 2012)

vulnerable host (the adversary is generally unknown), while intentional cyber-attacks are carried out against a specific target and are designed to exploit specific weaknesses of the target system (e.g., vulnerabilities in information systems, system procedures, controls, or implementation). In the latter case, the adversary can be an insider (internal cyber-attack), i.e., an individual with authorized access to the assets of the company (e.g., employee, contractor, business partner, vendor, etc.), or an outsider (external cyber-attack). Insiders are generally identified by an investigation, while outsiders generally claim the responsibility of the attacks. Therefore, adversarial causes include intentional malicious manipulations of the OT system (BPCS, SIS, or SCADA, see Fig. 1) or its accidental infection with malwares.

Examples from past events generated by adversarial causes (see Table 2) are the cyber-attack to the control system of the BTC (Baku-Tbilisi-Ceyhan) pipeline (Turkey) occurred in 2008 that generated a major explosion and other severe consequences; the Stuxnet attack to the Natanz nuclear power plant (Iran) in 2010 that was able to vary the spinning speed of thousands of centrifuges causing their damage and inducing the plant shutdown which prevented a plant disaster; the Triton attack that targeted the Saudi Arabian Oil Company (Saudi Arabia) in 2017 causing a serious near miss due to the infection and the malicious manipulation of the SIS (safety logics were reprogrammed); the accidental infection of several servers of the OT system of a Floating Production Offshore Platform (FPSO) in Chad in 2004 which caused server and communication failures without severe consequences (near miss).

Accidental causes consist in erroneous actions taken by individuals while executing their everyday responsibilities such as unaware personnel that do not follow cybersecurity policies. An example of past event initiated by an accidental cause that could have resulted in a serious cascading event is the near miss that took place in 2005 in a petrochemical plant (unknown location) where an IT security consultant carelessly accessed the OT system causing its failure (see Table 2). No severe consequences followed the accident.

The incident that took place in a petrochemical plant in UK in 2003 (see Table 2) is of particular interest, since it is an example of the combination of both adversarial and accidental causes: an employee accidentally infected his/her own laptop with SQL Slammer when installing a software (adversarial cause), then carelessly connected it to the Internet in violation of company policy (accidental cause) causing the malware outbreak and the shutdown of the BPCS. The prompt removal of the malware prevented more severe consequences.

A different category of events is triggered by structural causes, as failures of equipment, of environmental controls, or of software due to aging, resource depletion, or to other circumstances which deviate from expected operating parameters. Examples of past events initiated by structural causes (see Table 2) are the Washington Gasoline Pipeline major release (about 900,000 L of gasoline) in 1999 due to a malfunction of the SCADA system which inhibited the pipeline controllers, and the well-known Buncefield disaster (United Kingdom) in 2005 where the failure of the pumping shutdown system did not prevent the overfilling of a storage tank, initiating a cascading chain of events with catastrophic consequences.

Finally, environmental causes consist in natural disasters (e.g., fire, flood/tsunami, windstorm/tornado, hurricane, earthquake), man-made disasters (e.g., bombing), and failures of critical infrastructures on which the organization depends, but which are outside the control of the organization (e.g., electrical power). A well-known example that received international media coverage initiated by an environmental cause (see Table 2) is the Fukushima disaster, occurred in 2011, where a violent tsunami compromised the emergency power units, affecting the operation of the control rooms and the possibility to provide coolant water for the reactors, causing large-scale radiation leakage with tremendous consequences for plant employees, nearby citizens, and the local environment.

In conclusion, intentional cascading events are caused by intrusion of the OT system and consequent manipulation of the BPCS and/or SIS, or SCADA (adversarial cause), while unintentional cascading events (domino-like conventional cascading events) are caused by failure of OT components due to accidental causes, structural causes, or environmental causes.

4. Layers of protection (LOP) in chemical and process facilities

Process safety barriers are systematically applied in chemical and process facilities where hazardous materials are stored, processed, and transported, together with IT countermeasures, adopted due to the presence of a digitalized network structure such as that shown in Fig. 1. Overall, they constitute the Layers of Protection (LOP) needed to reduce the risk of accident propagation, including the risk of cascading events (Landucci, Necci, Antonioni, Argenti, & Cozzani, 2017; Mannan, 2012). Fig. 2 shows typical LOPs present in chemical and process facilities.

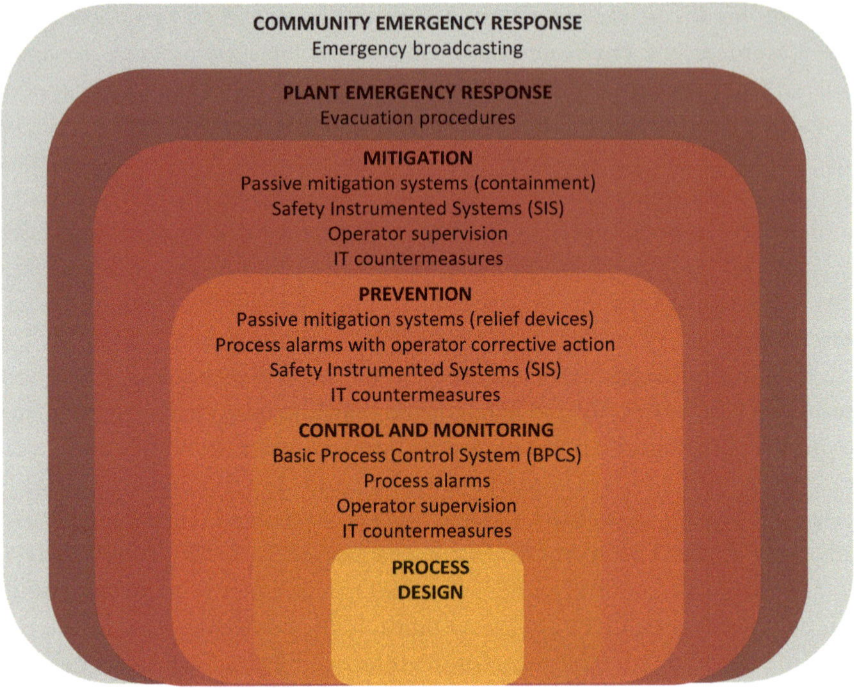

Fig. 2 Typical layers of protection (LOPs) in chemical and process facilities. *Adapted from Mannan, S. (2012).* Lees' loss prevention in the process industries: Hazard identification, assessment and control *(4th ed.). Elsevier, UK: Butterworth-Heinemann.*

Process safety barriers, which constitute the LOPs, may be classified according to CCPS (Center of Chemical Process Safety, 2001) in four categories:

- inherently safer design;
- passive protection systems;
- active protection systems;
- procedural and emergency measures.

The inherent safety approach is based on actions aimed to achieve process safety by a reduction or, possibly, by the elimination of the hazard through the adoption of inherently safer technologies (IST) (Cozzani, Tugnoli, & Salzano, 2009). Examples of possible application of IST are: the adoption of appropriate safety distances between the more hazardous process equipment items during the design phase and layout definition of a chemical and process facility; the reduction of the inventory in single equipment items or the reduction of the number of equipment items; the adoption of materials featuring less hazardous properties (Amyotte, Goraya, Hendershot, & Khan, 2007; Hendershot, 1997). This type of approach is extremely effective (Cozzani, Tugnoli, & Salzano, 2007) but its application is mostly limited to early design steps and suggested IST may conflict with other drivers of the design phase (costs, use of well-known technologies, standardization, etc.) (Landucci et al., 2017).

Passive protection systems are physical elements that do not require external activation to perform the protective action. They can be applied as prevention and/or mitigation barriers and are typically implemented in later detailed design stages. Examples of passive safety barriers are: pressure relief devices aimed at avoiding pressure build-up, such as PSVs (Pressure Safety Valves) and burst disks; heat resistant insulation on process equipment items to reduce the incoming heat flux due to fire; adoption of blast walls or mounds aimed at reducing the peak overpressure and impulse on the protected targets; adoption of catch basins surrounding process and storage equipment items in order to contain the spilled liquid and to avoid liquid spreading over a larger area (Tugnoli, Cozzani, Di Padova, Barbaresi, & Tallone, 2012; Tugnoli, Cozzani, Khan, & Amyotte, 2013).

Active protection systems require external automatic and/or manual activation and are aimed at either preventing or mitigating accident escalation and cascading effect. Examples are the logics which carry out automatic response actions on the process equipment items (Safety Instrumented Functions—SIFs), such as emergency shutdown (ESD) and emergency blowdown (EBD) logics. ESD and EBD systems are typically adopted to prevent cascading effects, reducing the escalation potential of the primary events: The ESD acts isolating the process equipment items, while the EBD acts depressurizing them, venting the hazardous inventory to a blowdown system as a flare (Iaiani, Tugnoli, Macini, & Cozzani, 2021; Landucci et al., 2017).

Finally, procedural and emergency measures may support the management and control of scenarios having an escalation potential by their integration with passive and/or active measures (Landucci et al., 2017; Mannan, 2012). Procedural and emergency measures include monitoring systems informing the operators of malfunctions or anomalies in process equipment items (e.g., alarms and warning lights), remote controls that allow carrying out corrective actions on the process equipment items from the control room or local panels (e.g., hand switches) and any other human-mediated action (Iaiani, Tugnoli, Bonvicini, & Cozzani, 2021b).

IT countermeasures are actions, devices, procedures, techniques, or other measures that reduce the vulnerability of an information system

(Paulsen & Byers, 2019). They can be grouped into seven categories based on their scope (International Society of Automation, International Electrotechnical Commission, 2018):

- identification and authentication control;
- use control;
- system integrity;
- data confidentiality;
- restricted data flow;
- timely response to events;
- resource availability.

Identification and authentication control provides for the identification and authentication of all users (humans, software processes and devices) before allowing them to access specific levels in the network structure shown in Fig. 1.

Use control provides for enforcing the assigned privileges of an authenticated user (human, software process or device) to perform the requested action on the IACS and monitors the use of these privileges.

System integrity provides for ensuring the integrity of the IACS to prevent unauthorized manipulation.

Data confidentiality provides for ensuring the confidentiality of information on communication channels and in data repositories to prevent unauthorized disclosure.

Restricted data flow provides for the segmentation of the network system via zones (grouping of logical or physical assets based upon risk or other criteria, such as criticality of assets, operational function, physical or logical location, required access or responsible organization) and conduits (logical grouping of communication channels that share common requirements connecting two or more zones) to limit the unnecessary flow of data.

Timely response to events provides for a response to security violations by notifying the proper authority, reporting required evidence of the violation and taking timely corrective actions when incidents are discovered.

Resource availability provides for ensuring the availability of the IACS against the degradation or denial of essential services.

5. Unintentional cascading effects in digitalized process systems

Unintentional cascading events in digitalized process systems are those caused by the failure of an OT component (e.g., a BPCS, SIS, or SCADA component) due to erroneous actions of personnel ("accidental cause" as defined in Section 3), to aging, to resource depletion, to other circumstances which exceed expected operating parameters ("structural cause" as defined in Section 3), or to external environmental events ("environmental cause" as defined in Section 3). In particular, the failure of the OT component generates a cause/effect chain that leads to the formation of adverse conditions (e.g., in terms of internal pressure and temperature) on plant equipment items, that cause the occurrence of the primary event, which, in case of consequence escalation, gives rise to an unintentional cascading event. Therefore, the failure of active and procedural safety barriers is an important contributor to unintentional cascading events in digitalized process systems, since it may play an important role in the extent and severity of the escalation of such events.

Each active and procedural safety barrier has a different action and a different availability, whose performance is typically assessed by standard techniques. In particular, the evaluation of process safety barrier performance in the specific framework of escalation prevention is aimed at quantifying barrier availability and barrier effectiveness (Landucci, Argenti, et al., 2015). Availability is defined as the probability of failure on demand (PFD) of a safety barrier: a PFD value of zero (0) means there is no probability of failure (i.e., the barrier is 100% dependable/guaranteed to properly perform when needed), while a PFD value of one (1) means it is completely undependable (i.e., guaranteed to fail when activated). Effectiveness is defined as the probability that the safety barrier, even if successfully activated, will be able to prevent the escalation.

The design and management of active and procedural safety barriers is addressed by functional safety, which is the part of the overall safety of plant and equipment items that depends on the correct operation of safety-related systems and other risk reduction measures such as BPCS, SIS, and SCADA systems. The general benchmark of good practice is standard IEC 61508 ("Functional safety of electrical/electronic/programmable electronic safety related systems"). This standard has been used as the basis for application-specific standards such as IEC 61511 (chemical and process industry), IEC 62061 (machinery), and IEC 61513 (nuclear power plants).

In particular, the standard specific for the chemical and process industry (standard IEC 61511 (International Electrotechnical Commission, 2017)), sets out several engineering and management requirements in order to define the safety lifecycle and the management of functional safety. The key principles of the safety lifecycle are to:

- use hazard and risk assessment to identify requirements and opportunities for risk reduction;

- allocate risk reduction to SIS or to other risk reduction measures (including instrumented systems providing safety functions of low / undefined safety integrity);
- specify the required function, integrity and other requirements of the SIS;
- design and implement the SIS to satisfy the safety requirements specification;
- install, commission and validate the SIS;
- operate, maintain and periodically proof-test the SIS;
- appropriately manage modifications of the SIS;
- decommission the SIS.

Based on the classification of the process safety barriers provided in Section 4, active and procedural safety barriers are automated or human-mediated actions which involve response by the IT and OT systems, while inherent and passive safety barriers provide their safety action independently of such systems: thus, while the success of the former is influenced by both the correct operation of the IT and OT systems and by the response of the operators in the control room (i.e., behavioral factors affecting human error), the latter are not affected by them. Therefore, inherent and passive safety barriers have a key role as well in the protection from unintentional cascading events, as they can prevent the occurrence of the primary event and/or they can mitigate its consequences in case they are able to handle the adverse conditions (e.g., of temperature and pressure) that arise as a result of the unavailability or ineffectiveness of active and procedural safety barriers. These adverse conditions are typically foreseen and considered in the design phase in the case of non-adversarial causes (e.g., see the design basis for pressure-relieving devices of API Recommended Practice 520), while they are not always foreseen in the case of adversarial attacks (e.g., cyber-attacks), since the adversaries can perform complex attack patterns within the IACS network structure (IT system and OT system) and give rise to non-expected adverse conditions (Iaiani, Tugnoli, Bonvicini, & Cozzani, 2021b).

IT countermeasures have a role as well, as they may influence the PFD of active and procedural safety barriers. For example, the network segmentation and restricted data flow, which are part of the more general defense-in-depth strategy, generally provide for the separation between different zones and conduits, e.g., between non-control system networks (IT system) and control system networks (OT system), allowing communication only through properly controlled and configured devices (e.g., firewalls, switches and routers). By this strategy, a potential outage of the corporate network would not affect the operability of the BPCS and the SIS. However, a trade-off is present. Although the division into zones and conduits of the network system may lead to the advantage discussed above, it may affect the operational reliability of the overall system as well. In fact, when interdependent subsystems are required to work together for maintaining regular operation, the introduction of excessive segregation may increase the PFDs.

6. Intentional cascading events in digitalized process systems

Intentional cascading events in digitalized process systems are those caused by both intentional and accidental cyber-attacks as defined in Section 3 (see "adversarial cause") which both imply an intrusion within the IACS network structure (IT and OT system, see Fig. 1).

Section 6.1 is thus aimed at pointing out the mechanisms of cyber-intrusion in IACS network structures (IT system and OT system) considering different types of adversaries (internal, external intentional, external accidental), while Section 6.2 describes the final impacts of the cyber-attacks after an adversary has gained access to the desired level within the IACS network structure and has launched the final hacking. The results of both the sections were obtained by Exploratory Data Analysis (EDA) of 82 cybersecurity-related incidents (CSIs in the following) occurred worldwide in the process industry and similar sectors (chemical, petrochemical, energy production, water/wastewater treatment sectors) in the last 50 years. Detailed information on the approach adopted and on the results obtained are reported elsewhere (Iaiani, Tugnoli, Bonvicini, & Cozzani, 2021a). Finally, Section 6.3 provides the description of two systematic and formally rigorous qualitative methodologies for cyber-risk identification that may be used to investigate case-specific attack patterns in terms of the set of manipulations of the BPCS and SIS that may originate critical events (with potential escalation of the consequences), thus gaining valuable insights on cyber-attacks to typical IACS network structures present in chemical and process plants. Actually, past incident analysis is a well consolidated tool used to support hazard identification in the safety management practice, allowing the recognition of lesson learnt and the definition of reference scenarios to consider as baseline information for authorities and practitioners to analyze more consistently the specific cases of concern. However, past incident analysis is not deemed suitable for cyber-hazard identification, as the mechanism of cyber-attacks strongly depends on the design of the OT system.

Therefore, a dedicated approach is needed for the analysis of the potential impacts that can arise from the malicious manipulation of the OT system, or in other words, to model the actions that the adversaries have to carry out once they have gained access to the desired level within the IACS network structure.

6.1 Mechanisms of cyber-intrusion in IACS network structures

Based on past incident analysis, footprinting, scanning, gaining access, escalating privileges and final hacking are the well-known steps for cyber-attacks to IT systems (Paulsen & Byers, 2019).

Footprinting and scanning allow an adversary to collect as much information as possible about the target system (e.g., reachable IP addresses, authentication mechanisms, network topology, etc.), and to identify the vulnerabilities which can be exploited in order to gain the access. The description of the recorded CSIs generally lacks of information on these steps, as the details are usually unknown and/or undisclosed in public reports. Even in the well-known case of Stuxnet (see Table 2) only hypotheses are available regarding how the network system of the Natanz nuclear power plant was violated (Appelbaum & Potras, 2013; Peter, 2013).

Gaining access to the IT or OT system consists in penetrating the network of the target system through its vulnerable access points. Different type of adversaries can access the network by different techniques. Intentional external adversaries obtain the necessary access information by footprinting and scanning steps. The analysis of eight CSIs with sufficient level of details from the available database reported by Iaiani and co-workers (Iaiani, Tugnoli, Bonvicini and Cozzani, 2021a) evidenced that different hacking techniques were applied in external attacks: brute force password-cracking, phishing, and trojan horses. Four entries of the database showed that gaining remote access can also involve physical actions on the IACS network structure, most commonly by the use of infected USB sticks that are connected to the network. Although this may be possible in specific cases, no recoded CSI features an unauthorized adversary physically accessing the hardware to infect the system or launch the attack (i.e., cyber-attack following a physical security breach).

On the other hand, accidental cyber-attacks penetrate the IT system when they are able to exploit its vulnerabilities (e.g., see the past incident occurred in UK in 2003 reported in Table 2).

As regards intentional internal adversaries, eight CSIs were documented in the database as caused by individuals related to the target organization (e.g., employees, contractors, etc.). These adversaries took advantage of their own credentials and knowledge of the system and were able to access at least some levels of the IACS network structure without the need of footprinting and scanning techniques. For example, a disgruntled employee accessed and disabled on purpose the leak detection system of three oil derricks by his own authentication credentials. As insiders usually have extensive knowledge of the process and of the plant, they are potentially a very critical category of adversaries.

Fig. 3A shows the overall results of the analysis of the recorded CSIs with concern to the type of adversary. Intentional and accidental cyber-attacks are found to be equally credible patterns (respectively 47 and 35 CSIs). The figure confirms that the external cyber-threat against chemical and process facilities is the more frequent reported. Nevertheless, the role of insiders is also evidenced as a possible threat (about 12% of the total CSIs recorded).

Escalation of the privileges consists in obtaining elevated access to resources that are normally protected from an application or user (e.g., admin, root, kernel resources), in order to manipulate more deeply the system (Paulsen & Byers, 2019). In the case of an attack to a system with a IACS network structure such the one shown in Fig. 1, escalation of privileges also includes gaining the access to sections of the network with a lower level. In particular, this step is necessary to infect the OT system after access is granted to the IT system. The network of the OT system is generally a trusted network, separated from the IT system by means of network security systems (e.g., firewalls, see Fig. 1) (Boudriga, 2010). However, six recorded CSIs confirm that a network design lacking of firewalls or with a wrong firewall configuration can result in the escalation of the infection, spreading from IT system to OT system. In case of the Stuxnet attack in 2010 (see Table 2), the escalation of the privileges used 0-day vulnerabilities (i.e., unknown computer-software vulnerabilities), allowing the bypass of network security systems (Falliere, O Murchu, & Chien, 2011).

The classification of the collected past cybersecurity-related incidents with respect to the type of system infected is reported in Fig. 3B. As shown in the figure, CSIs that affected the OT system resulted to be more than those that affected the IT system (46 and 36 respectively). As an attack to the OT system generally implies a former intrusion in the IT system, this result can be a consequence of a higher attention worldwide toward reporting attacks that resulted/may have resulted in physical effects in a

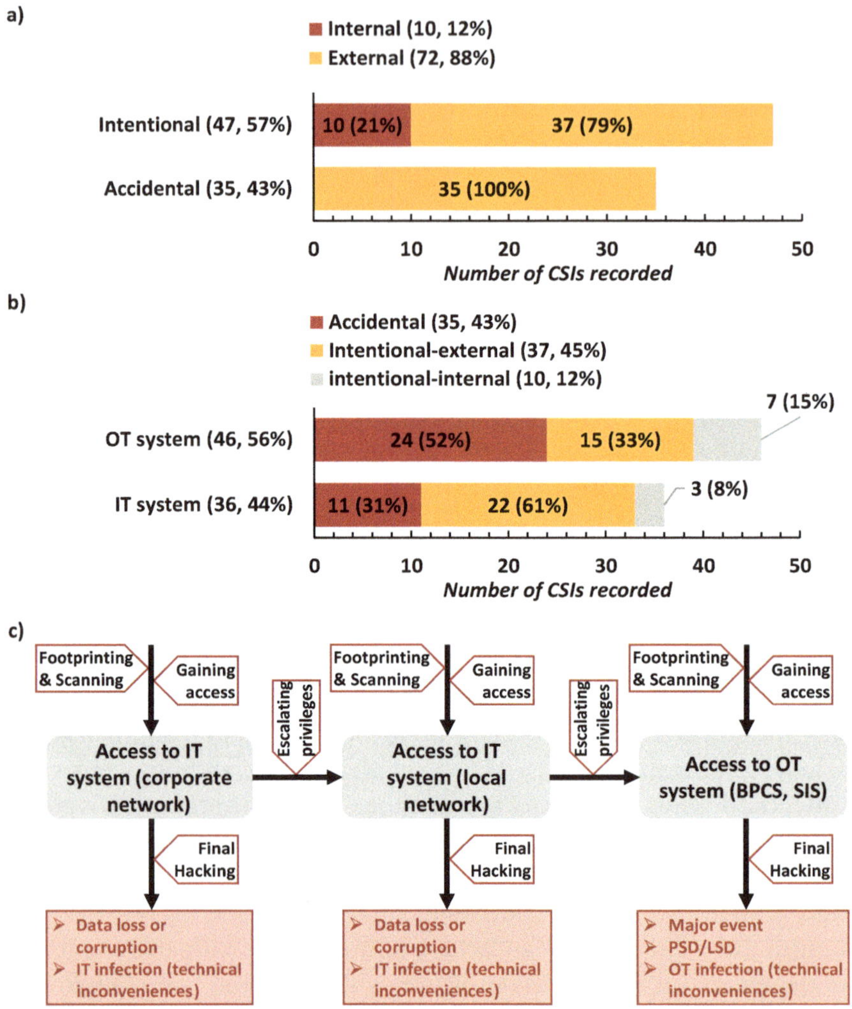

Fig. 3 Distribution of recorded CSIs based on the type of adversary (A) and the system affected (B). General steps of a cyber-attack to the typical IACS network structure present in a chemical and process facility (C).

facility. The CSIs reporting the detail of the OT subsystem affected identified as the ultimate target of the attack the control system (8 CSIs), the monitoring system (12 CSIs), and the automated safety and alarm system (3 CSIs). The targeted systems suggest the possibility for the adversary to tamper the physical remote-controlled devices on the plant by hacking the BPCS or SIS, but also to manipulate the information directed from the plant to the operators in the control room (e.g., values of process variables). In particular, the recorded CSIs evidence cases of information made unavailable (3 CSIs) or misreported (one CSI), potentially paving the way to trigger inappropriate or ineffective human control and correction actions. With respect to the possibility to cause automated safety devices to be ineffective, the recorded CSIs concerned deactivation of the leak detection system (3 CSIs), of the automated pressure relief valves (one CSI) and of the alarms (one CSI).

Fig. 3B also shows that an appreciable number of events classified as "accidental cyber-attacks" (i.e., 24 CSIs, 29% of the total incidents recorded), have indeed affected the OT system: therefore, the infection of the BPCS, SIS, or SCADA is proved to be possible also in this case. A wrong configuration of the IT and OT systems architecture, especially due to a low attention to security/cybersecurity issues, was tracked as cause of these incidents.

The final hacking is the last part of the cyber-attack, through which the impacts on the assets of the affected facility are originated. Fig. 3C shows the connection between the level of the IACS network structure infected and the impacts in selected entries of the database for which detailed description was available. No recorded CSI with infection of the IT system alone resulted in impacts on the process system, confirming that gaining access to the OT system is a distinctive feature required to achieve impacts on the physical plant in chemical and process facilities. The analysis of the 5 CSIs for which more detailed information about the final hacking was available revealed some attack techniques used by the adversaries in order to originate impacts on the assets of the target company: Denial of Service (adversary made intermittent the traffic between HMIs, PLCs and SCADA systems of a petrochemical facility), Man-in-the-Middle (adversary intercepted and modified signals between filed sensors and control rooms in order to carry out manipulation of process parameters), and data encryption (adversary launched a ransomware attack asking for payment for the release of the decryption key).

6.2 Impacts of cyber-attacks

Iaiani and co-workers (Iaiani, Tugnoli, Bonvicini and Cozzani, 2021a) carried out an analysis of past cybersecurity-related incidents considering the categories of impact defined in Table 3 (i.e., major event, PSD/LSD, OT infection, data loss or corruption, IT infection).

Table 3 Definitions of the impact categories considered (Iaiani, Tugnoli, Bonvicini, & Cozzani, 2021a).

Impact category	Definition
Major event	Hazardous material and/or energy release, explosion, fire, toxic dispersion, soil contamination, etc.
PSD/LSD	PSD (Process Shut Down) or LSD (Local Shut Down) originated either directly or by inducing an anomalous condition or abnormal mode of operation
OT infection	Infection of the OT system (e.g., infection of HIM workstations, OT servers, etc.)
Data loss or corruption	Theft and/or corruption of sensitive information regarding the company knowhow, employee data, process data (e.g., equipment data sheets, PFDs, P&IDs, historical data, etc.), economic data, etc.
IT infection	Infection of the IT system (e.g., infection of IT servers, PCs, etc.)

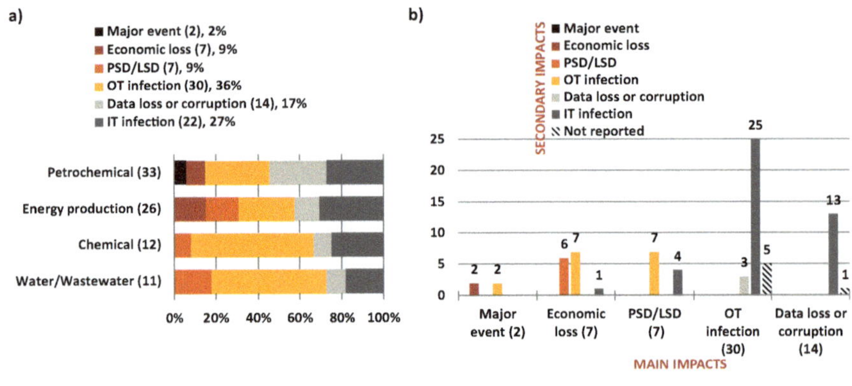

Fig. 4 (A) Distribution of the main impact categories of the CSIs recorded with respect to the industrial sectors considered in the study; (B) Secondary impacts recorded for the CSIs belonging to the same main impact category. *Adapted from Iaiani, M., Tugnoli, A., Bonvicini, S., & Cozzani, V. (2021a). Analysis of cybersecurity-related incidents in the process industry.* Reliability Engineering and System Safety, 209, *107485. https://doi.org/10.1016/j.ress.2021.107485.*

Fig. 4A reports the distribution of the "main impact" categories with respect to the industrial sectors considered (i.e., chemical, petrochemical, energy production, water/wastewater sectors). The "main impact" category is the one with the highest criticality (generally more than one impact category has been registered for each CSI).

The figure evidences that the number of recorded CSIs generally increases progressing from the "main impact" categories toward those having a lower criticality. In fact, the potentially most severe main impact (i.e., major event) is also the least frequent (only 2 CSIs recorded in the transportation of hydrocarbons by pipeline). The highest number of occurrences (36 CSIs) is reported for impacts requiring the infection of the IT system only: this can be explained by the presence of a higher number of safety/security barriers between the adversary and the target system when access to OT system is required.

Moreover, while the IT system of chemical and process facilities is generally similar to that of other business sectors, the OT system has specific design solutions: this means that an adversary aiming at affecting the OT system has to perform deeper and more difficult scanning and escalating phases rather than an adversary aiming at affecting the IT system alone.

With reference to Fig. 4A it can be observed that the distribution of the "main impact" categories with respect to the industrial sectors is roughly similar to the distribution of the total number of CSIs recorded for the sectors considered, with few exceptions. For instance, in the water/wastewater treatment sector, less major events and economic losses were experiences than in other sectors, due to the lower quantities of hazardous materials handled and to the lower revenues of the activity.

Fig. 4B reports the secondary impacts recorded for CSIs belonging to the same "main impact" category. No fatalities were registered in the two major events triggered by the malicious manipulation of the OT system. However, huge economic losses (due to production outage and cost for equipment repair) and environmental damages were reported in these incidents. In fact, in Fig. 4B, the categories "economic loss" and "OT infection" were associated with both the major events. Large economic losses were reported as the main impact in a significant number of cases. It should be noted that most of the incident reports provide only a verbal description of the economic losses: for example, expressions as "huge economic loss," "significant financial impact," "loss of production," "loss of revenue" are frequent. With reference to Fig. 4B it can be observed that this impact category is frequently associated with a local or process shutdown (PSD/LSD), and obviously with an infection of the OT system. In seven recorded CSIs the adversaries induced a local or process shutdown via remote manipulation: the totality of these events is associated with the infection of devices connected to the OT system.

Several cases of infection of the OT system (30 CSIs) did not result in major events, economic losses or PSDs/LSDs. The analysis of the descriptions of such incidents evidenced the following events: infection of the SCADA system (23 CSIs), short term loss of plant control (3 CSIs), malfunction of the detection system of dangerous substances (3 CSIs), remote manipulation of plant devices (1 CSI). Fig. 4B shows that for 3 CSIs a data theft or corruption was also reported, while for the majority of these incidents (25 CSIs) the infection of the IT system was registered. For 5 CSIs the secondary impacts were not reported.

Most of the CSIs, however, only affected the IT system (22 CSIs, see Fig. 4A). Data theft or corruption was recorded in 14 CSIs. These incidents featured sensitive data theft or data corruption. The majority of such attacks resulted also in technical inconveniences at IT level (e.g., server crashes).

In case of IT infection as the "main impact" category, an occurrence of technical inconveniences at IT level resulted to be the more frequent event recorded. Since it is the least severe impact class considered (lowest criticality), no secondary impacts were associated to these CSIs (therefore they are not reported in Fig. 4B).

6.3 Identification of potential scenarios deriving from cyber-attacks

Systematic and formally rigorous qualitative approaches to cyber-risk identification were developed in recent studies, the PHAROS and the POROS methodologies (Iaiani, Tugnoli, Bonvicini, & Cozzani, 2021b; Iaiani, Tugnoli, Macini, et al., 2021), to support the identification of scenarios deriving from cyber-attacks. These methodologies fill the gap previously present in the assessment of the actual link between malicious manipulations of the BPCS and SIS and the potential impacts that can be triggered on the physical plant, including cascading events.

6.3.1 The PHAROS and POROS methodologies

The PHAROS (Process Hazard Analysis of Remote manipulations through the cOntrol System) and the POROS (Process Operability analysis of Remote manipulations through the cOntrol System) are both aimed at supporting the identification of the specific set of manipulations of the BPCS and the SIS which may lead to adverse scenarios which can potentially trigger cascading events. However, PHAROS addresses major accident scenarios, while POROS addresses specifically operability issues. Overall, the information coming from PHAROS and POROS application support step

5.3 (determine consequences and impact), step 5.4 (determine unmitigated likelihood), step 5.8 (identify and evaluate existing countermeasures), and step 5.12 (identify additional cybersecurity countermeasures) of the detailed cybersecurity risk assessment proposed by ISA/IEC 62443, as well as the hazards and countermeasures identification steps of SVA/SRA methodologies.

PHAROS and POROS methodologies are based on a reverse-HazOp concept and require the application of 9 steps (see Fig. 5A) to be carried out by a team with knowledge of the process plant system, the logics of the BPCS and SIS, and the loss prevention system. No specific IT-OT skills are required in the team. The input information needed for the application of the methodologies is: the PFD (Process Flow Diagram) and the material balances, the P&ID (Piping and Instrumentation Diagram), the list of substances stored or handled and their hazardous properties, the operating conditions of each process unit, the logics of the BPCS and the SIS, and the data sheets of each process unit. The procedure is intended for application in the front-end design phase as well as in the security review of operating systems.

Fig. 5B shows the conceptual model of an attack as considered by the two methodologies. In particular, an adversary is intended to generate a Security Event (SE, e.g., release) through specific Mechanism of Actions (MAs, e.g., give rise to internal overpressure). In order to do this, he/she has to carry out Remote Manipulations (RMs, e.g., setpoint change) of the Manipulative Elements (MEs, e.g. controllers and their logics) so that Local Consequences (LCs, e.g. closing/opening) on the Remotely Manipulable Components (RMCs, e.g. automatic valves, pumps, compressors) of the physical plant take place, and to overcome the Inherent/Passive safeguards (IPSs, e.g. PSVs) and the Active/Procedural safeguards (APSs, e.g. PSD activation logics) which are present. Both PHAROS and POROS methodologies address the systematic identification of the specific elements described in the conceptual model of the attack.

In the following, the workflow of the methodologies is briefly described.

In step 1 (S1) the input information is collected.

In step 2 (S2) the plant is divided into nodes. In case of POROS, all the nodes are considered in the analysis (including the utilities), while only those in which hazardous substances are processed or stored are of concern in the PHAROS assessment (critical nodes, step S2.1). The division in process nodes may be carried out by a procedure similar to that applied in the HazOp study, as described by the IEC 61882 standard (International Electrotechnical Commission, 2016).

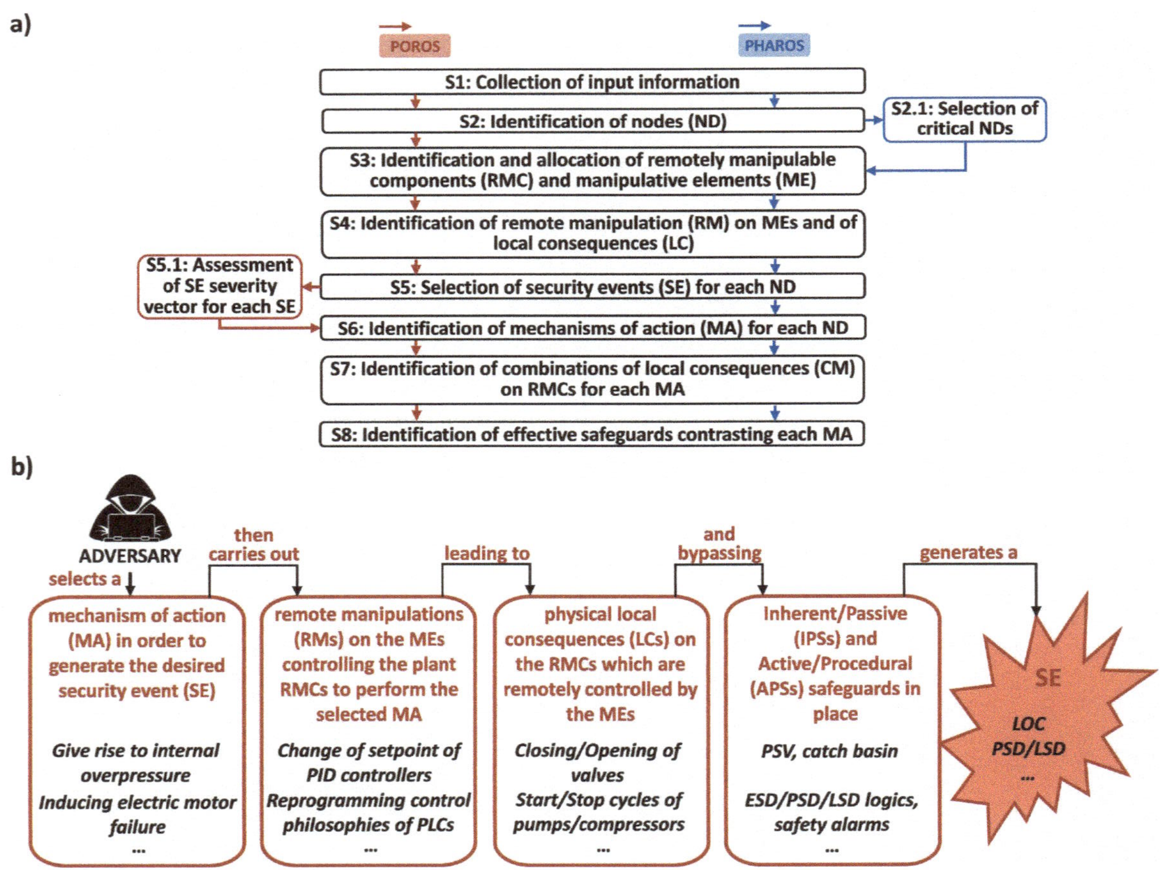

Fig. 5 (A) Flowchart of POROS (red arrows) and PHAROS (blue arrows) methodologies; (B) Core mechanism of POROS and PHAROS methodologies. *Adapted from Iaiani, M., Tugnoli, A., Bonvicini, S., & Cozzani, V. (2021b). Major accidents triggered by malicious manipulations of the control system in process facilities.* Safety Science, 134, *105043. doi:10.1016/j.ssci.2020.105043; Iaiani, M., Tugnoli, A., Macini, P., & Cozzani, V. (2021). Outage and asset damage triggered by malicious manipulation of the control system in process plants.* Reliability Engineering and System Safety, 213, *107685. https://doi.org/10.1016/j.ress.2021.*

Step 3 (S3) consists in the identification, based on a review of process documentation, of the RMCs, of the corresponding MEs, and in the allocation of RMCs to the selected nodes. The allocation of each RMC to one or more nodes is carried out based on the influence it can have on that node (ND). The following guidelines can support the allocation of RMCs: (i) if an RMC is located on a stream that directly connects process units belonging to two different nodes, it is allocated to both nodes; (ii) if an RMC is located on a stream internal to a node, it is allocated only to that node.

Step 4 (S4) consists in identifying, for each ME, all the possible RMs that can be carried out through an attack (e.g., the set-point change and the signal shut-down for a BPCS controller, or the function reprogramming for a SIS controller such as a PLC). Next, for each ME, the LCs on the corresponding RMC are identified (e.g., the increase/decrease in the opening degree of the controlled valve, the increase/decrease in the rotational speed of the controlled operating machine).

Step 5 (S5) consists in associating to each selected node the compatible SEs. In the POROS methodology, a SE is intended as any event which affect the operability and/or system integrity of the system analyzed (e.g., arrest/blockage of a piece of a equipment item, or product out of specification), while in the PHAROS methodology, the SEs are only hazardous material and/or energy releases (i.e., major events). Given the high number of SEs identified with POROS, a cut-off criterion based on severity vectors is suggested in order to focus the efforts on the improvement of prevention and mitigation measures on the worst-case SEs (step S5.1). Several well-known techniques can be used to identify SEs: HazOp analysis (International Electrotechnical Commission, 2016), what-if-analysis (Mannan, 2012), failure modes and effect analysis (FMEA) (International Electrotechnical Commission, 2018), MIMAH (Delvosalle, Fiévez, Pipart, & Debray, 2006), HazId (Mannan, 2012), DyPASI (Paltrinieri, Tugnoli, Buston, Wardman, & Cozzani, 2013), etc.

Step 6 (S6) consists in identifying all the MAs by means of which each SE can be initiated through an attack to the BPCS and SIS. Some MAs may require actions from a nearby node to occur (i.e., manipulation of RMCs belonging to a nearby node is required). In these cases, the information is propagated from a node to another, similarly to deviations propagating among different nodes in a traditional HazOp study.

Step 7 (S7) consists in identifying the combinations (CMs) of LCs from the allocated RMCs by means of which each MA in a node can be carried out. In general, not all the RMCs present in a node need to be manipulated in order to carry out a MA: in order to limit the number of CMs to the essential ones, a CM should be selected on the basis of the minimum set of manipulations required.

Step 8 (S8) consists in identifying, for each CM, the effective safeguards (i.e., process safety barriers) which are present in the node being analyzed (both IPSs and APSs). "Effective" means that the safeguard is able to contrast, directly or indirectly, the MA associated to a particular CM, avoiding the occurrence of the corresponding SEs. Effectiveness shall be checked case by case, as it also depends on design specifications of the barrier (e.g., matching or not the requirements from physical scenarios deriving from the attack).

Similarly to an HazOp study, once all the steps have been performed, a completeness check is required. It consists in verifying that all the nodes have been analyzed, and that all the MAs, by which the associated SE can be originated were developed as combinations of LCs on the relevant RMCs. Moreover, for the SEs (and corresponding MAs) that require actions from a nearby ND to occur, it must be verified that the additional SEs have been fully developed in the relevant nodes.

Examples of step-by-step application of PHAROS and POROS methodologies to onshore upstream Oil&Gas plants and offshore Oil&Gas platforms are reported by Iaiani, Tugnoli, Bonvicini, and Cozzani (2021b), Iaiani, Tugnoli, Landucci, and Cozzani (2020) and Iaiani, Tugnoli, Macini, et al. (2021).

6.3.2 Key findings obtained by the application of PHAROS and POROS methodologies

The application of PHAROS and POROS methodologies to representative case studies (an upstream Oil&Gas plant for the preliminary treatment of the crude oil and an offshore Oil&Gas platform for gas compression) highlighted that, when redundant active and procedural safety barriers are installed (see Section 4), as required by the conventional safety management of systems processing relevant amounts of hazardous materials, a successful cyber-attack has to infect both the BPCS and the SIS in order to give rise to security events such as major events and outage of important systems. Actually, the adversaries must not only tamper with the physical components of the plant (e.g., automated valves, pumps, compressors, etc.), but they must also bypass the active and procedural safety barriers in place. Moreover, it was evidenced that while the infection of a single system (e.g., the BPCS) can be sufficient to impart all the necessary manipulations originating security events, active and procedural safety barriers are mediated by both systems. For this reason, the separation (e.g., air gap) of the BPCS and the SIS in the IACS network architecture results to be an effective measure against both accidental and intentional cyber-attacks.

More in general, the manipulation of elements belonging to different zones may be required to cause a security event, making the attack more complex and, possibly, decreasing its likelihood. Thus, the division of the IACS network structure in smaller zones (network segmentation) can be identified as a good strategy to reduce the cyber-risk. Although the division into zones and conduits of the network system suggested by ISA/IEC 62443 leads to the advantages discussed above, it may affect the operational reliability of the overall system (similarly to what discussed in Section 5 for unintentional cascading events). In fact, when interdependent subsystems are required to work together for maintaining regular operation, the introduction of segmentation may increase the PFDs, reducing the system availability. This can be critical in case of the SIS system since it executes safety functions, where required performances of reliability shall be met (e.g., according to IEC 61511). Thus, a cost-benefit analysis of the separation of the network system into smaller zones and conduits, considering both the advantages in terms of system safety/security and the disadvantages in terms of system operational reliability is of crucial importance.

Passive and inherent safety barriers were found to have a key role in protecting the system from malicious manipulations of the OT system: since they are not electronic devices, they may not be remotely affected by the adversaries, representing an effective barrier whenever they are able to contrast the adverse conditions (e.g., in terms of internal pressure and temperature) that can arise as a result of the malicious manipulation of the BPCS and SIS. An appropriate sizing of such devices, considering the security cases and the critical events that can be identified through the application of PHAROS and POROS methodologies is therefore important to guarantee the protection of the process system from cascading events and other severe impacts.

7. Conclusions

A relevant number of cascading events with severe consequences on humans, assets, and the environment has been registered over the years in the chemical and process industry (CPI) due to the high degree of digitalization (adoption of IACSs - Industrial Automation and Control Systems) together with the presence of large amounts of hazardous materials processed and stored, making the assessment and management of such events an issue that the CPI can no longer disregard.

The specific causes of cascading events in digitalized process systems have been discussed and exemplified through past incidents. Intentional cascading events are caused by the intrusion of the OT (Operational Technology) system and the consequent manipulation of the BPCS (Basic Process Control System) and/or SIS (Safety Instrumented System) systems (adversarial cause), while unintentional cascading events (domino-like conventional cascading events) are caused by the failure of OT components, whether due to accidental causes (e.g., erroneous actions by personnel), structural causes (e.g., aging of components), or environmental causes (e.g., NaTech event).

Active and procedural safety barriers (e.g., emergency shutdown logics and alarm systems), as well as passive and inherent safety barriers (e.g., pressure relieving devices) turn out to have a different and, at the same time, important role in the prevention and mitigation of intentional and unintentional cascading events.

The analysis of 82 past cybersecurity-related incidents allowed gaining relevant insights on intentional cascading events with respect to the mechanisms of cyber intrusion in a typical IACS network structure (IT system and OT system) adopted by the CPI, on the type of adversary behind the cyber-attacks, and on the impacts that can be generated through them. Case-specific patterns of attacks (set of manipulations of the BPCS and SIS) that can cause a security event (including a cascading event) can be identified by the application of dedicated cyber-risk identification methods, such as the PHAROS and the POROS methodologies. The application of such methodologies to representative test cases allowed drawing indications on effective measures to reduce the risk of cascading events in digitalized process systems. In particular, the separation (e.g., air gap) of the BPCS and the SIS in the IACS network architecture resulted effective in preventing both accidental and intentional cyber-attacks. Passive and inherent safety barriers were found to have a key role in protecting the system from malicious manipulations of the OT system aimed at triggering cascading events.

In perspective, further work is needed to build on such results and to approach the specific design of robust safety barriers to prevent cascading events in digitalized process systems of chemical and process plants.

References

Abdolhamidzadeh, B., Abbasi, T., Rashtchian, D., & Abbasi, S. A. (2011). Domino effect in process-industry accidents—An inventory of past events and identification of some patterns. *Journal of Loss Prevention in the Process Industries*, *24*, 575–593. https://doi.org/10.1016/J.JLP.2010.06.013.

Amyotte, P. R., Goraya, A. U., Hendershot, D. C., & Khan, F. I. (2007). Incorporation of inherent safety principles in process safety management. *Process Safety Progress, 26*, 333–346. https://doi.org/10.1002/PRS.10217.

Appelbaum, J., & Potras, L. (2013). *The NSA and its willing helpers*. Edward Snowden Interview.

Bagster, D. F., & Pitblado, R. M. (1991). Estimation of domino incident frequencies—An approach. *Process Safety and Environment Protection, 69*, 195–199.

Boudriga, N. (2010). *Security of mobile communications*. Boca Raton: CRC Press.

eMARS database. (2021). (WWW Document) https://emars.jrc.ec.europa.eu/en/emars/content (accessed 12.23.20).

International Electrotechnical Commission (IEC). (2018). *IEC 60812: Failure modes and effects analysis (FMEA and FMECA)*.

International Electrotechnical Commission (IEC). (2017). *IEC 61511: Functional safety—Safety instrumented systems for the process industry sector*.

International Electrotechnical Commission (IEC). (2016). *IEC 61882: Hazard and operability studies (HAZOP studies)—Application guide*.

International Electrotechnical Commission (IEC). (2010). *IEC 61508: Functional safety of electrical/electronic/programmable electronic safety-related systems*.

International Society of Automation (ISA), International Electrotechnical Commission (IEC). (2018). *ISA/IEC 62443 series of standards: Industrial Automation and Control Systems Security*.

The *Repository of Industrial Security Incidents* (RISI) [WWW Document], (2015). URL https://www.risidata.com/Database (accessed 12.8.20).

Casal, J., & Darbra, R. M. (2013). 2. *Analysis of past accidents and relevant case-histories*. In *Domino effects in the process industries: Modelling, prevention and managing* (pp. 12–29). Elsevier. https://doi.org/10.1016/B978-0-444-54323-3.00002-6.

Center of Chemical Process Safety (CCPS). (2001). *Layer of protection analysis: Simplified process risk assessment*. New York, NY: American Institute of Chemical Engineers—Center of Chemical Process Safety.

Chen, C., Reniers, G., & Khakzad, N. (2019). Integrating safety and security resources to protect chemical industrial parks from man-made domino effects: A dynamic graph approach. *Reliability Engineering and System Safety, 191*, 106470. https://doi.org/10.1016/j.ress.2019.04.023.

Chen, C., Reniers, G., & Khakzad, N. (2020). Cost-benefit management of intentional domino effects in chemical industrial areas. *Process Safety and Environment Protection, 134*, 392–405. https://doi.org/10.1016/j.psep.2019.10.007.

Cherdantseva, Y., Burnap, P., Blyth, A., Eden, P., Jones, K., Soulsby, H., et al. (2016). A review of cyber security risk assessment methods for SCADA systems. *Computers & Security, 56*, 1–27. https://doi.org/10.1016/j.cose.2015.09.009.

Corallo, A., Lazoi, M., & Lezzi, M. (2020). Cybersecurity in the context of industry 4.0: A structured classification of critical assets and business impacts. *Computers in Industry, 114*, 103165. https://doi.org/10.1016/J.COMPIND.2019.103165.

Cozzani, V., Gubinelli, G., Antonioni, G., Spadoni, G., & Zanelli, S. (2005). The assessment of risk caused by domino effect in quantitative area risk analysis. *Journal of Hazardous Materials, 127*, 14–30. https://doi.org/10.1016/j.jhazmat.2005.07.003.

Cozzani, V., Gubinelli, G., & Salzano, E. (2006). Escalation thresholds in the assessment of domino accidental events. *Journal of Hazardous Materials, 129*, 1–21. https://doi.org/10.1016/j.jhazmat.2005.08.012.

Cozzani, V., & Salzano, E. (2004a). The quantitative assessment of domino effect caused by overpressure: Part II. Case studies. *Journal of Hazardous Materials, 107*, 81–94. https://doi.org/10.1016/j.jhazmat.2003.09.014.

Cozzani, V., & Salzano, E. (2004b). Threshold values for domino effects caused by blast wave interaction with process equipment. *Journal of Loss Prevention in the Process Industries*, *17*, 437–447. https://doi.org/10.1016/j.jlp.2004.08.003.

Cozzani, V., & Salzano, E. (2004c). The quantitative assessment of domino effects caused by overpressure: Part I. Probit models. *Journal of Hazardous Materials*, *107*, 67–80. https://doi.org/10.1016/J.JHAZMAT.2003.09.013.

Cozzani, V., Tugnoli, A., & Salzano, E. (2007). Prevention of domino effect: From active and passive strategies to inherently safer design. *Journal of Hazardous Materials*, *139*, 209–219. https://doi.org/10.1016/J.JHAZMAT.2006.06.041.

Cozzani, V., Tugnoli, A., & Salzano, E. (2009). The development of an inherent safety approach to the prevention of domino accidents. *Accident; Analysis and Prevention*, *41*, 1216–1227. https://doi.org/10.1016/j.aap.2008.06.002.

Crucitti, P., Latora, V., Marchiori, M., & Rapisarda, A. (2004). Error and attack tolerance of complex networks. *Physica A*, *340*, 388–394. https://doi.org/10.1016/j.physa.2004.04.031.

Darbra, R. M., Palacios, A., & Casal, J. (2010). Domino effect in chemical accidents: Main features and accident sequences. *Journal of Hazardous Materials*, *183*, 565–573. https://doi.org/10.1016/J.JHAZMAT.2010.07.061.

Delvosalle, C., (1996). Domino effect phenomena: Definition, overview and classification, in: European Seminar on Domino Effects. Leuven (B), pp. 5–10.

Delvosalle, C., Fiévez, C., Pipart, A., & Debray, B. (2006). ARAMIS project: A comprehensive methodology for the identification of reference accident scenarios in process industries. *Journal of Hazardous Materials*, *130*, 200–219. https://doi.org/10.1016/j.jhazmat.2005.07.005.

Falliere, N., O Murchu, L., & Chien, E. (2011). *W32.Stuxnet Dossier*.

Firoozjaei, M. D., Mahmoudyar, N., Baseri, Y., & Ghorbani, A. A. (2022). An evaluation framework for industrial control system cyber incidents. *International Journal of Critical Infrastructure Protection*, *36*, 100487. https://doi.org/10.1016/J.IJCIP.2021.100487.

Health and Safety Executive (HSE). (2021). *Electrical, control and instrumentation (E, C&I)—HSE online form [WWW Document]*. URL https://www.hse.gov.uk/eci/ (accessed 1.26.22).

Hendershot, D. C. (1997). Inherently safer chemical process design. *Journal of Loss Prevention in the Process Industries*, *10*, 151–157. https://doi.org/10.1016/S0950-4230(96)00055-1.

Iaiani, M., Casson Moreno, V., Reniers, G., Tugnoli, A., & Cozzani, V. (2021). Analysis of events involving the intentional release of hazardous substances from industrial facilities. *Reliability Engineering and System Safety*, *212*, 107593. https://doi.org/10.1016/J.RESS.2021.107593.

Iaiani, M., Tugnoli, A., Bonvicini, S., & Cozzani, V. (2021a). Analysis of cybersecurity-related incidents in the process industry. *Reliability Engineering and System Safety*, *209*, 107485. https://doi.org/10.1016/j.ress.2021.107485.

Iaiani, M., Tugnoli, A., Bonvicini, S., & Cozzani, V. (2021b). Major accidents triggered by malicious manipulations of the control system in process facilities. *Safety Science*, *134*, 105043. https://doi.org/10.1016/j.ssci.2020.105043.

Iaiani, M., Tugnoli, A., Landucci, G., & Cozzani, V. (2020). A systematic methodology for the identification of major accidents induced by malicious manipulation of the BPCS and the SIS in a process plant. *Chemical Engineering Transaction*, *82*, 319–324. https://doi.org/10.3303/CET2082054.

Iaiani, M., Tugnoli, A., Macini, P., & Cozzani, V. (2021). Outage and asset damage triggered by malicious manipulation of the control system in process plants. *Reliability Engineering and System Safety*, *213*, 107685. https://doi.org/10.1016/j.ress.2021.107685.

Johnson, B., Caban, D., Krotofil, M., Scali, D., Brubaker, N., & Glyer, C. (2017). *Attackers deploy new ICS attack framework "TRITON" and cause operational disruption to critical infrastructure [WWW document]*. Mandiant. URL https://www.mandiant.com/resources/attackers-deploy-new-ics-attack-framework-triton (accessed 1.24.22).

Khan, F. I., & Abbasi, S. A. (1998). Models for domino effect analysis in chemical process industries. *Process Safety Progress*, *17*, 107–123. https://doi.org/10.1002/PRS.680170207.

Kjaerland, M. (2006). A taxonomy and comparison of computer security incidents from the commercial and government sectors. *Computers & Security*, *25*, 522–538. https://doi.org/10.1016/j.cose.2006.08.004.

Kornecki, A., & Zalewski, J. (2010). Safety and security in industrial control. In *CSIIRW '10 proc. Sixth Annu. Work. Cyber Secur. Inf. Intell. Res* (pp. 1–4). https://doi.org/10.1145/1852666.1852754.

Kourniotis, S. P., Kiranoudis, C. T., & Markatos, N. C. (2000). Statistical analysis of domino chemical accidents. *Journal of Hazardous Materials*, *71*, 239–252. https://doi.org/10.1016/S0304-3894(99)00081-3.

Landucci, G., Argenti, F., Tugnoli, A., & Cozzani, V. (2015). Quantitative assessment of safety barrier performance in the prevention of domino scenarios triggered by fire. *Reliability Engineering and System Safety*, *143*, 30–43. https://doi.org/10.1016/j.ress.2015.03.023.

Landucci, G., Necci, A., Antonioni, G., Argenti, F., & Cozzani, V. (2017). Risk assessment of mitigated domino scenarios in process facilities. *Reliability Engineering and System Safety*, *160*, 37–53. https://doi.org/10.1016/j.ress.2016.11.023.

Landucci, G., & Reniers, G. (2019). Preface to special issue on quantitative security analysis of industrial facilities. *Reliability Engineering and System Safety*, *191*, 106611. https://doi.org/10.1016/j.ress.2019.106611.

Landucci, G., Reniers, G., Cozzani, V., & Salzano, E. (2015). Vulnerability of industrial facilities to attacks with improvised explosive devices aimed at triggering domino scenarios. *Reliability Engineering and System Safety*, *143*, 53–62. https://doi.org/10.1016/j.ress.2015.03.004.

Lee, R. M., Assante, M. J., & Conway, T. (2014). *ICS CP/PE (cyber-to-physical or process effects) case study paper - media report of the Baku-Tbilisi-Ceyhan (BTC) pipeline cyber attack.* (SANS ICS).

Lezzi, M., Lazoi, M., & Corallo, A. (2018). Cybersecurity for industry 4.0 in the current literature: A reference framework. *Computers in Industry*, *103*, 97–110. https://doi.org/10.1016/J.COMPIND.2018.09.004.

Mannan, S. (2012). *Lees' loss prevention in the process industries: Hazard identification, assessment and control* (4th ed.). Elsevier, UK: Butterworth-Heinemann.

Nicholson, A., Webber, S., Dyer, S., Patel, T., & Janicke, H. (2012). SCADA security in the light of cyber-warfare. *Computers & Security*, *31*(4), 418–436. https://doi.org/10.1016/j.cose.12.2002.009.

Ogie, R. (2017). Cyber security incidents on critical infrastructure and industrial networks. In *Proc. 9th Int. Conf. Comput. Autom. Eng* (pp. 254–258). https://doi.org/10.1145/3057039.3057076.

Paltrinieri, N., Dechy, N., Salzano, E., Wardman, M., & Cozzani, V. (2012). Lessons learned from Toulouse and Buncefield disasters: From risk analysis failures to the identification of atypical scenarios through a better knowledge management. *Risk Analysis*, *32*, 1404–1419. https://doi.org/10.1111/j.1539-6924.2011.01749.x.

Paltrinieri, N., Tugnoli, A., Buston, J., Wardman, M., & Cozzani, V. (2013). Dynamic procedure for atypical scenarios identification (DyPASI): A new systematic HAZID tool. *Journal of Loss Prevention in the Process Industries*, *26*, 683–695. https://doi.org/10.1016/j.jlp.2013.01.006.

Paulsen, C., & Byers, R. (2019). *NISTIR 7298 Rev. 3: Glossary of key information security terms*. https://doi.org/10.6028/NIST.IR.7298r3.

Peter, T. (2013). *Snowden confirms NSA created Stuxnet with Israeli aid.*

Rasmussen, K. (1996). *The Experience with the major accident reporting system from 1984 to 1993.* EUR 16341 EEN Luxembourg (L): Commission of the European Communities.

Reniers, G., & Cozzani, V. (2013). *Domino effects in the process industries: Modelling, prevention and managing, domino effects in the process industries: Modelling, prevention and managing*. Elsevier B.V. https://doi.org/10.1016/C2011-0-00004-2.

Reniers, G. L. L., & Dullaert, W. (2008). Knock-on accident prevention in a chemical cluster. *Expert Systems with Applications, 34*, 42–49. https://doi.org/10.1016/J.ESWA.2006.08.033.

Rennie, R., & Law, J. (2019). *A dictionary of physics*. https://doi.org/10.1093/acref/9780198821472.001.0001.

Srivastava, A., & Gupta, J. P. (2010). New methodologies for security risk assessment of oil and gas industry. *Process Safety and Environment Protection, 88*, 407–412. https://doi.org/10.1016/J.PSEP.2010.06.004.

Stouffer, K., Pillitteri, V., Lightman, S., Abrams, M., & Hahn, A. (2015). *NIST special publication 800–82 revision 2 guide to industrial control systems (ICS) security*. https://doi.org/10.6028/NIST.SP.800-82r2.

The American Nuclear Society special committee on Fukushima, (2012). FUKUSHIMA DAIICHI: ANS Committee Report.

Tugnoli, A., Cozzani, V., Di Padova, A., Barbaresi, T., & Tallone, F. (2012). Mitigation of fire damage and escalation by fireproofing: A risk-based strategy. *Reliability Engineering and System Safety, 105*, 25–35. https://doi.org/10.1016/j.ress.2011.11.002.

Tugnoli, A., Cozzani, V., Khan, F., & Amyotte, P. (2013). Managing domino effects from a design-based viewpoint. In *Domino effects in the process industries: Modelling, prevention and managing* (pp. 246–271). Elsevier. https://doi.org/10.1016/B978-0-444-54323-3.00012-9.

Wang, J., Jiang, C., & Qian, J. (2014). Robustness of internet under targeted attack: A cascading failure perspective. *Journal of Network and Computer Applications, 40*, 97–104. https://doi.org/10.1016/J.JNCA.2013.08.007.

Wang, J., Rong, L., Zhang, L., & Zhang, Z. (2008). Attack vulnerability of scale-free networks due to cascading failures. *Physica A: Statistical Mechanics and Its Applications, 387*, 6671–6678. https://doi.org/10.1016/J.PHYSA.2008.08.037.

Willems, E. (2011). Cyber-terrorism in the process industry. *Computer Fraud & Security, 2011*, 16–19. https://doi.org/10.1016/S1361-3723(11)70032-X.

CHAPTER ELEVEN

Uncertainty modeling in risk assessment of digitalized process systems

Mohammad Yazdi[a,c,*], Esmaeil Zarei[a], Sidum Adumene[b], Rouzbeh Abbassi[c], and Payam Rahnamayiezekavat[d]
[a]Centre for Risk, Integrity, and Safety Engineering (C-RISE), Faculty of Engineering and Applied Science, Memorial University of Newfoundland, St. John's, NL, Canada
[b]School of Ocean Technology, Marine Institute Memorial University of Newfoundland, St. John's, NL, Canada
[c]School of Engineering, Faculty of Science and Engineering, Macquarie University, Sydney, NSW, Australia
[d]School of Engineering, Design and Built Environment, Western Sydney University, Penrith, NSW, Australia
[*]Corresponding author: e-mail address: myazdi@mun.ca

Contents

1. Introduction

The digitalized process systems with its abundance of data have recently attracted system safety decision-makers (i.e., the risk analyst is meant who has to decide on what tool, model, or parameter value to use) as strategic contributors to increasing the efficiency, operational products, and system safety

Methods in Chemical Process Safety, Volume 6
ISSN 2468-6514
https://doi.org/10.1016/bs.mcps.2022.04.005

of process-based industrial sectors (Benson, Argyropoulos, Dimopoulos, Mikellidou, & Boustras, 2021; Lee, Cameron, & Hassall, 2019). The history of digitalized process systems dates back to more than 30 years ago by investigating neural networks to detect and diagnose the faults of process systems (Hoskins & Himmelblau, 1988; Li, Olson, & Chester, 1990; Sorsa, Koivo, & Koivisto, 1991). In fact, the frequencies of fatalities and injured people in the process-based sectors have reduced significantly since the last decades. However, the process-based accidents have not decreased in both number and consequence severities in the same period (Pitblado & Nelson, 2013). Thus, the digitalized process systems are essential and inevitable nowadays, aiming to parallel the critical priority of process system management to acquire a zero accident over time (Wang, 2021). Similar to process system safety, the digitalized process systems need to be integrated with the feasible and optimal design, control, and engineered system modeling supports. In addition, the digitalized process systems inherently assisted in having monitoring tools, high-tech computer simulators and software, human interactions, and on-time diagnosis (Hassall, 2015; Sultana, Andersen, & Haugen, 2019). Thus, it is believed that the digitalized process systems boost system safety performance by providing a safe and healthy environment for employees and maintaining operational assets from any potential hazardous events (Arzaghi et al., 2018; Golestani, Abbassi, Garaniya, Asadnia, & Khan, 2020; Romero et al., 2018). Some examples of digitalized process systems can be highlighted as early warnings (Baksh, Abbassi, Garaniya, & Khan, 2017; Chang, Khan, & Ahmed, 2011; Kalantarnia, Khan, & Hawboldt, 2010; Schmitz, Swuste, Reniers, & van Nunen, 2020), fault detection (Lou, Li, & Atoui, 2020; Wu & Zhao, 2021), and diagnosis (Amin, Khan, Ahmed, & Imtiaz, 2021a, 2021b) in process-based systems. The digitalization of the process systems characterizes the process abnormalities and deviations in the system (e.g., emergency or non-emergency shutdowns, failures, faults, errors, interruptions) (De Rademaeker, Suter, Pasman, & Fabiano, 2014; Hameed, Khan, & Ahmed, 2016; Kelly, 2011; Khan et al., 2016; Md Nor, Che Hassan, & Hussain, 2020; Yazdi, Adesina, Korhan, & Nikfar, 2019).

The techniques mentioned above can prevent the system from potential accidents or at least reduce risk levels. Besides, different types of information could be provided and updated subsequently, such as asset integrity management, operational and system reliability procedure, accidents/near-miss/mishap report investigation process (Liaw, 2019). For example, digitalization can be integrated with the hazard identification process

(Argyropoulos, Christolis, Nivolianitou, & Markatos, 2012), resilience (Patriarca, Falegnami, De Nicola, Villani, & Paltrinieri, 2019), and learning from past accidents.

However, up to the present day, the modernization of digitalization process-based systems within the decision-makers cognitive behaviors challenges the processing system with the different uncertainty aspects (subjective (Epistemic) or objective (Aleatory)). One of the main aspects of uncertainty challenges is the risk assessment of digitalized process systems—the uncertainty causes and further increases the inefficiency of the risk-based engineering perspective of the processing system (Aven, 2011, 2013, 2016; Flage, Aven, & Berner, 2018; Steen & Aven, 2011). Thus, the merits as mentioned earlier, of process systems' digitalization is negligible due to lack of good, safe operation process and unreliable decision-making purposes. In addition, the complicated process multivariate controlling system integrated with the products quality and operational conditions are then safety restricted.

This chapter reviews uncertainty modeling in risk assessment methods associated with the process systems' digitalization techniques. It is objected for interested readers to have insights regarding sources of uncertainties, distinguish the uncertainty contributors, and propagate uncertainty. This chapter also aims to develop advanced risk-based decision-making tools in system safety and reliability analysis methods applicable in digitalization process-based systems.

The organization of the present chapter is provided as the following. In Section 2, the term "uncertainty" is well defined. In Section 3, the category of uncertainty is comprehensively reviewed. In Section 4, the uncertainty contributors are explained and distinguished. The uncertainty propagation is briefly explained in Section 5. Finally, the conclusion, including current challenges and future direction of digitalization process-based systems, are presented.

2. The definition of uncertainty

The term *"uncertainty"* has been widely engaged within several existing state of arts explanations on risk evaluation, assessment, analysis, and management of process-based systems. By reviewing the published works, one can see that the term *"uncertainty"* is substituted with the term *"risk"* as equivalent meaning, or some authors indicated the term *"risk"* is somehow the" *uncertainty"* about the future occurrences. Considering the vast claims, different interpretations, and opposing interests of decision-makers about both terms' *"risk"* and *"uncertainty,"* in the present book chapter, we assume that the *"risk"* and *"uncertainty"* belong to the different concepts, and it should be distinguished adequately during the risk assessment process.

To have a clear understanding of *"uncertainty"* transfers the doubt/ hesitation or lack of information and knowledge about a decision. The *"risk"* is considered for the risk assessment process and decision-making purposes. In this regard, the decision-makers (assessors) must trust their decisions as accurately and reliably as possible. Thus, the decision-makers are willing to know how much the degree of their trust is for the obtained risk assessment outcomes based on their evaluations and analysis. Hence, this is the uncertainty that should be taken into account for the risk assessment outcomes.

Thus, one can define *"uncertainty"* as a measurement of confidence that decision-makers need to know in risk assessment outcomes. However, it should be highlighted that the *"uncertainty"* does not necessarily require quantification. The "*US National Research Council*" defined *"uncertainty"* as "*Lack or incompleteness of information. Quantitative uncertainty assessment attempts to analyze and describe the degree to which a calculated value may differ from the actual value; it sometimes uses probability distributions. Uncertainty depends on the quality, quantity, and relevance of data and the reliability and relevance of models and assumptions.* (National Research Council, 2009)".

Once we understand and distinguish what precisely the uncertainty is, it is required to categorize it by enabling decision-makers to model, treat, and propagate uncertainty in practice. In the next section, two categories of uncertainty are reviewed.

2.1 Uncertainty and risk assessment

Let us consider that there are a set of terms (parameters) as (i) uncertainty, (ii) risk assessment, (iii) input information (objective data, or subjective experts' opinions), (iv) risk-based model, and (v) decision-making. Fig. 1 presents the best performance of the risk-based decision-making tool (i.e., risk assessment methods) with the interconnection of the parameters (i) to (v). In fact, the picture should be considered as a loop. Once the new data becomes available, the procedure will be repeated. Additionally, it is not practical to measure the uncertainty of the whole risk assessment process. Such measurement is much more uncertain than risk estimation itself. Therefore, it is required to quantify the uncertainty, if possible,

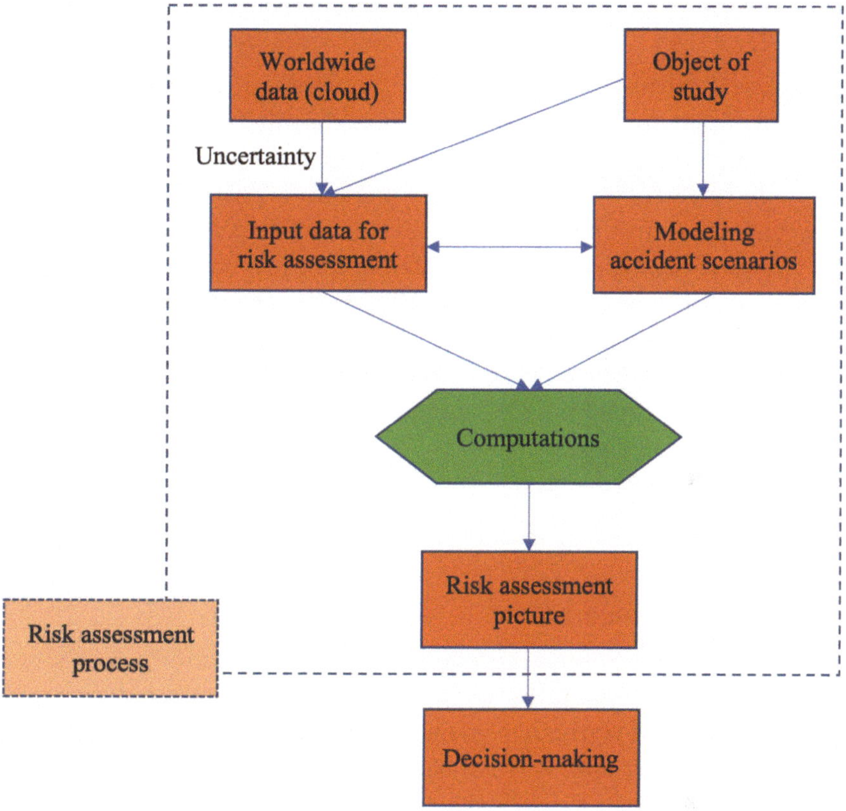

Fig. 1 The big picture of risk-based decision-making model.

in every single step of the risk assessment process systematically. Documenting the outcome makes the decision-makers confident about the results and the decisions made.

3. Reviewing the category of uncertainty

The uncertainty is categorized in (i) Aleatory and (ii) Epistemic, which refer to the rolling of dice (Latin), and knowledge (Greek), respectively (Ferson & Ginzburg, 1996; Rausand, 2011; Yazdi & Zarei, 2022).

3.1 Aleatory uncertainty

The source of Aleatory uncertainty is randomness and natural variability, and also named "*variability, random uncertainty, inherent uncertainty, and irreducible uncertainty.*" For example, the variation in the quality of products, direction of the wave. Mathematically, the Aleatory uncertainty is detected when a trial is conducted several times in the same circumstances, and the derived outcomes fluctuate every time. This simply means that increasing the number of trials does not necessarily decline the randomness and variability. However, it would provide more precise information regarding the probability distributions by explaining the outcomes' variabilities.

Let us assume that there is a failure in the digitalized process-based system like a fire and flame detector failure. The appropriate and early detection of installed detectors is needed to reduce the fire consequence in the open area. The time variability of wind direction and speed of detecting the failure or some response time should also be observed. Mathematically, we can find the best-fitted probability distributions for speed and wind direction. However, there is no chance that we could definitely say that the fire goes to a vulnerable asset (due to the detector failure). Meanwhile, the best probability distributions can be obtained to describe the situations.

3.2 Epistemic uncertainty

As mentioned earlier, epistemic uncertainty arises due to the lack of knowledge. Therefore, this can be reduced and further eliminated by gaining adequate knowledge about the understudy objective. It is also named "*ignorance, subjective uncertainty, knowledge-based uncertainty, phenomenological uncertainty, and reducible uncertainty*" (Aven, 2010; Martorell, Martón, Villamizar, Sánchez, & Carlos, 2014; Paté-Cornell, 1996; Yazdi, 2019a; Yazdi et al., 2021a) .

The bordering line between the two types of uncertainties has not set yet. Increasing knowledge regarding the objective of a study causes decision-makers to provide more information about the circumstances, and then the aleatory uncertainty is shortened. One can say that all the uncertainties would be epistemic ones at the end of a day (Gholamizadeh, Zarei, Omidvar, & Yazdi, 2022; Jiang et al., 2021).

Human error can be an appropriate example of epistemic uncertainty, which has declined over time. Since the last decades, the science to model and estimate human reliability has increased extensively, reducing epistemic uncertainty.

As the first rule of thumb, uncertainty is the absence of certainty. Therefore, distinguishing between epistemic and aleatory uncertainties is not always an important task. However, some researchers highlighted that it would be better to categorize uncertainty to understand better the risk assessment process (Anderson & Hattis, 1999; Bari & Park, 1989; Clemen & Winkler, 1999; Kiureghian and Ditlevsen, 2009; Parry, 1996).

As a second rule of thumb, decision-makers accuracy and confidence level about the risk assessment outcomes are the opposite of uncertainty. Hence, as much as the confidence level is high, the uncertainty is low.

3.2.1 Alternative representation of uncertainty

In this section, a brief review of four mathematical structures that can be utilized to represent epistemic uncertainty is provided, including (i) probability theory, (ii) evidence/Dempster-Shafer theory, (iii) possibility degree theory, and (iv) interval analysis.

3.2.1.1 Probability theory

Probability theory is based on the triple parameters as (X, D, p) for a x vector of uncertain variables $x_1, x_2, x_3, \ldots, x_n$, in which X is the set of possible values for x, X is preferably limited to subsets collection of X (i.e., in case $u \in D$, therefore, $u^c \in D$, and in case u_1, u_2, u_3 would be the countable sequence of D elements, therefore, $\bigcup_k u_k \in D$ and $\bigcap_k u_k \in D$, and finally, p is defined as the probability of D elements (i.e., $0 \leq p(u) \leq 1$, in case $u \in D$, $p(x) = 1$, and $p\left(\bigcup_k u_k\right) = \sum_k p(u_k)$ when u_1, u_2, u_3, ... would be the countable sequence non-interesting D elements.

In order to get more details regarding probability theory, an interested reader can refer to references (George, 1990; Parry & Winter, 1981).

3.2.1.2 Evidence/Dempster-Shafer theory

Evidence/Dempster-Shafer theory provides an uncertainty representation allowing more structure and specifications to decision-makers. Evidence/Dempster-Shafer theory is also based on triple parameters(X, D, m) for x vector of uncertain variables x_1, x_2, x_3, ..., x_n, in which X is the set of possible x values, D is the countable collection of X subsets, m is a function denoted for subset u of X satisfying $m(u) > 0$ in case $m(u) = 0$ if $u \notin D$, and $\sum_{u \in D} m(u) = 1$.

In the Evidence/Dempster-Shafer theory terminology, X is the sample space/interval set, D is the set of focal elements for both x and m, and $m(u)$ is the fundamental probability assignment connected with a subset u and x. Conceptually, the probability assignment $m(u)$ would provide a number of information measures named as "*credibility probability*," in which is connected with a subset of u of x. However, this cannot be decomposed into over u subsets.

In order to get more details regarding Evidence/Dempster-Shafer theory, an interested reader can refer to references (Ayoun & Smets, 2001; Lefevre, Colot, & Vannoorenberghe, 2002; Yager, 1987; Yazdi & Kabir, 2018; Zadeh, 1986; Zhang & Deng, 2018).

3.2.1.3 Possibility degree theory

Possibility theory can provide a reasonable representation of uncertainty that allows decision-makers to formulate the uncertainty properly. Possibility theory is based on parameters (X, r) for x vector of uncertain variables x_1, x_2, x_3, …, x_n, in which X is the set of possible x values, and r is the function defined X satisfying $0 \leq r(x) \leq 1$ for all $x \in X$, and $\mathbf{sup}\ \{r(x): x \in X\} = 1$. The function of r provides a measure for likelihood/confidence, assigned to every single element of x. The pair (X, r) is defined as a possibility space for x.

A value of $r(x) = 1$ shows that there would be unknown information reflecting the "occurrence" or "appropriateness" of the particular value of x included in X, and the value of $r(x) = 0$ denotes that the known information totally reflects the "occurrence" or "appropriateness" of x. In addition, enhancing the $r(x)$ value in intervals zero and 1 shows that there would be an increase in lack of information reflecting the "occurrence" or "appropriateness" of x. Besides, the $r(x) = 1$ denotes that x is totally possible in the sense of nothing in contract of x, satisfying $0 < r(x) < 1$. This condition means that x is possible; however, x may not be possible as $r(x)$ goes into the zero. In addition, $r(x) = 0$ indicates that x is known to be impossible.

In case, variables $x_1, x_2, x_3, \ldots, x_n$ have connected with possibility spaces (X_1, r_1), (X_2, r_2), …, (X_n, r_n), then the vector $x = [\,x_1, x_2, x_3, \ldots, x_n]$ has a connection possibility space (X, r), where $X = X_1 \times X_2 \times X_3 \times \ldots \times X_n$ and $r(x) = \min\{r_1(x_1), r_2(x_2), \ldots, r_n(x_n)\}$.

In order to get more details regarding Possibility degree theory, an interested reader can refer to the reference (Zadeh, 1999).

3.2.1.4 Interval analysis

Interval analysis is based on an assumption that a set $X = \{x : x = [x_1, x_1, \ldots, x_n]\}$ of possible values for x vector of variables x_1, x_1, …, x_n is known. However, there is no specific uncertainty structure with a set of x. Therefore, it is required to consider the "known" should be "true" or "appropriate" value for x contained with a set of x. In addition, the propagation of the individual x values in X within a function F results in the set of $y = \{y : x \in X \text{ and } y = F(x)\}$ of possible y values. This simply means that there is no uncertainty structure of X set, and there is no uncertainty structure for a set of y.

4. Uncertainty contributors

There are three uncertainty contributors for the risk assessment outcomes, including (i) model, (ii) parameter, and (iii) completeness uncertainty. In the risk assessment process, a decision-maker is looking to predict the upcoming outcomes and events. Thus, it apparently provides uncertainty as decision-makers cannot surely know about the state of the process-based system safety, such as the external factors influencing the system and how it will look. Each uncertainty contributor is provided briefly in the following since the condition mentioned above would impact the contributors.

4.1 Models uncertainty

Many models play a role in the risk assessment process, including stochastic, structural, accidental, human cognitive behavior, evacuation, and cloud dispersion models. These models are often a simplified version of actual case studies due to reducing the interest features by mathematical and analytical methods.

In fact, the decision-makers are free to select a method among numerous models. They only should care about the advantages and disadvantages of the models and how they can be appropriately fitted with the risk assessment of the problem. To reach the best model, decision-makers should know enough about objective study's problems' features and operational/technical characteristics (see Fig. 2).

It should be added that the availability of input data can significantly help decision-makers to select the best-fitted model for the risk assessment of the understudy objective. There are two probabilistic and deterministic models commonly utilized in the risk assessment process and describe the physical phenomena and life distributions, respectively. Some aspects are difficult to be modeled, such as but not limited to "*human behavior in critical situations, human reliability and human recovery of faulty situations, organizational factors,*

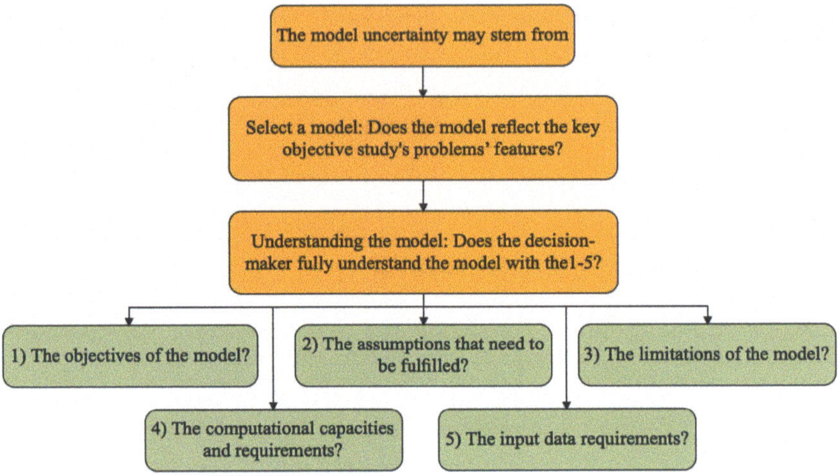

Fig. 2 The model uncertainty performance.

for example, the organization's safety culture, violations of rules and procedures, software functions, and software reliability, effects of maintenance and aging, hazards caused by modifications of the system, causes and factors that are not measurable, and deliberate hostile actions (Rausand, 2011)."

The existing methods that are widely used in literature, such as fault tree analysis (FTA) (Casamirra, Castiglia, Giardina, & Lombardo, 2009; Chiremsel, Nait Said, & Chiremsel, 2016; Kabir, Geok, Kumar, Yazdi, & Hossain, 2020; Shi, Wang, Zhang, Cheng, & Zhao, 2017; Yazdi, 2017; Yazdi, Korhan, & Daneshvar, 2018), event tree analysis (ETA) (de Ruijter & Guldenmund, 2015; Gholamizadeh et al., 2022; Mohammadfam & Zarei, 2015; Mohsendokht, 2017), risk matrix (Ahn & Chang, 2016; Markowski & Siuta, 2018, 2017; Yazdi, 2018), Bayesian Network (BN) (Adumene et al., 2021b; He et al., 2018; Nedjati, Yazdi, & Abbassi, 2021; Yazdi, Khan, Abbassi, & Rusli, 2020; Zarei, Yazdi, Abbassi, & Khan, 2019), Markov chain (Adumene et al., 2021a, 2020; Yazdi et al., 2021b, 2021c), failure mode and effect analysis (FMEA) (Daneshvar, Yazdi, & Adesina, 2020; Liu, 2016a; Liu, 2016b; Tsai et al., 2017; Weidl, Madsen, & Israelson, 2005; Mohammad Yazdi, 2019; Yazdi, Daneshvar, & Setareh, 2017), would inherently introduce uncertainty due to seeking simplification by decision-makers. The source of model uncertainty may be arisen from (i) the decision-makers putting wrong numbers and (ii) the decision-makers misinterpreting the input and output numbers because of low experience and knowledge.

4.2 Parameter's uncertainty

The parameter's uncertainty indicated the wide range of data that can be used in the risk assessment process. The parameters in risk assessment methods include but are not limited to "*failure rates for the various components and for the various failure modes, repair times and downtimes for the various components, intervals between proof tests, common cause failure rates, or β-factors, human error probabilities and effects of performance-influencing factors, wear parameters for deteriorating components, ignition probabilities for gas releases, frequencies for natural events* (e.g., *flooding, avalanche, lightning, and earth-quake*), *exposure data (who, where, for how long), activity data* (e.g., *production rate/volume, staffing*), *and process data actions* (Rausand & Haugen, 2020)." Fig. 3 illustrates the main important reason to highlight why data availability is uncertain. For example, in the digitalized process system, the data collected from the study's objective, e.g., the corresponding reliability information and how they would be related to the study's objective. The critical challenge is that many models in a real case are developing rapidly. Therefore, reliability data collection is inevitably old-fashioned because there is a necessary experience to predict the model parameters in the existing risk assessment methods.

According to the authors' understanding of parameters uncertainty and having support from references (Abuswer, Amyotte, Khan, & Morrison, 2013; Bae, Grandhi, & Canfield, 2004; Bari & Park, 1989; Cooke & Goossens, 2008; Flage & Aven, 2009; Hora, 1996; Zio & Apostolakis, 1996), it is an excellent idea that decision-makers ask themselves a set of questions when they decide to use data in the model, as "(1) *would the decision be different if the data were different?* (2) *would other data collection and research*

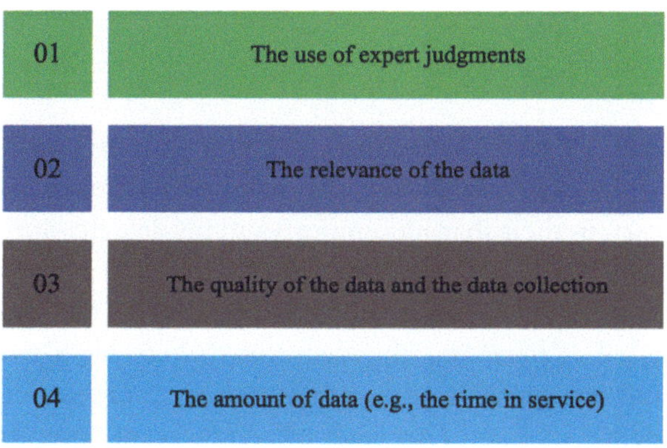

Fig. 3 The main reasons for data uncertainty.

probably lead to a different decision? (3) *how long will it take to collect the information,* (4) *how much would it cost, and* (5) *would the resulting decision be significantly different?"*

4.3 Completeness uncertainty

The completeness uncertainty is defined as the total quality of the risk assessment process, including its scope and objective, the capability of decision-makers, and the way of the risk assessment procedure. It is challenging to discriminate between the two uncertainties as model and completeness because the risk assessment model is not comprehensive and complete in a single word. Thus, in the digitalized proves system, completeness uncertainty is connected with the lack or missing hazard/failure and event identification present in the system.

The main objective of hazard/failure identification in the risk assessment procedure is to introduce a complete list of potential hazard/failure(s) and find out the critical to receiving further intervention actions to be at the acceptable level of risks (Hirayama, Shinozaki, Kasai, & Otaki, 2018; Yazdi, 2019b) or ALARP "*As low as reasonably practicable*" (Baybutt, 2014; Melchers, 2001; Pasman, Jung, Prem, Rogers, & Yang, 2009; Pike, Khan, & Amyotte, 2020). The available risk assessment methods are practical and efficient to improve digitalized process safety; however, it requires that the decision-makers have enough capability with sufficient knowledge about technical/operational and phenomena of objective. It should be added that different types of challenges have to be taken into account, such as: "*tightly coupled systems* (i.e., *systems where a failure in one item rapidly spreads to other items, new technology, or known technology used for new applications, systems with open boundaries to the environment, where events outside the system may lead to hazardous events within the system boundaries* (Rausand & Haugen, 2020)."

Pasman et al. (2009) discussed that the key contributor to the uncertainty is in defining the scenario during the risk assessment process, in which decision-makers, in their sharing opinions and model selection, had an inevitable impact on the assessment outcomes.

Moreover, the risk assessment process often has limited time and budget to be conducted. Therefore, the decision-makers have to select rapid-performing and straightforward methods, which there would no effort to do more in-depth analysis. This leads to a significant completeness uncertainty during the risk assessment process. This contributor's uncertainty is the most challenging task to be quantified.

The question is that we know why we should conduct an uncertain analysis, however, when decision-makers should perform it. Simply, the response is that first, an overview screening of computations within conservative bias predictions would show if there is any further investigation required or not. Second, when the impact and consequence of the risk outcomes are considerably high. On the opposite side, there is no need to analyze uncertainty when decision-makers highlight that the system's risk levels are in the acceptable area.

5. Uncertainty propagation

Uncertainty propagation assists decision-makers to have better insights regarding the impact of input variations on the risk assessment outcomes simultaneously. Two approaches can be used for this purpose, (i) "*Monte Carlo Simulation*" and (ii) "*Analytical technique*." Both approaches can help decision-makers formulate the risk assessment parameters mathematically, such as failure rate of component, probability of a top event in FTA, or dispersion fire and explosion model.

It should be added that uncertainty propagation is commonly used to analyze risk assessment outcomes according to the parameter's input data uncertainty (Jensen, Mayorga, Valdebenito, & Chen, 2020; Liu, Li, Zhao, & Song, 2021; Moges, Demissie, & Li, 2020; Shi et al., 2020; Yazdi & Kabir, 2017). Thus, it is not suggested to analyze the model uncertainty. A brief review of both methodologies is presented in the following.

5.1 Monte Carlo Simulation

To perform a "*Monte Carlo Simulation,*" the assessors need to have best-fitted probability distributions reflecting the decision-makers' knowledge about the uncertainty, for example, for each single input variable. In "*Monte Carlo Simulation,*" the set of samples is repeated for probability distributions of each input variable. The pseudo-random generator is then used to derive one value from each distribution. The simulator is also used to calculate the output variables. The process is then continued several times to determine many samples of output variables. Afterward, the derived samples can be drawn in histogram plots with specific mean, variance, and expected number characteristics.

The significant shortage of "*Monte Carlo Simulation*" is extensive and time-consuming for the complicated risk assessment models. In this regard, many searches have developed "*Monte Carlo Simulation*" in different domains to be much more effective and efficient in practice (Fazzolari & Tan, 2020; Jakab, Pázmándi, & Zagyvai, 2021; Rameback & Lindgren, 2018; Yang, 2021; Zhao & Hong, 2022). The advantages of "*Monte Carlo Simulation*" can also be highlighted: "it is easy to use and applied directly to the model; no one needs to develop surrogate models. For example, by Taylor expansion, one may select from different distributions for the input variables and test various alternatives" (Koc, Kazantzis, Nuttall, & Ma, 2013; Madani & Lund, 2011; Shekari, Khan, & Ahmed, 2015).

5.1.1 Copula solved by Monte Carlo simulation

In addition, an advanced extension of "*Monte Carlo Simulation*" is applied to Copula, known as a dependency function to create a joint multivariate distribution, in which one-dimensional marginal variable distributions (MDs) would be coupled. This is relevant as dependent variables in, e.g., fault tree configuration is another neglected aspect of risk assessments. Facilitating this point, though, Copulas have enough capability to have considerable flexibility in structurally characterizing and dependency degree capturing of those variables that are correlated. The Copula can construct a model considering more dependency or less due to the MD of the variables and could be different by a joint cumulative distribution. Moreover, the Copula is a robust and powerful tool to select a probability distribution, even in case of mistaken selection (D'Angelo & Weissfeld, 2013). Besides, the n-dimensional continuous multivariate random numbers as vector $x = (x_1, x_2, \ldots x_n)$ have this chance to be reformed in terms of n univariate marginal distributions $F_1(x_1)$, $F_2(x_2)$, ..., $F_n(x_n)$ and n-dimensional Copula function (CF) $\widetilde{C}$, in which defined in the following equations. The CF $\widetilde{C}[0,1]^d \to [0,1]$ represents univariate-marginal joint cumulative distributions $F_1(x_1)$, $F_2(x_2)$, ..., $F_n(x_n)$ into the joint distribution F (Hashemi, Khan, & Ahmed, 2016; Morales-Nápoles, Paprotny, Worm, Abspoel-Bukman, & Courage, 2017).

$$F(x_1, x_2, \ldots x_n) = \widetilde{C}(F_1(x_1), F_2(x_2), \ldots, F_n(x_n)) \tag{1}$$

In addition, the marginals have continual trends, $\widetilde{C}$ can be explained in the following equation:

$$\widetilde{C}(u_1, u_2, \ldots u_k) = F\left(F_1^{-1}(u_1), F_1^{-1}(u_2), \ldots, F_n^{-1}(u_n)\right) \tag{2}$$

where $F_i(x_i)$ for $i \in \{1, 2, \ldots n\}$ is the MD of x_i, $\widetilde{C}$ is based on the Copula function, and $u_i = F_i(x_i)$ for $i \in \{1, 2, \ldots n\}$. In a similar way, for the bivariate distribution, $F(x_1, x_2)$ of two-random variables could be presented underlying the context of CF and two-different marginal joint cumulative distribution as the following equation:

$$F(x_1, x_2) = \widetilde{C}(F_1(x_1), F_2(x_2), \theta) \tag{3}$$

in which, θ is signified as the CF parameter to obtain the causality between the two different variables x_1 and x_2, utilizing "*Pearson correlation coefficient*" and denoted as ρ. The parameter ρ can be therefore be obtained as the following equation:

$$\rho = \int_{-\infty}^{\infty} \int_{-\infty}^{\infty} \left(\frac{x_1 - \mu_{x_1}}{\sigma_{x_1}}\right) \left(\frac{x_2 - \mu_{x_2}}{\sigma_{x_2}}\right) f_1(x_1) f_2(x_2) \partial \widetilde{C} \left(\frac{F_1(x_1), F_2(x_2), \theta}{\partial F_1(x_1) \partial F_2(x_2) dx_1 dx_2}\right) \tag{4}$$

in which, μ_{x_1} and μ_{x_2} are the means with respect to x_1 and x_2. The σ_{x_1} and σ_{x_2} are the standard deviations of x_1 and x_2. In the same manner, the $f_1(x_1)$ and $f_2(x_2)$ are the marginal probability density functions (PDFs) of x_1 and x_2.

In order to properly use Copula in decision-making problems, it can be integrated into Bayesian Network. It should be highlighted that using Copula Bayesian Network is not only because it is full of Bayesian Network topology, but also it is because of its accuracy in measurement and fully considering the dependency between the variables in the network based on an existing database. With consideration of data availability from objective data or an elicitation process from decision-makers, CF can be obtained appropriately with two different aspects: (i) by MD to fit the variables' properties, and (ii) by using CF to model dependency structure. It is also essential to confirm the rationality of both the determined MD and CF. A brief explanation of determining the MD and CF is provided in the following.

5.1.1.1 Determining the marginal distribution

The most important task to determine the best-fitted MD for the variables is correctly describing a probability distribution. To handle this issue, three different types of marginal distributions can be considered: normal distribution, Beta distribution, and log-normal distribution. In Table 1, the aforementioned marginal distributions and their properties are provided. In order to evaluate the precision of marginal distributions, comparison tools like Akaike Information Criterion (AIC) (Akaike, 1974) presented in Eq. (5) are utilized. It can be concluded that the AIC with minimum value shows that MD is the best. In addition, AIC should be implemented to all the

Table 1 Three different types of MDs.

Name of distribution	PDF	Mean	Standard deviation
Normal	$\frac{1}{\sigma\sqrt{2\pi}}\exp\left(-\frac{(x-\mu)^2}{2\sigma^2}\right)$	μ	σ
Log-normal	$\frac{1}{x\sigma\sqrt{2\pi}}\exp\left(-\frac{(\log x-\mu)^2}{2\sigma^2}\right)$	$\exp\left(\mu+\frac{\sigma^2}{2}\right)$	$\sqrt{(\exp(\sigma^2)-1)\exp(2\mu+\sigma^2)}$
Beta	$\frac{(x-a)^{\alpha-1}(b-x)^{\beta-1}}{\beta(\alpha,\beta)(b-a)^{\alpha+\beta+1}}$	$\frac{\alpha}{\alpha+\beta}$	$\frac{\alpha\beta}{(\alpha+\beta)^2(\alpha+\beta+1)}$

variables in the network in accordance, and it is not required that the variables follow the equal MD.

$$AIC = -2 \times \log(\max likelihood) + 2 \times (number\ of\ parameters) \tag{5}$$

where the likelihood is the maximum value for the model.

5.1.1.2 Determining the Copula function

There are different types of Copula functions having different characteristics such as tail dependency, symmetry, etc. These Copula functions can fit the different models and generate an appropriate effect on the output's viability. In addition, the Gaussian normal Copula, which is one of the most essential and common Copula based on elliptical Copula, is presented in the following equation:

$$Co_{\rho}^{G}(x_1, x_2, \ldots x_n) = \phi_{\rho}\left(\phi^{-1}(x_1), \phi^{-1}(x_2), \ldots \phi^{-1}(x_n)\right) \tag{6}$$

The Gaussian normal Copula can be practically summarized by the following features: (i) it has n-dimensional generalization, (ii) it would be easy to structure the dependency even having an extensive uncertainty, and (iii) it is just for modeling bivariate distribution with lack of data (Pan, Zhang, Koh, & Deng, 2021). Therefore, Gaussian normal Copula is utilized in this work among existing marginal distributions with cumulative distribution function. The density function of Gaussian normal Copula is presented in Eq. (7). As a result, the main difference between Gaussian normal Copula and joint cumulative distribution function is that the Gaussian variables normal Copula are underlying the different types of the marginal cumulative distribution function, which better fit a complex system.

$$Co_{\rho}^{G}(x_1, x_2, \ldots x_n) = \frac{\partial Co_{\rho}^{G}(x_1, x_2, \ldots x_n)}{\partial x_1, \partial x_2, \ldots \partial x_n}$$
$$= |\rho|^{\left(-\frac{1}{2}\right)} \exp\left(-\frac{1}{2}\xi\left(\rho^{-1} - I\right)\right) \tag{7}$$

In the defined equations, the ρ is the n- "*order symmetric positive definite*" with the $diag(\rho) = 1$, ϕ_ρ is a standard "*multivariate normal distribution*" having correlation matrix ρ, ϕ^{-1} is the inverse of standard univariate-normal cumulative distribution function $\xi = (\phi^{-1}(x_1), \phi^{-1}(x_2), \ldots \phi^{-1}(x_n))$, and I is the unit matrix. Let us assume that the dimension of n is equal to be 2, the bivariate normal Copula can be determined from the following equation.

$$Co_{\rho_{12}}^{G}(u, v) = \int_{-\infty}^{\phi^{-1}(u)} \int_{-\infty}^{\phi^{-1}(v)} \frac{1}{2\pi\sqrt{1-\rho^2}} \exp\left(-\frac{s^2 + t^2 - 2\rho st}{2}\right) dsdt \tag{8}$$

where ρ_{12} is called the "correlation coefficient of the bivariate standard normal distribution".

Two different types of analysis are introduced to show that the proposed model can be effectively used in decision-making problems: (i) model validation, and (ii) correlation analysis.

5.1.1.3 Model validation

The Kolmogorov–Smirnov test could be performed to investigate the goodness of estimated best fitting marginal. The Kolmogorov–Smirnov test measures the distances between approximated and the empirical distributions, presented in Eq. (9). The test must indicate whether the two different distributions are considerably different. It should be added that the null hypothesis in which the data fall in the context of a particular distribution may be approved at a level of 0.5%, when $h = 0$ and P-value > 0.05.

$$D = \boldsymbol{Sup}_{-\infty < x < \infty} \left| F_{exp}(x) - F_{abs}(x) \right| \tag{9}$$

in which, F_{abs} is the empirical distributions data according to the data set, $F_{exp}(x)$ is the approximated distribution data, and ***Sup*** is a supremum of the distance measure.

Similarly, the empirical Copula (EC) depends on provided input data engaged in examining whether the Gaussian Copula is applicable for the best data fitting. Let us assume that (x_i, y_i) $(i = 1, 2, \ldots n)$ is a sample from (X, Y). The empirical distribution functions for X and Y can be presented by $F_n(x)$

and $G_n(x)$, respectively. According to this point, the empirical bivariate Copula is defined as the following equation:

$$\widehat{Co_n}(u,v) = \frac{1}{n}\sum_{i=1}^{n} I_{[F_n(x_i)\le u]} I_{[G_n(x_i)\le v]},\ u,v\in[0,1] \tag{10}$$

in which $I_{[,]}$ denotes the indicative representation of the function. When $F_n(x_i) \le u$, $I_{[F_n(xi)\le u]} = 1$; otherwise, $I_{[F_n(xi)\le u]} = 0$.

In addition, the other CFs could be compared with the empirical one according to the computation of Euclidean distance as the following equation.

$$d^2 = \sum_{i=1}^{n} \left| \widehat{Co_n}(u_i, v_i) - \widehat{Co}(u_i, v_i) \right|^2 \tag{11}$$

where $u_i = F_n(x_i)$, $v_i = G_n(x_i)$ $(i = 1, 2, \ldots n)$, $\widehat{Co_n}(u_i, v_i)$ is standing for the empirical Copula (EC), and $\widehat{Co}(u_i, v_i)$ is standing for the best-fitted Copula. As an example, two candidate bivariate Copulas, called Gaussian Copula $\widehat{Co}^{Ga}(u_i, v_i)$ and *t*-Copula $\widehat{Co}^{t}(u_i, v_i)$, can be selected to be fitted as Copula $\widehat{Co}(u_i, v_i)$ based on provided data. The distance with EC can be presented as d^2_{Ga} and d^2_t and is obtained from Eq. (11). If $d^2_{Ga} \le d^2_t$, it means in this case that the Gaussian Copula is better to be fitted from the data compared to the *t*-Copula.

5.1.1.4 Correlation analysis

This type of analysis provides an ability to measure the correlation degree between two variables or nodes, which shows the direction of how these two variables are changing relative to each other. The standard Pearson correlation coefficient measures the linear relation between two variables and does not consider the effect of other variables. However, un-controlled variables may have an effect on these two variables, which causes misleading outputs. To deal with this challenge, one can use the partial "correlation coefficient" as a measurement of relationships between these two variables under the influence of other variables in the network. As an example, variable z is related with the two variables of x and y, the partial correlation analysis of the x and y can be computed according to the standard correlation as presented in the following equation, in which the output is between zero and $|1|$, meaning that at zero there is no linear-relationship. The value of $+1$ denotes the highest linear relationship of variables changing in the same direction, and -1 in the opposite direction.

$$\gamma_{xy,z} = \frac{\gamma_{xy} - \gamma_{xz}\gamma_{yz}}{\sqrt{\left(1-\gamma_{xz}^2\right)\left(1-\gamma_{yz}^2\right)}} \tag{12}$$

where γ_{xz} denotes the correlation between two variables x and z, γ_{yz} denotes the correlation between the variables y and z, $\gamma_{xy,z}$ is the correlation between x and y, in which both of them are un-correlated with variable z.

In addition, Spearman's rank correlation can also be used to determine the relationships between the two variables. The implication of Spearman's rank correlation has similarities with the partial correlation coefficient. If the result is closer to +1 or −1, the relationships would be stronger. In addition, the result of zero shows that there is no relationship between the two variables. As seen from the following equation, Spearman's rank correlation is dependent on the ranking of data values of each variable rather than the existing data itself.

$$r_R = 1 - \frac{6\sum_i^n d_i^2}{n(n^2-1)} \tag{13}$$

where $d_i = rank(x_i) - rank(y_i)$ is the ranks' differentiations according to the element i, which is somehow the paired data set of x and y, and n denotes the number of data from the two x and y variables.

5.2 Analytical technique

A broad range of mathematical models has been utilized as a simple additive or multiplicative models in risk assessment (Cooke, 2014, 2016; Cooke, ElSaadany, & Huang, 2008). To seek simplicity, let us consider only the simple additive method here, described by a set of input variables ($X = X_1$, X_2, ..., X_n) and one output variable (Y). It is also considered that the "X" are independent random variables with a specific mean ($E(X_i)$) and variance ($Var\ (X_i)$). Thus, the idea behind this simple additive method is deriving ($E(Y)$) and variance ($Var\ (Y)$) of the output variables Y.

For the first output variables and the central limit theorem, it is known that the "Y_1" is usually distributed when the number of the input variables is high. Therefore, the obtained distributions can be used to determine the probabilistic characteristics of the "Y_1".

The multiplicative model on the same input variables can be reduced to a simple additive model by taking the proper logarithms actions (let us consider it "Y_2"). The central limit theorem (CLT) can also be used to claim that the "Y_2" is also normally distributed (lognormally).

As an example, $Y_2 = g(X) = \prod_{i=1}^{n} X_i$ with n number of variables is considered as a simple multiplicative model. This would change into the additive model by taking the log function, presented as $Y_2 = \ln(\prod_{i=1}^{n} X_i) = \sum_{i=1}^{n} ln\ X_i$.

Similarly, the CLT can be used to approve that the Y_2 is nearly normally distributed (i.e., lognormally distributed). In addition, the Taylor series may be used to extend the set of input variables as presented in Eq. (14):

$$Y = g(\mu) + \sum_{i=1}^{n} \left[\frac{\partial g(x)}{\partial X_i}\right]_{X=\mu} (X_i - \mu_i) + \frac{1}{2!} \sum_{i=1}^{n} \sum_{i=1}^{n} \left[\frac{\partial^2 g(X)}{\partial X_i \partial X_j}\right]_{X=\mu} \left[(X_i - \mu_i)\left(X_j - \mu_j\right)\right] + \Delta \tag{14}$$

The term Δ is a residual and includes the order of three and more. It should be highlighted that the derivatives are determined for the mean-nominal-value. In case, the value of $(X_i - \mu_i)$ is quite small; then, the higher-power would become much smaller. In this context, the residual value may be a minimal amount. The mean-nominal-value of Eq. (14) can be derived from Eq. (15) as the following:

$$E(Y) = g(\mu) + \sum_{i=1}^{n} \left[\frac{\partial g(x)}{\partial X_i}\right]_{X=\mu} E(X_i - \mu_i) + \frac{1}{2!} \sum_{i=1}^{n} \sum_{i=1}^{n} \left[\frac{\partial^2 g(X)}{\partial X_i \partial X_j}\right]_{X=\mu} E\left[(X_i - \mu_i)\left(X_j - \mu_j\right)\right] + E(\Delta) \tag{15}$$

In fact, the residual value for many $g(X)$ functions are negligible due to the following reasons: (i) it is assumed that the $X_1, X_2, \ldots, X_n$ are independent variables, thus the $Cov(Xi, Xj) = E[(X_i - \mu_i)(X_j - \mu_j)])$ for all $i \neq j$, and (ii) the mean-nominal-value is equal to μ_i.

The variance of X_i is denoted as $Var(X_i) = E[(X_i - \mu_i)(X_j - \mu_j)] = E(X_i - \mu_i)^2 = \sigma_i^2$, and Eq. (15) can be modified to Eq. (16) accordingly:

$$E(Y) \approx g(\mu) + \frac{1}{2} \sum_{i=1}^{n} \left[\frac{\partial^2 g(X)}{\partial X_i^2}\right]_{X=\mu} \sigma_i^2 \tag{16}$$

The variance of Y can be obtained using Eq. (16), in which the $g(\mu)$ is constant and should be eliminated. Therefore, the variance of Y is computed as:

$$Var(Y) \approx Var \sum_{i=1}^{n} \left[\frac{\partial g(x)}{\partial X_i}\right]_{X=\mu} (X_i - \mu_i) = \sum_{i=1}^{n} \left[\frac{\partial g(x)}{\partial X_i}\right]_{X=\mu}^2 \sigma_i^2 \tag{17}$$

6. Conclusion: Challenges and future direction

System safety and risk management have remained a key challenge for process-based industrial sectors. Considering the decades of decision-makers attention, the processing system safety tendencies and related plots have not followed the descending measure of personal lagging safety indicators. Many digitalized technologies are engaged to improve the process system safety.

However, the digitalized process systems are commonly on disparate platforms with a specific data set that is restricting and challenging due to uncertainty contributors. This chapter aimed to provide a picture of uncertainty modeling, treatment, and propagation in digitalized process systems.

Modeling the uncertainty in digitalized process systems would help decision-makers improve their risk-based operational decisions and control the processing system's potential hazards over time. In addition, digitalizing the process systems ensured onsite employees worked confidently and ensured that there would be a safe operational condition and environmental protection. The digitalized process systems are in the correct developing trend. Keeping the current maturity priority and allocating the most incredible attention to system safety can significantly reduce incidents. As an example, asset integrity failures can be minimized by implementing the digitalized corrosion models and risk-based inspection methods associated with the objective. The predictive maintenance and machine learning tools help decision-makers reach reliable solutions by handling uncertainties.

Additional work is also vital to improve the digitalized process systems with greater precision. Having the highest level of process system safety performance needs the priority of digitalization to be considered in an interdisciplinary strategic plan, which would be integrated with production scheduling, optimization, managing budget, and cost rather than focusing only on process safety.

Reference

Abuswer, M., Amyotte, P., Khan, F., & Morrison, L. (2013). An optimal level of dust explosion risk management: Framework and application. *Journal of Loss Prevention in the Process Industries*, *26*, 1530–1541. https://doi.org/10.1016/j.jlp.2013.08.018.

Adumene, S., Adedigba, S., Khan, F., & Zendehboudi, S. (2020). An integrated dynamic failure assessment model for offshore components under microbiologically influenced corrosion. *Ocean Engineering*, *218*, 108082. https://doi.org/10.1016/j.oceaneng.2020.108082.

Adumene, S., Khan, F., Adedigba, S., Zendehboudi, S., & Shiri, H. (2021a). Offshore pipeline integrity assessment considering material and parametric uncertainty. *Journal of Pipeline Science and Engineering*, *1*(3), 265–276. https://doi.org/10.1016/j.jpse.2021.09.005.

Adumene, S., Okwu, M., Yazdi, M., Afenyo, M., Islam, R., Orji, C. U., et al. (2021b). Dynamic logistics disruption risk model for offshore supply vessel operations in Arctic waters. *Maritime Transport Research*, *2*, 100039. https://doi.org/10.1016/j.martra.2021.100039.

Ahn, J., & Chang, D. (2016). Fuzzy-based HAZOP study for process industry. *Journal of Hazardous Materials*, *317*, 303–311. https://doi.org/10.1016/j.jhazmat.2016.05.096.

Akaike, H. (1974). A new look at the statistical model identification. *IEEE Transactions on Automatic Control*, *19*, 716–723. https://doi.org/10.1109/TAC.1974.1100705.

Amin, M. T., Khan, F., Ahmed, S., & Imtiaz, S. (2021a). Risk-based fault detection and diagnosis for nonlinear and non-Gaussian process systems using R-vine copula. *Process Safety and Environment Protection*, *150*, 123–136. https://doi.org/10.1016/j.psep.2021.04.010.

Amin, M. T., Khan, F., Ahmed, S., & Imtiaz, S. (2021b). A data-driven Bayesian network learning method for process fault diagnosis. *Process Safety and Environment Protection*, *150*, 110–122. https://doi.org/10.1016/j.psep.2021.04.004.

Anderson, E. L., & Hattis, D. (1999). A. Uncertainty and variability. *Risk Analysis*, *19*, 47–49. https://doi.org/10.1111/j.1539-6924.1999.tb00386.x.

Argyropoulos, C. D., Christolis, M. N., Nivolianitou, Z., & Markatos, N. C. (2012). A hazards assessment methodology for large liquid hydrocarbon fuel tanks. *Journal of Loss Prevention in the Process Industries*, *25*, 329–335. https://doi.org/10.1016/j.jlp.2011.12.003.

Arzaghi, E., Abaei, M. M., Abbassi, R., Garaniya, V., Binns, J., Chin, C., et al. (2018). A hierarchical Bayesian approach to modelling fate and transport of oil released from subsea pipelines. *Process Safety and Environment Protection*, *118*, 307–315. https://doi.org/10.1016/j.psep.2018.06.023.

Aven, T. (2010). On the need for restricting the probabilistic analysis in risk assessments to variability: Perspective. *Risk Analysis*, *30*, 354–360. https://doi.org/10.1111/j.1539-6924.2009.01314.x.

Aven, T. (2011). On some recent definitions and analysis frameworks for risk, vulnerability, and resilience. *Risk Analysis*, *31*, 515–522. https://doi.org/10.1111/j.1539-6924.2010.01528.x.

Aven, T. (2013). On how to Deal with deep uncertainties in a risk assessment and management context. *Risk Analysis*, *33*, 2082–2091. https://doi.org/10.1111/risa.12067.

Aven, T. (2016). Risk assessment and risk management: Review of recent advances on their foundation. *European Journal of Operational Research*, *253*, 1–13. https://doi.org/10.1016/j.ejor.2015.12.023.

Ayoun, A., & Smets, P. (2001). Data association in multi-target detection using the transferable belief model. *International Journal of Intelligence Systems*, *16*, 1167–1182. https://doi.org/10.1002/int.1054.

Bae, H. R., Grandhi, R. V., & Canfield, R. A. (2004). An approximation approach for uncertainty quantification using evidence theory. *Reliability Engineering and System Safety*, *86*, 215–225. https://doi.org/10.1016/j.ress.2004.01.011.

Baksh, A.-A., Abbassi, R., Garaniya, V., & Khan, F. (2017). A network based approach to envisage potential accidents in offshore process facilities. *Process Safety Progress*, *36*, 178–191. https://doi.org/10.1002/prs.11854.

Bari, R. A., & Park, C. K. (1989). Uncertainty characterization of data for probabilistic risk assessment. *Reliability Engineering and System Safety*, *26*, 163–172. https://doi.org/10.1016/0951-8320(89)90072-0.

Baybutt, P. (2014). The ALARP principle in process safety. *Process Safety Progress*, *33*, 36–40. https://doi.org/10.1002/prs.11599.

Benson, C., Argyropoulos, C. D., Dimopoulos, C., Mikellidou, C. V., & Boustras, G. (2021). Safety and risk analysis in digitalized process operations warning of possible deviating conditions in the process environment. *Process Safety and Environment Protection*, *149*, 750–757. https://doi.org/10.1016/j.psep.2021.02.039.

Casamirra, M., Castiglia, F., Giardina, M., & Lombardo, C. (2009). Safety studies of a hydrogen refuelling station: Determination of the occurrence frequency of the accidental scenarios. *International Journal of Hydrogen Energy*, *34*, 5846–5854. https://doi.org/10.1016/j.ijhydene.2009.01.096.

Chang, Y., Khan, F., & Ahmed, S. (2011). A risk-based approach to design warning system for processing facilities. *Process Safety and Environment Protection*, *89*, 310–316. https://doi.org/10.1016/j.psep.2011.06.003.

Chiremsel, Z., Nait Said, R., & Chiremsel, R. (2016). Probabilistic fault diagnosis of safety instrumented systems based on fault tree analysis and Bayesian network. *Journal of Failure Analysis and Prevention*, *16*, 747–760. https://doi.org/10.1007/s11668-016-0140-z.

Clemen, R. T., & Winkler, R. L. (1999). Combining probability Distributiond from experts in risk analysis. *Risk Analysis*, *19*, 155–156. https://doi.org/10.1023/A:1006917509560.

Cooke, R. M. (2014). *Validating expert judgment with the classical model* (pp. 191–212). Cha Springer. https://doi.org/10.1007/978-3-319-08551-7_10.

Cooke, R. M. (2016). Supplementary online material for cross validation of classical model for structured expert judgment. *Resources for the future* (pp. 1–35). Delft Univ. of Technology, Dept Mathematics. November 18.

Cooke, R. M., ElSaadany, S., & Huang, X. (2008). On the performance of social network and likelihood-based expert weighting schemes. *Reliability Engineering and System Safety*, *93*, 745–756. https://doi.org/10.1016/j.ress.2007.03.017.

Cooke, R. M., & Goossens, L. L. H. J. (2008). TU Delft expert judgment data base. *Reliability Engineering and System Safety*, *93*, 657–674. https://doi.org/10.1016/j.ress.2007.03.005.

D'Angelo, G. M., & Weissfeld, L. A. (2013). Application of copulas to improve covariance estimation for partial least squares. *Statistics in Medicine*, *32*, 685–696. https://doi.org/10.1002/sim.5533.

Daneshvar, S., Yazdi, M., & Adesina, K. A. (2020). Fuzzy smart failure modes and effects analysis to improve safety performance of syste Case study of an aircraft landing system. *Quality and Reliability Engineering International*, *36*, 890–909. https://doi.org/10.1002/qre.2607.

De Rademaeker, E., Suter, G., Pasman, H. J., & Fabiano, B. (2014). A review of the past, present and future of the European loss prevention and safety promotion in the process industries. *Process Safety and Environment Protection*, *92*, 280–291. https://doi.org/10.1016/j.psep.2014.03.007.

de Ruijter, A., & Guldenmund, F. (2015). The bowtie method: A review. *Safety Science*, *88*, 211–218. https://doi.org/10.1016/j.ssci.2016.03.001.

Der Kiureghian, A., & Ditlevsen, O. (2009). Aleatory or epistemic? Does it matter? *Structural Safety*, *31*, 105–112. https://doi.org/10.1016/j.strusafe.2008.06.020.

Fazzolari, F. A., & Tan, P. (2020). A linearised hybrid FE-SEA method for nonlinear dynamic systems excited by random and harmonic loadings. *Vibration*, *3*(3), 304–319. https://doi.org/10.3390/vibration3030021.

Ferson, S., & Ginzburg, L. R. (1996). Different methods are needed to propagate ignorance and variability. *Reliability Engineering and System Safety*, *54*, 133–144. https://doi.org/10.1016/S0951-8320(96)00071-3.

Flage, R., & Aven, T. (2009). On treatment of uncertainty in system planning. *Reliability Engineering and System Safety*, *94*, 884–890. https://doi.org/10.1016/j.ress.2008.09.011.

Flage, R., Aven, T., & Berner, C. L. (2018). A comparison between a probability bounds analysis and a subjective probability approach to express epistemic uncertainties in a risk assessment context – A simple illustrative example. *Reliability Engineering and System Safety*, *169*, 1–10. https://doi.org/10.1016/j.ress.2017.07.016.

George, A. (1990). The concept of probability in safety assessments of technological systems. *Science (80-.)*, *250*, 1359–1364. https://doi.org/10.1126/science.2255906.

Gholamizadeh, K., Zarei, E., Omidvar, M., & Yazdi, M. (2022). In M. Yazdi (Ed.), *Fuzzy sets theory and human reliability: Review, applications, and contributions BT - linguistic methods under fuzzy information in system safety and reliability analysis* (pp. 91–137). Cha Springer International Publishing. https://doi.org/10.1007/978-3-030-93352-4_5.

Golestani, N., Abbassi, R., Garaniya, V., Asadnia, M., & Khan, F. (2020). Human reliability assessment for complex physical operations in harsh operating conditions. *Process Safety and Environment Protection, 140*, 1–13. https://doi.org/10.1016/j.psep.2020.04.026.

Hameed, A., Khan, F., & Ahmed, S. (2016). A risk-based shutdown inspection and maintenance interval estimation considering human error. *Process Safety and Environment Protection, 100*, 9–21. https://doi.org/10.1016/j.psep.2015.11.011.

Hashemi, S. J., Khan, F., & Ahmed, S. (2016). Multivariate probabilistic safety analysis of process facilities using the copula Bayesian network model. *Computers and Chemical Engineering, 93*, 128–142. https://doi.org/10.1016/j.compchemeng.2016.06.011.

Hassall, M. (2015). Improving human control of hazards in industry. In *Proceeding 19th Trienn. Congr. IEA, Melb. 9-14 August 2015*.

He, R., Li, X., Chen, G., Wang, Y., Jiang, S., & Zhi, C. (2018). A quantitative risk analysis model considering uncertain information. *Process Safety and Environment Protection, 118*, 361–370. https://doi.org/10.1016/J.PSEP.2018.06.029.

Hirayama, M., Shinozaki, H., Kasai, N., & Otaki, T. (2018). Comparative risk study of hydrogen and gasoline dispensers for vehicles. *International Journal of Hydrogen Energy, 43*, 12584–12594. https://doi.org/10.1016/j.ijhydene.2018.05.003.

Hora, S. C. (1996). Aleatory and epistemic uncertainty in probability elicitation with an example from hazardous waste management. *Reliability Engineering and System Safety, 54*, 217–223. https://doi.org/10.1016/S0951-8320(96)00077-4.

Hoskins, J. C., & Himmelblau, D. M. (1988). Artificial neural network models of knowledge representation in chemical engineering. *Computers and Chemical Engineering, 12*, 881–890. https://doi.org/10.1016/0098-1354(88)87015-7.

Jakab, D., Pázmándi, T., & Zagyvai, P. (2021). Effects of the uncertainty contributions on the methods used for measurement uncertainty evaluation. *Applied Radiation and Isotopes, 173*, 109704. https://doi.org/10.1016/j.apradiso.2021.109704.

Jensen, H. A., Mayorga, F., Valdebenito, M., & Chen, J. (2020). An effective parametric model reduction technique for uncertainty propagation analysis in structural dynamics. *Reliability Engineering and System Safety, 195*, 106723. https://doi.org/10.1016/j.ress.2019.106723.

Jiang, G.-J., Chen, H.-X., Sun, H.-H., Yazdi, M., Nedjati, A., & Adesina, K. A. (2021). An improved multi-criteria emergency decision-making method in environmental disasters. *Soft Computing, 25*, 10351–10379. https://doi.org/10.1007/s00500-021-05826-x.

Kabir, S., Geok, T. K., Kumar, M., Yazdi, M., & Hossain, F. (2020). A method for temporal fault tree analysis using intuitionistic fuzzy set and expert elicitation. *IEEE Access, 8*, 980–996. https://doi.org/10.1109/ACCESS.2019.2961953.

Kalantarnia, M., Khan, F., & Hawboldt, K. (2010). Modelling of BP Texas City refinery accident using dynamic risk assessment approach. *Process Safety and Environment Protection, 88*, 191–199. https://doi.org/10.1016/j.psep.2010.01.004.

Kelly, B. D. (2011). Why process safety programs sometimes fail. *Process Safety Progress, 30*, 307–309. https://doi.org/10.1002/prs.10494.

Khan, F., Hashemi, S. J., Paltrinieri, N., Amyotte, P., Cozzani, V., & Reniers, G. (2016). Dynamic risk management: A contemporary approach to process safety management. *Current Opinion in Chemical Engineering, 14*, 9–17. https://doi.org/10.1016/j.coche.2016.07.006.

Koc, R., Kazantzis, N. K., Nuttall, W. J., & Ma, Y. H. (2013). An economic evaluation framework for membrane reactor modules in the presence of uncertainty: The case for process safety investment and risk reduction. *Journal of Loss Prevention in the Process Industries, 26*, 468–477. https://doi.org/10.1016/j.jlp.2012.07.012.

Lee, J., Cameron, I., & Hassall, M. (2019). Improving process safety: What roles for digitalization and industry 4.0? *Process Safety and Environment Protection*, *132*, 325–339. https://doi.org/10.1016/j.psep.2019.10.021.

Lefevre, E., Colot, O., & Vannoorenberghe, P. (2002). Belief function combination and conflict management. *Information Fusion*, *3*, 149–162. https://doi.org/10.1016/S1566-2535(02)00053-2.

Li, R., Olson, J. H., & Chester, D. L. (1990). Dynamic fault detection and diagnosis using neural networks. In *Vol. 2. Proceedings. 5th IEEE international symposium on intelligent control 1990* (pp. 1169–1174). https://doi.org/10.1109/ISIC.1990.128602.

Liaw, H.-J. (2019). Deficiencies frequently encountered in the management of process safety information. *Process Safety and Environment Protection*, *132*, 226–230. https://doi.org/10.1016/j.psep.2019.10.015.

Liu, H. (2016a). *FMEA using uncertainty theories and MCDM methods*. https://doi.org/10.1007/978-981-10-1466-6.

Liu, H. C. (2016b). *FMEA using uncertainty theories and MCDM methods, FMEA using uncertainty theories and MCDM methods*. https://doi.org/10.1007/978-981-10-1466-6.

Liu, Y., Li, L., Zhao, S., & Song, S. (2021). A global surrogate model technique based on principal component analysis and kriging for uncertainty propagation of dynamic systems. *Reliability Engineering and System Safety*, *207*, 107365. https://doi.org/10.1016/j.ress.2020.107365.

Lou, C., Li, X., & Atoui, M. A. (2020). Bayesian network based on an adaptive threshold scheme for fault detection and classification. *Industrial and Engineering Chemistry Research*, *59*, 15155–15164. https://doi.org/10.1021/acs.iecr.0c02762.

Madani, K., & Lund, J. R. (2011). A Monte-Carlo game theoretic approach for multi-criteria decision making under uncertainty. *Advances in Water Resources*, *34*, 607–616. https://doi.org/10.1016/j.advwatres.2011.02.009.

Markowski, A. S., & Siuta, D. (2017). Selection of representative accident scenarios for major industrial accidents. *Process Safety and Environment Protection*, *111*, 652–662. https://doi.org/10.1016/j.psep.2017.08.026.

Markowski, A. S., & Siuta, D. (2018). Fuzzy logic approach for identifying representative accident scenarios. *Journal of Loss Prevention in the Process Industries*, *56*, 414–423. https://doi.org/10.1016/j.jlp.2018.10.003.

Martorell, S., Martón, I., Villamizar, M., Sánchez, A. I., & Carlos, S. (2014). Evaluation of risk impact of changes to completion times addressing model and parameter uncertainties. *Reliability Engineering and System Safety*, *130*, 190–201. https://doi.org/10.1016/j.ress.2014.06.003.

Md Nor, N., Che Hassan, C. R., & Hussain, M. A. (2020). A review of data-driven fault detection and diagnosis methods: Applications in chemical process systems. *Reviews in Chemical Engineering*, *36*, 513–553. https://doi.org/10.1515/revce-2017-0069.

Melchers, R. E. (2001). On the ALARP approach to risk management. *Reliability Engineering and System Safety*, *71*, 201–208. https://doi.org/10.1016/S0951-8320(00)00096-X.

Moges, E., Demissie, Y., & Li, H. (2020). Uncertainty propagation in coupled hydrological models using winding stairs and null-space Monte Carlo methods. *Journal of Hydrology*, *589*, 125341. https://doi.org/10.1016/j.jhydrol.2020.125341.

Mohammadfam, I., & Zarei, E. (2015). Safety risk modeling and major accidents analysis of hydrogen and natural gas releases: A comprehensive risk analysis framework. *International Journal of Hydrogen Energy*, *40*, 13653–13663. https://doi.org/10.1016/j.ijhydene.2015.07.117.

Mohsendokht, M. (2017). Risk assessment of uranium hexafluoride release from a uranium conversion facility by using a fuzzy approach. *Journal of Loss Prevention in the Process Industries*, *45*, 217–228. https://doi.org/10.1016/j.jlp.2017.01.004.

Morales-Nápoles, O., Paprotny, D., Worm, D., Abspoel-Bukman, L., & Courage, W. (2017). Characterization of precipitation through copulas and expert judgement for risk assessment of infrastructure. *ASCE-ASME Journal of Risk and Uncertainty in Engineering Systems, Part A: Civil Engineering*, *3*, 04017012. https://doi.org/10.1061/ajrua6.0000914.

National Research Council. (2009). *Science and decisions: Advancing risk assessment*. Washington DC: National Academies Press.

Nedjati, A., Yazdi, M., & Abbassi, R. (2021). A sustainable perspective of optimal site selection of giant air - purifiers in large metropolitan areas. In *Environment, development and sustainability*. Netherlands: Springer. https://doi.org/10.1007/s10668-021-01807-0.

Pan, Y., Zhang, L., Koh, J., & Deng, Y. (2021). An adaptive decision making method with copula Bayesian network for location selection. *Information Sciences*, *544*, 56–77. https://doi.org/10.1016/j.ins.2020.07.063.

Parry, G. W. (1996). The characterization of uncertainty in probabilistic risk assessments of complex systems. *Reliability Engineering and System Safety*, *54*, 119–126. https://doi.org/10.1016/S0951-8320(96)00069-5.

Parry, G. W., & Winter, P. W. (1981). Characterization and evaluation of uncertainty in probabilistic risk analysis. *Nuclear Safety*, *22*, 28–42.

Pasman, H. J., Jung, S., Prem, K., Rogers, W. J., & Yang, X. (2009). Is risk analysis a useful tool for improving process safety? *Journal of Loss Prevention in the Process Industries*, *22*, 769–777. https://doi.org/10.1016/j.jlp.2009.08.001.

Paté-Cornell, M. E. (1996). Uncertainties in risk analysis: Six levels of treatment. *Reliability Engineering and System Safety*, *54*, 95–111. https://doi.org/10.1016/S0951-8320(96)00067-1.

Patriarca, R., Falegnami, A., De Nicola, A., Villani, M. L., & Paltrinieri, N. (2019). Serious games for industrial safety: An approach for developing resilience early warning indicators. *Safety Science*, *118*, 316–331. https://doi.org/10.1016/j.ssci.2019.05.031.

Pike, H., Khan, F., & Amyotte, P. (2020). Precautionary principle (PP) versus as low as reasonably practicable (ALARP): Which one to use and when. *Process Safety and Environment Protection*, *137*, 158–168. https://doi.org/10.1016/j.psep.2020.02.026.

Pitblado, R., & Nelson, W. R. (2013). Advanced safety barrier management with inclusion of human and organizational aspects. *Chemical Engineering Transactions*, *31*, 331–336. https://doi.org/10.3303/CET1331056.

Ramebäck, H., & Lindgren, P. (2018). Uncertainty evaluation in gamma spectrometric measurements: Uncertainty propagation versus Monte Carlo simulation. *Applied Radiation and Isotopes*, *142*, 71–76. https://doi.org/10.1016/j.apradiso.2018.09.024.

Rausand, M. (2011). *Risk assessment: Theory, methods, and applications*. Wiley.

Rausand, M., & Haugen, S. (2020). *Risk assessment: Theory, methods, and applications*. Wiley.

Romero, D., Mattsson, S., Fast-Berglund, Å., Wuest, T., Gorecky, D., & Stahre, J. (2018). Digitalizing occupational health, safety and productivity for the operator 4.0. In *Vol. 536. IFIP advances in information and communication technology* (pp. 473–481). https://doi.org/10.1007/978-3-319-99707-0_59.

Schmitz, P., Swuste, P., Reniers, G., & van Nunen, K. (2020). Mechanical integrity of process installations: Barrier alarm management based on bowties. *Process Safety and Environment Protection*, *138*, 139–147. https://doi.org/10.1016/j.psep.2020.03.009.

Shekari, E., Khan, F., & Ahmed, S. (2015). A predictive approach to fitness-for-service assessment of pitting corrosion. *International Journal of Pressure Vessels and Piping*, *137*, 13–21. https://doi.org/10.1016/j.ijpvp.2015.04.014.

Shi, S., Kuschmierz, R., Zhang, G., Lin, J., Czarske, J., & Qu, J. (2020). Modeling, quantification, and mitigation of uncertainty propagation in two-step roundness measurements. *Measurement*, *155*, 107530. https://doi.org/10.1016/j.measurement.2020.107530.

Shi, L., Wang, J., Zhang, G., Cheng, X., & Zhao, X. (2017). A risk assessment method to quantitatively investigate the methane explosion in underground coal mine. *Process Safety and Environment Protection*, *107*, 317–333. https://doi.org/10.1016/j.psep.2017.02.023.

Sorsa, T., Koivo, H. N., & Koivisto, H. (1991). Neural networks in process fault diagnosis. *IEEE Transactions on Systems, Man, and Cybernetics*, *21*, 815–825. https://doi.org/10.1109/21.108299.

Steen, R., & Aven, T. (2011). A risk perspective suitable for resilience engineering. *Safety Science*, *49*, 292–297. https://doi.org/10.1016/j.ssci.2010.09.003.

Sultana, S., Andersen, B. S., & Haugen, S. (2019). Identifying safety indicators for safety performance measurement using a system engineering approach. *Process Safety and Environment Protection*, *128*, 107–120. https://doi.org/10.1016/j.psep.2019.05.047.

Tsai, S. B., Zhou, J., Gao, Y., Wang, J., Li, G., Zheng, Y., et al. (2017). Combining FMEA with DEMATEL models to solve production process problems. *PLoS One*, *12*, 1–15. https://doi.org/10.1371/journal.pone.0183634.

Wang, B. (2021). Safety intelligence as an essential perspective for safety management in the era of safety 4.0: From a theoretical to a practical framework. *Process Safety and Environment Protection*, *148*, 189–199. https://doi.org/10.1016/j.psep.2020.10.008.

Weidl, G., Madsen, A. L., & Israelson, S. (2005). Applications of object-oriented Bayesian networks for condition monitoring, root cause analysis and decision support on operation of complex continuous processes. *Computers and Chemical Engineering*, *29*, 1996–2009. https://doi.org/10.1016/j.compchemeng.2005.05.005.

Wu, D., & Zhao, J. (2021). Process topology convolutional network model for chemical process fault diagnosis. *Process Safety and Environmental Protection*, *150*, 93–109. https://doi.org/10.1016/j.psep.2021.03.052.

Yager, R. R. (1987). On the dempster-Shafer framework and new combination rules. *Information Sciences*, *41*, 93–137. https://doi.org/10.1016/0020-0255(87)90007-7.

Yang, C. (2021). A novel uncertainty-oriented regularization method for load identification. *Mechanical Systems and Signal Processing*, *158*, 107774. https://doi.org/10.1016/j.ymssp.2021.107774.

Yazdi, M. (2017). Hybrid probabilistic risk assessment using fuzzy FTA and fuzzy AHP in a process industry. *Journal of Failure Analysis and Prevention*, *17*, 756–764. https://doi.org/10.1007/s11668-017-0305-4.

Yazdi, M. (2018). Risk assessment based on novel intuitionistic fuzzy-hybrid-modified TOPSIS approach. *Safety Science*, *110*, 438–448. https://doi.org/10.1016/j.ssci.2018.03.005.

Yazdi, M. (2019). Improving failure mode and effect analysis (FMEA) with consideration of uncertainty handling as an interactive approach. *International Journal on Interactive Design and Manufacturing*, *13*, 441–458. https://doi.org/10.1007/s12008-018-0496-2.

Yazdi, M. (2019a). Footprint of knowledge acquisition improvement in failure diagnosis analysis. *Quality and Reliability Engineering International*, *35*, 405–422. https://doi.org/10.1002/qre.2408.

Yazdi, M. (2019b). Introducing a heuristic approach to enhance the reliability of system safety assessment. *Quality and Reliability Engineering International*, *35*, 2612–2638. https://doi.org/10.1002/qre.2545.

Yazdi, M., Adesina, K. A., Korhan, O., & Nikfar, F. (2019). Learning from fire accident at Bouali Sina petrochemical complex plant. *Journal of Failure Analysis and Prevention*, *19*, 1517–1536. https://doi.org/10.1007/s11668-019-00769-w.

Yazdi, M., Daneshvar, S., & Setareh, H. (2017). An extension to fuzzy developed failure mode and effects analysis (FDFMEA) application for aircraft landing system. *Safety Science*, *98*, 113–123. https://doi.org/10.1016/j.ssci.2017.06.009.

Yazdi, M., Golilarz, N. A., Adesina, K. A., & Nedjati, A. (2021a). Probabilistic risk analysis of process systems considering epistemic and aleatory uncertainties: A comparison study. *International Journal of Uncertainty, Fuzziness and Knowledge-Based Systems*, *29*, 181–207. https://doi.org/10.1142/S0218488521500098.

Yazdi, M., & Kabir, S. (2017). A fuzzy Bayesian network approach for risk analysis in process industries. *Process Safety and Environment Protection, 111*, 507–519. https://doi.org/10.1016/j.psep.2017.08.015.

Yazdi, M., & Kabir, S. (2018). Fuzzy evidence theory and Bayesian networks for process systems risk analysis. *Human and Ecological Risk Assessment, 26*(1), 57–86. https://doi.org/10.1080/10807039.2018.1493679.

Yazdi, M., Khan, F., & Abbassi, R. (2021b). Microbiologically influenced corrosion (MIC) management using Bayesian inference. *Ocean Engineering, 226*, 108852. https://doi.org/10.1016/j.oceaneng.2021.108852.

Yazdi, M., Khan, F., & Abbassi, R. (2021c). Operational subsea pipeline assessment affected by multiple defects of microbiologically influenced corrosion. *Process Safety and Environment Protection, 158*, 159–171. https://doi.org/10.1016/j.psep.2021.11.032.

Yazdi, M., Khan, F., Abbassi, R., & Rusli, R. (2020). Improved DEMATEL methodology for effective safety management decision- making. *Safety Science, 127*, 104705. https://doi.org/10.1016/j.ssci.2020.104705.

Yazdi, M., Korhan, O., & Daneshvar, S. (2018). Application of fuzzy fault tree analysis based on modified fuzzy AHP and fuzzy TOPSIS for fire and explosion in process industry. *International Journal of Occupational Safety and Ergonomics, 0*, 1–18. https://doi.org/10.1080/10803548.2018.1454636.

Yazdi, M., & Zarei, E. (2022). In M. Yazdi (Ed.), *Step forward on how to treat linguistic terms in judgment in failure probability estimation BT - linguistic methods under fuzzy information in system safety and reliability analysis* (pp. 193–200). Cha Springer International Publishing. https://doi.org/10.1007/978-3-030-93352-4_10.

Zadeh, L. A. (1986). Simple view of the Dempster-Shafer theory of evidence and its implication for the rule of combination. *AI Magazine*, 7, 85–90. https://doi.org/10.1609/aimag.v7i2.542.

Zadeh, L. A. (1999). Fuzzy sets as a basis for a theory of possibility. *Fuzzy Sets and Systems, 100*, 9–34. https://doi.org/10.1016/S0165-0114(99)80004-9.

Zarei, E., Yazdi, M., Abbassi, R., & Khan, F. (2019). A hybrid model for human factor analysis in process accidents: FBN-HFACS. *Journal of Loss Prevention in the Process Industries, 57*, 142–155. https://doi.org/10.1016/j.jlp.2018.11.015.

Zhang, W., & Deng, Y. (2018). Combining conflicting evidence using the DEMATEL method. *Soft Computing, 23*, 8207–8216. https://doi.org/10.1007/s00500-018-3455-8.

Zhao, Q., & Hong, J. (2022). An analytical framework for local and global system kinematic reliability sensitivity of robotic manipulators. *Applied Mathematical Modelling, 102*, 331–350. https://doi.org/10.1016/j.apm.2021.09.021.

Zio, E., & Apostolakis, G. E. (1996). Two methods for the structured assessment of model uncertainty by experts in performance assessments of radioactive waste repositories. *Reliability Engineering and System Safety, 54*, 225–241. https://doi.org/10.1016/S0951-8320(96)00078-6.

CHAPTER TWELVE

Human factors in digitalized process operations

Rajagopalan Srinivasan[a,d,*], Babji Srinivasan[b,d], and Mohd Umair Iqbal[c]

[a]Department of Chemical Engineering, Indian Institute of Technology Madras, Chennai, India
[b]Department of Applied Mechanics, Indian Institute of Technology Madras, Chennai, India
[c]Department of Chemical Engineering, Indian Institute of Technology Gandhinagar, Gandhinagar, India
[d]American Express Lab for Data Analytics, Risk & Technology, Indian Institute of Technology Madras, Chennai, India
*Corresponding author: e-mail address: raj@iitm.ac.in

Contents

1. Introduction

Chemical and process industries are highly hazardous; occasionally, these hazards translate to catastrophic accidents. There have been significant improvements in safety methods resulting in safer processes. However, disasters continue to occur, shaking public confidence in the chemical process industries. Among the most significant contributors to accidents is human

Methods in Chemical Process Safety, Volume 6
ISSN 2468-6514
https://doi.org/10.1016/bs.mcps.2022.04.007

error—the root cause of 60–80% of accidents (Brauer, 2016). An examination of severe industrial incidents caused by hazardous chemicals from 2008 to 2018 indicates that more than 76.1% of them were caused by human error (Jung, Woo, & Kang, 2020). Thus, it is imperative to address the issue of human errors—which wholly or partially occur due to the way individuals interact with systems— to ensure safe and profitable operations in industries.

Addressing the issue of human errors has always been challenging, given the complexities and uncertainties associated with human behavior, along with the ever-evolving role of humans. Increased digitalization and sophistication have resulted in a change in the way operators interact with the system, and it can pose new safety issues. The requirement to acquire, segregate, and integrate information from multiple information sources and use them for maintaining optimal operating conditions can be cognitively challenging. An overabundance of information can result in increased workload and degrade operators' performance if they are not trained adequately. For instance, one of the causes of the Bayer CropScience accident (2008) was the incorporation of a new distributed control system (DCS) which posed new challenges to operators (CSB, 2008). The new DCS interface contained features intended to minimize human error, such as graphical display screens that simulated process flow and automated icons to display process variables. While the older system display and command entry was a spreadsheet, detailed process equipment displays in the new DCS were challenging to navigate. The requirement to move between multiple screens resulted in degraded operator awareness and response times during routine operations like troubleshooting alarms and starting a reaction. As a result, inadequate DCS checkout prevented operators from achieving correct operating conditions on the day of the accident (CSB, 2008). This clearly highlights the need for assessing the impact of new technologies on operator performance.

Digitization has resulted in the enhancement of autonomous process operation by using advanced methods for monitoring, modeling, optimization, and control (Khan, Amyotte, & Adedigba, 2021). However, there must be an appropriate trust calibration (Borenstein, Mahajan, Wagner, & Howard, 2020) between humans and automation. If trust is not enough, human intervention may risk the system. Conversely, if there is too much trust in an automated system, operators may lack sufficient time to act once the problem is identified. Problems can get complicated if people do not have enough training in the working of an automation system. The complications can be best exemplified by the crashes involving Boeing 737 MAX airplanes. Boeing had used a Maneuvering Characteristics Augmentation System (MCAS) software system to mask significant changes in the hardware configuration of their airplanes. However, the role of MCAS in compensating for design changes involving the repositioning of engines was not communicated to the pilots (Herkert, Borenstein, & Miller, 2020). These changes in the design could lead to the possibility of a nose-up stall during certain conditions (Glanz, Creswell, Kaplan, & Wichter, 2019). The MCAS system was designed to detect critical situations using an Angle of Attack (AOA) sensor and intervene in the event of an imminent stall. On the day of the crash, the failure of an Angle of Attack (AOA) sensor sent faulty data, which led to subsequent activation of MCAS, forcing the nose of the aircraft down several times (Ahmed, Glanz, & Beech, 2019; Gates, 2019). However, inadequate information and training about the upgraded software prevented the pilots from successfully taking back control of the airplane.

In the process industries, advanced control techniques such as Model Predictive Controller (MPC) have received limited operator acceptance compared to those of basic PI(D). This is despite MPC being a powerful tool for process control and offering numerous benefits such as optimization of multivariable control problems, ease of tuning, and incorporation of constraints (Mahmoud, 2016). A successful application requires that a controller should be perceived as simple and transparent by an operator (Hansen, Jespersen, Bram, & Yang, 2018). However, MPC is complex in nature. Its control actions are not easy to understand (Lindscheid, Bremer, Haßkerl, Tatulea-Codrean, & Engell, 2016), making it difficult for operators to monitor, diagnose, and control MPC (Forbes, Patwardhan, Hamadah, & Gopaluni, 2015). There have been several instances of the operator's lack of confidence in MPC due to difficulty in its understanding, because of which MPC is often disabled (Campos, Teixeira, Liporace, & Gomes, 2009; Lindscheid et al., 2016). In order to harness the benefits of an advanced controller, operators must be able to evaluate control actions (Hansen et al., 2018) so that errors are minimized when they have to take over. Thus, it is necessary to address challenges that are involved in human-automation interaction so that overall system performance is optimized and safety ensured.

In summary, digitalization and concomitant automation throw new challenges, especially in the manner people interact with and experience technology. If individuals are not given enough information about what the automation is doing, then human supervisors are disconnected from the functioning of the system. Such disconnects result in skill deterioration and loss of situational awareness. Adequate situational awareness and skills are critical, especially while dealing with abnormal situations which are often outside the purview of automated control systems. Further, automation does not merely replace tasks that people do; it changes the nature of tasks and invokes different cognitive responses, such as inattention when a machine is doing a job for an individual. Likewise, the emotional response of people towards technology may also change, such as the level of trust in automation. In addition to this, digitization also means a requirement for fewer people at workplaces which in turn can pose different challenges such as higher cognitive workload during emergency situations. Thus, it is imperative to account for and minimize human errors that arise due to automation. In this chapter, we discuss how the science of human factors can be used to minimize human errors. We also highlight the role of physiological measurements in providing additional insights into the cognitive aspects of human-machine interaction.

The rest of the chapter is organized as follows. Section 2 provides a detailed discussion on human factors and how their consideration is critical to optimal human-machine interactions. In Section 3, we provide a general discussion on human errors in the process industries and the various perspectives that exist. Next, in Section 4, we discuss several human reliability analysis techniques that can be used to quantify human failure so as to help in the minimization of human errors and improve safety. In Section 5, we highlight the role of various sensors such as eye-tracking, electroencephalography (EEG), galvanic skin resistance (GSR), and electrocardiography (ECG) in providing critical information about individual performances which cannot be extracted through traditional approaches. Finally, we provide conclusions in Section 5.

2. Human factors

Human factors (HF) is a scientific discipline that is concerned with the understanding of interactions of people with other elements of the work system in order to optimize their well-being and overall system performance (IEA, 2021). Within a work system, several types of human interactions are essential to achieve the system goal. The nature of these interactions is influenced by the characteristics of work system elements. The elements of a work system include the following (Edmonds, 2016).

(a) Work tasks: Functions that are required to be performed by the work system can be typically allocated either to machines or to people, or both. Human tasks typically can be grouped into job roles such as maintainer and operator. Performance during these tasks includes consideration of both physical elements as well as cognitive elements (psychological aspects). There has to be an optimal allocation of tasks between operators and automation.

(b) Work tools and Equipment: These elements are used by people to carry out allocated tasks. These can take any form, ranging from a simple spanner to standard operating procedures to complex control systems. Work tools and equipment have become technologically advanced; however, people have not evolved at the same pace (Friedman, 2017). Therefore, tools and equipment have to be designed to meet the needs of the end-user by taking into account human capabilities and limitations.

(c) Workspace and Work areas: It includes physical space in which people perform their tasks. Physical space serves two needs: provide sufficient area for people and the ability to reach and interact with equipment in a posture that is comfortable and unlikely to result in an injury. In addition to physical accommodation, the layout and arrangement of space have to be optimized to support the required functionality.

(d) Environmental context: Even when operators do not directly interact with the environment, it critically influences operator performance. HF commonly deals with environmental considerations that include vibration, noise, thermal comfort, and lighting. Other aspects of the environment could include hazards such as cold or hot surfaces. HF aims to consider two aspects which include protecting people from adverse effects of the environment and performance enhancement by optimal design of environmental conditions. The environment can affect a person's physical as well as psychological capability. For instance, noise can result in hearing loss and impact communication. Thermal conditions that can place excessive demand on the thermoregulation of operators can result in reduced performance. Similarly, adequate lighting conditions are required for operators to perform visual tasks. Adverse lighting conditions such as reflection and glare can have an adverse effect on performance and compromise safety.

(e) Organizational context: Several organizational factors have an impact on operator performance and safety. These include (i) organizational culture (shared attitudes, ways of behaving, and beliefs), (ii) management of change (a mismanaged change can have an adverse impact while a well-managed change has a positive impact), (iii) Staffing and workload (concerned with having an optimal number of people and optimally allocating people to roles for which they have adequate knowledge and skills. Required level of competence is achieved and maintained through proper training and experience), (iv) Fatigue (concerned with the way rosters are organized. Improper sleep and hours of work can adversely impact performance by exacerbating effects of fatigue), (v) safety-critical communication (methods of communication between transmitters and receivers can affect the reliability of communication and safe performance. Reliable communication is especially critical during situations such as shift handover and permit-to-work systems, which are prone to communication vulnerability), (vi) Psychological well-being (organizational factors such as shift size, training, rewards, and punishments can impact mental well-being, and the need to be optimized to maintain safe behaviors).

Human factors within the work system are not independent of each other; they represent a complex web of interactions with all aspects of the system. In the context of process industries, a basic model of HF can be represented as shown in Fig. 1 (Crowl, 2007). This model identifies three broad domains

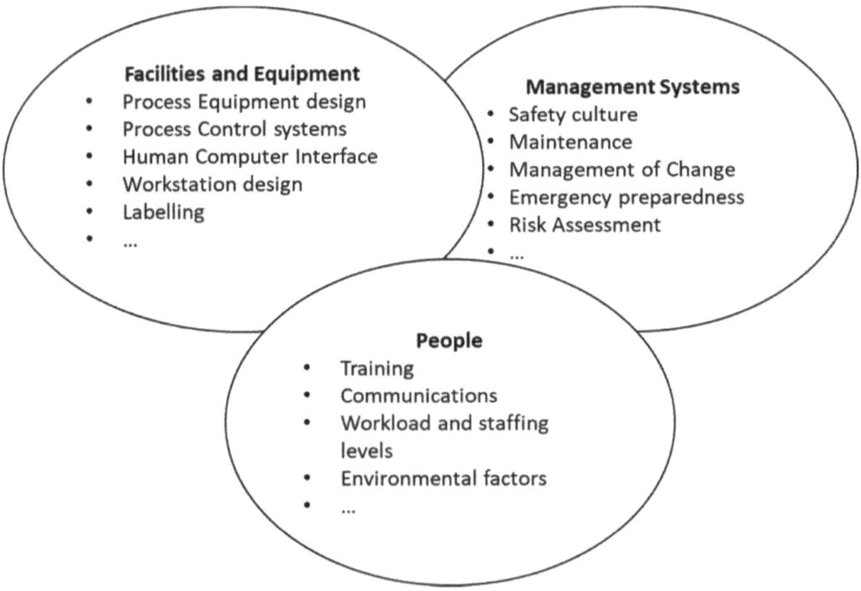

Fig. 1 A basic representation of human factors.

for HF—Facilities and Equipment, People, and Management Systems. The broad topics that are studied under each heading (Crowl, 2007) are also provided in Fig. 1. Typically, the Facilities and Equipment domain are concerned with physical characteristics, workspace design, maintenance of equipment, and reliability. People's domain deals with individual characteristics, skills, perceptions, and factors related to fatigue, stress, and fitness. The management systems domain is concerned with the framework under which operations are carried out and includes procedures, training, process safety-related work environment, and safety culture. All these domains overlap with each other with an ultimate aim to identify and mitigate sources and consequences of errors by fitting task and environment to the person rather than forcing significant adaptations in the person to perform a given task (Crowl, 2007).

Failure of HF has been one of the primary reasons for human errors leading to catastrophic accidents in highly hazardous industries. For instance, the lack of consideration of factors such as training and workforce competency, procedure deviations, and poor safety culture contributed to the Texas City Refinery explosion of 2005 (CSB, 2007). The accident occurred during the start-up of an isomerization unit. A raffinate splitter tower got overfilled and led to the release of flammables which caught fire and exploded. Procedures were not followed during start-up; operators were allowed to alter procedural steps without management of change analysis. Problems during start-up were so common that operators believed that deviations were necessary for an efficient start-up. BP has no shift communication policy, and formal shift turnover was not enforced between day board operators and outside operators. Instruments were misleading; the sight glass on the raffinate tower was unreadable since it was dirty, and a high alarm on the tower did not occur to alert operators. Likewise, the level indicator gave faulty readings due to miscalibration and suggested the level of liquid in the tower was declining. Day board operator was experiencing fatigue due to long 12h shifts for 29 consecutive days, which would have resulted in impaired performance due to operator fatigue. The Board operator did not recognize that feed was not being removed from the tower, and the tower was overfilling. The poor design of the control board made things worse; separate screens were used to display liquid in and liquid out of the unit (HSE, 2022b). Inadequate staffing and training levels also contributed to the accident; when the Day supervisor left the refinery, there was no one with experience in ISOM unit start-up to assist board operators

(Lees, 2012). The reader is referred to Crowl (2007) (p.12) for a summary of major accidents with HF-related issues.

In the context of highly hazardous industries, the UK Health and Safety Executive identified the top issues as human failures, HF in design, procedures, organizational culture, organizational change, staffing, training and competence, safety-critical communications, fatigue and shift work, and finally, maintenance, inspection, and testing. The application of HF principles can help minimize risk. The reader is referred to HSE (2022c) for more details.

HF checklists are typically utilized at various phases of a plant's lifecycle to increase safety. These checklists are used to investigate the nature of interaction among various work system components/elements, such as work tasks, work tools, workspaces/areas, environmental context, and organizational context (Edmonds, 2016). The fundamental goal of these approaches/analyses is to make the system safer by decreasing various hazards and lowering the chance of human errors. For example, checklists for the design of a control room may include lines of sight, space, and evacuation of the control room in order to reduce several ergonomic errors. It also takes into account HF for linked auxiliaries such as equipment rooms, office rooms, training rooms, and their layout, among other things. Similarly, checklists for HMI (Human Machine Interface) design in a control room are typically geared towards ensuring ease of interaction of people with computer interface by analyzing various aspects such as legibility of texts on an interface (e.g., font color, size), level of detail (too much or too little), and alarm status as different from normal plant operating conditions. Along with checklists, guidelines are provided to guarantee that HF is taken into account throughout the lifecycle of a plant. For example, guidelines are provided for training plant operators (depending on the nature of work), with a goal of guaranteeing a suitable level of training, regular performance assessment, and subsequently periodic training (if required). Guidelines are also offered to ensure that the organization adequately contributes towards improved performance and safety so that risk is minimized in a facility. Table 1 provides an example of an HF checklist for control room design and Human/Computer interface (HCI) (Crowl, 2007). The interested reader is referred to Crowl (2007) for checklists for process equipment design; process control systems; control center design; remote operations; facilities and workstation design; HCI; safe havens; labeling; training; communications; documentation design and use; environmental factors; workloads and staffing levels; shiftwork issues; etc.

Table 1 Human Factor Checklist for control center design and HMI (Crowl, 2007).

Control center design	Q1. Does the layout provide sufficient egress, access, and freedom of movement?
	Q.2. Is the layout conducive to clear lines of sight?
	Q.3 Is the design conducive to necessary communication?
	Q.4. Does the layout provide enough space for workers and their workstations?
	...
Human-computer interface	Q.1. Do screens use familiar metaphors (process functions, process structure, tasks, etc.)?
	Q.2. Is text size legible, i.e., is font size and type appropriate?
	Q.3. Is the information delivered in an appropriate manner (numeric value, tables, trends, etc.)?
	Q.4 Is there enough spacing between active areas of the touch screen?
	Q.5 Do screens provide only the information that is needed by the operator (no excessive/unnecessary details)? (Relevant to dynamic conditions)
	...

Minimizing risk is one of the primary needs of every industry, and when it comes to risk, the element of operator performance does contribute to the possibility of hazards transforming into accidents. This includes the potential of an operator making an error in a task or, alternatively, a poor design of equipment affecting operator performance. Nonetheless, humans are not always a source of hazard, but they are also defenses against system failures and consequent risk. In fact, at times, they are better at managing risk than automation, especially during abnormal situations. Designers cannot foresee all possibilities. While automated systems ensure consistency because they are based on logical arguments/rules, they cannot do anything in situations for which they are not programmed. For example, if X happens, then do Y. But if the program does not detect a value of X for which it is designed to perform a certain action, then the automated system fails. This is especially true during novel abnormal situations. Humans can perform better in such situations. For instance, Apollo 13, the US spaceflight, was launched on 11 April 1970 and suffered an oxygen tank explosion and rupture en route to the moon, threatening the lives of astronauts on board. There were no standard operating procedures to deal with such an abnormal situation. A decision was made to return to earth by using a lunar module. However, lithium hydroxide canisters that removed carbon dioxide in the lunar module would not last long, and afterward, the module would not support life. To make the lunar module habitable, astronauts used their knowledge and modified lithium canisters of the command module to fit them into the lunar module (Britannica, 2021). After this incident, components used in different modules were made to be compatible with each other so that modifications were not necessary.

The above example highlights that there is a need to proactively identify hazards by incorporating HF into risk assessment. The application of knowledge of HF to manage risk is based on two principles (Edmonds, 2016): (i) Identifying how human failures/errors can contribute to risk, and (ii) how that risk can be reduced, including how to best utilize people to reduce risk. These principles can be applied either at the task level or higher level and are often accounted for through human reliability analysis.

Next, we discuss how human errors can be accounted for in the process industries.

3. Human errors

A comprehensive framework for identifying and analyzing human factors has been a challenge for a long. Consequently, interventions could not be targeted appropriately to minimize human errors, nor could their effectiveness be objectively assessed. One of the early approaches to developing a framework to account for human error was the "Swiss Cheese" model (Reason, 1990). Holes in the cheese correspond to failed defenses and are a result of both active failures (error of omission or commission) and latent error (causal factor or contributing factor). As per this model, there are several layers of protection (depicted as cheese slices) for each hazard that prevent it from transforming into an accident. These protection layers correspond to people, work system, equipment and design, and organization. Drawing on Reason's Swiss Cheese model, the human factor analysis and classification system (HFACS) was developed by Shappell and Wiegmann (2000) as a broad human error framework. HFACS framework describes human errors at four different levels with an aim to understand underlying contributory factors that lead to an accident. The four levels include unsafe acts of operators that lead to accidents, preconditions for unsafe acts, unsafe supervision, and organizational influences.

3.1 Unsafe Acts

An overview of unsafe acts is shown in Fig. 2. These acts can be categorized into two broad levels: errors and violations (Reason, 1990). Errors are unintentional physical and/or mental activities of an operator that fail to achieve an intended goal. Violations, on the other hand, are willful actions on the part of the operator involving disregard for rules and regulations, compromising safety. For increasing granularity for effective accident prevention, errors are further classified into three sub-types, including skill-based, decision, and perceptual errors.

Skill-based errors usually occur in familiar and highly practiced operations that involve actions requiring little conscious thought. Examples of skill-based errors could be moving a switch up rather than down, taking a reading from a wrong instrument, omitting a checklist item, and missing a crucial step in a safety-critical procedure. These are often a result of a failure of attention and/or memory. Depending on whether skill-based errors include errors of commission or omission, failure modes are categorized into slips or lapses, respectively. Decision errors represent intentional behavior, as intended, but prove inappropriate or inadequate for the situation. These are a result of errors in judgment involving mental processes linked to planning, information acquisition, and communication. Decision errors are often categorized into procedural errors (rule-based mistakes), poor choices, and problem-solving errors (knowledge-based mistakes) (Rasmussen, 1982). An example of a rule-based mistake could be to ignore an alarm in a real emergency, following a history of false alarms. On the other hand, knowledge-based mistakes could occur in novel situations for which there

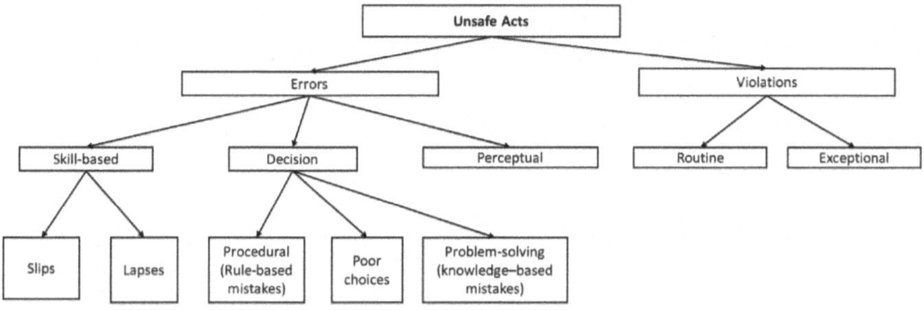

Fig. 2 An overview of unsafe acts.

are no pre-set rules or procedures, and operators rely on their knowledge and experience. An example of knowledge-based errors could be a misdiagnosis of process upset and consequently taking wrong control action(s). Perceptual errors typically occur when an operators' sensory input is degraded, and a decision is made based on faulty information.

Violations can be broadly classified into two types: routine violations and exceptional violations. Routine violations involve habitual actions that are often tolerated by enforcing agencies, such as resorting to shortcuts to increase production. Exceptional violations involve situations that do not represent an operators' typical behavior but are an isolated departure (Reason, 1990)—for instance, delaying emergency shut-down to prevent loss of production (HSE, 2022a).

3.2 Preconditions for unsafe acts

Focusing only on unsafe acts is akin to focusing only on the symptoms. To properly understand why errors occur, it is important to consider the conditions that lead to these errors. Accordingly, preconditions for unsafe acts are categorized into substandard conditions of operators, sub-standard practices of operators, and environmental factors. Sub-standard conditions of operators include adverse mental states (e.g., mental fatigue, distraction), adverse psychological states (e.g., medical illness, physical fatigue), and physical/mental limitations (e.g., insufficient reaction time, visual limitation). Sub-standard practices of operators include crew resource mismanagement (lack of effective communication and team coordination) and lack of personal readiness (ignoring rest requirements). Environmental factors include both physical environment and technological environment conditions that affect performance. Physical environmental includes factors such as heat, vibration, lighting, and weather condition. Technological environment refers to factors involving design and automation factors such as the design of equipment, design of interfaces, and layout. Fig. 3 provides the overall framework for understanding human error. A summary of the factors in unsafe acts is shown in Fig. 3A.

3.3 Unsafe supervision

Unsafe supervision can be categorized into four types: Inadequate supervision, planned inappropriate operations, failure to correct the problem, and supervisory violation (Shappell & Wiegmann, 2000). Inadequate supervision may include shortcomings in providing guidance, training

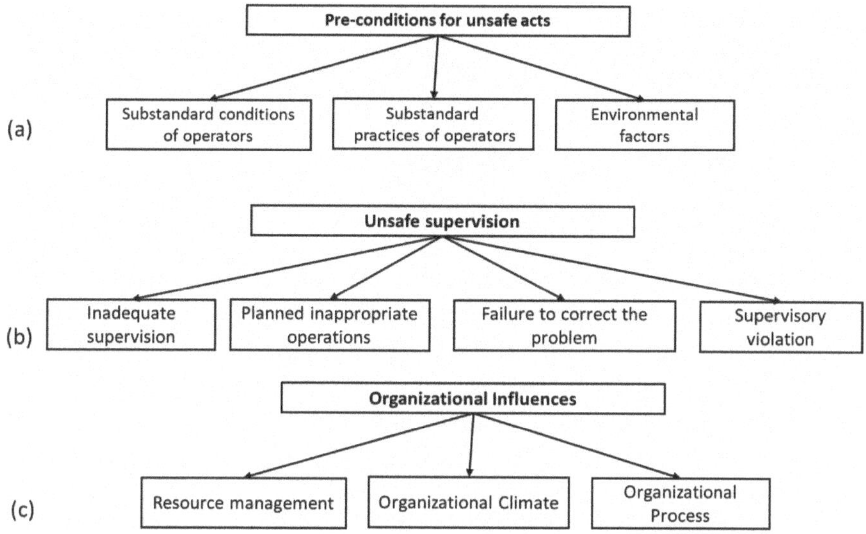

Fig. 3 An overview of the factors influencing human failure, including (A) unsafe acts, (B) unsafe supervision, and (C) organizational influences.

opportunities, leadership, motivation, and failure to track performances. Planned inappropriate operations refer to those operations that are unacceptable during routine operations but may be acceptable during emergencies (e.g., risk management). Failure to correct known problems refers to instances where deficiencies in equipment, individual, training, or related safety areas are known to the supervisor but are allowed to continue, for instance, not initiating corrective actions and failing to report unsafe tendencies. Supervisory violation refers to disregard for rules and regulations by a supervisor, e.g., authorizing unnecessary hazards and failure to enforce rules and regulations. An overview of what constitutes unsafe supervision is shown in Fig. 3B.

3.4 Organizational influences

These are categorized into three types, including resource management, organizational climate, and organizational process (as shown in Fig. 3C). It includes factors such as human resources, monetary and budget allocation for maintenance of organizational assets, organizational culture, organizational rules, procedures, and oversite. These have a significant impact on errors.

In summary, the classification of human errors along with their contributory factors allows us to foresee the variations in operator actions while they perform a task and subsequently predict how defenses can fail. We will next discuss various perspectives on human errors that have been followed in process industries.

3.5 Perspectives on human error

Various perspectives have been adopted in industrial systems to analyze, predict and reduce human error. The application of each perspective is based upon the underlying theory that led to the development of each perspective. A user of HF techniques must be acquainted with the rationale behind each perspective so as to provide appropriate recommendations/ error control strategies. Typically, there are four perspectives on human error. These include traditional safety engineering, human factors engineering/ergonomics, cognitive engineering, and sociotechnical systems perspectives (Embrey, Kontogiannis, & Green, 1994).

The traditional safety engineering perspective stresses that individual factors contribute to accidents and hence seeks improvement by modifying human behavior through motivational and disciplinary perspectives to error and accident reduction. Hence, the focus is on the individual rather than the system. Errors are primarily seen as being caused due to lack of discipline, lack of motivation to behave safely, or lack of knowledge of what constitutes safe behavior. These are assumed to give rise to "unsafe acts," which in combination with "unsafe situations" (e.g., toxic substances) cause accidents. Thus, the main area of application of this approach has been occupational safety which focuses on hazards that affect individual workers, rather than process safety which emphasizes major systems failures. Accordingly, it focuses on the elimination of hazards. Since hazards cannot always be eliminated (Srinivasan & Iqbal, 2017), this perspective promotes providing training in safe methods, appropriate guarding of hazards, personal protection equipment, and safety campaigns to improve motivation. Despite its success in some areas (occupational safety), the traditional perspective suffers from certain problems. Since it assumes that accidents are a result of individual factors, it does not encourage consideration of underlying causes or mechanisms of accidents (Embrey et al., 1994). Further, it is primarily suitable for manual tasks such as entry to closed spaces, lifting heavy objects, breaking-open process lines, etc., as the safe methods can be easily prescribed. It may not be suitable for minimizing the errors that involve decision-making. It does not place emphasis on contributory factors such as inadequate task design, inadequate procedures, and failures in communication. The introduction of computer-based control has drastically changed the role of operators to that of a decision-maker and problem solver in the event of emergencies or abnormal situations. In such situations, extensive support from the work system is also required, such as good process information presentation.

The human factors engineering (or ergonomics) perspective (HF/E) emphasizes that the main source of human error is a mismatch between human capabilities and the demands of the system. From this perspective, it is crucial that the design of the system considers the limited physical and mental characteristics of humans to avoid costly system failures. Therefore, this perspective seeks to improve the work situation (Lees, 2012) rather than focusing on correcting behavior. It thus requires a designer to consider the needs of the end-user, including factors such as workplace design, design of HMIs (e.g., control panels), design of physical environment (e.g., heat, lightning, and noise), and optimizing mental and physical workload on operators. By applications of design principles and standards, it seeks to minimize operator errors. In contrast to the traditional perspective of fitting a person to a job, it focuses on fitting the job to the person (Embrey et al., 1994). This perspective can be employed for occupational safety (similar to the traditional perspective) as well as for process safety and for manual and control room operations.

One of the main focuses of interest of the HF/E approach is the optimum design of HMIs. In CPI, the most extensively studied HMI is in control rooms. Even though in modern, highly digitized plants, the role of humans may be minimized, operator intervention is still critical, especially during unexpected events like abnormalities. Human interaction with HMI can occur in diverse ways such as monitoring, detection, prognosis, data entry, and control. Human errors can occur in each of these tasks and could be of the following nature: not reading the information correctly, failing to detect alarms, not responding or responding incorrectly, incorrect diagnosis, incorrect data entry, and incorrectly issuing instructions or not instructing at all. Human factors influence the likelihood of all these failure types. For instance, closely placing start buttons for pumps on a touch screen can lead to inadvertent activation of the wrong pump, causing a hazardous chemical spill. Likewise, a cluttered and complex screen can increase the likelihood of errors as well as response time. For the interested reader, a checklist of key issues for sensing, understanding, and using displays is provided in Crowl (2007) (Table 9–1). It should, however, be noted that this approach does not provide any systematic method for addressing and eliminating cognitive errors in diagnosis and problem-solving.

The cognitive systems engineering (CSE) perspective again accords primacy to work situations but also places significant emphasis on higher-level human functions (Lees, 2012), which are typically involved in CPI operations, such as problem-solving, decision making, and diagnosis. This perspective is based on the premise that rather than viewing a human as a passive black box akin to a hardware component—focusing primarily on information inputs and control action outputs—individuals should be seen from a cognitive psychology perspective. The cognitive perspective emphasizes the role of operators as they are involved in acquiring information, representing it internally, and processing it to guide their behavior. It involves a holistic assessment of how the design influences the information acquisition and, conversely, how operator mental models (understanding of the process) impact information acquisition and processing. It considers humans purposeful as their actions are influenced by future goals and objectives (e.g., maintaining a particular temperature profile or producing a given amount of product over several days). This perspective is particularly applicable to activities such as planning and dealing with abnormal situations (Embrey et al., 1994). Thus, this perspective is very relevant in the context of modern digitalized plants, where the role of humans is mostly that of monitoring, diagnosis, prognosis, and decision making. A detailed view of the cognitive aspects of human performance is provided in Reason (1990).

The sociotechnical systems perspective is based on the realization that human performance cannot be considered in isolation but is conditioned by social and cultural factors as well as management policies (Embrey et al., 1994). For instance, the mere existence of good procedures and designs does not ensure process safety. If a culture exists that encourages production and shortcuts over safety, it may promote unsafe behaviors. Therefore, a procedures design policy should include elements such as participation from end-users of procedures and design of procedures in accordance with sound and accepted HF principles. A sociotechnical system is a top-down approach that addresses the question of how management policies affect the likelihood of errors and are critical in any type of industrial setting—digitalized or not.

In summary, these perspectives try to encourage safe behavior (the traditional perspective), design systems to ensure a match between systems demands and human capabilities (the HF/E perspective), understand causes of errors (the cognitive modeling perspective), and finally, consider the impact of management policy and organizational culture. A summarized

Perspectives on human error		
Perspective	**Summary**	**Applicability**
Traditional safety engineering	• Individual factors contribute to accidents • Focus on individual rather than system	Occupational safety and manual operations
Human factors engineering HF/E	• Work situations such as design of HMIs impact human performance • Focus on fitting the job to person	Occupational/process safety, manual/control operation/ routine operations
Cognitive systems engineering	• Work situations and limitations in higher level human functions such as decision making can induce errors	Process safety, decision making, problem solving, and abnormal situation
Sociotechnical systems perspective	• Social and cultural factors, and management policies	Occupational/process safety, Policy aspects, safety culture

Fig. 4 An overview of the perspectives on human error.

view of these perspectives is provided in Fig. 4. Each of these perspectives is relevant depending on the nature of the task and the probable error type. Consideration of the traditional perspective is suitable for minimizing errors in manual operations (occupational safety). The HF/E perspective seeks to understand how elements of the work system such as workplace design, interface design, physical environment, and workload on the operator can impact their performance. The primary focus areas include occupational/process safety, manual/control operations, and routine operation (Embrey et al., 1994). The cognitive engineering approach takes special consideration of mental skills such as problem-solving and diagnosis. Its main areas of application include process safety, decision making/problem solving, and abnormal situations, which are often encountered in digitalized operations. In Section 4, we will highlight how critical can cognitive aspects be for operator performance by providing examples from diverse domains, including CPIs. However, the usage of HF/E and Cognitive engineering perspectives in CPIs is not extensive yet. Finally, the socio-technical approach is crucial for enhancing safety in general, irrespective of the nature of the task. Further, these perspectives focus on factors that need to be considered to reduce human error. However, for these strategies to be effective, the likelihood of human failure during the events needs to be quantified first. Such quantification will help identify events during an operation (or a task) that require a significant focus of human factor specialists and other assessors for devising error minimization strategies to enhance the reliability of the overall system. This is possible through human reliability analysis.

Next, we discuss how human errors can be systematically accounted for in order to get insights into the reliability of operations involving operator actions.

3.6 Human reliability analysis

Human reliability analysis (HRA), also sometimes referred to as human reliability assessment is the scientific approach used to assess the impact of human factors on risk. Primarily goals of HRA are achieved through its functions which include identifying what errors can occur (human error identification), what is the likelihood of these errors (error quantification), and finally, identifying ways by which the likelihood and consequences of errors can be reduced (error reduction) (Kirwan, 2017). A generic methodology that defines best practices for using HRA tools involves ten steps (Kirwan, 2008, 2017). These are:

1. Problem Scoping: This involves defining terms of reference for HRA, e.g., primary human interactions and functions that need to be assessed, what events may occur during an operation and what failures, etc. It is generally carried out by HF/ HRA specialists and safety assessors.
2. Task Analysis: It involves a detailed description of the task, including what an operator is supposed to do (actions, cognitive processes, or both) and resources available (displays, communication media, etc.) to achieve the task goal. The primary aim of this step is to create a detailed picture of human involvement as well as the information necessary to carry out those actions.
3. Human-Error Identification: It involves identifying what can go wrong at each step of a task from a human perspective. It may include identifying errors of omission (e.g., failure to notice an alarm), errors of commission (e.g., operating a valve in the wrong direction, taking action too late or too early), and extraneous acts (unrequired actions).
4. Representation: It involves integrating human contributions (both positive and negative) with other contributions (software, hardware, and environment) in a quantifiable format to allow overall system risk to be calculated. It provides a view of errors that can propagate through a system. Fault trees and event trees are commonly used for the purpose of representation in probabilistic safety assessment (PSA).

Operator-action even tree (OAET), similar in essence to the event tree, is more specific to HRA as it involves more emphasis on human operations.

5. Screening: It involves determining which errors are risk-sensitive and which errors have a negligible impact on overall risk to take a decision on the depth of analysis required. This can be done on the basis of pessimistic estimates of error probabilities.
6. Error Quantification: It involves using formal methods to derive human error probabilities (HEPs) for quantifying overall risk in the system.
7. Impact Assessment: It involves determining the relative contributions of individual errors to accident frequencies as well as overall impact.
8. Error Reduction: It involves specifying human factors and other safety requirements based on impact assessment and prior qualitative stages (task analysis, error identification, representations) to achieve an acceptable risk level by either minimizing HEPs or minimizing consequences.
9. Quality Assurance: It involves checking whether recommended error reduction measures are effective in practice
10. Documentation: It involves recording assumptions, calculations, results of analysis, and recommendations to inform designers as well as end-users.

It is important to mention that these steps are generic and do not necessarily always follow the same order. Next, we discuss various HRA methodologies that can be used by practitioners to improve safety and optimize system performance. Broadly speaking, HRA methodologies are categorized as first-generation and second-generation methods (Bell & Holroyd, 2009). These authors identified 35 HRA tools altogether, we shall here discuss a few prominent ones.

First-generation methods were developed to aid risk assessors in predicting and quantifying human error likelihood (Kim, 2001). However, humans are treated as akin to hardware components with a failure rate associated with them. These methods are atomistic in nature; they encourage the breaking of a task into component parts and then consider the impact of factors such as equipment design, time pressure, and stress to determine human error potential (HEP). First-generation methods focus on skill- and rule-based human actions (errors) but do not consider context, human organizational factors, and cognitive aspects. Examples of these methods include a technique for human error rate prediction

(THERP) (Swain & Guttmann, 1983). THERP involves carrying out a task analysis to identify human involvements, defining the sequence of events that need to be carried out for ensuring safety, and then quantifying these sequences using a database of human error probabilities (Kirwan, 2008). The database is contained in the THERP handbook (produced for the US Nuclear Regulatory Commission (NUREG/CR-1278)), which presents tabled entries of nominal human error probabilities (HEPs). The database contains error probabilities for tasks such as reading a display, responding to an alarm, and operating a joystick. These probabilities can be modified using performance shaping factors (PSFs) (such as time pressure, lightening, work hours, etc.) which affect operator performance. THERP also includes the calculation of dependence which means when the success or failure of the current action is related to the previous action or task. THERP is often criticized as being too fine-grained, i.e., it involves breaking a task into several sub-tasks, making the analysis cumbersome.

Another popular HRA technique is Human Error Assessment and Reduction Technique (HEART) (Williams, 1986). HEART has a more generic and smaller database compared to THERP, and it is applicable to a wide range of industrial settings. It is based on a set of 9 generic tasks developed on the basis of extensive psychological studies about how humans may fail. A few examples of generic tasks included in HEART are "Complex task requiring high levels of comprehension and skill" and "Routine highly-practiced, a rapid task involving relatively low level of skill." A generic HEP is associated with each generic task type. Error probability is for perfect conditions, but HEART is based on the premise that perfect conditions rarely exist and, therefore, provides a set of 38 error-producing conditions (EPCs) which affect generic HEP. For instance, high levels of stress increase error probability by a factor of 1.3. Operator inexperience has a weight of 3 in HEART. A few other examples of EPCs include unfamiliarity, shortage of time, ease of information assimilation, etc. It is worth mentioning that the selection and scale of EPCs are dependent on expert judgment. Following analysis, an overall HEP, as well as element-wise HEPs, are obtained to guide the formulation of recommendations, which are followed by an implementation to reduce the influence of EPCs. A schematic of the steps involved in applying HEART is provided in Fig. 5. We next present a simple case study demonstrating the application of the HEART technique (Kirwan, 1996).

Let us assume that we have to find human error probability (HEP) associated with a simple procedural task using HEART.

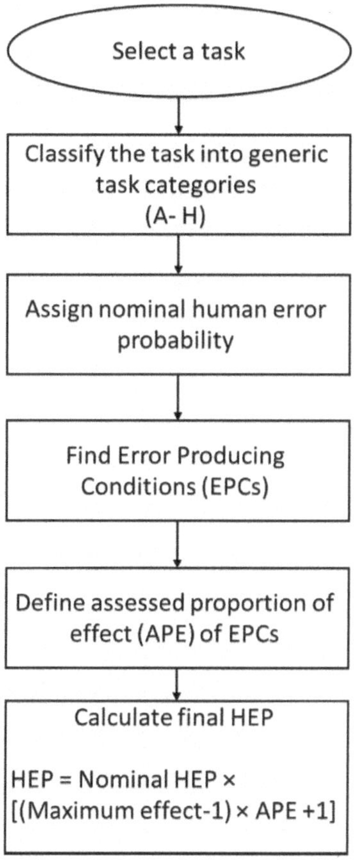

Fig. 5 Procedure for applying HEART for human failure quantification.

Step 1: Classify the task into one of the generic task categories (Williams, 1986), which are designated as A-H. These generic task categories are provided as a table in Williams (1986) along with respective proposed nominal HEPs. Allotting a task to a category is based on the assessor's judgment. Let us assume it is a procedural task with some checking; therefore, it is classified into category F.

Step 2: Assign a nominal human error probability to the task. For example, choose median HEP, which is 0.003 for this generic task type.

Step 3: Find out which Error Producing Condition(s) (EPCs or PSFs) affect the reliability of this task. Let us assume that there is time pressure. The corresponding EPC would be "A shortage of time for error detection and correction." Therefore, multiply nominal HEP with a factor of

11 (maximum predicted nominal amount by which unreliability might change, going from 'good' conditions to 'bad' possible).

Step 4: Define the extent of the negative effect of EPC on nominal HEP associated with the task, i.e., 'assessed proportion of effect.' Let's assume the assessor selects a proportion equivalent to 50%. If there are multiple EPCs, the assessed proportion of effect needs to be calculated for each EPC.

Step 5: Calculate the final human error probability as follows:

$$\begin{aligned} HEP &= Nominal\ HEP \times [(Maximum\ effect - 1) \times Assessed\ proportion\ of\ affect + 1] \\ &= 0.003 \times [(11 - 1) \times 0.50 + 1] \\ &= 0.003 \times 6 \\ &= 0.018 \end{aligned} \tag{1}$$

If another EPC, such as inexperience, existed (maximum effect = 3, assessed proportion of affect = 0.4), then calculations would be done as follows:

$$HEP = 0.003 \times 6 \times [(3 - 1) \times 0.4 + 1] = 0.0324$$

It is worth mentioning that THERP and HEART have also been considered suitable HRA methods for oil and gas Quantitative Risk Analysis (QRA) by the international association of oil and gas producers (IOGP) (IOGP, 2010) being generic. First-generation methods do not place emphasis on context and organizational factors.

Second-generation methods were developed to incorporate the effect of context (i.e., scenarios people find themselves in) in which a task is carried out and human organizational factors. Specifically, these focus on the causes of human errors rather than their frequency and place emphasis on interactions of factors leading to errors (Di Pasquale, Iannone, Miranda, & Riemma, 2013). These factors are called performance shaping factors (PSFs). In first-generation methods, PSFs relate only to operator characteristics while ignoring environment and task characteristics. Second-generation methods are marked by more focus on cognitive aspects (for an appropriate explanation of operator behavior) and qualitative assessment of human errors (Di Pasquale et al., 2013). Among the prominent second-generation methods, Cognitive Reliability and Error Analysis Method (CREAM) is preferred by HF practitioners because it is based on a well-established cognitive model and can explain why human errors occur (Hollnagel, 2003). For a review of various HRA techniques and their applications, the reader is referred to Ade and Peres (2022).

CREAM can be used to predict human error as well as to analyze and quantify it. CREAM allows analysts to identify those tasks, actions, or parts of work that require cognition and may be affected by variance in cognitive reliability. Subsequently, it helps provide an assessment of the consequences of human performance. It helps an analyst determine conditions that affect cognitive reliability and develop modifications to improve these conditions (Hollnagel, 1998). CREAM is based on a distinction between competence and control. This distinction is based on the context control model (COCOM), wherein competence includes a person's skills, and knowledge and competence are represented as running along a continuum where an individual has complete control to little or no control. The context is identified through a set of nine common performance conditions (CPCs). These include the adequacy of organization, working conditions, adequacy of man-machine interface, available time, time of day, adequacy of training and experience, etc. For each activity, a CPC level is assigned, e.g., for training adequacy descriptors could be improved, not significant, and reduced. These descriptors can also be represented in quantitative terms. Further, a classification scheme separates genotypes (causes) and phenotypes (manifestations/ error modes) to create a list of cognitive function failures. These phenotypes and genotypes are classified into consequents which have some general/specific antecedents. For example, for genotype "communication," the consequent may be "communication failure," while general and specific antecedents may be 'distraction' and 'noise' respectively. In a typical HRA process, CREAM can be useful in describing how an initiating event has the potential to lead to the occurrence of errors. An overview of the steps in using CREAM to determine the failure probability is shown in Fig. 6. The basic version of CREAM is applicable for screening an operation for detailed analysis. Next, we demonstrate, using an example, the application of the CREAM technique.

Consider a task requiring an operator to restart a furnace so that it reaches a temperature of 800 °C following a system trip. The steps required for determining the human failure probability associated with this task using the CREAM technique are explained below.

Step 1: Identify the list of activities that the operator needs to perform to achieve the end goal. Using task analysis, let the identified activities be:

(i) Prepare plant and services: This includes three steps (a) Ensure plant is in a ready state, (b) gas-oil supply is available, and (c) the oxygen system is working. These steps can be performed in any order.

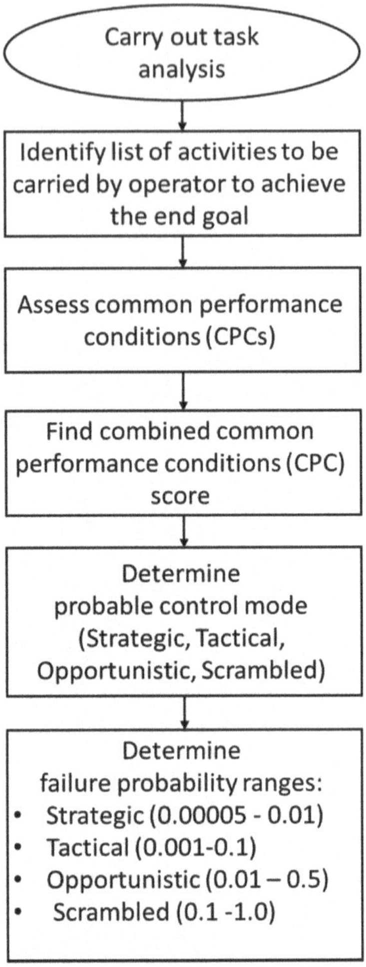

Fig. 6 Procedure for applying CREAM.

(ii) Start the air blower
(iii) Start the oil pump
(iv) Heat furnace to 800 °C while monitoring oxygen and temperature. Accordingly, the steps include increasing the set-point of the temperature controller as per instructions provided in the chart next to the controller, monitoring oxygen and temperature, and finally, switching to automatic mode when the desired temperature is attained.

Step 2: Assess common performance conditions (CPCs). These are assessed by experts who have proper information about the facility

Table 2 Common performance conditions for furnace start-up task (Hollnagel, 1998).

S. No.	CPC name	Descriptors	Effect on performance reliability (obtained from Tables present in CREAM)
1	Adequacy of organization	Very efficient/Efficient/ **Inefficient**/Deficient	Reduced
2	Working conditions	Advantageous/ **Compatible**/Incompatible/	Not Significant
3	Adequacy of MMI and operational support	Supportive/Adequate/ **Tolerable** / Inappropriate	Not Significant
4	Adequacy of procedures/ plans	Appropriate/Acceptable/ **Inappropriate**	Reduced
5	No. of simultaneous goals	Fewer than capacity/ **Matching current capacity**/More than capacity	Not Significant
6	Available time	**Adequate**/Temporarily inadequate/Continuously inadequate	Improved
7	Time of day	**Day-time (adjusted)**/ Night-time (unadjusted)	Not Significant
8	Adequacy of training and preparation	Adequate, high experience/ Adequate, limited experience/**Inadequate**	Reduced
9	Crew collaboration quality	Very efficient/ **Efficient**/Inefficient/ Deficient	Not Significant

and are not challenging. For the purpose of demonstration, we will assume conditions here. The 9 CPCs that exist in CREAM, along with their (assumed) descriptor label, are provided in Table 2.

Step 3: Find the combined CPC score, which is represented as a triplet $[\sum_{reduced}, \sum_{not\ significant}, \sum_{improved}]$. CREAM provides a mapping from descriptors to CPC scores: −1 for CPCs descriptors that result in reduced performance, 0 for descriptors that do not significantly influence performance, and + 1 for descriptors that increase performance.

From Table 2, and using information available in CREAM (Hollnagel, 1998), the CPC score is [3, 5, 1]. That means that 3 CPCs result in an increase in reduced human performance, 5 CPCs do not have any significant effect, and 1 CPC results in an increase in performance.

Step 4: Determine the Probable Control Mode so as to assess the effect of combined CPCs on human performance reliability. This can be obtained from the 2D plot ("Relation between CPC score and control modes"; Hollnagel, 1998) between $\sum_{\text{reduced reliability}}$ and $\sum_{\text{improved reliability}}$ present in the CREAM handbook (Hollnagel, 1998). From the plot, the control mode that the operator is expected to be in is an opportunist control mode. Alternatively, the control mode can also be obtained from the difference of $\sum_{\text{reduced reliability}}$ and $\sum_{\text{improved reliability}}$ as follows. If $\sum_{\text{reduced reliability}} - \sum_{\text{reduced reliability}} = -7\ to -4$, then the control mode is strategic. From−3 to 1, it is tactical; from 2 to 5, it is opportunist, and from 6 to 9, it is scrambled (He, Wang, Shen, & Huang, 2008).

Step 5: Determine general action probability for probable control mode. Since the mode is opportunist, failure probability ranges from 0.01 to 0.5. This failure probability range is quite high and corresponds well with assumptions (average conditions) regarding CPC descriptors. Since failure probability is unacceptably high, there is a need for continued analysis using extended-CREAM. The interested reader is referred to Bell and Holroyd (2009) and Hollnagel (1998) for a detailed description.

It is worth mentioning that determining the effect of the CPCs on human performance is not an easy task and is dealt with numerous uncertainties (Shirali, Hosseinzadeh, & Kalhori, 2019). The challenges exist due to the lack of availability of adequate data on CPCs, subjectivity involved, and unclear relationships (Akyuz & Celik, 2015; Ung, 2015). These can result in inter-assessor variability in failure rates. In fact, the lack of availability of objective data and consequent reliance on expert judgment to determine HEPs is one of the main limitations of current HRA approaches (Iqbal & Srinivasan, 2018).

In summary, the science of HF is critical to identifying elements of the system which may affect overall work system reliability and human performance. HRA helps obtain quantitative estimates of human error probabilities. However, such estimates are primarily based on information present in databanks and involve a great degree of human judgment. Judgments are prone to biases due to their subjective nature. As a result, there can be significant variability in the estimates of risk associated with an operation. There is a need for objectively and quantitatively assessing how cognition and context play a role in the performance and how this impacts the reliability of operations involving operator actions. Currently, the assessments are, at best qualitative in nature. The influence of cognition on overall system reliability is at best limited. With an increase in digitization in process plants, assessing cognitive aspects becomes crucial (Srinivasan, Srinivasan, Iqbal, Nemet, & Kravanja, 2019). For instance, the impact of task difficulty on cognitive performance (e.g., cognitive workload and fatigue) needs to be assessed quantitatively. Fortunately, this has been possible through advancements in wearable sensor technology (which allows us to tap on the potential of physiological measurements) and the parallel developments in the field of cognitive engineering.

4. Human error reduction in digitalized operations

Cognitive engineering has grown as a multi-disciplinary field that combines knowledge from cognitive science, human factors, human-computer interaction design, and systems engineering (Gersh, McKneely, & Remington, 2005)—that aims to understand mental workload, planning, decision-making process, and situational awareness in industrial settings (Parasuraman, Sheridan, & Wickens, 2008). With the increase in digitalization in industries, the cognitive aspects of performance have become more relevant than ever. An increase in the availability of information and the consequent interface management tasks often distract and burden the operator mentally (Porthin, Liinasuo, & Kling, 2020). There is a need to study how new interface designs influence operator performance. The requirement to segregate and integrate information to maintain situational awareness has also become challenging. In fact, the increased automation—a significant aspect of digitization—often pushes operators out of the loop with the system, affecting their situational awareness.

Depending on the nature of the task, humans need to acquire information, process it, and make decisions accordingly. Human working memory capacity is limited (Sweller, 1988), being able to hold only a few chunks of information (Miller, 1956). When the mental effort demanded by a task exceeds that of the human's mental capability, the likelihood of error increases due to an increase in cognitive workload (Mayer & Moreno, 2003). The relation between cognitive workload and human errors is shown in Fig. 7. Therefore, assessment of cognitive aspects of performance is critical to understanding human-machine interaction and optimizing it.

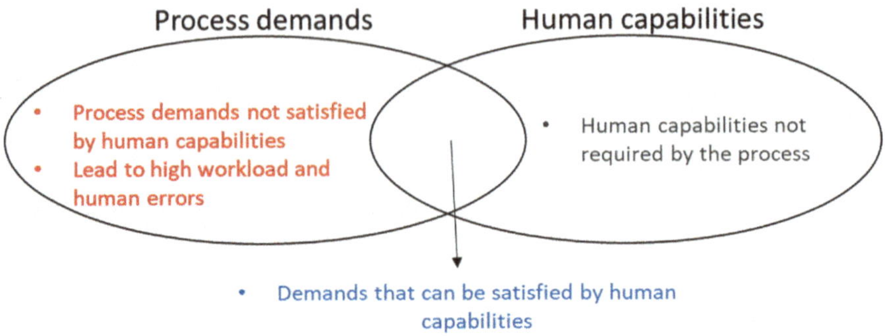

Fig. 7 Relation between cognitive workload and human errors.

For instance, cognitive workload significantly influences decision-making (Srinivasan et al., 2019; Zheng, Jiang, & Atkins, 2015). These aspects of performance, such as cognitive workload, are often intrinsic and not directly observable. Traditionally, these (especially workload) were accounted for through techniques such as National Aeronautics Space Administration Task Load Index, or NASA-TLX (Hart, 2006; Hart & Staveland, 1988), and Subjective Workload Assessment Technique, or SWAT (Reid & Nygren, 1988). A sample of the NASA-TLX questionnaire is provided in Table 3 (Hart, 2006). These techniques, however, are criticized for being subjective and hence prone to bias. These can also not be used to assess cognition in real-time. These shortcomings limit their utility in digitalized operations.

With advancements in sensor technology such as eye-tracking, electroencephalography (EEG), electrocardiography (ECG), and galvanic skin resistance (GSR), it has now been possible to obtain quantitative measures of cognitive aspects of performance based on physiological measurements. These measurements have the potential to provide an objective assessment of the capabilities of an operator in relation to the demands of the process, as shown in Fig. 7. Conversely, it has also been possible to assess the impact of design on human performance.

4.1 Cognitive performance evaluation

Here we provide a brief overview of sensor technologies that can enhance human performance and safety in the workplace.

4.1.1 Eye-tracking

Eye-tracking can be extensively used to understand the interaction between humans and equipment. Eye-tracking is based on the premise that the

Table 3 A summary of the NASA-TLX workload assessment technique (Hart, 2006).

S. No.	Parameter	Description	Likert scale Endpoints: High/Low
1	Mental Demand	How much mental and perceptual activity (e.g., thinking, calculating) was required in the task?	
2	Physical Demand	How physically demanding (e.g., pushing, pulling, activating) was the task?	
3	Temporal Demand	How much time pressure did you feel due to the pace of the task?	
4	Performance	How successful were you in accomplishing what you were asked to do?	
5	Effort	How hard did you have to work to accomplish your level of performance?	
6	Frustrations	How insecure, discouraged, irritated, stressed, and annoyed were you?	

movement of the eyes serves as a trace of attention allocation (Just & Carpenter, 1980). Eye-tracking involves tracking the movement of eyes with respect to the head of the person, which acts as a reference. It allows us to track the eye-gaze of a person using a video camera that captures images of eyes continuously (Holmqvist et al., 2011) to get a sense of how a person is acquiring information. For instance, during reading, the eye temporarily pauses on a word and then rapidly moves from one word to another word. Typically, eye movement is categorized into fixations and saccades. Fixations are pauses in eye movement during which a person acquires information, while a saccade is rapid-eye-movement between fixations. In the context of human-machine interaction, fixations provide critical insights into the relative importance of areas of information (AOIs). AOIs are the regions on the interface that are relevant to the task at hand. If the number and duration of fixations on a particular AOI is more, it means that information present on that AOI is more relevant to the task. There are several types of eye-trackers available in the market, which can be broadly classified into screen-based and wearable eye-trackers. Screen-based eye trackers are mounted on a screen to record the eye-gaze of a person interacting with the screen, while wearable eye trackers are commonly used when information has to be obtained from multiple sources, allowing freedom of movement (for instance, for a field operator).

4.1.2 Electroencephalography (EEG)

EEG involves the measurement of electrical activity generated by a population of neurons present in the brain by placing electrodes (sensors) on the scalp of a person. Fluctuations in voltage that EEG captures are a direct result of various biophysical processes that occur at the neuronal level depending upon the mental state of a person. Thus, it entails research into how electrical activity produced by the neuronal populations within the brain supports or implements diverse cognitive activities such as perception, memory, behavior monitoring and control, and emotions (Cohen, 2014). Since the brain is the center of cognition, EEG is one of the most important physiological parameters providing critical information about the mental state of a person. In addition, it is able to capture brain dynamics at a scale at which cognition occurs (Cohen, 2011), making it an effective tool to capture performance changes in real-time.

EEG signal is often studied in the frequency domain in terms of changes in various frequency bands. These frequency bands typically include delta (1–4 Hz), theta (4–7 Hz), alpha (8–12 Hz), beta (12–30 Hz), and gamma bands (30–80 Hz), and are commonly obtained using Fourier transformation of time series EEG signal. Each of these bands provides critical information about the mental state of a person; these states may represent the level of stress, fatigue, cognitive workload, and attention. For instance, the alpha band is often associated with arousal levels, with low alpha power being associated with high levels of arousals (Pfurtscheller, Stancak Jr, & Neuper, 1996). The beta band is associated with problem-solving and is sensitive to conscious focus (Abhang, Gawali, & Mehrotra, 2016). Similarly, delta activity can capture mental effort exerted by a person during different tasks and is associated with attention and internal mental processing (Harmony et al., 1996).

4.1.3 Galvanic skin response

Also called electrodermal activity (EDA), this refers to the change in the activity of sweat glands in response to the changes in emotional arousal of a person. An increase in arousal is often linked to an increase in skin conductance. GSR sensors capture the change in electrical activity resulting from a change in the activity of the sweat glands.

4.1.4 Electrocardiogram (ECG)

Electrocardiogram records the electrical signals generated by the heart. Heart rate variability (HRV) is one of the most commonly used metrics that has been used to assess various aspects of cognition, such as stress and fatigue. HRV captures variation in time between successive heartbeats and reflects the ability of the heart to adapt to changing circumstances (Rajendra Acharya, Paul Joseph, Kannathal, Lim, & Suri, 2006).

We will next discuss the potential applications of the physiological measurements obtained through these sensors for HF.

4.2 Human-machine interface assessment

The design of interfaces predominantly influences human performance. There are three main components of HMI, which include information displays, controls, and management of interaction. The way information is presented to the user requires consideration of visual, audio, and other tactile factors (Edmonds, 2016). This includes consideration of how the user detects information, choice of size, layout, coding, and grouping of information. The design of information displays is an important HF problem that has grown in importance due to the increasing information density resulting from digitalization (Lees, 2012). Thus, it requires significant focus on user requirements, attention to layout, and alarm management. If the design is not optimal, it will increase the chances of human errors and compromise safety. For instance, in a resin plant, a batch of resins from reactors was required to be discharged to blenders to achieve the desired viscosity through the addition of a solvent. The solvent was fed to the blenders using a solvent charging manifold. However, poor panel layout and labeling of the charging manifold led the floor operator to pump the solvent to Blender-21A instead of Blender-12A as directed by instructions (Embrey et al., 1994).

Eye-tracking can be utilized to assess the impact of HMI design on performance. For instance, Ha, Byon, Baek, and Seong (2016) studied the impact of faulty design (in level indicators) on nuclear power plant operators' performance during accident scenarios. The task required them to monitor a pilot NPP, understand the situation and report their diagnosis results. Performance was assessed using FIR (fixation to importance ratio of an information source), which captures attentional-resource effectiveness. Fixations on an AOI are necessary to acquire information. If there are any important AOIs and yet there are no fixations on them, it results in degradation of performance. For best effectiveness, FIR should have approached unity, but for faulty indicators, FIRs were much smaller than unity reflecting remarkably poor performance (Ha et al., 2016).

Chen and Li (2013) used eye-tracking to evaluate how interfaces with different information organizations (process layout vs. functional layout) influence human-computer interaction during emergencies in terms of difficulty of information abstraction, search efficiency, and workload. Even though they did not find significant differences in all evaluation measures, workload (as measured by pupil diameter) was more during interaction with the process layout compared to the functional layout.

Eye-tracking has also been used to automatically obtain the optimal layout design of control panels. Diego-Mas, Garzon-Leal, Poveda-Bautista, and Alcaide-Marzal (2019) used eye-tracking along with mouse tracking to identify the information elements (areas of interest on the control panel) that have a greater relationship between them. These AOIs included indicators, controls, and communication systems. The authors used an algorithm to modify the placement of various AOIs based on the relationship between them as well as their relative importance. They then carried out a comparative study to evaluate the usability of the new design with respect to the manually modified old design (based on user experience). The results revealed an improvement of 11% in the mean task duration and a reduction of 18% in the number of errors due to the modified interface. Thus, eye-tracking can be used to evaluate and improve the design of HMIs so as to optimize overall performance and decrease the chances of errors.

4.3 Expertise assessment

Operator expertise is critical for ensuring safety in a plant. In the process industries, a large number of accidents have occurred due to a lack of operator competence owing to poor training. Eye-tracking has been used in several safety-critical domains to assess the difference in expert and novice behavior during various tasks. For instance, eye-tracking has been used to find the difference in visual scan paths of expert and novice air-traffic controllers while they obtain information from the radar screen. Experts' scan path is usually characterized by a circular motion and fixations on altitudes of airplanes. On the other hand, novices do not have a specific pattern of information acquisition (Kang & Landry, 2014). Eye-tracking can also reveal how experts adjust their scan paths during complex situations. For instance, for complex tasks, the number of fixations by experts is fewer compared to novice air-traffic controllers (Lim et al., 2006). This indicates that experts are able to identify airplanes that are more critical for safety and distribute attention accordingly.

Information about differences in visual patterns can be used to provide targeted feedback to less-experienced personnel to increase their performance and minimize human errors. For instance, expert scan path treatment proved to be beneficial for quickly enhancing the performance of novice air traffic controllers during aircraft conflict detection tasks (Kang & Landry, 2014). Eye-tracking has been effectively utilized to identify differences in strategies of experts and novices across several safety-critical domains such as healthcare (Diaz-Piedra, Sanchez-Carrion, Rieiro, & Di Stasi, 2017) and driving (Pradhan, Fisher, & Pollatsek, 2005).

In the process industries, eye tracking has been used to identify differences in the expertise level of control room operators. Sharma, Bhavsar, Srinivasan, and Srinivasan (2016) assessed the information acquisition pattern of operators while they monitored a simulated ethanol production plant for any disturbances. The operators' task was to take corrective control actions in case of a disturbance in process variables. A majority of the operators were successfully able to control the disturbance. However, analysis of eye-gaze data revealed that the scan pattern of all the successful operators was not the same. A group of successful operators had eye-gaze patterns that resembled that of the failure group. Like failure group operators, they paid significantly lower attention to primary variables of interest, indicating a poor understanding of the process dynamics. The study highlights that consideration of cognition can provide more insights into performance rather than binary notions of success and failure. Gaze patterns can also be used to differentiate experts from novices. Bhavsar, Srinivasan, and Srinivasan (2017), in their study on control room operators, found that experts had lower gaze entropy compared to novices. Lower gaze entropy reflects a proper understanding of the process and directed attention to relevant areas of information. High gaze entropy is a result of poor understanding of the operator about which areas on a human-machine interface are relevant to control disturbance.

Liu et al. (2021) conducted experiments with nuclear power plant operators to identify signatures of expertise using electrodermal activity (EDA). They grouped operators on the basis of years of experience. Results of the study revealed that the operator group with more years of experience exhibited lower skin conductance (as obtained from EDA) while handling emergency conditions involving a fire accident. The authors argue that lower skin conductance in the more experienced group might be a result of lower mental stress than experience compared to less experienced operators.

Since these physiological measurements have the ability to identify distinct signatures in experts compared to novice operators, these have also been used to monitor training progress. Several studies have been carried out to assess training levels of control room operators using eye-tracking derived metrics. For instance, Iqbal, Srinivasan, and Srinivasan (2018) found that as operators learn to control a disturbance, their gaze entropy decreases. Similarly, a study on control room operators by Shahab, Srinivasan, and Srinivasan (2021b) reveals that fixation transition entropy decreases as operators learn the dynamics of the process with learning. Association metric (obtained from eye-gaze patterns) can be used to capture the operator's understanding of areas of interest on HMI that are relevant to a disturbance. When the operator develops an understanding of process dynamics, the association metric approaches unity (Shahab, Iqbal, Srinivasan, & Srinivasan, 2021a). Similarly, whether operators are adopting a pro-active monitoring strategy or not—critical for ensuring process safety—can be identified by the salience metric, which approaches unity during proactive monitoring. More recently, Shahab, Iqbal, Srinivasan, and Srinivasan (2022a, 2022b) demonstrated that mental models of operators could be captured using hidden Markov Modeling of eye-gaze and operator action data in a quantitative manner.

Training progress can also be captured using EEG. Recently, Iqbal, Shahab, Choudhary, Srinivasan, and Srinivasan (2021) used power spectral densities in the range of 1–30 Hz to assess the learning levels of control room operators. The study revealed that the fraction of low power clusters increased during a task with learning. The study also highlighted that assessing learning based on EEG power spectral densities reveals more information about the learning process rather than relying only on process-based metrics; EEG metrics can also capture the absence of learning when process-based metrics may suggest otherwise.

4.4 Workload and fatigue assessment

High cognitive workload and fatigue are known to be precursors to sub-optimal performance. A high workload is caused when the limited working memory is exerted beyond its capacity, either due to requirements of high information processing or the nature of the task. Therefore, there is a need to identify instances during which workload increases and the underlying reasons. Likewise, fatigue levels also need to be monitored during shifts so that the performance is not unacceptably degraded.

Eye-tracking-derived metrics have been utilized in several domains to quantify the level of workload and assess corresponding performance. For instance, when pilots faced highly challenging situations, they experienced higher levels of cognitive workload (Othman & Romli, 2016) compared to when disturbances were less challenging. The performance of pilots also degraded during higher workload levels. An increase in workload can influence monitoring strategy. Air traffic control officer (ATCo) participants with the task of monitoring air traffic and granting clearance changed their monitoring and control strategy (as measured by their gaze behavior) during high task load (Manske & Schier, 2015). They focused more on aircraft directly during low load conditions but paid more attention to radar display when the load increased. Eye-tracking measures have also been used to quantify the fatigue levels of drivers (Di Stasi et al., 2012). The study by Di Stasi et al. (2012) revealed that peak saccadic velocity (highest velocity achieved during saccades) is lesser in fatigued drivers. In addition, fatigue is also captured by an increase in saccadic duration, which is the time taken to complete a saccade.

Bongiorno, Bosurgi, Pellegrino, and Sollazzo (2017) used GSR along with eye-tracking to identify elements in the driver's visual frame that contributed to a high workload. These elements included both static objects (e.g., parked vehicles and road signs) as well as dynamic objects (e.g., moving vehicles and pedestrians). Various parameters were calculated from the GSR curve (which shows the Electrodermal activity measured in micro Seimens) such as GSR minimum, GSR maximum, latency, the difference in minimum and GSR values to obtain insights about workload experienced by the driver as well as recovery (signifying driver's resiliency). In general, an increase in GSR is related to high levels of mental workload.

Researchers have been using HRV to measure mental stress and other cognitive processes, which are known to influence HRV (Thayer, Åhs, Fredrikson, Sollers III, & Wager, 2012). For instance, HRV was used by Landi et al. (2018) to optimize performance in human-robot interactions. The authors utilized HRV to find out instances when human is under high workload so that robot can take over, thus managing the workload of operators by introducing assistive technologies. HRV has also been used to monitor fatigue and emotional states of drivers to prevent any near-miss accidents (De Nadai et al., 2016). The results of the study revealed that LF/HF ratio obtained from HRV increases during morning and evening hours, at crossroads, and at traffic signals. A higher LF/HF ratio indicates a higher mental load and an increased likelihood of errors.

In the process industries, eye-tracking has also been used for cognitive workload assessment. The study by Bhavsar, Srinivasan, and Srinivasan (2016) revealed that pupil diameter variation during an abnormal situation control task contains distinct signatures of the operator's ability to successfully manage abnormality. Their study revealed that once an operator is notified about an abnormality in a process through an alarm, their pupil diameter starts increasing, reflecting an increase in cognitive workload. If the operator takes the correct control action to control disturbance, then their pupil diameter starts decreasing. On the other hand, if the operator is confused about how to control disturbance, then pupil diameter continues to stay higher, reflecting poor performance (and higher workload) on the part of the operator. Das, Iqbal, Bhavsar, Srinivasan, and Srinivasan (2018) found that fixation duration and saccade duration can be used to identify the mental states of operators corresponding to inconsistencies (marked by a change in decisions) in their thought process. Fixation and saccade duration was found to increase during inconsistencies reflecting high cognitive workload and increased chances of errors.

EEG has also been extensively used in several domains for the assessment of cognitive aspects of performance. EEG can be used to assess fatigue levels of humans which degrades their performance. Long periods of sustained cognitive activity and time on task can result in cognitive fatigue (Warm, Parasuraman, & Matthews, 2008), which in turn induces disengagement, leaving operators ill-equipped to respond to unexpected events (Borragan Pedraz & Peigneux, 2016). EEG has the potential to quantify driver fatigue (Li, Su, & Lu, 2017); aircraft pilots' fatigue (Dehais et al., 2018), cognitive workload (Gentili et al., 2014), vigilance (Sauvet et al., 2014); changes in workload of marine pilots with task difficulty (Orlandi & Brooks, 2018); and visual attention of air-traffic controllers (Kamzanova, Kustubayeva, & Matthews, 2014). However, its use in the process industries is still limited.

In process industries, EEG has been used to understand operators' behavior during abnormal situations during which chances of error are maximum, as these often involve novel situations. The authors have studied the potential of EEG in assessing cognitive aspects of an operator's performance. It was found that the ratio of power spectral densities of theta and alpha waves obtained from raw EEG signals is higher for difficult disturbance rejection tasks compared to easier tasks (Iqbal, Srinivasan, & Srinivasan, 2019). Power spectral density of theta waves can capture a mismatch between the operator's expectations about process behavior and actual process behavior (Iqbal, Srinivasan, & Srinivasan, 2020). Whenever theta PSD increases, it reflects mismatch and consequent increase in cognitive workload, leading to the failure of the operator to control abnormalities in the process.

A schematic overview of how sensors can be used to improve human reliability during different operations is shown in Fig. 8. Both human performance and interface effectiveness need to be enhanced to achieve the

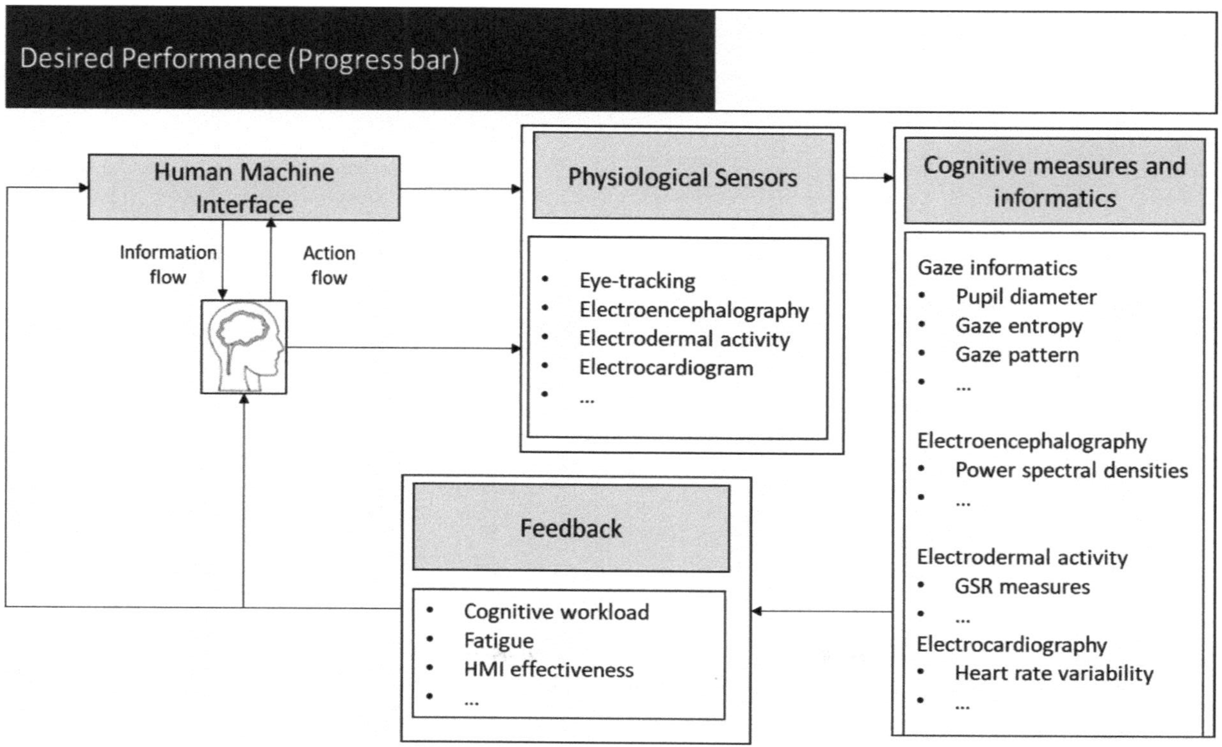

Fig. 8 Physiological measurements for error reduction.

desired performance for a task since they are interdependent. As shown in Fig. 8, during a typical task, sensor data can be recorded and then analyzed to identify shortcomings in HMIs and operator capabilities. These shortcomings can be used as a basis to improve the design and human capabilities to achieve the desired overall performance.

5. Summary

Human error is one of the main contributing factors to accidents in process industries, and there are several perspectives reported in the literature. In this chapter, we discussed the role of human factors in ensuring optimal system performance. Specifically, we highlighted how HRA could be utilized to account for human errors and understand human-machine interactions with an aim to improve human performance. Advancements in technology in the form of sophistication and digitization have shifted the role of humans more towards monitoring, supervision, diagnosis, and prognosis. These processes are cognitively challenging as they require operators to monitor numerous variables at the same time, aggregate information, and build an understanding of overall system functioning. Therefore, the need for assessing the impact of a huge information load on operator performance becomes imperative. We discussed the potential of physiological measurements in providing insights into cognitive aspects of performance such as workload, decision making, fatigue, and attention. We also highlighted how information from physiological measurements could be used to quantify and improve HF, especially in digitalized operations.

References

Abhang, P. A., Gawali, B., & Mehrotra, S. C. (2016). *Introduction to EEG-and speech-based emotion recognition*. Academic Press.

Ade, N., & Peres, S. C. (2022). A review of human reliability assessment methods for proposed application in quantitative risk analysis of offshore industries. *International Journal of Industrial Ergonomics*, *87*, 103238.

Ahmed, H., Glanz, J., & Beech, H. (2019). *Ethiopian airlines pilots followed Boeing's safety procedures before crash*. Report Shows. New York Times, April 4. https://www.nytimes.com/2019/04/04/world/asia/ethiopia-crash-boeing.html.

Akyuz, E., & Celik, M. (2015). Application of CREAM human reliability model to cargo loading process of LPG tankers. *Journal of Loss Prevention in the Process Industries*, *34*, 39–48.

Bell, J., & Holroyd, J. (2009). Review of human reliability assessment methods. *Health & Safety Laboratory*, *78*. Retrieved from: https://www.hse.gov.uk/research/rrpdf/rr679.pdf.

Bhavsar, P., Srinivasan, B., & Srinivasan, R. (2016). Pupillometry based real-time monitoring of operator's cognitive workload to prevent human error during abnormal situations. *Industrial & Engineering Chemistry Research, 55*(12), 3372–3382.

Bhavsar, P., Srinivasan, B., & Srinivasan, R. (2017). Quantifying situation awareness of control room operators using eye-gaze behavior. *Computers & Chemical Engineering, 106*, 191–201.

Bongiorno, N., Bosurgi, G., Pellegrino, O., & Sollazzo, G. (2017). How is driver's workload influenced by road environment? *Procedia Engineering, 187*, 5–13.

Borenstein, J., Mahajan, H. P., Wagner, A. R., & Howard, A. (2020). Trust and pediatric exoskeletons: A comparative study of clinician and parental perspectives. *IEEE Transactions on Technology and Society, 1*(2), 83–88.

Borragan Pedraz, G., & Peigneux, P. (2016). Behavioural bases and functional dynamics of cognitive fatigue. *Behavioural Bases and Functional Dynamics of Cognitive Fatigue (Doctorat)*. Sciences psychologiques et de l'éducation, Louvain.

Brauer, R. L. (2016). *Safety and health for engineers*. John Wiley & Sons.

Britannica, T. (Ed.). (2021, 13 December). *Apollo 13*, Encyclopedia Britannica. https://www.britannica.com/topic/Apollo-13-mission. (Accessed 1 March 2022).

Campos, M. C. M. M., Teixeira, H., Liporace, F., & Gomes, M. (2009). Challenges and problems with advanced control and optimization technologies. *IFAC Proceedings Volumes, 42*(11), 1–8.

Chen, K., & Li, Z. (2013). Evaluation of human-system interfaces with different information organization using an eye tracker. In *International Conference on Cross-Cultural Design* (pp. 288–295). Berlin, Heidelberg: Springer.

Cohen, M. X. (2011). It's about time. *Frontiers in Human Neuroscience, 5*, 1–15. Article 2.

Cohen, M. X. (2014). *Analyzing neural time series data: Theory and practice*. MIT press.

Crowl, D. A. (2007). *Human factors methods for improving performance in process industries*. John Wiley & Sons.

CSB. (2007). *Investigation report. Refinery Explosion and Fire*. Report No. 2005-04-I-TX.

CSB. (2008). *Investigation report. Pesticide Chemical Runaway Reaction Pressure Vessel Explosion*. Report No. 2008-08-I-WV.

Das, L., Iqbal, M. U., Bhavsar, P., Srinivasan, B., & Srinivasan, R. (2018). Toward preventing accidents in process industries by inferring cognitive state of control room operators through eye tracking. *ACS Sustainable Chemistry & Engineering, 6*(2), 2517–2528.

De Nadai, S., D'Incà, M., Parodi, F., Benza, M., Trotta, A., Zero, E., et al. (2016). Enhancing safety of transport by road by on-line monitoring of driver emotions. In *2016 11th System of Systems Engineering Conference (SoSE)* (pp. 1–4). IEEE.

Dehais, F., Dupres, A., Di Flumeri, G., Verdiere, K., Borghini, G., Babiloni, F., et al. (2018). Monitoring pilot's cognitive fatigue with engagement features in simulated and actual flight conditions using an hybrid fNIRS-EEG passive BCI. In *2018 IEEE International Conference on Systems, Man, and Cybernetics (SMC)* (pp. 544–549). IEEE.

Di Pasquale, V., Iannone, R., Miranda, S., & Riemma, S. (2013). An overview of human reliability analysis techniques in manufacturing operations. *Operations Management*, 221–240.

Di Stasi, L. L., Renner, R., Catena, A., Canas, J. J., Velichkovsky, B. M., & Pannasch, S. (2012). Towards a driver fatigue test based on the saccadic main sequence: A partial validation by subjective report data. *Transportation Research Part C: Emerging Technologies, 21*(1), 122–133.

Diaz-Piedra, C., Sanchez-Carrion, J. M., Rieiro, H., & Di Stasi, L. L. (2017). Gaze-based technology as a tool for surgical skills assessment and training in urology. *Urology, 107*, 26–30.

Diego-Mas, J. A., Garzon-Leal, D., Poveda-Bautista, R., & Alcaide-Marzal, J. (2019). User-interfaces layout optimization using eye-tracking, mouse movements and genetic algorithms. *Applied Ergonomics, 78*, 197–209.

Edmonds, J. (Ed.). (2016). *Human factors in the chemical and process industries: Making it work in practice*, Elsevier.

Embrey, D., Kontogiannis, T., & Green, M. (1994). *Guidelines for preventing human error in process safety*. New York: Center for Chemical Process Safety, American Institute for Chemical Engineers.

Forbes, M. G., Patwardhan, R. S., Hamadah, H., & Gopaluni, R. B. (2015). Model predictive control in industry: Challenges and opportunities. *IFAC-PapersOnLine, 48*(8), 531–538.

Friedman, T. L. (2017). *Thank you for being late: An optimist's guide to thriving in age of accelerations (Version 2.0, With a New Afterword)*. USA: Picador.

Gates, D. (2019). *Flawed analysis, failed oversight: How Boeing, FAA Certified Suspect 737 MAX Flight Control System*. Seattle Times, 17 March. https://www.seattletimes.com/business/boeing-aerospace/failed-certification-faa-missed-safety-issues-in- -737-max-system-implicated-in- -lion-air-crash/.

Gentili, R. J., Rietschel, J. C., Jaquess, K. J., Lo, L. C., Prevost, C. M., Miller, M. W., et al. (2014). Brain biomarkers based assessment of cognitive workload in pilots under various task demands. In *2014 36th Annual International Conference of IEEE Engineering in Medicine and Biology Society* (pp. 5860–5863). IEEE.

Gersh, J. R., McKneely, J. A., & Remington, R. W. (2005). Cognitive engineering: Understanding human interaction with complex systems. *Johns Hopkins APL Technical Digest, 26*(4), 377–382.

Glanz, J., Creswell, J., Kaplan, T., & Wichter, Z. (2019). *After a lion air 737 MAX crashed in October, questions about plane arose*. New York Times. https://www.nytimes.com/2019/02/03/world/asia/lion-air-plane-crash-pilots.html.

Ha, J. S., Byon, Y. J., Baek, J., & Seong, P. H. (2016). Method for inference of operators' thoughts from eye movement data in nuclear power plants. *Nuclear Engineering and Technology, 48*(1), 129–143.

Hansen, D. S., Jespersen, S., Bram, M. V., & Yang, Z. (2018). Human machine interface prototyping and application for advanced control of offshore topside separation processes. In *IECON 2018–44th Annual Conference of IEEE Industrial Electronics Society* (pp. 2341–2347). IEEE.

Harmony, T., Fernández, T., Silva, J., Bernal, J., Díaz-Comas, L., Reyes, A., et al. (1996). EEG delta activity: An indicator of attention to internal processing during performance of mental tasks. *International Journal of Psychophysiology, 24*(1–2), 161–171.

Hart, S. G. (2006). NASA-task load index (NASA-TLX); 20 years later. In *Vol. 50, No. 9. Proceedings of the human factors and ergonomics society annual meeting* (pp. 904–908). Sage CA: Los Angeles, CA: Sage Publications.

Hart, S. G., & Staveland, L. E. (1988). Development of NASA-TLX (task load index): Results of empirical and theoretical research. In P. A. Hancock, & N. Meshkati (Eds.), *Human mental workload* (pp. 139–193). North-Holland: Oxford.

He, X., Wang, Y., Shen, Z., & Huang, X. (2008). A simplified CREAM prospective quantification process and its application. *Reliability Engineering & System Safety, 93*(2), 298–306.

Herkert, J., Borenstein, J., & Miller, K. (2020). Boeing 737 MAX: Lessons for engineering ethics. *Science and Engineering Ethics, 26*(6), 2957–2974.

Hollnagel, E. (1998). *Cognitive reliability and error analysis method (CREAM)*. Elsevier.

Hollnagel, E. (2003). *Context, cognition and control. Co-operative process management: Cognition and information technology* (pp. 27–52). CRC Press.

Holmqvist, K., Nyström, M., Andersson, R., Dewhurst, R., Jarodzka, H., & Van de Weijer, J. (2011). *Eye tracking: A comprehensive guide to methods and measures*. OUP Oxford.

HSE. (2022a). *Human Failure Types*. https://www.hse.gov.uk/humanfactors/topics/types.pdf. (Last assessed: 26 March, 2022).

HSE. (2022b). *Texas City incident Human Factor aspects*. https://www.hse.gov.uk/chemicals/workshop/human-factors-09/texas-city-incident.pdf. (Last assessed: Feb 28, 2022).

HSE. (2022c). *Introducing key topics*. https://www.hse.gov.uk/humanfactors/top-ten.htm. (Last assessed: Feb 28,2022).

IEA. (2021). *Definitions, domains of specialization, systemic approach*. https://iea.cc/definition-and-domains-of-ergonomics/. (Accessed 19 February 2022).

IOGP. (2010). *Assessment risk data directory: Human factors in QRA* (p. 66). https://www.iogp.org/bookstore/product/risk-assessment-data-directory-human-factors-in-qra/.

Iqbal, M. U., Shahab, M. A., Choudhary, M., Srinivasan, B., & Srinivasan, R. (2021). Electroencephalography (EEG) based cognitive measures for evaluating effectiveness of operator training. *Process Safety and Environmental Protection, 150*, 51–67.

Iqbal, M. U., & Srinivasan, R. (2018). Simulator based performance metrics to estimate reliability of control room operators. *Journal of Loss Prevention in the Process Industries, 56*, 524–530.

Iqbal, M. U., Srinivasan, B., & Srinivasan, R. (2018). Towards obviating human errors in real-time through eye tracking. In *Vol. 43. Computer Aided Chemical Engineering* (p. 1189). 1194: Elsevier.

Iqbal, M. U., Srinivasan, B., & Srinivasan, R. (2019). Electroencephalogram based biomarkers for tracking cognitive workload of operators in process industries. In *Vol. 46. Computer Aided Chemical Engineering* (pp. 1393–1398). Elsevier.

Iqbal, M. U., Srinivasan, B., & Srinivasan, R. (2020). Dynamic assessment of control room operator's cognitive workload using electroencephalography (EEG). *Computers & Chemical Engineering, 141*, 106726.

Jung, S., Woo, J., & Kang, C. (2020). Analysis of severe industrial accidents caused by hazardous chemicals in South Korea from January 2008 to June 2018. *Safety Science, 124*, 104580.

Just, M. A., & Carpenter, P. A. (1980). A theory of reading: From eye fixations to comprehension. *Psychological Review, 87*(4), 329.

Kamzanova, A. T., Kustubayeva, A. M., & Matthews, G. (2014). Use of EEG workload indices for diagnostic monitoring of vigilance decrement. *Human Factors, 56*(6), 1136–1149.

Kang, Z., & Landry, S. J. (2014). Using scanpaths as a learning method for a conflict detection task of multiple target tracking. Human Factors, 56(6), 1150–1162.

Khan, F., Amyotte, P., & Adedigba, S. (2021). Process safety concerns in process system digitalization. *Education for Chemical Engineers, 34*, 33–46.

Kim, I. S. (2001). Human reliability analysis in man-machine interface design review. *Annals of Nuclear Energy, 28*, 1069–1081.

Kirwan, B. (1996). The validation of three human reliability quantification techniques—THERP, HEART and JHEDI: Part 1—technique descriptions and validation issues. Applied ergonomics, 27(6), 359–373.

Kirwan, B. (2008). Human reliability assessment. *Vol. 2. Encyclopedia of quantitative risk analysis and assessment*. Wiley.

Kirwan, B. (2017). *A guide to practical human reliability assessment*. CRC Press.

Landi, C. T., Villani, V., Ferraguti, F., Sabattini, L., Secchi, C., & Fantuzzi, C. (2018). Relieving operators' workload: Towards affective robotics in industrial scenarios. *Mechatronics, 54*, 144–154.

Lees, F. (2012). *Lees' loss prevention in process industries: Hazard identification, assessment and control*. Butterworth-Heinemann.

Li, R., Su, W., & Lu, Z. (2017). Physiological signal analysis for fatigue level of experienced and inexperienced drivers. *Traffic Injury Prevention, 18*(2), 139–144.

Lim, S. J., Shin, J. S., Jang, J. Y., Yoon, Y. S., Park, H. J., & Jung, H. C. (2006). Eyetracking and protocol analysis of expertise in air traffic control situation awareness. In: *5th International conference of the cognitive science*. British Columbia, Canada.

Lindscheid, C., Bremer, A., Haßkerl, D., Tatulea-Codrean, A., & Engell, S. (2016). A test environment to evaluate integration of operators in nonlinear model-predictive control of chemical processes. *IFAC-PapersOnLine*, *49*(32), 129–134.

Liu, Z., Zhao, Q., Zhang, L., Zhang, X., Fan, J., Wang, Q., et al. (2021). Quantitative evaluation on effect of experience under emergency situations in NPP Main control room based on multimodal data. *Nuclear Technology*, *207*(4), 575–581.

Mahmoud, M. S. (2016). *Microgrid: Advanced control methods and renewable energy system integration*. Elsevier.

Manske, P. G., & Schier, S. L. (2015). Visual scanning in an air traffic control tower–a simulation study. *Procedia Manufacturing*, *3*, 3274–3279.

Mayer, R. E., & Moreno, R. (2003). Nine ways to reduce cognitive load in multimedia learning. *Educational Psychologist*, *38*(1), 43–52.

Miller, G. A. (1956). The magical number seven, plus or minus two: Some limits on our capacity for processing information. *Psychological Review*, *63*(2), 81.

Orlandi, L., & Brooks, B. (2018). Measuring mental workload and physiological reactions in marine pilots: Building bridges towards redlines of performance. *Applied Ergonomics*, *69*, 74–92.

Othman, N., & Romli, F. I. (2016). Mental workload evaluation of pilots using pupil dilation. *International Review of Aerospace Engineering*, *9*(3), 80–84.

Parasuraman, R., Sheridan, T. B., & Wickens, C. D. (2008). Situation awareness, mental workload, and trust in automation: Viable, empirically supported cognitive engineering constructs. *Journal of Cognitive Engineering and Decision Making*, *2*(2), 140–160.

Pfurtscheller, G., Stancak, A., Jr., & Neuper, C. (1996). Event-related synchronization (ERS) in alpha band—An electrophysiological correlate of cortical idling: A review. *International Journal of Psychophysiology*, *24*(1–2), 39–46.

Porthin, M., Liinasuo, M., & Kling, T. (2020). Effects of digitalization of nuclear power plant control rooms on human reliability analysis–a review. *Reliability Engineering & System Safety*, *194*, 106415.

Pradhan, A. K., Fisher, D. L., & Pollatsek, A. (2005). Effects of PC-based training on novice drivers' risk awareness in a driving simulator. In *Proceedings of 3rd International Driving Symposium on Human Factors in Driver Assessment, Training and Vehicle Design*. Rockport, Maine: USA.

Rajendra Acharya, U., Paul Joseph, K., Kannathal, N., Lim, C. M., & Suri, J. S. (2006). Heart rate variability: A review. *Medical and Biological Engineering and Computing*, *44*(12), 1031–1051.

Rasmussen, J. (1982). Human errors. A taxonomy for describing human malfunction in industrial installations. *Journal of Occupational Accidents*, *4*(2–4), 311–333.

Reason, J. (1990). *Human error*. Cambridge University Press.

Reid, G. B., & Nygren, T. E. (1988). The subjective workload assessment technique: A scaling procedure for measuring mental workload. In *Vol. 52. Advances in psychology* (pp. 185–218). North-Holland.

Sauvet, F., Bougard, C., Coroenne, M., Lely, L., Van Beers, P., Elbaz, M., et al. (2014). In-flight automatic detection of vigilance states using a single EEG channel. *IEEE Transactions on Biomedical Engineering*, *61*(12), 2840–2847.

Shahab, M. A., Iqbal, M. U., Srinivasan, B., & Srinivasan, R. (2021a). Metrics for objectively assessing operator training using eye gaze patterns. *Process Safety and Environmental Protection*, *156*, 508–520.

Shahab, M. A., Iqbal, M. U., Srinivasan, B., & Srinivasan, R. (2022a). HMM-based models of control room operator's cognition during process abnormalities. 1. Formalism and model identification. *Journal of Loss Prevention in the Process Industries, 104748.*
Shahab, M. A., Iqbal, M. U., Srinivasan, B., & Srinivasan, R. (2022b). HMM-based models of control room operator's cognition during process abnormalities. 2. Application to operator training. *Journal of Loss Prevention in the Process Industries, 104749.*
Shahab, M. A., Srinivasan, B., & Srinivasan, R. (2021b). Evaluating control room operator training outcomes through eye gaze augmented multi-scale data. *Computer Aided Chemical Engineering, 50*, 1307–1312.
Shappell, S. A., & Wiegmann, D. A. (2000). *Human factors analysis and classification system—HFACS*. https://commons.erau.edu/publication/737/.
Sharma, C., Bhavsar, P., Srinivasan, B., & Srinivasan, R. (2016). Eye gaze movement studies of control room operators: A novel approach to improve process safety. *Computers & Chemical Engineering, 85*, 43–57.
Shirali, G. A., Hosseinzadeh, T., & Kalhori, S. R. N. (2019). Modifying a method for human reliability assessment based on CREAM-BN: A case study in control room of a petrochemical plant. *MethodsX, 6*, 300–315.
Srinivasan, R., & Iqbal, M. U. (2017). Inherently safe design. In *Handbook of safety principles* (pp. 386–396). John Wiley & Sons.
Srinivasan, R., Srinivasan, B., Iqbal, M. U., Nemet, A., & Kravanja, Z. (2019). Recent developments towards enhancing process safety: Inherent safety and cognitive engineering. *Computers & Chemical Engineering, 128*, 364–383.
Swain, A. D., & Guttmann, H. E. (1983). *Handbook of human-reliability analysis with emphasis on nuclear power plant applications. Final report.* (No. NUREG/CR-1278; SAND-80-0200). Sandia National Labs., Albuquerque, NM (USA).
Sweller, J. (1988). Cognitive load during problem solving: Effects on learning. *Cognitive Science, 12*(2), 257–285.
Thayer, J. F., Åhs, F., Fredrikson, M., Sollers, J. J., III, & Wager, T. D. (2012). A meta-analysis of heart rate variability and neuroimaging studies: Implications for heart rate variability as a marker of stress and health. *Neuroscience & Biobehavioral Reviews, 36*(2), 747–756.
Ung, S. T. (2015). A weighted CREAM model for maritime human reliability analysis. *Safety Science, 72*, 144–152.
Warm, J. S., Parasuraman, R., & Matthews, G. (2008). Vigilance requires hard mental work and is stressful. *Human Factors, 50*(3), 433–441.
Williams, J. C. (1986). A proposed method for assessing and reducing human error. In *Proc. 9th advances in reliability technology Symp*, Univ. of Bradford.
Zheng, B., Jiang, X., & Atkins, M. S. (2015). Detection of changes in surgical difficulty: Evidence from pupil responses. *Surgical Innovation, 22*(6), 629–635.

CHAPTER THIRTEEN

Safety assessment of complex socio-technical systems

Nicola Paltrinieri*
Norwegian University of Science and Technology – NTNU, Trondheim, Norway
*Corresponding author: e-mail address: nicola.paltrinieri@ntnu.no

Contents

1. Introduction

At the beginning of the 90s, Prof. Diekmann (1992) stated the following. "New analysis tools are emerging, which have the potential to allow complex risk analyses to be performed simply. These new tools, which are underpinned by decision analysis and, lately, expert-systems technology, may lead to powerful, yet simple, approaches to the representation of risky problems." Such an optimistic prediction on the future of risk analysis was also accompanied by the suggestion of a possible interdisciplinary direction. "Future approaches to risk analysis will certainly rely more on the advances being made in Artificial Intelligence (AI) and the cognitive sciences. New computer tools and knowledge-representation schemes will unquestionably lead to new techniques, insights and opportunities for risk analysis."

Methods in Chemical Process Safety, Volume 6
ISSN 2468-6514
https://doi.org/10.1016/bs.mcps.2022.05.004

In the same decade (1997), the Russian chess grandmaster Garry Kimovich Kasparov (former World Chess Champion, ranked world No. 1 from 1984 until his retirement in 2005) lost a chess game with the chess playing computer Deep Blue by IBM, which was an example of Good Old-Fashioned Artificial Intelligence (GOFAI) (Hsu, Campbell, & Hoane, 1995). On that game, Kasparov (2017) later stated the following. "Deep Blue was intelligent the way your programmable alarm clock is intelligent. Not that losing to a 10-million-dollar alarm clock made me feel any better."

Industrial risk analysis and safety management have tried to make use of AI, but they have unevenly progressed since the described events. They neither respected Diekmann's prediction (methodological gaps are still present (Paltrinieri, Comfort, & Reniers, 2019)), nor turned into "programmable-alarm-clock intelligence" thanks to the progressive refinement of machine learning models and the increase in available computing power (Goodfellow, Bengio, & Courville, 2016).

This contribution aims to outline what AI can bring to risk analysis and safety management of complex socio-technical systems by illustrating a series of examples (with emphasis on benefits and limitations) where AI techniques are used to continuously update the current status evaluation.

1.1 Artificial intelligence and machine learning

AI is intelligence demonstrated by machines and it is divided into sub-fields based on technical considerations, such as particular goals (e.g. "robotics" or "machine learning"), the use of particular tools ("logic" or artificial neural networks), or deep philosophical differences.

This contribution focuses on the sub-field of Machine Learning (ML). ML refers to techniques aiming to program computers to learn from experience (Samuel, 1959). Some of its models (e.g. deep learning) aim to simulate the learning model of the human brain (Goodfellow et al., 2016). Such models are composed of multiple processing layers to learn representations of data with multiple levels of abstraction.

A computer may be trained to assess risk for safety-critical industries such as Oil & Gas through these learning techniques. This would allow processing a large amount of information in the form of indicators from normal operations and past unwanted events (from mishaps to major accidents), which would be used for training. Due to the subjectivity of risk definition (Villa, Paltrinieri, Khan, & Cozzani, 2016), risk level cannot be assigned to each event with certainty and expert supervision is needed. Once the model has learned risk categorization, it uses its knowledge to evaluate real-time risk from the state of the monitored system.

1.2 Monitoring of early deviations and past events

Increasing attention has been dedicated to monitoring safety barrier performance through indicators, as a way to assess and control risk. Indicators may report a series of socio-technical factors: physical conditions of a plant (equipment pressure and temperature), number of failures of an equipment piece, maintenance backlog, number of emergency preparedness exercises run, amount of overtime worked, etc. A number of indicator typologies are theorized and used in literature (Øien, Utne, & Herrera, 2011). Øien et al. (2011) affirm that we can refer to risk indicators if: they provide numerical values (such as a number or a ratio); they are updated at regular intervals; they only cover some selected determinants of overall risk, in order to have a manageable set of them. That being said, the latter feature is quickly becoming outdated due to the extensive collection carried out in industry and the attempts to process large numbers of them (Paltrinieri & Reniers, 2017).

Øien et al. (2011), Paltrinieri, Landucci, Nelson, and Hauge (2016), Paltrinieri, Øien, and Cozzani (2012) and Landucci and Paltrinieri (2016) have produced several reviews on risk and barrier indicators. They show that definition and collection of risk indicators have become consolidated practices in "high-risk" sectors, such as the petroleum and chemical industries. For instance, the Norwegian Petroleum Safety Authority (PSA) requires indicators describing the technical performance of safety barriers within the Norwegian Oil & Gas industry since 1999 (PSA, 2016); while, the European directive "Seveso III" (European Parliament and Council, 2012) on the control of major-accident hazards involving dangerous substances suggests their use for sites handling hazardous substances (European Parliament and Council, 1982). Such trend towards definition and collection of higher numbers of indicators (Paltrinieri & Reniers, 2017) demonstrates the mentioned challenge on big data process for safety assessment.

2. Examples of AI-based prediction

Three examples of AI-based prediction for the safety assessment of complex socio-technical systems are described in the following.

2.1 Consequence class associated to a hazardous material release

ML techniques were applied by Tamascelli, Solini, Paltrinieri, and Cozzani (2022) to a database of past accidents with the purpose to simulate its application on the national databases managed by the Seveso-competent authorities. The dataset used is the Major hazard incident database (MHIDAS) (AEA Technology—Major Hazards Assessment Unit, 2003) launched by the UK Health and Safety Executive in 1986 and developed by AEA Technology until mid-1990s. The events included are based on public domain information sources and their characteristics are registered using keywords.

MHIDAS collects about 8972 hazardous events from 1916 to 1992, recorded by means of the set of items listed in Table 1. Some items use a taxonomy to systematically categorize the event.

The items listed in the upper part of Table 1 were considered as inputs to the ML models to predict the associated consequences—lower part of Table 1. The details of data pre-processing are explained elsewhere (Tamascelli et al., 2022). The study focused on the number of people killed and aimed to predict the occurrence of a hazardous event within one of the severity categories listed in Table 2 based on the considered inputs. Only categorical data are used.

2.2 Wellhead damage frequency in a drilling rig

To avoid potential damage during drilling operations for a new offshore Oil and gas well, a semi-submersible drilling unit should maintain the position above the wellhead (Fig. 1). This is particularly critical if the platform is in shallow waters, where small changes of position lead to higher riser (pipe connecting the platform to the subsea drilling system) angles. Exceeding physical inclination limits may result in damages to wellhead, Blowout Preventer (BOP—sealing the well) or Lower Marine Riser Package (LMRP—connecting riser and BOP) (Chen, Moan, & Verhoeven, 2008).

Platform positioning is maintained in an autonomous way (without mooring system) through the action of a set of thrusters controlled by the Dynamic Positioning (DP) system. Input for the DP system is provided by the position reference system (Differential Global Positioning System—DGPS and Hydroacoustic Position Reference—HPR), environmental sensors, gyrocompass, radar, and inclinometer (Chen et al., 2008).

Table 1 Set of items used to record hazardous events in MHIDAS (AEA Technology—Major Hazards Assessment Unit, 2003).

Items	Description	Category from taxonomy
Date	Date of the event	
Location	Location of event	
Substance	Substances involved in the event	X
Event type	Typology of event	X
Origin	Area of the plant and type of equipment from which the event started	X
Section	Plant section in which the event occurred	X
Quantity	Amount (ton) of released substance	
General causes	General causes the led to the event	X
Specific causes	Specific causes the led to the event	X
Evacuated	Number of people evacuated	
Consequences		
Damage	Economic damage to the property or production loss	
Injured	Number of people injured by the event	
Killed	Number of people killed by the event	

Specific keywords are used to describe some of the items.

Table 2 Severity categories considered by the study.

Severity categories	
0	**Event with no fatalities**
1–10	Event with a number of fatalities between 1 and 10
10–100	Event with a number of fatalities between 10 and 100

A Dynamic Positioning Operator (DPO) located in the Marine Control Room (MCR) is responsible for constant monitoring of DP panels and screens and carrying out emergency procedures if needed (Giddings, 2013). Platform position may be lost due to several reasons.

In this case study, Paltrinieri, Patriarca, Stefana, Brocal, and Reniers (2020) assume that the platform thrusters exercise propulsion towards a

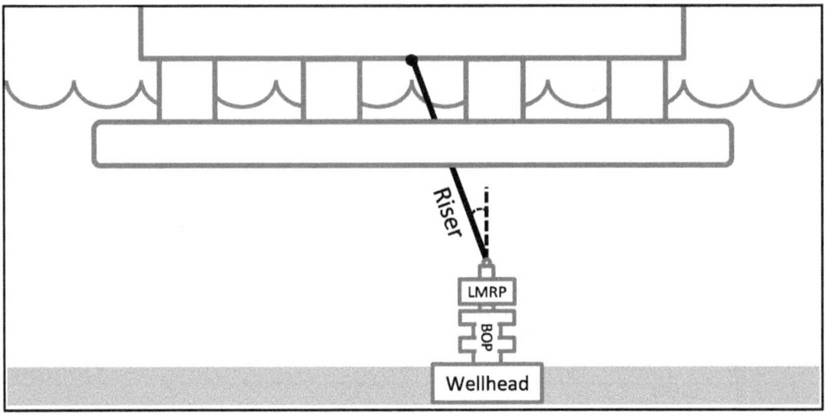

Fig. 1 Position of a semi-submersible drilling unit above the wellhead. *Adapted from Paltrinieri, N., Patriarca, R., Stefana, E., Brocal, F., & Reniers, G. (2020). Meta-learning for safety management.* Chemical Engineering Transactions, 83. *doi:10.3303/CET2082029.*

wrong direction, leading to a scenario of "drive-off." If the rig moves to an offset position, specific alarms turn on and suggest the DPO to stop the drive-off scenario by deactivating the thrusters and initiate the manual Emergency Disconnect Sequence (EDS) for the disconnection of the riser from the BOP. If the manual EDS ultimately fails, the automatic EDS activates at the ultimate position limit allowing for safe disconnection (Chen et al., 2008).

A number of works (Matteini, 2015; Paltrinieri et al., 2019; Paltrinieri, Massaiu, & Matteini, 2016) address the details of occurrence and development of drive-off scenarios. Relevant indicators are defined to assess the performance of socio-technical safety barriers and related systems. Examples of these indicators are the following.

- Thruster control failures in the last 3 months.
- Thruster monitoring sensors failures in the last 3 months.
- Simulator hours carried out by the DPO in the last 3 months.
- Inadequate DPO communication events in the last 3 months.
- Delays in DPO shifts in the last 3 months.
- Percentage of time in the last 3 months with more than one operator monitoring.

The simulations of their trends for a period of 30 years can be found in the literature (Paltrinieri et al., 2019). They are inspired to the typical bathtub curve for technical elements (Wang, Hsu, & Liu, 2002) and relevant expert judgment for the remaining elements.

As shown by Bucelli, Paltrinieri, and Landucci (2018), indicator values may be aggregated based on relative weights and hierarchical barrier models, in order to enable dynamic update of barrier failure probabilities. This can be used to update, in turn, occurrence frequencies of potential outcomes. Outcome frequencies are an expression of the scenario probability and, in turn, of the risk. If we assume that the other factors are constant, this represents a simplified risk model. However, Matteini (2015) points out a certain complexity within the hierarchical barrier model, which may be due to a tangled structure and an unclear approach to assign relative weights to single model elements. For this reason, a machine learning approach bypassing the construction of such hierarchies and aggregation rules is suggested by Paltrinieri et al. (2020b).

2.3 Alarm chattering in an ammonia plant

Alarm data from a section of an ammonia production process (Topsoe, 2020) are analyzed by Tamascelli, Paltrinieri, and Cozzani (2020). Due to the large quantity of hazardous substances stored and handled during normal activity, the plant has been classified as an "upper tier" Seveso III establishment. Extensive use of methane, hydrogen, and ammonia (anhydrous and aqueous solution) occurs in the plant section. Furthermore, due to the intrinsic properties of the processes involved—severe operating conditions (i.e., high pressure and high temperature) are often associated with corrosive substances. Additional information about ammonia production and the considered site can be found at: (Aika et al., 2012; Yara Italia, 2016).

The alarm database consists of alarm data collected during an observation period of more than 4 months. Each row of the database represents an alarm event (26,473 observations in total), and each column (36 in total) represents a piece of information (i.e. an "attribute") about the alarm. A list of the most meaningful attributes is presented in Table 3.

The Alarm Identifier (point 5. of the "Message" attribute) is a code that defines the alarm status. Examples of Alarm Identifiers are "HHH" (which means that the measured variable has exceeded the "high level" setpoint), "HTRP" (the measured variable has exceeded the "very high level" alarm setpoint and automatic block intervention procedures might be triggered), "IOP" (which indicates an instrumental failure or out-of-range measure), "LLL" and "LTRP" (same as "HHH" and "HTRP" but referring to a "low/very low level").

Table 3 Alarm database attributes.

Attribute	Meaning
Time stamp	Date and time (GMT) of the alarm event
Source	The source that triggered the alarm. It might be a measuring instrument or a PLC function
Jxxx	The safety interlock logic associated with the alarm
Message	The message that is shown to the operator contains the following five attributes:
	the Source;
	a concise description of the equipment involved;
	the safety interlock logic (Jxxx);
	the value and units of measures of the process variable;
	the Alarm Identifier (e.g. HHH, HTRP, LLL, LTRP, ACK, etc.)
Active time	Date and time (GMT) of the first alarm occurrence
Data value	The value of the process variable
Eng. unit	The units of measure of the process variable

According to Kondaveeti, Izadi, Shah, and Black (2010), an alarm event is uniquely identified by three attributes only: Time Stamp, Source, and Alarm Identifier (e.g. HHH, HTRP, LLL, LTRP, etc.). The combination of a "source" and an "alarm identifier" is called a "unique alarm".

More than 96% of the alarms registered in the database occurred within 1 month only when a considerable number of floods and chattering alarms must have occurred. In fact, only 10 alarm sources (out of 194 in total) were responsible for more than 80% of the alarms recorded.

Chattering alarms are alarms "that repeatedly transitions between active state and inactive state in a short period of time" (ANSI/ISA, 2016). Therefore, chattering alarms have the potential to produce a large count of alarms and reducing their number is a key step to improve the performance of the alarm system during alarm floods.

Kondaveeti et al. (2010) proposed a method for quantifying alarm chatter based on run lengths distributions. Although effective, this technique produces static results (i.e. chattering is quantified based on historical alarm data, but no conclusion can be drawn about the alarm's future behavior). This Chattering Index approach is modified by Tamascelli et al. (2020) to predict a chattering behavior by means of standard ML models.

3. Method

ML classification models were used for the three examples in Section 2. Moreover, comparison among different ML models is also beneficial. Results from multiple linear regression (MLR) were compared to the relatively more sophisticated deep neural network (DNN) models.

3.1 Multiple linear regression

A linear model, given a vector of inputs $X=(x_1, x_2, \ldots, x_p)$, predicts the output y (in this case an index for the risk R) via the following equation (Hastie, Tibshirani, & Friedman, 2009):

$$y = b_0 + \sum_{j=1}^{p} x_j w_j \sim R \tag{1}$$

where b_0 is the so-called bias and wj represents the model weights. This model needs then to be trained with a training set data to learn the weights of every provided input. Once the weights are known, the model can be used for prediction of y based on new inputs X.

3.2 Deep neural network

The deep learning model considered in this work is a feed-forward neural network, wherein connections between the units do not form a cycle (Svozil, Kvasnicka, & Pospichal, 1997). A linear model, such as MLR, would be restricted to linear functions, while a DNN model describes the target as a nonlinear function of the input features (Goodfellow et al., 2016). The DNN model can be described as a series of functional transformations associated to the model layers (Fig. 2).

The overall length of the chain gives the depth of the model. Specifically, the first network layer performs the following computation of the inputs $X=(x_1, x_2, \ldots, x_p)$:

$$a_i = b_i + \sum_{j=1}^{p} x_j w_{i,j} \tag{2}$$

with $i=1, \ldots, m$. where a_i, b_i and w_i are respectively defined as activation, bias and model weight.

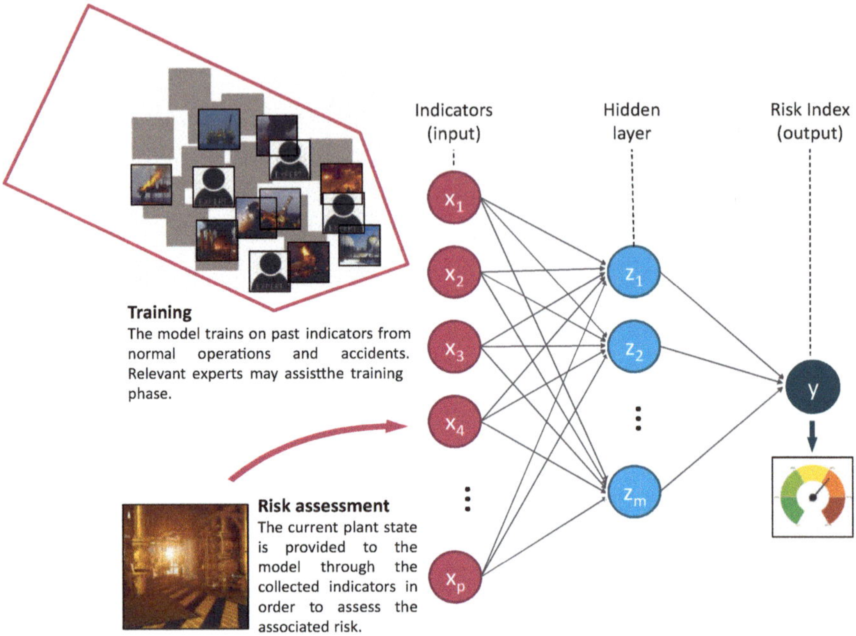

Fig. 2 Representation of deep learning approach for risk assessment. *Adapted from Paltrinieri, N., Comfort, L., & Reniers, G. (2019). Learning about risk: Machine learning for risk assessment.* Safety Science, 118, *475–486. doi:10.1016/j.ssci.2019.06.001.*

The activations are transformed by the activation function g within the hidden layer:

$$z_i = g(a_i) \tag{3}$$

where z_i is defined as hidden unit. The most used activation function is the sigmoid (Goodfellow et al., 2016). Fig. 2 shows only one hidden layer for the sake of simplicity, but there can be several.

The hidden units are combined to give the activations a_o of the output layer:

$$a_o = b_o + \sum_{j=1}^{m} z_j w_{o,j} \tag{4}$$

where a_o, b_o and w_o are activation, bias and model weight. Fig. 2 shows only one output for the sake of simplicity, but there can be several.

Finally, the activation function h is used to obtain the output y, which is an index for the risk R:

$$y = h(a_o) \sim R \tag{5}$$

Given a dataset of X and associated y, the model can be trained to minimize the final loss function in a supervised way, in order to predict y based on new inputs X.

3.3 Metrics

The performance of the used classification algorithms is assessed during the evaluation phase. As an example, consider a situation where accidents must be classified into two classes A or B. A positive prediction occurs when the model predicts the class A. Instead, a negative prediction occurs when the model predicts the class B. Whenever the model predicts the class of an object, there are four possible outcomes:

- TP = True Positive—i.e., predicted label = A, true label = A;
- TN = True Negative—i.e., predicted label = B, true label = B;
- FP = False Positive—i.e., predicted label = A, true label = B;
- FN = False Negative—i.e., predicted label = B, true label = A.

The sum of True Positives and True Negatives represents the number of correct predictions, while the sum of False Positives and False Negatives indicates the number of wrong predictions.

True Positives, True Negatives, False Positives, and False Negatives are used to obtain three performance indicators:

$$\text{Accuracy} = \frac{TP + TN}{TP + TN + FP + FN} \tag{6}$$

$$\text{Precision} = \frac{TP}{TP + FP} \tag{7}$$

$$\text{Recall} = \frac{TP}{TP + FN} \tag{8}$$

Accuracy represents the fraction of objects that have been correctly classified. Precision indicates the success rate of a positive prediction. Recall denotes the fraction of actual positives that have been correctly identified.

It is worth mentioning that metrics and indicators depend on the probability threshold used by the models to classify. For example, if the decision threshold is lowered, the model may produce more positive predictions. As a result, the Recall might increase, but the Precision might decrease (Scikit-learn.org, 2020). In fact, actions aimed at increasing Recall often lower the Precision, and vice-versa. A convenient means of displaying the effect of the decision threshold is the Precision-Recall curve—i.e., a plot where each point represents the couple Precision vs Recall at a specific decision threshold (Murphy, 2012). A convenient means of summarizing the information in the Precision-Recall curve is the area under the curve (AUC P-R) (Murphy, 2012), which takes values between 0 and 1. Being independent on the decision threshold, the AUC PR is considered a more comprehensive indicator of the model performance if compared with Accuracy, Precision, and Recall. In general, a large AUC P-R value indicates good performance (Scikit-learn.org, 2020).

4. Results and discussion

Table 4 summarizes all the results from the examples described in Section 2. The results from the two approaches used (MLR and DNN) are directly compared to identify the best predictive performance. MLR shows a higher number of higher values in green cells (9) if compared to DNN (6).

However, this overall result cannot convey the message that MLR performs better than DNN as "there ain't no such thing as a free lunch." In fact, in these examples, DNN was applied with default parameters (e.g. number of layers and nodes). In addition, DNN is relatively more sensitive to poor quality of data (Paltrinieri et al., 2019).

Table 4 reports all the metrics discussed in Section 3.3. If we exclusively focus on accuracy, we notice that the highest value (0.99) is obtained for both MLR and DNN predictions of the consequence class 10–100 fatalities

Table 4 Summary of results from the representative examples of machine learning application for safety purposes.

	Multi linear regression					**Deep neural network**				
	C: None	C: 1–10	C: 10–100	WDF	AC	C: None	C: 1–10	C: 10–100	WDF	AC
Accuracy	0.88	0.87	0.99	0.82	0.95	0.77	0.88	0.99	0.83	0.94
Precision	0.90	0.20	0.00	0.80	0.94	0.89	0.11	0.00	0.84	0.93
Recall	0.98	0.07	0.00	0.91	0.94	0.85	0.09	0.00	0.86	0.93
PR AUC	0.95	0.22	0.06	0.92	/	0.87	0.27	0.01	/	/

Two approaches are considered': Multi Linear Approach and Deep Neural Network. "C" stands for consequence class, "WDF" stands for wellhead damage frequency, "AC" stands for alarm chattering, and "PR AUC" stands for the area under the precision recall curve. Green and red cells respectively show higher and lower values when compared to the other approach.

associated to a hazardous substance release. However, accuracy alone is not informative if the problem involves the identification of rare classes—i.e., when the dataset is class imbalanced (Google, 2020).

Releases of hazardous substances with 10–100 fatalities are (fortunately) rare events as they represent about 1% of the records in the MHIDAS database. In this case, the models have learned that the result will be correct 99% of the times, if they predict that this kind of events never occurs. If the cost for a False Negative is higher than the cost for a False Positive (such as the case of a release of hazardous substance with 10–100 fatalities), Recall is the most meaningful metric. In this context, a good model must produce a high Recall, while a low precision might be considered acceptable and, to a certain extent, conservative.

The prediction of events with a relatively higher frequency and lower consequence (e.g., a release of a hazardous substance with no fatalities, an increase of wellhead damage frequency, or a chattering alarm) may instead benefit of a higher precision at the expense of the recall value.

For this reasons, rather than considering Precision and Recall individually, one may aggregate them into the so-called *F*-score (Chinchor, 1992), especially if the area under the precision recall curve indicates the potential of optimizing the model by tuning the decision threshold probability.

$$F_\beta = \left(1 + \beta^2\right) \bullet \frac{\text{Precision} \bullet \text{Recall}}{\left(\beta^2 \bullet \text{Precision}\right) + \text{Recall}} \tag{9}$$

where β is a non-negative real number. If $\beta = 1$, the score represents the harmonic mean between Precision and Recall (Han, Kamber, & Pei, 2012). If $\beta > 1$, the score is Recall oriented and, vice versa, if $\beta < 1$, the score is precision oriented (Sasaki, 2007).

Fig. 3 shows the precision recall curve used for the optimization of the model predicting the increase of wellhead damage frequency. Its area under the precision recall curve is relatively high (0.92), indicating potential for an improvement. In this case, we would like to increase the precision by tuning the decision probability threshold. For this reason, we select $\beta = 0.5$ and identify the maximum of the *F*-score for a threshold equal to 0.2 (Table 5), which translates into precision next to 1 (Fig. 3) and accuracy next to 0.8 (Table 5).

The results from these examples of AI-based prediction for safety purposes show how the techniques used require a deep understanding of their benefits, limitations, and application boundaries. For this reason, this

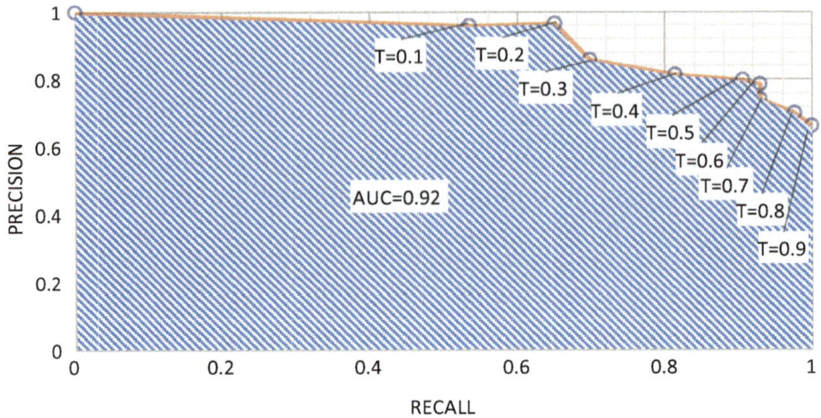

Fig. 3 Precision Recall curve used for the optimization of the model predicting the increase of wellhead damage frequency at various level of probability threshold, T.

Table 5 F-scores and accuracy values obtained by tuning the decision probability threshold.

Probability Threshold, *T*	$F_{0.5}$	**Accuracy**
0.1	0.827	0.734
0.2	0.881	0.797
0.3	0.820	0.772
0.4	0.814	0.797
0.5	0.816	0.823
0.6	0.810	0.823
0.7	0.772	0.785
0.8	0.759	0.742
0.9	0.722	0.710

contribution aims to convey the message that AI-based techniques must be considered as tools supporting and not substituting safety-related decision-making for socio-technical systems. Awareness and knowledge of these tools properties by the user is essential to effectively exploit their results. The role of the human as user of these tools is even more central than before. AI should not be intended as a way to replace the human, but only as an improved approach assisting the human. This is conform with the concept of trustworthy AI by the European Commission (EC's High Level Expert Group on AI, 2018) promoting explainable AI (XAI) human centrality by means of interpretability, info-besity (overload of information) avoidance, and transparency.

Once the principles of trustworthy AI and XAI are fully embraced, machine learning will have a vast potential in safety management, especially considering emerging variants of the traditional approaches described in this contribution, such as:

- *Transfer learning*, aiming at developing methods to exploit the knowledge gained in one task (i.e., the *source task*) to address a new task (i.e., *target task*).
- *Federated learning*, machine learning technique that trains an algorithm across multiple decentralized servers holding local data samples, without exchanging them.
- *Meta-learning*, focus on the learning model and its optimization towards new observations, in order to apprehend the emergence of unknown scenarios (e.g. unknown risks (Stefana & Paltrinieri, 2020, 2021)).

5. Conclusion

This contribution has illustrated examples of AI-based prediction used to continuously update the safety assessment of complex socio-technical systems. In fact, the examples refer to the impact prediction of a hazardous substance release in chemical industry, the wellhead damage frequency in offshore oil and gas and chattering alarms in ammonia production. The results can and must be read on different levels, carefully considering the available metrics based on the scenario addressed. This shows that we are not (and will not be in a near future) in a "no-brainer" condition in which the responsibility for human and system safety is entirely moved to the machine. At the same time, such advanced techniques have the concrete potential to provide a reliable support for critical decision making, guiding industry towards more risk-informed and safety-responsible planning.

References

AEA Technology—Major Hazards Assessment Unit. (2003). *MHIDAS—Major hazard incident data service. UK.*

Aika, K., Christiansen, L. J., Dybkjaer, I., Hansen, J. B., Nielsen, P. E. H., Nielsen, A., et al. (2012). *Ammonia: Catalysis and manufacture.* Springer Science & Business Media.

ANSI/ISA. (2016). *ANSI/ISA–18.2–2016 management of alarm systems for the process industries.* ANSI/ISA.

Bucelli, M., Paltrinieri, N., & Landucci, G. (2018). Integrated risk assessment for oil and gas installations in sensitive areas. *Ocean Engineering*, *150*, 377–390. https://doi.org/10.1016/J.OCEANENG.2017.12.035.

Chen, H., Moan, T., & Verhoeven, H. (2008). Safety of dynamic positioning operations on mobile offshore drilling units. *Reliability Engineering and System Safety*, *93*, 1072–1090. https://doi.org/10.1016/J.RESS.2007.04.003.

Chinchor, N. (1992). MUC-4 evaluation metrics. In *Proceedings of the 4th conference on message understanding, MUC4 '92. Association for Computational Linguistics, USA* (pp. 22–29). https://doi.org/10.3115/1072064.1072067.

Diekmann, E. J. (1992). Risk analysis: Lessons from artificial intelligence. *International Journal of Project Management, 10*, 75–80. https://doi.org/10.1016/0263-7863(92)90059-I.

EC's High Level Expert Group on AI. (2018). *Draft ethics guidelines for trustworthy AI. Brussels, Belgium.*

European Parliament and Council. (1982). Council directive 82/501/EEC of 24 June 1982 on the major-accident hazards of certain industrial activities. *Official Journal of the European Union, 1–18.*

European Parliament and Council. (2012). Directive 2012/18/EU of 4 July 2012 on the control of major-accident hazards involving dangerous substances, amending and subsequently repealing council directive 96/82/EC - Seveso III. *Official Journal of the European Union, 1–37.*

Giddings, I. C. (2013). *IMO guidelines for vessels with dynamic positioning systems.* Houston, TX, USA: Dynamic Positioning Conference.

Goodfellow, I. J., Bengio, Y., & Courville, A. (2016). *Deep learning.* Citeseer, Cambridge, MA, USA: The MIT Press.

Google. (2020). *Classification: Accuracy | Machine learning crash course.* (WWW Document).

Han, J., Kamber, M., & Pei, J. (2012). 8—Classification: Basic concepts. In J. Han, M. Kamber, & J. B. T.-D. M. Pei (Eds.), *The Morgan Kaufmann series in data management systems* (3rd ed., pp. 327–391). Boston: Morgan Kaufmann. https://doi.org/10.1016/B978-0-12-381479-1.00008-3.

Hastie, T., Tibshirani, R., & Friedman, J. (2009). Unsupervised learning. In *The elements of statistical learning* (pp. 485–585). Springer.

Hsu, F., Campbell, M. S., & Hoane, A. J., Jr. (1995). Deep blue system overview. In *Proceedings of the 9th International Conference on Supercomputing* (pp. 240–244).

Kasparov, G. (2017). *Deep thinking: Where machine intelligence ends and human creativity begins.* UK: Hachette.

Kondaveeti, S. R., Izadi, I., Shah, S. L., & Black, T. (2010). Graphical representation of industrial alarm data, IFAC proceedings volumes (IFAC-PapersOnline). *IFAC.* https://doi.org/10.3182/20100831-4-fr-2021.00033.

Landucci, G., & Paltrinieri, N. (2016). A methodology for frequency tailorization dedicated to the oil & gas sector. *Process Safety and Environment Protection, 104*, 123–141. https://doi.org/10.1016/j.psep.2016.08.012.

Matteini, A. (2015). *Human factors and dynamic risk analysis: A case-study in oil & gas drilling. Bologna, Italy.*

Murphy, K. P. (2012). *Machine learning: A probabilistic perspective, adaptive computation and machine learning.* Cambridge, MA, United States: MIT Press.

Øien, K., Utne, I. B., & Herrera, I. A. (2011). Building safety indicators: Part 1—Theoretical foundation. *Safety Science, 49*, 148–161. https://doi.org/10.1016/j.ssci.2010.05.012.

Paltrinieri, N., Comfort, L., & Reniers, G. (2019). Learning about risk: Machine learning for risk assessment. *Safety Science, 118*, 475–486. https://doi.org/10.1016/j.ssci.2019.06.001.

Paltrinieri, N., Landucci, G., Nelson, W. R., & Hauge, S. (2016). Proactive approaches of dynamic risk assessment based on indicators. In *Dynamic risk analysis in the chemical and petroleum industry: Evolution and interaction with parallel disciplines in the perspective of industrial application.* https://doi.org/10.1016/B978-0-12-803765-2.00006-8.

Paltrinieri, N., Massaiu, S., & Matteini, A. (2016). *Human reliability analysis in the petroleum industry: Tutorial and examples* (pp. 181–192). Butterworth-Heinemann. https://doi.org/10.1016/B978-0-12-803765-2.00015-9.

Paltrinieri, N., & Reniers, G. (2017). Dynamic risk analysis for Seveso sites. *Journal of Loss Prevention in the Process Industries, 49*, 111–119. https://doi.org/10.1016/j.jlp.2017.03.023.

Paltrinieri, N., Øien, K., & Cozzani, V. (2012). Assessment and comparison of two early warning indicator methods in the perspective of prevention of atypical accident scenarios. *Reliability Engineering and System Safety, 108*. https://doi.org/10.1016/j.ress.2012.06.017.

Paltrinieri, N., Patriarca, R., Stefana, E., Brocal, F., & Reniers, G. (2020). Meta-learning for safety management. *Chemical Engineering Transactions, 83*. https://doi.org/10.3303/CET2082029.

PSA. (2016). *Trends in risk level in the petroleum activity (RNNP) [WWW document]*. URL http://www.psa.no/about-rnnp/category911.html.

Samuel, A. L. (1959). Some studies in machine learning using the game of checkers. *IBM Journal of Research and Development, 3*, 210–229. https://doi.org/10.1147/rd.33.0210.

Sasaki, Y. (2007). *The truth of the F-measure* (p. 2007).

Scikit-learn.org. (2020). *Precision—Recall [WWW Document]*.

Stefana, E., & Paltrinieri, N. (2020). In P. Baraldi, F. Di Maio, & E. Zio (Eds.), *Meta-learning potential to assess uncertainties in dynamic risk management E-proceedings of the 30th European safety and reliability conference and 15th probabilistic safety assessment and management conference (ESREL2020 PSAM15)*. Research Publishing.

Stefana, E., & Paltrinieri, N. (2021). ProMetaUS: A proactive meta-learning uncertainty-based framework to select models for dynamic risk management. *Safety Science, 138*, 105238. https://doi.org/10.1016/j.ssci.2021.105238.

Svozil, D., Kvasnicka, V., & Pospichal, J. (1997). Introduction to multi-layer feed-forward neural networks. *Chemometrics and Intelligent Laboratory Systems, 39*, 43–62. https://doi.org/10.1016/S0169-7439(97)00061-0.

Tamascelli, N., Paltrinieri, N., & Cozzani, V. (2020). Predicting chattering alarms: A machine learning approach. *Computers and Chemical Engineering, 143*. https://doi.org/10.1016/j.compchemeng.2020.107122.

Tamascelli, N., Solini, R., Paltrinieri, N., & Cozzani, V. (2022). Learning from major accidents: A machine learning approach. *Computers and Chemical Engineering, 162*, 107786. https://doi.org/10.1016/J.COMPCHEMENG.2022.107786.

Topsoe, H. (2020). *Ammonia | NH3 | process | Haldor Topsoe [WWW document]*.

Villa, V., Paltrinieri, N., Khan, F., & Cozzani, V. (2016). Towards dynamic risk analysis: A review of the risk assessment approach and its limitations in the chemical process industry. *Safety Science, 89*. https://doi.org/10.1016/j.ssci.2016.06.002.

Wang, K. S., Hsu, F. S., & Liu, P. P. (2002). Modeling the bathtub shape hazard rate function in terms of reliability. *Reliability Engineering and System Safety, 75*, 397–406. https://doi.org/10.1016/S0951-8320(01)00124-7.

Yara Italia, S. P. A. (2016). *Relazione di riferimento della Yara Italia S.p.A.* (dello stabilimento di Ferrara).

CHAPTER FOURTEEN

Security of digitalized process systems

Ahmed Hamdy El-Kady, Syeda Halim, Hans Pasman*, and Faisal Khan*
Mary Kay O'Connor Process Safety Center, Artie McFerrin Department of Chemical Engineering, Texas A&M University, College Station, TX, United States
*Corresponding authors: e-mail address: hjpasman@gmail.com; fikhan@tamu.edu

Contents

1. Introduction

Digitalized process systems are composed of multiple components and subsystems that are embedded with advanced Industry 4.0 technologies so that each of these objects can sense, communicate, compute, and potentially

Methods in Chemical Process Safety, Volume 6
ISSN 2468-6514
https://doi.org/10.1016/bs.mcps.2022.05.005

actuate. They are susceptible to security threats on their physical as well as their cyber layer (Pasqualetti, Dörfler, & Bullo, 2013). This fact has been highlighted by the multiple reported cyber-physical benefit-driven or ideology-driven attacks on different process systems over the years. The Structured Query Language (SQL) "Slammer" worm that infected the Davis–Besse nuclear plant in 2003 was among the first attacks that targeted industrial control systems (ICS) and caused the plant process computer to be inaccessible for over 6h. However, the catastrophic consequences of highly targeted sabotages by powerful adversaries (e.g., nation states) on ICSs became clear after the Stuxnet attack in June 2010, which caused the failure of centrifuges at an Iranian nuclear plant. Stuxnet exploited multiple vulnerabilities which were unknown before they were exploited (Sadeghi, Wachsmann, & Waidner, 2015). It showed the urgency of an efficient approach to assess the impact of cyber threats against both the physical and the cyber parts of digitalized ICSs (Genge, Siaterlis, Fovino, & Masera, 2012; Hassanzadeh, Modi, & Mulchandani, 2015). Moreover, some vulnerabilities in digitalized systems can result in losing full control of these systems (Dunaka & McMillin, 2017; Storm, 2014). Such security sabotages can cause serious harm to the controlled physical entities and the people operating or utilizing them (Yong, Zhu, & Frazzoli, 2015).

2. Review of physical and cyber threats in digitalized process systems

The availability and integrity of digitalized process systems can not only be compromised by unintentional component failures, malfunctions and human errors, but also by intentional physical and cyber-attacks (Makhdoom, Abolhasan, Lipman, Liu, & Ni, 2018). However, if malicious attacks were confused with unintentional components failures, it will lead the wrong countermeasures during which the attacker may continue to eavesdrop to the system, manipulate data, or take control and damage the system (Dibaji, Hussain, & Ishii, 2022; Rahman, Mahmud, Oo, & Pota, 2016; Tabassum, Hossain, & Rahman, 2018).

2.1 Increasing vulnerabilities

Digitalized systems, due to their characteristics, are not only exposed to conventional (IT) threats, but also to new, IoT/CPS-specific attacks (Berger, Bürger, & Röglinger, 2020; He, Huang, Liu, & Zhang, 2021; Khaled, Ouchani, Tari, & Drira, 2020; Zografopoulos, Ospina, Liu, & Konstantinou, 2021).

Deep integration and interplay (increased interconnectivity and interdependency) between the heterogeneous and fragmented cyber and physical components increases the surface for malicious attacks (Aigner & Khelil, 2020; Friedberg, McLaughlin, Smith, Laverty, & Sezer, 2017; Guo, Gong, Njilla, & Kamhoua, 2018; Han, Li, Guo, & Tang, 2016; Hassanzadeh et al., 2015; Kure, Islam, & Razzaque, 2018; Oyewole & Jayaweera, 2020; Palaniswamy & McMillin, 2018; van Lier, 2017). Along with system complexity, the performance of digitalized process systems has also become limited by the computational (software) complexity; and the latter has increased significantly due to the high degree of cross-linking and decentralization required for efficient performance (Beg, Johnson, & Davoudi, 2017; Berger et al., 2020; Pan, Gusain, & Palensky, 2019; Roshandel, Dolatkhahi, Hosseinzadeh, & Davazdah-Emami, 2020; Xu, Liu, Tang, Zheng, & Zhao, 2012). Modern features of digitalized systems however remain vague and ambiguous to process engineers and risk analysts (Carreras Guzman, Wied, Kozine, & Lundteigen, 2020).

In a digitalized process system, various components communicate via wireless interfaces. Such interfaces can be weak links and open to the outside world (via internet), which negatively affect security of the controlled process systems (Li et al., 2020; Martin-Tricot, Eichler, & Berthomé, 2020; Yang et al., 2021). Remote control is the major advantage of a digitalized process, but it is a challenge to make sufficiently secure remote configurations. Additionally, outdated configurations are a major security vulnerability. Unsecured and weak configurations are clear and easy target for adversaries to introduce malicious configuration updates which can compromise a system's security and safety (Ulz, Pieber, Steger, Haas, & Matischek, 2018).

The increasing complexity, open nature, and immaturity of the systems lead to increasing uncertainties that results in security challenges (Gao, Peng, Chi, & Zhang, 2020; Sayed-Mouchaweh, 2018). More attack surface and security vulnerabilities in digitalized process systems can arise due to several device (component) limitations including: (i) physical limitations (Ding, Han, Xiang, Ge, & Zhang, 2018) (ii) limitations of sensing and data collection, and computation capabilities (iii) energy and resource limitations (Cherif, Sahnoun, Ramdane, & Nardjes, 2019).

2.2 Characteristics of cyber-physical attacks

The blended cyber-and-physical attacks (i.e., Physical-enabled cyber-attack or Cyber-enabled physical attack) typically have the following characteristics:

- Likely to cause sequential and cascading damage that have a destructive effect on the safety and security of the process system (Carreras Guzman et al., 2020; Hong, Chen, Liu, & Govindarasu, 2015; Lai, Illindala, & Subramaniam, 2019; Oyewole & Jayaweera, 2020; Wang & Liu, 2019; Zhang, Wang, Liang, Zhuang, & Xu, 2017).
- Dynamic in the sense that they change to adapt with the dynamic characteristics of the systems (El-Maliki & Seigne, 2016; Wang, Meng, Liu, Fu, & Cao, 2019).
- Stealthy with intelligent design that is capable of inflicting destructive damages while remaining undetected (Enăchescu, Sándor, & Genge, 2019). An example of stealthy attack is the insider attack where an attacker uses the inside knowledge to exploit high impact system vulnerabilities (Ning & Jiang, 2020).
- Persistent in the sense that they remain inside for a prolonged period of time. An example is the notorious advanced persistent threats (APT) attacks that use sophisticated hacking techniques to establish an illicit, long-term presence on a network in order to mine highly sensitive data with destructive consequences (Huang & Zhu, 2020; Tian et al., 2021).
- Simultaneous (Joint) such that an adversary attacks two or more parts of the system at the same time. An example of a joint attack is when the attacker prevents transmission of information from a physically attacked zone to mask the induced failures in the system (Hossain & Rahnamy-Naeini, 2019; Lai et al., 2019).

2.3 Examples of attacks

There exist four attack categories as shown in Fig. 1: physical attacks (in orange) on physical processes, *disclosure* attacks (in red) on sensor measurements by eavesdropping, *deception* attacks (in blue) similarly on sensor measurements by false injection, and finally *disruption* attacks (in purple) on control and actuation signals. Additionally, attackers also target computation and data storage, software, and even the attack detection algorithms.

2.3.1 Communication system attacks

Communication networks usually utilize unreliable routing protocols and weak RF-based wireless systems that make it easily accessed, interrupted and distorted (Chen, Hu, Yan, & Yu, 2019; Jahromi, Sakhnini, Karimpour, & Dehghantanha, 2019; Vaz, Nayak, & Nayak, 2020). An adversary relies on unsecured communication network to gain access to data in transit between devices. Attacks on information transmitted via

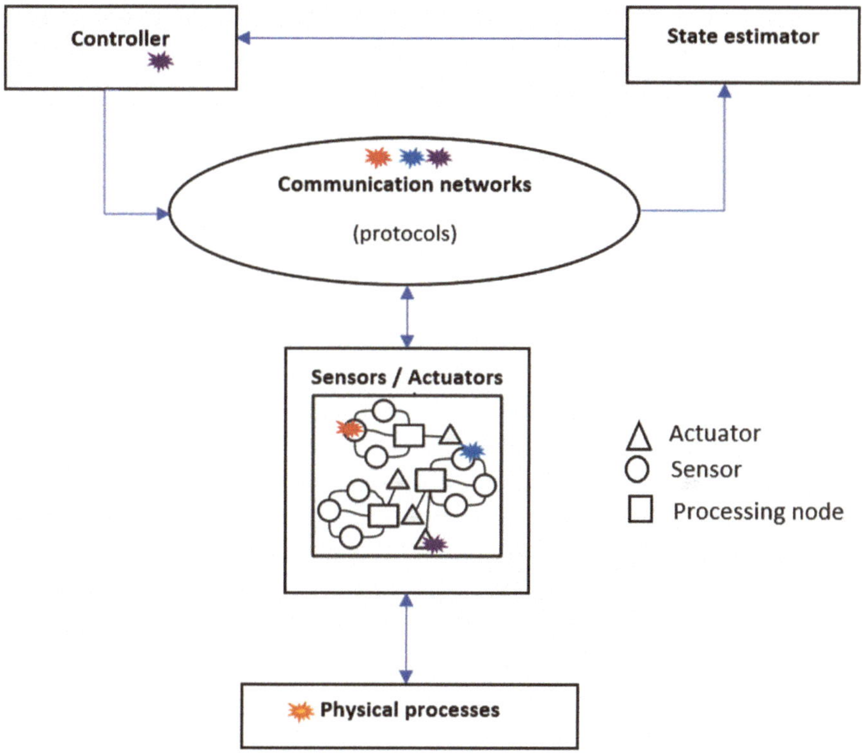

Fig. 1 A simple model of CPS showing physical attacks (in orange), disclosure attacks (in red), deception attacks (in blue), and disruption attacks (in purple).

communication networks may have a merely disclosive intent, however they usually carry out deceptive or disruptive goals by either modifying it or destroying it (Dibaji et al., 2022). The data gets replaced by a value of zero (disruptive) or value from a previous time step (deceptive) which causes data loss or delays, that can compromise system functionality and have catastrophic consequences (Khan & Tomić, 2020). Therefore, assuring security and reliability of communications is critical and challenging (Conti, Kaliyar, & Lal, 2019; Hou et al., 2020).

Types of communication attacks:

- Deceptive communication attacks:

System integrity "eavesdropping" attacks in which attackers intercept and modify critical data transmitted between two components of the system with a goal to deceive decision-making processes (Ntalampiras, 2016; Wei & Sikdar, 2019; Zhang, Wang, et al., 2017). Two main examples of deceptive communication attacks are: (i) Person in the *Middle Attack* (*PitM*) *attacks* where an adversary act like a middleman between critical components (e.g., sensor-to-controller (S-C) channel and the controller-to-actuator (C-A) channel) and change transmitted emergency information with normal process information (Rahman et al., 2016). (ii) *Replay attack* A special type of eavesdropping attack where an adversary records transmitted data for a while then maliciously resends recorded data to the receiver while hijacking the system by injecting malicious (exogenous) signals (Ding et al., 2018; Pasqualetti et al., 2013; Rahman et al., 2016).

- Disruptive communication attacks:

Most common type of communication disruption attacks is *denial-of-service* (DoS). The major target of DoS is to disrupt the availability of the network by preventing data from completely reaching or being transmitted to its destination (Ding, Han, Ge, & Wang, 2020). This is done by hijacking a critical transmission channel, flooding network with traffic, consuming memory resources, eliminating critical nodes, or injecting information that triggers a crash (Ashraf, 2018; Ding et al., 2018; Ledwaba & Venter, 2017; Rahman et al., 2016; Zhang, Wang, et al., 2017). DoS can lead to severer deterioration in the performance of the digitalized system in terms of process control (Wu, Sun, & Chen, 2016). A special type of DoS attack is *distributed denial-of-service* (DDoS) where multiple machines are operating together to flood targeted network and consume its resources (Du & Wang, 2019; Makhdoom et al., 2018). Two more examples of DoS attacks are *jamming attack* and *users' emulation attacks*.

More examples of communication attacks include *Masquerading attack*, *Sybil attack*, *Selective Forwarding attack*, and *Vampire attack*.

2.3.2 Sensor attacks

Malicious (possibly strategic) manipulation of the critical data sensed can cause system disruption and destabilization, increase running cost or, due to interconnectivities and interdependencies, destroy the whole process plant (Dunaka & McMillin, 2017; Essa, Al-Shoura, Al Nabulsi, Al-Ali, & Aloul, 2018; Satchidanandan & Kumar, 2017; Severson et al., 2020). The hacker can hijack sensors by having physical access to them, or through knowledge of the system infrastructure and network (Beg et al., 2017). Compromised sensors may report strategically fabricated measurements (by biasing true values with a non-zero value (Dibaji et al., 2022)) and therefore, assuring sensors veracity (fidelity) is critical (Krotofil & Cárdenas, 2013).

This is the most common (and most addressed in literature) sensor attack, termed *False data injection attack* (*FDIA*). However, it can target any component that deal with the measured data in the sensor-to-controller (S-C) channel including estimators, filters, and controllers by maliciously contaminating the system with undetected errors (Jiang, Dong, & Yin, 2020; Li, Lu, Wang, & Choo, 2017; Noorizadeh, Shakerpour, Meskin, Unal, & Khorasani, 2021; Rahman et al., 2016; Sandberg, 2018; Wang et al., 2019; Wu, Wang, Sun, & Xiong, 2019; Xue & Wu, 2020; Zhang & Ye, 2020).

To ensure and assess system resilience to sensor attacks, investigations should be undertaken, for example to identify the minimum number of sensors that can make the attack undetectable, if compromised (Chen, Kar, & Moura, 2015).

2.3.3 Actuator/controller attacks

Similar to sensor attacks, actuator/controller attacks aim at manipulating (altering) the control signals of a critical controller with strategically fabricated signals (by biasing true values with a non-zero value (Dibaji et al., 2022)). If the attacker is able to gain control and perform malicious actions, it can lead to unsafe control modes and shut down or failure of the process which result in tremendous cost to the system.

Examples of actuator/controller attack include: (i) *False Sequential Command Attack*, which aims at manipulating the control logic so that the order of the commands are altered, disrupting a required sequence in process operation (Rahman et al., 2016), (ii) *Interposition attacks on control systems*, which are stealthy and persistent attacks that transfer control to hijacked library functions (Bellettini & Rrushi, 2008), (iii) *Command disaggregation attack*, which targets sub-controllers (aggregators) like programmable logic controllers PLCs and generates malicious commands to harm the physical process (Xun, Zhu, Hu, Cui, & Zhang, 2017).

2.3.4 Other attacks

Other examples of attacks include attacks against computing system. A common example is *Memory-safety attacks*, which introduces violations that causes the process to crash or maliciously alter the operations. The conventional countermeasure is to abort and restart the system, however, the unavailability of the system leaves the control in critically unsafe state. More examples include *Cloud computing related attacks* (Bhattasali,

Chaki, & Chaki, 2013) and *Edge/fog computing-related attacks* (Hui, Zhou, Xu, & Lin, 2020; Roman, Zhou, & Lopez, 2013).

Another example is the malicious manipulation (exploitation) of vulnerabilities in software by targeted ransomware or malware (Al-Hawawreh, den Hartog, & Sitnikova, 2019; Huda et al., 2018). *Ransomware* has been used the last few years to extort considerable amounts of currency from many companies and institutions. A known example is the attack on the Colonial pipeline in the U.S., May 7, 2021 (e.g., Rivero, 2021). *Malware* can be a virus, worm, trojan, adware etc., that propagates in the cyber network and executes unauthorized actions in order to destroy the system. (Makhdoom et al., 2018). An example of malware attack is *Stuxnet attack* which is a worm that targets supervisory control and data acquisition (SCADA) systems to alter the operation of industrial processes (Genge et al., 2012; Hassanzadeh et al., 2015; Howser, 2014).

Finally, attacks can target the Machine Learning algorithms. ML has been recently deeply utilized for attack detection. which makes it a critical target for an adversary that seeks undetectability. Examples include *Causative poisoning attack* that maliciously manipulates training sets with purposely designed input to ML algorithms in order to cause a model to make a mistake in its predictions (misclassifications) and substantially decreases the algorithm's performance (Vähäkainu, Lehto, & Kariluoto, 2021).

2.4 Human factors

Human factors, i.e., unwanted behaviors of users and administrators and other technical staff, play a main role in the majority of cyber-incidents in digitalized process systems (Chowdhury, Adam, & Teubner, 2020). These behaviors include:

- Insecure collaboration between system components and humans for decision making (Peng et al., 2015)
- Lack of human trust in the digitalized system which mostly lead to hazardous human intervention (Howser, 2014)
- Insufficient/incorrect maintenance, testing, implementation of security policies, and installation of patches and updates.

Reasons include:

- High complexity and difficult human intervention (Chowdhury et al., 2020; Pinto, Gonçalves, Tovar, & Delsing, 2020).
- Complex socio-technical work

- Time pressure on human operator which result in a perceived limited time that leads to a faulty decision-making process (Chowdhury et al., 2020)
- Lack of training and testing
- Human carelessness and dissatisfaction

It is crucial to understand how these behaviors can critically impact the system and its components for safety and cybersecurity management in digitalized systems, i.e., enhanced safety and security (Kuznetsova & Konovalov, 2020). Then, this awareness shall be transferred to operators to increase cyber-security awareness and transform from a mere compliance to cyber-security culture (Chowdhury et al., 2020; Corallo, Lazoi, Lezzi, & Luperto, 2022). Additionally, tools that enable a strong cybersecurity culture include effective governance, training, testing, and collaboration. Authenticated access control may not only depend on a person's clearance but also to what level and for what reason, hence why, he/she wants access.

3. Review of physical and cyber defense/protection measures

After enumerating some of the security challenges for the application of digitalization in process systems, in this section some of the proposed countermeasures found in literature are listed, as seen in Fig. 2.

3.1 Attack prevention and information assurance

Information assurance encompasses all measures that protect and defend the use, storage and transmission of sensitive data to ensure their availability, integrity, authentication, confidentiality, and non-repudiation during acquisition and dissemination among the distributed components of digitalized systems (Conti et al., 2019; Elleuchi, Boujeleben, Abid, & BenSaleh, 2017; Gawanmeh & Alomari, 2018; Guan et al., 2017; Hou et al., 2020; Hui et al., 2020; Hussain et al., 2020; Khan & Byun, 2020; Manogaran, Alazab, Shakeel, & Hsu, 2021; Rathee, Balasaraswathi, Chandran, Gupta, & Boopathi, 2021; Tao et al., 2019; Zhang et al., 2021). A key to achieving information assurance and reducing risks of data breach is attack prevention (Pan & Zheng, 2020; Rahman et al., 2016). In this section, we will discuss some techniques used for attack prevention and information assurance:

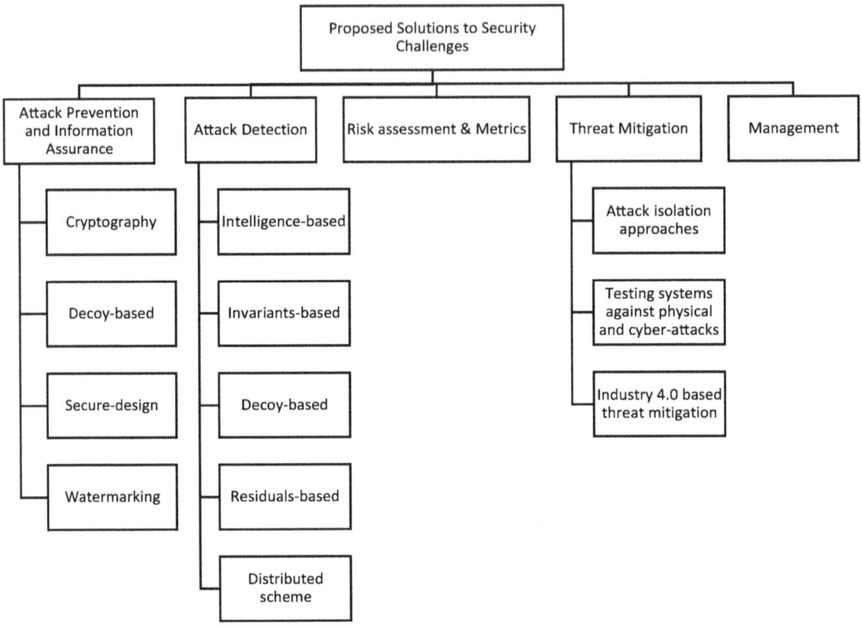

Fig. 2 The proposed solutions in literature for security and safety challenges.

3.1.1 Cryptography

It is the most critical countermeasure that includes authentication, authorization, and encryption of interconnected "communicating" critical components in the digitalized process to prevent third parties from accessing sensitive information (Abuhasel & Khan, 2020; Chen & Ng, 2017; Chen, Yang, & Chou, 2021; Falk & Fries, 2014; Prasath, Ramachandraiah, & Muthukumaran, 2021). Cryptography techniques include:

3.1.1.1 Identity-based schemes

Identity-based schemes is the simplest cryptography technique that secure control on critical components or access to critical information based on an identifier that is used by a server to generate an encrypting/decrypting public key (Elleuchi et al., 2017; Sani et al., 2019). Examples of identity-based cryptography techniques include:

3.1.1.1.1 Physically unclonable function Physically unclonable function (PUF) is a physical entity that is embodied in the system component to provide hardware security: cryptography and anti-counterfeiting. It exploits inherent randomness in the system to produce a physically defined unique response for any given input. The unique and unclonable response of the

PUF represent a unique identifier or a digital fingerprint of the system. Any malicious manipulation by an attacker will alter PUF's values and result in incorrect "fingerprint" leading to un-decryptable data and denied access. PUFs have widely been used for the purposes of data provenance, integrity, and identity management (Kanuparthi, Karri, & Addepalli, 2013).

3.1.1.1.2 Digital signature This mathematical scheme for cryptography generates a stamp (certificate) of authentication. In the context of digitalization, a digital signature is employed as an electronic identifier of a system component (device) to ensure its integrity and non-repudiation (Abuhasel & Khan, 2020). An example of digital signatures is 'hash signature'.

3.1.1.1.3 Attribute-based This is a special type of identity-based cryptography where the unique identifier is a set of attributes. It enables secured data acquisition. When a user possesses those attributes, he is considered legitimate and able to decrypt cyphertext Guan et al. (2017). The most common attribute-based cryptography technique is Rivest-Shamir-Adelman (RSA) (Abuhasel & Khan, 2020).

3.1.1.1.4 Token-based It is a cryptography protocol that verify identity of the user, then issue a trusted token that works for a specified time during which a legitimate user/element can access data and control components without having to re-prove its authenticity. It can be used to secure Supervisory Control and Data Acquisition (SCADA) system in digitalized processes (Yang et al., 2021).

- *Lightweight cryptography*: It is a cryptography technique that features a low computational complexity and, therefore, tailored for implementation in constrained environments such as digitalized systems (Hou et al., 2020; Hussain et al., 2020; Kouicem, Abdelmadjid, & Hicham, 2017). Constraints include device (component) constraint and limitations that include physical constraints (Ding et al., 2018), limitations of sensing and data collection, and computation capabilities and energy and resource limitations (Cherif et al., 2019).
- *Homomorphic cryptography*: In order to inhibit malicious access to data, homomorphic cryptography technique allows mathematical computations to be performed on encrypted data without needing to decrypt it according to the following steps (Dibaji et al., 2022):
  (1) Data in encrypted at the sensor (source)
  (2) Encrypted data is transmitted for computation and control
  (3) Computation is done on encrypted data

(4) Control signals remain encrypted and is transmitted to actuator
(5) Signals are decrypted only for actuation

- *Blockchain-based cryptography*: Since digitalized process is characterized by heterogeneity, the integration of blockchain techniques provide a powerful cryptography tool due to its decentralized nature. It leverages different cryptography techniques such as encryption keys and digital signatures for decentralized secure data sharing (Khan & Byun, 2020; Latif, Idrees, Ahmad, Zheng, & Zou, 2021; Manogaran et al., 2021; Rathee et al., 2021; Zhang et al., 2021).
- *Differential privacy (DP)*: It is a disclosure control technique for privacy that guarantees no critical process detail is disclosed or inferred even if there was a very high exposure to eavesdropping attacks on transmitted signals (Dibaji et al., 2022).
- *Other schemes*: Other cryptography schemes include chaos synchronization technology Hui et al. (2020) and distributed middleware (Tabassum et al., 2018).

3.1.2 Decoy-based

These are techniques that utilize deception to impede, mislead, and capture malicious entities. They present numerous vulnerabilities to entice the attacker while collecting enough data to prevent a critical attack and prosecute the attacker. Examples include *honeypot* (Du & Wang, 2019), and *moving target defense technique* (Abdelwahab, Lucia, & Youssef, 2020).

3.1.3 Secure design

In order to be proactive, it is critical to securely design the process system to prevent attackers from maliciously manipulating it. This can be done by:

- *Components physical limits*: Components such as sensors and actuators can be designed with physical limits to inhibit attackers (Giraldo, Kafash, Ruths, & Cardenas, 2020).
- *Secure wireless communication*: Current radio-based wireless communication has various limitations and vulnerabilities and easy to be interfered with. Visible light communication (VLC) can be an alternative since it is considered immune (resilient) towards inference and can be integrated with digitalization (Vaz et al., 2020) However, security issues arise regarding the open and unattended nature of VLC (Arfaoui et al., 2020). More research needs to be done to investigate secure techniques for communication in digitalized processes.

- *Reliable software*: Software is the backbone of any digitalized system that is responsible for control and organize communication networks. Hence, if compromised, it can lead to complete loss of control and destruction of the system. Reliability must be integrated in process digitalization software to achieve software assurance (Kornecki, Subramanian, & Zalewski, 2013; Xu et al., 2012). Additionally, continuous updates must be applied to ensure that software is free of zero-day vulnerabilities that can be exploited by adversaries. Some approaches to achieve a reliable software are Software Fault Tree Analysis (SFTA) and Fault Tolerant Software Architecture (Xu et al., 2012).

3.1.4 Watermarking

Digital watermarking is the process of inserting invisible information (a signal) into a component or digital transmitted data which can be detected by a computer algorithm to prove the authenticity and integrity of the component or the transmitted data (Dibaji et al., 2022). Dynamic watermarking gives one more layer of protection since each component of the system (sensors or actuator) can insert a random signal (noise) onto its own signal, and check measurements and control signals for consistency with the injected noise (Satchidanandan & Kumar, 2017). Finally, in order to efficiently control operation costs, a parsimonious watermarking policy can limit the injection of watermarks during sufficiently normal "attack-free" operations (Naha, Teixeira, Ahlén, & Dey, 2022).

3.2 Attack detection

Powerful detection techniques need to be in place to identify malicious attacks, in case prevention techniques are insufficient to completely prevent them. An attack detection method diagnoses attacks, usually by identifying unusual events that do not match expectations, then immediately informs the operator (authorized personnel) to promptly assess risk, and make informed mitigative decisions (Adepu & Mathur, 2017; Ding et al., 2018; Jahromi et al., 2019; Kosek, 2016; Minot, Sun, Nikovski, & Zhang, 2019; Ntalampiras, 2016; Pasqualetti et al., 2013; Wu et al., 2016; Xun et al., 2017). However, such process can be challenging and difficult to accomplish, leaving some attacks undetectable. These compromise the safety of the digitalized process (Jiang et al., 2020; Valencia, Alzate, Castro, Bayona, & García, 2019). In this section, we will discuss some techniques used for attack detection:

3.2.1 Intelligence-based

Intelligent attack detection leverages Machine Learning (ML) and Artificial intelligence (AI) techniques in order to identify and diagnose attacks (Colelli, Magri, Panzieri, & Pascucci, 2021; Ntalampiras, 2016; Otokwala, Petrovski, & Kalutarage, 2021; Xue & Wu, 2020; Yan, Mestha, & Abbaszadeh, 2019). ML techniques includes *deep learning* (Huda et al., 2018; Jahromi et al., 2019) and *artificial neural networks* (Enăchescu et al., 2019). AI techniques include what is called *Artificial Immune Systems* (*AIS*) which detect adversaries in a way that mimics the human dendritic cells' reaction to toxins or antigens (Pinto et al., 2020).

3.2.2 Invariants-based (specifications-based)

This technique for attack detection is based on specifying a set of invariants about the system that is discovered/derived (based on process dynamics) or simply encoded in the system component (Programmable Logic Controller (PLC)) during the initial phases (design). During operation, these invariants are used to build trust in the process system (validity check) and any deviation from them is considered an anomaly that indicate an adversarial input Adepu and Mathur (2017). Deriving and checking on invariants are not straightforward (Leopold, McMillin, Stiffler, & Lutes, 2020), this is done with the help of the following techniques:

- *Multiple Security Domain Nondeducibility* (*MSDND*): MSDND segregates the system into separate domains and replaces the notion of trust among those domains by invariant-based information assurance to detect any anomaly (Dunaka & McMillin, 2017; Palaniswamy & McMillin, 2018).
- *Reachability analysis*: Reachability analysis is conducted to generate the set of reachable states that a system can achieve in the normal state (reachability graph). This set is considered a system invariant (actual invariants). It is compared with observed reachable states (candidate invariants) and any mismatch can identify adversarial behavior (Beg et al., 2017).

3.2.3 Decoy-based

As described in the attack prevention section, decoy-based strategies can inhibit attacks by consuming the resources of an adversary on phony components and network connections (false vulnerabilities). In addition, decoy-based strategies such as honeypots can be effective tools to detect and identify attacks using their baiting abilities (Tian et al., 2021).

3.2.4 Residuals-based

Residual-based methods are the techniques that leverage inherent system noise to identify anomalies that results from any malicious attack. Examples include:

- *Kalman filter*: Kalman filtering is processing events (e.g., sensor measurements) that involve dynamic system random noise and inaccuracies to estimate and predict the security status of the system (a future projection of the probability of attacks) (Gokarn, Kulkarni, & Singh, 2017; Wang, Meng, et al., 2019). A residual-based recursive updating of a prior state prediction is done through subtracting estimated measurement from true measurement observed overtime.
- *Fingerprinting*: Residual-based fingerprinting synthesize a unique residual (or a fingerprint) of a component (e.g., sensor) using system noise as well as measurement noise. Any deviation from this unique fingerprint is indicative of adversarial input, and hence it is an effective attack detection technique (Hong, Yang, & Yu, 2020).

3.2.5 Distributed scheme

Centralized cloud or standalone attack detection approaches may be insufficient to handle the highly decentralized, heterogeneous, and resource limited digitalized systems. It is highly effective to have a somehow similarly distributed and collaborative attack detection scheme such as distributed intelligence schemes using fog computing (Li et al., 2017; Liu et al., 2020; Pasqualetti et al., 2013; Rahman et al., 2016).

3.3 Risk assessment

Due to the increased complexity, system intelligence, human factors, and uncertainties in the digitalized process systems, conducting security vulnerability analysis and risk management become highly challenging (Kure et al., 2018; Orojloo & Azgomi, 2017). Such analysis should:

- Investigate system vulnerabilities and how they can be exploited by an adversary (Zhang, Wang, et al., 2017). For example, identifying assets or operations (e.g., sensors, actuator, intellectual property, or process) that could be affected by malicious cyber-attack
- Assess the physical damage inflicted by cyber-attacks and estimate and prioritize their individual risk (Friedberg et al., 2017; Guo et al., 2018; Khan, McLaughlin, Laverty, & Sezer, 2017), and
- Examine the possibility of cascading damage caused by cyber-attacks (Zhao, Li, & Wang, 2020).

Exploitation of system vulnerabilities by adversaries through malicious cyber-attacks can introduce various risks including disruption of critical system availability, degradation of performance, unsafe operation mode (threatening human lives), and, due to interconnectivity and interdependency, complete destruction of the whole process system (Ding et al., 2020; Gokarn et al., 2017; Kure et al., 2018; Sadeghi et al., 2015; Tripathi, Singh, Tripathi, & Chaturvedi, 2021; van Lier, 2017).

In a digitalized process, risk assessment (RA) must consider interdependencies among various internal and external components and be performed automatically (MacDonald et al., 2013; Vu, Tippenhauer, Chen, Nicol, & Kalbarczyk, 2014). A simple cyber-risk estimation methodology can follow this concept:

$$\begin{aligned}\text{Data breach risk} &= \text{Likelihood of data breach} \times \text{Information Value}\\ &= (\text{Threat} \times \text{Vulnerability}) \times \text{Information Value}\end{aligned}$$

Common cyber risk assessment techniques utilize:

- *Testbeds* (Genge et al., 2012; Hong et al., 2015; Tantawy, Abdelwahed, Erradi, & Shaban, 2020),
- *Analytic Hierarchy Process* ((*AHP*)) (Zhao et al., 2020),
- *Multiple information factors* (Liu et al., 2017),
- *Failure Modes and Effects Analysis* ((*FMEA*)) (Zúñiga, Baleia, Fernandes, & Branco, 2020),
- *Petrinets/Bayesian Network* (Tripathi et al., 2021; Zhang et al., 2017), and
- *Semantic modeling* (Aigner & Khelil, 2020).

Details on these techniques are in Section 4 of this chapter (Approaches for intrusion modeling). Finally, security metrics are quantifiable and objective and hence, they can be insightful to the risk assessment process. Those metrics include:

- Mean-Time-to-Security-Failure (MTTSF): a useful life of an asset operates before it fails (Mitchell & Chen, 2015; Orojloo & Azgomi, 2017)
- Mean-Time-to-Detect (MTTI) and Mean-Time-to-Respond (MTTR)
- Mean-Time to implement security patches
- Number of users with unlimited access
- Number of open communication ports during a period of time

3.4 Threat mitigation

In the context of cybersecurity, threat mitigation are the strategies taken for containing the attack by isolating it and limiting its impact (Kure et al., 2018;

Saad, Faddel, Youssef, & Mohammed, 2020; Zahid, Inayat, Daneva, & Mehmood, 2021). As cyber-attacks are becoming more sophisticated and cyber defenses are proving insufficient, it is becoming clear that management approaches should not be about eliminating risk, but about the capability of having a resilient system with consistent performance, despite adversarial attacks (Cheng & Chow, 2020; Ding et al., 2018; Gokarn et al., 2017; Huang, Zhou, Yang, & Qin, 2015; Kure et al., 2018; Noorizadeh et al., 2021; Pasqualetti et al., 2013; Saad et al., 2020). A resilient system is characterized by secure estimation as well as secure control, i.e., correctly assessing system state and giving appropriate actuation when communicated data are corrupted (Ding et al., 2020; Krotofil & Cárdenas, 2013; Satchidanandan & Kumar, 2016; Wu et al., 2016; Yong et al., 2015). This is done by means of attack isolation, mitigation, and compensation and making sure that there are no degradation or loss of performance of the digitalized system (e.g., modeling CPS performance under different cyber-attacks and testing for alternatives such as having redundant components) (Ashraf, 2018; Lakshminarayana, Karachiwala, Teng, Tan, & Yau, 2019; Mahmoud & Hamdan, 2019; Minot et al., 2019; Pasqualetti et al., 2013; Sandberg, 2018; Severson et al., 2020; Wang, Ruan, et al., 2019; Wu et al., 2019; Yong et al., 2015). El-Halwagi et al. (2020) discussed twelve, Industry 4.0 based, strategies to achieve resilient physical infrastructures and CPSs. Secure communication is the main path to secure estimation and control and subsequently resilient systems (Pan & Zheng, 2020; Sani et al., 2019). Unfortunately, this process is extremely challenging due to the increased system complexity and interdependency in today's industry (Gräßler, Bodden, Pottebaum, Geismann, & Roesmann, 2020). Approaches for threat mitigation include:

3.4.1 Attack isolation approaches

Most attack isolation approaches are zoning-based where access is restricted to various zones depending on the applied security policy. When a zone is intruded, other surrounding zones cooperate to diagnose the attack category, specify which component is compromised (to be isolated), and finally, determine the appropriate defense or mitigation strategies (Hossain & Rahnamy-Naeini, 2019; Yang, Zhou, Tian, & An, 2020).

3.4.2 Testing systems against physical and cyber-attacks

This approach aims at identifying all feasible attacks to the system and experimenting with various attack scenarios to determine the set of appropriate mitigative actions as well as recovery strategies. This is accomplished by:

1. Reinforcement learning (El Hariri, Harmon, Habib, Youssef, & Mohammed, 2017; Khaled et al., 2020).
2. Design Space Exploration (DSE) (Gressl, Steger, & Neffe, 2021).
3. Attack gains and cost principle (AGCP) (Wang & Liu, 2019).
4. Reachability restoration strategies (Alcaraz & Lopez, 2016).

3.4.3 Industry 4.0 based threat mitigation

This includes utilizing Industry 4.0 emerging technologies such as:

1. Fog/edge computing: as mentioned in the section of distributed scheme for attack detection, fog/edge computing can more efficiently handle the computation in decentralized, heterogeneous, and resource limited digitalized systems (Roman et al., 2013). In fog/edge computing, distributed objects collaborate with each other in a dynamic way to provide more appropriate threat mitigation strategies.
2. Digital Twin: a virtual representation of the process that digitally replicates the data, the processes, the operation states, and the lifecycle of the asset and hence, helps to understand the impact of cyber threats and how to mitigate them (Saad et al., 2020).

3.5 Cyber security management in the process industry: Current progress and future directions

(Authors: Alessandro Tugnoli, and Valerio Cozzani)

A management system is the strategy by which organizations control the interrelated parts of their business in order to gain their objectives, in this case cybersecurity. The application of dedicated approaches is currently not the more frequently adopted approach for the management of cybersecurity threats in the process industry. Generic cybersecurity management system approaches are usually exploited, originally developed either for IT systems (e.g., computer networks) or for generic industrial automation and control systems (IACS), also named Operating Technology systems or OT systems. The current methodologies dedicated to security vulnerability or risk assessment in process plant (e.g. the VAM-CF methodology (Jaeger, 2002), the CCPS methodology (American Institute of Chemical Engineers, Center of Chemical Process Safety (AIChE-CCPS), 2003) and API RP 780 methodology (API, 2013)) recognize as a threat, attacks to the control system. However, all these methodologies are mostly geared toward physical security issues, providing little support to the identification and management of cyber-threats.

Ylönen et al. (2022) classified three main types of threats that cyber-security risk management in a process facility would need to address:

1. attacks of the IT system of the company (i.e., generic computer network of the company, not dedicated to process control and supervision). These attacks usually aim at compromising the system or at acquiring sensitive data/information;
2. attacks to the OT system (e.g., comprising Basic Process Control System (BPCS) or the Safety Instrumented System (SIS)), and aimed at loss of production (e.g., production shutdown or product out of specification);
3. attacks to the OT system and aimed to generate a major event (e.g., loss of containment of hazardous material or explosion).

A more extensive description of IT and OT systems and of the potential threats to them can be found in chapter "Risk of cascading effects in digitalized process systems" by Iaiani et al. of this volume.

Attacks to the IT system aimed to acquire or compromise sensitive data/information are common to any computer network, independently of the fact it belongs to a process plant or not. Therefore, they are the "traditional" object of cyber-security management and can be effectively addressed by the current best practices on information security management, such as those provided by the ISO/IEC 27000 series of standards (ISO/IEC, 2018). The standards of the ISO/IEC 27000 series describe the basic features of an Information Security Management System (ISMS) aimed at developing and implementing a framework for managing the security of the information assets of an organization (e.g., financial information, intellectual property, employee details, information entrusted to them by customers or third parties). The classical steps of identification of information assets and their associated information security requirements, information security risk assessment, implementation of relevant controls measures to unacceptable risks, and monitoring and improvement are cyclically applied to establish and maintain an ISMS. More specifically ISO/IEC 27005 describes guidelines for information security risk management, which implements the general structure of the Risk Management process defined in ISO 31000 (Fig. 3A) (ISO, 2009).

However, the ISO/IEC 27000 series of standards, due to its general structure, does not provide any specific support to the management of security issues of the OT system of process industry. Attacks to the OT system

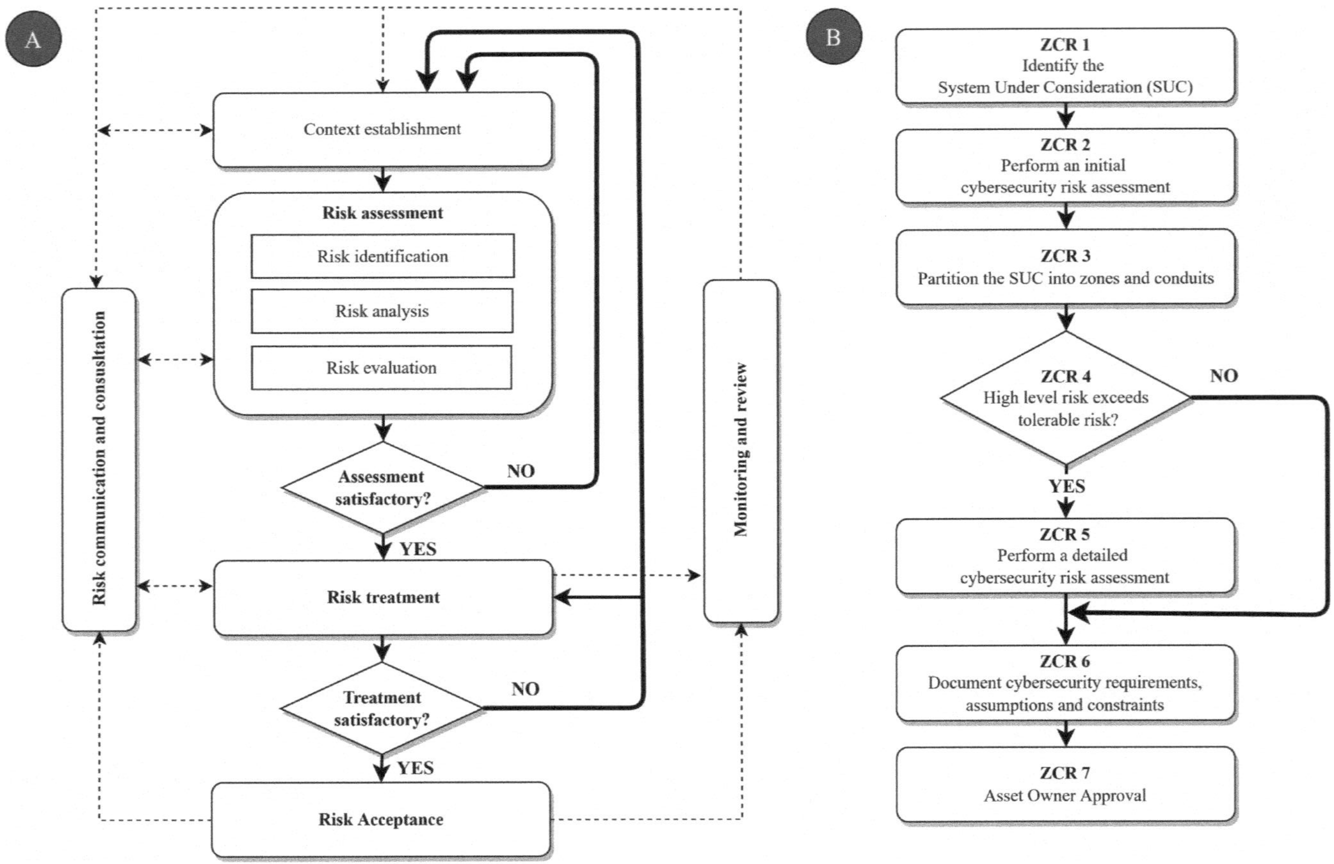

Fig. 3 (A) Security Risk Management Process drawn after a description in ISO 21005. (B) Workflow of Security risk assessment and system design drawn after a description in ISA/IEC 62443.

require a management specific to the IACS, as the one addressed by ISA/IEC 62443 series of standards or by NIST Guide to Industrial Control Systems (ICS) Security (ISA/IEC, 2018; NIST, 2015). In particular, these standards require the evaluation of all the impacts (including those on the physical process plant) that can result from intentional malicious attacks to the OT system in order to evaluate the actual level of cyber risk of a facility, implementing proper cybersecurity countermeasures for its reduction. ISA/IEC 62443 proposes (in part 3-2) a security risk assessment and system design procedure for IACS (Fig. 3B). This procedure divides the IACS into zones (i.e., groups of logical or physical assets) connected by conduits (i.e., communication channels) and assesses the cybersecurity risk and the security level to be provided for each zone and conduit.

Health and Safety Executive, in UK, recently proposed an operational guidance (OG86) on 'Cyber Security for Industrial Automation and Control Systems (IACS)' (HSE, 2017). The approach was aimed at identifying the required cyber security countermeasures to address low levels of cyber security risk in major accident hazard installations. This guidance is aimed at HSE inspectors, but provides also support to operators in achieving compliance with regulations describing the basic elements that need to be assessed and managed. It is mostly based on current standards and regulatory cyber-security requirements for critical infrastructures.

The analysis of security aspects is also explicitly recalled by the latest versions of the standards for the assessment of functional safety (IEC 61508 and IEC 61511), recognizing the importance of protecting the Safety Instrumented System from cyber-attacks (see also Chapter 10 of this volume) (IEC, 2010, 2018). However, also in this case, no specific hazard assessment technique for cyber-threats is provided or referenced.

Standards as the ISA/IEC 62443 are quite scant in providing support to the identification of adverse outcomes, and practitioners usually rely on semi-quantitative matrices or scoring approaches. More sophisticated approaches to support this phase include cyber Bow-Tie approaches (Abdo, Kaouk, Flaus, & Masse, 2018), cyber Process Hazard Analysis approaches (Cusimano & Rostick, 2018), reverse-HazOp approaches for attacks to BPCS and SIS (Iaiani, Tugnoli, Bonvicini, & Cozzani, 2021a; Iaiani, Tugnoli, Macini, & Cozzani, 2021), and cause-effect matrix approaches (Hashimoto et al., 2013).

In particular, such methods fall short when the OT infection is not aimed at information access or interference with production (e.g., causing a shutdown of production), as in the case of an attacker aiming at causing a major event, e.g. a massive release of a hazardous substance (Iaiani, Tugnoli, Bonvicini, & Cozzani, 2021b). This is because these types of attacks involve not only an electronic system, but also the physical system of the process plant: understanding the dynamic response of the process system to induced perturbations and the role of physical safety devices (e.g., pressure relief valves) involves competencies and analysis procedures that are quite different from those required to assess a computer system.

Current practice resolves the lack of specific assessment procedures on the physical part of the system by introducing simplified assumptions: for example, considering the worst-case consequences as expected from the safety assessments (PHA, process hazard analysis). However, the impacts and consequences that can be induced by a cyber-attack are potentially different from those considered in the safety study (e.g. some abnormal states of the plant cannot be induced through the BPCS and the SIS).

Hence, a management system of cyber-threats aimed at prevention of causation of major events through infection of the OT system is expected to obtain the greater benefits from the integration of safety and security management (Ylönen et al., 2022). Major events characterization and modeling traditionally belongs to the area of expertise of process safety and safety management (Mannan, 2012). Methods and tools used for the prediction of expected consequences (i.e., physical effect modeling for major accident scenarios, damage models, calculation of risk indexes) developed in the context of safety risk analysis are directly applicable also to the major events induced by malicious attacks. However, the identification of the potential major accidents (i.e., characterization of the accident scenario in terms of mode of loss of containment, operating conditions before release, inventory released, etc.) cannot be achieved with the identification techniques normally used in the safety domain (e.g., HazOp, FMEA, HEA), since those focus on events originated from random failures or human error.

A systematic and formally rigorous methodology, PHAROS (Process Hazard Analysis of Remote manipulations through the control System), aimed at the identification of major scenarios that possibly have originated from malicious manipulations, which may lead to major accidents, was developed within the framework of the 4STER project (Iaiani, Tugnoli, Macini, & Cozzani, 2021). PHAROS proposes a team-based review, similar to HazOp, that identifies the specific set of manipulations of the BPCS and the SIS, which may lead to major accident scenarios. The method also allows to assess the protection requirements for the physical safeguards on the plant and the basis for improved design for cybersecurity of the network system (division into zones and conduits as for ISA/IEC 62443).

3.6 Approaches for intrusion modeling

As discussed in the previous section, intrusion modelling is an inherent part of attack prevention, attack detection, risk assessment, as well as threat mitigation (Ding et al., 2018; Khan et al., 2017; Ledwaba & Venter, 2017; Pan et al., 2019; Wu et al., 2016). Actually, model-based techniques are the most used techniques for these purposes. They include combinatorial (e.g., attack tree and Bayesian networks) and state space-based (e.g., Petri net and Markov chain) modeling (Zahid et al., 2021). Modeling digitalized process can be either for the purpose of safe and secure designing and engineering of the system or for the purpose of analyzing its response to adversarial input (Mitchell & Chen, 2015). Hence, appropriate digitalization models must represent all system components and the interdependencies involved in the system (Gräßler et al., 2020; Zografopoulos et al., 2021). Similarly, risk analysis models must represent all relevant parameters, including risk prioritization criteria and the intentionality and resourcefulness of adversaries (Rios Insua et al., 2021).

In this section we are briefly discussing some of the modelling-based techniques:

3.6.1 Safe experimentation with attacks

Safe and secure experimenting with malicious attacks is enabled by the following:

1. Testbeds

They mimic the cyber-physical systems, processes, their response to physical and cyber-attacks, as well as attack behavior and threat propagation with means of emulation and simulation (Al-Hawawreh et al., 2019; Deloglos, Elks, & Tantawy, 2020; Genge et al., 2012; Hong et al., 2015; Noorizadeh et al., 2021; Tantawy et al., 2020)

2. A game theoretic approach

Game theory allows modelling of cyber incidents as an interaction between players i.e., attackers and system operators. (In Appendix: principles, background and why game theory contributes, is summarized). Interactions can be cooperative (between system components) or opposing (system component and an attacker). Each player has its strategies to achieve a specific goal or target. For example, an attacker targets maximizing its influence in the system/network to hijack control and cause destroy the system, while system operators will defend the system and minimize damage (Sanjab & Saad, 2016; Wadhawan & Neuman, 2016). Musman and Turner (2018) utilized the game theory concepts, taking into account system interconnectivity, to identify attack paths, optimal defense to minimize risk, and possible attacker reactions to defense strategies. Finally, stochastic (Bayesian) game theory can predict the possibility attacks as well as possible paths that they can follow against digitalized systems (Guo et al., 2018; Huang, Zhou, Qin, & Tu, 2019; Liu, Zhang, Zhu, Tan, & Yin, 2021).

3.6.2 Dynamic model

Dynamic attack models investigate how the attack and the response evolve over time by means of modelling system dynamics and the interaction between different cyber and physical components (Cheng & Chow, 2020; Sayed-Mouchaweh, 2018; Wang, Meng, et al., 2019). This is critical to analyze the attack/fault propagation (Guo, Tan, & Wang, 2020) and cascading failure dynamics (Gao et al., 2020; Shen, Gao, & Peng, 2020). Enabling concepts for these purposes are:

1. Markovian process (Oyewole & Jayaweera, 2020)
2. Percolation theory (Han et al., 2016).

3.6.3 Hybrid attack model

Hybrid attack models investigate the simultaneous multiple attack types, their malicious impact, and approaches to defend against them and mitigate their consequences. Such hybrid modelling is critical, since multiple attacks are believed to cause confusion to control and manipulate system defenses (Tu, Xia, Chi, & Chen, 2020).

3.6.4 Risk management model

By the means of a risk management model, various risks and related uncertainties are identified, quantified, and prioritized then measures are applied to prevent or mitigate their impact. Examples of approaches used for these modelling purposes include:

1. Analytic Hierarchy Process (AHP): involves assessing scales rather than measures. So, it can prioritize parameters that are hard to measure such as risk (Zhao et al., 2020).
2. Failure Modes and Effects Analysis (FMEA): prioritizes failures depending on factors such as easy detection, severe consequence, and frequency of occurrence (Zúñiga et al., 2020).
3. Petri-net: models and analyzes interactions in the process systems and involved risks based on graphical and mathematical approaches (Tripathi et al., 2021).

3.6.5 Reliability model

System reliability is the probability that it will correctly and consistently perform all operations (tasks) for which it was designed and built, in a timely manner, and in a specified environment regardless of the presence of accidents, attacks, or failures. It encompasses system performance, availability, and resilience (Bernieri & Pascucci, 2019; Lin et al., 2020; Zúñiga et al., 2020). Ensuring reliability (availability and resilience) of digitalized processes is very challenging process due to the following factors:

- Increased attack surface due to vulnerable cyber infrastructure
- Defects and faults in system architecture and cyber components (Lei, Chen, Butler-Purry, & Singh, 2018).
- Short release cycle which implies system immaturity
- Adverse environments with high failure risk (Kopsidas, Cruzat, & Ni, 2021).
- Sequential or cascading failure is a major bottleneck to network reliability (Gao, Peng, & Chi, 2021). Due to interconnectivity and interdependency, any exploitation in one component of the system (e.g., hijacking one sensor) triggers a sequence of cascading effects (Alcaraz & Lopez, 2016; Anand et al., 2020; Han, Guo, Ma, & Song, 2018; Li, Sun, & Su, 2021; Liu et al., 2017; Shen et al., 2020; Sun, Liu, & Deng, 2020; Xing, 2020; Zhao, Bai, Liang, Ma, & Deng, 2018). An example of cascading failures in the digitalized systems is the recent worldwide blackouts in power systems.
- Complexity of the environment of digitalized systems (with the large number of connected objects via IoT) (Kouicem et al., 2017).
- Finally, uncertainties associated with the system (Gao et al., 2020).

Modelling and evaluating system reliability is critical to estimate as well as maximize maintenance free lifetime of the digitalized process system. This is done through techniques that leverage:

1. Markovian process Lin et al. (2020) Kopsidas et al. (2021)
2. Monte Carlo Kopsidas et al. (2021)
3. Fault tree Sun et al. (2020)
4. Non-intrusive, deterministic fault-injection tests (Fröhlich, Frtunikj, Rothbauer, & Stückjürgen, 2016), and
5. Temporal causal diagrams (TCDs) (Chhokra et al., 2018).

4. Conclusions

As the foregoing shows, cybersecurity aspects are numerous and diverse. Unfortunately, in the rush to build and implement IoT devices, security and privacy have not been adequately addressed. This created Internet of Vulnerabilities (IoV) in which IoT devices are vulnerable to cyber-attacks, which threaten the security and privacy of users and systems (Croteau et al., 2017; Premnath & Haas, 2015; Upadhyay et al., 2020).

Digitalized systems are typically characterized by heterogeneity and dependencies, complexity and ambiguity, unattended nature, increased machine intelligence, autonomous reconfiguration decisions and compensatory actions, short release cycles, uncertainties, as well as several constraints and limitations. Major security issues and challenges related to the application of digitalization in process systems include cyber-physical attacks, unsecured remote configuration, reliability challenges in the presence of sequential or cascading failure or persistent adversary.

Recent literature focuses on (i) Intelligent-based anomaly detection, (ii) Emulation-based risk assessment, (iii) Blockchain-enabled technologies in addressing confidentiality issues, (iv) Decoy-based attack prevention (v) Secure communication protocols for resilient control.

Optimal solution is most likely a combination of multiple layers of defense and control (i.e., authentication and authorization), enough to withstand the attacks. Safety and security engineers need to consider the following:

- Vulnerabilities that arise from digitalized process characteristics, especially the integration of cyber and physical domains. It needs to consider variables of physical system, connectivity and dependency, and evolution of scenarios.
- Zero-day vulnerabilities are likely to be present in the system. The critical goal of safety and security analysis is to identify and eliminate them before an adversary can exploit them.
- Cyber-risk assessment helps in discovering the weaknesses of process design. While it is extremely challenging, results of a process-aware security assessment serve as an important strategy to improve resilience against intentional disturbances.
- System and attack model assumptions need to be reviewed to make an effective scheme against expected intentional and unintentional disturbances.
- Effect of simultaneous multiple disturbances on the performance of the system need to be studied.
- Integrated system safety and security standards are be developed instead of the multiple component standards.

Further, with respect to cyber risk analysis an interesting perspective paper by Scala, Reilly, Goethals, and Cukier (2019) draws attention to the Five Hard Problems of Cybersecurity defined by the Science of Security (SoS) initiative at the U.S. National Security Agency.

Finally, as recommended by Khan, Amyotte, and Adedigba (2021), incorporating digital process safety as part of the engineering curriculum is critical. It will address the process industry's need of digital solutions, increase awareness and reinforce the concept of making process safety learning a conscious choice.

Appendix: Principles of game theory and why it can contribute to security

Author: Chao Chen

Game theory is a mathematical tool for studying and optimizing the effect of strategic policies and measures among two or more players or opponents (Myerson, 1997). It is widely used in all fields of social science such as economics, psychology, and politics, and natural science such as biology, logic, and computer science. In such opponent situations, game theory is one of the best mathematical tools for decision-making (Chen, Yang, & Chou, 2021; Wang, Wang, Liu, Huang, & Xie, 2016). It originated from two-person zero-sum games, in which each player's payoffs are exactly balanced by those of other players. Now, game theory has been developed also into types, such as non-zero-sum games, cooperative games, and non-cooperative games. To apply game theory to the security of digitalized process systems, we need to select a type of game according to the available information, the move sequence of players, and cooperative or competitive relations between players.

Main elements of game theory

A game mainly consists of three elements: players, strategies, and payoffs. With the assumptions of "common knowledge" and "rationality," different game solutions (e.g., equilibrium) can be obtained.

(1) Players

Players are strategic decision-makers who have their own goals and preferences involved in a game. A game with two players is called a two-player game while one with three or more players is called a multi-player game. The players can be rational or irrational while normally they are assumed to be rational. If the players are rational, they know well about their own goals and interests and expect to maximize their own payoffs. In a non-cooperative game, there are usually an attacker and a defender. In terms of security of digital safety, there are the defender who aims to protect the system, and the attacker who expects to attack the digital system to obtain his/her own goal.

(2) Strategy set

A strategy set is a predetermined plan of actions that a player may take given the set of circumstances that might arise in the game. In each move of a player, he/she takes an action. By the application of game theory in security, we, as defenders, always expect to find an optimal action in the strategy set to maximize our payoffs. Players can select their actions simultaneously or sequentially. The former is called a simultaneous game while the latter is called a sequential game.

(3) Payoffs

Payoff refers to the amount of satisfaction, that a player derives from actions. They represent the player's interests, goals, aims, and preferences in the game. Pay-off units are arbitrary, loss is negative, gain is positive. In our case, defender pay-off is clearly success in warding off an attack. Every combination of strategies of all players in a game has a final payoff and a rational player determines his/her strategy based on the payoffs. As shown in the overly known example tree representation of Fig. A.1, both the defender and attacker have two strategies. The defender can select defense or no defense while the attacker can select attack and no attack. In that case, there are four combinations of strategies and each one has a payoff for each player.

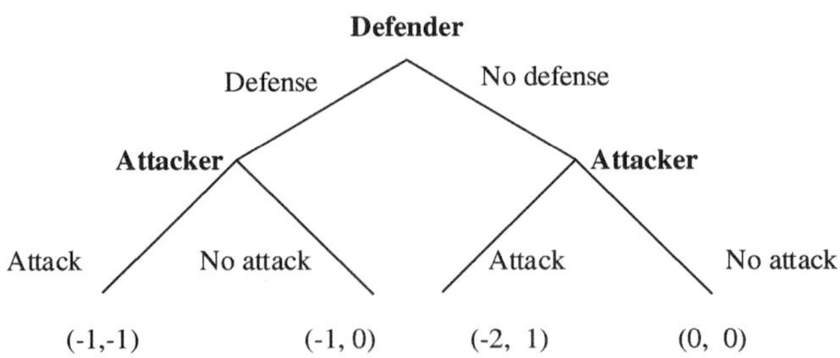

Fig. A.1 A game-theoretical model with two players from a defender's point of view.

Equilibrium

A solution to a game is called an equilibrium. In decision-making based on game theory, there are two types of widely used equilibriums: Nash Equilibrium and Stackelberg Equilibrium.

(1) Nash equilibrium

Nash equilibrium is the most widely used solution for a non-cooperative game such as the security game with a defending and an attacking party. A Nash equilibrium is a set of players' strategies in which each player's strategy in the equilibrium is the best response to the opposing player's strategies. In a non-cooperative game, players cannot communicate with each other, but their payoffs are determined by their own strategies. In a Nash equilibrium, each player is assumed to know the equilibrium strategies of the other player, and the player cannot increase his/her payoff by only changing his/her own strategy. The Nash equilibrium is a solution for players to find out the most optimal strategy in order to obtain the highest payoff. Take the prisoner's dilemma (Poundstone, 1993) as an example, as shown in Table A.1. Two members of the same criminal organization are arrested and locked in separately. If both prisoners choose to stay silent, they get payoffs (−1, −1), which is optimal for both. However, if one prisoner changes his/her strategy and speaks up, he/she would obtain 0 from −1 and the opponent −10. As a result, both of them may select betrays since if one changes his/her selection, he/she will lose his/her payoff (from −5 to −10) (Nash, 1951).

Table A.1 The payoffs of the prisoner's dilemma, example outcome.

	Prisoner 2	
Prisoner 1	**Stays silent**	**Betrays**
Stays silent	−1, −1	−10, 0
Betrays	0, −10	−5, −5

(2) Stackelberg equilibrium

Stackelberg game is a common sequential and not simultaneous game in decision making in which one player moves first and the other follows. The former is called a leader while the latter is called a follower. In a Stackelberg game, since the leader goes first, the follower can optimize his/her strategy according to the leader's strategy. Besides, the leader also knows that the follower can obtain the information about his/her strategy, therefore the leader can also optimize his/her strategy accordingly. Stackelberg equilibrium is an equilibrium for two-player games in which one player is a leader and the other player is a follower. If the follower breaks ties in favor of the leader, it is a strong Stackelberg equilibrium (SSE), otherwise, it is a weak Stackelberg equilibrium (WSE). Taking the inspection game as an example. In such game, an inspector chooses whether to inspect or not, and the inspectee chooses whether to cheat or not (for an extensive treatment, see e.g., Avenhaus, Von Stengel, & Zamir, 2001). If it is a simultaneous game, there is a single unique Nash equilibrium in this game. For example, the inspection probability is 0.1, and the cheating probability is 0.2. If the inspector goes first to be the leader in a Stackelberg game, any strategy that inspects with a probability of at least 0.1 will make not cheating the best strategy for the inspectee. There is an SSE in the game for the inspector to inspect with a probability of 0.1. As shown in Table A.2, the expected payoff is −1, 0, in case of no cheat which is much higher than the simultaneous game with a payoff of −2.

Table A.2 The payoffs of an example inspection game.

	Inspectee	
Inspector	**Cheat**	**No cheat**
Inspection	−6, −9	−1, 0
No inspection	−10, 1	0, 0

In the security domain, the defender is usually assumed to move first after which the attacker moves (Chen, Reniers, & Khakzad, 2020; Rezazadeh, Zhang, Reniers, Khakzad, & Cozzani, 2017). As a result, in principle, an equilibrium can be achieved between the leader and the follower, resulting in an optimal defense strategy.

Other types of games

We already mentioned zero-sum/non-zero-sum games, cooperative/non-cooperative games, simultaneous/sequential games, but there are also perfect information/imperfect information games, complete information/incomplete information games, discrete/continuous games, etc. Different types may be applied to different problems (see, Cox, 2009; Elkind & Rothe, 2016; Owen, 2013; Rezazadeh, Talarico, Reniers, Cozzani, & Zhang, 2019; Talarico, Reniers, Sörensen, & Springael, 2015; Zhang & Reniers, 2016).

(1) Perfect information/imperfect information games
In a perfect information game, Players know all of the previous strategies (actions) of all other players when they move. Alternatively, if one or more players do not know all information about other players' actions when they make a decision, it is an imperfect information game. Because it is difficult for all players in a game to know the exact actions of the others, imperfect information games seem the more suitable for security issues (Mirman, 1989).

(2) Complete information/incomplete information games
A complete information game represents the game that all information such as strategy sets and payoff functions of the game are open to all players. In a complete information game, individual players are fully aware of all other players' strategy sets, and payoff functions before they decide. If one or more players do not know all the information, it is an incomplete information game (Baye, Kovenock, & De Vries, 1996).

(3) Discrete/continuous games
In a discrete game, there is a finite number of players, moves, events, and outcomes while continuous games allow players to choose a strategy from continuous strategy sets (infinite) (Park, Kim, Kim, Jung, & Park, 2001).

A game theory framework of security of digital systems

As game theory consists of many different game theoretical models that may be used in the security of digital systems, for obtaining optimal defense strategies. The attackers aim to damage the digital system while the defender expects to protect the digital system to ensure its security of the digital system. As a result, game theory may be an ideal tool for decision-making on digital system security. Fig. A.2 shows the flowchart of a game theory framework for digital system security.

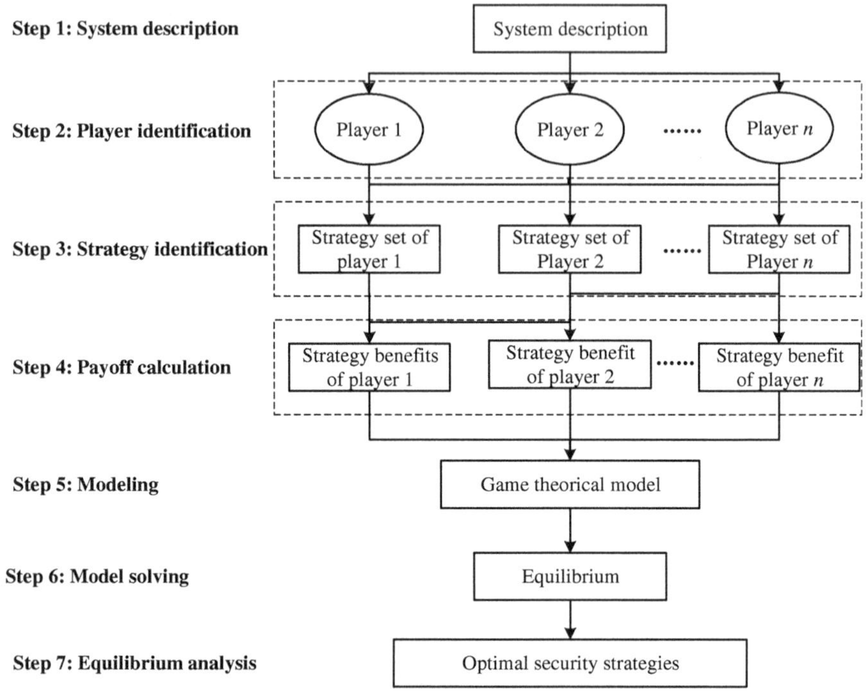

Fig. A.2 A game theory framework of security of digital systems.

As shown in Fig. A.2, the framework consists of seven steps: (1) system description, (2) player identification, (3) strategy identification, (4) payoff calculation, (5) modeling, (6) model solving, and (7) Equilibrium analysis.

(1) System description

In the process industry, different digital systems may be used for obtaining different functions. Therefore, the first step is to characterize the digital system, finding out the characteristics of the system such as equipment, software, and network used in this system as well as the process controlled by the system.

(2) Player identification

The second step is to identify the players in terms of the security of digital systems according to the characteristics of the system. In this step, possible threats and the attractiveness of the digital system should be identified. The main threats (attackers) should be considered as players. The threat is considered to be any indication, circumstance, or event with the potential to damage the digital system. The attackers may be criminals (e.g. white collar, cyber, organized, opportunists), activists (pressure groups, single-issue zealots), terrorists (international or domestic), and disgruntled personnel (API, 2013).

(3) Strategy identification
Strategy identification aims to identify possible strategies of defenders and attackers based on the characteristics of all players. The identification of strategies of attackers is the key task of this step for obtaining the attack objective, attack types, attack weapons, etc.

(4) Payoff calculation
The payoffs of the defender and attackers should be calculated under different combinations of defense and attack strategies. The payoff for defenders is the avoided losses while the payoff for attackers should be based on their motivations and objectives. The losses caused by the attack are always regarded as the payoff for attackers in chemical security. Since fire, explosion, toxic gas release and domino effect events may be triggered by intentional attacks in the process industry, the losses caused by possible major accidents should also be considered.

(5) Modeling
According to the results of system description, player identification, strategy identification, and payoff calculation, we need to choose the suitable game-theoretical models. Then the model parameters should be determined according to the results of steps 1–4 and the established model.

(6) Model solving
According to the theoretical model and the parameters of the model, we need to analyze the equilibrium of the model. According to the model types, one of the two equilibriums (Nash Equilibrium and Stackelberg Equilibrium) may be obtained.

(7) Equilibrium analysis
A pure equilibrium in which each player adopts their strictly dominant strategy may be obtained by solving the game theoretical model. If the equilibrium is not a pure strategy, a mixed equilibrium should be further analyzed to obtain the optimal prevention strategy according to the model characteristics. For instance, the defense resources may be divided according to the ratio of the resource in the mixed strategy.

According to the above steps, we illustrate the application of game theory in the security of digital systems, to obtain the optimal defense strategy accounting for the interactions among strategies of rational defenders and players.

References

Abdelwahab, A., Lucia, W., & Youssef, A. (2020). In *Decoy-based moving target defense against cyber-physical attacks on smart grid Paper presented at the 2020 IEEE Electric Power and Energy Conference (EPEC)*.

Abdo, H., Kaouk, M., Flaus, J. M., & Masse, F. (2018). A safety/security risk analysis approach of industrial control systems: A cyber bowtie—Combining new version of attack tree with bowtie analysis. *Computers & Security, 72*, 175–195. https://doi.org/10.1016/j.cose.2017.09.004.

Abuhasel, K. A., & Khan, M. A. (2020). A secure industrial internet of things (IIoT) framework for resource management in smart manufacturing. *IEEE Access, 8*, 117354–117364.

Adepu, S., & Mathur, A. (2017). In *From design to invariants: Detecting attacks on cyber physical systems Paper presented at the 2017 IEEE International Conference on Software Quality, Reliability and Security Companion (QRS-C)*.

Aigner, A., & Khelil, A. (2020). In *A Semantic Model-based Security Engineering Framework for Cyber-Physical Systems Paper presented at the 2020 IEEE 19th International Conference on Trust, Security and Privacy in Computing and Communications (TrustCom)*.

Alcaraz, C., & Lopez, J. (2016). In *Safeguarding structural controllability in cyber-physical control systems Paper presented at the European Symposium on Research in Computer Security*.

Al-Hawawreh, M., den Hartog, F., & Sitnikova, E. (2019). Targeted ransomware: A new cyber threat to edge system of brownfield industrial Internet of Things. *IEEE Internet of Things Journal, 6*(4), 7137–7151.

American Institute of Chemical Engineers, Center of Chemical Process Safety (AIChE-CCPS). (2003). *Guidelines for analysing and managing the security vulnerabilities of fixed chemical sites. American Institute of Chemical Engineers*. New York: Center of Chemical Process Safety.

Anand, P., Singh, Y., Selwal, A., Singh, P. K., Felseghi, R. A., & Raboaca, M. S. (2020). IoVT: Internet of vulnerable things? Threat architecture, attack surfaces, and vulnerabilities in internet of things and its applications towards smart grids. *Energies, 13*(18), 4813.

API. (2013). *ANSI/API 780 standard—Security risk assessment methodology for the petroleum and petrochemical industry*. New York: American Petroleum Institute.

Arfaoui, M. A., Soltani, M. D., Tavakkolnia, I., Ghrayeb, A., Safari, M., Assi, C. M., et al. (2020). Physical layer security for visible light communication systems: A survey. *IEEE Communications Surveys & Tutorials, 22*(3), 1887–1908. https://doi.org/10.1109/COMST.2020.2988615.

Ashraf, U. (2018). PROSE–Proactive resilience in internet of things: Targeted attacks and countermeasures. *IEEE Sensors Journal, 18*(24), 10049–10057.

Avenhaus, R., Von Stengel, B., & Zamir, S. (2001). *Inspection games*. London School of Economics. http://www.maths.lse.ac.uk/Personal/stengel/TEXTE/insp.pdf.

Baye, M. R., Kovenock, D., & De Vries, C. G. (1996). The all-pay auction with complete information. *Economic Theory, 8*, 291–305.

Beg, O. A., Johnson, T. T., & Davoudi, A. (2017). Detection of false-data injection attacks in cyber-physical DC microgrids. *IEEE Transactions on Industrial Informatics, 13*(5), 2693–2703.

Bellettini, C., & Rrushi, J. L. (2008). In *A product machine model for anomaly detection of interposition attacks on cyber-physical systems Paper presented at the IFIP International Information Security Conference*.

Berger, S., Bürger, O., & Röglinger, M. (2020). Attacks on the industrial internet of things–Development of a multi-layer taxonomy. *Computers & Security, 93*, 101790.

Bernieri, G., & Pascucci, F. (2019). Improving security in industrial internet of things: A distributed intrusion detection methodology. In *Security and privacy trends in the industrial Internet of Things* (pp. 161–179). Springer.

Bhattasali, T., Chaki, R., & Chaki, N. (2013). In *Study of security issues in pervasive environment of next generation internet of things Paper presented at the IFIP International Conference on Computer Information Systems and Industrial Management.*

Carreras Guzman, N. H., Wied, M., Kozine, I., & Lundteigen, M. A. (2020). Conceptualizing the key features of cyber-physical systems in a multi-layered representation for safety and security analysis. *Systems Engineering, 23*(2), 189–210.

Chen, H., Hu, M., Yan, H., & Yu, P. (2019). In *Research on industrial internet of things security architecture and protection strategy Paper presented at the 2019 International conference on virtual reality and intelligent systems (ICVRIS).*

Chen, Y., Kar, S., & Moura, J. M. (2015). In *Cyber-physical systems: Dynamic sensor attacks and strong observability Paper presented at the 2015 IEEE International Conference on Acoustics, Speech and Signal Processing (ICASSP).*

Chen, G., & Ng, W. S. (2017). In *An efficient authorization framework for securing industrial Internet of Things Paper presented at the TENCON 2017-2017 IEEE Region 10 Conference.*

Chen, C., Reniers, G., & Khakzad, N. (2020). Cost-benefit management of intentional domino effects in chemical industrial areas. *Process Safety and Environmental Protection, 134*, 392–405.

Chen, H. C., Yang, W. J., & Chou, C. L. (2021). An online cognitive authentication and trust evaluation application programming interface for cognitive security gateway based on distributed massive Internet of Things network. *Concurrency and Computation: Practice and Experience, 33*(19), e6128.

Cheng, Z., & Chow, M.-Y. (2020). In *Reputation-based collaborative distributed energy management system framework for cyber-physical microgrids: Resilience against profit-driven attacks Paper presented at the 2020 IEEE Power & Energy Society Innovative Smart Grid Technologies Conference (ISGT).*

Cherif, B., Sahnoun, Z., Ramdane, M., & Nardjes, B. (2019). In *Internet of Things: Security Between Challenges and Attacks Paper presented at the International Conference on Machine Learning for Networking.*

Chhokra, A., Mahadevan, N., Dubey, A., Hasan, S., Balasubramanian, D., & Karsai, G. (2018). Hierarchical reasoning about faults in cyber-physical energy systems using temporal causal diagrams. *International Journal of Prognostics and Health Management, 9*. https://doi.org/10.36001/ijphm.2018.v9i1.2669.

Chowdhury, N. H., Adam, M. T., & Teubner, T. (2020). Time pressure in human cybersecurity behavior: Theoretical framework and countermeasures. *Computers & Security, 97*, 101931.

Colelli, R., Magri, F., Panzieri, S., & Pascucci, F. (2021). In *Anomaly-based intrusion detection system for cyber-physical system security Paper presented at the 2021 29th Mediterranean Conference on Control and Automation (MED).*

Conti, M., Kaliyar, P., & Lal, C. (2019). Secure machine to machine communication in industrial Internet of Things. In *Security and privacy trends in the industrial Internet of Things* (pp. 199–219). Springer.

Corallo, A., Lazoi, M., Lezzi, M., & Luperto, A. (2022). Cybersecurity awareness in the context of the industrial Internet of Things: A systematic literature review. *Computers in Industry, 137*, 103614.

Cox, J. L. A. (2009). Game theory and risk analysis. *Risk Analysis, 29*, 1062–1068.

Croteau, B., Krishnankutty, D., Robucci, R., Patel, C., Banerjee, N., Kiriakidis, K., et al. (2017). In *Cross-level detection of sensor-based deception attacks on cyber-physical systems Paper presented at the 2017 IEEE 7th Annual International Conference on CYBER Technology in Automation, Control, and Intelligent Systems (CYBER).*

Cusimano, J., & Rostick, P. (2018). In *If it isn't secure, it isn't safe: Incorporating cybersecurity into process safety AIChE Spring Meet Glob Congr Process Saf.*

Deloglos, C., Elks, C., & Tantawy, A. (2020). In *An Attacker Modeling Framework for the Assessment of Cyber-Physical Systems Security Paper presented at the International Conference on Computer Safety, Reliability, and Security.*

Dibaji, S. M., Hussain, A., & Ishii, H. (2022). A tutorial on security and privacy challenges in CPS. In H. Ishii, & Q. Zhu (Eds.), *Security and Resilience of Control Systems: Theory and Applications* (pp. 121–146). Cham: Springer International Publishing.

Ding, D., Han, Q.-L., Ge, X., & Wang, J. (2020). Secure state estimation and control of cyber-physical systems: A survey. *IEEE Transactions on Systems, Man, and Cybernetics: Systems, 51*(1), 176–190.

Ding, D., Han, Q.-L., Xiang, Y., Ge, X., & Zhang, X.-M. (2018). A survey on security control and attack detection for industrial cyber-physical systems. *Neurocomputing, 275*, 1674–1683.

Du, M., & Wang, K. (2019). An SDN-enabled pseudo-honeypot strategy for distributed denial of service attacks in industrial Internet of Things. *IEEE Transactions on Industrial Informatics, 16*(1), 648–657.

Dunaka, P. R., & McMillin, B. (2017). In *Cyber-physical security of a chemical plant Paper presented at the 2017 IEEE 18th International Symposium on High Assurance Systems Engineering (HASE).*

El Hariri, M., Harmon, E., Habib, H., Youssef, T., & Mohammed, O. (2017). In *A targeted attack for enhancing resiliency of intelligent intrusion detection modules in energy cyber physical systems Paper presented at the 2017 19th International Conference on Intelligent System Application to Power Systems (ISAP).*

El-Halwagi, M. M., Sengupta, D., Pistikopoulos, E. N., Sammons, J., Eljack, F., & Kazi, M.-K. (2020). Disaster-resilient design of manufacturing facilities through process integration: Principal strategies, perspectives, and research challenges. *Frontiers in Sustainability, 1.* https://doi.org/10.3389/frsus.2020.595961.

Elkind, E., & Rothe, J. (2016). *Cooperative game theory, Economics and computation* (pp. 135–193). Springer.

Elleuchi, M., Boujeleben, M., Abid, M., & BenSaleh, M. S. (2017). In *Securing RPL-based Internet of Things applied for water pipeline monitoring Paper presented at the 2017 25th International Conference on Software, Telecommunications and Computer Networks (SoftCOM).*

El-Maliki, T., & Seigne, J.-M. (2016). In *Efficient security adaptation framework for internet of things Paper presented at the 2016 International Conference on Computational Science and Computational Intelligence (CSCI).*

Enăchescu, C., Sándor, H., & Genge, B. (2019). A multi-model-based approach to detect cyber stealth attacks in industrial Internet of Things. In *Paper presented at the 2019 International Conference on Software, Telecommunications and Computer Networks (SoftCOM).*

Essa, A., Al-Shoura, T., Al Nabulsi, A., Al-Ali, A., & Aloul, F. (2018). In *Cyber physical sensors system security: Threats, vulnerabilities, and solutions Paper presented at the 2018 2nd International Conference on Smart Grid and Smart Cities (ICSGSC).*

Falk, R., & Fries, S. (2014). Managed certificate whitelisting–a basis for internet of things security in industrial automation applications. *Securware, 2014*, 178.

Friedberg, I., McLaughlin, K., Smith, P., Laverty, D., & Sezer, S. (2017). STPA-SafeSec: Safety and security analysis for cyber-physical systems. *Journal of Information Security and Applications, 34*, 183–196.

Fröhlich, J., Frtunikj, J., Rothbauer, S., & Stückjürgen, C. (2016). *Testing safety properties of cyber-physical systems with non-intrusive fault injection—An industrial case study, Cham.*

Gao, X., Peng, M., & Chi, K. T. (2021). Cascading failure analysis of cyber physical power systems considering routing strategy. *IEEE Transactions on Circuits and Systems II: Express Briefs.*

Gao, X., Peng, M., Chi, K. T., & Zhang, H. (2020). A stochastic model of cascading failure dynamics in cyber-physical power systems. *IEEE Systems Journal, 14*(3), 4626–4637.

Gawanmeh, A., & Alomari, A. (2018). In *Taxonomy analysis of security aspects in cyber physical systems applications Paper presented at the 2018 IEEE International Conference on Communications Workshops (ICC Workshops).*
Genge, B., Siaterlis, C., Fovino, I. N., & Masera, M. (2012). A cyber-physical experimentation environment for the security analysis of networked industrial control systems. *Computers & Electrical Engineering, 38*(5), 1146–1161.
Giraldo, J., Kafash, S. H., Ruths, J., & Cardenas, A. A. (2020). In *DARIA: Designing actuators to resist arbitrary attacks against cyber-physical systems Paper presented at the 2020 IEEE European Symposium on Security and Privacy (EuroS&P).*
Gokarn, V., Kulkarni, V., & Singh, P. (2017). In *Enhancing cyber physical system security via anomaly detection using behaviour analysis Paper presented at the 2017 International Conference on Wireless Communications, Signal Processing and Networking (WiSPNET).*
Gräßler, I., Bodden, E., Pottebaum, J., Geismann, J., & Roesmann, D. (2020). Security-oriented fault-tolerance in systems engineering: a conceptual threat modelling approach for cyber-physical production systems. In *Advanced, contemporary control* (pp. 1458–1469). Springer.
Gressl, L., Steger, C., & Neffe, U. (2021). Design space exploration for secure IoT devices and cyber-physical systems. *ACM Transactions on Embedded Computing Systems (TECS), 20*(4), 1–24.
Guan, Z., Li, J., Wu, L., Zhang, Y., Wu, J., & Du, X. (2017). Achieving efficient and secure data acquisition for cloud-supported internet of things in smart grid. *IEEE Internet of Things Journal, 4*(6), 1934–1944.
Guo, Y., Gong, Y., Njilla, L. L., & Kamhoua, C. A. (2018). In *A stochastic game approach to cyber-physical security with applications to smart grid Paper presented at the IEEE INFOCOM 2018-IEEE Conference on Computer Communications Workshops (INFOCOM WKSHPS).*
Guo, X., Tan, Y., & Wang, F. (2020). Modeling and fault propagation analysis of cyber–physical power system. *Energies, 13*(3), 539.
Han, Y., Guo, C., Ma, S., & Song, D. (2018). Modeling cascading failures and mitigation strategies in PMU based cyber-physical power systems. *Journal of Modern Power Systems and Clean Energy, 6*(5), 944–957.
Han, Y., Li, Z., Guo, C., & Tang, Y. (2016). In *Improved percolation theory incorporating power flow analysis to model cascading failures in cyber-physical power system Paper presented at the 2016 IEEE Power and Energy Society General Meeting (PESGM).*
Hashimoto, Y., Toyoshima, T., Yogo, S., Koike, M., Hamaguchi, T., Jing, S., et al. (2013). Safety securing approach against cyber-attacks for process control system. *Computers & Chemical Engineering, 57,* 181–186. https://doi.org/10.1016/j.compchemeng.2013.04.019.
Hassanzadeh, A., Modi, S., & Mulchandani, S. (2015). In *Towards effective security control assignment in the Industrial Internet of Things Paper presented at the 2015 IEEE 2nd World Forum on Internet of Things (WF-IoT).*
He, H., Huang, S., Liu, Y., & Zhang, T. (2021). A tri-level optimization model for power grid defense with the consideration of post-allocated DGs against coordinated cyber-physical attacks. *International Journal of Electrical Power & Energy Systems, 130,* 106903.
Health Safety Executive (HSE). (2017). *Cyber Security for Industrial Automation and Control Systems (IACS) (OG86).* https://www.hse.gov.uk/foi/internalops/og/og-0086.pdf.
Hong, J., Chen, Y., Liu, C.-C., & Govindarasu, M. (2015). Cyber-physical security testbed for substations in a power grid. In *Cyber physical systems approach to smart electric power grid* (pp. 261–301). Springer.
Hong, Z., Yang, C., & Yu, L. (2020). R-print: A system residuals-based fingerprinting for attack detection in industrial cyber-physical systems. *IEEE Transactions on Industrial Electronics.*

Hossain, M. J., & Rahnamy-Naeini, M. (2019). Line failure detection from PMU data after a joint cyber-physical attack. In *Paper presented at the 2019 IEEE Power & Energy Society General Meeting (PESGM)*.

Hou, R., Ren, G., Zhou, C., Yue, H., Liu, H., & Liu, J. (2020). Analysis and research on network security and privacy security in ubiquitous electricity Internet of Things. *Computer Communications*, *158*, 64–72.

Howser, G. W. (2014). *Multiple security domain nondeducibility in cyber-physical systems*. Missouri University of Science and Technology.

Huang, K., Zhou, C., Qin, Y., & Tu, W. (2019). A game-theoretic approach to cross-layer security decision-making in industrial cyber-physical systems. *IEEE Transactions on Industrial Electronics*, *67*(3), 2371–2379.

Huang, S., Zhou, C.-J., Yang, S.-H., & Qin, Y.-Q. (2015). Cyber-physical system security for networked industrial processes. *International Journal of Automation and Computing*, *12*(6), 567–578.

Huang, L., & Zhu, Q. (2020). A dynamic games approach to proactive defense strategies against advanced persistent threats in cyber-physical systems. *Computers & Security*, *89*, 101660.

Huda, S., Miah, S., Yearwood, J., Alyahya, S., Al-Dossari, H., & Doss, R. (2018). A malicious threat detection model for cloud assisted internet of things (CoT) based industrial control system (ICS) networks using deep belief network. *Journal of Parallel and Distributed Computing*, *120*, 23–31.

Hui, H., Zhou, C., Xu, S., & Lin, F. (2020). A novel secure data transmission scheme in industrial internet of things. *China Communications*, *17*(1), 73–88.

Hussain, S., Ullah, I., Khattak, H., Adnan, M., Kumari, S., Ullah, S. S., et al. (2020). A lightweight and formally secure certificate based signcryption with proxy re-encryption (CBSRE) for Internet of Things enabled smart grid. *IEEE Access*, *8*, 93230–93248.

Iaiani, M., Tugnoli, A., Bonvicini, S., & Cozzani, V. (2021a). Major accidents triggered by malicious manipulations of the control system in process facilities. *Safety Science*, *134*, 105043. https://doi.org/10.1016/j.ssci.2020.105043.

Iaiani, M., Tugnoli, A., Bonvicini, S., & Cozzani, V. (2021b). Analysis of cybersecurity-related incidents in the process industry. *Reliability Engineering & System Safety*, *209*, 107485. https://doi.org/10.1016/j.ress.2021.107485.

Iaiani, M., Tugnoli, A., Macini, P., & Cozzani, V. (2021). Outage and asset damage triggered by malicious manipulation of the control system in process plants. *Reliability Engineering & System Safety*, *213*, 107685. https://doi.org/10.1016/j.ress.2021.107685.

International Electrotechnical Commission (IEC). (2010). *IEC 61508: Functional safety of electrical/electronic/programmable electronic safety-related systems*. International Electrotechnical Commission.

International Electrotechnical Commission (IEC). (2018). *IEC 61511: Functional safety—Safety instrumented systems for the process industry sector*. International Electrotechnical Commission.

International Organization for Standardization (ISO). (2009). *31000:2009: Risk management—Principles and guidelines*. International Organization for Standardization.

International Organization for Standardization, International Electrotechnical Commission (ISO-IEC). (2018). *ISO/IEC 27000 series of standards: Information technology—Security techniques - Information security management systems*. International Organization for Standardization, International Electrotechnical Commission.

International Society of Automation, International Electrotechnical Commission (ISA/IEC). (2018). *ISA/IEC 62443 series of standards: Industrial Automation and Control Systems Security*. International Society of Automation, International Electrotechnical Commission.

Jaeger, D. (2002). Vulnerability assessment methodology for chemical facilities (VAM-CF). *ACS Chemical Health & Safety*, *9*(6), 15–19. https://doi.org/10.1016/S1074-9098(02)00389-1.

Jahromi, A. N., Sakhnini, J., Karimpour, H., & Dehghantanha, A. (2019). In *A deep unsupervised representation learning approach for effective cyber-physical attack detection and identification on highly imbalanced data Paper presented at the Proceedings of the 29th Annual International Conference on Computer Science and Software Engineering.*

Jiang, Y., Dong, J., & Yin, S. (2020). In *Improving the safety of distributed cyber-physical systems against false data injection attack by establishing interconnections Paper presented at the IECON 2020 The 46th Annual Conference of the IEEE Industrial Electronics Society.*

Kanuparthi, A., Karri, R., & Addepalli, S. (2013). In *Hardware and embedded security in the context of internet of things Paper presented at the Proceedings of the 2013 ACM workshop on Security, privacy & dependability for cyber vehicles.*

Khaled, A., Ouchani, S., Tari, Z., & Drira, K. (2020). Assessing the severity of smart attacks in industrial cyber-physical systems. *ACM Transactions on Cyber-Physical Systems*, *5*(1), 1–28.

Khan, F., Amyotte, P., & Adedigba, S. (2021). Process safety concerns in process system digitalization. *Education for Chemical Engineers*, *34*, 33–46.

Khan, P. W., & Byun, Y. (2020). A blockchain-based secure image encryption scheme for the industrial Internet of Things. *Entropy*, *22*(2), 175.

Khan, R., McLaughlin, K., Laverty, D., & Sezer, S. (2017). In *STRIDE-based threat modeling for cyber-physical systems Paper presented at the 2017 IEEE PES Innovative Smart Grid Technologies Conference Europe (ISGT-Europe).*

Khan, M. T., & Tomić, I. (2020). Securing industrial cyber–Physical systems: A run-time multilayer monitoring. *IEEE Transactions on Industrial Informatics*, *17*(9), 6251–6259.

Kopsidas, K., Cruzat, C., & Ni, M. (2021). *Cyber-physical reliability of dynamic line rating ICT failures in OHL networks*. IET Generation, Transmission & Distribution.

Kornecki, A. J., Subramanian, N., & Zalewski, J. (2013). In *Studying interrelationships of safety and security for software assurance in cyber-physical systems: Approach based on bayesian belief networks Paper presented at the 2013 Federated Conference on Computer Science and Information Systems.*

Kosek, A. M. (2016). In *Contextual anomaly detection for cyber-physical security in smart grids based on an artificial neural network model Paper presented at the 2016 Joint Workshop on Cyber-Physical Security and Resilience in Smart Grids (CPSR-SG).*

Kouicem, D. E., Abdelmadjid, B., & Hicham, L. (2017). In *Distributed fine-grained secure control of smart actuators in Internet of Things Paper presented at the 2017 IEEE International Symposium on Parallel and Distributed Processing with Applications and 2017 IEEE International Conference on Ubiquitous Computing and Communications (ISPA/IUCC).*

Krotofil, M., & Cárdenas, A. A. (2013). In *Resilience of process control systems to cyber-physical attacks Paper presented at the Nordic Conference on Secure IT Systems.*

Kure, H. I., Islam, S., & Razzaque, M. A. (2018). An integrated cyber security risk management approach for a cyber-physical system. *Applied Sciences*, *8*(6), 898.

Kuznetsova, N., & Konovalov, Y. (2020). In *Enhancing the safety of energy systems functioning at their digitalization Paper presented at the E3S Web of Conferences.*

Lai, K., Illindala, M., & Subramaniam, K. (2019). A tri-level optimization model to mitigate coordinated attacks on electric power systems in a cyber-physical environment. *Applied Energy*, *235*, 204–218.

Lakshminarayana, S., Karachiwala, J. S., Teng, T. Z., Tan, R., & Yau, D. K. (2019). Performance and resilience of cyber-physical control systems with reactive attack mitigation. *IEEE Transactions on Smart Grid*, *10*(6), 6640–6654.

Latif, S., Idrees, Z., Ahmad, J., Zheng, L., & Zou, Z. (2021). A blockchain-based architecture for secure and trustworthy operations in the industrial Internet of Things. *Journal of Industrial Information Integration*, *21*, 100190.

Ledwaba, L., & Venter, H. (2017). In *A threat-vulnerability based risk analysis model for cyber physical system security Paper presented at the Proceedings of the 50th Hawaii International Conference on System Sciences.*

Lei, H., Chen, B., Butler-Purry, K. L., & Singh, C. (2018). In *Security and reliability perspectives in cyber-physical smart grids Paper presented at the 2018 IEEE Innovative Smart Grid Technologies-Asia (ISGT Asia).*

Leopold, J., McMillin, B., Stiffler, R., & Lutes, N. (2020). In *Comparison of design-centric and data-centric methods for distributed attack detection in cyber-physical systems Paper presented at the International Conference on Critical Infrastructure Protection.*

Li, Z., Liu, Y., Liu, D., Zhang, N., Lu, D., & Huang, X. (2020). In *A security defense model for ubiquitous electric internet of things based on game theory Paper presented at the 2020 IEEE 4th Conference on Energy Internet and Energy System Integration (EI2).*

Li, B., Lu, R., Wang, W., & Choo, K.-K. R. (2017). Distributed host-based collaborative detection for false data injection attacks in smart grid cyber-physical system. *Journal of Parallel and Distributed Computing, 103*, 32–41.

Li, J., Sun, C., & Su, Q. (2021). Analysis of cascading failures of power cyber-physical systems considering false data injection attacks. *Global Energy Interconnection, 4*(2), 204–213.

Lin, D., Liu, Q., Li, Z., Zeng, G., Wang, Z., Yu, T., et al. (2020). Elaborate reliability evaluation of cyber physical distribution systems considering fault location, isolation and supply restoration process. *IEEE Access, 8*, 128574–128590.

Liu, Y., Lu, D., Deng, L., Bai, T., Hou, K., & Zeng, Y. (2017). Risk assessment for the cascading failure of electric cyber-physical system considering multiple information factors. *IET Cyber-Physical Systems: Theory & Applications, 2*(4), 155–160.

Liu, J., Zhang, W., Ma, T., Tang, Z., Xie, Y., Gui, W., et al. (2020). Toward security monitoring of industrial cyber-physical systems via hierarchically distributed intrusion detection. *Expert Systems with Applications, 158*, 113578.

Liu, X., Zhang, J., Zhu, P., Tan, Q., & Yin, W. (2021). Quantitative cyber-physical security analysis methodology for industrial control systems based on incomplete information Bayesian game. *Computers & Security, 102*, 102138.

MacDonald, D., Clements, S. L., Patrick, S. W., Perkins, C., Muller, G., Lancaster, M. J., et al. (2013). In *Cyber/physical security vulnerability assessment integration Paper presented at the 2013 IEEE PES Innovative Smart Grid Technologies Conference (ISGT).*

Mahmoud, M. S., & Hamdan, M. M. (2019). Improved control of cyber-physical systems subject to cyber and physical attacks. *Cyber-Physical Systems, 5*(3), 173–190.

Makhdoom, I., Abolhasan, M., Lipman, J., Liu, R. P., & Ni, W. (2018). Anatomy of threats to the internet of things. *IEEE Communications Surveys & Tutorials, 21*(2), 1636–1675.

Mannan, S. (2012). *Lees' loss prevention in the process industries: Hazard identification, assessment and control* (4th ed.). Elsevier, UK: Butterworth-Heinemann.

Manogaran, G., Alazab, M., Shakeel, P. M., & Hsu, C.-H. (2021). Blockchain assisted secure data sharing model for Internet of Things based smart industries. *IEEE Transactions on Reliability.*

Martin-Tricot, F., Eichler, C., & Berthomé, P. (2020). In *An enrolment gateway for data security in heterogeneous industrial Internet of Things Paper presented at the 2020 IEEE 29th International Conference on Enabling Technologies: Infrastructure for Collaborative Enterprises (WETICE).*

Minot, A., Sun, H., Nikovski, D., & Zhang, J. (2019). In *Distributed estimation and detection of cyber-physical attacks in power systems Paper presented at the 2019 IEEE International Conference on Communications Workshops (ICC Workshops).*

Mirman, L. J. (1989). *Perfect information, game theory* (pp. 194–198). Springer.

Mitchell, R., & Chen, R. (2015). Modeling and analysis of attacks and counter defense mechanisms for cyber physical systems. *IEEE Transactions on Reliability, 65*(1), 350–358.

Musman, S., & Turner, A. J. (2018). A game oriented approach to minimizing cybersecurity risk. *Safety and Security Studies, 27*.
Myerson, R. B. (1997). *Game theory: Analysis of conflict*. Harvard University Press.
Naha, A., Teixeira, A., Ahlén, A., & Dey, S. (2022). Quickest detection of deception attacks on cyber-physical systems with a parsimonious watermarking policy. *arXiv*. 2201.09389.
Nash, J. (1951). Non-cooperative games. *Annals of Mathematics, 286–295*.
National Institute of Standards and Technology (NIST). (2015). *Guide to Industrial Control Systems (ICS) Security*. NIST. https://doi.org/10.6028/NIST.SP.800-82r2.
Ning, X., & Jiang, J. (2020). In the mind of an insider attacker on cyber-physical systems and how not being fooled. *IET Cyber-Physical Systems: Theory & Applications, 5*(2), 153–161.
Noorizadeh, M., Shakerpour, M., Meskin, N., Unal, D., & Khorasani, K. (2021). A cyber-security methodology for a cyber-physical industrial control system testbed. *IEEE Access, 9*, 16239–16253.
Ntalampiras, S. (2016). Automatic identification of integrity attacks in cyber-physical systems. *Expert Systems with Applications, 58*, 164–173.
Orojloo, H., & Azgomi, M. A. (2017). A game-theoretic approach to model and quantify the security of cyber-physical systems. *Computers in Industry, 88*, 44–57.
Otokwala, U., Petrovski, A., & Kalutarage, H. (2021). *Effective detection of cyber attack in a cyber-physical power grid system*.
Owen, G. (2013). *Game theory*. Emerald Group Publishing.
Oyewole, P. A., & Jayaweera, D. (2020). Power system security with cyber-physical power system operation. *IEEE Access, 8*, 179970–179982.
Palaniswamy, P., & McMillin, B. (2018). In *Cyber-physical security of an electric microgrid Paper presented at the 2018 IEEE 23rd Pacific Rim International Symposium on Dependable Computing (PRDC)*.
Pan, K., Gusain, D., & Palensky, P. (2019). In *Modelica-supported attack impact evaluation in cyber physical energy system Paper presented at the 2019 IEEE 19th International Symposium on High Assurance Systems Engineering (HASE)*.
Pan, X., & Zheng, Z. (2020). In *Cooperative distributed antenna systems based secure communications for industrial Internet of Things Paper presented at the 2020 IEEE 20th International Conference on Communication Technology (ICCT)*.
Park, J.-B., Kim, B. H., Kim, J.-H., Jung, M.-H., & Park, J.-K. (2001). A continuous strategy game for power transactions analysis in competitive electricity markets. *IEEE Transactions on Power Systems, 16*, 847–855.
Pasqualetti, F., Dörfler, F., & Bullo, F. (2013). Attack detection and identification in cyber-physical systems. *IEEE Transactions on Automatic Control, 58*(11), 2715–2729.
Peng, Y., Wang, Y., Xiang, C., Liu, X., Wen, Z., Chen, D., et al. (2015). In *Cyber-physical attack-oriented Industrial Control Systems (ICS) modeling, analysis and experiment environment Paper presented at the 2015 International Conference on Intelligent Information Hiding and Multimedia Signal Processing (IIH-MSP)*.
Pinto, R., Gonçalves, G., Tovar, E., & Delsing, J. (2020). In *Attack detection in cyber-physical production systems using the deterministic dendritic cell algorithm Paper presented at the 2020 25th IEEE International Conference on Emerging Technologies and Factory Automation (ETFA)*.
Poundstone, W. (1993). *Prisoner's Dilemma*. New York: Ancor books, ISBN:0-385-41580-X.
Prasath, J., Ramachandraiah, U., & Muthukumaran, G. (2021). Modified hardware security algorithms for process industries using Internet of Things. *Journal of Applied Security Research, 16*(1), 127–140.
Premnath, S. N., & Haas, Z. J. (2015). Security and privacy in the internet-of-things under time-and-budget-limited adversary model. *IEEE Wireless Communications Letters, 4*(3), 277–280.

Rahman, M. S., Mahmud, M. A., Oo, A. M. T., & Pota, H. R. (2016). Multi-agent approach for enhancing security of protection schemes in cyber-physical energy systems. *IEEE Transactions on Industrial Informatics*, *13*(2), 436–447.

Rathee, G., Balasaraswathi, M., Chandran, K. P., Gupta, S. D., & Boopathi, C. (2021). A secure IoT sensors communication in industry 4.0 using blockchain technology. *Journal of ambient intelligence and humanized computing*, *12*(1), 533–545.

Rezazadeh, A., Talarico, L., Reniers, G., Cozzani, V., & Zhang, L. (2019). Applying game theory for securing oil and gas pipelines against terrorism. *Reliability Engineering and System Safety*, *191*, 106140.

Rezazadeh, A., Zhang, L., Reniers, G., Khakzad, N., & Cozzani, V. (2017). Optimal patrol scheduling of hazardous pipelines using game theory. *Process Safety and Environmental Protection*, *109*, 242–256.

Rios Insua, D., Couce-Vieira, A., Rubio, J. A., Pieters, W., Labunets, K., Rasines, G., et al. (2021). An adversarial risk analysis framework for cybersecurity. *Risk Analysis*, *41*(1), 16–36.

Rivero, N. (2021). *Hacking collective DarkSide are state-sanctioned pirates*. Quartz (publication). Archived from the original on May 12, 2021. Retrieved April, 2022.

Roman, R., Zhou, J., & Lopez, J. (2013). On the features and challenges of security and privacy in distributed internet of things. *Computer Networks*, *57*(10), 2266–2279.

Roshandel, E., Dolatkhahi, F.-A., Hosseinzadeh, A., & Davazdah-Emami, H. (2020). Cyber–physical energy system toward safety in smart grids. *Journal of Supercomputing*, *76*(11).

Saad, A., Faddel, S., Youssef, T., & Mohammed, O. A. (2020). On the implementation of IoT-based digital twin for networked microgrids resiliency against cyber attacks. *IEEE Transactions on Smart Grid*, *11*(6), 5138–5150.

Sadeghi, A.-R., Wachsmann, C., & Waidner, M. (2015). In *Security and privacy challenges in industrial internet of things Paper presented at the 2015 52nd ACM/EDAC/IEEE Design Automation Conference (DAC)*.

Sandberg, H. (2018). In *Control Theory for Practical Cyber-Physical Security Paper presented at the Proceedings of the 4th ACM Workshop on Cyber-Physical System Security*.

Sani, A. S., Yuan, D., Jin, J., Gao, L., Yu, S., & Dong, Z. Y. (2019). Cyber security framework for Internet of Things-based Energy Internet. *Future Generation Computer Systems*, *93*, 849–859.

Sanjab, A., & Saad, W. (2016). In *On bounded rationality in cyber-physical systems security: Game-theoretic analysis with application to smart grid protection Paper presented at the 2016 Joint Workshop on Cyber-Physical Security and Resilience in Smart Grids (CPSR-SG)*.

Satchidanandan, B., & Kumar, P. R. (2016). In *Secure control of networked cyber-physical systems Paper presented at the 2016 IEEE 55th Conference on Decision and Control (CDC)*.

Satchidanandan, B., & Kumar, P. (2017). In *Defending cyber-physical systems from sensor attacks Paper presented at the International Conference on Communication Systems and Networks*.

Sayed-Mouchaweh, M. (2018). *Diagnosability, security and safety of hybrid dynamic and cyber-physical systems*. Springer.

Scala, N. M., Reilly, A. C., Goethals, P. L., & Cukier, M. (2019). Risk and the five hard problems of cybersecurity. *Risk Analysis*, *39*(10), 2119–2126.

Severson, T. A., Croteau, B., Rodríguez-Seda, E. J., Kiriakidis, K., Robucci, R., & Patel, C. (2020). A resilient framework for sensor-based attacks on cyber–physical systems using trust-based consensus and self-triggered control. *Control Engineering Practice*, *101*, 104509.

Shen, M., Gao, X., & Peng, M. (2020). In *Cascading failure dynamic of cyber-physical power system considering malware attacks Paper presented at the Journal of Physics: Conference Series*.

Storm, D. (2014). *Hackers exploit SCADA holes to take full control of critical infrastructure.* Retrieved from https://www.computerworld.com/article/2475789/hackers-exploit-scada-holes-to-take-full-control-of-critical-infrastructure.html.

Sun, X., Liu, Y., & Deng, L. (2020). Reliability assessment of cyber-physical distribution network based on the fault tree. *Renewable Energy, 155*, 1411–1424.

Tabassum, T., Hossain, S. A., & Rahman, M. A. (2018). In *A secure key management technique through distributed middleware for the Internet of Things Paper presented at the Science and Information Conference.*

Talarico, L., Reniers, G., Sörensen, K., & Springael, J. (2015). MISTRAL: A game-theoretical model to allocate security measures in a multi-modal chemical transportation network with adaptive adversaries. *Reliability Engineering & System Safety, 138*, 105–114.

Tantawy, A., Abdelwahed, S., Erradi, A., & Shaban, K. (2020). Model-based risk assessment for cyber physical systems security. *Computers & Security, 96*, 101864.

Tao, H., Bhuiyan, M. Z. A., Rahman, M. A., Wang, T., Wu, J., Salih, S. Q., et al. (2019). TrustData: Trustworthy and secured data collection for event detection in industrial cyber-physical system. *IEEE Transactions on Industrial Informatics, 16*(5), 3311–3321.

Tian, W., Du, M., Ji, X., Liu, G., Dai, Y., & Han, Z. (2021). Honeypot detection strategy against advanced persistent threats in industrial Internet of Things: A prospect theoretic game. *IEEE Internet of Things Journal.*

Tripathi, D., Singh, L. K., Tripathi, A. K., & Chaturvedi, A. (2021). Model based security verification of cyber-physical system based on petrinet: A case study of nuclear power plant. *Annals of Nuclear Energy, 159*, 108306.

Tu, H., Xia, Y., Chi, K. T., & Chen, X. (2020). A hybrid cyber attack model for cyber-physical power systems. *IEEE Access, 8*, 114876–114883.

Ulz, T., Pieber, T., Steger, C., Haas, S., & Matischek, R. (2018). In *Secured remote configuration approach for industrial cyber-physical systems Paper presented at the 2018 IEEE Industrial Cyber-Physical Systems (ICPS).*

Upadhyay, S., Kumar, S., Dutta, S., Srivastava, A. K., Mondal, A. K., & Kaundal, V. (2020). A comprehensive review on the issues related to the data security of Internet of Things (IoT) devices. *Intelligent Communication, Control and Devices, 727–734.*

Vähäkainu, P., Lehto, M., & Kariluoto, A. (2021). *Adversarial poisoning attack's impact on prediction functionality of ML-based feedback loop system in cyber-physical context* (pp. 373–381). Reading: Academic Conferences International Limited. XIII, XV.

Valencia, C. M. P., Alzate, R. E., Castro, D. M., Bayona, A. F., & García, D. R. (2019). In *Detection and isolation of DoS and integrity attacks in Cyber-Physical Microgrid System Paper presented at the 2019 IEEE 4th Colombian Conference on Automatic Control (CCAC).*

van Lier, B. (2017). In *The industrial internet of things and cyber security: An ecological and systemic perspective on security in digital industrial ecosystems Paper presented at the 2017 21st International Conference on System Theory, Control and Computing (ICSTCC).*

Vaz, A. C., Nayak, C. G., & Nayak, D. (2020). Role of visible light communication in enhancing the safety of cyber-physical systems. In *Advances in Control Instrumentation Systems* (pp. 183–192). Springer.

Vu, A. H., Tippenhauer, N. O., Chen, B., Nicol, D. M., & Kalbarczyk, Z. (2014). In *Cybersage: A tool for automatic security assessment of cyber-physical systems Paper presented at the International Conference on Quantitative Evaluation of Systems.*

Wadhawan, Y., & Neuman, C. (2016). In *Defending cyber-physical attacks on oil pipeline systems: a game-theoretic approach Paper presented at the Proceedings of the 1st International Workshop on AI for Privacy and Security.*

Wang, Y., & Liu, Y. (2019). Deducing cascading failures caused by cyberattacks based on attack gains and cost principle in cyber-physical power systems. *Journal of Modern Power Systems and Clean Energy*, 7(6), 1450–1460.

Wang, H., Meng, A., Liu, Y., Fu, X., & Cao, G. (2019). Unscented Kalman Filter based interval state estimation of cyber physical energy system for detection of dynamic attack. *Energy*, *188*, 116036.

Wang, H., Ruan, J., Zhou, B., Li, C., Wu, Q., Raza, M. Q., et al. (2019). Dynamic data injection attack detection of cyber physical power systems with uncertainties. *IEEE Transactions on Industrial Informatics*, *15*(10), 5505–5518.

Wang, Y., Wang, Y., Liu, J., Huang, Z., & Xie, P. (2016). In *A survey of game theoretic methods for cyber security 2016 IEEE First International Conference on Data Science in Cyberspace (DSC)* (pp. 631–636).

Wei, X., & Sikdar, B. (2019). In *Impact of GPS time spoofing attacks on cyber physical systems Paper presented at the 2019 IEEE international conference on industrial technology (ICIT)*.

Wu, G., Sun, J., & Chen, J. (2016). A survey on the security of cyber-physical systems. *Control Theory and Technology*, *14*(1), 2–10.

Wu, G., Wang, G., Sun, J., & Xiong, L. (2019). Optimal switching attacks and countermeasures in cyber-physical systems. *IEEE Transactions on Systems, Man, and Cybernetics: Systems*.

Xing, L. (2020). Cascading failures in internet of things: Review and perspectives on reliability and resilience. *IEEE Internet of Things Journal*, *8*(1), 44–64.

Xu, T., Liu, Z., Tang, T., Zheng, W., & Zhao, L. (2012). In *Component based design of fault tolerant devices in cyber physical system Paper presented at the 2012 IEEE 15th International Symposium on Object/Component/Service-Oriented Real-Time Distributed Computing Workshops*.

Xue, W., & Wu, T. (2020). Active learning-based XGBoost for cyber physical system against generic AC false data injection attacks. *IEEE Access*, *8*, 144575–144584.

Xun, P., Zhu, P.-D., Hu, Y.-F., Cui, P.-S., & Zhang, Y. (2017). Command disaggregation attack and mitigation in industrial internet of things. *Sensors*, *17*(10), 2408.

Yan, W., Mestha, L. K., & Abbaszadeh, M. (2019). Attack detection for securing cyber physical systems. *IEEE Internet of Things Journal*, *6*(5), 8471–8481.

Yang, Y.-S., Lee, S.-H., Chen, W.-C., Yang, C.-S., Huang, Y.-M., & Hou, T.-W. (2021). TTAS: Trusted token authentication service of securing SCADA network in energy management system for Industrial Internet of Things. *Sensors*, *21*(8), 2685.

Yang, J., Zhou, C., Tian, Y.-C., & An, C. (2020). A zoning-based secure control approach against actuator attacks in industrial cyber-physical systems. *IEEE Transactions on Industrial Electronics*, *68*(3), 2637–2647.

Ylönen, M., Tugnoli, A., Oliva, G., Heikkilä, J., Nissilä, M., Iaiani, M., et al. (2022). Integrated management of safety and security in Seveso sites - sociotechnical perspectives. *Safety Science*, *151*, 105741. https://doi.org/10.1016/j.ssci.2022.105741.

Yong, S. Z., Zhu, M., & Frazzoli, E. (2015). In *Resilient state estimation against switching attacks on stochastic cyber-physical systems Paper presented at the 2015 54th IEEE Conference on Decision and Control (CDC)*.

Zahid, M., Inayat, I., Daneva, M., & Mehmood, Z. (2021). Security risks in cyber physical systems—A systematic mapping study. *Journal of Software: Evolution and Process*, e2346.

Zhang, Q., Li, Y., Wang, R., Liu, L., Tan, Y.a., & Hu, J. (2021). Data security sharing model based on privacy protection for blockchain-enabled industrial Internet of Things. *International Journal of Intelligent Systems*, *36*(1), 94–111.

Zhang, L., & Reniers, G. (2016). A game-theoretical model to improve process plant protection from terrorist attacks. *Risk Analysis*, *36*, 2285–2297.

Zhang, T., Wang, Y., Liang, X., Zhuang, Z., & Xu, W. (2017). In *Cyber attacks in cyber-physical power systems: A case study with GPRS-based SCADA systems Paper presented at the 2017 29th Chinese control and decision conference (CCDC).*

Zhang, T.-Y., & Ye, D. (2020). False data injection attacks with complete stealthiness in cyber–physical systems: A self-generated approach. *Automatica*, *120*, 109117.

Zhang, Q., Zhou, C., Tian, Y.-C., Xiong, N., Qin, Y., & Hu, B. (2017). A fuzzy probability Bayesian network approach for dynamic cybersecurity risk assessment in industrial control systems. *IEEE Transactions on Industrial Informatics*, *14*(6), 2497–2506.

Zhao, Y., Bai, M., Liang, Y., Ma, J., & Deng, P. (2018). In *Fault modeling and simulation based on cyber physical system in complex distribution network Paper presented at the 2018 China International Conference on Electricity Distribution (CICED).*

Zhao, A., Li, M., & Wang, Q. (2020). In *Cyber security risk identification in power Internet of Things system Paper presented at the 2020 IEEE 4th Conference on Energy Internet and Energy System Integration (EI2).*

Zografopoulos, I., Ospina, J., Liu, X., & Konstantinou, C. (2021). Cyber-physical energy systems security: Threat modeling, risk assessment, resources, metrics, and case studies. *IEEE Access*, *9*, 29775–29818.

Zúñiga, A. A., Baleia, A., Fernandes, J., & Branco, P. J. D. C. (2020). Classical failure modes and effects analysis in the context of smart grid cyber-physical systems. *Energies*, *13*(5), 1215.

CHAPTER FIFTEEN

Integrated dynamic risk management in process plants

Mohammed Taleb-Berrouane[a,*] and Hans Pasman[b]
[a]Centre for Risk, Integrity and Safety Engineering (C-RISE), Faculty of Engineering & Applied Science, Memorial University of Newfoundland, St. John's, NL, Canada
[b]TEES Mary Kay O'Connor Process Safety Center, Texas A&M University System, College Station, TX, United States
*Corresponding author: e-mail address: m.talebberrouane@mun.ca

Contents

1. General introduction

This chapter will exhibit some other aspects of dynamic risk assessment and management than have been treated in chapter "Dynamic operational risk assessment in process safety management" by Li et al. Motivation

Methods in Chemical Process Safety, Volume 6
ISSN 2468-6514
https://doi.org/10.1016/bs.mcps.2022.05.006

to develop methodology for this approach is among other that every time a catastrophic accident happens, the risk analysis practice is criticized for not being effective, proactive or dynamic enough, to capture or predict these accidents. Since the 1980s, the science of risk analysis has evolved by virtue of the progress in both hardware and software technologies along with the maturation and expansion of the field of risk analysis. With the world entering in the last decade the era of big data and The Fourth Industrial Revolution, also known as "Industry 4.0," within its wake "Safety 4.0," the field of risk analysis is moving toward a fully integrated dynamic risk analysis (IDRA) practice as a basis for management decision-making.

As opposed to the conventional risk analysis, the models used for IDRA are flexible enough to be constantly calibrated and revised based on new, and in some cases, real-time data, as pieces of evidence adapting the models to the fast-changing processes.

The world integrated is defined by the online Cambridge dictionary as: "to combine two or more things in order to become more effective." This integration requires a high level of adaptability between the techniques or tools that are used for this type of risk analysis. Where the world "dynamic" is defined by the same dictionary as "continuously changing or developing." The concept of dynamic risk analysis was first introduced at the Massachusetts Institute of Technology in the work of Dister LeRoy Deoss (Deoss, 1989) on developing a simulation model for dynamic system availability analysis as part of his thesis in nuclear engineering.

Section 2 will be dedicated to dynamic risk assessment, Section 3 to safety and security and Section to the contribution loss functions can provide.

2. Real-time risk assessment of process operations

2.1 Introduction

The technological development in the 20th and early 21st centuries has generated a massive number of new technologies applied to the way the process plants are now designed, built and operating. The Third Industrial Revolution, also known as "Industry 3.0" started in 1970s by introducing simple forms of automations with the intent of intensifying the production. Industry 3.0 was described as the digital revolution mainly because of the extensive use of the programmable logic controllers (PLC) in most forms of manufacturing activities. Currently most of the process plants around the world are still operating at this level (i.e., Industry 3.0). Besides productivity, this automation also increased the reliability and safety of the assets.

The concept of "Industry 4.0" was first used in Germany in 2011. It was part of the high-tech strategy of the German government, which promotes the computerization of manufacturing. Industry 4.0 consists of generating data making use of the large diversity of developed sensors and other observational techniques, linking or merging the digital data streams of the physical assets with the advanced digital technologies such as Internet of Things (IoT), and treating the data to extract information with techniques, such as Artificial Intelligence (AI), Machine Learning (ML), and cloud computing. This combination or interconnection allows for real-time communication, analysis, decision making and taking actions (Dima, 2021).

2.2 Overview and recent developments

In a condensed fashion in Table 1 a number of pioneering papers is summarized based on the tool used to describe the probabilistic dynamic process of risk assessment.

2.3 Operation data and data analytics

Operational risk analysis plays a primary role in safety-critical industries such as oil and gas exploration, pipeline transportations, and downstream operations. This essential task is facing a series of challenges due to increased complexity and volume of generated data (i.e., big data processing and analysis). Current tools used by the process industries are centralized and offline which makes it less convenient, hard to access and communicate, and slow in response to emergency situations and dangerous escalations. The emerging technologies such as cloud-based solutions are now getting bridged from the research work in quantitative risk analysis to applied software tools accessible to the industrial end-users. This is made possible due to the recent advancements in cloud computing power provided by information technology giants such as Google (i.e., Google Cloud Platform) and Amazon (i.e., Amazon Web Services). These tools will help engineers and executives to track current and predictive industrial risks such as toxic releases, fire hazards and explosions due to loss of containment. Real-time process analytics are generated, visualized, and processed to predict dangerous escalations. Fig. 1 presents a data analytics framework applicable to the process industries. Fig. 2 presents a view on how data analytics lifecycle can be integrated with cloud services for process applications.

Data analytics is becoming harder and harder to neglect in the process industry for decision-making, the analytics process can be defined by a

Table 1 Major contributions in dynamic risk analysis of process facilities.

Tool	Description	Related works
Bayesian network	*Information used*: Process history, accident precursor data, alarm databases, expert judgment data *Updating mechanism*: Bayesian updating mechanism is used to update the prior beliefs about the accidents by incorporating new information from system. *Advantages*: Capability to handle uncertainty, multi-state variables, complex causal relationships, and sequentially dependent failures. – Simple graphical interface founded on strong mathematical base (Bayes' theorem) *Disadvantages*: – High computational burden to construct conditional probability tables – Inability to model complex dependencies and transitional mechanism among variables – Application of deterministic and/or normally distributed probabilities	Meel and Seider (2006, 2008), Pariyani, Seider, Oktem, and Masoud (2012a), Pariyani, Seider, Oktem, and Masoud (2012b), Vinnem et al. (2012), Taleb-Berrouane, Khan, Hawboldt, Eckert, and Skovhus (2018), Taleb-Berrouane, Sterrahmane, Mehdaoui, and Lounis (2017), Taleb-Berrrouane (2019), Kalantarnia, Khan, and Hawboldt (2009), Khakzad (2019), and Khakzad, Khan, and Amyotte (2012b)
Bow-tie analysis	*Information used*: Accident precursor data; new identified scenarios, expert judgment data *Updating mechanism*: Continuous safety-related information retrieval is integrated with conventional bow-tie analysis to dynamically estimate the risk. *Advantages*: – Simple and practical implementation – Graphical structure easy to read and modify *Disadvantages*: – Limited capability to handle uncertainty, multistate variables and dependent – failures due to application of simple Boolean functions in bow-tie analysis – Application of deterministic probability values – Inability to model complex dependencies among variables	Paltrinieri, Tugnoli, Buston, Wardman, and Cozzani (2013a), Paltrinieri, Khan, Amyotte, and Cozzani (2014), Paltrinieri, Tugnoli, Buston, Wardman, and Cozzani (2013b), Pasman and Rogers (2014), Taleb-Berrouane and Lounis (2016), Taleb-Berrouane, Khan, and Hawboldt (2021), Khakzad, Khan, and Amyotte (2012a), and Villa, Paltrinieri, Khan, and Cozzani (2016)

Stochastic Petri nets	*Information used*: Process history, time series data, accident precursor data, equipment failure datasets, expert judgment data *Updating mechanism*: Updating may require complex structure changes. However, updating rates is easy and can be done in real time. *Advantages*: – Capability to handle stochastic behavior, uncertainty, multi-state variables, complex causal relationships, and sequentially dependent failures. – Can model concurrency, conflict management, synchronization, and resources sharing. *Disadvantages*: – High computational burden to extract certain transitional rates – Difficult to track large models – May lead to state explosion if the complexity and the size of the system increase – Need to be coupled with Monte Carlo simulation to provide accurate results	Taleb-Berrouane, Khan, and Kamil (2019), Taleb-Berrouane and Khan (n.d.), Taleb-Berrouane, Khan, and Lounis (2016), and Zhou & Reniers, 2016, 2017
Principal component analysis (PCA)	*Information used*: Process history *Updating mechanism*: Probability of a fault is calculated based on the PCA filtered score, and the severity of the fault is a weighted average of the consequences of each variable in the score. *Advantages*: – Multivariate techniques – Takes advantage of the correlation between process input and output – Extracts latent features from high dimensional data *Disadvantages*: – Relies on linear models and assumes Gaussian noise – Unable to model complex dependencies among variables – Most applications require the process model	Jiang and Yan (2015), Ge and Song (2013), and Zadakbar, Imtiaz, and Khan (2013)

Continued

Table 1 Major contributions in dynamic risk analysis of process facilities.—cont'd

Tool	Description	Related works
Loss functions	*Information used*: Deviation of key process characteristics from target values *Updating mechanism*: Loss functions are used to relate process deviations to economic losses. *Advantages*: – Provides a mechanism for real-time loss modeling – Promotes continuous improvement of process safety through proactive loss minimization *Disadvantages*: – Selection of a proper loss function could be difficult for data-scarce processes	Hashemi, Ahmed, and Khan (2014), Ali (2015), and Pan and Chen (2013)
Risk barometer	*Information used*: Technical, operational and organizational indicators assessing deviation from optimal condition *Updating mechanism*: Indicators are used to assess safety barrier performance on a regular basis, which in turn allows for continuous assessment of overall risk variation. *Advantages*: – Technical indicators are integrated with proactive operational/organizational indicators in order to assess early deviations potentially leading to unwanted events – Degrading safety barriers leading to critical risk increase may be identified and their improvement may be prioritized *Disadvantages*: – It is case-specific and partially based on expert judgment from operators. – Relies on linear models – It is based on relevant indicators whose collection may be irregular. Such irregularities may compromise the overall risk assessment	Paltrinieri, Scarponi, Khan, and Hauge (2014), Paltrinieri and Hokstad (2014), and Villa et al. (2016)

Adapted from Khan, F., Hashemi, S. J., Paltrinieri, N., Amyotte, P., Cozzani, V., & Reniers, G. (2016). Dynamic risk management: A contemporary approach to process safety management. In Current opinion in chemical engineering (Vol. 14, pp. 9–17). Elsevier Ltd. https://doi.org/10.1016/j.coche.2016.07.006; Taleb-Berrouane, M., Khan, F., & Amyotte, P. (2020). Bayesian stochastic petri nets (BSPN) - A new modelling tool for dynamic safety and reliability analysis. Reliability Engineering and System Safety, 193. https://doi.org/10.1016/j.ress.2019.106587; Kamil, M. Z., Taleb-Berrouane, M., Khan, F., & Amyotte, P. (2021). Data-driven operational failure likelihood model for microbiologically influenced corrosion. Process Safety and Environmental Protection, 153, 472–485. https://doi.org/10.1016/j.psep.2021.07.040.

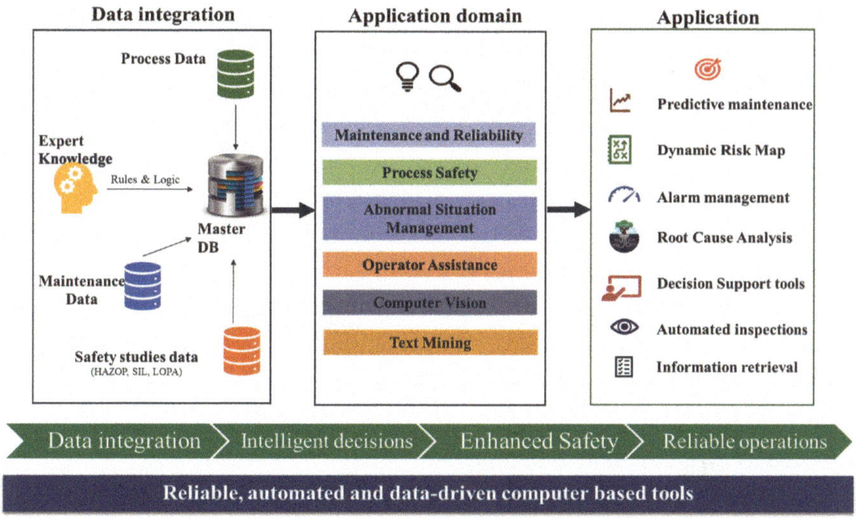

Fig. 1 Data analytics framework (Goel, Jain, Pasman, Pistikopoulos, & Datta, 2020).

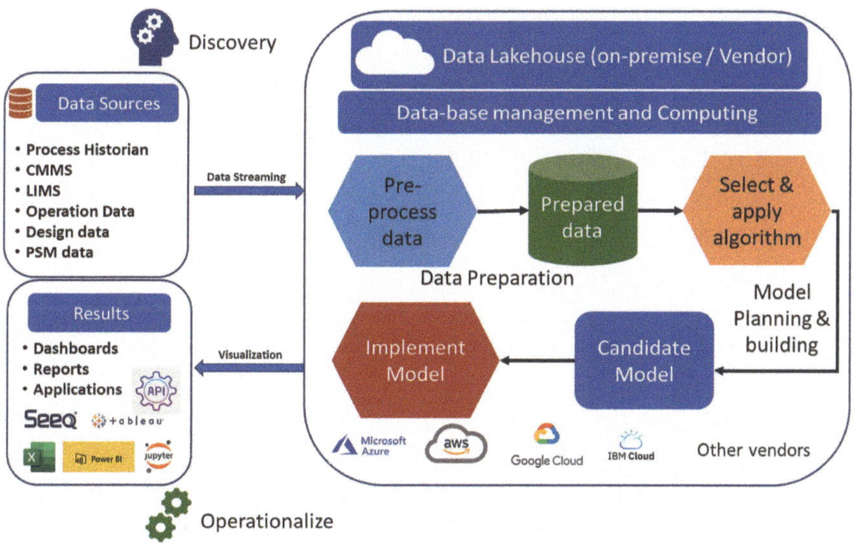

Fig. 2 Data analytics lifecycle integrated with cloud services (Goel et al., 2020).

systems approach with six key stages as shown in Fig. 2. Moreover, the tasks can be addressed in multiple stages at the same time. The main stages of this lifecycle include.

1. Discovery: define goals and plans to achieve the goals, in other words, define the business requirement.

2. Data Preparation: define data requirements, implement methods involving collection, processing, and cleaning of data.
3. Model planning & building: perform exploratory analysis on data, which may include transforming data (eliminating noise, removal of dirty data, reducing skewness, dealing with missing data, and so on), aggregation, integration, and data scrubbing. Identify the best model for the data and execute the model to ensure the model fits the data.
4. Operationalize: communicate (within the organization) or publish (to outside entities) the results and implement the solution in a live operational environment, monitor results and update any stage actions to get desired results (Goel et al., 2020).

2.4 Data processing methods

In previous chapters of this book Volume 6 various aspects of data treatment and information extraction are described, which will not repeated here. It suffices to state that this field of method development is still expanding. On the level of method details many variations appear to be possible and beneficial for certain applications, so that the most useful and stable methods will emerge in due time. The stage of writing reviews has already arrived as shown by references in previous chapters, e.g., chapters "Data-driven approaches: Use of digitized operational data in process safety" by Bai et al.; "Industry 4.0 based process data analytics platform" by Wanasinghe et al.; "Statistical approaches and artificial neural networks for process monitoring" by Alauddin et al.

2.5 Applications to process systems

The complexity of modern process facilities has significantly increased. This complexity has led to a substantial risk of failure leading to casualties, damage to assets, and air, soil, and marine pollution such as the 2010 Macando well blowout in the Gulf of Mexico. To handle this complexity and prevent accidents, research in process safety and risk analysis, known as risk engineering (Mannan, 2014.; Modarres, 2006), has evolved from qualitative and semi-quantitative assessments to dynamic risk analysis (Khakzad, Khan, & Amyotte, 2013) and artificial intelligence (AI)-powered safety models (Adedigba, Khan, & Yang, 2017; Gheraibia, Kabir, Aslansefat, Sorokos, & Papadopoulos, 2019). Current software solutions have not yet adopted these novel techniques resulting from academic research. For instance, the software tool PHA Pro (PHA Pro, 2020) developed by

Sphera is an offline software that can run conventional risk analysis techniques such as Hazard and operability study (HAZOP) and Layer of Protection Analysis (LOPA) without any dynamic features such as time dependencies and integration of changes in process data. Another widely used software for risk and root cause analysis is TapRoot (TapRoot - Root Cause Analysis, n.d.). This software is offline and focuses on human error and performance. The analysis using TapRoot is very static and does not capture the complexity of the control and safety systems in place. Multianalytic risk visualization tool—MARV (Multi-Analytic Risk Visualization Tool - MARV™, n.d.) developed by DNV-GL is a powerful tool for real-time risk assessment using Bayesian networks. Besides its high-cost and complexity, this tool is dedicated to risk assessment in pipelines systems only and cannot be applied to other process equipment.

In the modern era of digital twin (Fei et al., 2018), efficient risk analysis practice requires access to real-time data and large historical datasets, robust computing power, and dynamic risk models. Using machine learning, emerging dynamic risk models can predict the risk and potential escalation scenarios and help in decision making.

Recently, several research works made attempts to implement machine learning for risk analysis. The most recent work by Goel et al. (Goel et al., 2020) at the Mary Kay O'Connor Process Safety Center at Texas A&M University demonstrated the use of machine learning on four different case studies: (1) failure prediction of a turbo engine and predictive maintenance modeling using deep learning, (2) Natural Language Processing (NLP) for mining US OSHA incident data from text, (3) barrier assessment for dynamic risk mapping (DRA), and (4) correlation development for sustainability indicators. The four studies have shown that machine learning can significantly improve the outcome of risk analysis and still work is required for further testing and optimization. In another work at the University of Alberta, Kurian et al. (Kurian, Sattari, Lefsrud, & Ma, 2020) used machine learning and keyword analytics to analyze 15,000 incident reports and extract useful information to improve safety performance.

3. Safety vs security risk

3.1 Introductory remarks

This section will be on process safety and security aspects of integrated risk management. In case of security, one distinguishes physical and cyber security, where in the former attackers are near or inside a facility, in the latter attacker are remote using the internet. In several languages there are no separate words for safety and security, which may lead to confusion as is clearly true in case of Kukfisz, Kuczy'nska, Piec, and Szykuła-Piec (2022). This will be due to the many similarities between those concepts of safety and security as both can only be measured via risk of being breached, but there exist also fundamental differences, which in a general sense have been analyzed with respect to various general aspects by a range of authors in a FonCSI book edited by Bieder and Pettersen Gould (2020). In relation with process plants, a similarity is that in case deviations, irregularities or failures take place that disrupt the safety or security situation, risks evolve leading to major hazards. Main difference between the two is human act, which in the case of safety breach is non-existent or accidental and certainly non-intentional, while in case of a security incident external or internal threats by malicious act occur with the purpose to intentionally disrupt the course of the process and to damage the plant. This also characterizes the difference in risk assessment methodology. In case of safety risk failure rate of people, plant and procedures play an important role, whereas security risk is fundamentally antagonistic of character, in which frequency does not figure but in which attack capability costs and protection capability costs are optimized given assets of both parties are analyzed by game theory-based methods. Whereas safety in industry has been strived for many years already, security is a relatively new concept as threats from human actors is in fact from the start of this century.

Over the various decades after the Second World War, and in fact already starting before, how to foster safety was thought about. In particular in the 1950s and later, tools have been developed to analyze systematically how accidents can occur. It started with Heinrich's (1941) falling domino stone model, but a first step in the direction of risk analysis was Watson's (1961) fault tree analysis, and an early most comprehensive accident investigation tool was Management Oversight and Risk Tree (MORT) developed by Johnson (1973). An evolutionary process of growing insight to improve safety followed as illustrated in Fig. 3. Reniers and Khakzad (2017) present a time line of safety progress in the chemical industry and distinguish in the development even three "safety revolutions," the last one taking place at present. Today, with big data & analytics we are indeed seeing a major leap in technology forward, indicated as Industry 4.0 to which is associated Safety 4.0 which is also enabled by the same techniques of analyzing acquired data presented in this MCPS volume 6.

Security in the sense of physical threat to process installations became an important topic in the United States after the 2001 terrorist attacks on the

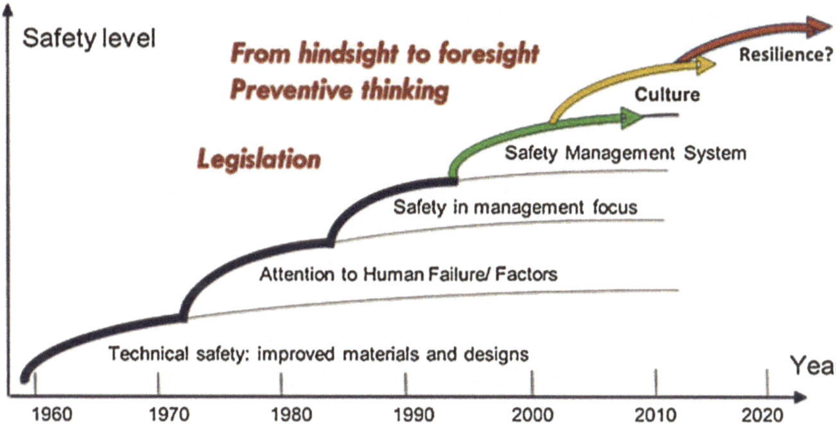

Fig. 3 Growing insight leading to improved safety levels in the process industry (Pasman & Fabiano, 2021).

WTC buildings in New York. It led to new regulation, e.g., the Critical Infrastructures Protection Act (2001), already initiated under the Clinton presidency, and the Chemical Facility Anti-Terrorism Standards (CFATS) Act (2014) (CFATS, 2014). In particular, the latter meant quite an effort for the process industry with respect to analyzing vulnerabilities and installing adequate preventative and protective measures; for details, see Pasman (2017). For transport of hazardous materials there is the United States Federal regulation addressing safety and security: 49 CFR PART 172 Subpart I—Safety and Security, while for LNG (liquified natural gas) additional regulation has been created with 49 CFR PART 193, Subpart J—Security. Obviously, this is all to protect for physical threats.

More recently, awareness of Cyber threats increased after the known Stuxnet attack on the Iranian nuclear uranium enrichment facilities in 2010. Iaiani, Casson Moreno, Reniers, Tugnoli, and Cozzani (2021) developed a data base and presented a more complete overview of intentionally caused industrial incidents. Attackers are terrorists and hackers, but also saboteurs, disgruntled employees, thieves and vandals. Physical security attacks are frequently targeted on storage tanks and pipelines, while cyber-security attacks often try to damage safety instrumented systems. In 2013, after the disastrous accident of ammonium nitrate detonation in West, Texas, President Obama issued an Executive order "Improving Chemical Facility Safety and Security" to screen all regulation with respect to hazardous materials and assess its adequacy (see Obama, 2013).

Regarding cyber-physical security, an extensive review of the challenges and opportunities in relation to digitalization of processes and consequences for process safety is given in the previous chapter. In this chapter, we may have some overlaps with the previous but the emphasis will be on tools to determine the vulnerability of plant to both safety and security threats. To reduce the risk of process disruption, work has been focusing on what tools will achieve the goal how safety and security can be combined. In the following, we shall go in more detail based on a limited literature search based on the keyword combination "safety and security," resulting in 30 papers on a broad spectrum of aspects. As mentioned much literature has been reviewed in the previous chapter, here a few early books on the topic, such as Nolan (2015) and Reniers, Khakzad, and Van Gelder (2018) will be reviewed as well as a number of very recent papers. Most of the latter are on approaches to solve the problem, methods and models; the categorization is not sharp as approaches lead to methods while models are part of methods. There are too a few papers on conceptual, societal, cultural, policy and management system aspects.

3.2 The non-technical aspects of safety & security

Most comprehensive is the book edited by Bieder and Pettersen Gould (2020) on the interrelations between safety and security. The Contents, reproduced in Table 2, presents a fair impression of the topics treated in the book. In the last chapter, editors Bieder and Pettersen Gould (2020) address the required research. It first concludes that safety and security are coming together and mixed in the management of hazards. However, besides the similarities there are differences, although risk is the overarching framework for management. One difference is strikingly expressed in the heading of a section: "In safety we trust, in security we distrust." Due to the similarities, it is tempting to apply the conventional safety tools also to the new security threat. This was particularly the early-on thought, but as we shall see later in the recent papers a considerable number of new methods appeared to be necessary to cope with the complexity and the interactions of components.

The chapter continues by mentioning that as society over the years increasingly requires adequate risk control, which trend is characterized as the "risk society," organizational safety and security will have to share a research agenda. The chapter ends with stating that if a global approach would be developed to organizational safety and security it will be needed to think in different scales. With more interconnections and critical infrastructure an individual act can have worldwide consequences. Also, little

Table 2 Contents Bieder and Pettersen Gould (2020).

Chapter		Page
1	**Safety and Security: The Challenges of Bringing Them Together**	1
	Kenneth Pettersen Gould and Corinne Bieder	
2	The Concepts of Risk, Safety, and Security: A Fundamental Exploration and Understanding of Similarities and Differences	9
	Peter J. Blokland and Genserik L. Reniers	
3	Safety and Security Are Two Sides of the Same Coin	17
	Nancy Leveson	
4	Safety Versus Security in Aviation	29
	Heinz Wipf	
5	Security and Safety Culture—Dual or Distinct Phenomena?	43
	Sissel H. Jore	
6	User Safety and Security Experience: Innovation Through	53
	Design-Inspired Methods in Airports	
	Ivano Bongiovanni	
7	Divergence of Safety and Security	63
	David J. Brooks and Michael Coole	
8	Doing Safety ... and then Security: Mixing Operational Challenges—Preparing to Be Surprised	75
	Todd R. La Porte	
9	Safety and Security: Managerial Tensions and Synergies	87
	Paul R. Schulman	
10	The Interface of Safety and Security; The Workplace	97
	George Boustras	
11	Exploring the Interrelations Between Safety and Security:	105
	Research and Management Challenges	
	Corinne Bieder and Kenneth Pettersen Gould	

has been done to anticipate the unexpected and hidden effects of safety and security on organizations due to the risk society trend. For this also research can help to uncover what global trends will bring in this respect.

Already mentioned before, Reniers and Khakzad (2017) exposed an historical evolution of safety by distinguishing three safety revolutions in 1900/10 with "safety first" movement, 1960/70 with technical safety, barriers thinking, the first risk assessment and management activities, and this time 2020/30, where we just had resilience adopted as concept and system thinking and are now with CHESS[a] heading for strategic and pro-active cooperation/high transparency, and further into the future after 2030 a focus on education, security integration and safety innovations. Here, the comment can be made that the 2030 mentioned topics seem to have started already.

Hydyrov (2017) considers the world of multinational companies and the interests and right of workers and users, and the management systems that form the core of protection of those. Further it is discussed how parallel the management systems of safety and security will run and how these are supported by corporate governance. Interests of various stakeholders should be balanced. Security is obviously used here in the context of secured information system in relation to workers and users. Hence, although the words are not used it seems having to do in this paper with privacy protection.

Glesner, Van Oudheusden, Turcanu, and Fallon (2020) review safety and security cultures. They first recall the emerging of safety culture after the Chernobyl nuclear power plant disaster. The authors distinguish functionalist and interpretative perspectives of safety culture. Functionalist is managerial and top-down, while interpretative is how an organization is and something does. Then, the development of security culture is described. They observe that there is an asymmetrical treatment of security culture in relation to safety culture. Co-production and Actor-Network Theory (ANT) concepts are called in to close that gap. Co-production refers to the interaction of science and society, while ANT is defined by constantly shifting networks of relationships. In addressing the gaps applying these two concepts, the following aspects are considered: (1) Security dimension of safety devices: Labelling of hazardous materials; (2) Safety dimensions of the security principle: The "four eyes" principle; (3) Combining co-production and ANT to the analysis of safety and security cultures.

[a] In case an acronym is not directly or near-by explained, see list of acronyms at the end.

Boustras and Waring (2020) discuss safety and security at societal level, where these are being addressed as discrete and separate entities at governmental, corporate and professional organizations. Authors describe conceptual issues, such as the terminology, and further the links, interfaces and interactions between safety and security domains. They conclude that safety and security are similar as risk domains but differ in management objects. As a result of the separateness of the two, the impact on the public and on occupational health and safety of workers of security threats, such as global warming consequences, terrorism, radicalization, hate crimes, white collar crime, cyber-crime, economic crime, and trafficking, has not obtained the attention it deserves. Therefore, guiding management standards should be adopted and legislation requiring effective security policies be developed to damp the current imbalance.

Popov (OECD, 2018) makes the point of the need for the European Union to enable a general nuclear power plant security obligation bypassing the sovereignty of the Member states, just as the practice is in case of safety, although that has never been formally accepted.

3.3 Review papers on safety & security

Chen, Li, Reniers, and Yang (2021) reviewed specifically incidents of oil and gas pipelines. Bibliometric analysis of 598 papers (1970–2019) was performed on oil and gas pipeline intentional or unintentional damage and risk of leakage. Causes comprised corrosion, lack of integrity management, absence of risk assessment, security, and failure analysis. Most frequently used research methods are quantitative risk assessment (QRA), making use of fuzzy theory to estimate event probability values and Bayesian network simulating causation structure enabling integration and final risk value. Emphasis has been on technical safety issues but more attention should be given to security breach related incident prevention and also to management and culture influences. Other research items would be integration of environmental risk assessment, oil spill cleaning technology, carbon footprint reduction, system resilience, in summary enhancing sustainability.

Gou et al. (2022) conducted a broader, more general review by means of the web of science (WoS) and identified 1091 papers on safety and security research in the realm of system dynamics from 2000 to 2020. Most represented journals were IEEE Transactions on Power System, Safety Science, Energy Policy, System Dynamics Review, and Reliability

Engineering & System Safety. Safety and security have significant differences domains but are very similar in risk analysis. Current hot topics are safety management, energy security, resource utilization, and industry applications. Current frontiers are system safety and reliability, energy consumption and security, power system dynamics and stability.

3.4 General approach papers and some threat aspects

Kriaa, Pietre-Cambacedes, Bouissou, and Halgand (2015) reviewed approaches to the security analysis of industrial control systems. After dealing with terminology, the interdependencies of safety and security were described. Although there are many commonalities, as the authors express it, the differences appear in tools, standards and the way of risk assessment. To understand the interactions between the two, consider the situation that to be safe puts a condition of security or the other way around, they can reinforce each other, be in conflict, or be independent. In an appendix, standards are summarized: *safety* IEC 61508, 13 (nuclear) and 11 (process industry), and *security* IEC 62443 automation and control), IEC 15408 (IT), ISO/IEC 27001 (information security management systems) and ISO/IEC 27005 (risk management—with commonalities with ISO/IEC 15026 safety standard for system and software integrity), and IEC 62645 (nuclear instrumentation and control). The paper analyzes the life cycle of development, safety management and information security management alignment key points.

Calling it approaches, Kriaa et al. (2015) also review methods: FMEA, extended FT, BDMP, SPN, SEFT, BBN, UML, CHASSIS (with various diagram types), CARISMA, SysML, MBSE, CRUTIAL, TVRA, STPA-Sec. The approaches show that safety and security should no longer be treated separately.

Another, very ambitious approach paper (Wu, 2021) is addressing the severe threats due to the so-called dark functions of vulnerabilities and backdoors (either by code loopholes or artificially embedded ones). Endogenous security problems cannot be completely eliminated. The solution is by introducing endogenous security mechanisms and characteristics based on dynamic heterogeneous redundancy (DHR) architecture. Author proposes that DHR can deal with traditional reliability problems and nontraditional network security problems. DHR is a further development of DRS the dissimilar redundant structure, which consists of a number of parallel not the same executor elements. DHR combines the multimodal decision-based scheduling mechanism and the multidimensional dynamic reconstruction robust-control mechanism. All signals pass a multimodal decision element. In DHR the executors are reconfigurable and the system has a feedback control element that receives an input control parameter and a decision parameter. The latter is further influenced by a signal from the mimic decision element. This is based on the cyberspace mimic defense, which is introduced by Guo, Wu, Zhang, and Wu (2020). It is claimed that DHR becomes an enabling technology for a new generation of software and hardware products, managing cyberspace security issues at the source.

Busquim e Silva, Piqueira, Cruz, and Marques (2021) performed a simulation to see the risks of unauthorized software changing real-world system behaviors. This particular type of security breach is a major and critical infrastructure safety and security issue. Two different cyberattacks were simulated on the Asherah Nuclear Power Plant Simulator. The interaction of safety and security and the dynamics of interacting instrumentation and control (I&C) and operational technology (OT) was clearly seen. The paper provides many details.

3.5 Safety: Security risk assessment methods

In the following brief summaries of papers with an emphasis on methods will be presented in table form for a better overview and to save space. Often, it is not easy to distinguish whether the emphasis is on the development of a method or a model. In many papers, a method centers around a model. However, may be arbitrarily we made a distinction and in Table 3 the method papers are summarized.

3.6 Safety: Security interaction analysis models

For models holds the same as for methods. Vulnerability assessment is emphasized. Papers are summarized in Table 5.

3.7 Safety vs security concluding remarks

Safety and security assessment including their interactions focuses on vulnerability assessment. It is a lively field of research spreading over many aspects form social and cultural to model and method development and application studies. Also, a strong accelerating advancement is noticeable between 2015 and 2022. The problems to assure safety and simultaneously security are complex and can only be approximated by merging various methods. Various rather different approaches are attempted.

Table 3 Summary of papers describing *methods* in consecutive years.

Nolan (2015)	This book describes in quite detail conventional methods of safety and physical security risk assessment. Such as Preliminary Hazard Assessment, What-If method (Appendix D with long checklists), and HAZOP (Appendix E many Guide words). The acronym PHA means also Process Hazard Analysis. This analysis is mandated by the US Occupational Health and Safety Administration (OSHA) for the process industry for credible accidental and security release scenarios. Also consequence of failures need to be determined and together with hazard identification it produces a Security Vulnerability Assessment (SVA). Further extensions are Computer HAZOP (CHAZOP), Electricity HAZOP (EHazop), and the known Bow-tie, LOPA, SIL analysis, and Fishbone causation. In case of cybersecurity there is CSVA and the Ring of Protection, ROPA similar to LOPA, ANSI/ISA-TR99.03.01. Guidance for the functioning of an analysis team is given in detail. Also needed management support and responsibilities get attention. Applications are shown. Review procedures and worksheets are described in much detail, as well as reporting and handling recommendations, and cost and schedule procedures. The book is practically oriented and popular as it appears already in a few years as 4th edition
Reniers et al. (2018)	This book on security risk assessment provides a quite comprehensive overview of the US legislation on physical security for the process industry and relevant assessment methods. The contents of this book is presented in Table 4
Hirsch, Gummersbach, and Bartsch (2016)	Is a Physical Separation of the SIS necessary? IEC61508 and 11 now require cyber risk analysis. Between safety and security are similarities and differences (security Standard IEC 62443). Options are air-gapped safety instrumented system (SIS), interfaced SIS, and zone 1 and 2 integration possibilities (SA-TR84.00.09-2013). Zone 1 means SIS and basic process control system (BPCS) are integrated providing greater communication between those systems and higher-level systems in the architecture. (No duplication of equipment is needed; cheaper). Zone 2 is similar, but SIS is not directly connected to the internet. A security risk assessment is required, also for the gap solution
Song, Khan, and Yang (2018)	The paper stresses the need of various risk integration and proposes an influence diagram (BN + decision + utility nodes); effect of measures; risk assessment/decision-making; An illustrative example is given of overfilling a gasoline storage tank. Probability values are obtained from literature and expert judgment

Reniers, Landucci, and Khakzad (2020)	Where process safety risk management has both consequence and likelihood of an unwanted event in focus, in case of physical security risk it is vulnerability assessment. On the other hand, security risk management has many other questions to solve. A number of risk management aspects, principles, approaches and models are compared
Kraidi, Shah, Matipa, and Borthwick (2021)	The paper describes the development of a risk mitigation method (RMM) for safety & security of oil pipelines in Iraq. Four risk management steps are distinguished: (1) risk identification and registration; (2) analysis; (3) RMM response; (4) monitoring & control. 30 Different risk factors were identified, among which security ones (Third-Party Disruption) and 12 RMMs. Expert judgment based on a questionnaire and a Likert scale triangular fuzzy number risk levels were estimated enabling risk ranking
George and Renjith (2021)	These authors provided a detailed description of BN applications in process risk assessment, including software platforms, validation, and sensitivity/criticality analysis. "pre-BN" assessment techniques of safety and physical and cyber security risk assessment are compared with application of BNs. A comparison with artificial neural networks was made and noted that BN includes causation and ANN only data. Petri net offers the possibility of dynamic modeling and feedbacks overcoming BN limitations
Ji et al. (2021)	This paper focuses on industrial control systems (ICS) or supervisory control and data acquisition (SCADA) systems forming so called Cyber Physical System (CPS). Systems become more complex, which also makes that safety and security become more interwoven. Risks should be assessed and managed, though, in an integrated way. First, attack route root cause models (ARMs) were collected from literature. Then, Physical (PSPR) safety prevention route and Cyber (CSPR) security prevention routes are analyzed, also safety critical variable analysis (SCVA) is proposed and all integrated in harmonized bowtie. Various attack types were simulated: DDOS, malware, phishing, etc., but as single attacks. The case was a coal-to-methanol plant. Future studies will treat simultaneous attacks
Kumar, Narra, Kela, and Singh (2022)	Condition-based monitoring is spreading in the industry. Attack-Fault Maintenance Tree (AFMT) is proposed as model to guide maintenance, repairs, and inspections, including safety and security dependencies, and answering reliability and cost questions. AFMT basic events are Basic Attack Steps (BAS, malicious) and Basic Component Failure (BCF, accidental); What-If scenario analysis drives. Reactive vs pro-active maintenance modules are built in. Security metrics are encoded in UPPAAL query language. Case study: oil pipeline in which based on sensor data a SCADA via a remote terminal unit (RTU) controlling valves and pumps, pressure will be kept below a threshold. Tool is available at http://afmt-simulator.com/

Table 4 Contents of Reniers et al. (2018).

Contents
Preface—v
List of Contributors—xi
1. Introduction—1
2. American legislation and regulatory measures: a lesson for Europe?—6
Hans Pasman
3. Security vulnerability analysis:
protecting process plants from physical and cyber threats—31
Paul Baybutt
4. Security risk assessment: Some techniques—75
Shailendra Bajpai, J. P. Gupta
5. A methodology for the evaluation of attractiveness with respect to external
acts of interference dedicated to the chemical and process industry—93
Gabriele Landucci, Francesca Argenti, Valerio Cozzani
6. Applying game theory for adversarial risk analysis in chemical plants—110
Laobing Zhang, Genserik Reniers
7. Dynamic security assessment: benefits and limitations—131
Nicola Paltrinieri, Cecilia Haskins
8. Security vulnerability assessment:
A review of Bayesian network approaches—142
Nima Khakzad
9. OR methods to enhance security in the chemical process industry—162
Luca Talarico, Genserik Reniers
10. Conclusions—185
Index—189

Table 5 Summary of papers describing *models* in consecutive years.

Kriaa, Bouissou, and Laarouchi (2019)	Authors propose for safety and security joint risk analysis for Industrial control systems (ICS), e.g., SCADA, S-cube (SCADA safety and security). Communications are TCP/IP based. Model requirements are Domain Specific Languages (DSLs), Cyber Security Modeling Language (CySeMol) and Multihost, multistage Vulnerability AnaLysis (MulVAl). Generic DSLs are Figaro and AltaRica; specific for electronics is Architecture Analysis & Design Language (AADL), which can generate FMEA and FT. ICSs monitor real-time and therefore need to have high availability and to be fault-tolerant. S-cube KB (Knowledge Base on Figaro language) contains many ICS component models exchanging data flows in the system architecture. Attacks are simulated and vulnerabilities identified. Safety metrics are TTF, MTTF, CCFs; Security metrics are TTS, MTTS via a Markovian TTC model; joining yields the Common Vulnerability Scoring System (CVSS). A tool chain is built. A case is demonstrated. S Cube KB result can improve the architecture
Song, Khan, and Yang (2019)	As in their PSP'19 paper presented under methods, the authors stress integration. Here, it is by applying BN. Case is an oil storage tank fire directly converted in BN from a FT taking into account CCFs. Probabilities were estimated by experts and historical data. A dynamic feature can be included
Shin et al. (2021)	STPA safety and security vulnerability analysis is compared with FMVEA, EFT, SAHARA, and CHASSIS on a NPP condensate water testbed. STPA SafSec is applied on the testbed to generate hazard scenarios, security analysis, and mitigation strategy. STPA SafSec is more complicated in use, user-unfriendly, not quantitative, but more effective to find interactions of security and safety than the alternatives, and hence, more useful
Carreras Guzman, Kozine, and Lundteigen (2021), Carreras Guzman, Zhang, Xie, and Glomsrud (2021)	Cyber security threats are now key factors in complex accident scenarios in CPSs. Overlap between safety and security analysis is still a loosely defined domain without theories and methods. Due to limitations earlier methods, UFoI-E (Uncontrolled Flows of Information and Energy) with causality, CPS master and CyPHASS, (extended bowtie, scenario ontologies, steps scheme, validation cases) has been developed. Case study is performed on an ICS. Limitation is the reductionist approach leading to missing interactions between parts

Continued

Table 5 Summary of papers describing *models* in consecutive years.—cont'd

Carreras Guzman, Kozine, and Lundteigen (2021), Carreras Guzman, Zhang, et al. (2021)	A comparison is made between STPA and the Uncontrolled Flows of Information and Energy UFoI S&S co-analysis based on a tailored CPS-master diagram. UFoI-E consists of fundamental theory for causality, a CPS model and a CyPHASS harm builder. A case study was performed on the ReVolt autonomous vessel prototype analyzed by an UFoI-E team (1) and a STPA team (2). Completeness and effort are evaluation criteria. Completeness was for both about 68%. Efforts (1) 45 h vs (2) 50. Completeness of risk scenarios compared to design specific causes is about the same, but scenarios identified are different. Combination of methods is the solution.
Zhang, Feng, Shi, and Srinivasan (2021)	Authors wrote a survey on Multi-Agent Systems (MASs) with independent sensing, communication, computing, and decision-making capabilities. Safety and security were analyzed and a synthesis made arising from physical faults and cyber-attacks. Results were gathered on fault estimation, detection, diagnosis and fault-tolerant-control, as well as cyber-attack detection and secure control of MASs subject to typical DDOS and Deception attack. Combination with Replay and Stealth attacks makes it more difficult. With AI at attacker side mutations will be possible to avoid detection. To counter one shall resort to game-theory Also, MASs may crash when subject to simultaneous physical and cyber threats. MASs are safety-critical; hence, malicious signal detection and secure consensus should be in finite-time. Focus of research is on detection and only little defense
Ferrari and Teixeira (2021)	This book contains much technically interesting information. In Table 6 the book table of contents is reproduced
Chen, Reniers, and Yang (2022)	This Springer book is asset vulnerability oriented, and contains eight sections: (1) Safety and security of domino effects in the process industry: the state of the art, (2) Dynamic risk assessment of fire induced domino effects, (3) Dynamic assessment of vapor cloud explosion induced domino effects, (4) Risk assessment of coupled hazardous scenarios, (5) Integrated safety and security management to tackle domino effects, (6) An economic approach for domino effect management, (7) A Resilience based approach for the prevention and mitigation of domino, (8) Conclusions and Future research.

Viksnin, Marinenkov, and Chuprov (2022)	CPS safety and security assurance is a challenging research problem for the "Smart" city and other concepts. Technical faults and malicious attack agents can disrupt. Each agent has a reputation indicator to show its trustworthiness. Mixed strategies Game theory is proposed to calculate Reputation and Trust-based indicator values. Agents with different strategies produce pay-off matrix. The case study consists of a group of unmanned vehicles to pass conflict-free an intersection. Attack scenarios are created by transmitting some incorrect data. Applying game theory increased the number of detected intruders
Kolb, Nicoletti, Peppelman, and Stoelinga (2022)	The paper surveys 14 model-based formalisms (methods) for risk assessment, among other. FT/AttackT, SAHARA, CHASSIS, and STAMP. It describes their principles and the interaction dependency of safety-security. The formalisms are tested with the lock-door example (fire but the emergency exit door is closed due to burglary). The majority of formalisms are FT/AttackT based, no novel models seem to have been introduced yet, safety-security interactions are still not well modeled, no novel metrics are proposed, no large case studies are conducted, various formalisms model safety-security interaction differently. These findings motivate future research. Here, the comment can be made that meanwhile from other 2022 papers appears that these needs are being addressed
Alanen et al. (2022)	Authors applied RAMSS (Reliability, Availability, Maintainability for Safety and Security) ontology as a model-based, data-driven, support for engineering mission-critical systems harmonizing basic concepts between dependability of safety and security to develop STA (security threat analysis) of ICS. The model is demonstrated on a nuclear fuel pool cooling control system. Ontology (ISO/IEC 19,763–1, 2015) is applied as proposed for HAZOP by Single, Schmidt, and Denecke (2020) in OWL (Ontology Web Language) (or UML) to enable automated reasoning mechanism. Model-based systems engineering (MBSE) is followed resulting in a RAMSS risk assessment ontology and related STA data model. The method is demonstrated on a nuclear fuel pool cooling control system

Table 6 Contents of Ferrari and Teixeira (2021)

1. Introduction to the Book—1
Riccardo M. G. Ferrari and André M. H. Teixeira
2. Fault Tolerance in Networked Control Systems by Flexible
Task Assignment—vv—9
Kai Schenk and Jan Lunze
3. Resilient Control Under Denial-of-Service: Results
and Research Directions—41
Claudio De Persis and Pietro Tesi
4. Stealthy False Data Injection Attacks in Feedback Systems
Revisited—61
Henrik Sandberg
5. Detection of Attacks in Cyber-Physical Systems: Theory
and Applications—79
Vaibhav Katewa, Cheng-Zong Bai, Vijay Gupta, and Fabio Pasqualetti
6. Security Metrics for Control Systems—99
André M. H. Teixeira
7. The Secure State Estimation Problem—123
Yasser Shoukry and Paulo Tabuada
8. Active Detection Against Replay Attack: A Survey on Watermark
Design for Cyber-Physical Systems—145
Hanxiao Liu, Yilin Mo, and Karl Henrik Johansson
9. Detection of Cyber-Attacks: A Multiplicative Watermarking
Scheme—173
Riccardo M. G. Ferrari and André M. H. Teixeira
10. Differentially Private Anomaly Detection for Interconnected
Systems—203
Riccardo M. G. Ferrari, Kwassi H. Degue, and Jerome Le Ny
11. Remote State Estimation in the Presence of an Eavesdropper—231
Alex S. Leong, Daniel E. Quevedo, Daniel Dolz, and Subhrakanti Dey

Table 6 Contents of Ferrari and Teixeira (2021)—cont'd

12. Secure Networked Control Systems Design Using
Semi-homomorphic Encryption—257
Yankai Lin, Farhad Farokhi, Iman Shames, and Dragan Nešić
13. Deception-as-Defense Framework for Cyber-Physical Systems—287
Muhammed O. Sayin and Tamer Başar
14. Cyber-Risk: Cyber-Physical Systems Versus Information
Technology Systems—319
Carlos Barreto, Galina Schwartz, and Alvaro A. Cardenas
15. Cyber-Insurance—347
Carlos Barreto, Galina Schwartz, and Alvaro A. Cardenas
16. Concluding Remarks and Future Outlook—377
Riccardo M. G. Ferrari

3.7.1 Acronyms in use in Section 3

AI Artificial Intelligence
BBN, also BN Bayesian (Belief) Networks
BDMP Boolean logic Driven Markov Processes
Bowtie is no acronym but the combination of a fault and an event tree connected by a critical (release) event.
CCF Common Cause Failure
CHASSIS combined harm assessment for safety and security for information systems
CHESS Cluster-thinking and intensified cooperation, High transparency and efficient inspection, Education, training and learning, Security integration, Safety innovation and dynamic risk assessments
CPS Cyber-Physical System
CRUTIAL CRitical Utility InfrastructurAL resilience
CyPHASS Cyber-Physical Harm Analysis for Safety and Security
EFT Extended Fault Tree
EMF Eclipse Modeling Framework
FMEA Failure Mode and Effect Analysis
FMVEA Failure Mode Vulnerability and Effect Analysis
FT Fault Tree
HAZOP Hazard and Operability study
LOPA Layer of Protection Analysis
MBSE Model-Based System Engineering
MTTF Mean Time to Failure
MTTS Mean Time to Success exploiting a vulnerability

NPP Nuclear Power Plant
SAHARA Security-Aware Hazard Analysis and Risk Assessment
SCADA Supervisory Control and Data Acquisition
SIL Safety Integrity Level
SIS Safety Instrumented System
SPN Stochastic Petri Nets
STAMP System-Theoretic Accident Model and Processes
STPA-sec System-Theoretic Process Analysis - Security
SysML System Modeling Language
TCP/IP Transmission Control Protocol/Internet Protocol
TTC Time to Compromise
TTF Time to Failure
TTS Time to Success exploiting a vulnerability
UFoI Uncontrolled Flows of Information and Energy
UML Unified Modeling Language has been building with EMF a tool in the EU CARISMA project.
UPPAAL Uppsala—Aalborg universities software product: Integrated tool environment for modeling, validation and verification of real-time systems modeled as networks of timed automata, extended with data types (bounded integers, arrays, etc.)

4. Economics of risk reduction; applications of LFs

The goal of implementing safety measures in operations is to reduce existing and potential or future risks. Preventive measures aim at reducing the likelihood of occurrence while the mitigative measures aim at reducing the consequences of an accident.

Operational safety and risk reduction measures come at a cost. Direct and indirect costs involved in it are assessed by managers who are mandated to meet financial targets dictated by stakeholders. These measures and accident scenarios are inherently hypothetical, uncertain in nature, and subjected to scarce if not missing data. Given these circumstances, managers found themselves in a dilemma searching for conclusive data to support their decision to invest or not invest in a particular risk reduction measure when recommended. This is where techniques such as loss functions can be used to model the financial loss resulting from the deviations from the target values.

4.1 Overview and recent developments

Loss functions were first introduced by Genichi Taguchi, a Japanese statistician who wanted to give a graphical depiction of loss in a finance. This was latter known as Taguchi's loss function. The concept that

Table 7 Major contributions and applications of loss functions.

Domain	Related work
Economics and business analysis	Spiring (1993), Kapur and Cho (1996), Leung and Spiring (2004), and Kulkarni (2008)
Process safety and risk analysis	Sii, Ruxton, and Wang (2001), Pan and Chen (2012), Ali, Aslam, and Kazmi (2013), Yu (2017), and Abimbola and Khan (2018)
Reliability assessment	Murphy, Tsui, and Allen (2005), Dey (2012), and Afenyo, Jiang, and Ng (2021)
Logistics and supply chain management	Kulkarni (2008, 2008) and Ordoobadi (2009)

Taguchi introduced was against the common concept of quality where a product either has the desired quality (i.e., product approved) or does not have it (i.e., product rejected). Taguchi considered that the quality gets reduced progressively with respect to the amount of deviation from the target value. Therefore, there is a cost for any finite deviation (Hashemi et al., 2014) (Table 7).

4.2 Applications of loss functions

The application of loss functions in process safety assessment improves the performance of chemical process systems against various disturbances. A disturbance in a chemical process system causes deviations to its process parameters and consequently lead to potential losses. To investigate the effects of process deviations on operational safety, Hashemi, Ahmed, and Khan (2014) proposed the application of loss functions to monitor process safety performance and improve the safety of these process plants. The authors laid out the development of loss functions into four stages:

- In the first stage, identification of the target value and how to represent the deviation from it. Thus, these loss function models aim to estimate the deviations of the predicted results from the target values. They are often used in classification and regression.
- The second stage focuses on modeling of the deviation from a target value and the definition of estimated maximum loss. The maximum loss is first introduced in loss functions to represent the worst situation, in which a maximum acceptable deviation occurs. These types of loss functions are often used in quality management.

- In the third stage, the occurrence of a random deviation is considered and modeled using different types of probability distributions. Between the target value and the maximum loss value, the random changes of values between these two points are connected to a certain type of probability distribution.
- The fourth stage investigates the use of multiple variables in loss functions while only single variable is considered in the previous three stages. Probability distributions are used to model the inherent feature of each variable and the dependency between the variables was considered (Khan, Wang, & Yang, 2016).

The square and absolute functions are simple ways to quantify losses associated with deviations from the predetermined target values. Traditionally, losses are quantified as squared error, absolute error, weighted and zero–one losses (Leung & Spiring, 2004):

- **Squared error loss function**

The squared function is a type of non-negative function, widely used in determining distance from two values. Consequently, the squared error loss function is used to describe the deviation from the target value.

$$L(y) = (y - T)^{2.}$$

- **Absolute error loss function**

Another type of widely used non-negative function is the absolute function. It is used to describe the deviation between current and target values.

$$L(y) = |\mathrm{y} - \mathrm{T}|$$

- **Weighted loss function**

The weighted loss is a supplement to squared error loss, when there are more than one current and target values. Different weights are assigned to the target values. The weighted loss functions are in the formulism as below.

$$L(y) = \Sigma w_i (y_i - T_i)^2$$

- **Zero–one loss function**

The zero–one loss function is a frequently used loss function in statistics and decision theory.

$$L(y) = 0, if\ y = T$$
$$L(y) = 1, if\ y \neq T$$

The zero–one loss function is used to describe the classification results: when the classification gives the right result, the loss is zero; otherwise, it is one.

The above-mentioned loss functions describe the distance from the target value, but in many cases, there is an upper limit of the loss, i.e., the estimated maximum loss. To better indicate the loss and consider the maximum estimated loss, polynomial-based loss functions were developed to quantify losses and the maximum loss.

4.3 Loss quantification of a process system

The use of loss functions to estimate the consequence of a process at an end process scenario is demonstrated. A loss function usually describes the deviation from only few variables; while there are more variables need to be considered in the large system. As a result, the loss of process is decomposed into different variables groups. Within these variables' groups, the approximate loss functions are applied to determine the deviations and calculate these losses accordingly. Then, with properly designed strategy, these local losses will compound and represent the losses of the whole process. A chemical process system has many units. Since different types of units are to be monitored in a process, appropriate types of inverted probability distribution functions need to be selected to estimate the potential losses:

- For units only requiring the upper boundary of an operating variable: the inverted normal loss function is used.
- For units requiring both the upper and lower boundaries of an operating variable: the inverted Beta loss function is used.
- For units requiring multivariate monitoring: the multivariate inverted loss function is used.

Using these loss functions, the losses at an end process scenario can be estimated. The losses are estimated continuously, with every state of deviations from the target values.

In a process system, the key variable is denoted as $X(t)$. Among all these monitored units, m of them are unique variable described ones, that is, there is only one key variable applied to monitor the unit. At the same time, there are other n units where more than one key variable is monitored, furthermore, the numbers of monitored key variables were not all the same in these units. In this situation, the estimated loss will be replaced by the maximum value from all calculated losses. As a result, the loss function of system will be formulated as follow (Khan, Wang, & Yang, 2016):

$$L(X(t^*)) = \sum_{i=1}^{m} L_i(x_i(t^*)) + \sum_{i=1}^{n} \max\left(L_{j,k}\left(x_{j,k}(t^*)\right)\right)$$

In this formulation, t^* denotes certain moment during process operating, $X(t^*)$ denotes all variables values at that moment. $L_i(x_i(t^*))$ represents the loss of unit with only one monitored variable, and $L_{j,\ k}(x_{j,\ k}(t^*))$ represent the loss of unit with k monitored variables.

5. Progress and future opportunities

Risk assessment integration encompasses inclusion of dynamics, process safety and security assessment, and possible loss estimation in relation to risk reducing measures installed. From the foregoing can be concluded that progress in this field is stormy. Hence, it is not predictable where it will end. One thing is sure, namely, precision and accurateness of risk predictions of process operations real time and online will be realized in the future. It will affect the whole socio-technical system and it will support the resiliency of enterprises.

Interaction between security attacks and process safety can be complex. Analysis to prevent and protect requires also new methods.

The field of accident loss functions related to process industry is rather young and will undoubtedly get more also more attention in the future, where it can be expected to combine with resilience assessment.

References

Abimbola, M., & Khan, F. (2018). Dynamic blowout risk analysis using loss functions. *Risk Analysis*, *38*(2), 255–271.

Adedigba, S. A., Khan, F., & Yang, M. (2017). Dynamic failure analysis of process systems using neural networks. *Process Safety and Environmental Protection*, *111*, 529–543. https://doi.org/10.1016/j.psep.2017.08.005.

Afenyo, M., Jiang, C., & Ng, A. K. (2021). A Bayesian-loss function model for assessing marine liability regime for ship-source spills. *SSRN Electronic Journal*, *3994309*. https://doi.org/10.2139/ssrn.3994309.

Alanen, J., Linnosmaa, J., Malm, N., Papakonstantinou, T., Ahonen, T., Heikkilä, E., et al. (2022). Hybrid ontology for safety, security, and dependability risk assessments and security threat analysis (STA) method for industrial control systems. *Reliability Engineering and System Safety*, *220*, 108270. https://doi.org/10.1016/j.ress.2021.108270.

Ali, S. (2015). Mixture of the inverse Rayleigh distribution: Properties and estimation in a Bayesian framework. *Applied Mathematical Modelling*, *39*(2), 515–530. https://doi.org/10.1016/j.apm.2014.05.039.

Ali, S., Aslam, M., & Kazmi, S. M. A. (2013). A study of the effect of the loss function on bayes estimate, posterior risk and hazard function for Lindley distribution. *Applied Mathematical Modelling*, *37*(8), 6068–6078.

Bieder, C., & Pettersen Gould, K. (Eds.). (2020). *The coupling of safety and security: Exploring interrelations in theory and practice* FonCSI (Foundation for an Industrial Safety Culture, Toulouse) coordinated Springer. Open access https://doi.org/10.1007/978-3-030-47229-0.

Boustras, G., & Waring, A. (2020). Towards a reconceptualization of safety and security, their interactions, and policy requirements in a 21st century context. *Safety Science, 132*, 104950. https://doi.org/10.1016/j.ssci.2020.104950.

Busquim e Silva, R. A., Piqueira, J. R. C., Cruz, J. J., & Marques, R. P. (2021). Cybersecurity assessment framework for digital Interface between safety and security at nuclear power plants. *International Journal of Critical Infrastructure Protection, 34*, 100453. https://doi.org/10.1016/j.ijcip.2021.100453.

Carreras Guzman, N. H., Kozine, I., & Lundteigen, M. A. (2021). An integrated safety and security analysis for cyber-physical harm scenarios. *Safety Science, 144*, 105458. https://doi.org/10.1016/j.ssci.2021.105458.

Carreras Guzman, N. H., Zhang, J., Xie, J., & Glomsrud, J. A. (2021). A comparative study of STPA-extension and the UFoI-E method for safety and security co-analysis. *Reliability Engineering and System Safety, 211*, 107633. https://doi.org/10.1016/j.ress.2021.107633.

CFATS. (2014). *Title XXI – Chemical facility anti–terrorism standards, public law 113–254.* December 18, 2014. 128 STAT. 2898. https://www.congress.gov/113/plaws/publ254/PLAW-113publ254.pdf.

Chen, C., Li, C., Reniers, G., & Yang, F. (2021). Safety and security of oil and gas pipeline transportation: A systematic analysis of research trends and future needs using WoS. *Journal of Cleaner Production, 279*, 123583. https://doi.org/10.1016/j.jclepro.2020.123583.

Chen, C., Reniers, G., & Yang, M. (2022). *Integrating safety and security management to protect chemical industrial areas from domino effects*. Springer Series in Reliability Engineering. ISBN 978-3-030-88910-4, ISBN 978-3-030-88911-1 (eBook). https://doi.org/10.1007/978-3-030-88911-1.

Deoss, D. L., Jr. (1989). *A simulation model for dynamic system availability analysis.* Massachusetts Institute of Technology.

Dey, S. (2012). Bayesian estimation of the parameter and reliability function of an inverse Rayleigh distribution. *Malaysian Journal of Mathematical Sciences, 6*(1), 113–124.

Dima, A. (2021). *Short history of manufacturing: From industry 1.0 to industry 4.0.* https://kfactory.eu/short-history-of-manufacturing-from-industry-1-0-to-industry-4-0/.

Fei, T., Cheng, J., Qi, Q., Zhang, M., Zhang, H., & Sui, F. (2018). Digital twin-driven product design, manufacturing and service with big data. *The International Journal of Advanced Manufacturing Technology, 94*(9–12), 3563–3576.

Ferrari, R. M. G., & Teixeira, A. M. H. (2021). Safety, security and privacy for cyber-physical systems. In *Lecture notes in control and information sciences 486*, Springer, ISBN:978-3-030-65047-6. https://doi.org/https://doi.org/10.1007/978-3-030-65048-3.

Ge, Z., & Song, Z. (2013). Distributed PCA model for plant-wide process monitoring. *Industrial and Engineering Chemistry Research, 52*(5), 1947–1957. https://doi.org/10.1021/ie301945s.

George, P. G., & Renjith, V. R. (2021). Evolution of safety and security risk assessment methodologies towards the use of Bayesian networks in process industries. *Process Safety and Environmental Protection, 149*, 758–775. https://doi.org/https://doi.org/10.1016/j.psep.2021.03.031.

Gheraibia, Y., Kabir, S., Aslansefat, K., Sorokos, I., & Papadopoulos, Y. (2019). Safety + ai: A novel approach to update safety models using artificial intelligence. *IEEE Access, 7*, 135855–135869. https://doi.org/10.1109/ACCESS.2019.2941566.

Glesner, C., Van Oudheusden, M., Turcanu, C., & Fallon, C. (2020). Bringing symmetry between and within safety and security cultures in highrisk organizations. *Safety Science, 132*, 104950. https://doi.org/10.1016/j.ssci.2020.104950.

Goel, P., Jain, P., Pasman, H. J., Pistikopoulos, E. N., & Datta, A. (2020). Integration of data analytics with cloud services for safer process systems, application examples and implementation challenges. *Journal of Loss Prevention in the Process Industries, 104316.* https://doi.org/10.1016/j.jlp.2020.104316.

Gou, X., Liu, H., Qiang, Y., Lang, Z., Wang, H., Ye, D., et al. (2022). In-depth analysis on safety and security research based on system dynamics: A bibliometric mapping approach-based study. *Safety Science, 147*, 105617. https://doi.org/10.1016/j.ssci.2021.105617.

Guo, W., Wu, Z., Zhang, F., & Wu, J. (2020). Scheduling sequence control method based on sliding window in cyberspace mimic defense. *IEEE Access, 8*, 1517–1533. https://doi.org/10.1109/ACCESS.2019.2961644.

Hashemi, S. J., Ahmed, S., & Khan, F. I. (2014). Risk-based operational performance analysis using loss functions. *Chemical Engineering Science, 116*, 99–108. https://doi.org/10.1016/j.ces.2014.04.042.

Heinrich, H. (1941). *Industrial accident prevention, A scientific approach* (2nd ed.). London: McGraw-Hill Book Company.

Hirsch, V, Gummersbach, J.-L., & Bartsch, T. (2016). Safety & Security - Is a Physically Separation of the SIS Necessary? *Chemical Engineering Transactions, 48*, 673–678. https://doi.org/10.3303/CET1648113.

Hydyrov, A. (2017). Governance responsibilities and security and safety management systems: Does corporate governance support security and safety management systems? *Contemporary Readings in Law and Social Justice, 9*(2), 360–365, (ISSN 1948-9137, eISSN 2162-2752).

Iaiani, M., Casson Moreno, V., Reniers, G., Tugnoli, A., & Cozzani, V. (2021). Analysis of events involving the intentional release of hazardous substances from industrial facilities. *Reliability Engineering and System Safety, 212*, 107593. https://doi.org/10.1016/j.ress.2021.107593.

Ji, Z., Yang, S., Cao, Y., Wang, Y., Zhou, C., Yue, L., et al. (2021). Harmonizing safety and security risk analysis and prevention in cyber-physical systems. *Process Safety and Environmental Protection, 148*, 1279–1291. https://doi.org/10.1016/j.psep.2021.03.004.

Jiang, Q., & Yan, X. (2015). Nonlinear plant-wide process monitoring using MI-spectral clustering and Bayesian inference-based multiblock KPCA. *Journal of Process Control, 32*, 38–50. https://doi.org/10.1016/j.jprocont.2015.04.014.

Johnson, W. G. (1973). *The management oversight and risk tree – MORT.* (Including Systems Developed by the Idaho Operations Office and Aerojet Nuclear Company. U.S. Atomic Energy Commission SAN 821–822 UC-41).

Kalantarnia, M., Khan, F., & Hawboldt, K. (2009). Dynamic risk assessment using failure assessment and Bayesian theory. *Journal of Loss Prevention in the Process Industries, 22*(5), 600–606. https://doi.org/10.1016/j.jlp.2009.04.006.

Kapur, K. C., & Cho, B. R. (1996). Economic design of the specification region for multiple quality characteristics. *IIE Transactions, 28*(3), 237–248.

Khakzad, N. (2019). Modeling wildfire spread in wildland-industrial interfaces using dynamic Bayesian network. Reliability Engineering and System Safety, 189(March), 165–176. https://doi.org/10.1016/j.ress.2019.04.006.

Khakzad, N., Khan, F., & Amyotte, P. (2012a). Dynamic risk analysis using bow-tie approach. *Reliability Engineering & System Safety, 104*, 36–44.

Khakzad, N., Khan, F., & Amyotte, P. (2012b). Dynamic safety analysis of process systems by mapping bow-tie into Bayesian network. *Process Safety and Environmental Protection, 91*(1–2), 46–53. https://doi.org/10.1016/j.psep.2012.01.005.

Khakzad, N., Khan, F., & Amyotte, P. (2013). Quantitative risk analysis of offshore drilling operations: A Bayesian approach. *Safety Science, 57*, 108–117. https://doi.org/10.1016/j.ssci.2013.01.022.

Khan, F., Wang, H., & Yang, M. (2016). Application of loss functions in process economic risk assessment. *Chemical Engineering Research and Design, 111*, 371–386.

Kolb, C., Nicoletti, S. M., Peppelman, M., & Stoelinga, M. (2022). Model-based safety and security co-analysis: Survey and identification of gaps. *ArXiv*. https://doi.org/10.48550/arXiv.2106.06272.

Kraidi, L., Shah, R., Matipa, W., & Borthwick, F. (2021). An investigation of mitigating the safety and security risks allied with oil and gas pipeline projects. *Journal of Pipeline Science and Engineering, 1*, 349–359. https://doi.org/10.1016/j.jpse.2021.08.002.

Kriaa, S., Bouissou, M., & Laarouchi, Y. (2019). A new safety and security risk analysis framework for industrial control systems. *Proceedings of the Institution of Mechanical Engineers, Part O: Journal of Risk and Reliability, 233*(2), 151–174. https://doi.org/10.1177/1748006x18765885. hal-02274615. SAGE Publications, In press.

Kriaa, S., Pietre-Cambacedes, L., Bouissou, M., & Halgand, Y. (2015). A survey of approaches combining safety and security for industrial control systems. *Reliability Engineering and System Safety, 139*, 156–178. https://doi.org/10.1016/j.ress.2015.02.008.

Kukfisz, B., Kuczy'nska, A., Piec, R., & Szykuła-Piec, B. (2022). Research on the safety and security distance of above-ground liquefied gas storage tanks and dispensers. *International Journal of Environmental Research and Public Health, 19*, 839–850. https://doi.org/10.3390/ijerph19020839.

Kulkarni, S. S. (2008). Loss-based quality costs and inventory planning: General models and insights. *European Journal of Operational Research, 188*(2), 428–449.

Kumar, R., Narra, B., Kela, R., & Singh, S. (2022). AFMT: Maintaining the safety-security of industrial control systems. *Computers in Industry, 136*, 103584. https://doi.org/10.1016/j.compind.2021.103584.

Kurian, D., Sattari, F., Lefsrud, L., & Ma, Y. (2020). Using machine learning and keyword analysis to analyze incidents and reduce risk in oil sands operations. *Safety Science, 130*(May), 104873. https://doi.org/10.1016/j.ssci.2020.104873.

Leung, B. P., & Spiring, F. A. (2004). Some properties of the family of inverted probability loss functions. *Quality Technology & Quantitative Management, 1*(1), 125–147.

Mannan, S. (2014). *Hazard identification, assessment and control lees ' process safety essentials* (1st ed.). Elsevier Inc., ISBN:978-1-85617-776-4.

MARV. Multi-analytic risk visualization tool - MARV™. (n.d.). https://www.dnvgl.com/oilgas/pipeline-evaluation-portal/multi-analytic-risk-visualization-tool-marv.html.

Meel, A., & Seider, W. D. (2006). Plant-specific dynamic failure assessment using Bayesian theory. *Chemical Engineering Science, 61*(21), 7036–7056. https://doi.org/10.1016/j.ces.2006.07.007.

Meel, A., & Seider, W. D. (2008). Real-time risk analysis of safety systems. *Computers and Chemical Engineering, 32*(4–5), 827–840. https://doi.org/10.1016/j.compchemeng.2007.03.006.

Modarres, M. (2006). *Risk analysis in engineering: Techniques, tools, and trends*. CRC press, (ISBN-13: 978-1574447941).

Murphy, T. E., Tsui, K. L., & Allen, J. K. (2005). A review of robust design methods for multiple responses. *Research in Engineering Design, 15*(4), 201–215.

Nolan, D. P. (2015). *Safety and security review for the process industries; application of HAZOP, PHA, what-IF and SVA reviews* (4th Ed). Elsevier, ISBN:978-0-323-32295-9.

Obama, B. (2013). *Improving chemical facility safety and security*. (Executive Order 13650 of August 1, 2013. In Federal Register /Vol. 78, No. 152 /Wednesday, August 7, 2013 / Presidential Documents 48029).

OECD. (2018). Euratom competence in the areas of nuclear security and nuclear safety: An impossible parallel? Author Popov, A. Nuclear Law Bulletin No. 101/VOL. 2018/2, ISSN 1609-7378 https://inis.iaea.org/collection/NCLCollectionStore/_Public/50/020/50020742.pdf?r=1A.

Ordoobadi, S. (2009). Application of Taguchi loss functions for supplier selection. *Supply Chain Management: An International Journal, 14*(1), 22–30. https://doi.org/10.1108/13598540910927278.

Paltrinieri, N., & Hokstad, P. (2014). Dynamic risk assessment: Development of a basic structure. In T. Novakowski, et al. (Eds.), *Safety and reliability: Methodology and applications* (pp. 1421–1428). CRC Press. (ISBN-13: 978-1138026810).

Paltrinieri, N., Khan, F., Amyotte, P., & Cozzani, V. (2014). Dynamic approach to risk management: Application to the Hoeganaes metal dust accidents. *Process Safety and Environmental Protection, 92*(6), 669–679. https://doi.org/10.1016/j.psep.2013.11.008.

Paltrinieri, N., Scarponi, G. E., Khan, F., & Hauge, S. (2014). Addressing dynamic risk in the petroleum industry by means of innovative analysis solutions. *Chemical Engineering Transactions, 36*, 451–456. https://doi.org/10.3303/CET1436076.

Paltrinieri, N., Tugnoli, A., Buston, J., Wardman, M., & Cozzani, V. (2013a). Dynamic procedure for atypical scenarios identification (DyPASI): A new systematic HAZID tool. *Journal of Loss Prevention in the Process Industries, 26*(4), 683–695. https://doi.org/10.1016/j.jlp.2013.01.006.

Paltrinieri, N., Tugnoli, A., Buston, J., Wardman, M., & Cozzani, V. (2013b). DyPASI methodology: From information retrieval to integration of HAZID process. *Chemical Engineering Transactions, 32*, 433–438. https://doi.org/10.3303/CET1332073.

Pan, J. N., & Chen, S. C. (2012). A new approach for assessing the correlated risk. *Industrial Management & Data Systems, 112*, 1348–1365.

Pan, J. N., & Chen, S. C. (2013). A loss-function based approach for evaluating reliability improvement of an engineering design. *Expert Systems with Applications, 40*(14), 5703–5708. https://doi.org/10.1016/j.eswa.2013.04.032.

Pariyani, A., Seider, W. D., Oktem, U. G., & Masoud, S. (2012a). Dynamic risk analysis using alarm databases to improve process safety and product quality: Part I—Data compaction. *AIChE Journal, 59*(4), 215–228. https://doi.org/10.1002/aic.

Pariyani, A., Seider, W. D., Oktem, U. G., & Masoud, S. (2012b). Dynamic risk analysis using alarm databases to improve process safety and product quality: Part II—Bayesian analysis. *AIChE Journal, 59*(4), 215–228. https://doi.org/10.1002/aic.

Pasman, H. J. (2017). American legislation and regulatory measures: A lesson for Europe? In G. Reniers, N. Khakzad, & P. Van Gelder (Eds.), *Security risk assessment in the chemical and process industry, chapter 2* (pp. 6–30). De Gruyter Inc. https://doi.org/https://doi.org/10.1515/9783110499087-002.

Pasman, H. J., & Fabiano, B. (2021). The Delft 1974 and 2019 European loss prevention symposia: Highlights and an impression of process safety evolutionary changes from the 1st to the 16th LPS. *Process Safety and Environmental Protection, 147*, 80–91. https://doi.org/10.1016/j.psep.2020.09.024.

Pasman, H., & Rogers, W. (2014). How can we use the information provided by process safety performance indicators? Possibilities and limitations. *Journal of Loss Prevention in the Process Industries, 30*(1), 197–206. https://doi.org/10.1016/j.jlp.2013.06.001.

PHA Pro. (2020). https://sphera.com/pha-pro/.

Reniers, G., Khakzad, N., & Van Gelder, P. (Eds.). (2018). *Security Risk Assessment in the Chemical and Process Industry* De Gruyter, Inc. https://doi.org/10.1515/9783110499087-002.

Reniers, G., & Khakzad, N. (2017). Revolutionizing safety and security in the chemical and process industry: Applying the CHESS concept. *Journal of Integrated Security Science, 1*(1), 2–15.

Reniers, G., Landucci, G., & Khakzad, N. (2020). What safety models and principles can be adapted and used in security science? *Journal of Loss Prevention in the Process Industries, 64*, 104068. https://doi.org/10.1016/j.jlp.2020.104068.

Shin, J., Choi, J.-G., Lee, J.-W., Lee, C.-K., Song, J.-G., & Son, J.-Y. (2021). Application of STPA-SafeSec for a cyber-attack impact analysis of NPPs with a condensate water system test-bed. *Nuclear Engineering and Technology*, *53*, 3319–3326. https://doi.org/10.1016/j.net.2021.04.031.

Sii, H. S., Ruxton, T., & Wang, J. (2001). Taguchi concepts and their applications in marine and offshore safety studies. *Journal of Engineering Design*, *12*(4), 331–358.

Single, J. I., Schmidt, J., & Denecke, J. (2020). Ontology-based computer aid for the automation of HAZOP studies. *Journal of Loss Prevention in the Process Industries*, *68*, 104321. https://doi.org/10.1016/j.jlp.2020.104321.

Song, G., Khan, F. I., & Yang, M. (2018). Integrated risk management of hazardous processing facilities. *Process Safety Progress*, *48*(1), 42–50. https://doi.org/10.1002/prs.11978.

Song, G., Khan, F. I., & Yang, M. (2019). Probabilistic assessment of integrated safety and security related abnormal events: A case of chemical plants. *Safety Science*, *113*, 115–125. https://doi.org/10.1016/j.ssci.2018.11.004.

Spiring, F. A. (1993). The reflected normal loss function. *The Canadian Journal of Statistics*, *21*(3), 321–330.

Taleb- Berrouane, M., & Lounis, Z. (2016). Safety assessment of flare systems by fault tree analysis. *Journal of Chemical Technology and Metallurgy*, *51*(2) (ISSN 1314-7471).

Taleb-Berrouane M., Khan, F. (n.d.). Bayesian stochastic petri nets (BSPN) - A hybrid formalism for dynamic safety analysis. 1–32.

Taleb-Berrouane, M., Khan, F., & Hawboldt, K. (2021). Corrosion risk assessment using adaptive bow-tie (ABT) analysis. *Reliability Engineering & System Safety*, *214*(May), 107731. https://doi.org/10.1016/j.ress.2021.107731.

Taleb-Berrouane, M., Khan, F., & Kamil, M. Z. (2019). Dynamic RAMS analysis using advanced probabilistic approach. *Chemical Engineering Transactions*, 77, 241–246. https://doi.org/10.3303/CET1977041.

Taleb-Berrouane, M., Khan, F., & Lounis, Z. (2016). Availability analysis of safety critical systems using advanced fault tree and stochastic petri net formalisms. *Journal of Loss Prevention in the Process Industries*, *44*, 193–203. https://doi.org/10.1016/j.jlp.2016.09.007.

Taleb-Berrouane, M., Khan, F., Hawboldt, K., Eckert, R., & Skovhus, T. L. (2018). Model for microbiologically influenced corrosion potential assessment for the oil and gas industry and gas industry. *Corrosion Engineering, Science and Technology*, *53*(5), 378–392. https://doi.org/10.1080/1478422X.2018.1483221.

Taleb-Berrouane, M., Sterrahmane, A., Mehdaoui, D., & Lounis, Z. (2017). Emergency response plan assessment using Bayesian belief networks. In *3rd Workshop and Symposium on Safety and Integrity Management of Operations in Harsh Environments (C-RISE3)*. https://www.mun.ca/engineering/crise/workshops/crise3/CRISE3_Program_Final.pdf.

Taleb-Berrrouane, M. (2019). *Dynamic corrosion risk assessment in the oil and gas production and processing facility (issue October)*. Memorial University of Newfoundland. (https://research.library.mun.ca/14349/).

TapRoot - Root Cause Analysis. (n.d.). https://www.taproot.com/.

Viksnin, I. I., Marinenkov, E. D., & Chuprov, S. S. (2022). A game theory approach for communication security and safety assurance in cyber-physical systems with reputation and trust-based mechanisms. *Scientific and Technical Journal of Information Technologies, Mechanics and Optics*, *22*(1), 47–59. https://doi.org/10.17586/2226-1494-2022-22-1-47-59.

Villa, V., Paltrinieri, N., Khan, F., & Cozzani, V. (2016). Towards dynamic risk analysis: A review of the risk assessment approach and its limitations in the chemical process industry. *Safety Science*, *89*, 77–93. https://doi.org/10.1016/j.ssci.2016.06.002.

Vinnem, J. E., Bye, R., Gran, B. A., Kongsvik, T., Nyheim, O. M., Okstad, E. H., et al. (2012). Risk modelling of maintenance work on major process equipment on offshore petroleum installations. *Journal of Loss Prevention in the Process Industries*, *25*(2), 274–292. https://doi.org/10.1016/j.jlp.2011.11.001.

Watson, H. A. (1961). *Launch control safety study, section VII, vol 1*. Murray Hill, NJ: Bell Labs.

Wu, J. (2021). Cyberspace endogenous safety and security. *Engineering*. https://doi.org/10.1016/j.eng.2021.05.015.

Yu, H. (2017). Dynamic risk assessment of complex process operations based on a novel synthesis of soft-sensing and loss function. *Process Safety and Environmental Protection*, *105*, 1–11.

Zadakbar, O., Imtiaz, S., & Khan, F. (2013). Dynamic risk assessment and fault detection using principal component analysis. *Process Safety Progress*, *32*(4), 365–375. https://doi.org/10.1002/prs.11609.

Zhang, D., Feng, G., Shi, Y., & Srinivasan, D. (2021). Physical safety and cyber security analysis of multi-agent systems: A survey of recent advances. *IEEE/CAA Journal of Automatica Sinica*, *8*(2), 319–333. https://doi.org/10.1109/JAS.2021.1003820.

Zhou, J., & Reniers, G. (2016). Petri-net based simulation analysis for emergency response to multiple simultaneous large-scale fires. *Journal of Loss Prevention in the Process Industries*, *40*, 554–562. https://doi.org/10.1016/j.jlp.2016.01.026.

Zhou, J., & Reniers, G. (2017). Petri-net based cascading effect analysis of vapor cloud explosions. *Journal of Loss Prevention in the Process Industries*, *48*, 118–125. https://doi.org/10.1016/j.jlp.2017.04.017.

Further reading

Kamil, M. Z., Taleb-Berrouane, M., Khan, F., & Amyotte, P. (2021). Data-driven operational failure likelihood model for microbiologically influenced corrosion. *Process Safety and Environmental Protection*, *153*, 472–485. https://doi.org/10.1016/j.psep.2021.07.040.

Khan, F., Hashemi, S. J., Paltrinieri, N., Amyotte, P., Cozzani, V., & Reniers, G. (2016). Dynamic risk management: A contemporary approach to process safety management. In *Vol. 14. Current opinion in chemical engineering* (pp. 9–17). Elsevier Ltd. https://doi.org/10.1016/j.coche.2016.07.006.

Legal Information Institute. (2001). *Critical infrastructures protection act of 2001. 42 U. S. Code § 5195c – Critical infrastructures protection*. https://www.law.cornell.edu/uscode/text/42/5195c.

Taleb-Berrouane, M., Khan, F., & Amyotte, P. (2020). Bayesian stochastic petri nets (BSPN) - A new modelling tool for dynamic safety and reliability analysis. *Reliability Engineering and System Safety*, *193*. https://doi.org/10.1016/j.ress.2019.106587.

CHAPTER SIXTEEN

Use of digital twins for process safety management

Arvind Keprate* and Nikhil Bagalkot*
Department of Mechanical, Electronics and Chemical Engineering, Oslo Metropolitan University, Norway
*Corresponding authors: e-mail address: arvindke@oslomet.no; nikhilba@oslomet.no

Contents

1. Introduction to digital twin

In the past two decades, the concept of Digital Twin (DT) has found numerous applications across different industries such as aerospace, manufacturing, automotive, healthcare, process, etc. The concept was first introduced in the field of aircraft and aerospace as an information mirroring model for spacecraft state simulation and to obtain accurate data for decision-making assistance. The first definition of the DT was postulated by Grieves as a "*Physical product in real space, virtual product in virtual space and the connection of data and information that ties the two spaces together.*" Ever since then, numerous definitions of DT have existed in the literature arising particularly due to the usage of DTs in different industries. In context to defense and aerospace industries DT is defined as "*an integrated multi-physics, multiscale, probabilistic simulation of an as-built system, enabled by Digital Thread, that uses the best available models, sensor information, and input data to mirror and predict activities/performance over the life of its corresponding*

Methods in Chemical Process Safety, Volume 6
ISSN 2468-6514
https://doi.org/10.1016/bs.mcps.2022.05.001

physical twin." (West & Blackburn, 2017). The main focus in the aforementioned definition was on core concepts of structural mechanics, material science, and estimating the long-term behavior of aircrafts. However, for the manufacturing industry, besides system performance prediction, DT also ensures continuous information transmission amongst various phases of product lifecycle, based on computer-aided simulations (Haag & Anderl, 2018). The main goal of using DT in the process industry is for online operational monitoring, predictive maintenance, leakage detections, dynamic process optimization, hazard identification, and operator training (Lee, Cameron, & Hassall, 2019). Thus, in context to process industry DT can be defined as "*as a virtual representation of a physical asset enabled through data and simulators for real-time prediction, optimization, monitoring, controlling, and improved decision making*" (Rasheed, San, & Kvamsdal, 2020). Irrespective of which industry DTs are used, the three main pillars of the DT are, physical system, virtual representation of the physical system (this could be 3D CAD model, ontologies or Simulink/Simscape/Modellica model) and a digital thread (continuous stream of data transfer between physical and virtual system). A common architecture of the DT is shown in Fig. 1.

The concept of DT can be classified based upon its application area and also on how the dataflow occurs between physical system and its virtual representation. Based on the application area the different types of DTs are (Grieves & Vickers, 2017):

1. Digital Twin Prototype (DTP): A DTP consists of the informational sets (such as 3D models, Bills of Materials, Bills of Processes, etc.) which is imperative for describing and producing a physical version that emulates the virtual version.
2. Digital Twin Instance (DTI): A DT that represents its physical counterpart throughout its life cycle and is linked to a particular physical asset with the help of continuous digital thread (historical data, live data from sensors, tests and inspection).
3. Digital Twin Aggregate (DTA): An aggregation of all the DTIs, of a particular system or system of systems.
4. Digital Twin Environment: It refers to fully integrated, multidisciplinary physics application space for operating a DT for predictive (to estimate the future behavior) or interrogative (to understand the historical or current state) purposes.

Based upon how the dataflow occurs between physical system and its virtual representation, distinction can be made between digital model, digital

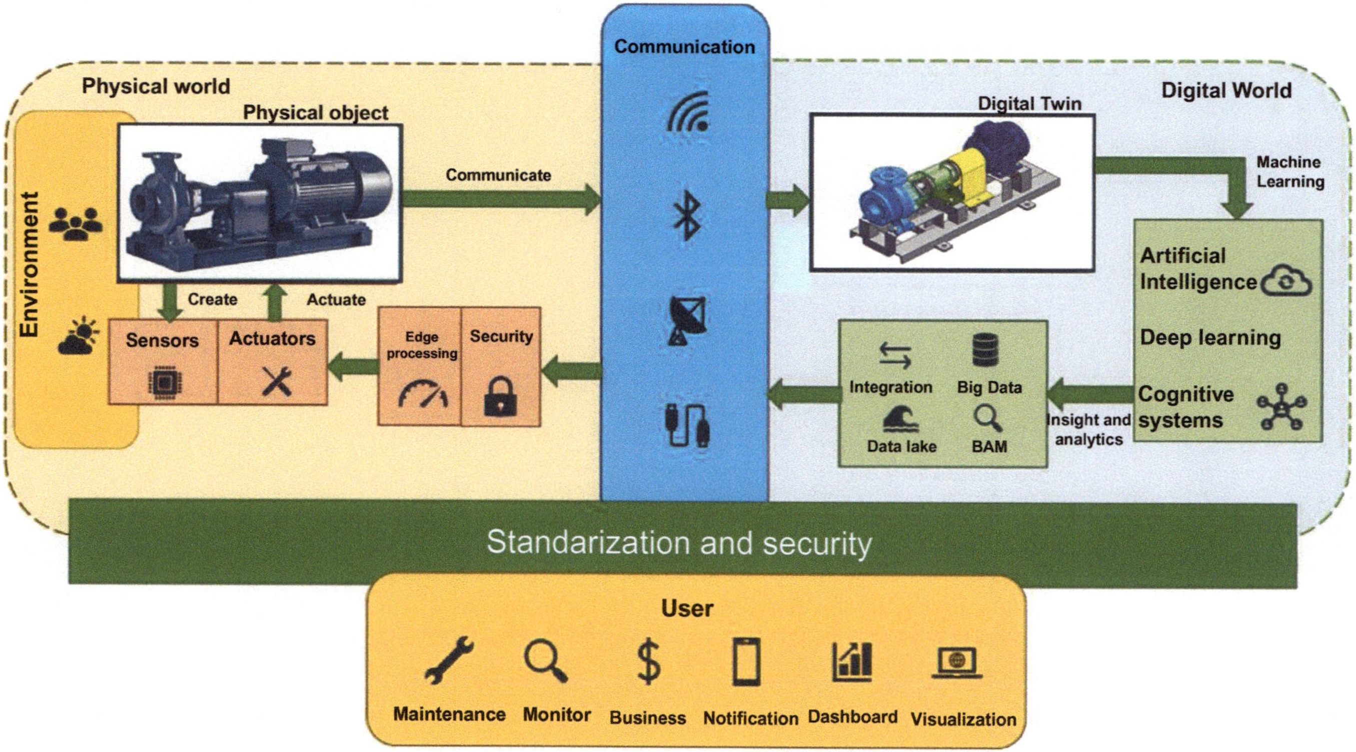

Fig. 1 Common architecture of a digital twin. *Adapted from Campos-Ferreira, A., Lozoya-Santos, J. J., Vargas-Martínez, A., Mendoza, R., & Morales-Menéndez, R. (2019). Digital twin applications: A review. In Memorias del Congreso Nacional de Control Automático (pp. 606–611). Puebla, Mexico: Asociación de México de Control Automático.*

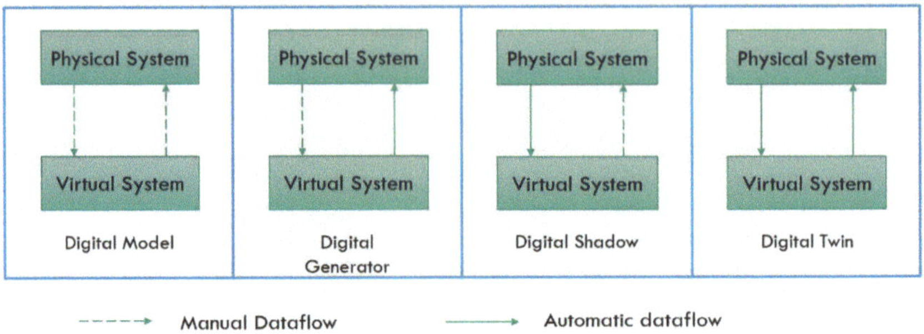

Fig. 2 Difference between concepts related to digital twin. *Adapted from Tekinerdogan, B., & Verdouw, C. (2020). Systems architecture design pattern catalog for developing digital twins.* Sensors, 20*(18), 5103. https://doi.org/10.3390/s20185103.*

generator, digital shadow and a digital twin, as highlighted in Fig. 2. For the digital model, the data integration between the physical asset and its virtual representation is not automatic which implies any change either in physical asset or in a virtual model must be executed manually by analyst/engineer. A bit more advanced concept in terms of data integration is a digital shadow which allows unidirectional automatic information flow from the physical world to the virtual world. For, e.g., a system where sensors measure information from the physical model and transfer signals to the virtual model is an example of the Digital Shadow. Irrespective of whether information flows are continuous or discretized, as long as it is automatic, the integration level can be determined as a digital shadow (Botín-Sanabria et al., 2022). As can be seen from Fig. 2, for a DT, interaction and convergence are its two key aspects, meaning that autonomous bi-directional data exchange between physical system and virtual system is a must. For a DT, the information flowing from the virtual world to the physical asset will instruct actuators to perform certain operation which will change the physical asset in some way. On the other hand, the data from the sensors on the physical asset generally changes the virtual twin automatically in such a way that it accurately represents the current state and the future evolution of its physical counterpart. Thus, a DT is something alive that changes, improves and evolves while maintaining the comparison between physical and virtual space (Botín-Sanabria et al., 2022).

In order to avoid confusion between DT and several other related concepts such as model, simulations, Cyber physical systems (CPSs) and Internet of things (IoT) it is important to create a clear distinction between these concepts. Engineering models (such as mathematical models, statistical models, 3D CAD, data driven, etc.) generally consists of many idealizations and approximations and are used to understand the underlying physical phenomenon/system. For example, Paris law is an empirical crack growth model which is used in different industries in order to calculate remaining fatigue life of the component. Simulations on the other hand primarily involve computer-based models which are used during the design phase to perform what-if analysis for the underlying physical system. Simulation technology has changed considerably from single discipline software (1985) to CAD/CAM integrated multi-disciplinary software (2000) as shown in Fig. 3. Nevertheless, simulation results depend upon the input given by the analyst and hence it is a static virtual representation of the physical system. DTs overcome this gap by coupling simulation modeling with IoT platform, and thus the input to the DT is the real-time data captured with the help of sensors on the physical system and thus enabling the analyst to virtually observe the operation of the physical system. Thus, simulation, models, IoT platform are combined to form the DT in order to enhance the value across the entire life cycle of the product. It must be mentioned there that many real assets (such as Wind Turbines) are cyber-physical systems, meaning that they consist of both the physical/ hardware elements (such as Blades, tower, Nacelle, generator) and cyber/ software elements (such as control system, etc.). While building a DT of a CPS, the physical elements are represented by a virtual model (which are approximations of reality as discussed above), while the cyber elements, can be included directly in the DT, which can then be used for modeling, simulating, and optimizing the CPS. The DTs can be used during entire life cycle of the system, for example during design phase they can be used for virtual testing, while during manufacturing they can simulate the production process. From the process safety management point of view, application of

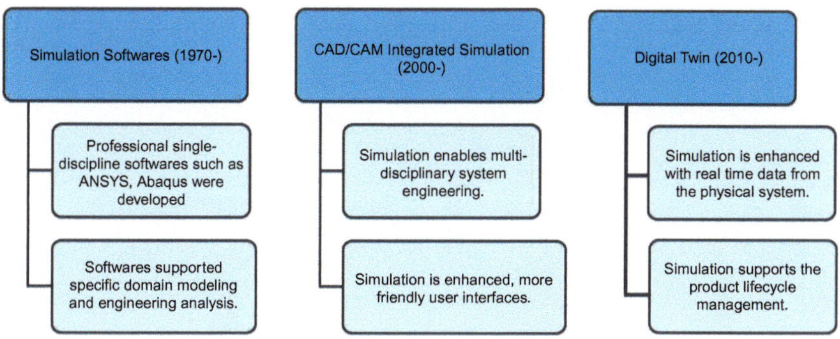

Fig. 3 Difference between simulation and a digital twin.

DT during the operational phase, offers the maximum benefit as they can be utilized to continuously reflect the past, present and future health of the system/asset, which in turn can be used to formulate inspection and maintenance plans. Besides this, DTs also allow analyst to perform complex and safety critical simulations, in order to calculate the dynamic risk, before such an event happens in real life. The operators can then take necessary decisions in order to mitigate the dynamic risk estimated by the DT, and thus maintain process safety at the industry regulated levels.

2. Process safety management of small-bore piping

On a typical process facility, the process piping can be categorized into mainline piping and Small Bore Piping (SBP), as shown in Fig.4. The SBP is a process piping having diameter of 2 in. or less which typically branches off the main process piping and is often used as relief lines, nozzles, instrumentation ports and drains. The connection of SBP with the mainline piping is called as Small-Bore Connection (SBC). In the recent years the refining industry has had a history of failures occurring in Small-bore Connections (SBCs) (Baggett, 2019). Likewise, both the petroleum and maritime industry also report significant SBP failures leading to Hydrocarbon Release (HCR). Although all HCRs emanating from failed pipework are of concern, the main danger comes from gaseous releases, as the gas cloud produced has the potential to quickly spread across the OOG platform. For instance, a gas release from a 6-mm diameter hole in a piping system operating at about 150 bar (i.e., 15 MPa) has the potential

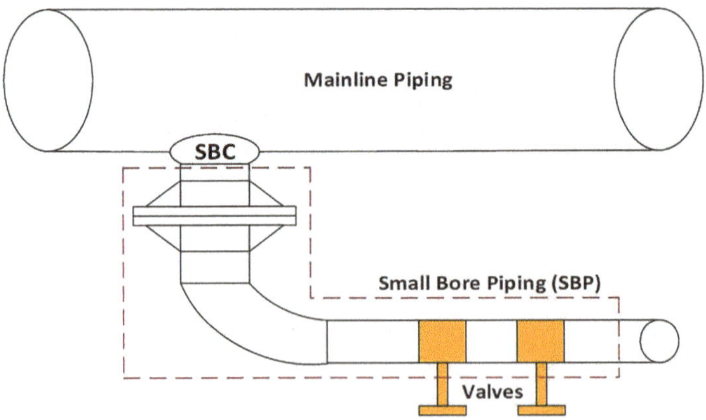

Fig. 4 Schematic of typical process piping.

to turn into a major HCR (i.e., greater than 300 kg) in approximately eight minutes (EI, 2013). If ignited, HCR can escalate into a major accident, resulting in significant economic losses and environmental damage, while causing a serious threat to the lives of personnel. Different degradation mechanisms (such as corrosion, erosion, stress corrosion cracking, corrosion under insulation and fatigue, etc.) are responsible for diminishing the health state of SBP, nevertheless fatigue (emanating due to vibrations) is attributed as the most prominent deterioration phenomenon causing failure of SBP as depicted in Fig. 5.

In a process piping system, cyclic loads emanate due to the excitation of the mainline piping carrying the hydrocarbons. This consequently leads to the magnified vibrations in small bore piping and ultimately causing the fatigue failure of the SBCs leading to HCR. Hence, from process safety perspective it is necessary to comprehend the chain of actions causing the fatigue failure of the SBCs. Fig. 6 depicts the failure chain due to the Vibration Induced Fatigue (VIF). It can be seen from Fig. 6, that the excitation forces responsible for vibrating the mainline piping are classified as steady state or transient. The former occurs during the normal operating conditions and emanates due to phenomenon such as flow-induced

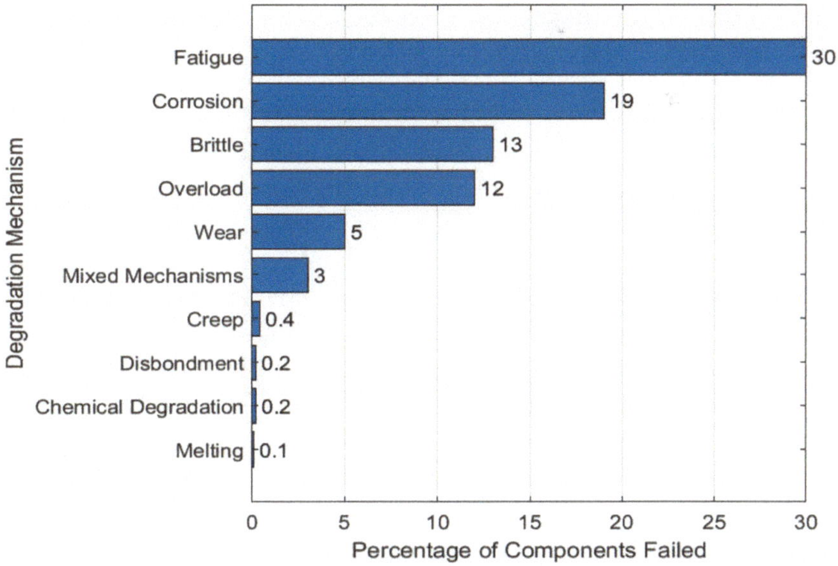

Fig. 5 Most dominant degradation Mechanisms. *Adapted from DNV, (2020). https://www.dnvgl.com/oilgas/laboratories-testsites/article/tubes-and-piping-are-the-most-failureprone-components.html.*

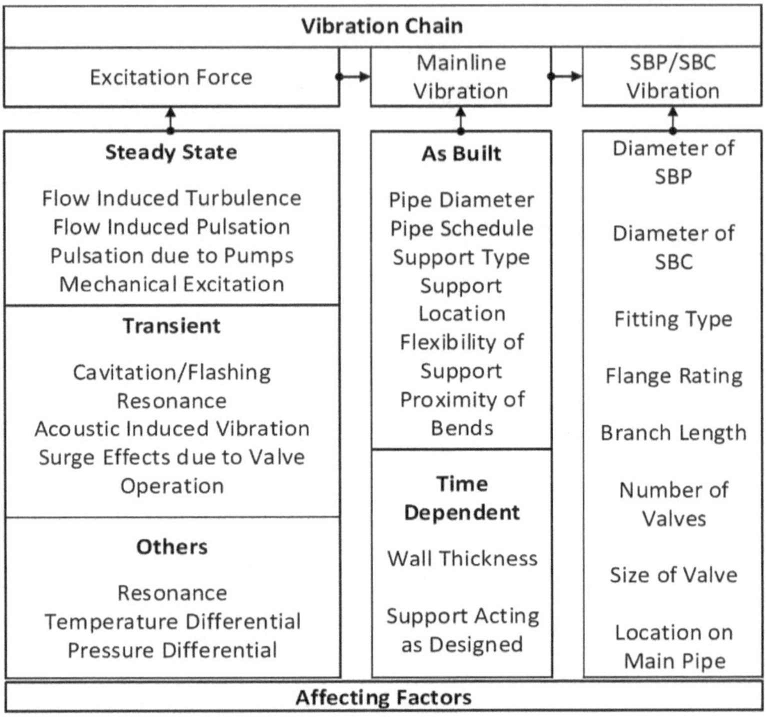

Fig. 6 Vibration induced fatigue failure chain. *Adapted from Swindell, R. (2003). Vibration fatigue in process pipework: A risk-based assessment methodology. Southampton, UK: Bureau Veritas.*

turbulence, flow-induced pulsations, pulsations from pumps/compressors, etc. (EI Guidelines, 2007). On the contrary the latter originate during events such as changing operating conditions, additional units coming online, normal start-up and shutdown, emergency shutdown, valve operation, etc. (HSE-OTR 028, 2002). In order to mitigate the likelihood of the HCR from the process piping on the offshore platforms, it is necessary to deeply understand the excitation mechanisms depicted in Fig. 6. The taxonomy of the fatigue originating in process piping is illustrated in Fig. 7. Process piping especially near rotary equipment is subjected to High Cycle Fatigue (HCF) and VIF is the major reason for it's failure. Although piping vibration is seldomly detected visually, nevertheless, expert knowledge (from standards such as EI Guidelines and from practicing engineers) coupled with condition monitoring data can help to frame effective inspection and maintenance (I&M) strategies for preventing unwanted piping breakdown due to fatigue. In order to cater for better understanding of the fatigue loading of SBP, EI

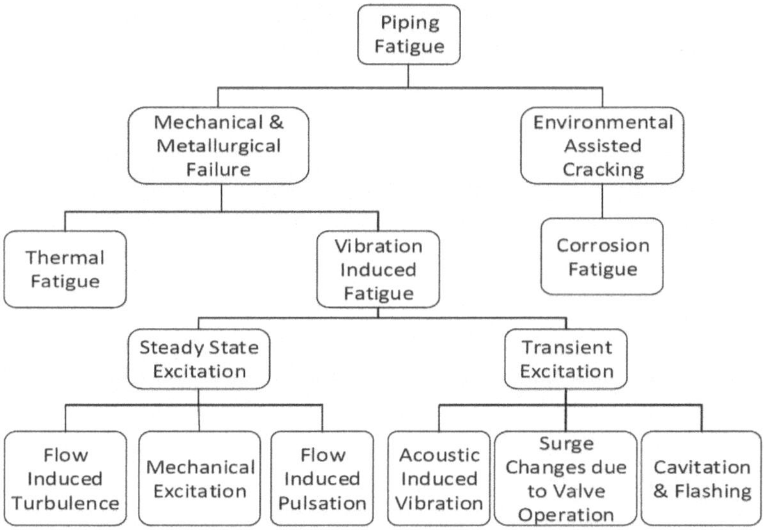

Fig. 7 Taxonomy of the piping fatigue.

guidelines suggest a more detailed analysis of SBP using advanced simulation techniques, such as Computational Fluid Dynamics (CFD), Finite Element Analysis (FEA), as these techniques allow, analyst to perform the dynamic simulation of pressure, temperature and flow characteristics in process facilities, which in turn are used to compute risk and reliability of the SBP (EI, 2007).

3. Digital twin for small bore piping

As discussed, a DT is the most realistic representation of a physical asset, which incorporates models and all necessary available information (operational, organizational, and technical) of the asset. Some of the areas where a DT can be used effectively in the process industry is for the early anomaly detection of production process parameters, prognostics and health management (PHM) of assets, support for decision-making on process safety management (PSM) (El Saddik, 2018; Tao, Zhang, & Nee, 2019). Lately, in the process industry the concept of DT is being applied for predictive maintenance of the various assets such as pumps, heat exchangers, etc. (Min, Lu, Liu, Su, & Wang, 2019). However, when it comes to usage of DT technology for SBP, very little work has been done in this direction although SBP is the most commonly failed equipment. The primary challenge for the implementation of the DT for PHM of SBP is due to lack of sensor data and the current industry practice of selecting fatigue critical inspection location based on ad-hoc basis. The lack of sensor data stems from the fact that on a typical offshore platform there could be more than 100,000 SBPs, hence placing a physical sensor for all these SBPs would be practically impossible. Thus, if a DT needs to be built for the SBP, then it is vital to forge a connection between the real-world observations (for example from experiments, inspections) and the information obtained from the virtual sensors (data driven models) on the virtual model (CAD model or Simulink model). An analogy that fits this scenario is the one given by Allen and Tildesley (1987) in the book Computer Simulations of Liquids, where they demonstrate how coupling experiments (and experience), with theory and computer simulations, can enhance the up-to-date knowledge of a real system (see Fig. 8). Similar connections between real-world observations and data-driven models are equally relevant while making a DT for the SBP, so that an analyst/inspection engineer can get an up-to date health state of the asset which could then be used to estimate risk and frame I&M strategies. Hence, while building a DT for the SBP, it is necessary to combine the expert knowledge (either given in the standards or available from the practicing engineers), the data (coming from inspection, experiments) and the virtual models.

Modeling methods can either be physics-based (they try to capture the actual physics of the process, e.g., CFD, FEA) or data-driven models (they try to find trend based upon the data, e.g., machine learning, deep learning). It is a common practice in process industry to perform risk analysis of

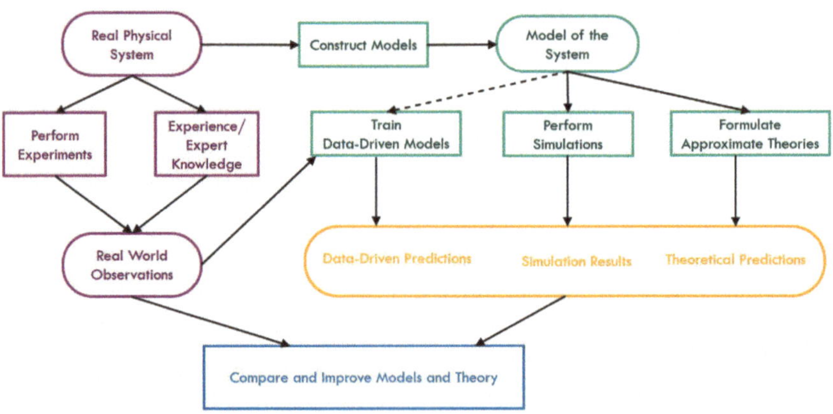

Fig. 8 Coupling real world observations with information from virtual model. *Adapted from Allen, M.P., & Tildesley, D.J. (1987). Computer simulation of liquids. Oxford Scholarship.*

engineering assets by leveraging the results from CFD and FEA techniques owing to its ability to simulate the physical phenomenon with great accuracy. For example, using CFD dispersion modeling, Fiates and Vianna (2016) performed risk assessment of gas leakage on a typical offshore topside plant. Kumaresh, Fossan, Venkatraman, and Jensen (2016) calculated transient ignition probability of gas leaks in enclosure using CFD, Li et al. (2019) presented a CFD based approach to describe the behavior of underwater gas release and dispersion from subsea gas pipelines leak. However, as argued by Keprate, Ratnayake, and Sankararaman (2019) the higher accuracy of the advanced dynamic simulation techniques is achieved at the expense of the greater time required for simulations to run, which makes these techniques computationally expensive and time-consuming. In order to overcome the aforementioned shortcomings of CFD and FEA, recently engineering community has resorted to surrogate models (SMs) as they are faster than the main simulation code (i.e., FEA, CFD), while still being usefully accurate (Forrester, Sobester, & Keane, 2008). Many researchers such as Forrester, Bresslof, and Keane (2006), Moore, Jetmarova, and Gebreslassie (2016), Wilson, Wakes, and Mayo (2017), Loy, Rangaiah, and Lakshminarayan (2018), Jeon, Yang, Kang, Na, and Lee (2019), and Keprate, Ratnayake, and Sankararaman (2017a, 2017b), Bagalkot, Keprate, and Orderloøkken (2021) have used SMs to replace CFD simulations for various engineering applications and demonstrated the accuracy and reduced computational expense of SMs. Thus, while building a DT for the SBP, a hybrid method which consists of physical modeling (using CFD and FEA) and data driven modeling (consisting of machine learning models) can be used. Furthermore, when the DT is to be used for risk assessment of safety critical equipment, it is necessary to account for uncertainty in the physical phenomenon and thus combine the above two modeling methods with the probabilistic graphical models such as Bayesian Networks. Fig. 9 demonstrates the components used for building DT for estimating the RFL for the SBP. As can be seen the process flow parameters such as pressure, flow velocity serves as an input to the DT model. In reality the value of these parameters can be continuously communicated to the DT using principles of IoT, however, for the sake of illustration, in the case study, authors shall use discrete values of these parameters. The next step is to build a virtual sensor for stress estimation using CFD and FEA coupled with Machine learning models. The input to the virtual sensor is pressure and flow velocity and the output is stress value at the SBC. Thereafter, the calculated stress acts as one of the inputs (along with initial

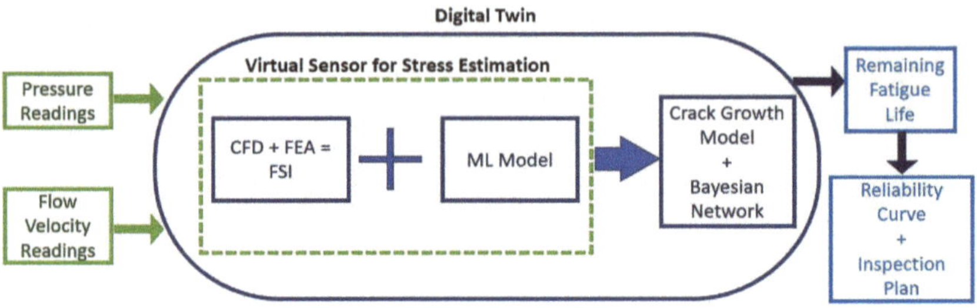

Fig. 9 Digital twin framework for estimating RFL of SBP.

crack size, critical crack size and material parameters) to the empirical crack growth model (Paris law) which is then coupled with a Bayesian Network model to estimate the probability density function (PDF) of the RFL of the SBP. Finally, the Cumulative Density Function (CDF) is used to establish the reliability curve which in turn is used to frame inspection intervals for SBP.

4. Illustrative case study

4.1 General

For this case study, a DT for estimating the Remaining Fatigue Life (RFL) of SBP shall be developed using hybrid approach which includes physics based modeling (captured by CFD/FEA and analytical crack growth model), the surrogate models (using machine learning models) and logical models (Bayesian Newtork). AISI Type 304 Stainless Steel (SUS 304) having yield strength of 215 MPa, is chosen as the material of the process piping. A semi-elliptical surface crack (having crack depth of *a*, and crack length of *2c*) is assumed to be initiated at the weld toe of the SBC, which shall propagate under the influence of the stress acting at the weld toe interface. *A total of four small bore sizes have been used for a fixed size of main pipe. For a 10-in. (Nominal 80S) size main pipe, four small bore size 0.5-in., 1-in., 1.5-in., and 2-in. (all Nominal 80S) have been used as shown in Table 1.*

The first step in creation of the DT is to create a virtual representation of the piping system and use this virtual model for FEA and CFD analysis to estimate the stress acting on the SBC. Since CFD and FEA simulations are computationally expensive, so a surrogate model is used as a virtual model, for estimating stress. The detailed methodology for stress prediction using CFD, FEA, and ML approach as shown in Fig. 3 is discussed. ANSYS software is used to perform Fluid Structure Interaction (FSI) by coupling CFD

Table 1 Details of the size of the main pipe and small bore.

DN (in.)	DN (mm)	OD (mm)	ID (mm)	Thickness (mm)
0.5 (small-bore)	15	21.3	13.84	3.73
1 (small-bore)	25	33.4	24.3	4.55
1.5 (small-bore)	40	48.3	38.14	5.08
2 (small-bore)	50	60.3	49.22	5.54
10 (main pipe)	250	273.1	247.7	12.7

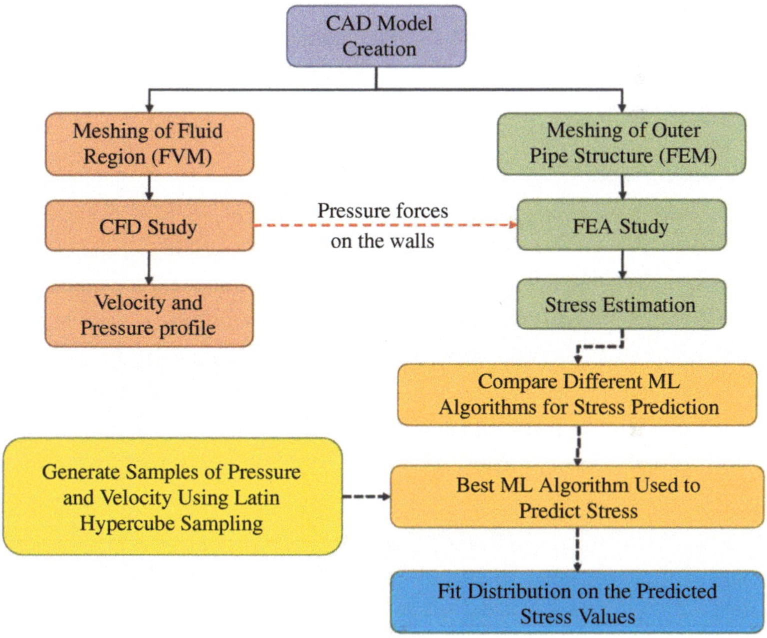

Fig. 10 Methodology used for stress estimation (using virtual sensor).

and FEA analysis. The details of each step in the methodology shall be presented in the upcoming sections. As shown in Fig. 10, first the CFD analysis shall be carried out, the results of this analysis are then transferred to a FE analysis tool, which will estimate the stress distribution on the weld toe of the SBP. Thereafter 19 different Machine Learning (ML) algorithm are trained, tested and validated using the data generated from the CFD/FEA analysis. The most accurate ML algorithm (Optimizable Gaussian Process Regression in our case) is then utilized to predict value of the stress for

10,000 samples (of pressure and velocity variables) generated using Latin Hypercubic Sampling (LHS). Thereafter, different distributions are fitted on the predicted maximum stress value and the Akaike Information Criterion (AIC) value is calculated to determine the best distribution fit for the stress acting on SBP.

4.2 Physics based modeling

4.2.1 Creation of the model

A single small-bore attached to the main pipe configuration is used to carry out the analysis in the current study. The dimensions of the main pipe and small-bore is adopted from Energy Institute (EI Guidelines, 2007). *Fig. 11 shows the 3D CAD model with a small-bore and main pipe used to carry out both CFD and finite element (FE) analysis. The 3D CAD model is segmented into a fluid region representing the fluid flow and a solid region representing the pipe body.*

4.2.2 Model meshing

For creating high fidelity physics based model using CFD/FEA, meshing plays a very important role. Too coarse mesh can lead to lowering the accuracy of the model, while too fine mesh can significantly increase the computation time. Thus a mesh sensitivity study was performed for this work. The 3D model shown in Fig. 11 was meshed using ANSYS Meshing Tool. Since there are two modes of simulation CFD and FEA, there will be two types of meshing finite volume and finite element. For the CFD analysis a finite volume mesh was created on the fluid flow region. Fig. 12B shows the mesh distribution at the junction of small bore and main pipe, the distribution of inflation layers as shown in Fig. 12C, along with mesh details. The mesh size was selected after a comprehensive mesh sensitivity study. The size of the first layer of the inflation was based on the $Y+<1$ criteria, a total of 12 layers of inflation with growth rate of 1.15 was used for CFD meshing. The minimum face size of the mesh was selected to be $8^{*}10^{-5}$ m

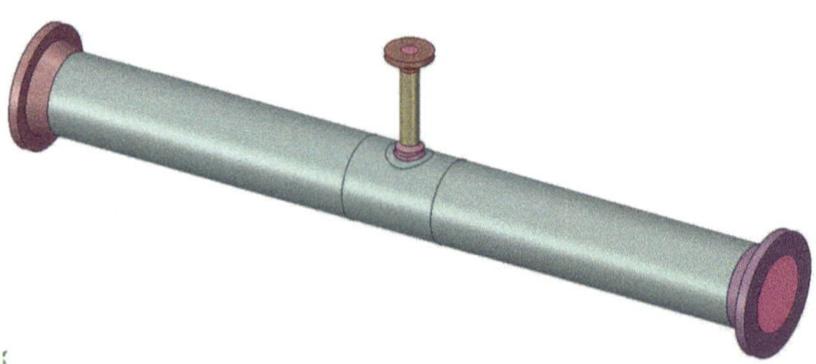

Fig. 11 CAD model of mainline piping and SBP.

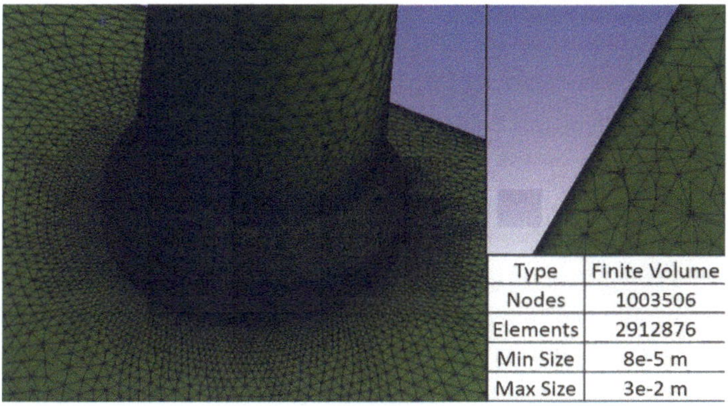

Type	Finite Volume
Nodes	1003506
Elements	2912876
Min Size	8e-5 m
Max Size	3e-2 m

Fig. 12 Mesh distribution and mesh details for CFD analysis.

Type	Finite Element
Nodes	251912
Elements	107911
Max Size	3e-2 m
Material	SUS 304

Fig. 13 Mesh distribution and mesh details for FE analysis.

and max face size was restricted to $2^{*}10^{-3}$m, the increment in the size is based on the growth rate of 1.15. *For the region represented by the structure of the main pipe and small bore, a finite element mesh is implemented. The nodes of the mesh on the pipes will be able to transfer the pressure loads arising from fluid flow and impact and transfer is as stresses within the pipe structure. The element distribution at the main pipe and small bore is shown in Fig. 13. Similar to CFD mesh the mesh size for stress analysis was selected after a comprehensive mesh sensitivity analysis. The supports were provided at the either ends of the main pipe.*

4.2.3 CFD and FEA analysis

After the model is meshed, ANSYS FLUENT 2021 R2 is used to carry out a comprehensive CFD analysis. For each size of bore the CFD analysis is carried out range of operational pressure and velocity. Due to limited time availability for illustration purposes, the analysis was carried out for four operating pressures 5, 15, 25, and 35 bar, for each of these operating pressures, the analysis was carried out for 7 operating flow rates, the flow rates are represented as velocity (keeping the inlet density and area constant) in the range of 1.5–8.4 m/s. From the boundary condition perspective, a velocity inlet with inlet gauge pressure is used as inlet boundary condition, and pressure outlet is implemented at the exit to the main pipe. The small bore acts a dead-leg and has no flow exiting it. Compressible air is used for the simulation with Peng Robinson model. A two-equation k-epsilon, with scalable wall function turbulence model option was implemented. The pressure force applied by the fluid on the internal walls of the pipe calculated by the FLUENT software is transferred to the FEA software, to estimate the mechanical stress on the structure arising due to the pressure loads. *The software is setup to estimate the Equivalent Von-Mises stress at the junction of main pipe and small bore. The stress on the internal and external wall of the SBP is estimated. The boundary condition includes fixed support at the end of the main pipe, and gravity effect, the influence of the weight of equipment at the end of small bore like check valves, pressure or temperature sensors, was neglected.*

All the results of the simulation are of similar nature except from change in magnitude, therefore as a representative case the FSI (CFD and FEA) results for the case study of 1.0-in. small bore at fixed pressure of 5 bar and at three different flow rates will be presented in detail. For the rest of the cases the output FSI result (Von-Mises stress) will be reported, which will later be used to build the surrogate model for stress prediction.

Fig. 14 shows the velocity vector at the junction of the main pipe and small-bore for 5 bar (1 in. small bore, operational velocity 8.5 m/s). Fig. 14A is along a central plane parallel to the flow direction (positive x-axis) and Fig. 14B is along the plane perpendicular to the flow direction. The velocity vector gives a clear picture of where the stress might get induced in the pipe due to the flow. As can be seen a major turbulence region is created in the vicinity of junction, this turbulence gives rise to region of pressure fluctuations or uneven pressures as shown in Fig. 15.

Fig. 16 shows the Von-Mises stress estimated by the ANSYS Mechanical software using the pressure loads from the ANSYS FLUENT software for flow condition mentioned in description of above CFD results. Fig. 16A and B show the stress distribution at the external and internal surface of the junction respectively and flow direction is in the positive X-direction. It may be observed that the maximum stress concentration, as well as stress distribution (regions of low and high stresses) stress is at

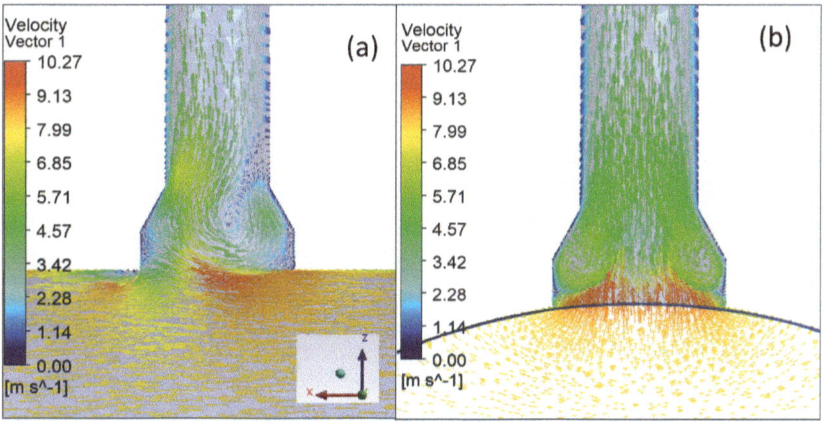

Fig. 14 Velocity vector at the SBC.

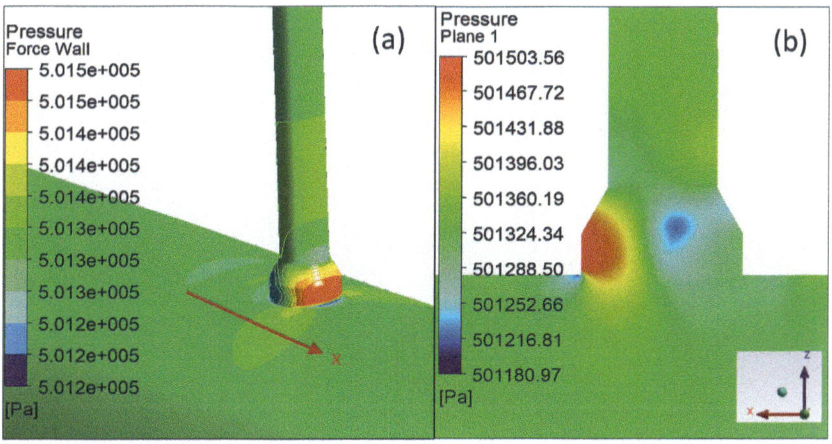

Fig. 15 Pressure concentration at the SBC.

SBC. The observations made in Fig. 16 is supported from the flow and force results shown in Figs. 14 and Fig. 15. Furthermore, the stress is slightly higher on the inner diameter of the pipe compared to outer diameter. Similarly, for a fixed size, the larger the flow rate, the greater is the stress at the junction of small bore and main pipe as can be seen in Fig. 16.

Although FEA/CFD is a good approach for emulating the physical phenomenon, but is can be seen that it is practically impossible to estimate stress at SBC for different values of parameters (pressure and volume flow rate). Thus, we shall employ surrogate model which once trained and tested can act as a replacement to the computationally expensive and/or time-consuming CFD/FEA simulations, without compromising the accuracy of the output.

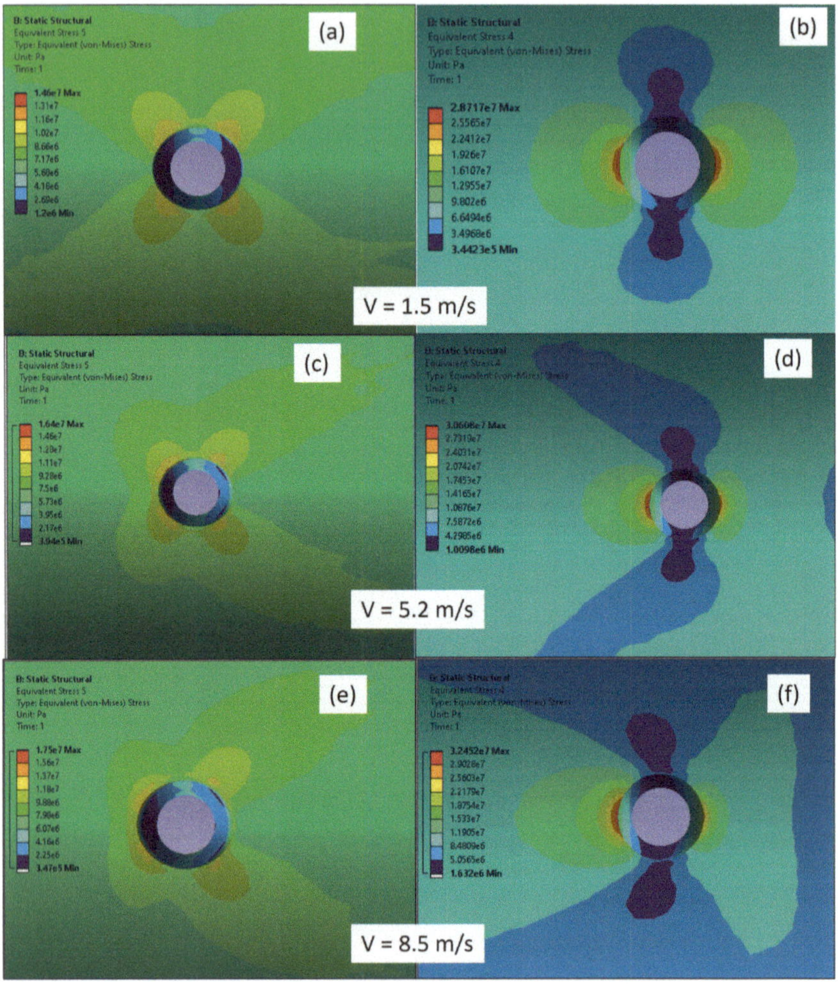

Fig. 16 Stress distribution at SBC at different velocity (1-in. small bore, 5 bar).

4.3 Surrogate modeling

The most commonly used SMs in the engineering domain are linear regression, Tree based methods, Ensemble methods, Boosting methods, Support Vector Machine and GPR. The mathematical background and theory of the various SMs used in this manuscript are discussed briefly (Keprate, Ratnayake, & Sankararaman, 2017a, 2017b). The data generated from CFD/FSI simulations was used for training, testing and validation of these SMs. Eighty-Four data points are used to train the aforementioned SMs, while 28 data points are used as for the validation purpose. Nineteen different SMs (consisting of linear regression, Tree based methods, Ensemble

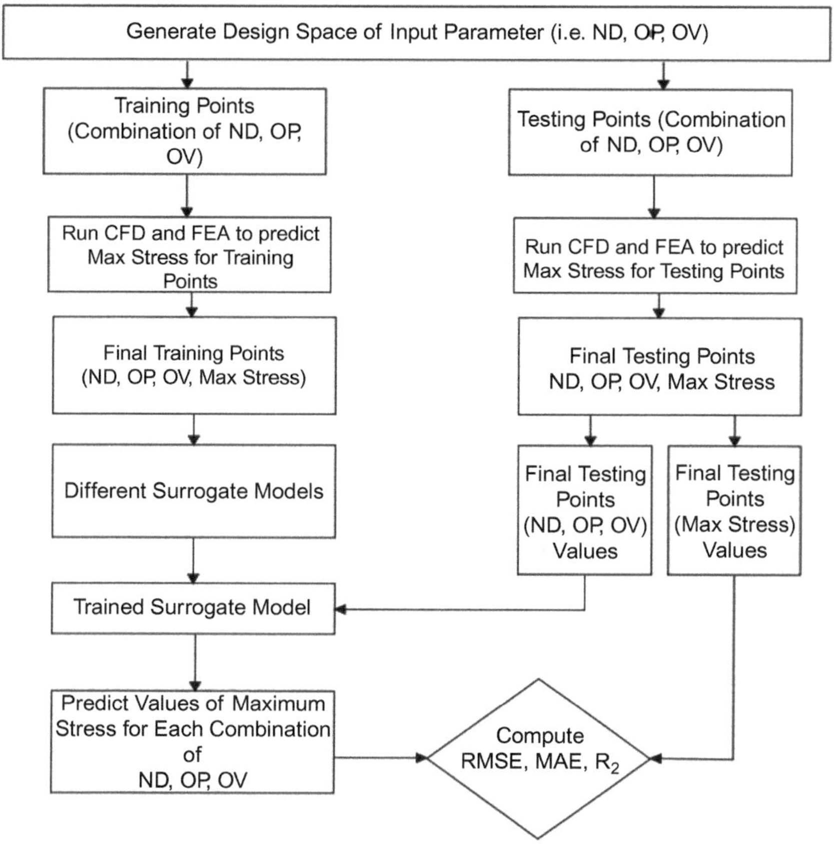

Fig. 17 Flowchart to build surrogate models for maximum stress prediction.

methods, Boosting methods, Support Vector Machine and GPR) are evaluated using 4 metrics, namely, Root Mean Square Error (RMSE), Maximum Absolute Error (MAE), Coefficient of Determination (R^2) and training time. Thereafter, using the flowchart shown in Fig. 17, different SMs are constructed to predict the value of the Maximum Stress. K-fold (10-folds in our case) cross validation is used during the training of the surrogate models. Afterward, the most accurate SM is further optimized using Bayesian Optimization to further improve its accuracy. Finally, the optimized SM is trained on validation dataset and the predicted values of maximum stress are compared to the values obtained from CFD and FEA.

In order to compare the performance of the SMs, four metrics, namely, Root Mean Square Error (RMSE), Maximum Absolute Error (MAE), coefficient-of-determination (R^2) and training time are used. Mathematically, these are written as:

$$RMSE = \left(\sum_{i=1}^{n} (y_i - \widehat{y}_i)^2 \right)^{0.5}$$
$$MAE = \max\left(|y_1 - \widehat{y}_1|, |y_2 - \widehat{y}_2|, .., |y_n - \widehat{y}_n|\right) \tag{1}$$
$$R^2 = 1 - \frac{SS_{Error}}{SS_{Total}}$$

Generally, the most efficient algorithm should have minimum value of RMSE, MAE and training time, and the value of R^2 should be closer to 1. The comparison of different SMs is presented in Table 2 and it can be seen that Ensemble Bagged trees is the worst performing algorithm while, Gaussian Process Regression (GPR) having a squared exponential kernel function is the best performing algorithm.

Table 2 Comparison of different surrogate models.

Surrogate Model	*RMSE (MPa)*	*MAE (MPa)*	R^2	Training Time (s)
Linear Regression	7.91	5.87	0.9	2.41
Linear Regression with Interaction	3.99	5.71	0.97	0.50
Robust Linear Regression	7.97	7.87	0.89	0.58
Stepwise Linear Regression	3.99	3.03	0.97	2.06
Fine Tree Algorithm	13.99	11.26	0.68	0.69
Medium Tree Algorithm	16.81	12.86	0.53	0.17
Coarse Tree Algorithm	19.01	14.85	0.4	0.17
Ensemble Boosted Trees	7.54	4.9	0.91	2.05
Ensemble Bagged Trees	24.41	19.76	0.01	1.65
Linear SVM	8.00	5.63	0.89	0.84
Quadratic SVM	3.08	2.46	0.98	0.30
Cubic SVM	2.21	1.83	0.99	0.24
Fine Gaussian SVM	18.17	13.86	0.45	0.29
Medium Gaussian SVM	5.51	3.65	0.95	0.21
Coarse Gaussian SVM	7.67	5.03	0.90	0.28
Rational Quadratic GPR	2.02	1.40	0.99	0.98
Squared Exponential GPR	2.01	1.39	0.99	0.46
Matern 5/2 GPR	2.04	1.42	0.99	0.53
Exponential GPR	3.31	2.12	0.98	0.46

It is further possible to enhance the accuracy of GPR, by performing Bayesian optimization. In total 30 iterations are performed across the entire hyperparameter space (such as Basis Function, kernel function, kernel scale, etc.) of the GPR to arrive at the optimized parameters (which are constant basis function, non-isotropic Matérn 5/2 kernel function, 34.89 kernel scale). The value of evaluation metrics for optimized GPR are RMSE of 1.81 MPa, MAE of 1.28 MPa and R^2 of 0.99. However, the training time increased to 102.99 s, which is logical as 30 iterations were performed with different values of hyperparameters, to arrive at the optimized parameter values. Furthermore, the plot comparing the stress values predicted by CFD/FEA and GPR are shown in Fig. 18, in which true response represents the former stress values while the predicted response represents the stress value estimated by optimized GPR. It is seen that for most of observations GPR predicts accurately the value of stress. However, it is vital to check the optimized GPR performance on the validation dataset. As can be seen from Fig. 19 and also inferred from the value of RMSE which is 3.012 MPa, and R^2 which is 0.974, thus performance of optimized GPR is good even on validation dataset. The, trained optimized GPR shall now serve as a virtual/soft sensor which can be used to predict maximum stress for SBP for the given flow conditions.

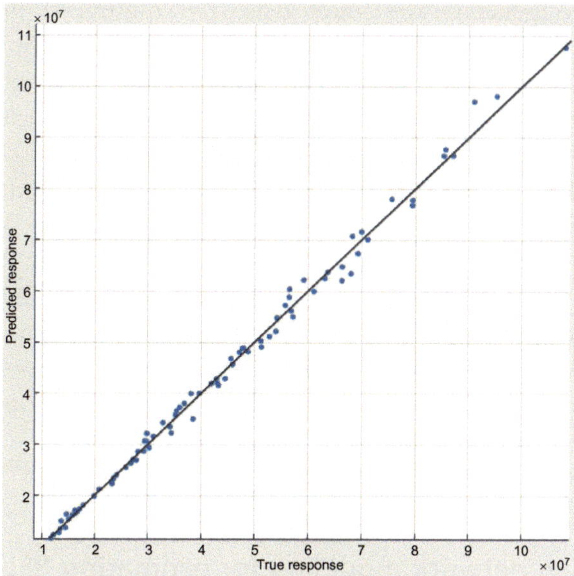

Fig. 18 True vs predicted stress values for training dataset.

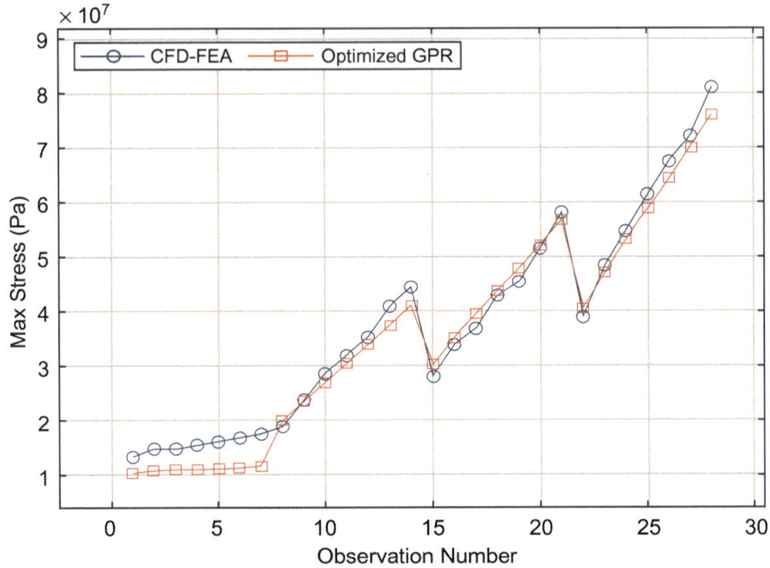

Fig. 19 Maximum stress values predicted for validation dataset.

If the flow data and pressure readings are obtained in real time from the process facility, then they can serve as in out data to the developed soft sensor. However, if it is not possible then we can utilize various sampling techniques to generate the input data. For illustrative purpose, Latin Hypercubic Sampling (LHS) is utilized to obtain 10,000 samples of pressure variable (between 5 bar and 35 bar), and of velocity variable between 1.5 and 8.7 m/s, which act as input to the virtual sensor, which in turn estimates the value of max stress for these 10,000 samples for a 2 in. (50 mm Nominal Diameter) SBP. Thereafter, different distributions are fitted on predicted maximum stress value. The two best fitted distributions are lognormal (log location = 3.37 and log scale = 0.388) and 2-parameter Weibull distribution (scale = 35.23, shape r = 2.86) as can be seen in Fig. 20. The Akaike information criterion (AIC) value was calculated to determine of the quality of the distribution fit. Generally, the distribution with the low AIC value is usually the preferred distribution. Amongst the two distributions, lognormal distribution was the best fit for the given, as its AIC value was lower than Weibull distribution (AIC_LN = 76,939, while AIC_Wbl = 77,113).

4.4 Bayesian network modeling for estimating RFL

The next step is to use probabilistic graphical model, i.e., Bayesian Network (BN) to build a DT for estimating the RFL of SBP. The main component of

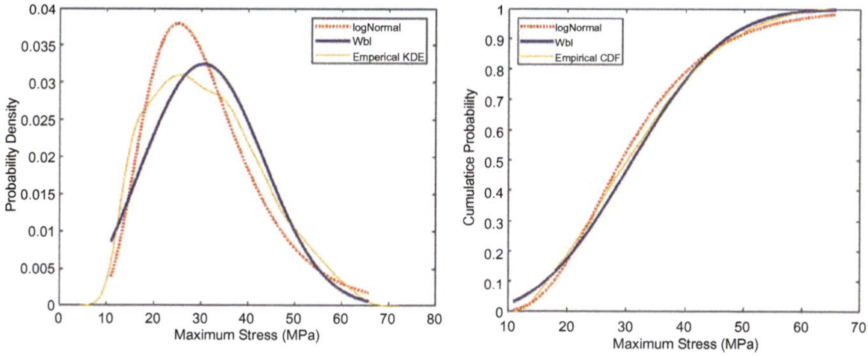

Fig. 20 Distribution fitting on stress values estimated from the virtual sensor.

the BN is the empirical Paris crack growth law, which captures the physics behind the crack propagation. Paris law is stated as:

$$\frac{da}{dN} = C(\Delta K)^m \text{ for} \Delta K_{th} \leq \Delta K \leq K_{mat} \tag{2}$$

where a is the crack size in m, N is the number of cycles, $\frac{da}{dN}$ is the increment in the crack growth rate in m/cycle, ΔK is the stress intensity factor (SIF) range in MPa$\sqrt{m}$, ΔK_{th} is the threshold value of the SIF, $K_{mat}$is the material fracture toughness, and m and C are the material parameters. The range of the SIF depends upon the parameters such as the range of the remote stress ($\Delta\sigma$), crack size (a) and the geometric function (Y). The number of fatigue loading cycles (RFL) required to grow a crack from the (ICS), a_i, to the final crack size, a_N, can be determined by integrating Eq. (3).

$$RFL = \frac{a_N^{1-\frac{m}{2}} - a_i^{1-\frac{m}{2}}}{C\left(1 - \frac{m}{2}\right)\left(\Delta\sigma\ Y\ \sqrt{\pi}\right)^m} \tag{3}$$

The various parameters of the Paris model are considered as random variables, and are represented by a suitable probability distribution and associated parameters as shown in Table 3. For our analysis we consider the uncertainty only in three parameters namely, initial crack size (a_i), material parameter (C) and the remote stress range ($\Delta\sigma$), while other parameters (critical crack size and material parameter (m)) are treated as non-random variables. The value of initial crack size is calculated using Equivalent Initial Flaw Size (EIFS) concept given in Sankararaman, Ling, and Mahadevan (2011), while the value of parameter C is taken from the literature. Furthermore, since Paris model is an empirical law, hence it is also subjected to uncertainty which is represented by a modeling error ε_r whose value in practice is

Table 3 Uncertainty quantification for RFL estimation.

Uncertainty source	Random variable	Parameter value
Initial crack size	a_0 (mm)	LN (0.47, 0.0333)
Material parameter	C	LN (−21.9, 0.3708)
Material parameter	m	3.5
Remote stress range	$\Delta\sigma$ (MPa)	LN (3.37, 0.388)
Critical crack size	a_c(mm)	12.7
Geometric function	Y	0.952

obtained from the experiments. However, in this chapter for the sake of illustration, ε_r is represented by a 5% Gaussian white noise. The value of remote stress range ($\Delta\sigma$) is obtained from the distribution fitting to the data obtained from the virtual sensor developed earlier in the chapter, while the value of geometric function Y is calculated using suitable industrial standards. All the yellow color nodes are input to the BN, while orange color node represents the crack size evolution with time. The BN can finally generate the distribution of RFL (using 10,000 samples) which is shown in Fig. 21. It must be mentioned here that for the sale of illustration, authors are manually assigning values to the various parameters of the BN, however, in real world implementation of the DT, the entire data input can be automated.

A closer look at the Probability Density Function (PDF) of the RFL tells us that the mean value of the RFL equals to 2.7492e06 cycles, while the 95% confidence interval (CI) is calculated as [6.1557e04 and 1.68e07] cycles. The estimated RFL can be used to estimate the reliability (in the form of reliability curve as shown in Fig. 22) of SBP, which in turn can be used to frame the future inspection intervals. In this method the operator sets the target reliability depending upon the limit states and the safety class. For example, if we consider fatigue limit state and high safety class then the target reliability level for the piping system should be 0.9999. From Fig. 23, the corresponding number of cycles for a target reliability of 0.9999 approximately equals to 20,000 cycles (as shown by red colored line on the figure). Thus, the first inspection of 2 in. SBP must be done after 20,000 cycles. The same DT concept can be used to frame inspection intervals of different sizes of SBPs (0.5-in., 1 in., etc.) and in this way the inspection plan of SBPs for entire process plant can be generated in a short period

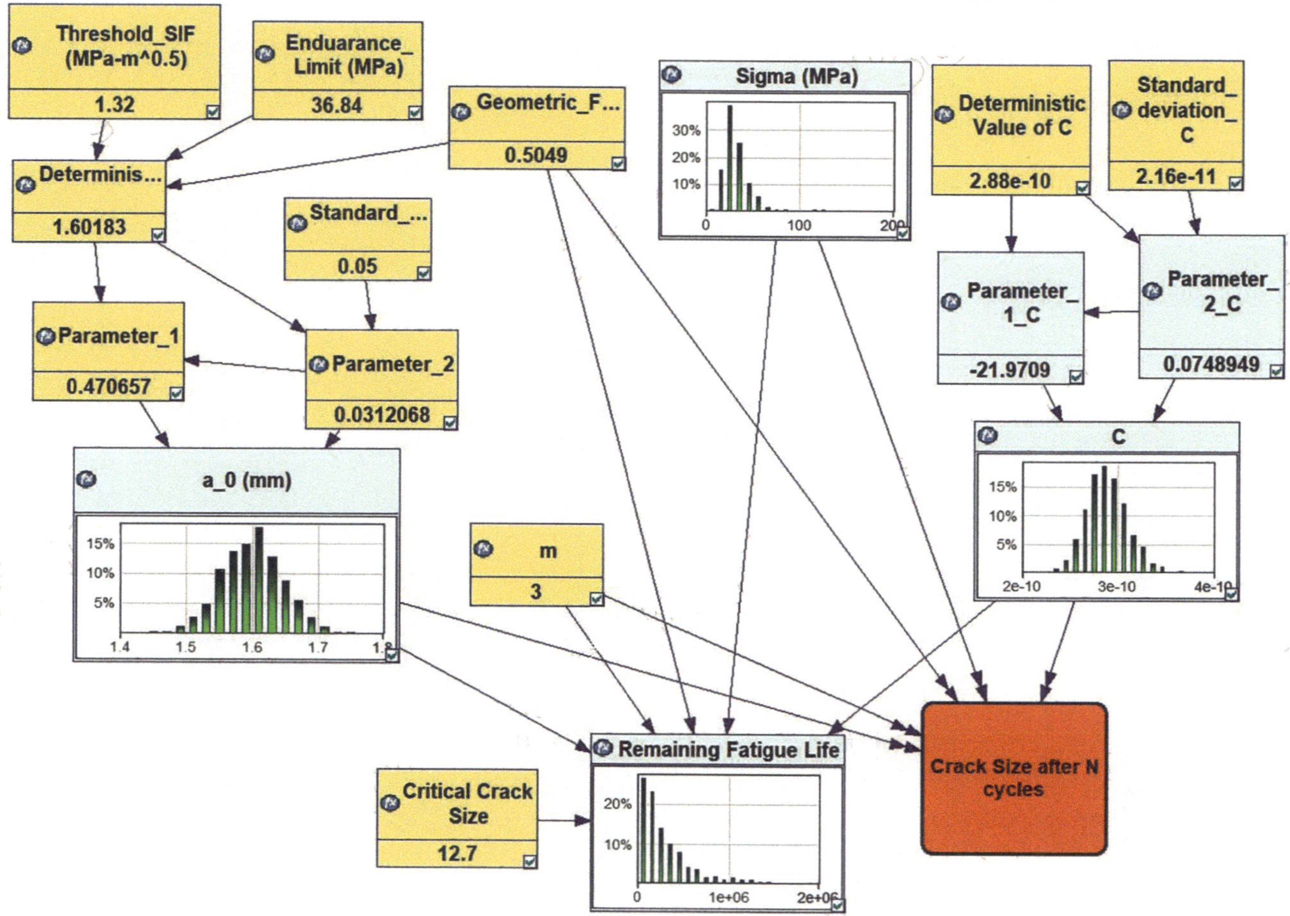

Fig. 21 Bayesian network enabled DT for RFL estimation of SBP.

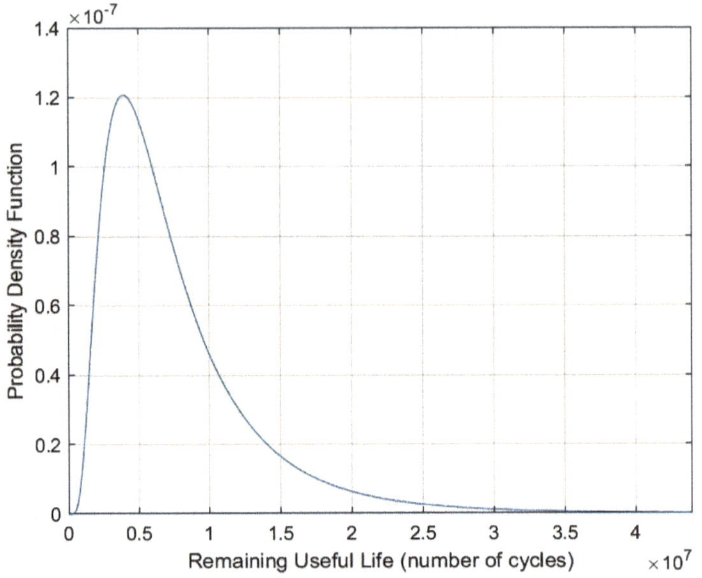

Fig. 22 Probability density function of RFL of SBP (10,000 samples).

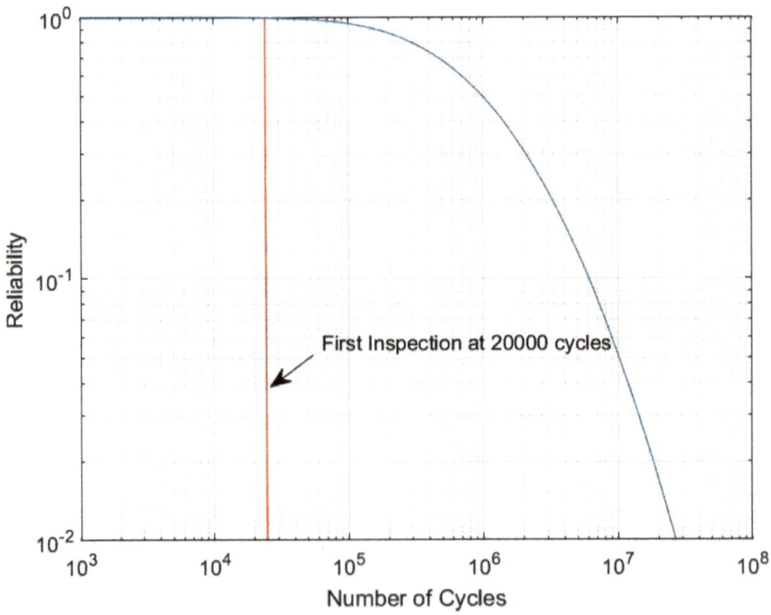

Fig. 23 Reliability curve for the 2 in. SBP (10,000 samples).

of time. Such a proactive strategy would help inspection engineers to timely inspect the SBP and thus help in preventing unwanted breakdown due to fatigue, thus enhancing process safety.

5. Conclusion

In this chapter, a Digital Twin Framework for estimating the Remaining Fatigue Life of Small Bore Piping was developed. Due to large number of SBPs on a typical process facility, the problem of placing a physical sernsor was also discussed. As a solution to the aofrementioned problem, a virtual sensor for utilizing CFD/FEA and Surroagte model was developed. The input to the virtual sensor were process parameters (i.e., presure and flow velocity) while maxium Von-Mises stress at Small Bore Connection (SBC) was the output. The unceratinty in the estimated stress was catered by fitting the most suitable probability disribution (logormal distrubtion for our case) to the stress value. Thereafter, crack propagation in the piping was modeled using Probabilistic Paris Law, which had input parameters such as intial crack size, critical crack size, material parameters and sress. A Bayesian Newtork was developed for representing the probabilistic crack growth phenomenon and estimating the RFL. Finally, reliability curve coupledwith threshold reliability value, was utilized to arrive at the inspection plan of the SBP. In this chapter the input values to the virtual ensors were entered manually for the sake of demonstration, howver, if the DT is deployed on cloud, it is also possible to direlty feed real time process parameters int the DT and estimate RFL. Likewise, more input process parameters such as tempterature, pH, etc. can aslo act as input to the DT. Although the DT framework has been deeloped for RFL estimation of SBPonly, it can easily be extended to other assets on a praocess faiclity and also for other degradation mechanisms such as corssion, erosion, etc.

References

HSE-OTR 028. (2002). *Offshore hydrocarbon releases statistics and analysis*. HID Statistics Report by Health Safety Executive, UK.

Allen, M. P., & Tildesley, D. J. (1987). *Computer simulation of liquids*. Oxford Scholarship.

Bagalkot, N., Keprate, A., & Orderløøkken, R. (2021). Combining computational fluid dynamics and gradient boosting Regressor for predicting force distribution on horizontal axis wind turbine. *Vibration*, *4*(1), 248–262.

Baggett, R. T. (2019). Inspect your primary process small bore piping the right way, using profile radiography. *Inspectioneering Journal*, *25*(2), 34–37.

Botín-Sanabria, D. M., Mihaita, A.-S., Peimbert-García, R. E., Ramírez-Moreno, M. A., Ramírez-Mendoza, R. A., & Lozoya-Santos, J.d. J. (2022). Digital twin technology challenges and applications: A comprehensive review. *Remote Sensing, 14*, 1335.

El Saddik, A. (2018). Digital twins: The convergence of multimedia technologies. *IEEE Multimedia, 25*(2), 87–92.

Fiates, J., & Vianna, S. S. (2016). Numerical modelling of gas dispersion using OpenFOAM. *Process Safety and Environmental Protection, 104*, 277–293.

Forrester, A. I., Bresslof, N. W., & Keane, A. J. (2006). Optimization using surrogate models and partially converged computational fluid dynamics simulations. *Proceedings of the Royal Society A: Mathematical, Physical and Engineering Sciences, 462*(2071), 2177–2204.

Forrester, A. I. J., Sobester, A., & Keane, A. J. (2008). *Engineering design via surrogate modelling*. UK: Wiley. ISBN: 9780470060681.

Grieves, M. W., & Vickers, J. H. (2017). Digital twin: Mitigating unpredictable, undesirable emergent behavior in complex systems. In *Transdisciplinary perspectives on complex systems* (pp. 85–113). Springer International Publishing. https://doi.org/10.1007/978-3-319-38756-7_4.

EI Guideline. (2013). *Guidelines for the design, installation and management of small bore tubing assemblies*. UK: The Energy Institute London.

EI Guidelines. (2007). *Guidelines for the avoidance of vibration induced fatigue failure in process pipework*. UK: The Energy Institute London.

Haag, S., & Anderl, R. (2018). Digital twin—Proof of concept. *Manufacturing Letters, 15*, 64–66.

Jeon, K., Yang, S., Kang, D., Na, J., & Lee, W. B. (2019). Development of surrogate model using CFD and deep neural networks to optimize gas detector layout. *Korean Journal of Chemical Engineering, 36*(3), 325–332.

Keprate, A., Ratnayake, R. M. C., & Sankararaman, S. (2017a). Adaptive Gaussian process regression as an alternative to FEM for prediction of stress intensity factor to assess fatigue degradation in offshore pipeline. *International Journal of Pressure Vessels and Piping, 153*, 45–58.

Keprate, A., Ratnayake, R. M. C., & Sankararaman, S. (2017b). Comparing different metamodelling approaches to predict stress intensity factor of a semi-elliptic crack. In *International conference on offshore mechanics and arctic engineering*. American Society of Mechanical Engineers.

Keprate, A., Ratnayake, R. M. C., & Sankararaman, S. (2019). Experimental validation of the adaptive Gaussian process regression model used for prediction of stress intensity factor as an alternative to finite element method. *Journal of Offshore Mechanics and Arctic Engineering, 141*(2), 021606. https://doi.org/10.1115/1.4041457.

Kumaresh, G. R., Fossan, I., Venkatraman, M. M., Gilarhus, K. E., & Jensen, S. (2016). CFD-based transient ignition probability modeling of gas leaks in enclosures. *International Journal of Computational Methods, 13*(02), 1640006. https://doi.org/10.1142/s0219876216400065.

Lee, J., Cameron, I., & Hassall, M. (2019). Improving process safety: What roles for digitalization and Industry4.0? *Process Safety and Environmental Protection, 132*, 325–339.

Loy, Y., Rangaiah, G. P., & Lakshminarayan, S. (2018). Surrogate modelling of net radiation flux from pool fires in a hydrocarbon storage facility. *Process Safety and Environmental Protection, 114*, 296–309.

Min, Q., Lu, Y., Liu, Z., Su, C., & Wang, B. (2019). Machine learning based digital twin framework for production optimization in petrochemical industry. *International Journal of Information Management, 49*, 502–519.

Moore, W. O. W., Jetmarova, H. M., Gebreslassie, M., Tabor, G. R., Belmnot, M. R., & Savic, D. A. (2016). Comparison of multiple surrogates for 3D CFD model in tidal farm optimisation. *Procedia Engineering*, *154*, 1132–1139.

Rasheed, A., San, O., & Kvamsdal, T. (2020). Digital twin: Values, challenges and enablers from a modeling perspective. *IEEE Access*, *8*, 21980–22012.

Sankararaman, S., Ling, Y., & Mahadevan, S. (2011). Uncertainty quantification and model validation of fatigue crack growth prediction. *Engineering Fracture Mechanics*, *78*(7), 1487–1504.

Tao, F., Zhang, M., & Nee, A. Y. C. (2019). Digital twin driven PHM. In F. Tao, M. Zhang, & A. Y. C. Nee (Eds.), *Digital twin driven smart manufacturing* Academic Press.

West, T., & Blackburn, M. (2017). Is digital thread/digital twin affordable? A systemic assessment of the cost of DoD's latest Manhattan project. Complex adaptive systems. *Procedia Computer Science*, *114*, 47–56.

Wilson, B., Wakes, S., & Mayo, M. (2017). Surrogate modeling a computational fluid dynamics-based wind turbine wake simulation using machine learning. In *IEEE symposium series on computational intelligence (SSCI)*.

CHAPTER SEVENTEEN

Resilience analysis of digitalized process systems

Rioshar Yarveisy[a,*], Hao Sun[b,c], Ming Yang[b,*], and Hans Pasman[d,*]
[a]C-RISE, Memorial University of Newfoundland, St. John's, NL, Canada
[b]Safety and Security Science Section, Department of Values, Technology, and Innovation, Faculty of Technology, Policy, and Management, Delft University of Technology, Delft, The Netherlands
[c]College of Mechanical and Electronic Engineering, China University of Petroleum (East China), Qingdao, China
[d]Mary Kay O'Connor Process Safety Center, Artie McFerrin Department of Chemical Engineering, Texas A&M University, College Station, TX, United States
*Corresponding authors: e-mail address: ryarveisy@mun.ca; m.yang-1@tudelft.nl; hjpasman@gmail.com

Contents

1. What is resilience?

The relatively constant rate of using the word resilience until the 1970s and 80s has experienced a steep rise since the early 2000s. Resilience has become a common term in many fields such as ecology, psychology, economy, sociology, politology, and engineering. Even though resilience is no more an obscure term, it is not apparent enough either. A clear definition of resilience is the first step in understanding the concept, implications, and scope.

Methods in Chemical Process Safety, Volume 6
ISSN 2468-6514
https://doi.org/10.1016/bs.mcps.2022.05.002

1.1 Resilience definitions

Originally, resilience was defined as the "act of rebounding" (derived from the Latin verb resilire meaning "to jump back" or "recoil," which comes from the verb salire itself, which means "to leap") and later on associated with "springing back" or "elasticity" in mechanics. However, Holling (1973) first coined the term resilience as a metaphor meaning the persistence of relationships in a system and a measure of its ability to absorb change. Holling believed persistence or extinction in the context of ecology is the result of system resilience. This definition has influenced later attempts to establish the essence of resilience.

Resilience Engineering (RE) is a concept concerned with promoting resilient performance to achieve higher levels of safety through developing its four cornerstones, i.e., respond, monitor, anticipate, and learn. The evolution of RE definitions observed in the works of Hollnagel provides valuable insights into the progress of resilience as a paradigm for safety.

The first definition (Hollnagel, 2006, p. 16) states: "The essence of resilience is, therefore, the intrinsic ability of an organization (system) to maintain or regain a dynamically stable state, which allows it to continue operations after a major mishap and/or in the presence of a continuous stress."

The second definition by Hollnagel (2008, p. xii) and Hollnagel (2009b, p. 117) suggests a resilient system is defined by "its ability to effectively adjust its functioning before or following changes and disturbances so that it can continue its functioning after a disruption or a major mishap and in the presence of continuous stresses."

The third working definition describes resilience as "The intrinsic ability of a system to adjust its functioning before, during or following changes and disturbances so that it can sustain required operations under both expected and unexpected conditions." (Hollnagel, 2011).

From the above definitions of RE, two progressing aspects stand out: (1) extension of the time horizon of the resilient action from following a disruption to prior and during the adverse event. This progression reflects the importance of anticipation and the need to gain foresight about possible threats in the future. (2) The progression of the limited scope of threats from major mishaps and continuous stress to the inclusive level of expected and unexpected conditions. This attempt to include a wide-ranging spectrum of threats emphasizes the critical role of resilience in dealing with uncertainties and overcoming unexpected blows as effectively as possible.

Supply chain resilience is recognized and studied as the opposite of vulnerability. In supply chains, vulnerability may be defined as being at risk or prone to damage (Peck, 2006). Barroso, Machado, and Cruz Machado (2009) define vulnerability as the "incapacity of the supply chain to react to the disturbances at a given moment." Like other fields, supply chain resilience literature lacks consensus on its definition (Tukamuhabwa, Stevenson, Busby, & Zorzini, 2015). Nevertheless, Ponis and Koronis (2012) provide a comprehensive definition for supply chain resilience as "the ability to proactively plan and design the supply chain network for anticipating unexpected disruptive events, respond adaptively to disruptions while maintaining control over structure and function and transcending to a post robust state of operations, if possible a more favorable one than that prior to the event, thus gaining a competitive advantage."

Duchek (2020) suggests the expectations from ***organizational resilience*** could be divided into three points of view; (1) resilience as a reactive ability to resist adversities and/or recover to a state of normalcy; (2) resilience as an adaption, integration, and reconfiguration strategy which goes beyond the scope of maintenance and recovery; (3) resilience as proactive readiness, to assure organizational survival and growth in adverse conditions.

The first group could be divided into three subcategories: (a) "precursor resilience" (Boin & van Eeten, 2013), focused on absorption of adverse impacts. (b) Focused on recovery, "recovery resilience" (Boin & van Eeten, 2013) and the ability to "bounce back to a state of normalcy." (c) Resilience as the combined effect of absorption and recovery "the organizational capacity to absorb the impact and recover from the actual occurrence of an extreme weather event" (Linnenluecke & Griffiths, 2012) The second group extends the scope of resilience further by incorporating adaption to the impact of adversities into the first group's perspective of organizational robustness. According to Hamel and Valikangas (2003), this attribute enables organizations to "dynamically reinvent" themselves. In this context, Lengnick-Hall, Beck, and Lengnick-Hall (2011) define resilience as "a firm's ability to effectively absorb, develop situation-specific responses, and ultimately engage in transformative activities to capitalize on disruptive surprises that potentially threaten organization survival." The third group extends the scope by including the requirements for anticipation, implying the need for proactive adaption by predicting and preventing potential hazards. Prevention in this context does not mean avoidance but rather the capacity to act more timely and vigilantly. Anticipation is made up of the ability to observe internal and external organizations, prepare for unexpected events, and capacity to identify emerging threats (Duchek, 2020).

The combined points of view explored through the above definitions culminate in the idea of a system: prepared to face familiar and unknown adversities with undetermined severity; able to withstand their adverse effects; continue operations in an impaired but stable state; ready to recover to improved operational qualities following the disturbance; potentially improve its performance compared to its pre-disruption state by taking advantage of the adversities. Although the reviewed definitions suggest a relatively unified point of view on what constitutes resilience, the contrasting feature is the definitions of adversity and success. The ambiguity in experienced adversity and resilient performance could cause challenges when dealing with individual systems. It is helpful to adapt resilience definitions suitable for specific systems based on their structure. As Ayyub (2014) suggests, operational resilience definitions should enable quantification, i.e., adopting an appropriate definition is a requisite for effective resilience analysis. Hence, adopting a proper definition that envelopes the adversity types and resilience capacities involved with digitized chemical process industries could prove advantageous.

It is essential to visualize the chemical process industries as multiple technical subsystems nested in multi-layered organizations, constituting vital global infrastructures that serve numerous supply chains. The supply chain, organizational, and RE definitions of resilience, display a lack of detail when discussing the system's resilience performance and the hazards. When concerned with digitized industries, in addition to pre-existing operational hazards, rapid technology integration comes along with a range of unprecedented hazards. In addition to increased human interaction with the system, e.g., focussed on change in design and operational control, unknown implications of new technology, and issues with software reliability, increased vulnerability to external threats is a concern. As examples of the latter, cyberattacks on Iran's central gasoline rationing system (BBC, 2021) and the ransomware attack on Colonial Pipeline (Kelly & Resnick-ault, 2021) show how limited surprise attacks could impact state-wide functions. Hence, inclusive resilience definitions developed by policymakers with critical infrastructure in mind could be a proper foundation for defining the digitized chemical industries.

The ASCE (2021) defines resilience of critical infrastructure as "The ability to plan, prepare for, mitigate, and adapt to changing conditions from hazards to enable rapid recovery of physical, social, economic, and ecological infrastructure. Improving resilience before or following a hazard event should engage physical infrastructure and social systems with adaptive capacity to ensure a rapid return to functionality, accounting for interdependencies within and across all sectors."

Ayyub (2014) provides an inclusive definition of system resilience by building upon US President's PPD-21 (2013) "Resilience notionally means the ability to prepare for and adapt to changing conditions and withstand and recover rapidly from disruptions. Resilience includes the ability to withstand and recover from disturbances of the deliberate attack types, accidents, or naturally occurring threats or incidents. The resilience of a system's function can be measured based on the persistence of a corresponding functional performance under uncertainty in the face of disturbances."

The above definition states the need for *anticipation, absorption, adaption*, and *recovery* from the impact of both internal and external threats, whether they are deliberate or not. Although this definition is not concerned with safety alone, it includes the elements required for the safety of digitized process systems.

2. Why seek resiliency?

The conventional approach to safety assumes that unwanted outcomes, i.e., accidents and incidents, can be described by decomposing the system and finding the failing component causal chains. Among others, Hollnagel (2014) discusses that the assumptions and approach of conventional safety are no more valid, and a new point of view should be adopted concerning socio-technical systems. He argues that the deficiency of preventive safety approaches are due to the unprecedented rate of technology development, changes to ever-growing organizational structures, and the increasingly complex relationship between humans and automation. Additionally, potential consequences of industrial failures increase as technology advances, decreasing the tolerance toward accidents. Hence, both the technical aspects and the public pressure are incentives for improving safety by incorporating resilience into organizations.

According to Marsh (2020), the analysis of the 100 major accidents in the hydrocarbon industry reveals no discernible reduction in the number or cost of such events since 1979. Suggesting advances in safety, even though effective, may not be sufficient. Jarvis and Goddard (2017) explore the causes of the 100 major onshore oil, gas, and petrochemical losses from 1996 to 2015. This study implicates the primary system failure causes as:

- Mechanical failures for 43% of losses, of which 70% are due to corrosion
- Operations, practices, and procedures for 25%
- Control of work 21%
- Process hazard analysis for 4% directly and 60% of non-mechanical failures as the secondary cause
- Management of change for 4%
- Safety-critical systems for 3%

The above statistics illustrate the impact of management system failures on accident occurrence. Moreover, nearly 60% of mechanical failures were due to management issues, such as inadequate inspection programs. Lastly, hazard identification is implicated in more than half of losses due to management failures. This section focuses on two aspects of industrial environments resilience could improve before discussing the challenges and benefits of digitization.

2.1 The organizational aspect

Even after repeated major incidents such as Bhopal (1984), Piper Alpha (1988), BP Texas City Refinery (2005), Deepwater Horizon (2010), and West Fertilizer Plant explosion (2013) where safety culture is implicated as one of the root causes, disregarding safe behavior to benefit other mostly short-term gains is still common. The House Committee of Transportation & Infrastructure's report on Boeing 737 MAX (2020, p. 14) (Final Committee Report, 2020) makes a note of this issue in the Aviation Safety Organization (AVS) by stating, "many believe that AVS senior leaders are overly concerned with achieving the business-oriented outcomes of industry stakeholders and are not held accountable for safety-related decisions." In an earlier instance, the official investigation into the Columbia disaster states, "In the Board's view, NASA's organizational culture had as much to do with this accident as the External Tank foam" (Columbia Accident Investigation Board, 2003).

These instances spanning close to four decades underline the necessity of developing the high consequence industries around safety.

API recommended practice 754 (API, 2010) categorizes the safety indicators in the refining and petrochemical industries into four groups. Tier-1 and Tier-2 events signify lagging indicators, i.e., loss of containment with greater and lesser consequences, respectively. On the other hand, Tier-3 and Tier-4 are challenges to safety systems and dents to operating discipline & safety management performance, hence countable as leading indicators. Mannan, Reyes-Valdes, Jain, Tamim, and Ahammad (2016) regard dependence on lagging indicators (conventional safety practices) over leading indicators (safety culture) as a "*risky practice.*" Emphasizing that sustainable safety cannot be achieved without a strong safety culture.

Dinh, Pasman, Gao, and Mannan (2012) highlight the importance and the influence of safety culture on organizational performance. The authors point out how a positive safety culture starts from management and subsequently takes over. The authors separate resilience engineering as an essential component of process safety. From an earlier point of view, Olive, O'Connor, and Mannan (2006) present resilience as a "feature of a good safety culture." They discuss that commitment to safety, inclination toward resilience, and flexibility with an attitude of constant vigilance are indicators of a good safety culture. Olive et al. (2006) call attention to the importance of separating safety climate from safety culture by exploring the interval between the Challenger space shuttle explosion (1986) and the Columbia disaster (2003). They discuss how the Challenger explosion created a climate supporting safety at NASA, yet the underlying organizational culture did not prioritize safety, leading to the Columbia disaster.

The risk-oriented approach of conventional safety does not guarantee the methodical merging of organizational values with safety practices. Hence, risk management exists as a feature of the organization, not its essence. Sequentially, safety culture should be integrated separately to promote safe behaviors, leading to a potential duality between safety practices and organizational values. This potential duality of safety culture and climate emphasizes the necessity of administering a structured safety-oriented framework capable of managing the complexities of socio-technical systems. This goal could be achieved through a paradigm shift to resilience. Resilience promoting concepts effectively blend safety culture and practice into one concept. They hypothesize that safety is the outcome of mindful and safety-oriented interaction with the system based on deep knowledge about the operation and its variations.

2.2 Technical factors

Advances in science and engineering have resulted in increasingly complex socio-technical systems. These complexities could result in deviation from the safe operation and cause incidents or accidents. One drawback of conventional safety approaches is focusing on resultant outcomes where cause and effect are linear. Due to the influence of organizational, human, and financial factors on operations in complex socio-technical systems, the causes are elusive and non-linear. In other words, due to multi-layered organizational structures and increased human interaction with the system, the causes of accidents are not clear. Hence, one drawback of safety is the failure to identify hazards due to a flawed understanding of the system and the people who interact with it. These hazards occur as emergent conditions resulting from the interaction between system components under specific conditions. These particular conditions occur due to stochastic variations in system parameters or performance variability of human and organizational factors (Hollnagel, 2017).

Relying solely on risk assessment to analyze what could happen may result in catastrophic ramifications even when utilizing quantitative and probabilistic methods. Firstly, imagining and foreseeing all possible hazardous scenarios is unachievable. Secondly, PRA, probabilistic risk analysis, requires probabilistic modeling techniques such as fault trees, event trees, or Bayesian networks. The issue with the probabilistic approaches is assigning probabilities to events as if they are mutually exclusive or collectively exhaustive. Assuming exclusivity of initial events requires that underlying events do not occur simultaneously. This assumption overlooks that these events may have a common cause unnoticed by the practitioners. Besides, the assumption of exhaustivity, for instance, in event trees, ignores the possibility of events or conditions unknown to the practitioner. Additionally, PRA's reliance on historical data diminishes its ability to consider rare events effectively. Moreover, the fast-paced technology development and integration in the industries will inevitably cause unprecedented hazardous conditions. Furthermore, the data on the novel hazards such as deliberate attacks and the impact of climate change by extreme weather effects and flooding is very limited. This lack of data further impairs probabilistic results used in quantitative risk assessments.

Another flawed aspect is the approach to considering human factors. Leveson (2016) points out that the statistics on human error are inconsistent with reality, as assigning blame to operators is a convenient stopping point in the causal analysis. The author suggests: (1) the data misrepresents human intervention as they do not reflect the times where operator action is successful. (2) The blame might be assigned due to unrealistic expectations. (3) The operators might intervene in conditions outside the design boundaries of the system. (4) It is easy to come up with correct solutions after the fact but acting correctly in the heat of the moment before control is lost is very difficult. Traditional safety fails to capture the impact of underlying causes such as system design, inconsistencies in mental models, and societal factors.

Lastly, the limited time horizon considered when assessing the risks ignores the dynamic nature of systems. In conventional risk assessment, the hazards and their impact are studied. Meanwhile, underlying organizational, economic, and human factors might have contributed to the root cause over long periods. In the Texas City Refinery explosion with 15 killed, 180 injured, and more than $2 billion in compensation, management believed deviation from start-up procedures caused the accident. Meanwhile, various investigations (CSB, 2007) revealed causes, including low-quality equipment and instrumentation forcing the operators to deviate, weak safety culture, human factors, and mismanagement of change as the root causes of the accident.

From the above, it can be concluded that risk assessment deficiencies are either due to a disability to identify hazards, determine the causes, or access to data. These three issues could be summarized as uncertainty about the socio-technical system. Some challenges posed by some uncertainties may be reduced by diligent study of the system, environment, and people. On the other hand, the practically irreducible uncertainties, such as tightly coupled non-linear component interactions (Perrow, 2000) and dysfunctional ones (Leveson, 2016), will inevitably cause unexpected accidents. Resilience-oriented concepts promise survival by increasing the readiness to deal with such occasions through resilient system capacities. Hence, in cases where avoidance is impossible, operational flexibility and effective resource allocation will serve operational continuity and speedy recovery. However, resilience analysis and integrating strengthening measures presuppose that thorough risk assessments have been performed (Pasman, Kottawar, & Jain, 2020). Shadow side may be possible misuse and lack of precaution for short-term gain (Ale, Hartford, & Slater, 2021).

Lee, Cameron, and Hassall (2019) suggest that digitization of chemical process industries could improve the mechanical issues by introducing early warning signaling, corrosion monitoring, remote sensing, increased connectivity, application of predictive models relying on real-time data, and machine learning. Moreover, they suggest that digital systems could improve work processes, risk assessment, operator interfaces, and alarm management systems.

Introducing new technology could improve safety culture by controlling procedural compliance and effective training. Implementing sensing technologies, real-time data analysis, and integrating dynamic system models into digital twins could promote safe operation. Goal-oriented data collection, e.g., on key performance indicators, can reduce uncertainties and improve predictive models. Meanwhile, the overlooked aspect is the system's increased complexity, which inevitably leads to hazardous conditions on its own or due to wrong interventions by the operators due to their deficient mental image of the system's processes. Furthermore, the rapid introduction of technologies without rigorous training alienates the operators from the foundation of the systems. This shows that even though digitization could improve our knowledge of system functions in some aspects, it also decreases our understanding by introducing additional system layers. Hence, the need for resilience-oriented operations remains intact while conventional safety's effectiveness further diminishes.

3. How to attain resilience?

This section divides the literature on resilience conceptualization into two hierarchical categories: (1) resilience promoting concepts that encourage systems thinking focused on safe operation. (2) Resilience frameworks developed based on the new systems thinking. This hierarchy reflects the logical progression from resilience concept development to implementation.

3.1 Resilience promoting concepts

Vaezie, Peterson, and Bourne (2019) define *High Reliability Organization (HRO)* as an "organization that experiences fewer than anticipated accidents or events of harm, despite operating in highly complex, high-risk environments." HROs achieve reliable performance by pursuing two approaches, prevention (anticipation) and resilience (containment) (Schulman, 2004). Known too is the promotional work by Weick and Sutcliffe (2011) Prevention relies on the capacity of members to anticipate the failures, identify scenarios that may lead to their occurrence, and devise procedures to avoid them. HROs, parallel to their preventive measure, develop capacities to detect, isolate, and recover from unavoidable failures caused by the uncertainties they cannot foresee. Roberts, Madsen, Desai, and van Stralen (2005) highlight several properties of HROs as diligent technology and procedure design, continuous improvement, frequent processes audit and analysis, well-trained personnel, and continuous training.

Resilience Engineering (RE) aims to put forward a new way of thinking about safety. Hollnagel (2008) identifies four cornerstones for resilient systems: (1) the ability to *respond* to both normal and rare disruptions and threats. (2) The ability to flexibly *monitor* both the system's own performance and outside threats. (3) The ability to *anticipate* adversities and their consequences in the medium to long-term. (4) The ability to *learn*. According to psychologist and human factor specialist Hollnagel (2011), resilience engineering aims to increase safety by increasing instances where things go right rather than decreasing the things that go wrong. This goal is achievable by observing and understanding how socio-technical systems normally perform, which provides enough information on how things go wrong. This observation and understanding are achieved by developing the four cornerstones.

Hollnagel (2014), Safety-I and Safety-II, puts forward another concept concerned with improving the state of safety through changing the point of view about how it should be achieved. In contrast to Safety-I, where the goal is to ensure as little as possible goes wrong, Safety-II focuses on assuring as much as possible goes right. Hollnagel points out that the increasing complexity results in increasing interactions. Due to not being fully aware of how work gets done, one cannot be sure how things might go wrong. To resolve this issue, adding flexibility to procedures will result in resilience when facing variations. Hollnagel defines Safety-II in congress with resilience as "the ability to succeed under expected and unexpected conditions alike so that the number of intended and acceptable outcomes ... is as high as possible." (Hollnagel, 2014, p. 134). The weakness of this approach has been made clear by the objections from a safety engineering point of view formulated by Leveson (2020).

From the above, it could be concluded that resilient systems are established through:

(1) diligent technology development and integration; (2) thorough procedure design; (3) in-depth study of system functions; (4) mindful system operation and procedural compliance; (5) recognition of immediate and long-term threats; (6) careful observation of the system and environment; (7) the flexibility to adapt to adverse conditions; (8) focus on safe system operation.

The digitized process industry possesses the tools required to adopt resilience-oriented safety-promoting concepts. Safety culture, risk management programs, probabilistic risk assessment, dynamic simulation techniques, digital twins, increased connectivity, sensing, data collection, and object-oriented interface design are some examples of such resources. The challenging aspect is re-organizing and repurposing available resources to accommodate resilience-oriented safety management. Resilience building frameworks discussed below are examples of how this task could be approached.

3.2 Resilience building frameworks

The literature on resilience-building frameworks is varied and numerous. This section looks into a general resilience-building framework and two field-specific frameworks concerned with resilience building in the chemical process industries. These frameworks provide procedures and strategies to improve system safety through resilience.

First, the socio-technical system (STS) analysis approaches: Following Rasmussen's original idea (Rasmussen, 1997), Leveson (2016) developed a socio-technical system-theoretic accident model and processes (STAMP) and derived from this, the *system-theoretic process analysis (STPA)* method to find any lack of control on the various hierarchical levels of an STS from the top regulator down to the physical operating process. In principle, all control loops can be checked on deficiencies being control action not provided, control action unsafe, too early or too late or out of sequence, or control action stopped too early or applied too long. Due to the complexity of many systems, control loops come in large numbers.

Hollnagel developed the *Functional Resonance Analysis Method (FRAM)*, mimicking a network of connected system functions as an accident analysis and risk assessment method (Hollnagel, 2004), which later evolved into its current form (Hollnagel, 2012) Because internal aspect details of the six function elements can be ignored, the method is rather convenient in use. FRAM is formulated based on four principles:

- *Equivalence of failure and success*, i.e., things go wrong the same way they go very right.
- The *approximate adjustments* reflect the variability in organizational and human performance. The successful operation of socio-technical systems requires flexibility. The actions resulting from flexible adjustments are approximate as there is insufficient time or resources to explore every action thoroughly.
- *Emergence* articulates that not all hazardous conditions are products of causal relations. Rather, interaction or combination of specific conditions cause emerging conditions that leave no clear trace for causal analysis. Hence, the event is no more resultant rather emergent.
- *Functional resonance* is the detectable signal that originates from the inadvertent interaction of multiple signals due to their approximate

adjustment. Functional resonance aims to highlight the somewhat predictable performance of people and organizations due to approximate adjustments as these adjustments follow recognizable patterns rather than being random.

Jain, Pasman, Waldram, Pistikopoulos, and Mannan (2018) put forward the *Process Resilience analysis Framework (PRAF)* as a risk assessment and management approach. PRAF utilizes systems thinking to include organizational and human factors, in addition to purely technical aspects in risk management. Jain, Pasman, Waldram, Pistikopoulos, and Mannan (2018) define the phases of process resilience as avoidance, survival, and recovery and put forward a holistic quantitative approach by suggesting using resilience metrics for quantification based on system performance. PRAF integrates data-driven methods, statistical approaches, Bayesian analysis, and dynamic simulation to provide an effective methodology. Considering the importance of profitability and feasibility in industrial environments, the authors incorporate cost-effectiveness as an objective function.

Jain, Rogers, Pasman, Keim, & Mannan, 2018; Jain, Rogers, Pasman, & Mannan, 2018) developed as part of PRAF the *Resilience-based Integrated Process Systems Hazard Analysis (RIPSHA)* as a bi-layered framework for resilience-oriented hazard identification. RIPSHA divides the chemical process industries into two layers: the management system and the plant system. The management system layer comprises safety culture and leadership, operational discipline, and process safety system. The plant system layer includes equipment, procedures, and human hazards. The system requirements for resilient operation are early detection, error-tolerant design, plasticity, and recoverability which correspond to monitor, anticipate, respond, and learn. When dealing with the physical system, the authors exploit the familiarity and widespread use of the HAZOP and enhance its framework to include various states of operation, procedures, and human factors to reflect the impact of emerging conditions and performance variability in their hazard assessment.

The general frameworks such as Leveson's STAMP/STPA (Leveson, 2004) and Hollnagel's FRAM suggest adopting a systems thinking point of view and integrating behavioral aspects. They are more concerned with the inner workings of the systems and interactions. On the other hand, field-specific frameworks focus on the system of interest and provide tailored approaches to analyze resilience or integrate it into existing technical or organizational structures. They are focused on their understanding of the system and try to include previously overlooked or mismanaged factors in their risk management programs.

When examining the prospects of digitized process systems, it is evident that progress in this direction will inevitably give rise to two hazard types, threats posed directly by the integrated technology and conditions resulting from interaction with the technology. Hazard identification when dealing with such novel conditions requires involving parties with expertise in the IT infrastructure, hardware, and software, in addition to personnel involved with the process systems. Although the mentioned methods can fulfill this new requirement for hazard identification in principle, it could become a pitfall leading to future challenges or beneficial for integrating resilience into the system. Taking advantage of the need for hazard identification to develop software, interfaces, and procedures according to real-life expertise with resilience in mind could prevent hazardous conditions.

In contrast to the potentially detrimental impact of technology integration, typical digital enablers, such as data collection, procedural compliance monitoring, and connectivity could escalate resilience implementation through quantification. Many resilience frameworks do not provide quantified measures, or if they do, the units are not tangible. Generic resilience metrics quantify resilience relying on the system's state and phases of resilient performance following a disruption. Goal-oriented data collection from the organization's technical and societal components could help assess system resilience and evaluate the efficacy of resilience-building strategies.

4. Resilience assessment methods

As mentioned, resilience is composed of four main capacities: anticipation, absorption, adaptation, and restoration. Anticipation is that the system denoises the process data and extracts valuable information (e.g., using statistic methods, machine learning, and artificial intelligence) and predicts the operating trend of the system based on the data to predict and prevent internal and/or external disruptions. Besides, anticipation consists of the ability to prepare for unexpected events and the capacity to identify emerging threats (Duchek, 2020). Besides recognizing signals as a looming threat, it also requires convincing top management to initiate action (Pasman, 2021). Absorption refers to the capacity of a system to withstand and resist disruptions and mitigate the impact of disruptions on a system. It determines the minimal performance of a system after disruptions. Adaptation is the capacity of a system to recover its performance without external maintenance activities. Restoration indicates the capacity of a system to restore its performance within external maintenance activities. To quantify and measure resilience, various methods and metrics are developed by peer researchers.

4.1 Resilience assessment methods

As discussed above, various methods have been developed to assess resilience. These types of methods are divided into four categories to introduce and describe them:

- Indicator/Index-based methods
- Statistical methods
- System performance-based methods
- Simulation-based methods

4.1.1 Indicator/Index-based methods

Indicator/Index-based approaches are a convenient tool for resilience assessment in engineering systems. They are usually utilized in the preliminary design and evaluation of complex systems (Kammouh, Gardoni, & Cimellaro, 2020). The general operation steps of this type of method are:

- (i) Determining main indicators which contribute to resilience;
- (ii) Constructing the quantitative relationship between resilience and its indicators;
- (iii) Assessing resilience through quantifying the main indicators.

Wu, Hou, and Chen (2021) developed an indicator-based method, which includes two indicators, i.e., traffic efficiency and safety performance, to assess the resilience of transportation networks during the post-earthquake long-term recovery period.

Kammouh et al. (2020) presented an indicator-based method, using three indicators, i.e., vulnerability, robustness, and recoverability, to quantify the resilience of engineering systems. Each indicator has its own sub-indexes. A dynamic Bayesian network represents the logic and structure relationship between resilience and these three indicators and sub-indexes, as shown in Fig. 1 and Table 1. Their quantitative relationships are demonstrated by conditional probability tables (CPTs). After that, the system's resilience can be determined.

Zhang, Xu, and Dui (2020) proposed an indicator-based method, including the node and edge indicators, to measure the resilience of network systems. The primary step of the proposed method is constructing matrices for node resilience and edge resilience based on the characteristics of nodes

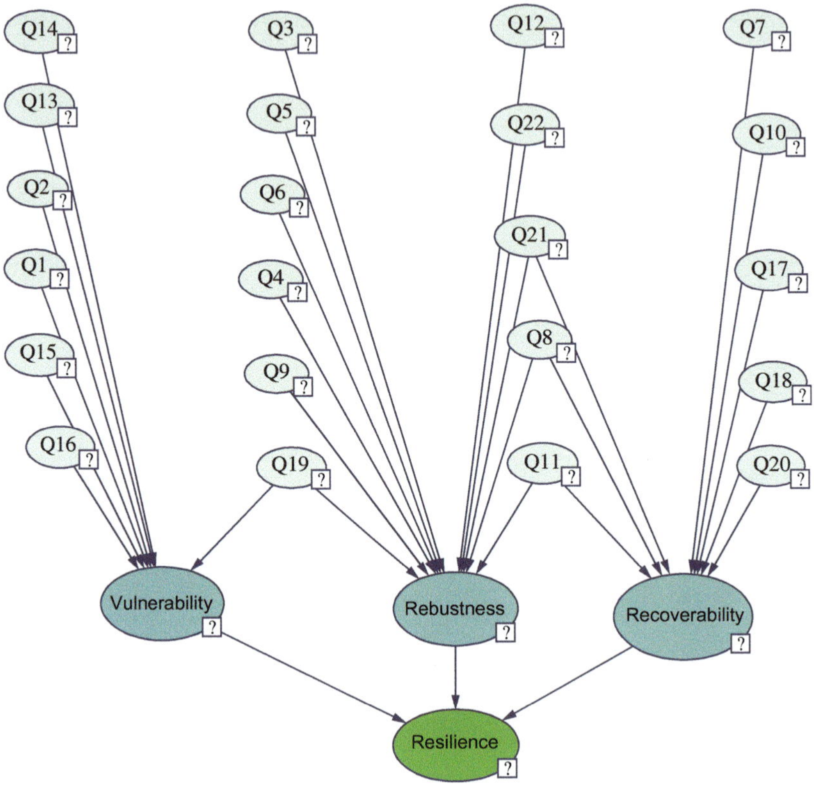

Fig. 1 The Bayesian network. *Adapted from Kammouh, O., Gardoni, P., Cimellaro, G.P. (2020). Probabilistic framework to evaluate the resilience of engineering systems using Bayesian and dynamic Bayesian networks.* Reliability Engineering and System Safety, 198.

and edges during failure. The network system resilience is then quantified by calculating these two types of matrices.

Bragatto, Vairo, Milazzo, and Fabiano (2021) developed an indicator-based approach, including four main indicators, i.e., anticipate, monitor, react, and learn, to quantify the resilience of chemical process systems. Analytic Hierarchy Process (AHP) is used to determine the weights assigned to each indicator and sub-indexes.

Tong, Yang, and Zinetullina (2020) came up with an indicator-based method by using a dynamic Bayesian network to represent the relationship between four main indicators, i.e., absorption, adaptation, restoration, and learning, to quantify the resilience of chemical process systems. According to the process system, the main contributors of each indicator are identified. For example, absorption consists of two sub-indexes, redundancy and

Table 1 The indexes of the developed Bayesian network (ddr means disaster risk reduction).

Indexes	Descriptions	Indexes	Descriptions
Q1	National policy and legal framework for ddr	Q12	Country-wide public awareness strategy
Q2	Dedicated and adequate resources	Q13	Ddr is an integral objective of policies
Q3	Community participation and decentralization	Q14	Social development policies implementation
Q4	National multisectoral platform for ddr	Q15	Economic and productive sectorial policies
Q5	National and local risk assessment	Q16	Planning and management incorporate ddr
Q6	Monitoring systems to disseminate data	Q17	ddr measures embedded in recovery
Q7	Early warning widespread systems	Q18	Assessment procedures of disaster impact
Q8	Transboundary and regional risk assessment	Q19	Policy and mechanisms for disaster risk management
Q9	Information on disasters accessible at all levels	Q20	Disaster preparedness plans and contingency plans
Q10	Education on ddr and recovery	Q21	Financial reserves and contingency mechanisms
Q11	Cost-benefit analysis and research methods for RA	Q22	Exchange information

robustness. These two sub-indexes together determine the absorption capacity of the process system.

As mentioned above, it can be seen that various indicators are proposed to evaluate resilience in different fields and domains. In digitalized process system, the prevailing view is that resilience is composed of four main capacities, i.e., anticipation, absorption, adaptation, and restoration. Researchers focused on the latter three when quantifying system resilience using an indicator-based method, while scant attention has been paid to the anticipation capacity. This research gap should be filled in future work.

4.1.2 *Statistical methods*

Statistic-based approaches employ historical databases and/or monitoring data to determine critical information of system key variables to quantify resilience.

The general operational steps of this type of statistical method are:

- (i) Ascertaining information and data of system key variables from the database;
- (ii) Building the mathematical relationship between resilience and key variables;
- (iii) Quantifying resilience based on the database and the developed relationship.

Shen, Cassottana, and Tang (2018) proposed a statistic-based approach to determine the resilience of a power grid. The presented methodology considers three critical variables, i.e., the time between disruptions, the performance loss of each disruption, and the time needed for recovery, and integrates them to measure the power grid resilience.

Ji et al. (2016) analyzed data from four regions of New York during Super Storm Sandy and daily operations to determine the relationship between resilience and infrastructural failures and recoveries and cost.

Ilbeigi (2019) proposed a statistic process control to monitor the topological features of the transportation network over time to detect unusual patterns caused by extreme events and quantify transportation systems' resilience.

Nelson, Johnson, Fahy, and Hansen (2020) presented a statistic-based method to assess the resilience of grid-connected microgrids. In this study, survivability is the key variable expressed as the probability of a microgrid satisfying critical load requirements during an islanding event. The information and data came from an asset-level reliability database.

4.1.3 *System performance-based methods*

System performance-based approaches have been widely used in various fields (e.g., infrastructure systems, process systems, engineering systems, lifeline networks, et cetera) to assess resilience. In this method, the key is to determine the system's performance response function (PRF). Various metrics have been proposed to measure performance, such as reliability, availability, functionality, product quality, productivity, et cetera. Although the indicators are different, the common purpose is to determine the PRF of the system. The ratio of integral value under the PRF to the integral value under total performance is employed to represent the resilience of systems. Therefore, the main steps of measuring system resilience are:

- Determining system performance response function (PRF);
- Utilizing the integral under the performance curve to express the system resilience.

According to Hollnagel's ETTO principle (Hollnagel, 2009a), a balance between accuracy and efficiency should be reached. To examine this balance, a cost-benefit analysis is required to justify the investment when seeking resilience. A quantified resilience metric with palpable units and relatable to familiar subjects such as reliability and risk can assess resilience implementation efficacy. Hosseini, Barker, and Ramirez-Marquez (2016) entitle the means to quantify resilience based on system performance regardless of its structure to compare resilient performance across different systems with a similar logic as "*generic resilience metrics*." Cook and Nemeth (2006) regard resilient performance as the empirical evidence of resilience. Resilient performance can be monitored based on the system's state. This information can be translated into resilience curves. Some metrics use these curves, such as the examples shown in Fig. 2, to quantify resilience.

Yarveisy et al. (2020) developed their metrics based on the logical relation among three resilience capacities, i.e., absorptive (*Ab*), adaptive (*Ad*), and restorative (*Res*), as shown in Eq. (1) and Fig. 2 (b). The authors include the impact of failure and restoration profiles, degradation, and various restoration targets in their metrics. The proposed metrics measure resilience by comparing the failure and recovery curves with their respective ideal profiles.

$$RE = Ab \cup (Ad \cap \mathrm{Res}) = Ab + Ad \times \mathrm{Res} - Ab \times Ad \times \mathrm{Res} \quad (1)$$

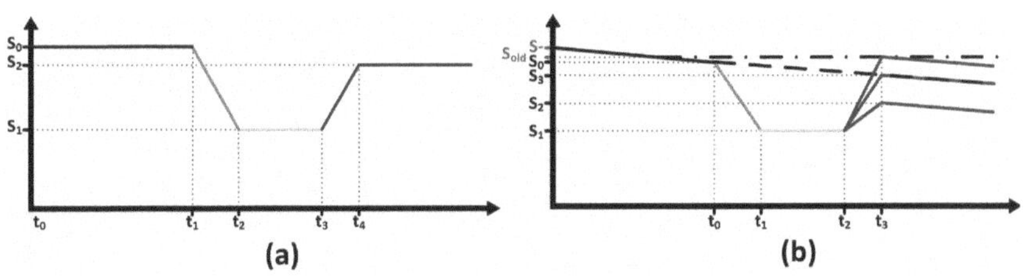

Fig. 2 Post disruption resilient performance conceptualizations. *Adapted from (A) Henry, D., & Emmanuel Ramirez-Marquez, J. (2012). Generic metrics and quantitative approaches for system resilience as a function of time.* Reliability Engineering & System Safety, 99, 114–122 *and (B) Yarveisy, R., Gao, C., & Khan, F. (2020). A simple yet robust resilience assessment metrics.* Reliability Engineering & System Safety, 197.

Bruneau et al. (2003) define resilient performance as a failure event following an abrupt disruption and recovery. The authors assess resilient performance as the area under the resilience curve as shown in Eq. (2), where S(t) represents performance quality.

$$R = \int_{t_0}^{t_1} [100 - S(t)]dt \tag{2}$$

Henry and Emmanuel Ramirez-Marquez (2012) specify the resilient performance as shown in Fig. 2A, based on an initial state pre-disruption (S_0), a failure phase following the disruption, impaired operation (S_1), and a recovery phase followed by a stable post-recovery state (S_2). The metric proposed by the authors, Eq. (3), measures resilience based on the achieved post-recovery state (S_2).

$$Я\left(t_r \middle| e_j\right) = \frac{F\left(t_4 \middle| e_j\right) - F\left(t_2 \middle| e_j\right)}{F(t_1) - F\left(t_2 \middle| e_j\right)} \tag{3}$$

Ayyub (2014) provides a more inclusive metric by accounting for degradation and various recovery and failure profiles. Resilience is measured using recovery efficacy, absorptive capacity, and overall duration of the resilient performance based on the area under the curve. Eq. (4) below shows the resilience metrics, and Eq. (5) illustrates how failure and recovery efforts are assessed individually.

$$\text{Resilience (Re)} = \frac{t_1 + F(t_2 - t_1) + R(t_3 - t_2)}{t_3 - t_1} \tag{4}$$

$$\text{Failure } (F) = \frac{\int_{t_1}^{t_3} f dt}{\int_{t_1}^{t_3} S dt}, \text{ Recovery } (R) = \frac{\int_{t_1}^{t_3} r dt}{\int_{t_1}^{t_3} S dt} \tag{5}$$

Sun, Wang, Yang, and Reniers (2021) proposed a performance-based approach to assess the resilience of safety barriers of a chemical process system. In their study, availability is used to indicate the system performance. The ratio of residual performance after disruptions to initial performance is used to represent the absorption capacity of the system; The ratio of the recovery value of the system performance to the total performance value of the corresponding period is employed to represent the adaptation and restoration capability of the system.

Cincotta, Khakzad, Cozzani, and Reniers (2019) presented a performance-based approach to quantify the resilience of process plants to optimize the firefighting strategies. The ratio of the number of intact tanks at time *t* to the total number of tanks represents the PRF. Based on the obtained PRF, the system resilience can be determined.

Cai, Zhang, Wang, et al. (2021) proposed a performance-based method to assess the resilience of engineering systems under different disruptions. The presented external disaster model is used to calculate the failure rate of each component under different disruptions. The obtained is inputted into the Markov and DBN models to determine PRF. After this, the system resilience under various disasters can be quantified.

Based on resilience principles and resilience indicator metrics within the PRAF framework, process outputs, such as PVC manufacturing (Jain, Chakraborty, Pistikopoulos, & Mannan, 2018) and reactor cooling systems (Jain, Diangelakis, Pistikopoulos, & Mannan, 2019), can be optimized by managing uncertainties and avoiding possible upsets. This was realized by applying various methods, such as Bayesian and sensitivity analysis. Ahead, both lagging and leading resilience indicators, technical and social for the avoidance, survival and recovery phases were identified by a survey of process industry experts and academia based on a Likert-type questionnaire (Jain, Mentzer, & Mannan, 2018). Metrics are determined by indicator monitoring.

4.1.4 Simulation-based methods

Simulation-based approaches aim to develop simulations to obtain the key data of system variables to quantify resilience. The general operation steps of this type of method are:

- Determining the key variables in systems;
- Developing a relationship between variables and resilience;
- Designing simulations to obtain the data of key variables;
- Quantifying system resilience based on resilience metrics and the obtained data.

Zhang, Cai, Liu, et al. (2021) proposed a simulation-based method by combining the finite element model and DBN model to assess the resilience of a subsea wellhead connector. The finite element model is utilized to simulate structure performance state and export simulation data. The obtained data represents the relationship between external load and each component's performance.

Senkel, Bode, and Schmitz (2021) developed a simulation-based approach to assess the resilience of integrated energy systems and evaluate the improvement measures. In this study, the gas pressure, enthalpy flow rate, room temperature, heat flow rate, frequency, and electric power were obtained by the simulation. As a result, these simulation results were combined with the proposed resilience metric to quantify the system resilience.

Lohmer, Bugert, and Lasch (2020) presented a simulation-based approach to quantify the resilience of a complex supply network under disruption conditions. Based on this, the proposed method is used to investigate indications of the exact functioning mechanisms of the resilience alterations.

Moosavi and Hosseini (2021) developed a simulation-based method to assess the impact of resilience strategies in a multi-stage supply chain during a pandemic. The system resilience was deduced from the simulation. Two resilience strategies and their impact on supply chain resilience and financial performance were simulated.

Patriarca, Paolis, Costantino, and Gravio (2021) developed a simulation-based method by formulating a simulation model to obtain the reliability of each component of a hot water generation plant for hospitals. The calculated data was then used to represent the system performance. After that, the system performance can be determined by integrating the performance curve and proposed resilience metric.

Resilience metrics quantify it by relying on various techniques and utilizing different system states and resilient capacities. The important aspect is accomodating a broad spectrum of performance indicators that reflect resilient performance. In addition to resilience assessment and cost-benefit analysis, metrics can provide valuable insights by evaluating dynamic simulation results. Generic metrics can assess process resilience based on system performance data. This capacity turns them into convenient tools for performance analysis and design based on simulation results.

4.2 The application of the four types of methods

4.2.1 Indicator/Index-based methods

According to the aforementioned main steps of indicator-based methods, the first step is to determine the main resilience indicators. Tong et al. (2020) proposed a four indicators-based approach using the DBN model to quantify the system resilience. The indicator/Index-based methods are

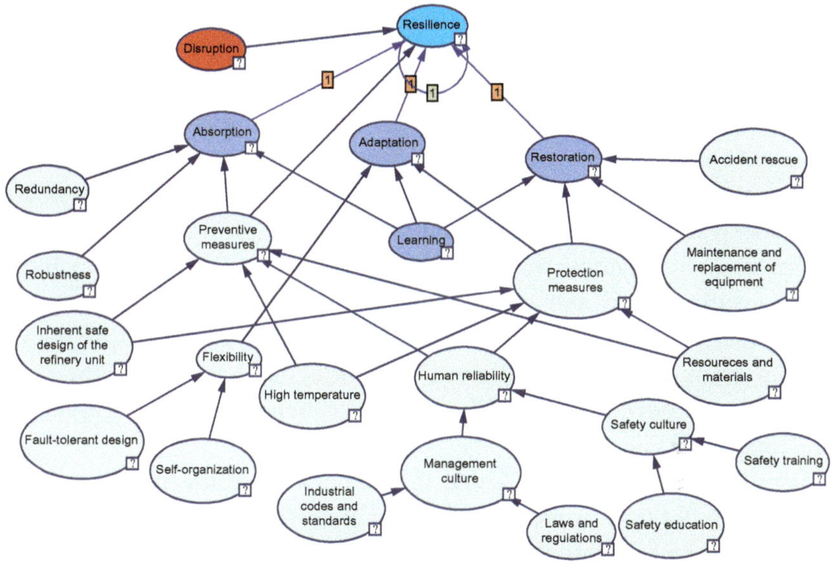

Fig. 3 DBN model the Chevron Richmond refinery crude unit. *Adapted from Tong, Q., Yang, M., Zinetullina, A. (2020). A Dynamic Bayesian Network-based approach to Resilience Assessment of Engineered Systems.* Journal of Loss Prevention in the Process Industries, 65.

demonstrated based on their study. There are four main indicators, absorption, adaptation, restoration, and learning, and each indicator possesses different sub-indexes, which can be seen in Fig. 3.

- Absorption: Redundancy plays a key role in enhancing absorption capacity. For example, It can provide redundant components to reduce the impact of disruptions on the system. For a 2 out of 1 parallel device, one device fails due to external disruptions, and then another one can ensure the system's normal operation. Besides, the robustness and preventive measures can also improve absorption capacity. Moreover, the learning capacity can provide essential information and experience to reinforce the absorption capacity.
- Adaptation: Flexibility, which comprises fault-tolerant design and self-organization, refers to the ability of a system to adapt itself to changes and external disruptions. The more flexible the system is, the more adaptable the system itself will be. In addition, protection measures are also an essential element. For instance, emergency shutdown is designed and employed to mitigate the consequence of accidents.
- Restoration: External maintenance and rescue activities are critical measures to increase the restoration capacity. These activities are utilized to

recover the function of failed components. Besides, the learning capacity and protection measures are used to mitigate the consequence caused by disruptions.

- Learning: Learning capacity provides essential information and experience to intensify the absorption, adaptation, and restoration capacity.

After determining the main indicators and corresponding sub-indexes, the second step is ascertaining the quantitative relationship between resilience and its main indicators. In this case study, the conditional probabilities, which historical data and expert judgment can determine, are employed to represent the quantitative relationship of each DBN node. Once the relationship among each node of the DBN model is determined, the system resilience can be quantified. The resilience behavior changes over time are depicted in Fig. 4.

It can be seen from Fig. 4 that in the beginning, the system resilience dropped rapidly until it reached the minimal value of 0.36 at 10 min. The minimal value of the system resilience depends on the absorption capacity. The stronger the absorption capacity, the greater the minimal value. After that, with the capacity for adaptation and restoration, the system resilience increased gradually. The stable value of the system resilience is 0.98, which is lower than the initial resilience value of 1. This is owing to the relatively

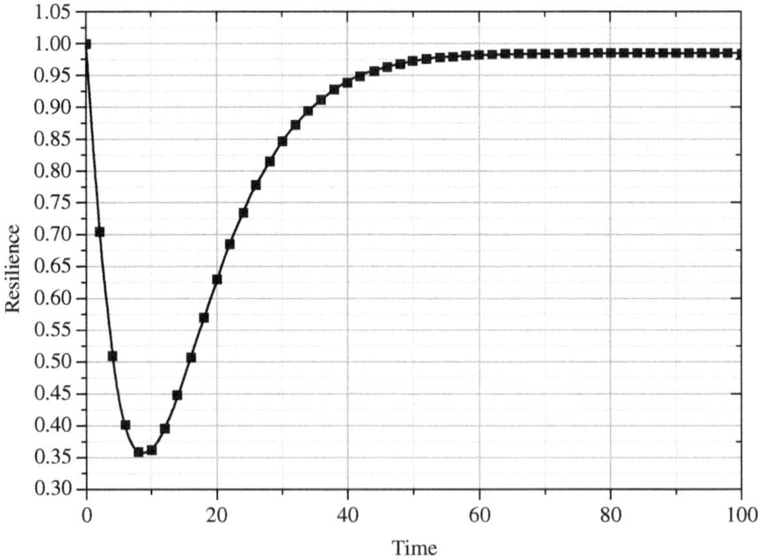

Fig. 4 DBN model the Chevron Richmond refinery crude unit.

lower adaptation and restoration capacity. If the adaptation and restoration capacity is strong enough, the stable value of the system resilience can exceed the initial resilience 1.

4.2.2 Statistical methods

Statistical methods are based on the historical database and/or monitoring data of key variables of the system. Ilbeigi (2019) proposed a statistical method to assess the resilience of transportation networks. The *closeness centrality* (mean distance from a network vertex to other vertices) and the *average pace* (normalized travel time) are key variables in transportation networks. Process control methods are then proposed to detect anomalous patterns. The system's resilience can be determined by the mathematical relationship between resilience and key variables.

Firstly, the key variables should be determined. The *closeness centrality* is considered one of the critical variables and used to represent the topological features of transportation networks, expressed as Eq. (6). The overall closeness centrality of a network at time t is considered the average closeness over all vertices in the network at time t, illustrated in Eq. (7). Besides, in a transportation network, the average travel time from two neighboring vertices is also an important variable in determining the weights of its connecting edges. Therefore, the *average pace* indicates the normalized travel time, shown in Eq. (8). The *average pace* is obtained from the historical data, shown in Fig. 5.

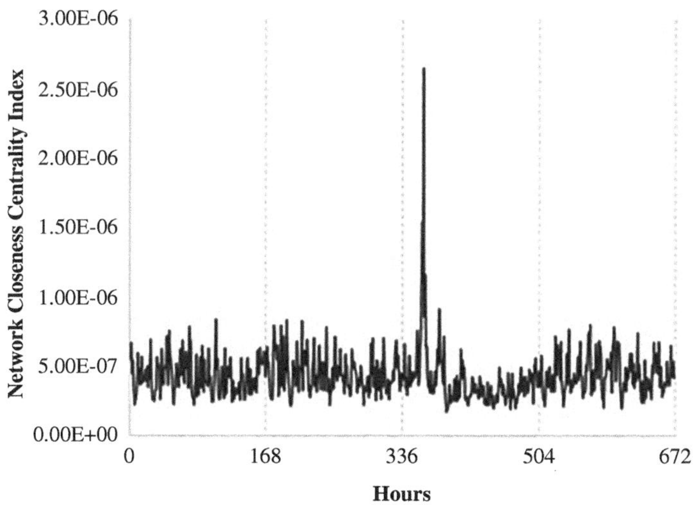

Fig. 5 New York City transportation network closeness centrality index from October 15, 2012, to November 12, 2012.

$$C_{it} = (n-1)\frac{1}{\sum_j d_{ijt}} \tag{6}$$

$$C_{Nt} = \frac{1}{n}\sum_i C_{it} \tag{7}$$

$$P(i,j,t) = \frac{\sum T_{ijt}}{\sum L_{ij}} \tag{8}$$

where n is the number of vertices in the network, d_{ijt} refers to the shortest distance from vertex i to vertex j at time t, T_{ijt} is the travel time from vertex i to j at time t, L_{ij} is the distance from i to j, $P(i,j,t)$ stands for average pace.

Second, statistical monitoring of quantified network characteristics through time uses process control methods to detect anomalous patterns, deviations, and recovery points. In this step, the forward and backward process monitors are developed to detect the deviation point and recovery point, shown in Fig. 6.

After that, to estimate the severity of the extreme event's impact, the quantified network features are compared to their values before and after the disturbance period based on the determined data. Finally, the average closeness centrality index of the transportation network during the disturbance period is compared to the average closeness indices before and after the disturbance to analyze the system resilience. The consequence can be seen in Fig. 7. This method provides a way to help decision-makers determine which parts of the network are more vulnerable and more likely to fail.

4.2.3 System performance-based methods

System performance-based methods have been proposed to measure system resilience in various fields. To assess the resilience of a safety barrier system in chemical process systems, Sun et al. (2021) proposed a performance-based approach using the DBN model to quantify the safety barrier system resilience. The availability of the system is used to represent the system performance. The DBN model ascertained the performance curves under different conditions (i.e., normal and disrupted conditions). After that, a novel resilience metric is proposed to quantify system resilience based on the determined performance.

Following the system performance-based methods mentioned above, the first step is determining the system performance response function (PRF). In this case study, availability represents the system performance, as shown in Figs 8. $f(T_1)$ refers to the availability curve of the degradation

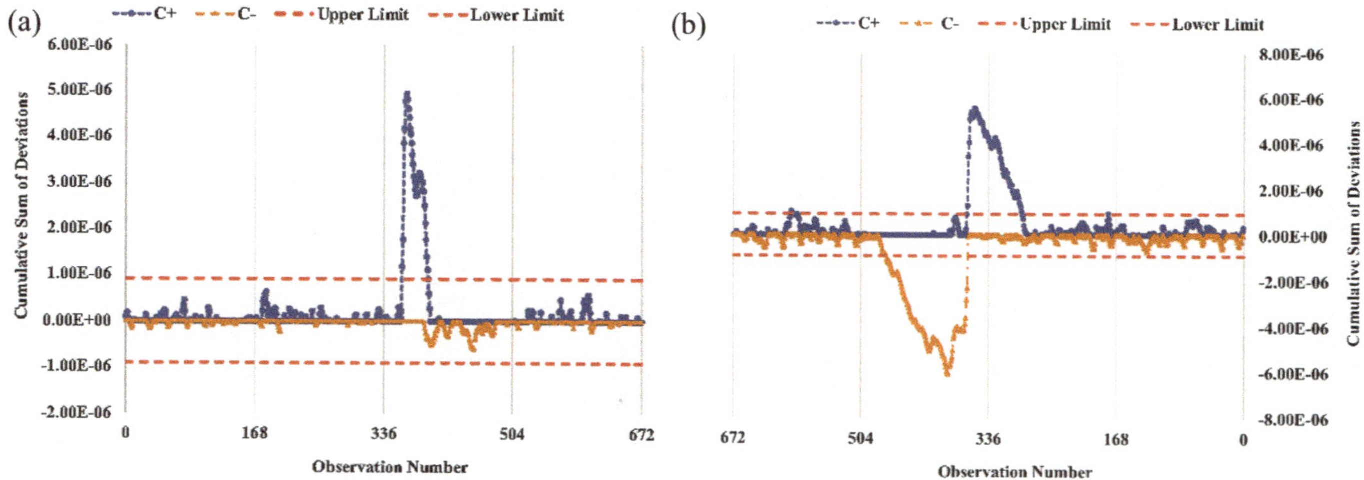

Fig. 6 (A) Forward control, (B) backward control.

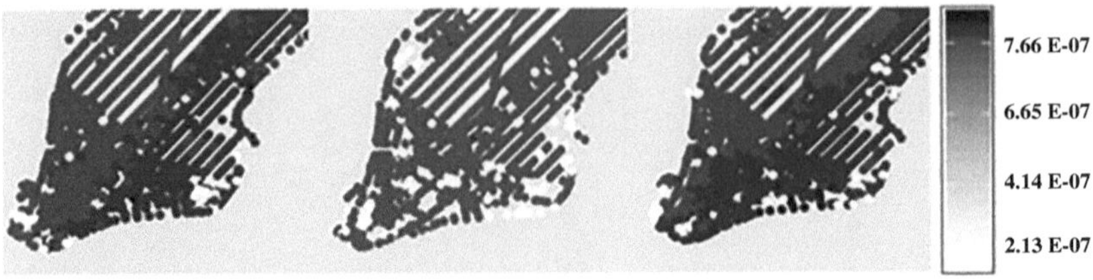

Fig. 7 Closeness centrality index in lower Manhattan before, during, and after disruptions.

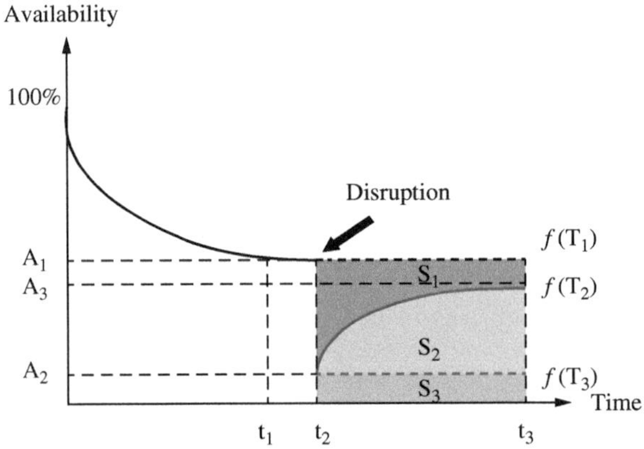

Fig. 8 Availability curve of the system. *Adapted from Sun, H., Wang, H.Q., Yang, M., Reniers, G. (2021). Resilience-based approach to safety barrier performance assessment in process facilities.* Journal of Loss Prevention in the Process Industries, 73.

process under normal conditions. $f(T_2)$ is the availability curve after the disruption. The DBN model determines $f(T1)$ and $f(T2)$. A resilience metric is then proposed to measure the system resilience based on $f(T_1)$ and $f(T_2)$.

The absorption capacity can be presented as Eq. (9), where A_b refers to absorption capacity.

$$A_b = \frac{A_2}{A_1} \tag{9}$$

The function of system availability during the adaptation and restoration phase is $f(T_2)$. The total amount of availability under normal conditions is shown in Eq. (10), where S represents the total system availability.

$$S = \int_{t_2}^{t_3} f(T_1)dt \tag{10}$$

The loss of system availability is S_1. The availability recovery amount of the system during the adaptation and restoration period is S_2, which can be expressed as Eq. (11).

$$S_2 = \int_{t2}^{t3} f(T_2) - f(T_3)dt \tag{11}$$

The expression of system resilience R is represented in Eq. (12):

$$R = \frac{\int_{t_2}^{t_3} f(T_2) - f(T_3)dt}{\int_{t_2}^{t_3} f(T_1)dt} + A_b \tag{12}$$

The system resilience can be determined after determining the PRF and resilience metric.

In this case study, the escalation prevention barrier (EPB) of the Chevron Richmond refinery crude unit is utilized to illustrate system performance-based methods. The DBN model is shown in Fig. 9. Historical data and

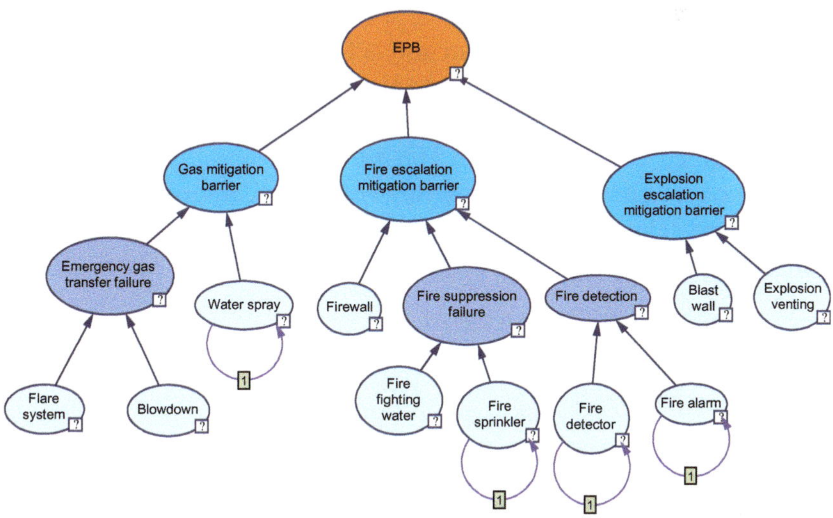

Fig. 9 DBN model for EPB dependencies. *Adapted from Sun, H., Wang, H.Q., Yang, M., Reniers, G. (2021). Resilience-based approach to safety barrier performance assessment in process facilities.* Journal of Loss Prevention in the Process Industries, 73.

expert judgment can determine prior, conditional, and transition probabilities. The specific process of obtaining data can be seen in Sun et al. (2021).

The PRF is depicted in Fig. 10 (left) according to the DBN model. Based on Fig. 17–10 (left) and the proposed resilience metric, the system's resilience is illustrated in Fig. 10 (right). It can be seen from Fig. 10 (left) that the minimal performance of EPB is 0.642. According to Eq. (9), the absorption capacity is 0.656. Combining the PRF with Eq. (12), the system resilience behavior is determined, and the final resilience is 0.852.

4.2.4 Simulation-based methods

The core of simulation-based methods is using simulation software and/or pseudo-code to obtain the information and data of key variables. After that, the system resilience can be measured according to the proposed resilience metric. Patriarca et al. (2021) proposed a simulation-based approach to assess the resilience of an engineered system. Their work is used to illustrate simulation-based methods.

The first step is identifying key variables. In this case, the reliability of a component is viewed as the key variable. Therefore, the main task in the next step is to design a simulation to calculate the reliability of each component. Fig. 11 is the process of the proposed methodology, including the simulation model and resilience metric. The result of reliability for each component is illustrated in Fig. 12.

The performance response function (PRF) can be ascertained based on the reliability of each component, which is shown in Fig. 13. Once the PRF is obtained, the proposed resilience metric can assess the system resilience. The determined PRF is shown in Fig. 14.

In this case, resilience comprises three capacities, absorption, adaptation, and recovery. The corresponding metrics of each capacities capacity and resilience metric are expressed as follows.

$$Ab = \frac{F'_{I1}}{F_0} \tag{13}$$

$$Ad = 1 - \left(\frac{T_{I2} - T_{I1}}{T_f - T_0}\right) \tag{14}$$

$$\text{Rec} = \frac{\text{Arctan}\left[\frac{F_f - F'_{I2}}{\left(\frac{T_f - T_{I2}}{T_f - T_0}\right)}\right]}{90} \times \left(\frac{T_{I2} - T_0}{T_f - T_0}\right) \tag{15}$$

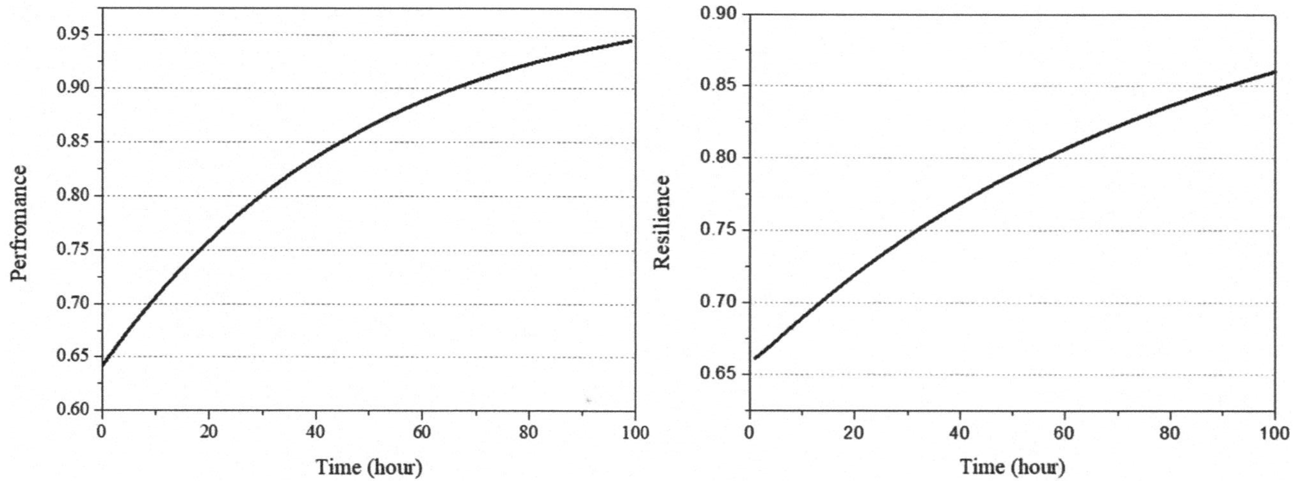

Fig. 10 (left) The performance response function of EPB, (right) The resilience behavior of EPB.

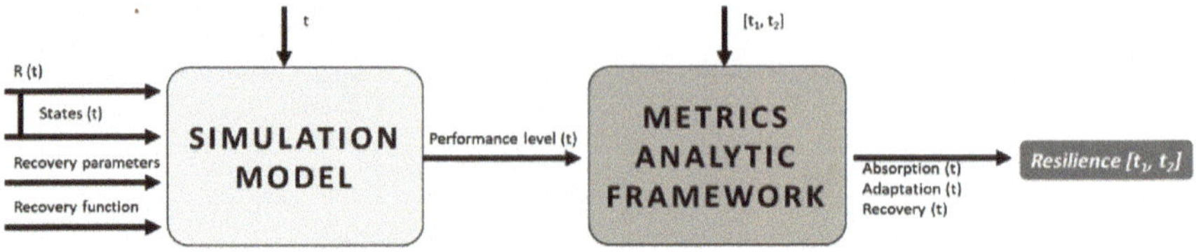

Fig. 11 The simulation procedure of the proposed methodology. *Adapted from Patriarca, R., Paolis, A.D., Costantino, F., Gravio, G.D. (2021). Simulation model for simple yet robust resilience assessment metrics for engineered systems.* Reliability Engineering and System Safety, 209.

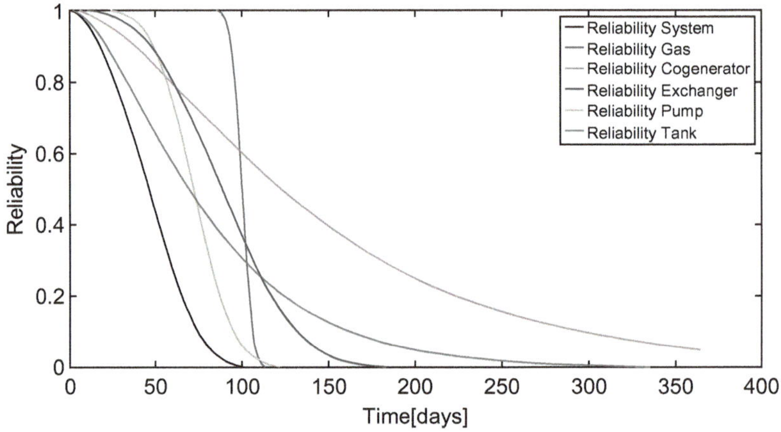

Fig. 12 The simulation result of each component and the system.

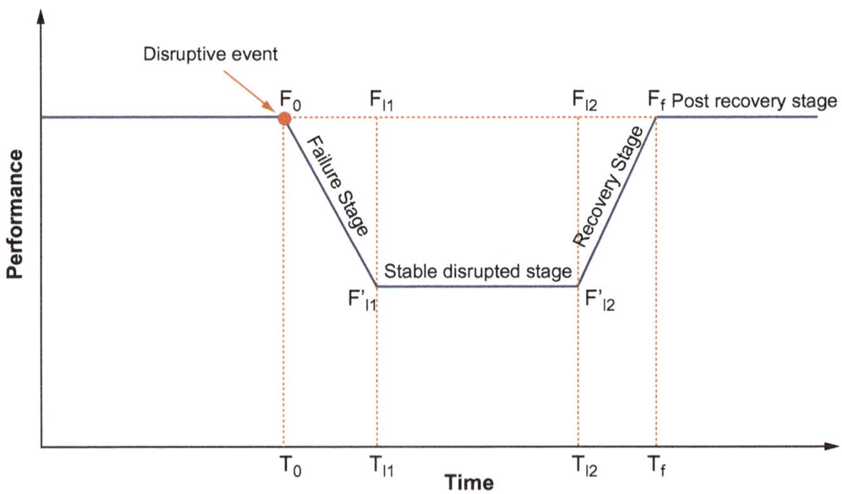

Fig. 13 The PRF over time.

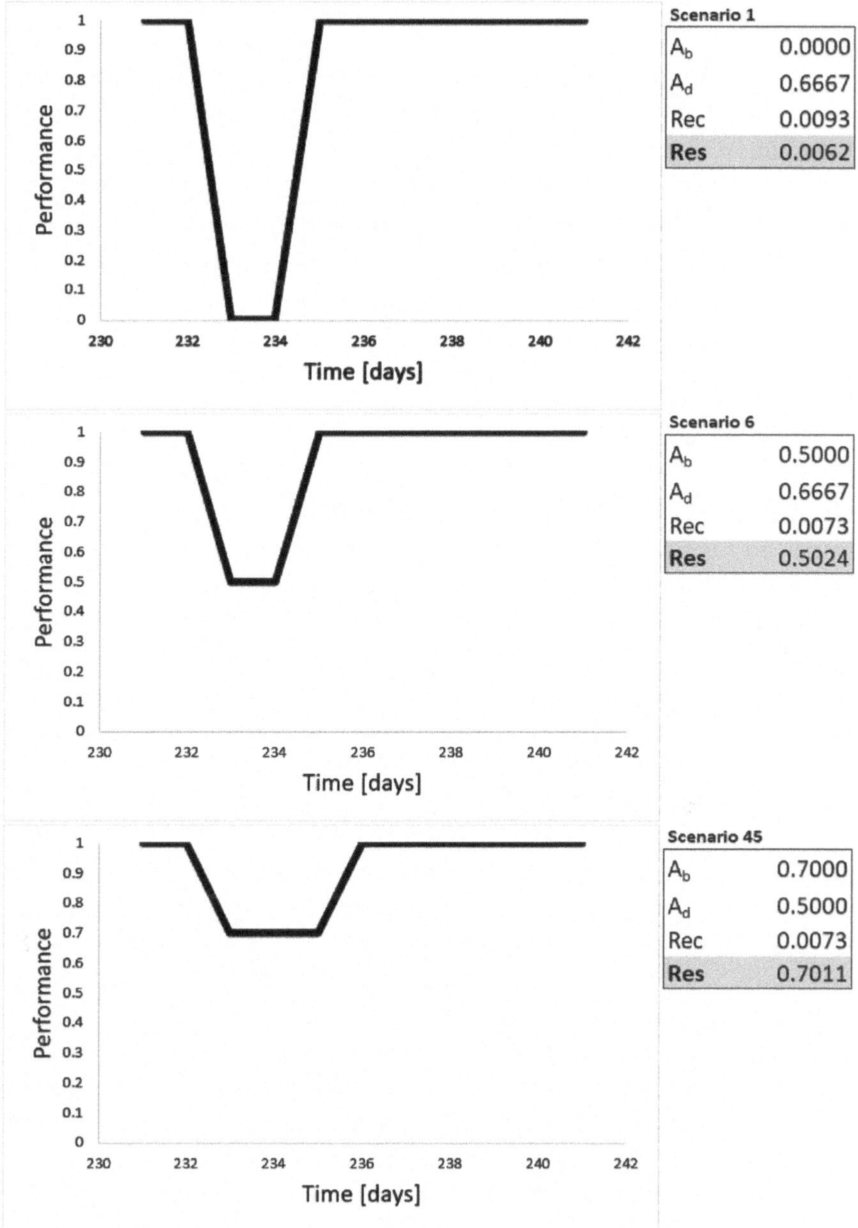

Fig. 14 The PRF over time and corresponding resilience.

$$\mathrm{Re} = Ab + Ad \times \mathrm{Re}\,c - Ab \times Ad \times \mathrm{Re}\,c \tag{16}$$

where *Ab* represents the absorption capacity, *Ad* stands for adaptation capacity, *Rec* indicates recovery capacity, and *Re* in Eq. (16) is resilience. The remaining parameters of Eqs. (13), (14), and (15) can be seen in Fig. 11.

According to the reliability of each component, the PRF can be determined, which is shown in Fig. 14. Based on the PRF and proposed resilience metric, the system resilience can be measured and demonstrated in Fig. 14.

5. Conclusion

Process systems store and process many hazardous materials, which may cause casualties, property damage, and ecological pollution. In Industry 4.0, sensing technologies, real-time data processing, and system state prediction are employed to boost performance, reduce uncertainties, promote safe operation, and support decision-making. To enhance system safety, digital technologies provide data that enable developing knowledge about possible risk patterns while data are introduced, improving safety culture by controlling procedural compliance and effective training. Besides, sensing technologies, real-time data processing, and system state prediction are employed to reduce uncertainties, promote safe operation, and support decision-making. However, every coin has two sides. While the introduction of digital technologies enhances automation, process data visualization, fault detection, diagnosis, alarm management, et cetera, it also makes the system and its subsystems much more interdependent, interacted, and complex. Due to the interdependency, interaction, and complexity among human-technical-organizational factors, the root causes of incidents and/or accidents are harder to identify. Since conventional risk assessment is usually causal-based, this may make conventional risk assessment methods inappropriate for analyzing and predicting a complete risk. New system-based methods, such as STPA and FRAM, will improve results but, in the end, depend still on human imagination, which is fallible. Therefore, and for perfect storm situations and maybe a rare black swan, resilience thinking needs to be considered.

Resilience as a novel paradigm has drawn wide attention. Various metrics and methods are proposed to assess and measure the resilience in different systems (e.g., infrastructure systems, supply chains, transportation systems, power networks, process systems, engineering systems, et cetera). Although there is no universally accepted resilience assessment framework, the works that have been done led to increased clarity about what resilience is and how it can be assessed. The developed resilience assessment methods are divided into four categories: indicator/index-based approaches, statistical approaches, system performance-based approaches, and simulation-based approaches. Examples are provided to illustrate the methods of each category. Designing a resilient system is an ideal practice to help process systems deal with uncertainties and enhance the system's ability to handle uncertain disruptions.

Future work should pay more attention to clarifying the concept of resilience and developing a potential unified approach for resilience assessment.

References

Ale, B. J. M., Hartford, D. N. D., & Slater, D. H. (2021). Prevention, precaution and resilience: Are they worth the cost? *Safety Science*, *140*(105), 271. https://doi.org/10.1016/j.ssci.2021.105271.

API. (2010). *API RP 754: Process safety performance indicators for the refining and petrochemical industries*. American Petroleum Institute.

ASCE. (2021). *ASCE Policy statement 518 - Unified definitions for critical infrastructure resilience*. American Society of Civil Engineers. https://www.asce.org/advocacy/policy-statements/ps518—unified-definitions-for-critical-infrastructure-resilience.

Ayyub, B. M. (2014). Systems resilience for multihazard environments: definition, metrics, and valuation for decision making. *Risk Analysis*, *34*(2), 340–355. https://doi.org/10.1111/risa.12093.

Barroso, A. P., Machado, V. H., & Cruz Machado, V. (2009). Identifying vulnerabilities in the supply chain. In *2009 IEEE International Conference on Industrial Engineering and Engineering Management*. https://doi.org/10.1109/IEEM.2009.5373062.

BBC. (2021). *Iran blames foreign country for cyberattack on petrol stations*. BBC NEWS. https://www.bbc.com/news/world-middle-east-59062907.

Boin, A., & van Eeten, M. J. G. (2013). The resilient organization. *Public Management Review*, *15*(3). https://doi.org/10.1080/14719037.2013.769856.

Bragatto, P., Vairo, T., Milazzo, M. F., & Fabiano, B. (2021). The impact of the COVID-19 pandemic on the safety management in Italian Seveso industries. *Journal of Loss Prevention in the Process Industries*, *70*, 104393.

Bruneau, M., Chang, S. E., Eguchi, R. T., Lee, G. C., O'Rourke, T. D., Reinhorn, A. M., et al. (2003). A framework to quantitatively assess and enhance the seismic resilience of communities. *Earthquake Spectra*, *19*(4), 733–752. https://doi.org/10.1193/1.1623497.

Cai, B. P., Zhang, Y. P., Wang, H. F., et al. (2021). Resilience evaluation methodology of engineering systems with dynamic-Bayesian-network-based degradation and maintenance. *Reliability Engineering and System Safety*, *209*(107), 464. https://doi.org/10.1016/j.ress.2021.107464.

Cincotta, S., Khakzad, N., Cozzani, V., & Reniers, G. (2019). Resilience-based optimal firefighting to prevent domino effects in process plants. *Journal of Loss Prevention in the Process Industries*, *58*, 82–89. https://doi.org/10.1016/j.jlp.2019.02.004.

Columbia Accident Investigation Board. (2003). *Columbia accident investigation board report*.

Cook, R. I., & Nemeth, C. (2006). Taking things in one's stride: Cognitive features of two resilient performances. In E. Hollnagel, D. D. Woods, & N. Leveson (Eds.), *Resilience engineering: Concepts and percepts* (pp. 205–222). Ashgate Publishing, Ltd.

CSB. (2007). *Investigation report: Refinery explosion and Fire, Bp Texas City, Texas, March 23, 2005, Report No. 2005-04-I-TX*.

Dinh, L. T. T., Pasman, H., Gao, X., & Mannan, M. S. (2012). Resilience engineering of industrial processes: Principles and contributing factors. *Journal of Loss Prevention in the Process Industries, 25*(2), 233–241. https://doi.org/10.1016/j.jlp.2011.09.003.

Duchek, S. (2020). Organizational resilience: a capability-based conceptualization. *Business Research, 13*(1). https://doi.org/10.1007/s40685-019-0085-7.

Final Committee Report. (2020). *The design, development & certification of the Boeing 737 Max*. House Committee on Transportation and Infrastructure.

Hamel, G., & Valikangas, L. (2003). *The quest for resilience*. Harvard Business Review. https://hbr.org/2003/09/the-quest-for-resilience.

Henry, D., & Emmanuel Ramirez-Marquez, J. (2012). Generic metrics and quantitative approaches for system resilience as a function of time. *Reliability Engineering & System Safety, 99*, 114–122. https://doi.org/10.1016/j.ress.2011.09.002.

Holling, C. S. (1973). Resilience and Stability of Ecological Systems. *Annual Review of Ecology and Systematics, 4*(1). https://doi.org/10.1146/annurev.es.04.110173.000245.

Hollnagel, E. (2004). *Barriers and accident prevention*. Ashgate Publishing, Ltd.

Hollnagel, E. (2006). Resilience - the Challenge of the Unstable. In E. Hollnagel, D. D. Woods, & N. Leveson (Eds.), *Resilience engineering: Cocnepts and percepts* (pp. 9–17). Ashgate Publishing, Ltd.

Hollnagel, E. (2008). Resilience Engineering in a Nutshell. In E. Hollnagel, C. P. Nemeth, & S. Dekker (Eds.), *Resilience engineering perspectives, Volume 1: Remaining sensitive to the possibility of failure* (pp. xi–xiv). CRC Press.

Hollnagel, E. (2009a). *The ETTO principle: Efficiency-thoroughness trade-off: Why things that go right sometimes go wrong*. Ashgate Publishing, Ltd.

Hollnagel, E. (2009b). The four cornerstones of resilience engineering. In C. P. Nemeth, E. Hollnagel, & S. Dekker (Eds.), *Resilience engineering perspectives, Volume 2: Preparation and restoration* (pp. 117–133). CRC Press.

Hollnagel, E. (2011). The Scope of Resilience Engineering. In E. Hollnagel, J. Paries, D. D. Woods, & J. Wreathall (Eds.), *Resilience engineering in practice: A guidebook* Ashgate Publishing, Ltd (pp. xxix–xxxix).

Hollnagel, E. (2012). *FRAM: The functional resonance analysis method: Modeling complex socio-technical systems*. Ashgate Publishing, Ltd.

Hollnagel, E. (2014). *Safety-I and Safety-II: The past and future of safety management*. Ltd: Ashgate Publishing.

Hollnagel, E. (2017). *FRAM: The functional resonance analysis method*. CRC Press. https://doi.org/10.1201/9781315255071.

Hosseini, S., Barker, K., & Ramirez-Marquez, J. E. (2016). A review of definitions and measures of system resilience. *Reliability Engineering & System Safety, 145*, 47–61. https://doi.org/10.1016/j.ress.2015.08.006.

Ilbeigi, M. (2019). Statistic process control for analyzing resilience of transportation networks. *International Journal of Disaster Risk Reduction, 33*, 155–161. https://doi.org/10.1016/j.ijdrr.2018.10.002.

Jain, P., Chakraborty, A., Pistikopoulos, E. N., & Mannan, M. S. (2018). Resilience-Based Process Upset Event Prediction Analysis for Uncertainty Management Using Bayesian Deep Learning: Application to a Polyvinyl Chloride Process System. *Industrial and Engineering Chemistry Research, 57*, 14822–14836. https://doi.org/10.1021/acs.iecr.8b01069.

Jain, P., Diangelakis, N. A., Pistikopoulos, E. N., & Mannan, M. S. (2019). Process resilience based upset events prediction analysis: Application to a batch reactor. *Journal of Loss Prevention in the Process Industries, 62*. https://doi.org/10.1016/j.jlp.2019.103957.

Jain, P., Mentzer, R., & Mannan, M. S. (2018). Resilience metrics for improved process-risk decision making: survey, analysis and application. *Safety Science, 108*, 13–28. https://doi.org/10.1016/j.ssci.2018.04.012.

Jain, P., Pasman, H. J., Waldram, S., Pistikopoulos, E. N., & Mannan, M. S. (2018). Process Resilience Analysis Framework (PRAF): A systems approach for improved risk and safety management. *Journal of Loss Prevention in the Process Industries*, *53*, 61–73. https://doi.org/10.1016/j.jlp.2017.08.006.

Jain, P., Rogers, W. J., Pasman, H. J., Keim, K. K., & Mannan, M. S. (2018). A Resilience-based Integrated Process Systems Hazard Analysis (RIPSHA) approach: Part I plant system layer. *Process Safety and Environmental Protection*, *116*, 92–105. https://doi.org/10.1016/j.psep.2018.01.016.

Jain, P., Rogers, W. J., Pasman, H. J., & Mannan, M. S. (2018). A resilience-based integrated process systems hazard analysis (RIPSHA) approach: Part II management system layer. *Process Safety and Environmental Protection*, *118*, 115–124. https://doi.org/10.1016/j.psep.2018.06.037.

Jarvis, R., & Goddard, A. (2017). An analysis of common causes of major losses in the onshore oil, gas & petrochemical industries. *Loss Prevention Bulletin*, *255*.

Ji, C., Wei, Y., Mei, H., Calzada, J., Carey, M., Church, S., et al. (2016). Large-scale data analysis of power grid resilience across multiple us service regions. *Nature Energy*, *2016*(1), 16052.

Kammouh, O., Gardoni, P., & Cimellaro, G. P. (2020). Probabilistic framework to evaluate the resilience of engineering systems using Bayesian and dynamic Bayesian networks. *Reliability Engineering and System Safety*, *198*. https://doi.org/10.1016/j.ress.2020.106813.

Kelly, S., & Resnick-ault, J. (2021). *One password allowed hackers to disrupt Colonial Pipeline, CEO tells senators*. Reuters. https://www.reuters.com/business/colonial-pipeline-ceo-tells-senate-cyber-defenses-were-compromised-ahead-hack-2021-06-08/.

Lee, J., Cameron, I., & Hassall, M. (2019). Improving process safety: What roles for Digitalization and Industry 4.0? *Process Safety and Environmental Protection*, *132*, 325–339. https://doi.org/10.1016/j.psep.2019.10.021.

Lengnick-Hall, C. A., Beck, T. E., & Lengnick-Hall, M. L. (2011). Developing a capacity for organizational resilience through strategic human resource management. *Human Resource Management Review*, *21*(3). https://doi.org/10.1016/j.hrmr.2010.07.001.

Leveson, N. (2004). A new accident model for engineering safer systems. *Safety Science*, *42*(4), 237–270. https://doi.org/10.1016/S0925-7535(03)00047-X.

Leveson, N. G. (2016). *Engineering a safer world: Systems thinking applied to safety*. The MIT Press.

Leveson, N. G. (2020). *Safety III: A systems approach to safety and resilience*. MIT Systems Engineering Lab. Available online: http://sunnyday.mit.edu.

Linnenluecke, M. K., & Griffiths, A. (2012). Assessing organizational resilience to climate and weather extremes: complexities and methodological pathways. *Climatic Change*, *113*(3–4). https://doi.org/10.1007/s10584-011-0380-6.

Lohmer, J., Bugert, N., & Lasch, R. (2020). Analysis of resilience strategies and ripple effect in blockchain-coordinated supply chains: An agent-based simulation study. *International Journal of Production Economies*, *228*. https://doi.org/10.1016/j.ijpe.2020.107882.

Mannan, M. S., Reyes-Valdes, O., Jain, P., Tamim, N., & Ahammad, M. (2016). The Evolution of Process Safety: Current Status and Future Direction. *Annual Review of Chemical and Biomolecular Engineering*, 7, 135–162. https://doi.org/10.1146/annurev-chembioeng-080615-033640.

Marsh. (2020). *100 Largest Losses in the Hydrocarbon Industry: 1974–2019*. https://www.marsh.com/us/industries/energy-and-power/insights/100-largest-losses-in-the-hydrocarbon-industry.html.

Moosavi, J., & Hosseini, S. (2021). Simulation-based assessment of supply chain with consideration of recovery strategies in the COVID-19 pandemic context. *Computers and Industrial Engineering*, *160*. https://doi.org/10.1016/j.cie.2021.107593.

Nelson, J., Johnson, N. G., Fahy, K., & Hansen, T. A. (2020). Statistical development of microgrid resilience during island operations. *Applied Energy*, *279*(115), 724. https://doi.org/10.1016/j.apenergy.2020.115724.

Olive, C., O'Connor, T. M., & Mannan, M. S. (2006). Relationship of safety culture and process safety. *Journal of Hazardous Materials*, *130*(1–2), 133–140. https://doi.org/10.1016/j.jhazmat.2005.07.043.

Pasman, H. J. (2021). Early warning signals noticed, but management doesn't act adequately or not at all: a brief analysis and direction of possible improvement. *Journal of Loss Prevention in the Process Industries*, *70*(104), 272. https://doi.org/10.1016/j.jlp.2020.104272.

Pasman, H., Kottawar, K., & Jain, P. (2020). Resilience of Process Plant: What, Why, and How Resilience Can Improve Safety and Sustainability. *Sustainability*, *12*(15), 6152. https://doi.org/10.3390/su12156152.

Patriarca, R., Paolis, A. D., Costantino, F., & Gravio, G. D. (2021). Simulation model for simple yet robust resilience assessment metrics for engineered systems. *Reliability Engineering and System Safety*, *209*. https://doi.org/10.1016/j.ress.2021.107467.

Peck, H. (2006). Reconciling supply chain vulnerability, risk and supply chain management. *International Journal of Logistics Research and Applications*, *9*(2). https://doi.org/10.1080/13675560600673578.

Perrow, C. (2000). *Normal Accidents*. Princeton University Press. https://doi.org/10.1515/9781400828494.

Ponis, S. T., & Koronis, E. (2012). Supply Chain Resilience: Definition Of Concept And Its Formative Elements. *Journal of Applied Business Research (JABR)*, *28*(5). https://doi.org/10.19030/jabr.v28i5.7234.

PPD. (2013). *Presidential policy directive (PPD-21) – Critical infrastructure security and resilience*. The White House. https://obamawhitehouse.archives.gov/the-press-office/2013/02/12/presidential-policy-directive-critical-infrastructure-security-and-resil.

Rasmussen, J. (1997). Risk management in a dynamic society: a modeling problem. *Safety Science*, *27*(2–3), 183–213.

Roberts, K. H., Madsen, P., Desai, V., & van Stralen, D. (2005). A case of the birth and death of a high reliability healthcare organization. *BMJ Quality and Safety*, *14*(3), 216–220.

Schulman, P. R. (2004). General attributes of safe organizations. *BMJ Quality and Safety*, *13*(suppl 2). ii39–ii44.

Senkel, A., Bode, C., & Schmitz, G. (2021). Quantification of the resilience of integrated energy systems using dynamic simulation. *Reliability Engineering and Sysytem Safety*, *209*. https://doi.org/10.1016/j.ress.2021.107447.

Shen, L. J., Cassottana, B., & Tang, L. C. (2018). Statistical trend test for resilience of power systems. *Reliability Engineering and Sysytem Safety*, *177*, 138–147. https://doi.org/10.1016/j.ress.2018.05.006.

Sun, H., Wang, H. Q., Yang, M., & Reniers, G. (2021). Resilience-based approach to safety barrier performance assessment in process facilities. *Journal of Loss Prevention in the Process Industries*, *73*. https://doi.org/10.1016/j.jlp.2021.104599.

Tong, Q., Yang, M., & Zinetullina, A. (2020). A Dynamic Bayesian Network-based approach to Resilience Assessment of Engineered Systems. *Journal of Loss Prevention in the Process Industries*, *65*. https://doi.org/10.1016/j.jlp.2020.104152.

Tukamuhabwa, B. R., Stevenson, M., Busby, J., & Zorzini, M. (2015). Supply chain resilience: definition, review and theoretical foundations for further study. *International Journal of Production Research*, *53*(18). https://doi.org/10.1080/00207543.2015.1037934.

Vaezie, S., Peterson, K., & Bourne, D. (2019). Implementation of high reliability organization principles. *Journal of Patient Safety*.

Weick, K. E., & Sutcliffe, K. M. (2011). *Managing the unexpected: Resilient performance in an age of uncertainty. Vol. 8*. John Wiley & Sons.

Wu, Y. Y., Hou, G. Y., & Chen, S. R. (2021). Post-earthquake resilience assessment and long-term restoration prioritization of transportation network. *Reliability Engineering and System Safety, 211*. https://doi.org/10.1016/j.ress.2021.107612.
Yarveisy, R., Gao, C., & Khan, F. (2020). A simple yet robust resilience assessment metrics. *Reliability Engineering & System Safety, 197*. https://doi.org/10.1016/j.ress.2020.106810.
Zhang, Y. P., Cai, B. P., Liu, Y. L., et al. (2021). Resilience assessment approach of mechanical structure combining finite element models and dynamic Bayesian networks. *Reliability Engineering and System Safety, 216*. https://doi.org/10.1016/j.ress,2021.108043.
Zhang, C., Xu, X., & Dui, H. Y. (2020). Resilience measure of network system by node and edge indicators. *Reliability Engineering and System Safety, 202*. https://doi.org/10.1016/j.ress,2021.107035.

CHAPTER EIGHTEEN

Risk assessment in Industry 4.0

Md Tanjin Amin[a] and Faisal Khan[b,*]
[a]Centre for Risk, Integrity and Safety Engineering (C-RISE), Faculty of Engineering & Applied Science, Memorial University of Newfoundland, St. John's, NL, Canada
[b]Mary Kay O'Connor Process Safety Center, Artie McFerrin Department of Chemical Engineering, Texas A&M University, College Station, TX, United States
*Corresponding author: e-mail address: fikhan@tamu.edu

Contents

1. Introduction

1.1 Background

The ever-growing world population, together with technological and social advancements, continually press the process industries to produce more chemicals, energy, food, and other daily necessities under more restrictive conditions. This thrust leads to increased size and complexity of chemical processing plants (Bakshi, 1998; Khan & Abbasi, 1999a). This enhancement is beneficial from the demand and production perspectives. Unfortunately, at the same time, it brings a few demerits, as well. For instance, the earlier days' simple process operations can no longer be kept simple—these need to be significantly modified to cope with the emerged conditions. As a result, the process operations are becoming more sophisticated and complex (Khan, Amyotte, & Amin, 2020). For example, closed loop control has become a mainstream practice in modern process plants, which may cause added loss magnitude due to an accident since a process upset gets the potential to affect more parts of a process plant (Albalawi, Durand, Alanqar, & Christofides, 2018; Amin, Khan, Ahmed, & Imtiaz, 2021b).

Methods in Chemical Process Safety, Volume 6
ISSN 2468-6514
https://doi.org/10.1016/bs.mcps.2022.05.003

Process safety is a common term that is often used to explain the methods for hazard identification and analysis, risk assessment and management. Hazard identification is the first step of process safety analysis. It is brainstorming about what can go wrong in a process system and then finding all the possible hazards. Risk assessment and management are the other crucial steps of safety analysis. Risk is the measure of safety; usually, safety cannot be measured directly. It is estimated from the level of risk. Safety is inversely proportional to risk (Khan et al., 2020). Risk measures how bad an accident can be and how frequently it can happen. Quantitatively, it is the product of the probability of a failure and the severity of that failure (CCPS, 2007).

The first component of risk is primarily concerned with quantifying accident probability, while the second one focuses on measuring the loss in financial terms. Probability quantification has gained much popularity from an engineering perspective since it is often difficult to quantify the loss, and fixed values with different consequences are generally considered (Amin, Khan, Ahmed, & Imtiaz, 2020; Yu, Khan, & Garaniya, 2015). However, loss functions can successfully be applied for dynamic severity estimation (Hashemi, Ahmed, & Khan, 2014, 2015). Probability estimates are even more uncertain.

Risk management uses risk as a benchmark to manage risk by assessment, evaluation, and decision-making for process safety performance improvement. Athough strongly related, process safety differs from occupational safety. The focus of process safety is on preventing and minimizing fire, explosion, and hazardous material release. On the other hand, occupational safety focuses on workplace hazards such as trips, slips, and falls.

1.2 Historical perspective of process safety

The evolution of process safety is closely related to the occurrence of major accidents. Although process safety was in industrial practice from the dawn of industrialization, the documented research started in the 1920s (Amin, Khan, & Amyotte, 2019; Klein, 2009). Significant accidents like the Flixborough explosion in North Lincolnshire, England (1974), the Seveso disaster in Seveso, Italy (1976), the Union Carbide plant gas leakage in Bhopal, India (1984), the Piper Alpha offshore disaster (1988), the Phillips 66 chemical plant vapor cloud explosion in Texas, USA (1989), and the Exxon Valdez crude oil spill in Alaska, USA (1989) paved the way for process safety research.

Professor Kletz's contribution at the beginning of process safety analysis is noteworthy. Professor Kletz and his colleagues from the Heavy Organic Chemicals Division of Imperial Chemical Industries (ICI) are credited for developing Hazard and Operability studies (HAZOP), one of the earliest and most powerful methods for identifying potential hazards in a process system (Gowland, 2012; Khan, Rathnayaka, & Ahmed, 2015).

HAZOP is the application of a formal and systematic examination procedure to assess the potential malfunctioning of equipment and its consequences (Dunjó, Fthenakis, Vílchez, & Arnaldos, 2010). Initially, it was applied to a new phenol plant (Kletz, 2001). However, it can be and is actually applied to existing plants, as well. The HAZOP study starts with breaking down a process into several manageable units using the process flow diagrams (PFDs) and piping and instrumentation diagrams (P&IDs). These units are called nodes. A process parameter is chosen for each node, and a guideword is applied to the chosen process parameter to suggest possible deviation from setpoints or design intents. The possible causes and associated consequences of any deviation are evaluated.

Inherent safety principles were also proposed by Professor Kletz during his time at ICI. Instead of relying on added safety devices, inherent safety primarily focuses on modifying existing safety technologies with the aim of eliminating or minimizing hazards from industries. Minimization, substitution, moderation, and simplification are the building blocks of an inherently safer design. The first task is to reduce the quantity of hazardous substances and replace the materials with less hazardous ones. If possible, the state of a material is also changed to lower its hazardous nature. Finally, the process design is simplified to enhance the safety features (Khan & Amyotte, 2003). Table 1 shows how inherent safety can be applied to a highly hazardous process plant.

Table 1 Inherent safety design example (Khan et al., 2020).

Principle	Action
Minimize	Use the chemicals in a lesser quantity.
Substitute	Replace the chemical with a less hazardous one if it serves the purpose.
Moderate	Use the material in a form that provides the least hazard. Ambient conditions may need to be modified to ensure that the hazard strength has been reduced.
Simplify	Simplify the process flow diagram and equipment.

The other methodologies, such as the failure mode and effect analysis (FMEA), also contributed significantly to process risk analysis. FMEA was initially developed for military purposes. However, it proved its significance and became widespread in many engineering fields. Several tools, such as the fault tree analysis (FTA), event tree analysis (ETA), bow tie (BT), Bayesian network (BN), layer of protection analysis (LOPA), Petri net (PN), analytic hierarchy process (AHP), and fuzzy theory (FT), have emerged and advanced the pace of process safety research. Fig. 1 shows the major tools with their evolution timeline. Interested readers are referred to (de Ruijter & Guldenmund, 2015; Dunjó et al., 2010; Ishizaka & Labib, 2011; Kabir & Papadopoulos, 2018; Lee, Grosh, Tillman and Lie, 1985; Markowski, 2012; Mentes & Helvacioglu, 2011; Pekalski, Zevenbergen, Lemkowitz, & Pasman, 2005; Ruijters & Stoelinga, 2015; Weber, Medina-Oliva, Simon, & Iung, 2012) for more information of these tools.

Contributions from global bodies such as the American Institute of Chemical Engineers (AIChE), Canadian Chemical Producers' Association (CCPA), European Federation of Chemical Engineering (EFCE), US Environmental Protection Agency (EPA), and UK Health and Safety Executive (HSE) are noteworthy, as well. They introduced regulations and suggestions to operate the process to avoid catastrophes. Additionally, accident databases such as the EU Major Accident Reporting System (eMARS), French ARIA, Dutch Factsonline, AIChE-CCPS Process Safety Incident Database (PSID), Japanese AIST's Relational Information System for Chemical Accidents Database (RISCAD), Pondicherry

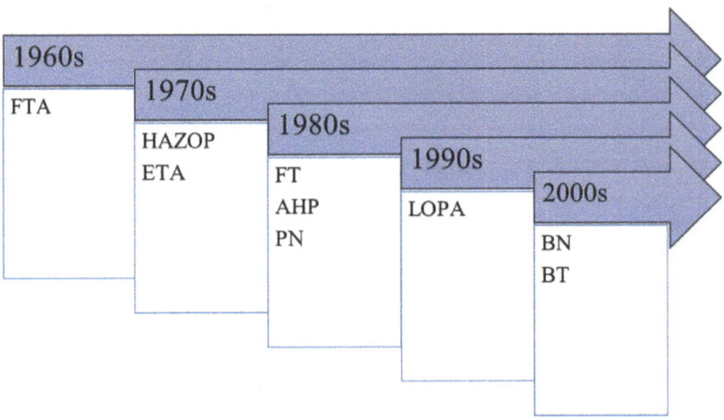

Fig. 1 Evolution of major process safety analysis tools. *Modified from Amin, M.T., Khan, F., & Amyotte, P. (2019). A bibliometric review of process safety and risk analysis.* Process Safety and Environmental Protection, *126.*

University Process-Industry Accident Database (PUPAD), and Sintef's Offshore Blowout Database are maintained. These databases help to promote safety awareness. Also, these make the accident histories and corresponding consequence data available, which are of vital importance since it is relatively easier to determine from retrospective analysis what causes led to an incident rather than predict what can go wrong that will lead to an incident in the future (Halim, Yu, Escobar, & Quddus, 2020). It is also easier to focus on technical issues that can lead to a failure.

In the earlier stages, safety analysis was utterly qualitative. For instance, HAZOP is a purely qualitative method. Indeed it can guide us to safe and unsafe conditions. However, it cannot tell how much unsafe if a deviation is encountered. Likewise, in earlier days, most risk assessment methods were qualitative or semi-quantitative. For example, LOPA, one of the most widely used methods for simplified risk assessment given a certain initiating event, is a semi-quantitative technique; it can give a quantitative result. However, due to assumptions and uncertainty, it is often inaccurate or vague from an engineering perspective (Guo, Khan, & Imtiaz, 2019).

Several methodologies have been proposed for quantitative risk assessment (Antonioni, Spadoni, & Cozzani, 2009; Khan & Abbasi, 1998; Khan & Abbasi, 1999b). The major benefit of these quantitative methods is their ability to provide an accurate result considering logical relations among process parameters and safety barriers. Nonetheless, the conventional risk assessment methods are static; these are insensitive to adaptive plant changes and process conditions. Hence, in reality assessing risk dynamically is a more viable mean (Villa, Paltrinieri, Khan, & Cozzani, 2016).

The FTA, ETA, BT, and their derivatives are widely used for dynamic quantitative risk assessment (Ferdous, Khan, Sadiq, Amyotte, & Veitch, 2013; Islam, Khan, & Venkatesan, 2017; Khakzad, Khan, & Amyotte, 2013). However, these tools cannot handle uncertainty. Also, these are not ideal for capturing dependency among multiple states and process variables or events. The use of BN helped to overcome these limitations. In the current era, the BN and its derivatives are the ideal tools for dynamic risk assessment and management (Amin, Khan, & Imtiaz, 2019; Bhandari, Arzaghi, Abbassi, Garaniya, & Khan, 2016; Wu, Zhang, Zheng, Liu, & Lunteigen, 2016).

1.3 Industry 4.0: A pathway to digital transformation

From the dawn of civilization, human beings have tried to adopt technologies to both survive and make life easier. However, the economy was primarily agriculture-based. The first stimulus towards industrialization was the invention of the steam engine (Wrigley, 2013). It helped to shift the agrarian economies to industry-centric ones. As a result, a rapid societal change happened around the 1760s, termed the industrial revolution, aka the first industrial revolution. Mechanization was the most significant change during this revolution. Coal became a key energy source, and massive coal extraction started worldwide (Clark & Jacks, 2007). There were accidents, e.g., boiler explosions, due to less emphasis on practising safe operating conditions. Process safety was mostly ignored during this period. However, occupational safety principles were practised in several plants.

The second industrial revolution started at the end of the 19th century and continued till the beginning of the 20th century. Human beings saw mass production during this period. Oil and gas emerged as new sources of energy (Kennedy, 2020); electricity empowered civilization. The laboratory experiments were implemented in industries. The invention of the aeroplane and chemical fertilizers increased the pace of development. Process safety was implemented during this period (Klein, 2009). However, documented research on process safety analysis was yet to be started (Amin, Khan, & Amyotte, 2019).

The third industrial revolution started at the beginning of the 1950s, and it brought a massive change thanks to the development of electronics towards digital technologies (Cooper & Kaplinsky, 2005). Enhanced use of computers and digital solutions became mainstream during this period. However, the first impetus towards the third industrial revolution was at the beginning of the 20th century when Lieutenant John Dumaresq of the Royal Navy invented the Dumaresq, one of the earliest analogue computers to handle a ship's fire control problem (Schleihauf, 2001). The FERMIAC and MONIAC are the two other notable analogue computers based on mechanical or electro-mechanical signals.

Sensor technologies advanced dramatically in the 1950s (Gungor & Hancke, 2009). Although process industries showed a late adoption, sensors became a common data collection instrument in the 1970s. This trend gradually replaced the conventional practice of an analogue data collection system with a digital one. This shift was the steppingstone for digitalization and digital transformation in process industries.

Process operations embraced mass digitalization—the analogue data were converted to digital data, and software was used to process these data for meaningful results. Also, the advancements in control and automation technologies gave a better way to ensure process safety since the minor

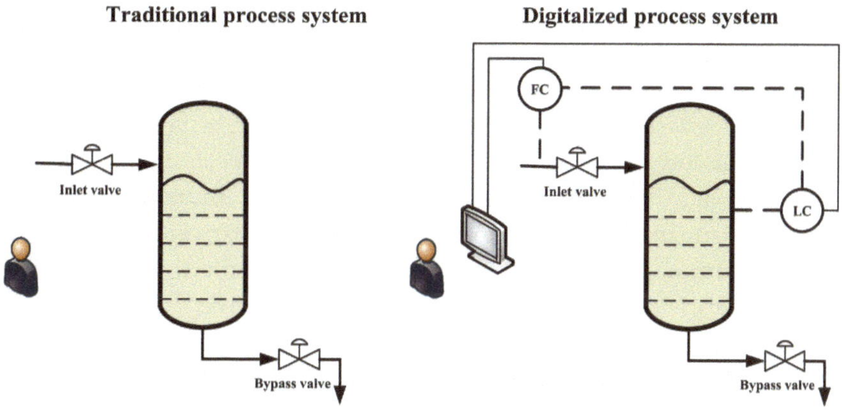

Fig. 2 Conventional vs digitalized process system.

process deviations could be mitigated using process controllers. At the same time, sensors enabled collecting data digitally.

Fig. 2 compares a traditional and a digitalized process system from the control and monitoring perspectives; a tank filling system has been considered. In earlier days, operator(s) used to monitor the level of the tank to protect against any potential overfilling. The complete control and monitoring mechanisms were manually done. On the other hand, with the advancement of sensor technologies and control and automation, this practice has been shifted to automated control (Moshgbar & Hammond, 2010). A sensor can monitor the level inside the tank, and the inlet flow controller can use this information to increase or decrease the flow rate to maintain the level, thereby protecting overflow.

As sensors became widespread, these were installed in many parts of process plants. For the first time, it became possible to collect data from places that were once unthinkable. Many redundant sensors were installed too. Soon, process plants were transformed into a source of many sensors and high-dimensional process data. These posed two challenges.

The first one was related to process monitoring or fault detection and diagnosis (FDD). In earlier days, one or two keys variables needed to be monitored. Hence, univariate control charts, such as the Shewhart, exponentially weighted moving average (EWMA), and cumulative sum (CUSUM) charts were predominant (Amin, Imtiaz and Khan, 2018b; Amin, Khan, Imtiaz, & Ahmed, 2019). These charts were easier to implement. Although these gave slower fault detection due to higher threshold values and the impact of ignoring multivariate correlation, fault diagnosis was pretty easier.

The second challenge came from the number of sensors. Due to variables' correlation, a deviation in a process variable can cause deviation in other variables, resulting in alarm flooding. This made finding the actual bad actor a severe challenge. During a process upset, a significant amount of time was lost to finding the true root cause (Abele, Anic, Gutmann, Folmer, et al., 2013; Ahmed, Izadi, Chen, Joe, & Burton, 2013; Goel, Datta, & Mannan, 2017). This might lead to unnecessary process downtime and (sometimes) significant loss due to a lack of timely response (Amin et al., 2021b).

The third industrial revolution brought a rapid change towards process monitoring. Multivariate process monitoring or fault detection and diagnosis (FDD) was rapidly adopted to tackle the enormous number of process variables. Multivariate tools can detect the faults earlier than their univariate counterparts since these consider the correlation among variables (Venkatasubramanian, Rengaswamy, Yin, & Kavuri, 2003b). However, fault diagnosis remained a challenge, as the earlier methods mostly focused on fault detection.

In the mid of the 2000s' first decade, multivariate fault diagnosis got significant attention. Multivariate contribution plots were developed; these contribution plots could identify the potential root cause(s). Nonetheless, an accurate root cause diagnosis remained a challenge since multivariate contribution plots often pointed towards a group of variables as a probable cause of a fault. To overcome this issue, many hybrid methods were developed (Amin, 2021; Amin, Khan, Ahmed, & Imtiaz, 2021a; Li, Qin, & Yuan, 2016; Vedam & Venkatasubramanian, 1999). These integrated tools could provide a better fault diagnosis result and improve process safety significantly.

The conventional qualitative risk assessment methods experienced a shift too. Probabilistic risk assessment (PRA), also known as Quantitative Risk Analysis (QRA), was introduced using the FTA (Haasl, 1965). Later, a consequence analysis tool: the ETA, was also used. Bow-tie was used for visualizing cause and consequences simultaneously (de Ruijter & Guldenmund, 2015). Many software were developed for supporting process digitalization (Khan & Abbasi, 1999a, 1999d, 1999c).

The research on process safety analysis saw significant growth during this period. At first, safety was deemed a purely technological issue, and more emphasis was given to analyzing technical aspects of safety. However, it was proved to be an improper concept. Especially, the association of human factors in several catastrophic accidents made the researchers and regulators think about the role of humans in process safety analysis (Hale & Hovden, 1998).

Therefore, methodologies were developed to quantify human error probability (Khan, Amyotte, & DiMattia, 2006; Kirwan, 1987). Later, it was realized that integrated safety management and culture were crucial for preventing accidents and enhancing safety (Wen, Khan, Amin, & Halim, 2022).

Industry 4.0 is the fourth and the latest industrial revolution—a new era in manufacturing using information and communication technology (ICT)-driven technologies (Culot, Nassimbeni, Orzes, & Sartor, 2020). This revolution is based on digital transformation and the enhanced use of artificial intelligence (Ghobakhloo, 2020; Pandey, Rastogi, Kainkaryam, Bhattacharya, & Saputelli, 2020). Digital transformation connects different business departments; it has the potential to bring the world under an umbrella. Digital transformation exceeds the technical aspects only. It includes business strategies too.

Digital transformation benefits a plant by enhancing safety, smoothness and rapidness of day to day operation and process optimization, and overall profitability (Devold, Graven, & Halvorsrød, 2017). Despite being common terms in most industries, digitalization and digital transformation were not emphasized in chemical industries to a level that these should have been focused. However, the situation has changed in recent years. The COVID-19 pandemic worked as a boon behind it. The policymakers have realized that there is no alternative to digital transformation to keep pace with the situation since it enables to operate a process plant more efficiently with minimum to no physical presence at the facilities.

The safety technologies improved over time to second these four industrial revolutions, resulting in a lower frequency of catastrophic accidents. However, major accidents such as the Caribbean Petroleum Company (CAPECO) refining tank explosion and fire (2009), Macondo well blowout (2010), Tianjin port explosion (2015), and Beirut ammonium nitrate explosion (2020) kept occurring, indicating the lack of the required level of safety that we should be satisfied (Amin, Imtiaz, & Khan, 2018a; Mannan, Reyes-Valdes, Jain, Tamim, & Ahammad, 2016). Risk assessment and management remain the key to ensuring process systems are being operated within prescribed limits with an adequate safety level.

2. Risk assessment and Industry 4.0

As we are in a new industrial revolutionary era, the critical areas of process safety research have enlarged. Along with the conventional areas (i.e., risk analysis, risk management, safety analysis, inherent safety, quantitative risk assessment, safety culture, process monitoring, and consequence modeling), the new areas, such as the big data, internet of things (IoT), cyber-physical systems (CPS), and digital twins are getting higher importance. Proper risk assessment in Industry 4.0 framework is crucial for next-generation safety system design.

Industry 4.0's largest benefit is data availability. Today, in principle, all operational processes can be monitored by indicators, also the human and socially associated performance quantities, such as economics, safety performance, and even culture. Industry 4.0 has offered several advantages, such as data-driven process optimization, remote process monitoring and control, barrier management, condition-based maintenance, and supply chain management (Amin et al., 2021a; Centobelli, Cerchione, & Ertz, 2020; Giunta, Bernasconi, Giro, & Cesari, 2020). Digital twin and digitized design data are of great help to identify hazards and to train operators (Lee, Cameron, & Hassall, 2019; Lee, Cameron, & Hassall, 2022). However, digitalization on a large scale poses some new threats to industries, as well. For instance, dependence on a large volume of datasets is essential in Industry 4.0. The security of these datasets is of vital importance since any data loss or modification can adversely affect intelligent monitoring technologies and plant safety.

On many occasions, this data theft is done using cyber means. Since the departments are interconnected through the internet, it is possible to get access to a process plant's system from anywhere. Unfortunately, due to changing geopolitical conditions, cyber crimes are on the increase. Hence, cybersecurity has become a prime concern in this Industry 4.0 era. Cyber threats have the same or even higher potential than the conventional threat to process safety. Compared to the other three industrial revolutions, cyber crime is the most notable added risk factor to process safety in Industry 4.0.

Catastrophic accidents are rare and often considered a Black Swan. However, it must not be a valid reason to wait for an accident to happen and decide about the required safety measures. The major question is—how we can fight against something that we do not know and are not aware of? We may not have proper models for these unknown events; nonetheless, with continuous learning, we may improve our understanding of these unseens. Paltrinieri, Landucci, Nelson, and Hauge (2016) viewed a tertiary event as a combination of single events. Therefore, learning about these standalone events may help us to get a better prediction of the final event. Also, acting against these events may lower the probability of a catastrophic accident. We will improve in foresight but will never be totally certain.

Hence, if a large value is at stake, it will be recommendable to invest in resilience assessment and resilience strengthening measures.

Conventionally, logical predictive models, such as the ETA, FTA, and BT, are used for accident analysis and developing strategies to prevent them. Safety barriers play a pivotal role in the overall safety performance. Safety barriers can be defined as the technical, operational and organizational elements that are designed or used to reduce the possibility of a hazard or an error or an accident (Paltrinieri, Comfort, & Reniers, 2019). Also, these can be used to reduce the magnitude of an abnormal situation.

Several researchers have stressed the importance of monitoring and assessing the performance of safety barriers through indicators. These indicators may report distinct information about a process plant. Physical parameters (e.g. flow rate, pressure, and temperature), part failure data, maintenance history, human response in an emergency situation, etc., are a few examples of such indicators. These performance indicators are often used for risk assessment in process industries (Landucci & Paltrinieri, 2016; Leveson, 2015; Øien, Utne, & Herrera, 2011; Podgórski, 2015).

One of the major drawbacks of logical failure models is their inability to predict the probability of an initial event; they need to be provided with the real-time initiating event probability to get a dynamic risk measure. Also, these tools cannot provide any information about fault detection and diagnosis (FDD), which are of great importance from operational smoothness, unnecessary downtime prevention, and safety perspectives. In fact, proper fault management is a prerequisite to risk assessment and abnormal situation management (Arunthavanathan, Khan, Ahmed, & Imtiaz, 2021).

Process FDD has received attention over the past few decades. The conventional FDD methods can be divided into three categories: quantitative model-based, qualitative model-based, and process history or data-based methods (Venkatasubramanian, Rengaswamy, Yin, & Kavuri, 2003a). The data-based methods have got special attention in Industry 4.0 since managing data and using them for a meaningful predictive outcome is crucial for implementing the latest industrial revolution.

The data-based methods can be segregated into statistical and machine learning-based methods. The principal component analysis (PCA), independent component analysis (ICA), partial least squares (PLS), and Gaussian mixture model (GMM) are the major statistical process monitoring (SPM) tools (Ge, Song, & Gao, 2013). On the contrary, the support vector machine (SVM) and artificial neural networks (ANN) are the dominant members of machine learning-based process monitoring techniques.

Compared to the SVM or ANN, statistical tools typically require fewer data since these can perform without knowing in-depth fault information. Nevertheless, SPM techniques suffer from accurate diagnosis feature (Amin et al., 2018b; Venkatasubramanian et al., 2003b). Usually, the diagnostic performance of SVM or ANN is superior to the PCA or ICA. However, none of the tools mentioned here is readily usable for quantitative risk assessment. For instance, PCA can provide a measure of deterministic fault assessment using T^2 or Q statistics. Similarly, ANNs or SVM can predict a class's probability rather than the probability of a system failure.

Efforts have been made to estimate dynamic risk using data-based tools (Yu, Khan, & Garaniya, 2016; Zadakbar, Imtiaz, & Khan, 2012). These tools use PCA scores or dynamic loading from self-organizing map (SOM) for risk assessment and fault detection. However, in real industrial scenarios, risk assessment is incomplete without incorporating safety barriers' performance evaluation. Indeed, we can get a snapshot of process safety without information on safety barriers; nevertheless, it will be an incomplete and numerically inaccurate safety assessment. Therefore, several methods have been developed for bridging this gap (Amin et al., 2020, 2021b; Yu et al., 2015).

The discussed above progress on process safety includes the unintentional failure causes. However, safety can be infringed due to deliberate reasons, as well. These are typically considered under security analysis. Though process industries suffered security-related issues from the 1960s, these intentional causes were less emphasized before the 9/11 attack (Iaiani, Moreno, Reniers, Tugnoli, & Cozzani, 2021; Iaiani, Tugnoli, Macini, & Cozzani, 2021). After the terroristic attack on the Twin Towers in the USA, process safety practitioners, regulatory bodies, and legislators reacted quickly to include security risk assessment for chemical plants as an inevitable practice.

With the rapid use of cyber-physical systems (CPS) in chemical plants, security threats are no longer considered lightly. The nature of the industrial internet of things (IIoT) also demands rigorous protection against cyberattacks. Since internet connections interconnect all the departments in a plant, it is easier to affect the whole plant by taking control of a smaller part. This control can be gained remotely due to a vulnerable security system.

Opposed power policies of countries is another critical factor. An Iranian nuclear power plant suffered an attack where the reactor temperature was modified using a malicious computer program that was injected into the controller exploiting cyber vulnerabilities (Kerr, Rollins, & Theohary, 2011). In many places of the world, terrorists were able to damage an industrial facility or (in some cases) take control of the entire plant through physical attacks. For instance, the Islamic State in Iraq and Syria (ISIS) attacked on the Iraqi Baiji oil refinery (BBC, 2014).

Process security risk analysis has received significant attention in the past two decades. The earlier methods for security risk assessment were mostly qualitative. Baybutt (2002) proposed a framework for security threats and vulnerability analysis. Baybutt and Ready (2003) presented a list of countermeasures to prevent or neutralize several attack scenarios. The authors suggested the industries to implement these steps for proper security enhancement.

Lou, Muthusamy, and Huang (2003) proposed a security zone classification-based semi-quantitative approach. The risk index was computed by observing the continuous process variables. The works by Bajpai & Gupta (2005, 2007) are the other earlier semi-quantitative approaches. Although these works have given a foundation for process security risk assessment and advanced the field notably, these cannot provide an accurate risk estimate.

To relax the limitations of qualitative and semi-quantitative methods, several quantitative methods have been proposed. van Staalduinen, Khan, and Gadag (2016) proposed a probabilistic predictive security vulnerability assessment methodology, SVAPP, using ETA, FTA, and BN. Argenti, Landucci, Reniers, and Cozzani (2018) proposed a BN-based security vulnerability assessment methodology. BN enabled capturing interactions among attack impact vectors and resistance of process units that combinedly determine an attack's success likelihood.

Attractiveness is a crucial factor for an asset to be targeted by terrorists since one of the primary motivations of attackers is to cause havoc and gain significant media attention (Bajpai & Gupta, 2005). Argenti, Landucci, Spadoni, and Cozzani (2015) proposed a semi-quantitative approach for attractiveness estimation of process facilities. Fakhravar, Khakzad, Reniers, and Cozzani (2017) developed a BN-based security assessment framework for gas pipelines using the attractiveness index proposed by Argenti et al. (2015).

Song, Khan, and Yang (2018) proposed a framework for several intrusion scenario modeling using BN. Nonetheless, this work is only applicable to physical attacks. The other BN-based methods for process security risk assessment can be found in the works by Landucci, Argenti, Cozzani, and Reniers (2017), van Staalduinen, Khan, Gadag, and Reniers (2017), and Misuri, Khakzad, Reniers, and Cozzani (2019), just to name a few.

3. Conclusions and future directions

Although process safety is a matured topic, it is yet to be adequately addressed in terms of Industry 4.0. It is still in the lap of the future how process safety will develop; the hope is of course it will improve. The interaction of security and safety as a result of cyber security breaches inducing physical safety problems will need new vulnerability analysis methods.

The latest industrial revolution has brought some unique features that must be considered while developing the current risk assessment frameworks. As Aven and Krohn (2014) suggested that our understanding of risk is continuously evolving due to the flourished knowledge level and understanding of risk assessment boundaries. Conventionally, risk used to be measured in two dimensions: the probability of failure and the severity of consequence. The authors proposed adding knowledge as another dimension. The risk should be higher where more uncertainty prevails. Delimiting uncertainty is crucial for sound decision making.

Although it is questionable how we can quantify knowledge, it is straightforward that known knowledge should be incorporated in risk assessment. However, the key question is—are we adequately addressing what we know? For instance, Industry 4.0 has culminated in the need for security-related risk assessment, and process industries have suffered from deliberate attacks. These attacks have been triggered using physical and cyber means; also, we have seen occasions when cyberattacks have been used to disable the safety system or modify the control systems (Iaiani, Moreno, et al., 2021; Iaiani, Tugnoli, et al., 2021; Moreno, Reniers, Salzano, & Cozzani, 2018).

Most of the current security analysis methodologies are developed considering either physical attacks or cyberattacks only. However, these attack modes can be triggered simultaneously, or cyberattacks can be used to command the conventional process safety parameters (Iaiani, Moreno, et al., 2021; Iaiani, Tugnoli, et al., 2021). Few efforts have been made to capture both the intentional and unintentional failure causes in process risk assessment (Aven, 2007; Ayyub, McGill, & Kaminskiy, 2007; Bernatik, Senovsky, & Pitt, 2011; Ji, Yang, Cao, Wang, et al., 2021; Song, Khan, & Yang, 2019). These works suffer from including every aspect of safety and security in a unified framework. Therefore, a holistic framework is much needed that can handle safety and security-related factors. It may result in a complex model for risk assessment; however, it will capture the available knowledge we have.

Machine learning algorithms are quite impressive in knowledge acquisition and representation. An interesting study was presented by Matteini (2015). The author suggested adding human knowledge and experience as input to the deep neural networks (DNNs). The output can be several indexes. For instance, Matteini (2015) considered risk increase or decrease as outputs. The results suggested that the proposed DNN-based method could outperform the performance of the conventional machine learning algorithms. Wu and Zhao (2021) showed how the integration of knowledge with a convolutional neural network (CNN) could improve fault diagnosis results. These works may not provide purely quantitative risk assessment; however, these will provide stimulus to future methodology development.

Hegde and Rokseth (2020) presented a review of the application of machine learning tools for engineering risk assessment. Though the scope of this review was beyond chemical process safety, the findings were significant. The authors found that machine learning had been successfully applied to both textual and numerical data analysis. Since many databases record accident history in textual form, it will be pretty interesting to investigate how we can incorporate such data for quantitative risk assessment. Developing a causal pattern among the accident histories using natural language processing (NLP) can be one of the promising avenues to explore.

Risk assessment provides additional information for safety analysis. Many authors have suggested using risk as an indicator of fault detection (Amin et al., 2021b; Yu et al., 2015; Zadakbar et al., 2012). However, controllers need to be risk-informed. Risk-based smart controller design will be an interesting field. Industry 4.0 requires autonomous fault diagnosis and maintenance action. Regrettably, these are yet to be implemented in many chemical plants (Arunthavanathan, Khan, Ahmed, & Imtiaz, 2022). Hence, these two aspects need to be automated for proper risk management in smart industries.

References

Abele, L., Anic, M., Gutmann, T., Folmer, J., … Vogel-Heuser, B. (2013). Combining knowledge modeling and machine learning for alarm root cause analysis. *IFAC Proceedings Volumes*, *46*(9), 1843–1848. https://doi.org/10.3182/20130619-3-RU-3018.00057.

Ahmed, K., Izadi, I., Chen, T., Joe, D., & Burton, T. (2013). Similarity analysis of industrial alarm flood data. *IEEE Transactions on Automation Science and Engineering*, *10*(2), 452–457. https://doi.org/10.1109/TASE.2012.2230627.

Albalawi, F., Durand, H., Alanqar, A., & Christofides, P. D. (2018). Achieving operational process safety via model predictive control. *Journal of Loss Prevention in the Process Industries*, *53*, 74–88. https://doi.org/10.1016/j.jlp.2016.11.021.

Amin, M. T. (2021). An integrated methodology for fault detection, root cause diagnosis, and propagation pathway analysis in chemical process systems. *Cleaner Engineering and Technology*, *4*, 100187. https://doi.org/10.1016/j.clet.2021.100187.

Amin, M. T., Imtiaz, S., & Khan, F. (2018a). Dynamic availability assessment of safety critical systems using a dynamic Bayesian network. *Reliability Engineering & System Safety, 178*, 108–117. https://doi.org/10.1016/j.ress.2018.05.017.

Amin, M. T., Imtiaz, S., & Khan, F. (2018b). Process system fault detection and diagnosis using a hybrid technique. *Chemical Engineering Science, 189*, 191–211. https://doi.org/10.1016/j.ces.2018.05.045.

Amin, M. T., Khan, F., Ahmed, S., & Imtiaz, S. (2020). A novel data-driven methodology for fault detection and dynamic risk assessment. *Canadian Journal of Chemical Engineering, 98*(11), 2397–2416. https://doi.org/10.1002/cjce.23760.

Amin, M. T., Khan, F., Ahmed, S., & Imtiaz, S. (2021a). A data-driven Bayesian network learning method for process fault diagnosis. *Process Safety and Environmental Protection, 150*, 110–122. https://doi.org/10.1016/j.psep.2021.04.004.

Amin, M. T., Khan, F., Ahmed, S., & Imtiaz, S. (2021b). Risk-based fault detection and diagnosis for nonlinear and non-Gaussian process systems using R-vine copula. *Process Safety and Environmental Protection, 150*, 123–136. https://doi.org/10.1016/j.psep.2021.04.010.

Amin, M. T., Khan, F., & Amyotte, P. (2019). A bibliometric review of process safety and risk analysis. *Process Safety and Environmental Protection, 126*. https://doi.org/10.1016/j.psep.2019.04.015.

Amin, M. T., Khan, F., & Imtiaz, S. (2019). Fault detection and pathway analysis using a dynamic Bayesian network. *Chemical Engineering Science, 195*, 777–790. https://doi.org/10.1016/j.ces.2018.10.024.

Amin, M. T., Khan, F., Imtiaz, S., & Ahmed, S. (2019). Robust process monitoring methodology for detection and diagnosis of unobservable faults. *Industrial and Engineering Chemistry Research, 58*(41). https://doi.org/10.1021/acs.iecr.9b03406.

Antonioni, G., Spadoni, G., & Cozzani, V. (2009). Application of domino effect quantitative risk assessment to an extended industrial area. *Journal of Loss Prevention in the Process Industries, 22*(5), 614–624. https://doi.org/10.1016/j.jlp.2009.02.012.

Argenti, F., Landucci, G., Reniers, G., & Cozzani, V. (2018). Vulnerability assessment of chemical facilities to intentional attacks based on Bayesian network. *Reliability Engineering and System Safety, 169*, 515–530. https://doi.org/10.1016/j.ress.2017.09.023.

Argenti, F., Landucci, G., Spadoni, G., & Cozzani, V. (2015). The assessment of the attractiveness of process facilities to terrorist attacks. *Safety Science*, 77, 169–181. https://doi.org/10.1016/j.ssci.2015.02.013.

Arunthavanathan, R., Khan, F., Ahmed, S., & Imtiaz, S. (2021). An analysis of process fault diagnosis methods from safety perspectives. *Computers & Chemical Engineering, 145*, 107197. https://doi.org/10.1016/j.compchemeng.2020.107197.

Arunthavanathan, R., Khan, F., Ahmed, S., & Imtiaz, S. (2022). Autonomous fault diagnosis and root cause analysis for the processing system using one-class SVM and NN permutation algorithm. *Industrial & Engineering Chemistry Research*. https://doi.org/10.1021/ACS.IECR.1C02731.

Aven, T. (2007). A unified framework for risk and vulnerability analysis covering both safety and security. *Reliability Engineering and System Safety, 92*(6), 745–754. https://doi.org/10.1016/j.ress.2006.03.008.

Aven, T., & Krohn, B. S. (2014). A new perspective on how to understand, assess and manage risk and the unforeseen. *Reliability Engineering & System Safety, 121*, 1–10. https://doi.org/10.1016/j.ress.2013.07.005.

Ayyub, B. M., McGill, W. L., & Kaminskiy, M. (2007). Critical asset and portfolio risk analysis: An all-hazards framework. *Risk Analysis*, 27(4), 789–801. https://doi.org/10.1111/j.1539-6924.2007.00911.x.

Bajpai, S., & Gupta, J. P. (2005). Site security for chemical process industries. *Journal of Loss Prevention in the Process Industries, 18*(4–6), 301–309. https://doi.org/10.1016/j.jlp.2005.06.011.

Bajpai, S., & Gupta, J. P. (2007). Securing oil and gas infrastructure. *Journal of Petroleum Science and Engineering*, *55*(1–2), 174–186. https://doi.org/10.1016/j.petrol.2006.04.007.
Bakshi, B. R. (1998). Multiscale PCA with application to multivariate statistical process monitoring. *AICHE Journal*, *44*(7), 1596–1610. https://doi.org/10.1002/aic.690440712.
Baybutt, P. (2002). Assessing risks from threats to process plants: Threat and vulnerability analysis. *Process Safety Progress*, *21*(4), 269–275. https://doi.org/10.1002/prs.680210403.
Baybutt, P., & Ready, V. (2003). Strategies for protecting process plants against terrorism, sabotage and other criminal acts. *Homeland Defence Journal*, *2*(1), 1–7.
BBC. (2014). *Iraq crisis: Battle grips vital Baiji oil refinery*. https://www.bbc.com/news/world-middle-east-27897648 (last accessed: September 27, 2021).
Bernatik, A., Senovsky, P., & Pitt, M. (2011). LNG as a potential alternative fuel - safety and security of storage facilities. *Journal of Loss Prevention in the Process Industries*, *24*(1), 19–24. https://doi.org/10.1016/j.jlp.2010.08.003.
Bhandari, J., Arzaghi, E., Abbassi, R., Garaniya, V., & Khan, F. (2016). Dynamic risk-based maintenance for offshore processing facility. *Process Safety Progress*. https://doi.org/10.1002/prs.11829.
CCPS. (2007). *Guidelines for risk based process safety*. Center for Chemical Process Safety and John Wiley & Sons, Inc.
Centobelli, P., Cerchione, R., & Ertz, M. (2020). Agile supply chain management: Where did it come from and where will it go in the era of digital transformation? *Industrial Marketing Management*, *90*, 324–345. https://doi.org/10.1016/j.indmarman.2020.07.011.
Clark, G., & Jacks, D. (2007). Coal and the industrial revolution, 1700–1869. *European Review of Economic History*, *11*(1), 39–72.
Cooper, C., & Kaplinsky, R. (2005). *Technology and development in the third industrial revolution*. Routledge.
Culot, G., Nassimbeni, G., Orzes, G., & Sartor, M. (2020). Behind the definition of Industry 4.0: Analysis and open questions. *International Journal of Production Economics*, *226*, 107617. https://doi.org/10.1016/j.ijpe.2020.107617.
de Ruijter, A., & Guldenmund, F. (2015). The bowtie method: A review. *Safety Science*, *88*, 211–218. https://doi.org/10.1016/j.ssci.2016.03.001.
Devold, H., Graven, T., & Halvorsrød, S. O. (2017). Digitalization of oil and gas facilities reduce cost and improve maintenance operations. *Offshore Technology Conference*. https://doi.org/10.4043/27788-MS.
Dunjó, J., Fthenakis, V., Vílchez, J. A., & Arnaldos, J. (2010). Hazard and operability (HAZOP) analysis. A literature review. *Journal of Hazardous Materials*, *173*(1–3), 19–32. https://doi.org/10.1016/j.jhazmat.2009.08.076.
Fakhravar, D., Khakzad, N., Reniers, G., & Cozzani, V. (2017). Security vulnerability assessment of gas pipelines using discrete-time Bayesian network. *Process Safety and Environmental Protection*, *111*, 714–725. https://doi.org/10.1016/j.psep.2017.08.036.
Ferdous, R., Khan, F., Sadiq, R., Amyotte, P., & Veitch, B. (2013). Analyzing system safety and risks under uncertainty using a bow-tie diagram: An innovative approach. *Process Safety and Environmental Protection*, *91*(1–2), 1–18. https://doi.org/10.1016/j.psep.2011.08.010.
Ge, Z., Song, Z., & Gao, F. (2013). Review of recent research on data-based process monitoring. *Industrial & Engineering Chemistry Research*, *52*(10), 3543–3562. https://doi.org/10.1021/ie302069q.
Ghobakhloo, M. (2020). Industry 4.0, digitization, and opportunities for sustainability. *Journal of Cleaner Production*, *252*, 119869. https://doi.org/10.1016/j.jclepro.2019.119869.
Giunta, G., Bernasconi, G., Giro, R. A., & Cesari, S. (2020). Digital transformation of historical data for advanced predictive maintenance. *Abu Dhabi International Petroleum Exhibition & Conference*. https://doi.org/10.2118/202906-MS.

Goel, P., Datta, A., & Mannan, M. S. (2017). Industrial alarm systems: Challenges and opportunities. *Journal of Loss Prevention in the Process Industries*, *50*(Part A), 23–36. https://doi.org/10.1016/j.jlp.2017.09.001.

Gowland, R. (2012). A journey into process safety with Trevor Kletz. *Journal of Loss Prevention in the Process Industries*, *25*(5), 768–769. https://doi.org/10.1016/j.jlp.2012.03.012.

Gungor, V. C., & Hancke, G. P. (2009). Industrial wireless sensor networks: Challenges, design principles, and technical approaches. *IEEE Transactions on Industrial Electronics*, *56*(10), 4258–4265. https://doi.org/10.1109/TIE.2009.2015754.

Guo, C., Khan, F., & Imtiaz, S. (2019). Copula-based Bayesian network model for process system risk assessment. *Process Safety and Environmental Protection*, *123*, 317–326. https://doi.org/10.1016/j.psep.2019.01.022.

Haasl, D. F. (1965). Advanced concepts in fault tree analysis(fault tree construction and probability evaluation techniques used in safety evaluation of rocket launching device). *System Safety Symposium, Seattle, Wash*.

Hale, A. R., & Hovden, J. (1998). Management and culture: The third age of safety. A review of approaches to organizational aspects of safety, health and environment. In *Occupational injury: Risk, prevention and intervention* (pp. 129–165). Taylor & Francis London.

Halim, S. Z., Yu, M., Escobar, H., & Quddus, N. (2020). Towards a causal model from pipeline incident data analysis. *Process Safety and Environmental Protection*, *143*, 348–360. https://doi.org/10.1016/j.psep.2020.06.047.

Hashemi, S. J., Ahmed, S., & Khan, F. (2014). Loss functions and their applications in process safety assessment. *Process Safety Progress*. https://doi.org/10.1002/prs.11659.

Hashemi, S. J., Ahmed, S., & Khan, F. I. (2015). Loss scenario analysis and loss aggregation for process facilities. *Chemical Engineering Science*, *128*, 119–129. https://doi.org/10.1016/j.ces.2015.01.061.

Hegde, J., & Rokseth, B. (2020). Applications of machine learning methods for engineering risk assessment–a review. *Safety Science*, *122*, 104492. https://doi.org/10.1016/j.ssci.2019.09.015.

Iaiani, M., Moreno, V. C., Reniers, G., Tugnoli, A., & Cozzani, V. (2021). Analysis of events involving the intentional release of hazardous substances from industrial facilities. *Reliability Engineering and System Safety*, *212*, 107593. https://doi.org/10.1016/j.ress.2021.107593.

Iaiani, M., Tugnoli, A., Macini, P., & Cozzani, V. (2021). Outage and asset damage triggered by malicious manipulation of the control system in process plants. *Reliability Engineering & System Safety*, *213*, 107685. https://doi.org/10.1016/j.ress.2021.107685.

Ishizaka, A., & Labib, A. (2011). Review of the main developments in the analytic hierarchy process. *Expert Systems with Applications*, *38*(11), 14336–14345. https://doi.org/10.1016/j.eswa.2011.04.143.

Islam, R., Khan, F., & Venkatesan, R. (2017). Real time risk analysis of kick detection: Testing and validation. *Reliability Engineering and System Safety*, *161*, 25–37. https://doi.org/10.1016/j.ress.2016.12.014.

Ji, Z., Yang, S. H., Cao, Y., Wang, Y., … Zhang, Y. (2021). Harmonizing safety and security risk analysis and prevention in cyber-physical systems. *Process Safety and Environmental Protection*, *148*, 1279–1291. https://doi.org/10.1016/j.psep.2021.03.004.

Kabir, S., & Papadopoulos, Y. (2018). A review of applications of fuzzy sets to safety and reliability engineering. *International Journal of Approximate Reasoning*, *100*, 29–55. https://doi.org/10.1016/j.ijar.2018.05.005.

Kennedy, C. (2020). The energy embodied in the first and second industrial revolutions. *Journal of Industrial Ecology*, *24*(4), 887–898. https://doi.org/10.1111/jiec.12994.

Kerr, P. K., Rollins, J., & Theohary, C. A. (2011). The stuxnet computer worm: Harbinger of an emerging warfare capability. In *The Stuxnet Computer Worm and Industrial Control System Security* (pp. 1–10). Congressional Research Service.

Khakzad, N., Khan, F., & Amyotte, P. (2013). Dynamic safety analysis of process systems by mapping bow-tie into Bayesian network. *Process Safety and Environmental Protection*, *91*(1–2), 46–53. https://doi.org/10.1016/j.psep.2012.01.005.

Khan, F., & Abbasi, S. A. (1998). DOMIFFECT (DOMIno eFFECT): User-friendly software for domino effect analysis. *Environmental Modelling and Software*, *13*(2), 163–177. https://doi.org/10.1016/S1364-8152(98)00018-8.

Khan, F. I., & Abbasi, S. A. (1999a). HAZDIG: A new software package for assessing the risks of accidental release of toxic chemicals. *Journal of Loss Prevention in the Process Industries*, *12*(2), 167–181. https://doi.org/10.1016/S0950-4230(97)00043-0.

Khan, F. I., & Abbasi, S. A. (1999b). Major accidents in process industries and an analysis of causes and consequences. *Journal of Loss Prevention in the Process Industries*, *12*(5), 361–378. https://doi.org/10.1016/S0950-4230(98)00062-X.

Khan, F. I., & Abbasi, S. A. (1999c). MAXCRED: A new software package for rapid risk assessment in chemical process industries. *Environmental Modelling & Software*, *14*(1), 11–25.

Khan, F. I., & Abbasi, S. A. (1999d). PROFAT: A user friendly system for probabilistic fault tree analysis. *Process Safety Progress*, *18*(1), 42–49. https://doi.org/10.1002/prs.680180109.

Khan, F. I., & Amyotte, P. R. (2003). How to make inherent safety practice a reality. *Canadian Journal of Chemical Engineering*, *81*(1), 2–16. https://doi.org/10.1002/cjce.5450810101.

Khan, F. I., Amyotte, P. R., & Amin, M. T. (2020). Advanced methods of risk assessment and management: An overview. In *Methods in Chemical Process Safety* (pp. 1–34). Elsevier. https://doi.org/10.1016/bs.mcps.2020.03.002.

Khan, F. I., Amyotte, P. R., & DiMattia, D. G. (2006). HEPI: A new tool for human error probability calculation for offshore operation. *Safety Science*, *44*, 313–334. https://doi.org/10.1016/j.ssci.2005.10.008.

Khan, F., Rathnayaka, S., & Ahmed, S. (2015). Methods and models in process safety and risk management: Past, present and future. *Process Safety and Environmental Protection*, *98*, 116–147. https://doi.org/10.1016/j.psep.2015.07.005.

Kirwan, B. (1987). Human reliability analysis of an offshore emergency blowdown system. *Applied Ergonomics*, *18*(1), 23–33. https://doi.org/10.1016/0003-6870(87)90067-6.

Klein, J. A. (2009). Two centuries of process safety at DuPont. *Process Safety Progress*, *28*(2), 114–122. https://doi.org/10.1002/prs.10309.

Kletz, T. A. (2001). *HAZOP and HAZAN: Identifying and assessing process industry hazards*. IChemE.

Landucci, G., Argenti, F., Cozzani, V., & Reniers, G. (2017). Assessment of attack likelihood to support security risk assessment studies for chemical facilities. *Process Safety and Environmental Protection*, *110*, 102–114. https://doi.org/10.1016/j.psep.2017.06.019.

Landucci, G., & Paltrinieri, N. (2016). Dynamic evaluation of risk: From safety indicators to proactive techniques. *Chemical Engineering Transactions*, *53*, 169–174. https://doi.org/10.3303/CET1653029.

Lee, J., Cameron, I., & Hassall, M. (2019). Improving process safety: What roles for Digitalization and Industry 4.0? *Process Safety and Environmental Protection*, *132*, 325–339. https://doi.org/10.1016/j.psep.2019.10.021.

Lee, J., Cameron, I., & Hassall, M. (2022). Information needs and challenges in future process safety. *Digital Chemical Engineering*. https://doi.org/10.1016/j.dche.2022.100017.

Lee, W. S., Grosh, D. L., Tillman, F. A., & Lie, C. H. (1985). Fault tree analysis, methods, and applications—A review. *IEEE Transactions on Reliability*, *R-34*(3), 194–203. https://doi.org/10.1109/TR.1985.5222114.

Leveson, N. (2015). A systems approach to risk management through leading safety indicators. *Reliability Engineering & System Safety*, *136*, 17–34. https://doi.org/10.1016/j.ress.2014.10.008.

Li, G., Qin, S. J., & Yuan, T. (2016). Data-driven root cause diagnosis of faults in process industries. *Chemometrics and Intelligent Laboratory Systems, 159*, 1–11. https://doi.org/10.1016/j.chemolab.2016.09.006.

Lou, H. H., Muthusamy, R., & Huang, Y. (2003). Process security assessment: Operational space classification and process security index. *Process Safety and Environmental Protection: Transactions of the Institution of Chemical Engineers, Part B, 81*(6), 418–429. https://doi.org/10.1205/095758203770866593.

Mannan, M. S., Reyes-Valdes, O., Jain, P., Tamim, N., & Ahammad, M. (2016). The evolution of process safety: Current status and future direction. *Annual Review of Chemical and Biomolecular Engineering*, 7, 135–162. https://doi.org/10.1146/annurev-chembioeng-080615-033640.

Markowski, A. S. (2012). A review of layer of protection analysis techniques for oil and gas industry. *International Journal of Oil, Gas and Coal Technology, 5*(1), 66–79. https://doi.org/10.1504/IJOGCT.2012.044178.

Matteini, A. (2015). *Human factors and dynamic risk analysis: A case-study in oil&gas drilling*. Italy: Bologna.

Mentes, A., & Helvacioglu, I. H. (2011). Review of fuzzy set theory applications in safety assessment for marine and offshore industries. In *ASME 2011 30th international conference on ocean, offshore and Arctic engineering* (pp. 875–884).

Misuri, A., Khakzad, N., Reniers, G., & Cozzani, V. (2019). A Bayesian network methodology for optimal security management of critical infrastructures. *Reliability Engineering & System Safety, 191*, 106112. https://doi.org/10.1016/j.ress.2018.03.028.

Moreno, V. C., Reniers, G., Salzano, E., & Cozzani, V. (2018). Analysis of physical and cyber security-related events in the chemical and process industry. *Process Safety and Environmental Protection, 116*, 621–631. https://doi.org/10.1016/j.psep.2018.03.026.

Moshgbar, M., & Hammond, S. (2010). Advanced process control. *The Quality Assurance Journal, 13*(3–4), 62–66. https://doi.org/10.1002/qaj.472.

Øien, K., Utne, I. B., & Herrera, I. A. (2011). Building safety indicators: Part 1–theoretical foundation. *Safety Science, 49*(2), 148–161. https://doi.org/10.1016/j.ssci.2010.05.012.

Paltrinieri, N., Comfort, L., & Reniers, G. (2019). Learning about risk: Machine learning for risk assessment. *Safety Science, 118*, 475–486. https://doi.org/10.1016/j.ssci.2019.06.001.

Paltrinieri, N., Landucci, G., Nelson, W. R., & Hauge, S. (2016). Proactive approaches of dynamic risk assessment based on indicators. In *Dynamic risk analysis in the chemical and petroleum industry* (pp. 63–73). Elsevier. https://doi.org/10.1016/B978-0-12-803765-2.00006-8.

Pandey, Y. N., Rastogi, A., Kainkaryam, S., Bhattacharya, S., & Saputelli, L. (2020). *Machine learning in the oil and gas industry*. Apress. https://doi.org/10.1007/978-1-4842-6094-4.

Pekalski, A. A., Zevenbergen, J. F., Lemkowitz, S. M., & Pasman, H. J. (2005). A review of explosion prevention and protection systems suitable as ultimate layer of protection in chemical process installations. *Process Safety and Environmental Protection, 83*(1), 1–17. https://doi.org/10.1205/psep.04023.

Podgórski, D. (2015). Measuring operational performance of OSH management system–A demonstration of AHP-based selection of leading key performance indicators. *Safety Science, 73*, 146–166. https://doi.org/10.1016/j.ssci.2014.11.018.

Ruijters, E., & Stoelinga, M. (2015). Fault tree analysis: A survey of the state-of-the-art in modeling, analysis and tools. *Computer Science Review, 15–16*, 29–62. https://doi.org/10.1016/j.cosrev.2015.03.001.

Schleihauf, W. (2001). The Dumaresq and the Dreyer: Part III. *Warship International, 38*(3), 221–233.

Song, G., Khan, F., & Yang, M. (2018). Security assessment of process facilities – Intrusion modeling. *Process Safety and Environmental Protection, 117*, 639–650. https://doi.org/10.1016/j.psep.2018.05.019.

Song, G., Khan, F., & Yang, M. (2019). Probabilistic assessment of integrated safety and security related abnormal events: A case of chemical plants. *Safety Science, 113*, 115–125. https://doi.org/10.1016/j.ssci.2018.11.004.

van Staalduinen, M. A., Khan, F., & Gadag, V. (2016). SVAPP methodology: A predictive security vulnerability assessment modeling method. *Journal of Loss Prevention in the Process Industries, 43*, 397–413. https://doi.org/10.1016/j.jlp.2016.06.017.

van Staalduinen, M. A., Khan, F., Gadag, V., & Reniers, G. (2017). Functional quantitative security risk analysis (QSRA) to assist in protecting critical process infrastructure. *Reliability Engineering & System Safety, 157*, 23–34. https://doi.org/10.1016/j.ress.2016.08.014.

Vedam, H., & Venkatasubramanian, V. (1999). PCA-SDG based process monitoring and fault diagnosis. *Control Engineering Practice*, 7(7), 903–917. https://doi.org/10.1016/S0967-0661(99)00040-4.

Venkatasubramanian, V., Rengaswamy, R., Yin, K., & Kavuri, S. N. (2003a). A review of process fault detection and diagnosis, part I quantitative model-based methods. *Computers & Chemical Engineering, 27*(3), 293–311. https://doi.org/10.1016/S0098-1354(02)00160-6.

Venkatasubramanian, V., Rengaswamy, R., Yin, K., & Kavuri, S. N. (2003b). A review of process fault detection and diagnosis part III: Process history based methods. *Computers and Chemical Engineering, Vol. 27*(3), 293–311. https://doi.org/10.1016/S0098-1354(02)00160-6.

Villa, V., Paltrinieri, N., Khan, F., & Cozzani, V. (2016). Towards dynamic risk analysis: A review of the risk assessment approach and its limitations in the chemical process industry. *Safety Science, 89*, 77–93. https://doi.org/10.1016/j.ssci.2016.06.002.

Weber, P., Medina-Oliva, G., Simon, C., & Iung, B. (2012). Overview on Bayesian networks applications for dependability, risk analysis and maintenance areas. *Engineering Applications of Artificial Intelligence, 25*(4), 671–682. https://doi.org/10.1016/j.engappai.2010.06.002.

Wen, H., Khan, F., Amin, M. T., & Halim, S. Z. (2022). Myths and misconceptions of data-driven methods: Applications to process safety analysis. *Computers & Chemical Engineering, 158*, 107639. https://doi.org/10.1016/j.compchemeng.2021.107639.

Wrigley, E. A. (2013). Energy and the English industrial revolution. *Philosophical Transactions of the Royal Society A: Mathematical, Physical and Engineering Sciences, 371*(1986), 20110568. https://doi.org/10.1098/rsta.2011.0568.

Wu, S., Zhang, L., Zheng, W., Liu, Y., & Lunteigen, M. A. (2016). A DBN-based risk assessment model for prediction and diagnosis of offshore drilling incidents. *Journal of Natural Gas Science and Engineering*. https://doi.org/10.1016/j.jngse.2016.06.054.

Wu, D., & Zhao, J. (2021). Process topology convolutional network model for chemical process fault diagnosis. *Process Safety and Environmental Protection, 150*, 93–109. https://doi.org/10.1016/j.psep.2021.03.052.

Yu, H., Khan, F., & Garaniya, V. (2015). Risk-based fault detection using self-organizing map. *Reliability Engineering and System Safety, 139*(82–96), 82–96. https://doi.org/10.1016/j.ress.2015.02.011.

Yu, H., Khan, F., & Garaniya, V. (2016). Risk-based process system monitoring using self-organizing map integrated with loss functions. *Canadian Journal of Chemical Engineering, 94*(7), 1295–1307. https://doi.org/10.1002/cjce.22480.

Zadakbar, O., Imtiaz, S., & Khan, F. (2012). Dynamic risk assessment and fault detection using principal component analysis. *Industrial & Engineering Chemistry Research, 52*, 809–816. https://doi.org/10.1021/ie202880w.

Kolibri 360 GmbH
Ferdinand-Jühlke-Straße 7
99095 Erfurt, Deutschland
produktsicherheit@kolibri360.de